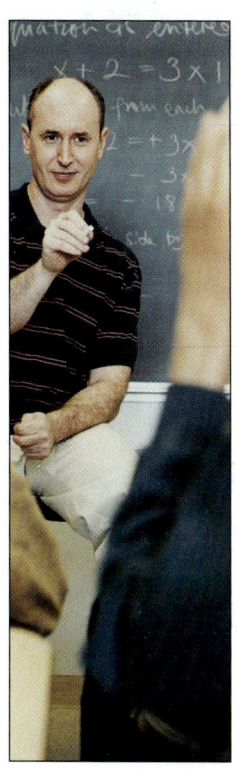

The first person to invent a car that runs on water…

… may be sitting right in your classroom! Every one of your students has the potential to make a difference. And realizing that potential starts right here, in your course.

When students succeed in your course—when they stay on-task and make the breakthrough that turns confusion into confidence—they are empowered to realize the possibilities for greatness that lie within each of them. We know your goal is to create an environment where students reach their full potential and experience the exhilaration of academic success that will last them a lifetime. *WileyPLUS* can help you reach that goal.

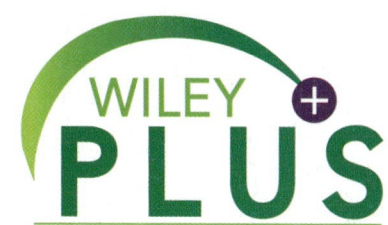

WileyPLUS is an online suite of resources—including the complete text—that will help your students:

- come to class better prepared for your lectures
- get immediate feedback and context-sensitive help on assignments and quizzes
- track their progress throughout the course

D0207355

"I just wanted to say how much this program helped me in studying… I was able to actually see my mistakes and correct them. … I really think that other students should have the chance to use *WileyPLUS*."

Ashlee Krisko, *Oakland University*

www.wiley.com/college/wileyplus

80% of students surveyed said it improved their understanding of the material.

FOR INSTRUCTORS

Wiley**PLUS** is built around the activities you perform in your class each day. With Wiley**PLUS** you can:

Prepare & Present

Create outstanding class presentations using a wealth of resources such as Power-erPoint™ slides, image galleries, interactive simulations, and more. You can even add materials you have created yourself.

Create Assignments

Automate the assigning and grading of homework or quizzes by using the provided question banks, or by writing your own.

Track Student Progress

Keep track of your students' progress and analyze individual and overall class results.

Now Available with WebCT and Blackboard!

"It has been a great help, and I believe it has helped me to achieve a better grade."

Michael Morris,
Columbia Basin College

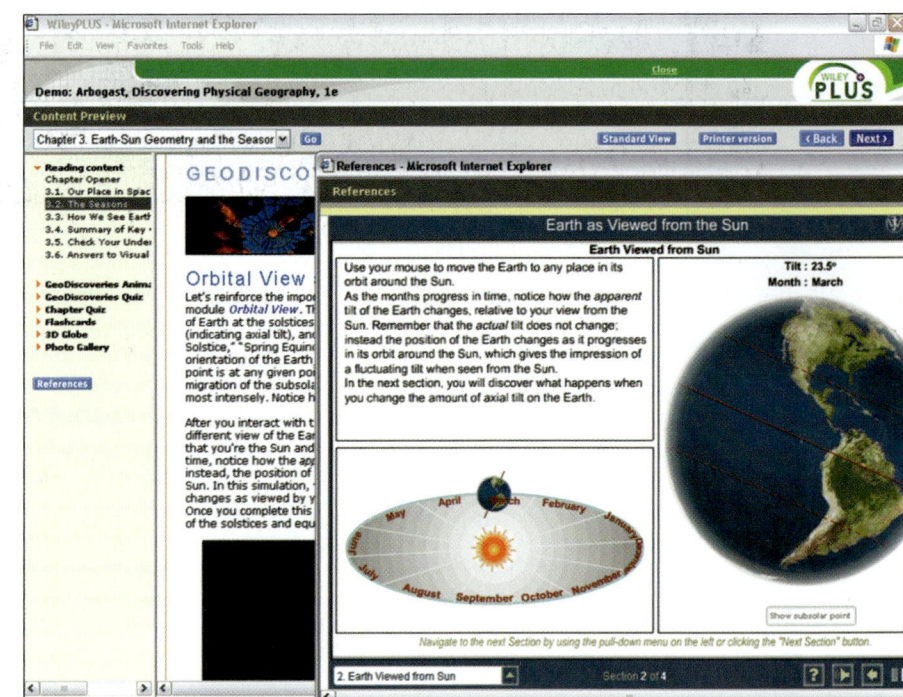

FOR STUDENTS

You have the potential to make a difference!

WileyPLUS is a powerful online system packed with features to help you make the most of your potential and get the best grade you can!

With Wiley**PLUS** you get:

- A complete online version of your text and other study resources.

- Problem-solving help, instant grading, and feedback on your homework and quizzes.

- The ability to track your progress and grades throughout the term.

For more information on what *WileyPLUS* can do to help you and your students reach their potential, please visit www.wiley.com/college/*wileyplus*.

76% of students surveyed said it made them better prepared for tests. *

*Based on a survey of 972 student users of *WileyPLUS*

DISCOVERING
PHYSICAL GEOGRAPHY

THE WILEY BICENTENNIAL—KNOWLEDGE FOR GENERATIONS

*E*ach generation has its unique needs and aspirations. When Charles Wiley first opened his small printing shop in lower Manhattan in 1807, it was a generation of boundless potential searching for an identity. And we were there, helping to define a new American literary tradition. Over half a century later, in the midst of the Second Industrial Revolution, it was a generation focused on building the future. Once again, we were there, supplying the critical scientific, technical, and engineering knowledge that helped frame the world. Throughout the 20th Century, and into the new millennium, nations began to reach out beyond their own borders and a new international community was born. Wiley was there, expanding its operations around the world to enable a global exchange of ideas, opinions, and know-how.

For 200 years, Wiley has been an integral part of each generation's journey, enabling the flow of information and understanding necessary to meet their needs and fulfill their aspirations. Today, bold new technologies are changing the way we live and learn. Wiley will be there, providing you the must-have knowledge you need to imagine new worlds, new possibilities, and new opportunities.

Generations come and go, but you can always count on Wiley to provide you the knowledge you need, when and where you need it!

WILLIAM J. PESCE
PRESIDENT AND CHIEF EXECUTIVE OFFICER

PETER BOOTH WILEY
CHAIRMAN OF THE BOARD

DISCOVERING PHYSICAL GEOGRAPHY

Alan F. Arbogast | *Michigan State University*

BICENTENNIAL
1807
WILEY
2007
BICENTENNIAL

John Wiley & Sons, Inc.

VICE PRESIDENT AND PUBLISHER	Jay O'Callaghan
EXECUTIVE EDITOR	Ryan Flahive
DEVELOPMENT EDITORS	Suzanne Thibodeau; David Chelton
MARKETING MANAGER	Emily Streutker
MARKETING ASSISTANT	Nicole Ferrato
MEDIA EDITOR	Lynn Pearlman
ASSISTANT EDITOR	Laura Kelleher
SENIOR PRODUCTION EDITOR	Lisa Wojcik
EDITORIAL ASSISTANT	Courtney Nelson
SENIOR DESIGNER	Harry Nolan
SENIOR PHOTO EDITOR	Jennifer MacMillan
COVER PHOTO	Arnul Husmo/Stone/Getty Images
PHOTO RESEARCHERS	Elyse Rieder; Ramón Rivera-Moret
SENIOR ILLUSTRATION EDITOR	Sandra Rigby
ANNIVERSARY LOGO DESIGN	Richard Pacifico

This book was set in Quark by Prepare and printed and bound by Von Hoffmann Press. This book is printed on acid free paper. ∞

To order books or for customer service please, call 1-800-CALL WILEY (225-5945).

Arbogast, Alan F.
Discovering physical geography / Alan F. Arbogast.
   p. cm.
   ISBN-13: 978-0-471-43860-1 (pbk.)
   ISBN-10: 0-471-43860-X (pbk.)
1. Physical geography–Textbooks. I. Title.
GB54.5.A73 2007
910'.02–dc22
                              2006034737

Printed in the United States of America

10 9 8 7 6 5 4 3

For Jennifer, Hannah, and Rosie

ABOUT THE AUTHOR

Alan F. Arbogast is an Associate Professor of Geography at Michigan State University in East Lansing, Michigan. His research focuses on Holocene landscape evolution of eolian, coastal, and fluvial environments in the Great Lakes region. He spends most of his research time studying the formation of coastal sand dunes along the eastern shore of Lake Michigan. His research has been funded by NASA, the National Science Foundation, and the state of Michigan.

Alan teaches a variety of courses at Michigan State, with the majority related to physical geography and geomorphology. He is a member of the Association of American Geographers, the Geological Society of America, the American Quaternary Association, and the International Quaternary Association. Alan is married to Jennifer and has two daughters, Hannah and Rosie.

Alan sits on a sand dune along Lake Michigan with his daughters Hannah (left) and Rosie (right).

Introduction to Physical Geography is a high-enrollment course at most universities. The usual goal of this course is to help students understand the Earth as a natural system and how various processes on the planet operate globally over space and time. Given the interactions that occur among these natural processes on Earth, physical geography requires integration of many different topics. For example, students are expected to understand how seasonal Earth–Sun relationships affect atmospheric circulation, which in turn influences the distribution of vegetation. In addition, physical geography is an applied discipline that can inform decisions about important environmental issues such as global warming, earthquake hazards, coastal erosion in populated areas, soil degradation, and deforestation.

Why a New Textbook?

Although a variety of textbooks teach physical geography at the introductory level, each presents the subject matter in essentially the same way; that is, text supplemented by diagrams and photographs. While this method is fundamentally effective, especially for highly motivated and scientifically inclined students, it is generally static and does not provide much sense of dimension or time. It also does not effectively illustrate the interactivity of various processes. Streams, for example, are dynamic features that *move*. The nature of this movement depends upon a variety of factors, such as precipitation, slope, and position in the drainage basin, to name a few. Stream movement is not only confined to channel flow, but also occurs horizontally in association with floodplain formation. Traditional texts show these processes as flow lines that are meant to depict movement or migration. Such a two-dimensional approach does not fully facilitate learning because it does not give students opportunities to *interact* with the various Earth processes. The recent inclusion of accompanying multimedia in other textbooks has addressed the static nature of traditional texts to some extent, but the multimedia material is generally inconsistently applied, poorly integrated with the text, and usually presented as an afterthought at the ends of chapters.

As a result of these limitations, many students find traditional texts boring and irrelevant because they do not engage *them*. This disillusionment is magnified by the fact that many students have daily experience with highly visual materials—magazines, websites, and interactive multimedia (video/computer games)—that draw them into a virtual landscape. As a result, I believe that many students, particularly those not sci-

entifically inclined, would learn best from a visually oriented textbook that is fully integrated with interactive multimedia. This approach will also increase the relevance of environmental issues to students because it will engage them in a way that makes them more aware of the outcomes of human interactions with various Earth processes.

Discovering Physical Geography: A Visually Oriented, Interactive Approach

Everyone associated with geography is fully aware that geographical literacy in the U.S. is very poor. This poor comprehension exists not only with respect to the fundamental issue of locating places, but also in understanding the age and processes associated with physical landscapes. Because many students enroll in a physical geography course only to fulfill a general education requirement in natural science, they frequently have little enthusiasm for the subject. Given this general apathy, students often focus on short-term learning to pass exams and forget much of the subject after the course ends.

Discovering Physical Geography confronts the related issues of student apathy and geographic literacy by offering students rich graphics and striking photos that depict physical processes and the natural variability of the landscape in memorable ways. The quality and breadth of the illustrations will spark students' interest and help them see the relevance of physical geography to their daily lives.

The illustrations are accompanied by a dynamic tool, *GeoDiscoveries*, an interactive, web-based multimedia resource. *GeoDiscoveries* consists of a variety of animations and simulations that will allow students to visualize and manipulate many of the factors associated with geographical processes and see the results over time and space. The multimedia will enhance students' learning as they participate more closely with geographical processes and will reinforce the integrative nature of the discipline by showing related variables in motion. This form of active learning will, in turn, help promote long-term retention of the material. The multimedia is fully integrated within the chapter text in distinct sections that direct students to the related multimedia on the website and explain to students what they should expect to learn by interacting with it.

Each *GeoDiscoveries* module on the website also includes a variety of self-assessment questions for students. Students can use these questions to test their understanding of topics, or instructors can assign them as homework. Such questions allow both students and instructors to assess learning. They also provide the foundation for exam questions that are independent of class lectures. The *GeoDiscoveries* modules should motivate more students to interact with the textbook and media because they will more readily see their connection with the course.

In short, my goal has been to create a textbook with rich graphics, fully integrated multimedia, and a variety of self-assessment tools that will engage the interest of *all* students enrolled in physical geography, not just those who are scientifically inclined or have a background in Earth science. To accomplish this goal, I provide students with vivid mental images that will help them reconstruct their knowledge for the rest of their lives in a way that substantially improves their geographical literacy. To draw the average non-majors into the book, the text is written in a direct, conversational style that connects important concepts with what students may experience. After all, most people are fundamentally interested in physical geography—otherwise, they wouldn't want to visit the Grand Canyon, the Rocky Mountains, or the Bahamas, or even to watch the Weather Channel. I want to show students the relevance of physical geography to their everyday lives, as well as increase their understanding of its integrative nature, so that they may better appreciate (for example) seasonal changes, weather patterns, or even a drive across the Great Plains into the Rocky Mountain rainshadow.

Special Features of the Text

To help students navigate their way through the book and better appreciate the nature and scope of physical geography, the chapters include a number of special and innovative features:

- *What a Geographer Sees*—This feature presents a photo and explains how a geographer would interpret the physical landscape seen in the photo. The goal of this feature is to make students realize that there is more to the physical landscape than meets the eye, which will hopefully spark their interest in what they see around them.
- *Amazing Places*—These photo essays depict striking or unusual geographic places and connect them to the chapter content. The goal of this feature is to make students more aware of how physical geography shapes the wide variety of fascinating places in the world.
- *GeoDiscoveries Multimedia*—Multimedia modules in every chapter explain to students what they can expect to see and learn as they interact with the *GeoDiscoveries* simulations and animations on the text's website. The website media also include a variety of self-assessment questions for students.
- *Visual Concept Check*—To provide students with a means of self-testing within the flow of chapter content, this feature offers a scenario with an illustration and questions to test students' understanding of key chapter concepts. Answers to the visual concept checks appear at the

end of the chapter.
- *Key Concepts to Remember*—This feature is an interim summary that appears after specific sections of the chapter to help students check their comprehension of the key concepts covered.
- *Locator Maps with Photographs*—Photographs of non-U.S. sites are accompanied by a small map indicating the location of the site shown.
- *Glossary*—Key terms are set in boldface type in the text and defined at the foot of the page for easy recognition and reference.
- *The Big Picture*—This concluding section uses a photograph to explain how the concepts learned in the chapter relate to the overall scope of physical geography, using an illustration to demonstrate the relationship.
- *Summary of Key Concepts*—The main points of the chapter are summarized.
- *Check Your Understanding*—Self-assessment questions at the end of the chapter allow students to test their comprehension.

Acknowledgments

A project of this scope naturally required the help and support of a number of people. I would first like to thank Michigan State University and the friendly people at John Wiley & Sons, Inc., for their faith in my vision and for giving me the freedom to see it through. In particular, Ryan Flahive deserves special thanks for recruiting me, backing me up many times, making a number of good things happen along the way, and generally being a good friend. Lynn Pearlman, David Shaw, and Dr. Lawrence McGlinn at SUNY New Paltz did an outstanding job with media development and integration. Major thanks also to Emily Streutker, who oversaw the evolution of a creative marketing strategy, and to the Wiley sales reps who enthusiastically carried it out.

In the context of the nuts and bolts of the book's evolution, I wish to specifically thank my two development editors, David Chelton and Suzanne Thibodeau. David helped me a lot in the middle phase of the effort and taught me how to write a textbook. Suzanne worked with me in the beginning and final phases, and I simply do not have the words to convey my gratitude for her support and guidance. She became a great friend who counseled and consoled me in countless other ways. Thank you, Suzanne!

Many talented people helped me achieve my goal of a visually oriented book. For the striking photographs, I am especially indebted to Jennifer MacMillan for her excellent judgment; I very much enjoyed working with her and appreciated her help immensely. My thanks also go to Elyse Rieder for her superb photo research skills; to Sandra Rigby for the rich illustration program; and to Harry Nolan, Hope Miller, and Maddy Lesure for the beautiful design that supports all our efforts. For the expert coordination of all the players involved in producing the book, I am extremely grateful to Lisa Wojcik.

I also want to extend my thanks to the many geographers

and scientists who are doing the research that needs to be done in this complex time of Earth history. Their research is crucial to a better understanding of our world and for the development of sustainable land-use practices. Many members of this team served as reviewers for this book and I want to thank them for taking the time out of their very busy schedules to help out. In particular I wish to thank Dr. Miriam H. Hill at Jacksonville State University for her thoughtful and thorough review of the line art. Other reviewers include:

John All, *Western Kentucky University*
John Anderton, *Northern Michigan University*
Jake Armour, *University of North Carolina, Charlotte*
Barbara Batterson-Rossi, *Cuyamaca College (California)*
Kevin Baumann, *Indiana State University*
Sheryl Beach, *George Mason University*
Jason Blackburn, *Baton Rouge Community College*
Greg Bohr, *California Polytechnic State University, San Luis Obispo*
Margaret Boorstein, *Long Island University, C.W. Post*
William Budke, *Ventura College (California)*
Michaele Ann Buell, *Northwest Arkansas Community College*
David Cairns, *Texas A&M University*
Tom Carlson, *University of Washington, Tacoma*
Nicole Cerveny, *Mesa Community College*
Philip Chaney, *Auburn University*
Richard Cooker, *Kutztown University (Pennsylvania)*
Ron Crawford, *University of Alaska, Anchorage*
James Davis, *University of Utah*
Lisa DeChano, *Western Michigan University*
Carol DeLong, *Victor Valley College (California)*
Mike DeVivo, *Grand Rapids Community College*
Lynn Fielding, *El Camino College*
Donald Friend, *Minnesota State University, Mankato*
Colleen Garrity, *SUNY Geneseo*
Alan Gaugert, *Glendale Community College (Arizona)*
David Goldblum, *University of Wisconsin, Whitewater*
John Greene, *University of Oklahoma, Norman*
Duane Griffin, *Bucknell University*
Chris Groves, *Western Kentucky University*
Curt Holder, *University of Colorado, Colorado Springs*
David Harms Holt, *Miami University of Ohio*
Michael Holtzclaw, *Central Oregon Community College*
Robert Hordon, *Rutgers University*
Hixiong (Shawn) Hu, *East Stroudsburg University (Pennsylvania)*
Peter Jacobs, *University of Wisconsin*
Scott Jeffrey, *Community College of Baltimore County, Catonsville*
Kris Jones, *Long Beach City College*
Stacy Jorgenson, *University of Hawaii, Manoa*
Theron Josephson, *Brigham Young University, Idaho*
Trudy Kavanagh, *University of Wisconsin, Oshkosh*
Ryan Kelly, *Lexington Community College*
Joseph Kerski, *U.S. Geological Survey*
Eric Keys, *Arizona State University*
Marti Klein, *Saddleback College (California)*

Dafna Kohn, *Mt. San Antonio College (California)*
Jean Kowal, *University of Wisconsin, Parkside*
Jack Kranz, *California State University, Northridge*
Barry Kronenfeld, *George Mason University*
Steve LaDochy, *California State University, Los Angeles*
Jeff Lee, *Texas Tech University*
Elena Lioubimtseva, *Grand Valley State University (Michigan)*
James Lowry, *Stephen F. Austin State University*
David Lyons, *Century College (Minnesota)*
Michael Madsen, *Brigham Young University, Idaho*
Emmanuel Mbobi, *Kent State University, Stark*
Christine McMichael, *Morehead State University*
Beverly Meyer, *Oklahoma Panhandle State University*
Peter Mires, *Eastern Shore Community College (Virginia)*
Laurie Molina, *Florida State University*
Christoper Murphy, *Community College of Philadelphia*
Steven Namikas, *Louisiana State University*
Andrew Oliphant, *San Francisco State University*
Darren Parnell, *Kutztown University (Pennsylvania)*
Charlie Parson, *Bemidji State University (Minnesota)*
Brooks Pearson, *University of West Georgia*
Robert Pinker, *Johnson County Community College (Kansas)*
Kevin Price, *University of Kansas*
David Privette, *Central Piedmont Community College (North Carolina)*
Carl Reese, *University of Southern Mississippi*
David Sallee, *University of North Texas*
Peter Scull, *Colgate University*
Glenn Sebastian, *University of Southern Alabama*
John Sharp, *SUNY New Paltz*
Wendy Shaw, *Southern Illinois University, Edwardsville*
Andrew Shears, *Kent State University*
Binita Sinha, *Diablo Valley College (California)*
Brent Skeeter, *Salisbury University (Maryland)*
Thomas Small, *Frostburg State University (Maryland)*
Lee Stocks, *Kent State University*
Sam Stutsman, *University of South Alabama*
Aondover Tarhule, *University of Oklahoma, Norman*
Roosmarijn Tarhule-Lips, *University of Oklahoma, Norman*
Thomas Terich, *Western Washington University*
Donald Thieme, *Georgia Perimeter College*
U. Sunday Tim, *Iowa State University*
Erika Trigoso, *University of Arizona*
Alice Turkington, *University of Kentucky, Lexington*
Lensyl Urbano, *University of Memphis*
David Welk, *State Center Community College District, Clovis Center (California)*
Forrest Wilkerson, *Minnesota State University, Mankato*
Dennis Williams, *Southern Nazarene University (Oklahoma)*
Joy Wolf, *University of Wisconsin, Parkside*
Lin Wu, *California State Polytechnic University, Pomona*
Ken Yanow, *Southwestern College, Chula Vista*
Brent Yarnal, *Pennsylvania State University*
Hengchun Ye, *California State University, Los Angeles*
Zhongbo Yu, *University of Nevada, Las Vegas*
Guoqing Zhou, *Old Dominion University*
Yu Zhou, *Bowling Green State University*

Tongxin Zhu, *University of Minnesota, Duluth*
Charles Zinser, *SUNY Plattsburgh*
Craig ZumBrunnen, *University of Washington*

Lastly, and most importantly, I wish to express my heart-felt thanks to my family for their support. My wife, Jennifer, is a wonderful and beautiful person who has enriched my life in ways that I cannot describe. She urged me to write this book in the early phases and has patiently supported me through the many ups and downs along the way. My children, Hannah and Rosie, have consistently given me their affection throughout this process and I love them dearly. I hope this effort inspires them in some way to reach for things that may seem at first glance to be beyond their reach.

Alan F. Arbogast

Teaching and Learning Resources

Discovering Physical Geography is supported by a comprehensive package of teaching and learning resources that includes an extensive selection of print, visual, and electronic materials. Additional information about these resources, including prices and ISBNs for ordering, can be obtained by contacting John Wiley & Sons. You can find your local representative and his or her contact information using our "rep locator" website: www.wiley.com/college/rep.

RESOURCES THAT HELP TEACHERS TEACH

LECTURE LAUNCHER VIDEOS AND ANIMATIONS

This new collection of video and animation lecture launchers allows instructors to provide a visual context for key concepts, ideas, and terms that they plan to introduce during their lectures. These videos are available in a DVD format that is optimized for in-class presentation as well as in a streaming format that is available online. Streaming videos will also be made available to students in the context of WileyPLUS assignments that can be graded online and added to the WileyPLUS instructor gradebook.

INSTRUCTOR'S WEBSITE (WWW.WILEY.COM/COLLEGE/ARBOGAST)

This comprehensive website includes numerous resources to help instructors enhance their current presentations, create new presentations, and employ our pre-made PowerPoint presentations. These resources include:

- **Image Gallery.** Online electronic files for the line illustrations and maps in the text, which the instructor can customize for presenting in class (for example, in handouts, overhead transparencies, or PowerPoints).
- A complete collection of **PowerPoint presentations** available in beautifully rendered, 4-color format. The presentations have been sized and edited for maximum effectiveness in large lecture halls.

- A comprehensive electronic **Test Bank** with multiple-choice, fill-in, matching, and essay questions that is distributed via the secure Instructor's website and can be saved into all major word-processing programs.
- A comprehensive collection of **animations and videos.**

INSTRUCTOR'S MANUAL AND TEST BANK

The Instructor's Manual includes a print version of the Test Bank, chapter overviews and outlines, and lecture suggestions. The Test Bank contains multiple-choice, fill-in, matching, and essay questions. Both the Instructor's Manual and Test Bank are also distributed via the secure Instructor's website as electronic files and can be saved into all major word-processing programs.

WILEY FACULTY NETWORK

This peer-to-peer network of faculty is ready to support your use of online course management tools and discipline-specific software/learning systems in the classroom. The Wiley Faculty Network will help you apply innovative classroom techniques, implement software packages, tailor the technology experience to the needs of each individual class, and provide you with virtual training sessions led by faculty for faculty. For more information, go to www.Faculty ResourceNetwork.com.

COURSE MANAGEMENT

Online course management assets are available to accompany *Discovering Physical Geography.*

RESOURCES THAT HELP STUDENTS LEARN

STUDENT COMPANION WEBSITE (WWW.WILEY.COM/COLLEGE/ARBOGAST)

This easy-to-use and student-focused website helps reinforce and illustrate key concepts from the text. It also provides interactive media content that helps students prepare for tests and improve their grades. This website provides additional resources that complement the textbook and enhance students' understanding of Physical Geography:

- **Interactive Animations and Exercises** allow students to explore key concepts from the text.
- **Flashcards** offer an excellent way to drill and practice key concepts, ideas, and terms from the text.
- *GeoDiscoveries* **Modules** allow students to explore key concepts in greater depth using videos, animations, and interactive exercises.
- **Chapter Review Quizzes** provide immediate feedback to true/false, multiple-choice, and short-answer questions.
- **Interactive Drag-and-Drop Exercises** challenge students to correctly label important illustrations from the textbook.
- **Annotated Web Links** put useful electronic resources into context.
- The **Learning Styles Survey** will help students identify their unique way of understanding and processing information.
- With the **Study Skills Chart**, students can create a customized learning plan that best suits their personal learning style.

A GUIDE TO THE FEATURES

WHAT A GEOGRAPHER SEES—This feature presents a photo and explains how a geographer would interpret the physical landscape seen in the photo. The goal of this feature is to make students realize that there is more to the physical landscape than meets the eye, which will hopefully spark their interest in what they see around them.

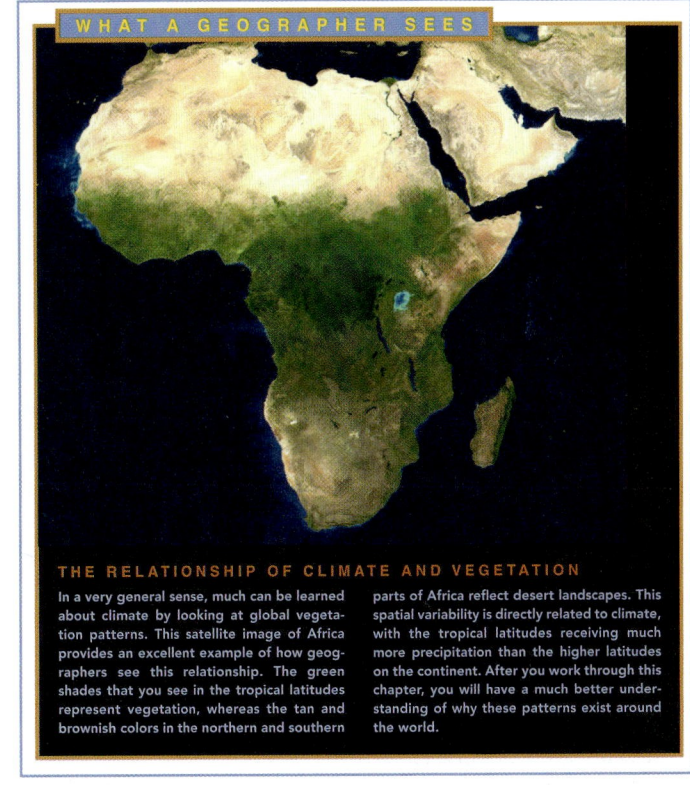

WHAT A GEOGRAPHER SEES

THE RELATIONSHIP OF CLIMATE AND VEGETATION

In a very general sense, much can be learned about climate by looking at global vegetation patterns. This satellite image of Africa provides an excellent example of how geographers see this relationship. The green shades that you see in the tropical latitudes represent vegetation, whereas the tan and brownish colors in the northern and southern parts of Africa reflect desert landscapes. This spatial variability is directly related to climate, with the tropical latitudes receiving much more precipitation than the higher latitudes on the continent. After you work through this chapter, you will have a much better understanding of why these patterns exist around the world.

AMAZING PLACES—These photo essays depict striking or unusual geographic places and connect them to the chapter content. The goal of this feature is to make students more aware of how physical geography shapes the wide variety of fascinating places in the world.

AMAZING PLACES

ATACAMA DESERT

The driest place on Earth is the Atacama Desert on the west coast of South America. This region is on the northern margin of the cold midlatitude desert climate (BWk) and extends about 1000 km (600 mi) from the southern border of Peru into northern Chile. The Atacama Desert is considered to be the driest place on Earth because there are areas within the desert that have never received any measurable precipitation in recorded history.

A typical scene in the Atacama Desert. Note the very sparse vegetation in this extremely dry climate.

Several interrelated factors explain the lack of precipitation in the Atacama Desert. The region lies on the west side of the Andes Mountains, which are a significant orographic barrier to the moisture-laden easterly winds that blow across equatorial South America. Thus, the Atacama Desert lies in the rainshadow of the Andes, where descending air under high pressure compresses and warms such that precipitation is rare. In addition to being in the Andean rainshadow, the Atacama Desert lies immediately east of the cold Humboldt Current (also called the Peru Current). Given the low temperature of this current, the air above it is stable, with very little convection. In short, the Atacama Desert is so dry because it lies in the lee of a major mountain range and to the east of a cold ocean current.

This map shows the geographic relationship of the Atacama Desert, the Humboldt Current, and the Andes Mountains.

GEODISCOVERIES MULTIMEDIA—Multimedia modules in every chapter explain to students what they can expect to see and learn as they interact with the *GeoDiscoveries* simulations and animations on the text's website. The website media also include a variety of self-assessment questions for students.

KEY CONCEPTS TO REMEMBER—This feature is an interim summary that appears after specific sections of the chapter to help students check their comprehension of the key concepts covered.

KEY CONCEPTS TO REMEMBER ABOUT THE VARIABLES THAT INFLUENCE LARGE-SCALE WINDS

1. Air generally flows from high pressure to low pressure.

2. Large-scale atmospheric circulation is caused by the unequal heating of the tropics and poles. The process of air flow begins at the Equator because of convection.

3. The speed of air flow is determined by the pressure gradient force. The steeper the gradient, the stronger the winds.

4. The Coriolis force causes air moving toward the Equator to be deflected to the right in the Northern Hemisphere and to the left in the Southern Hemisphere.

5. Features on the Earth's surface, such as mountains, forests, and buildings, create a frictional force that acts opposite to the wind's direction.

6. The combined effect of the pressure gradient force, Coriolis force, and frictional force causes a spiral motion of air in both low and high pressure systems.

VISUAL CONCEPT CHECK—To provide students with a means of self-testing within the flow of chapter content, this feature offers a scenario with an illustration and questions to test students' understanding of key chapter concepts. Answers to the visual concept checks appear at the end of the chapter.

VISUAL CONCEPT CHECK 8.2

This Doppler radar image shows a line of thunderstorms along a cold front in Kentucky and Tennessee. Which one of the following choices best explains what occurred along the front?

a) The warmest air temperatures were northwest of the front.

b) The cold front caused mT air to rapidly uplift along the front.

c) The air southeast of the front consisted of cP air.

d) The most humid air was located northwest of the front.

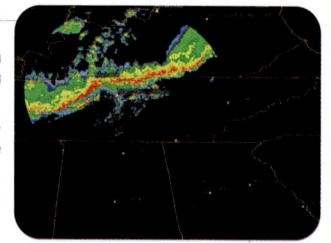

ANSWERS TO VISUAL CONCEPT CHECKS

Visual Concept Check 8.1

The answer is *d*. The cold front extends from western Michgan to Central Texas. You can tell the location of the cold front because of a sharp temperature difference between the east and west sides of the frontal boundary.

Visual Concept Check 8.2

The answer is *b*. The cold front caused mT air to rapidly uplift. Remember that a cold front plows aggressively into relatively warm, moist air. This interaction causes the mT air to lift rapidly.

Visual Concept Check 8.3

One would expect Hurricane Floyd to move to the west because the prevailing winds are easterly at low latitudes. At some point, the hurricane would begin to interact with the midlatitude westerlies, causing the hurricane to veer northeastward. In fact, Hurricane Floyd migrated in just this manner, moving westward toward Florida. Just before it reached Florida, however, it veered to the northeast and just grazed the mainland of the United States.

LOCATOR MAPS WITH PHOTOGRAPHS—
Photographs of non-U.S. sites are accompanied by a small map indicating the location of the site shown.

THE BIG PICTURE—This concluding section uses a photograph to explain how the concepts learned in the chapter relate to the overall scope of physical geography, using an illustration to demonstrate the relationship.

GLOSSARY—Key terms are set in boldface type in the text and defined at the foot of the page for easy recognition and reference.

Fronts

We have just seen that large bodies of air with distinct characteristics originate over specific places and then move over parts of the North American continent. These air masses have definable boundaries at the surface of the Earth, which are called *fronts*. The dominant front in the Northern Hemisphere midlatitudes is the polar front, which, if you recall, is the boundary between cold, dry (cP) air to the north and warmer, moist (mT) air to the south. These contrasting air masses flow parallel to one another along a **stationary front** most of the time. Sometimes, however, atmospheric conditions arise that cause one particular air mass to advance into another along distinct frontal boundaries. It's helpful to think of these advancing air masses as being analogous to a military force that is advancing into a particular area, with the lead columns of the force being the front. This part of the chapter focuses on these atmospheric boundaries and the types of precipitation that occur along them as air is uplifted. Recall from Chapter 7 that this type of uplift is called *frontal uplift*.

Warm Fronts **Warm fronts** occur in places where warm air advances into relatively cool air. This interaction causes the warmer air to slowly slide over the top of the underlying cooler air (Figure 8.2). This process of gradual overriding of cooler air actually occurs in the upper atmosphere ahead of the surface warm front and causes the lifting mT air to cool adiabatically. As the air cools, clouds form, beginning with high-level cirrus clouds at the deepest penetration of upper level warm air

Stationary front *A boundary where contrasting air masses are flowing parallel to one another.*

Warm front *A frontal boundary where warm air is advancing into relatively cool air. This front is typically associated with slow, steady precipitation.*

The Big Picture

Now that we've discussed how the factors of insolation and net radiation influence air temperature, we can examine how various atmospheric processes operate, such as wind patterns and the development of storm systems. A good place to begin this discussion is Chapter 6, which focuses on atmospheric air pressure and circulation and the way these concepts relate to Earth/Sun relationships and global temperature patterns. This image is a nice example of this relationship. Notice the stream of clouds across the image. This line of clouds is associated with an atmospheric feature called the jet stream, which is a band of strong winds that exists in the lower stratosphere. These winds can be particularly strong in the midlatitudes when large differences exist with respect to temperature at higher and lower latitudes. This pattern, as well as a variety of others related to the flow of air in the atmosphere, will be discussed in Chapter 6. As in previous chapters, be sure to consider how all of these concepts are interrelated and how they relate to previous discussions.

Summary of Key Concepts

1. The atmosphere serves as a protective shield that filters the potentially deadly effects of ultraviolet radiation from the Sun, allowing mostly visible and infrared wavelengths to reach the Earth.

2. The lower part of the atmosphere is warmed primarily due to the greenhouse effect. This warming occurs because variable gases in the atmosphere, such as carbon dioxide and water vapor (as well as others), trap longwave radiation that is emitted by the Earth.

3. Insolation refers to the amount of solar radiation received at the top of the atmosphere. From this point, radiation follows several paths. Approximately 50% of all radiation reaches the Earth, with some arriving directly and some indirectly. The remaining radiation is either absorbed in the atmosphere, scattered by dust, or reflected directly back into space.

4. Albedo refers to the reflectivity of surfaces on Earth. Snowy surfaces have the highest albedo, whereas darker surfaces such as roads and oceans have the lowest albedo. Overall, the Earth reflects about 31% of all incoming solar radiation through albedo.

5. The global radiation budget refers to the overall balance of incoming versus outgoing radiation. The Earth receives shortwave radiation from the Sun. This radiation is either absorbed by the atmosphere, the various Earth surfaces, or reflected and scattered back to the surface or back into space. The radiation that reaches the Earth's surface is absorbed and then re-radiated as longwave radiation. The overall radiation budget must be in balance, otherwise the Earth would become progressively hotter or colder.

Check Your Understanding

1. Why is air pressure greater at low altitudes rather than high altitudes?

2. Which direction does air flow at the surface, from cyclone to anticyclone, or the reverse?

3. Which place is more likely to be a zone of convection, the ITCZ or STH? Why does this pattern occur?

4. Why is the polar front an important feature in the midlatitudes?

5. Why are Rossby waves more likely to form during the winter months than in the summer months?

6. How does Earth–Sun geometry influence the position of atmospheric pressure systems?

7. How is a land–sea circulatory system like a tropical Hadley cell?

8. How are land–sea breezes and mountain breezes similar? What conditions are required for these circulatory systems to develop?

9. Within the Atlantic gyre, what is the primary purpose of the Gulf Stream?

10. In the equatorial Pacific, the normal circulatory flow of surface waters is easterly. Why does this flow pattern normally occur?

BRIEF CONTENTS

CONTENTS

INTRODUCTION TO PHYSICAL GEOGRAPHY

I want to welcome you to this introductory textbook about physical geography. Physical geography is an exciting scientific discipline that is central to our understanding of the Earth and how it functions. Every region of the planet is part of a delicately balanced natural system that has distinctive physical characteristics, including the type of climate, weather, vegetation, soils, and landforms. These characteristics are the result of processes that, in some cases, have been operating for billions of years. In other instances, significant changes occur over the

This view of Oregon's Mt. Hood illustrates many geographical processes, including the way air flows in the atmosphere, how water is stored in the hydrosphere, the role of climate and its impact on vegetation, and the way that landscapes evolve over time.

course of a single day. Our first goal in this book is to examine these physical processes and resulting geographical patterns on the Earth. Second, we will examine the various ways that these processes affect our lives and, in turn, how we impact them. In this opening chapter we will outline the topics discussed in this book and place them in the context of the overall discipline of geography. Then we'll discuss the various components and features of the book and how they will assist with your learning.

The Scope of Geography

When people are asked to define geography, the most common answer is that it's all about the locations of countries, capital cities, rivers, and oceans. Although the location of places is certainly a part of geography, the field encompasses *much* more. In fact, you may be surprised to learn that whatever you like to do and wherever you live, geography is very much a part of your everyday life.

Geography is an ancient discipline that examines the spatial attributes of the Earth's surface and how they differ from one place to another. Derived from the Greek words for "*Earth description*," the concept of geography has likely been a central part of the human experience for tens of thousands of years. It's easy to imagine, for example, that prehistoric hunters and gatherers were intimately aware of their surroundings, including the location and character of forests, streams, lakes, berry patches, animal herds, and competing groups of people. In short, an understanding of local geography would have been absolutely essential to sustain life. It would also have been necessary to be able to describe those patterns for future generations so they, in turn, could thrive.

So, for thousands of years at least, geography was a descriptive discipline that focused on the generalized location and character of places and things. Slowly, however, geography became an academic discipline with numerous specialized subfields. Scientists became experts in areas such as geology, meteorology, ecology, and human cultural differences. Interest in geography grew especially between the 15th and 19th centuries when explorers like Christopher Columbus, Ferdinand Magellan, James Cook, Charles Darwin, and Lewis and Clark began to investigate parts of the world that were previously unknown to people of European descent (including Americans). These explorers, as well as many others, brought detailed descriptions of exotic places and animals to a keenly interested public. The new knowledge and perspectives gained from this time were a major driving force for the development of the modern world.

The trend toward specialization in geography has continued to the present time. Most geographers consider themselves first of all to be either a physical or human specialist. Within these two broad fields are a range of geographical subdisciplines, as shown in Figure 1.1a. Although each of these subfields has a unique focus, such as soils or agricultural land use, you should realize that geographers draw from many of these subfields when they analyze any particular geographical pattern. For example, to fully understand human settlement patterns in Africa, it is important to consider the interaction of subfields such as climatology, soils, and vegetation (biogeography), to name a few. In turn, to understand the nature of soils in any given place, you must consider the effects of climatology, vegetation, geomorphology, and perhaps even regional cultural practices (Figure 1.1b).

Although we are concerned with physical geography in this book, it is important to know that all subfields of geography share a common methodology that makes them part of the overall discipline. Geographers use a method known as **spatial analysis**, which attempts to explain patterns or distributions of specific variables over physical space. Most geographers are fundamentally interested in knowing two things: *Where* and *why*? In other words, geographers examine why a specific environmental or cultural variable in a specific place or region has one particular set of characteristics, whereas the same variable someplace else has different characteristics.

For example, consider for a moment the concept of cultural diversity. The Middle East is a region of the Earth that is culturally diverse, with many different religious sects, dialects, and tribal identities. Although this diver-

Spatial analysis *A method of analyzing data that specifically includes information about the location of places and their defining characteristics.*

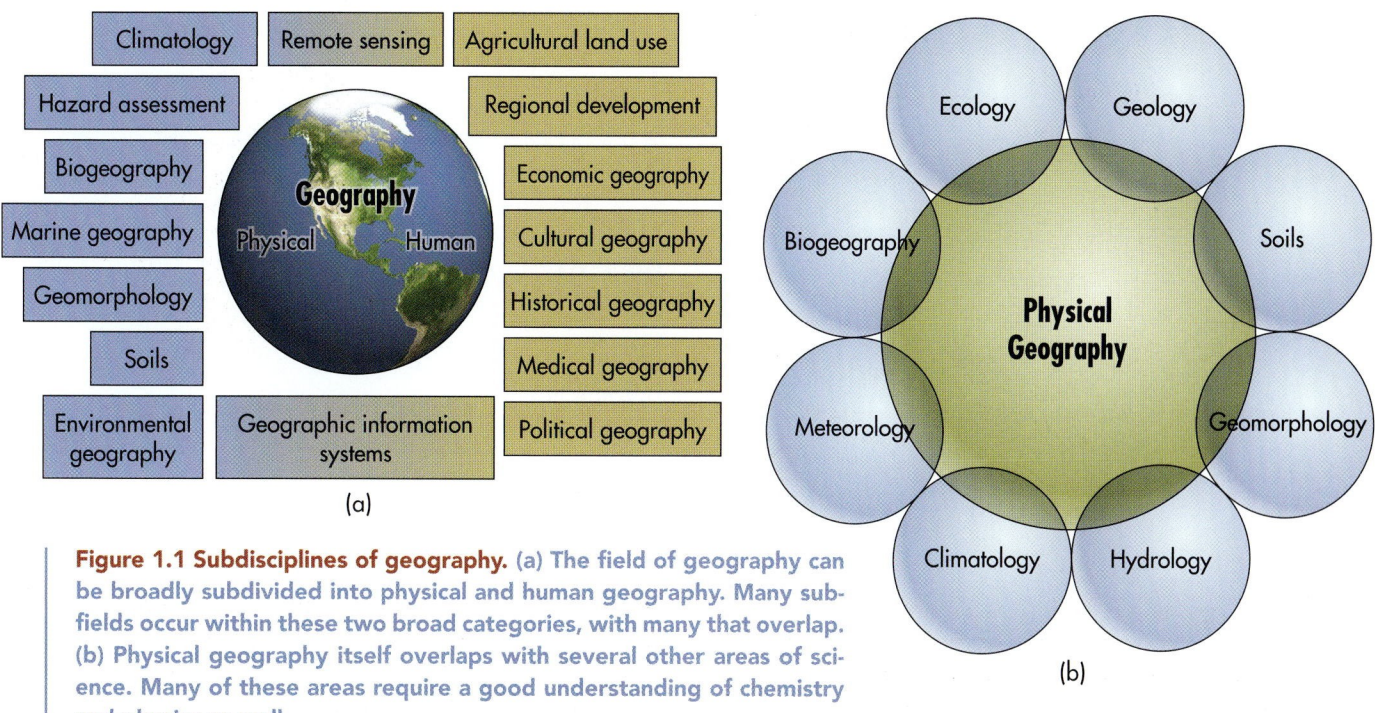

Figure 1.1 Subdisciplines of geography. (a) The field of geography can be broadly subdivided into physical and human geography. Many sub-fields occur within these two broad categories, with many that overlap. (b) Physical geography itself overlaps with several other areas of science. Many of these areas require a good understanding of chemistry and physics as well.

sity has produced a rich heritage, it has also resulted in a great deal of conflict between and among various cultural groups. A cultural geographer can examine the spatial distribution of the many groups in the area (in other words, *where are they?*), as well as study why people in one place differ in, say, their political or religious views, from people someplace else. The geographer might look for similarities (such as language) among groups across physical space, which might explain why certain people align themselves politically with others. In the course of this study, the geographer would have integrated several variables into one picture, including language, religion, history, and climate. Such a study might contribute to an understanding of why people differ across this region and why sources of conflict remain.

Defining Physical Geography

Now that we have examined the basic nature of geography as a broad field, let's focus on a more comprehensive definition of physical geography. You probably already have some kind of an interest in physical geography, whether you know it or not. For example, have you ever wondered why violent storms occur? You may know that tornadoes frequently occur in the central United States, especially in the springtime (Figure 1.2a). Do you know why? Maybe you wonder why large mountains are found in Washington but not in Texas (Figure 1.2b). Perhaps you've heard about the Sahara Desert and wonder why it's so dry there and why some of it is covered with sand dunes (Figure 1.2c). Like many people, you might enjoy

the seashore and wonder why nice beaches form in some places (Figure 1.2d) but not in others. If you've asked yourself questions like these, then you're probably interested in physical geography at some level. The fact is, most people are; they just don't realize it.

Simply stated, **physical geography** involves the spatial analysis of the various physical components and natural processes of the Earth. Some examples of the Earth's physical components are air, water, rocks, vegetation, and soil. The term **process** broadly refers to a series of actions that can be measured and which produce a predictable end result. In physical geography, these processes are fundamentally products of the energy that flows from the Sun to the Earth in the form of solar radiation. Once this energy reaches the Earth, it then flows from one place to another on the planet in various forms. Some examples of natural processes directly related to the flow of solar radiation (Figure 1.3) are the circulation of the atmosphere (Chapter 7), the distribution of vegetation (Chapter 10), the formation of soils (Chapter 11), and the movement of water in the air, in streams, and collecting in lakes (Chapters 5 and 16). As you'll see throughout the book, many processes behave in an interconnected way as *natural systems* where one environmental variable has a direct impact on another.

Physical geography *Spatial analysis of the physical components and natural processes that combine to form the environment.*

Process *A naturally occurring series of events or reactions that can be measured and which result in predictable outcomes.*

Figure 1.2 Some elements of physical geography. (a) A tornado in the central United States (see Chapter 8). (b) The Olympic Mountains overlooking Puget Sound in Washington (see Chapter 13). (c) Sand dunes in the Sahara Desert (see Chapters 9 and 18). (d) Coastline at Big Sur in northern California (see Chapter 19).

(a)

(b)

(c)

(d)

(a)

(b)

(c)

(d)

Figure 1.3 Some energy flows on Earth. (a) The Earth receives its energy from the Sun in the form of solar radiation (see Chapters 3 and 4). (b) The atmosphere circulates energy around the Earth, as can be seen in this stream of clouds (see Chapter 7). (c) Some energy is transferred when water flows from the atmosphere to the Earth as rain (see Chapter 5). (d) Some of the energy on the surface of the Earth is transferred through the flow of water (see Chapter 16).

To see an example of how environmental variables relate to one another, let's imagine that you are interested in the spatial distribution of rivers in the United States. One way to see the geographical concentration of rivers in the country is with a map showing the location of gauging stations, which are places where the U.S. Geological Survey continuously monitors the quantity of water in the streams (Figure 1.4). Note that the eastern part of the country contains many more gauging stations than the interior west in places like Nevada and Utah. This pattern reflects the fact that the eastern part of the U.S. includes more streams than the interior west.

The question a geographer would ask about this is: *Why* are there more rivers in the eastern U.S. than in the western part of the country? This question leads us to examine the interconnected nature of processes associated with water movement and its spatial distribution in rivers. The simple reason for this geographical distribution is that more precipitation falls in the eastern part of the U.S. than in the western states. Later you will study why this geographical variability occurs, but for now it is sufficient to say that it exists because the atmosphere over the eastern U.S. typically contains more water than in the interior parts of the western states. Thus, more water flows from

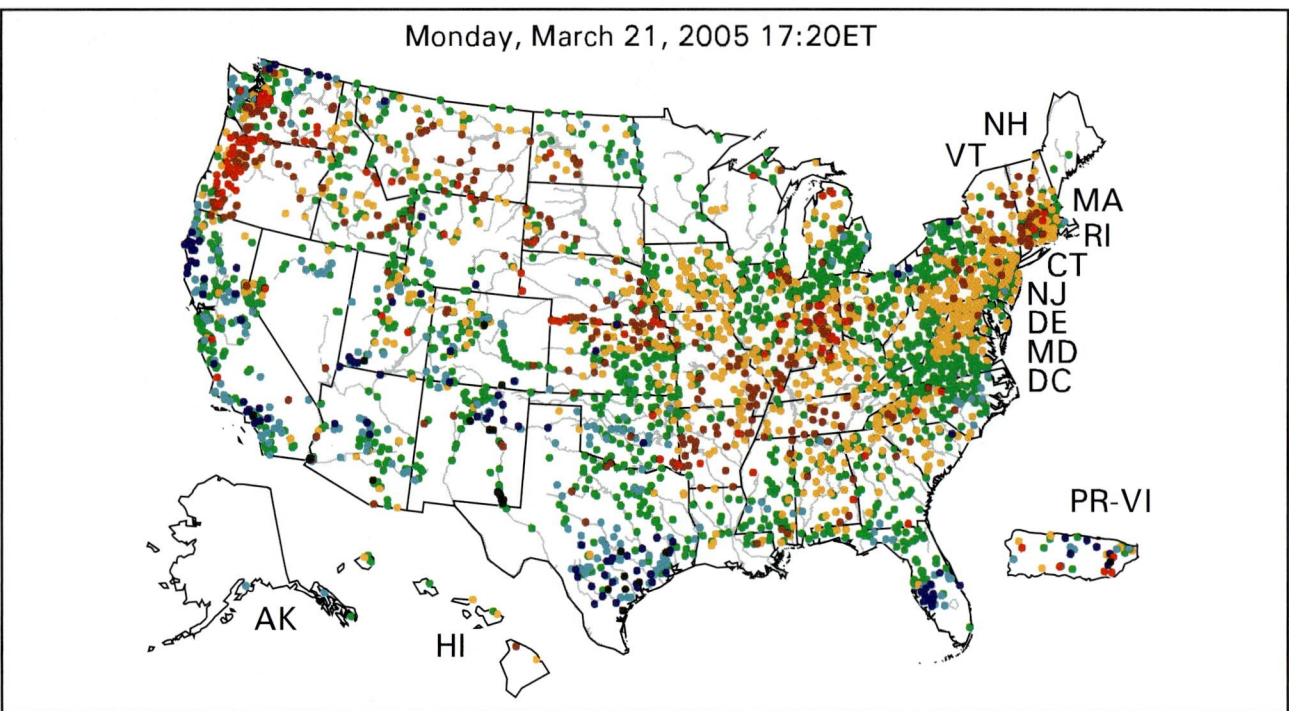

Monday, March 21, 2005 17:20ET

NH
VT
MA
RI
CT
NJ
DE
MD
DC
PR-VI
AK
HI

Figure 1.4 Location of stream gauging stations in the United States. The dense concentration of stations in the eastern half of the country reflects the fact that there are many more streams in the eastern U.S. than in the interior western states. The map is color coded to illustrate the amount of water in any given stream on March 21, 2005, relative to its average flow on that day in past years. For example, red dots reflect very low water levels, whereas dark blue/purple dots mean that stream flow was relatively high on that day.

the atmosphere to the ground as precipitation in the eastern U.S. than in the western states. Some of this water flows directly across the Earth's surface into streams. A great deal of it slowly absorbs into the ground where it is steadily released into streams. As a result of these interconnected processes, the eastern U.S. contains more streams than the western part of the country.

The Earth's Four Spheres

As you can imagine, a huge number of component/process combinations exist for geographical study. In physical geography, these various combinations can be grouped into the four "spheres" on Earth (Figure 1.5):

1. **Atmosphere**—The atmosphere is the gaseous shell that surrounds the Earth. This sphere is composed of many critical components essential to life, such as oxygen, carbon, water, and nitrogen, to name a few, that flow around the Earth.

2. **Lithosphere**—The lithosphere is the solid part of the Earth, including soil and minerals. A good example of process in this sphere is the way in which water, minerals, and organic matter flow in the outermost layer of the Earth to form soil. This sphere provides the habitat and nutrients for many life forms.

3. **Hydrosphere**—The hydrosphere is the part of the Earth where water, in all its forms (solid ice, liquid water, and gaseous water vapor), flows and is stored. This sphere is absolutely critical to life and is one that humans regularly interact with—for example, through irrigation and navigation.

4. **Biosphere**—The biosphere is the living portion of the Earth and includes all of the plants and animals (including humans) on the planet. Various components of this sphere regularly flow from one place to another, both on a seasonal basis and through human intervention. Humans interact with this sphere in a wide variety of ways, with agriculture being a very prominent example.

Atmosphere *The gaseous shell that surrounds the Earth.*

Lithosphere *A layer of solid, brittle rock that comprises the outer 70 km (44 mi) of the Earth.*

Hydrosphere *The part of the Earth where water, in all its forms, flows and is stored.*

Biosphere *The portion of the Earth and atmosphere that supports life.*

Atmosphere

Hydrosphere

Earth

Lithosphere

Biosphere

Figure 1.5 The four Earth spheres. Each sphere encompasses a major component of the Earth's natural environment.

These four spheres overlap to form the natural environment that makes Earth a unique place within our solar system. Physical geography examines the spatial variation within these spheres, the interrelationships among them, and the manner in which components flow from one sphere to another. This discipline is a science because research is conducted within the framework of the scientific method, which is the systematic pursuit of knowledge through the recognition of a problem, the formulation of hypotheses, and the testing of hypotheses through the collection of data by measurement, observation, and experiment. Among many other things, physical geographers collect data from the atmosphere, rocks, soils, ice cores, satellite images, the Earth's magnetic field, and even other planets (Figure 1.6). In order to understand how geographical patterns develop, it is essential that scientists understand physical laws and have the ability to mathematically analyze and compare them.

Many of the scientific analyses associated with physical geography are driven by the growing impact that human activities have within and among the Earth spheres.

Given the nature of human impact on the natural environment, physical geography is at the forefront of research on many environmental issues that face the world today. These issues include (to name a very few):

- **Global climate change**—Human activities are increasing the levels of carbon dioxide in the atmosphere. Abundant evidence suggests that this relationship may be contributing significantly to global climate change.

- **Deforestation**—The clearing of the tropical rainforests is occurring at an alarming rate, leading to soil erosion, loss of wildlife habitat, and species extinctions.

- **Farmland loss**—Due to increasing global population, farmland is being converted to zones of economic development and residential housing. This loss of farmland is resulting in more intensive farming of agricultural soils still in use, which increases the risk of soil erosion and pollution due to extensive application of fertilizers and pesticides.

(a)

(b)

(c)

(d)

Figure 1.6 Examples of collecting scientific data about the Earth.
(a) Many space shuttle missions are designed to obtain measurements about the atmosphere, oceans, and the distribution and character of plants, among many other things. (b) To learn about the behavior of streams in the past, scientists study the type of sediment deposited by the stream through time. (c) One way to learn about past climate changes on Earth is to obtain samples of ancient ice on the Greenland and Antarctic ice caps. (d) New methods of surveying enable scientists to obtain accurate measurements about elevation and location.

- **Natural hazards**—Hazards occur when extreme events result in danger to humans. Examples of natural hazards include hurricanes, tornadoes, flooding, earthquakes, and volcanoes. This is a particularly important area of geographical study because as global population grows, increasing numbers of people are moving into areas that are susceptible to extreme natural events.

A recent natural disaster in the U.S. vividly illustrates the integrated nature of physical geography and the critical role that geographers play with respect to solving real-world problems. As you probably know, Hurricane Katrina devastated the Gulf Coast in the summer of 2005, causing billions of dollars of damage, much of it occurring when the city of New Orleans was extensively flooded. Before the storm reached land, geographers were at the forefront of the effort to monitor the storm's path and predict where the most significant damage would occur. Once the storm passed, geographers began conducting research on the impact that the hurricane had on a variety of issues, including the shape of the coastline, distribution of wetlands, and the regional economy, to name just a few. These studies have profound implications for future environmental decisions, politics, and economic development in the Gulf region.

Organization of This Book

The chapters in this book are organized to provide you with a good understanding of the basic concepts associated with physical geography. They contain information that ranges in scale from global to local, which will allow you to better grasp your place both in the world and even your neighborhood. Chapter 2 will focus on the various kinds of tools that geographers use in their work, such as maps, remote sensing, and geographical information systems. Chapters 3 through 5 center on our relationship with the Sun (Chapter 3), the way we receive solar radiation (Chapter 4), and how those interactions relate to temperature (Chapter 5). The processes discussed in these chapters will prepare you for the topics we'll cover in the rest of the book. Chapters 6 through 9 revolve around the atmosphere, including the way that air circulates within it (Chapter 6), precipitation processes (Chapter 7), weather systems (Chapter 8), and global climate patterns (Chapter 9). We then examine the influence of the atmosphere and its interactions with Earth's other spheres by focusing on plant geography in Chapter 10 and soils in Chapter 11.

The last eight chapters of the book deal mainly with the lithosphere and hydrosphere. Chapter 12 describes the Earth's internal structure, rock cycle, and geologic time. This discussion leads directly into Chapter 13, which focuses on the lithosphere and tectonic landforms. From there, we'll turn our attention in Chapter 14 to the way that rocks weather and how sediment moves through mass wasting processes. Chapters 15 and 16 discuss the way that water moves on the Earth and how it is stored within it. Chapter 15 focuses specifically on groundwater processes and the formation of landforms such as caves. In Chapter 16 we look at how water flows across the surface in streams and the landforms that result. The last three chapters of the book are devoted to specific geomorphic processes and the resulting landforms, including glaciers (Chapter 17), eolian (wind) processes and arid landscapes (Chapter 18), and coastal regions (Chapter 19).

Exploring Cause-and-Effect Relationships

As you work through these chapters, you will constantly see the effects of interactions among the four Earth spheres. To understand how such interactions work, consider the following question, which encompasses elements of the atmosphere and the hydrosphere:

Imagine that an extensive drought, one lasting 10 years, occurred in the eastern United States. How would the quantity of water in rivers in the region be affected?

The answer is pretty clear: the quantity of water in rivers would decrease. The reason for this drop is that a significant drought would result in less water flowing from the atmosphere (as precipitation) to the Earth. As a result less water would then be available to flow across the surface in rivers. In addition, the quantity of water stored in the ground would decrease, which would also reduce river levels because a great deal of water in streams is derived from the ground. Potential further impacts of this drought could be that some forms of vegetation might become less common or that the likelihood of fire would increase.

This book relies heavily on exploring these kinds of cause-and-effect relationships in a variety of ways. One way in which we present this material is through the traditional use of text accompanied by photographs, diagrams, and tables. Each chapter contains detailed discussions that connect important concepts to events that you may have experienced. Each chapter also ends with a visual preview of the next chapter specifically designed to show how one chapter relates to the next.

GeoDiscoveries: An Interactive Tool

These cause-and-effect discussions in the text are accompanied by graphics and photographs, as well as a more dynamic tool—interactive digital modules called *GeoDiscoveries*. These modules consist of a variety of animations and simulations that allow you to visualize and manipulate many of the factors associated with geographical processes and see the results over time and space. The animations and simulations will enhance your learning as you participate more closely with geographical processes and will reinforce the interactive nature of the discipline by showing related variables in motion. The media are integrated entirely within the chapter text as distinct sections that explain what you should expect to learn by interacting with them.

Here's a good example of how the digital modules may enhance your learning. As you may already know, rivers are bodies of water that flow from one place to another. This concept is described in great detail in Chapter 16. The discussion in Chapter 16 is accompanied by a variety of diagrams that illustrate how water flows, using flow lines and arrows embedded within them. It is also supplemented by several *GeoDiscoveries* modules that are accessed on the text's accompanying website. One of the modules in that chapter shows how streams snake across the river valley in a process called *meandering*. Through this process, the geographical position of streams actually moves through time. The *GeoDiscoveries* module shows this process in animated form, which will enable you to comprehend it better. Have a look now to see what these modules are like.

STREAM MEANDERING

An excellent way to get a preliminary feel for the interactive media in this book is by examining one of the animations. A simple one shows how streams move across the landscape. Go to the *GeoDiscoveries* website and select the **Stream Meandering** module. This animation allows you to visualize the way that streams move across the landscape through the process of meandering. The animation illustrates how meander bends form over time in an initially straight river and also shows the formation of a feature called an oxbow

lake. Throughout the book, you'll frequently encounter *GeoDiscoveries* modules like this one that will give you a better idea of how geographical processes actually move. Some of the modules will be simple ones like this, which require only that you carefully watch them. Others are simulations that allow you to manipulate variables to see how outcomes change. Regardless of the type of media, each module will contain a series of questions that can be used to test your understanding of the concepts after you watch the animation or simulation.

Focus on Geographical Literacy

In addition to improving your overall understanding of physical geography, a secondary goal of this book is to enhance your geographical literacy. It's common knowledge that the overall geographical literacy of average Americans is very poor. How many Americans can identify, for example, the countries of the Middle East, where so much of our national focus is presently centered? This illiteracy not only extends to the world at large, but applies to the U.S. as well. In a recent poll, for example, Americans chose more than 30 different states (of the 50 in the country) as being the state of New York.

Geographical literacy also involves knowing where distinct physical regions exist on Earth, such as the Sahara Desert or the Himalaya Mountains, and why they exist. In this context, you will see that discussions include maps of the places described. You'll also find a number of *Amazing Places* features throughout the book. These features focus on places that are geographically unique, such as the Grand Canyon or Death Valley, which specifically relate to the material discussed in the chapter.

I also hope that your *visual* geographical literacy will improve by using this text. In other words, this book is designed to sharpen *your* eye so that when *you* see things on the landscape, which you may have previously ignored, you might better appreciate them and why they exist. Two features—*What a Geographer Sees* and *Visual Concept Check*—are specifically designed to help you improve this aspect of your geographical literacy. The *What a Geographer Sees* feature allows you to see the physical landscape through the eyes of a geographer. In other words, you will see the kinds of visual cues that geographers use to interpret why physical patterns exist, which will hopefully train your eye to look for similar visual cues. The *Visual Concept Check* features are placed after key topic discussions so you can test your understanding of those topics immediately after encountering them.

Physical Geography Is Interesting, Exciting, and Relevant to Your Life

As with any new endeavor you pursue, you can expect that improving your understanding of physical geography may be difficult at times. Many students initially avoid this subject because they feel that geography is boring, or they are intimidated by science, or see no relevance of geography to their lives. If you genuinely give this subject a chance, however, you will see that physical geography is indeed relevant to *your* everyday life and, most of all, is interesting and even exciting. At a fundamental level, how else can you explain the popularity of weather and nature programs, national and state parks, the travel industry, mountains, or beautiful coastlines? Why do people go on exotic vacations if not, in part, to enjoy the uniqueness of the physical landscape in new places? With a greater understanding of physical geography, you will appreciate those trips more. You may even appreciate the immediate world around you more.

In addition, with an understanding of physical geography you will be able to make *informed* decisions when you are confronted with important environmental issues in your lifetime. Thus, you will become a better citizen, one who is capable of protecting the best interests of your family and community. For instance, at some point in the future you may be confronted with a choice of where to place a landfill in your city or town. In order to make the most informed decision, one that perhaps ensures the safety of your drinking water, it will be important to understand the geology of the site, the character of soils, and the way water moves through the ground and is stored within it.

It's also possible that you may even decide after using this text that you want to become a physical geographer, as many people have. A number of excellent, well-paying jobs can be obtained with a specialization in physical geography, including environmental analysts, cultural resource managers, conservation agents, teachers, meteorologists, and landscape architects to name a very few. If you decide to purse such a career, you'll find that collecting, analyzing, and reporting about geographical data is very rewarding. The Earth is a beautiful and complex place. Regardless of whether you are a geographer or not, you are about to understand it better. Enjoy the ride!

The Big Picture

It's now time to move more deeply into the nature of physical geography. We begin this discussion with an overview of the tools that physical geographers use to obtain measurements about the Earth. This is important because it will give you a feel for how scientists came to understand how the Earth functions. An excellent example of a tool that physical geographers use is satellite remote sensing. The basic premise of this data-gathering method is that scientists can view the Earth from space with satellites and monitor changes that occur across the planet. The image shown here is a satellite image of the Baltimore, Maryland area. It is an infrared image, which means that it's measuring energy that you can't see with your naked eye. The gray areas are urban areas associated with the Baltimore metropolitan area, whereas the red zones are places where vegetation is especially dense, such as in a park or in suburban zones. Black is water (part of Chesapeake Bay). The next chapter will describe the different kinds of tools that geographers use and will provide examples of the data that can be collected.

CHAPTER TWO

THE GEOGRAPHERS' TOOLS

Before you begin to study the various concepts associated with physical geography, it's

important to examine the tools that geographers use to locate places and to gather and display

spatial information. Some of these tools, such as maps, are probably familiar because you've

used them yourself. Other aids, such as remote sensing, Geographic Information Systems (GIS),

Geographic technology allows us to view landscapes in unique and insightful ways. This image combines a satellite image with a digital elevation model to show the relationship of urban areas (in gray shades) and topography near Los Angeles, California. Of particular interest is the nearby location of the San Andreas Fault, which is the linear feature that angles from the upper center of the image to the lower right.

and Global Positioning Systems (GPS), are relatively new and not well understood by most

people. This chapter focuses on these tools, with the goal of providing a better understanding of

how geographers do their work. The material in this chapter will also help you understand how

much of the information presented in this text was obtained and what it means.

The Geographic Grid

One of the most basic things that geographers consider is the location of places on Earth. The simplest way that location can be determined is by a relative comparison to some other place. For example, a geographer might identify Los Angeles as being south of San Francisco, or that Africa is on the eastern side of the Atlantic Ocean. You've no doubt made such a comparison yourself, perhaps by indicating that you live on the east or west side of town or that your favorite park is in the eastern part of a state.

Although relative location provides an accurate depiction of place compared to some other location, such a statement is very general. In most circumstances, geographers desire to identify the absolute location of a place, which is most effectively done using the geographic grid. As you probably know, a grid is typically a network of intersecting lines. If the Earth were flat, then it would be easy to construct a grid similar to a sheet a graph paper using the basic directions: north, south, east, and west (Figure 2.1). In this imaginary situation, the grid lines would extend from these four directions and intersect at various places on the grid. With this grid in place, it would be easy to choose a place of origin and then determine the distance from this reference point to your location by measuring along the nearest grid lines, using some unit of measurement such as miles or kilometers.

However, the Earth is not flat, but instead has a curved surface. In order to account for the Earth's shape when locating places, the grid must consist of a series of intersecting circles, with one set extending north and south and the other set east and west. Let's begin by examining the circles that are oriented in a north/south direction. These circles converge at the North and South Poles on their way to the other side of the Earth. These circles are called **great circles** because they are the largest circles that can be drawn on a sphere (Figure 2.2a).

Great circles   *Circles that pass through the center of the Earth that divide the planet into equal halves.*

A good analogy of great circles in the Earth's grid system is the set of circles that separate orange segments after the fruit is peeled. If you examine a peeled orange closely, you'll see that these great circles all converge on the top and bottom of the fruit and bisect it into many segments.

Great circles can be drawn that bisect the Earth in a myriad of ways (Figure 2.2a). The common feature of each of these circles is they have the center of the Earth as their center and they bisect the Earth into two equal halves. Great circles are also important because their outlines reflect the shortest distance between any two points on the planet. This concept is highly relevant to international travel because pilots plot their courses by great circle routes to reach destinations as quickly as possible and save on fuel expenditures. If you plan to travel from New York City to Moscow, for example, you would probably be inclined to think that the shortest route is to fly due east across the Atlantic Ocean. But, in fact, the great circle route from New York to Moscow passes north over Greenland.

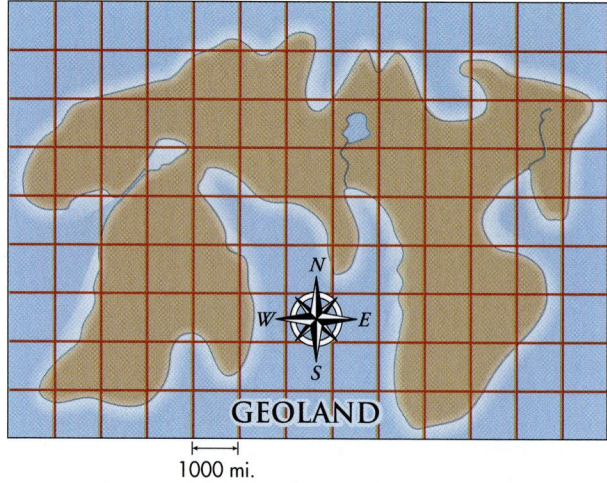

1000 mi.

Figure 2.1 Hypothetical flat land and grid. If the Earth were shaped in this way, you could determine your location by measuring the distance from the nearest grid lines.

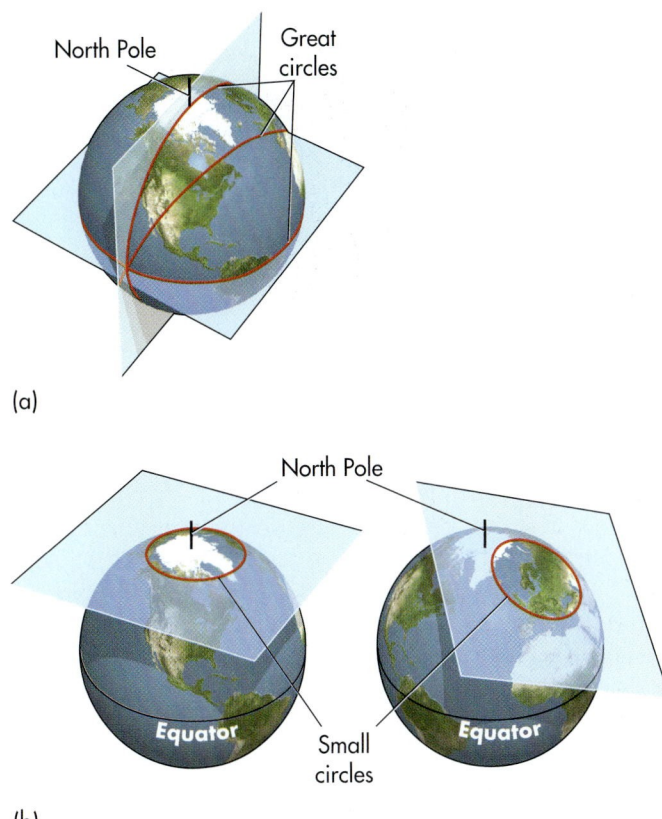

(a)

(b)

Figure 2.2 Great and small circles. (a) Examples of great circles on Earth. (b) Examples of small circles on Earth.

With respect to the Earth's grid system, the most important group of great circles is that which converge at the North and South Poles. Along with this group of circles, there is another corresponding set of circles with outlines that extend east and west. These outlines are always parallel to one another and thus never meet. The largest of these parallel circles (the Equator) lies halfway between the two poles and is a great circle because its outline corresponds with the maximum circumference of the Earth. North and south of the Equator, however, the parallel circles have smaller diameters and are known as **small circles** (Fig. 2.2b).

To visualize this sequence of small circles, imagine what happens when you slice a round loaf of bread. First, the slices are generally parallel to one another as you work across the loaf. Second, the circumference of each slice is progressively smaller as you move away from the center of the loaf, which is the great circle. As with great circles, small circles can be aligned in any direction on the globe. Geographers use this network of north–south intersecting great circles and east–west parallel small circles to create a grid on the Earth's surface (Figure 2.3) that allows them to determine any location within about 30 m (100 ft). This system is known as the latitude/longitude coordinate system. Let's now see how the latitude/longitude system works.

Latitude

Latitude is the portion of the grid system that uses the parallel set of circles with outlines extending east and west. Using these circles, latitude measures location

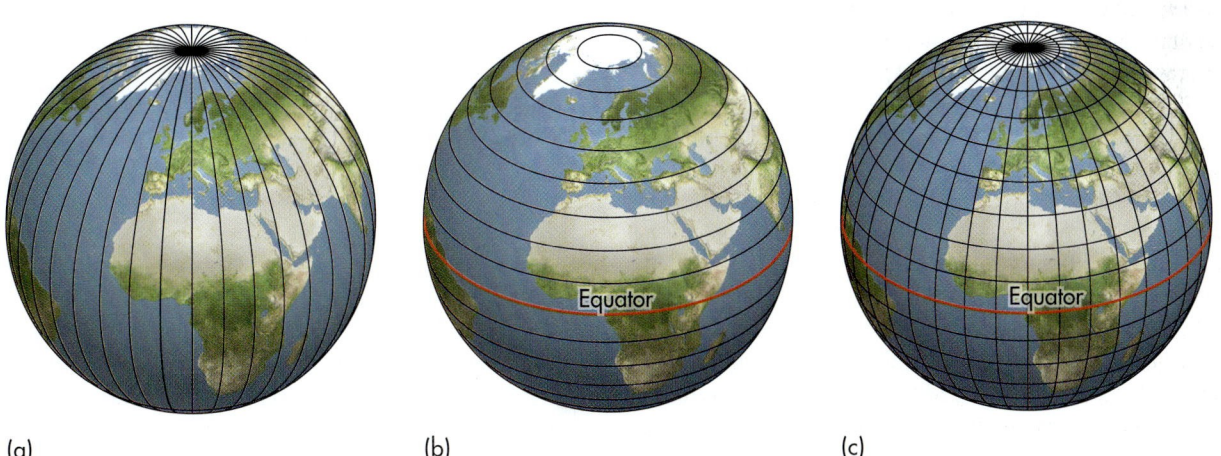

(a)   (b)   (c)

Figure 2.3 Orientation of great and small circles. Together, these circles form Earth's grid system. (a) All the great circles that are oriented north and south intersect at the poles. (b) Except for the great circle at the Equator, all circles oriented east and west on Earth are small circles. (c) Overlay of great and small circles to form the Earth's grid system.

Small circles *Circles that intersect the Earth's surface and that do not pass through the center of the planet.*

Latitude *The part of the Earth's grid system that determines location north and south of the Equator.*

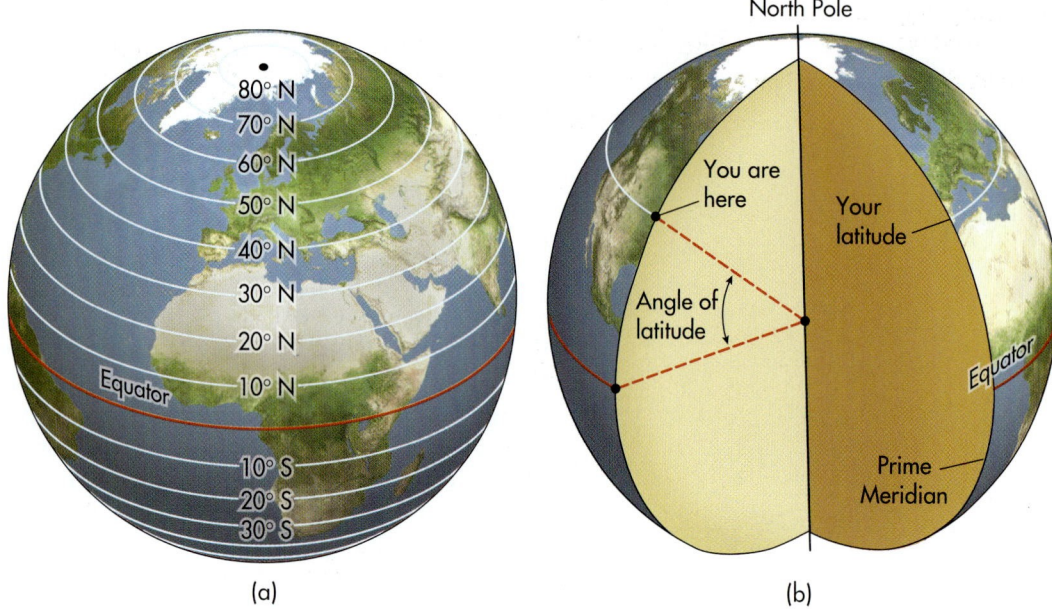

(a)   (b)

Figure 2.4 Determining latitude. (a) The reference point for latitude is the Equator, which divides the Earth into Northern and Southern Hemispheres. Lati-tude ranges from 0° to 90° N and S. (b) Latitude is cal-culated by simply measuring the angle of a point rela-tive to the Equator from the center of the Earth.

north and south of a reference point. The natural refer-ence point for latitude is the great circle that bisects the middle of the Earth at its maximum east/west circumfer-ence. This reference circle is called the **Equator** (Fig-ure 2.4a). The portion of the Earth north of the Equator is known as the **Northern Hemisphere**, whereas the area south of the Equator is the **Southern Hemi-sphere.** Latitude is determined using simple geometry by measuring the angle of any point on the Earth's sur-face north or south of the Equator. This calculation is ac-complished by projecting two lines from the center of the Earth to its surface. One of these lines always ex-tends from the center of the Earth to the surface along the Equator. The other line extends to the north or south from the center of the Earth directly to the location in question, depending upon whether it's in the northern or Southern Hemisphere, respectively (Figure 2.4b). We can then calculate the latitude of the location from the geometric arc between the two intersecting lines.

In this system the Equator has a value of 0° latitude. Latitudes close to the Equator have small geometric arcs from the Earth's center and, therefore, have relatively low latitude designations such as 5° or 10° N or S (see Figure 2.4a). Localities progressively farther north (N) or south (S) from the Equator have progressively greater angles from the Earth's center and, therefore, have progressively larger latitude designations. The maximum angle is per-pendicular (90°) to the center of the equatorial great cir-cle. In the Northern Hemisphere, this angle corresponds to the North Pole (90° N), whereas it correlates with the South Pole (90° S) in the Southern Hemisphere.

As far as distance is concerned, degrees of latitude are separated by about 110 km (69 mil) from each other, regardless of location. In other words, you can specify a location on the Earth's surface only within 110 km north or south at the level of a degree. It is possible to more closely locate places by subdividing degrees of latitude into smaller units. One way to more specifically identify locations is to progressively subdivide degrees of latitude into 60 minutes (′) and each minute into 60 seconds (″). An example of such a latitude designation would be a lo-cation in the Northern Hemisphere that has latitude of 39° 52′ 46″ N. At this level of detail, you can determine loca-tion within 30.5 m (100 ft) on the Earth's surface.

Another way to designate places with a higher level of precision is simply to convert latitude designations contain-

Equator   *The great circle that lies halfway between the North and South Poles.*

Northern Hemisphere   *The half of the Earth that lies north of the Equator.*

Southern Hemisphere   *The half of the Earth that lies south of the Equator.*

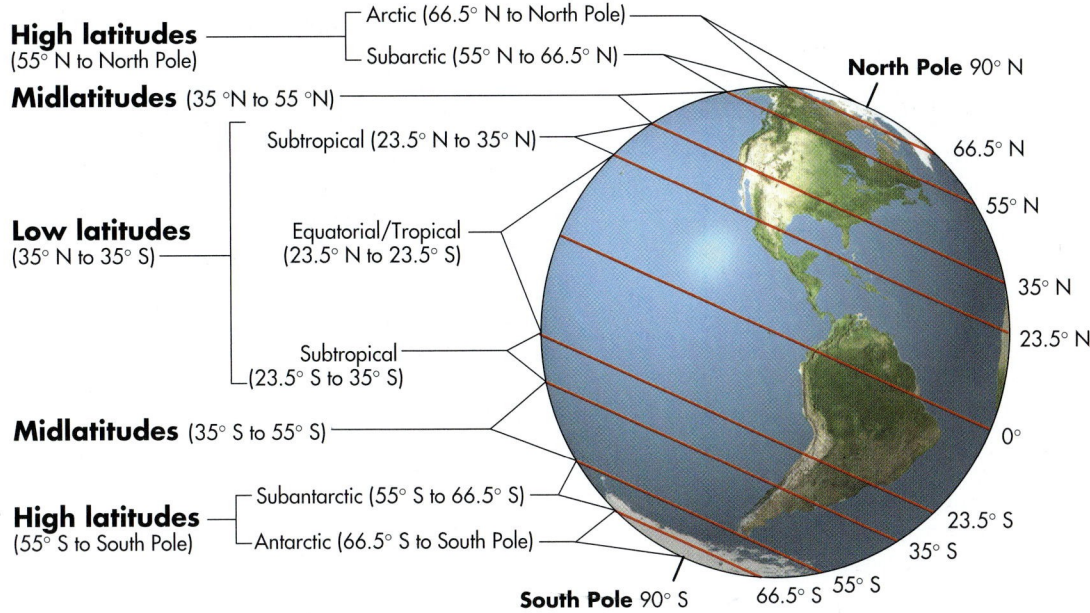

Figure 2.5 Important lines of latitude on Earth and general latitude zones. These zones of latitude often have distinctive climate, vegetation, and soil characteristics.

ing minutes and seconds to full decimal notations. This conversion is done with a simple equation that would, for example, change the previous latitude of 39° 52′ 46″ N to 39.8794° using a full decimal notation. The use of full decimal notations is very common and is most frequently used to designate such important lines of latitude as the Tropic of Cancer (23.5° N), Tropic of Capricorn (23.5° S), Arctic Circle (66.5° N), and the Antarctic Circle (66.5° S) (Figure 2.5). These lines of latitude will be discussed in more detail in the next chapter, but for now, suffice it to say that they are related to Earth–Sun geometry and the way in which the Sun's rays strike the Earth on a seasonal basis.

Lines of latitude are also called **parallels** because they are parallel to the Equator and each other. The value of any given line of latitude remains the same no matter the location, whether it's on one side of the Earth (such as in China)

or the other (such as in the United States). A common misconception that many people have is that lines of latitude must indicate direction east and west because that is the direction that these lines follow. To remember how the system really works, think of degrees of latitude as being numbered sequentially north and south of the Equator.

Although you can certainly think of lines of latitude in individual terms to locate specific places, it's also useful to think of them collectively by region. In that context, lines of latitude can be lumped into nine geographic zones (see Figure 2.5). These nine regions can be grouped further into three general zones (Table 2.1) in which many distinct physical processes occur. The **low latitudes** range from about 35° N to 35° S of the Equator and are generally the warmest part of the Earth. The **midlatitudes** range from about 35° to 55° latitude, in

TABLE 2.1	General Zones of Latitude	
Latitude Zone	Range	General Characteristics
Low latitudes	About 35° N–35° S	Very warm region of the Earth with generally consistent weather over the course of the year
Midlatitudes	About 35°–55° in both hemispheres	Seasonal weather with warm/hot summers and cool/cold winters
High latitudes	From about 55° to 90° in both hemispheres	Very cold region of the Earth with very short summers

Parallels   *Lines of latitude.*

Low latitudes   *The zone of latitude that lies between about 35° N and 35° S.*

Midlatitudes   *The zone of latitude that lies between about 35° and 55° in both hemispheres.*

both hemispheres, and are typically regions of the Earth that experience highly variable weather over the course of the year. Finally, the **high latitudes** range from 55° to 90° latitude, again in both hemispheres, and are generally the coldest places on Earth.

KEY CONCEPTS TO REMEMBER ABOUT LATITUDE

1. Lines of latitude run exactly east and west; they are also called parallels because they always parallel the Equator.

2. Lines of latitude are determined by measuring the geometric arc between two lines projected from the center of the Earth to the surface at the Equator and at the location in question. The reference point for latitude designations is the Equator (or 0°); latitude designations extend to 90° N and S.

3. Lines of latitude never intersect with one another and therefore have the same value over the entire circle.

4. Zones of latitude can be grouped into nine geographic zones. These nine can be further grouped into three general categories: the low, middle, and high latitudes.

Longitude

The previous discussion demonstrated how latitude is used to identify location north and south of the Equator. Now we turn to how east and west locations are determined with **longitude.** The foundation of the longitude system is the system of great circles that pass through the poles. Half of such a great circle is a spherically shaped outline that connects each pole. Each of these half circles is called a meridian of longitude, or simply a **meridian.** In contrast to the latitude system, which has the Equator as the natural reference point because it is the only great circle of the parallels, the easterly and westerly longitude calculations require an arbitrary reference meridian be-

Figure 2.6 The Prime Meridian. This line of longitude at the Old Royal Observatory in Greenwich, England, is the reference meridian for the longitude system.

cause all meridians are the half outlines of the great circles that pass through the poles. The arbitrary reference meridian for longitude is the **Prime Meridian,** which passes through Greenwich, England at the Old Royal Observatory (Figure 2.6). This meridian was chosen as the reference point of longitude at an 1884 conference in Washington, D.C.

High latitudes   *The zone of latitude that lies from about 55° to 90° in both hemispheres.*

Longitude   *The part of the Earth's grid system that determines location east and west of the Prime Meridian.*

Meridians   *Lines of longitude.*

Prime Meridian   *The arbitrary reference point for longitude that passes through Greenwich, England.*

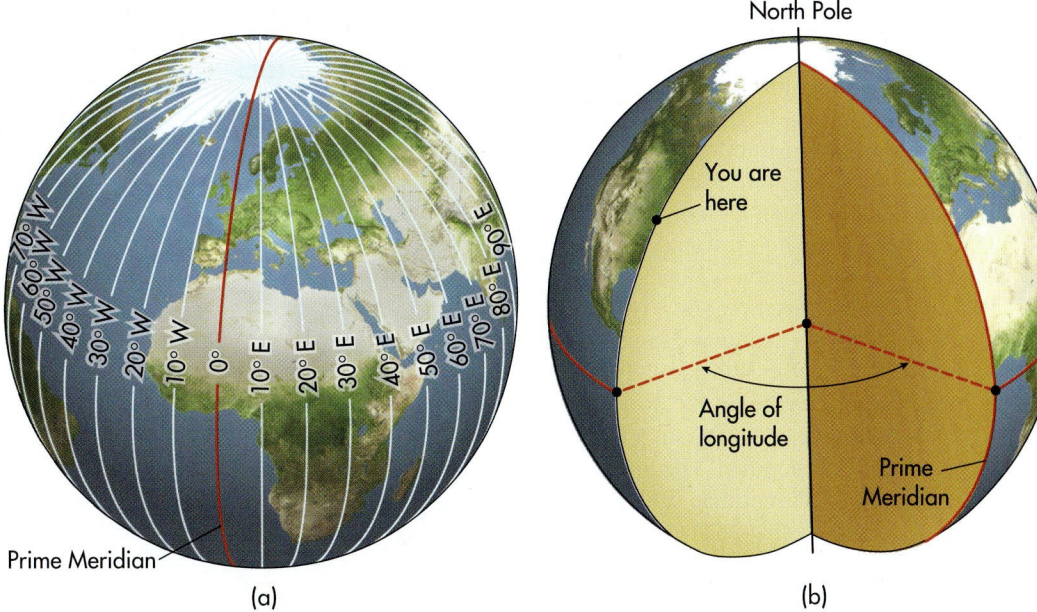

Figure 2.7 Calculating longitude. (a) The longitude grid consists of a series of great circles that converge at the poles and that are measured east and west of the Prime Meridian. (b) Longitude is calculated on the angle of arcs measured from the center of the Earth and relative to the Prime Meridian.

As with the Equator, the Prime Meridian has a value of 0°, and all points north and south along that meridian are at 0° longitude. The geometric basis for determining longitude is that a circle contains 360°. Think of the circle (the Earth) as divided into eastern and western halves by the Prime Meridian. The longitude grid consists of a series of meridians that extend 180° east and west of the Prime Meridian to complete the 360° circle. Using Figure 2.7 as a guide, imagine you are floating in space and looking at a section of the Earth at the Equator. Longitude is calculated by measuring the arc formed by two lines projected from the Earth's center at the Equator. One line extends to the Prime Meridian, the other to the meridian of the location in question. The angle of the arc formed by the two lines determines the longitude. Calculated this way, smaller geometric arcs from the Prime Meridian have lower longitude designations, and greater arcs have higher designations (see Figure 2.7b). As with latitude, degrees of longitude can be subdivided into minutes and seconds or be given full decimal notations.

A primary difference between longitude and latitude is that while lines of latitude are always parallel to one another, lines of longitude converge at the poles (Figure 2.7a). Thus, degrees of longitude are about 110 km (69 mi) apart at the Equator and progressively lessen until they meet at the poles. As with latitude, a common misconception that many nongeographers have about longitude is that meridians are used to determine location

north and south because that is the direction in which the semicircle outlines extend. Instead, remember that longitude determines location east and west of the Prime Meridian.

KEY CONCEPTS TO REMEMBER ABOUT LONGITUDE

1. Lines of longitude run north and south and are also called meridians.

2. Longitude is determined by measuring the geometric arc between two lines projected from the center of the Earth to the surface at the Equator and to the meridian in question. The reference point for longitude is the Prime Meridian (or 0°), which divides the Earth into eastern and western halves.

3. Meridians are located east and west of the Prime Meridian and extend 180° E and 180° W to complete a full circle.

4. The distance between meridians is greatest at the Equator, but decreases steadily until converging at the poles.

Using the Geographic Grid

Now that we have established the fundamental north–south and east–west grids through latitude and longitude, we can overlay the two subsystems to form a complete grid network that encompasses the Earth (Figure 2.8). Thus, we can describe any location in terms of its position north or south of the Equator and east or west of the Prime Meridian. Imagine, for example, that you wanted to find Washington, D.C. in the grid; Washington is located at 39° N (latitude), 77° W (longitude). To see how you would locate Washington within the grid, examine Figure 2.8 again. Beginning with latitude, find the 38th parallel north of the Equator to identify the location in the Northern Hemisphere. Once this is accomplished, locate the Prime Meridian and arc west to the 77th meridian. The intersection of the 38th parallel and the 77th meridian is the basic position of Washington, D.C. Notice that you can first find the latitude or you can first find the longitude; the order doesn't matter, you wind up at the same place either way. However, remember that by convention latitude is always given before longitude.

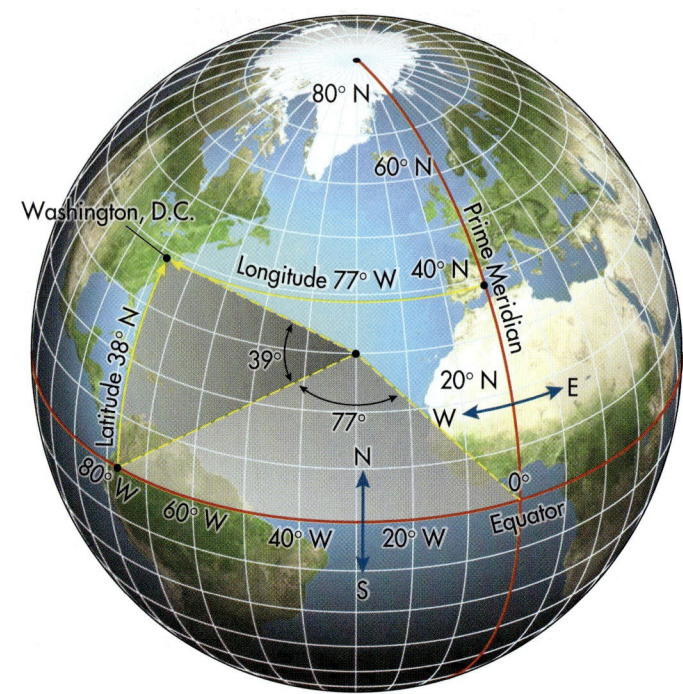

Figure 2.8 Determining latitude and longitude of Washington, D.C. Using the grid network we can locate Washington, D.C. at approximately 39° N, 77° W.

VISUAL CONCEPT CHECK 2.1

Beautiful natural features are scattered around the world on all seven continents. Locate the following places on a globe or map of the world and determine the country in which each one is located.

a) Mont Blanc, 45.50° N, 6.52° E
b) Lake Chad, 13.30° N, 14.00° E
c) Valle de Luna (Valley of the Moon), 23.50° S, 69.00° W

(a)

(b)

(c)

USING THE GEOGRAPHIC GRID

In order to help you fully comprehend how the complete geographic grid works, go to the *GeoDiscoveries* website and select the module **Using the Geographic Grid**. In this simulation you will be able to select a number of places on Earth and determine their latitude and longitude. This, in turn, should help you better understand where locations are in physical space. Once you complete the module, be sure to answer the questions at the end of the module to test your understanding of the geographic grid.

Maps—The Basic Tool of Geographers

Now that we've discussed the basics of the Earth's grid system, we can investigate the various tools that geographers use to portray spatial information within the grid. Most everyone knows that the primary tool associated with geography is the map. Figure 2.9 shows a typical thematic map and is one with which you're probably familiar. This map shows the U.S. and includes such geographic information as state boundaries, location of major cities, the course of major rivers, and vegetation patterns. You probably also know that a map such as this one contains a scale, a compass arrow pointing north, and a legend that defines major categories within the map, such as population density, the area covered by cities, and the relative size of highways, or the location of

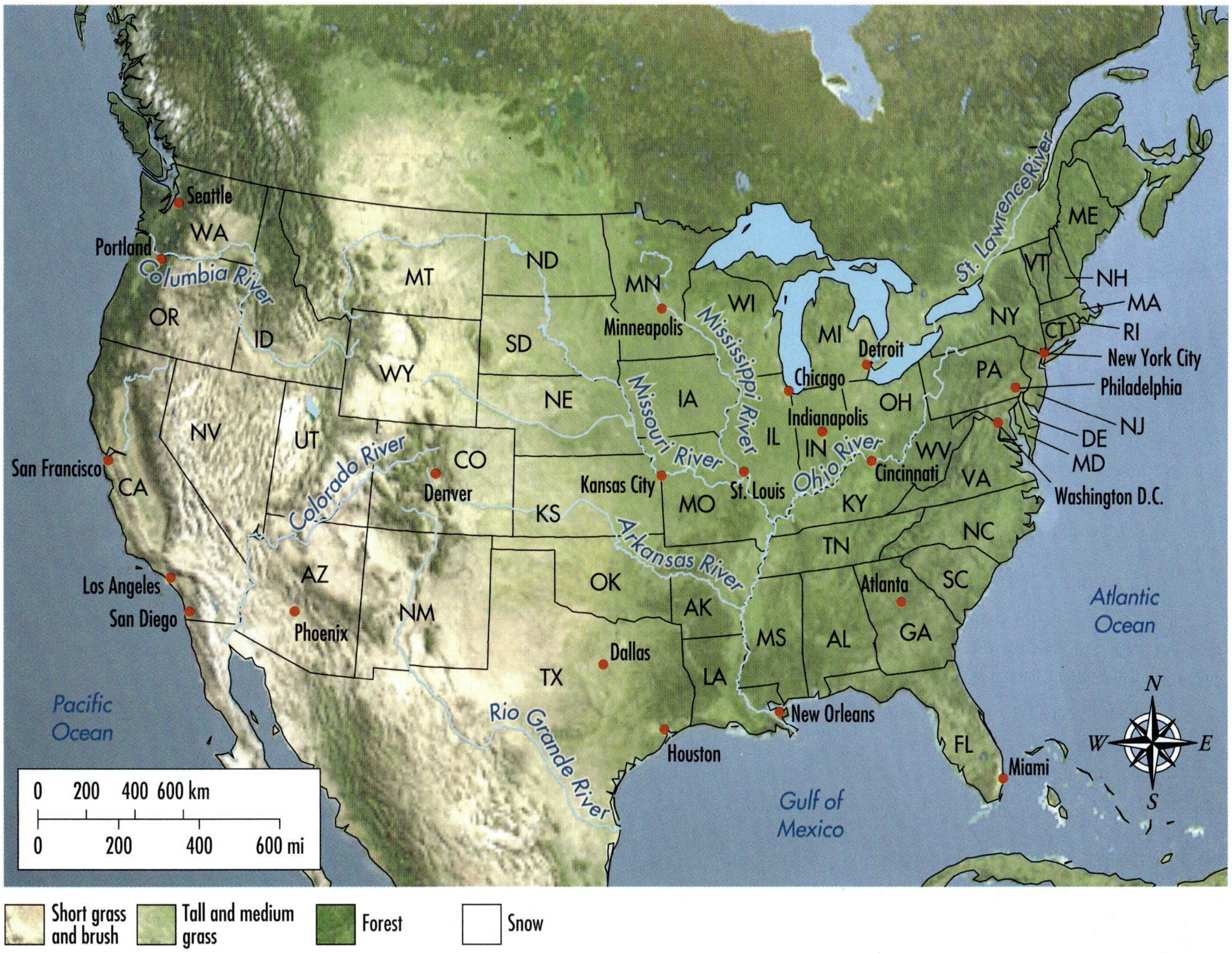

Figure 2.9 A basic thematic map. This map shows the location of some important features within the U.S., including states, major rivers, and cities. Note the legend that reflects general types of vegetation.

On the upper slopes of K2, the air is so thin that breathing is difficult. Climbers usually take oxygen tanks with them to recover from shortness of breath during the climb.

BIG BUT HARD TO REACH

Knowing the latitude and longitude of a place does-n't mean it's easy to get to. Some of the largest features on Earth are well known but hard to reach.

The mountain K2, also known as Mt. Godwin Austen, lies on the border of China and India, 35.53° N, 76.30° E. It is the second tallest mountain in the world (8611 m or 28,253 ft). However, it can be reached only by arduous hiking for many days from the nearest populated town. The first successful ascent was not until 1954.

The tallest uninterrupted waterfall in the world occurs along the Churun River in Venezuela, a tributary of the Orinoco River, 5.44° N, 62.27° W. However, it wasn't seen by anyone outside the local area until 1933, when it was sighted by American aviator Jimmie Angel. It is now known as Angel Falls and is 979 m (3212 ft) tall.

The largest freestanding rock in the world rises 348 m (1142 ft) above the Simpson Desert in Australia, 25.15° S, 136.00° E. Known at one time as Ayers Rock, it is now called by its Aboriginal name of

mountain ranges. A wide variety of additional information can be portrayed on maps, including the geographic distribution and thickness of rock units, temperature, precipitation patterns, intensity of solar radiation, vegetation, and soils. Much of the information in this text will be presented to you in map form, so in order to navigate through this book thoroughly, and to be successful in the course you are taking, you must first understand basic cartography.

Cartography is the subdiscipline of geography that focuses on the many ways to display spatial information so that it can be used and understood efficiently. The essence of cartography is the manner in which the Earth is portrayed in a usable fashion. There are a variety of ways to accomplish this task. The most visually accurate and complete way to illustrate the Earth is through the use of a globe (Figure 2.10), which depicts locations on the three-dimensional representation of the Earth. Globes can be cumbersome to use, however, and often do not illustrate the level of detail that's desired. Instead, geographers frequently rely on two-dimensional maps to show various geographic locations and attributes. Although you've probably used maps at some point in your life, perhaps through your travels, it's useful to review what they really represent.

What Is a Map?

1. A map is a two-dimensional (flat) representation of the whole Earth or a specific region of the Earth.

2. A map is a generalized view of an area, as seen from above, that is reduced in size.

3. A map is a tool that is used to depict spatial information and to analyze spatial relationships.

Cartography *The design and production of maps.*

Angel Falls is so remote that tourists must fly to the area by helicopter and then hike or take a canoe trip to reach it.

Uluru. The rock is made of sandstone and is remarkable for its vivid red color, which changes with the angle of the Sun. Uluru is in the middle of a desert—the nearest town, Alice Springs, is 280 miles away.

Uluru is located close to the exact center of Australia. Several native Aboriginal tribes consider the rock sacred and have carved pictures into shallow caves around the base.

Figure 2.10 Illustration of a globe. (a) A globe portrays geographic information about the Earth in a three-dimensional fashion. (b) Compare the image of the globe with this picture of the Earth taken from space. This particular Earth view focuses on Africa.

(a)

(b)

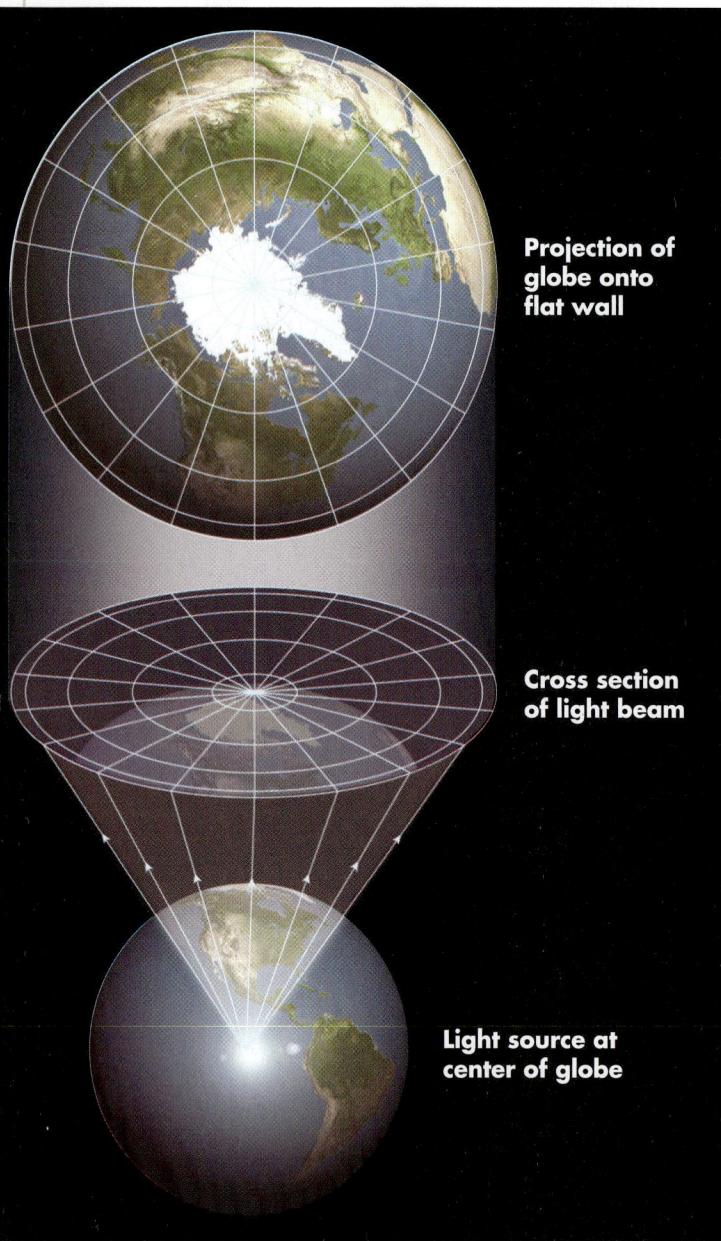

Projection of globe onto flat wall

Cross section of light beam

Light source at center of globe

Figure 2.11 The projection theory. Light from a source projects the information on the curved surface of the Earth onto a plane, forming a flat map.

Map Projections

The most important step in the cartographic process is the presentation of three-dimensional information of the Earth on a two-dimensional map. In order to create a map, locations on the roughly spherical surface of the Earth must somehow be transferred to a flat surface, such as paper or a computer screen. This is done through the process of **map projection.**

A simple way to visualize how projection works is as follows: Imagine a light source passing through a curved translucent surface that has an image on it, pro-

jecting the image from the curved surface onto a flat wall. Figure 2.11 shows this concept in some detail. Light from the source (the light bulb) streams out toward the wall in straight lines. Before it reaches the wall, however, the light passes through the spherical surface of the Earth. As the light beams move through this spherical surface, imagine that they carry with them the collective information from the Earth's surface through which they just passed. When the light beams reach the wall, this information is then displayed for you to see. Although the image was originally curved on the Earth, it becomes flat on the wall. Theoretically, the process of map projection works in the same manner.

While projection serves the purpose of creating a flat map that is potentially more usable than a globe, a negative by-product results because it is not possible to flatten a sphere (such as the Earth) without some distortion of the image. Look at what happens in Figure 2.12 if a lime is sliced in two so that two hemispheres are created. If either of the two hemispheres is flattened on a table, it covers a larger area than it did when it was intact. Also notice that the edges of the flattened hemisphere are severely distorted, in this case they're split. In fact, the only way to make the hemisphere become perfectly flat is to either stretch or cut the edges.

Maps contain the same kind of distortion because spatial information from the curved surface of the Earth is essentially spread over a flat surface. Because of this distortion, a variety of map projections have been developed,

Figure 2.12 Why distortion occurs. On the left is the semicircular (undistorted) half of a lime. Once the center of the lime is pressed to the table to make it flat, distortion occurs at the edges. Note that the area covered by the flattened part of the lime is larger than that beneath the uncut part of the lime.

Map projection *The representation of the three-dimensional Earth on a two-dimensional surface.*

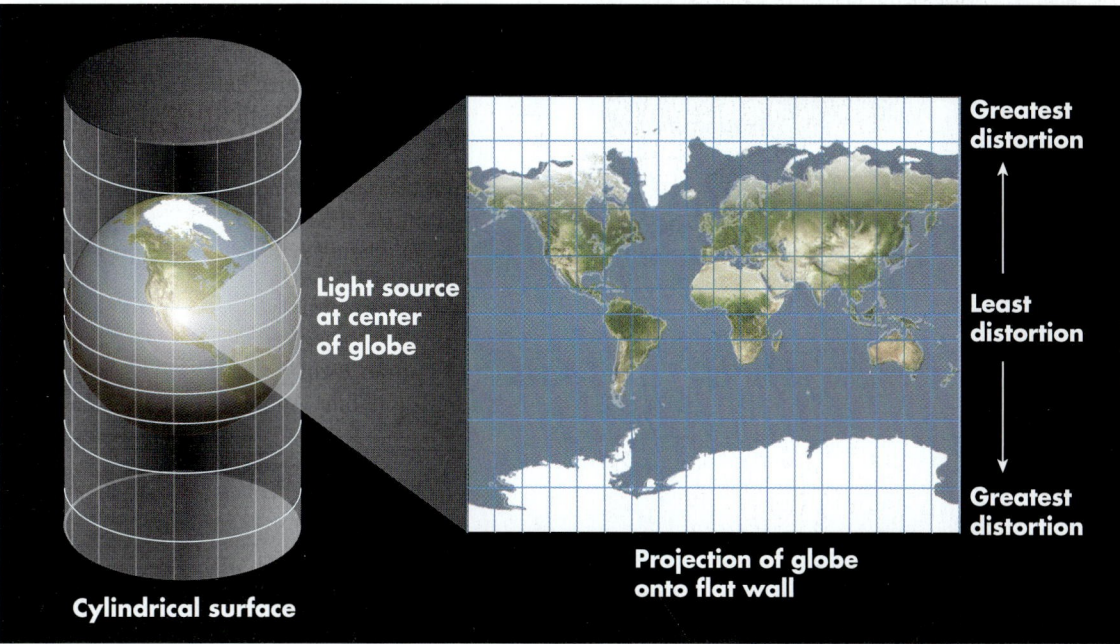

Figure 2.13 A conformal projection. Imagine a cylindrical surface surrounding the globe with a source of light at the center of the globe. The surface is unwound to produce a flat map. The regions of greatest distortion occur at the top and bottom of the map.

with each kind presenting location and distortion in a different way that can preserve either shape or size but not both. Geographers are intimately aware of the strengths and weaknesses of each kind of projection with respect to spatial analysis and select the type of map they want to use accordingly. Although it's not necessary at this introductory level for you to understand maps as thoroughly as a geographer does, you must know something of the properties that result from the various projections that are used to create them. We'll focus on this issue by examining the two basic kinds of projections: conformal and equivalent.

Conformal (or True Shape) A **conformal projection** is a map that maintains the correct shape of features on the Earth, but distorts their relative size to one another. A conformal projection is constructed (in concept) by placing a translucent cylinder over the globe and shining a light centered within the globe outward so that an image is transferred to the cylinder (Figure 2.13). If the cylinder is then removed from the globe, and cut so that it could be laid flat, a map would result.

One of the most common conformal projections is the Mercator projection (Figure 2.14), which was developed in 1568 by the Flemish geographer and mathematician Gerhardus Mercator. This projection is used fre-

Conformal projection *A map that maintains the correct shape of features on the Earth but distorts their relative size to one another.*

quently because it correctly shows the relative shape of features on the globe. As an example of this accuracy, look at Greenland and the Americas in Figure 2.14 and notice that the actual shapes of these places have been maintained. You might ask yourself, why would this be useful? Imagine that you wanted to compare the shape of coastlines between Greenland and the United States. The only way to accomplish this task accurately would be to use a conformal projection such as the Mercator, which would preserve the correct shape of the coastline. Thus, you can make an accurate comparison between these places.

Remember, however, that some spatial attribute must be distorted to keep the shape accurate. In the case of the Mercator projection the distorted feature is size. Notice how lines of latitude and longitude are at right angles to each other because meridians don't converge at the poles, and that the distance between parallels increases with distance from the Equator. This increase occurs because the higher latitudes are geometrically stretched more during the projection process than are the lower latitudes. Another way to look at this type of distortion is to realize the map in Figure 2.14 was stretched much like the lime in Figure 2.12.

You can see this distortion particularly well by looking at Greenland and Antarctica in Figure 2.14. Notice that Antarctica covers almost a quarter of the map and is significantly larger than Africa. In reality, however, the Antarctic continent is less than half the size of Africa (8.7 million km^2 [3.4 million mi^2] vs. 18.8 million km^2

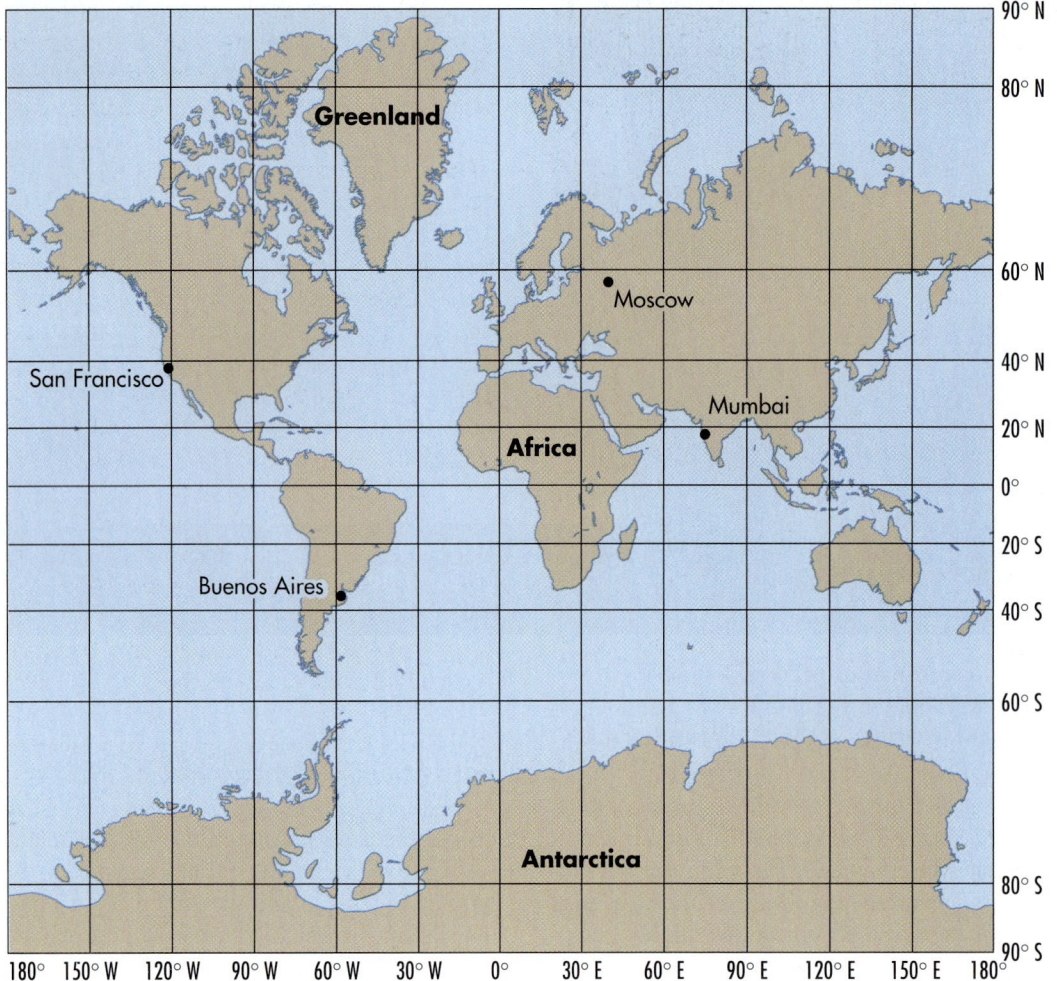

Figure 2.14 A Mercator projection. This map is a conformal projection that maintains shape at the expense of size at high latitudes. Accordingly, scale increases from low to high latitudes.

[7.3 million mi²]). The size of Greenland is also distorted, which you can see through comparison with the United States. Although Greenland appears to be significantly larger than the U.S. on the map, in reality Greenland is about 840,000 square miles in size, while the U.S. covers an area of about 3.7 million square miles—more than four times the size of Greenland.

Equivalent Projection: Once you know how the angular relationships between places compare through use of a conformal projection, you might want to see how their relative size differs. In order to accomplish this visualization you would need to create an **equivalent projection,** which accurately portrays size features throughout the map. This map can be constructed (in concept) by placing a cone over a portion of the Earth, with the sharp point over the North Pole (Figure 2.15). In this instance

---

Equivalent projection *A map projection that accurately portrays size features throughout the map.*

---

light is projected upward through the cone, which is then laid flat to form the map.

One kind of equivalent projection that results from this projection process is an Albers equal-area projection (see Figure 2.15), which was developed in 1805 by the German geographer H. C. Albers. Notice how this map, which focuses on North America, compares with the Mercator projection in Figure 2.14. One primary difference is that the lines of latitude are now curved. Given that these particular lines are actually curved on Earth, this kind of map represents a more accurate presentation of the geographic grid than a Mercator projection.

Another difference lies in the respective shapes of land masses, which can be nicely seen again at Greenland. Remember that in the Mercator projection (Figure 2.14) the shape of Greenland was accurately portrayed but the size was distorted. How does Greenland now look in Figure 2.15? Notice that Greenland looks squatty in the Albers projection compared to the Mercator projection. Why does this difference exist? Remember that a map projection can preserve only one spatial variable at a time. In the case of

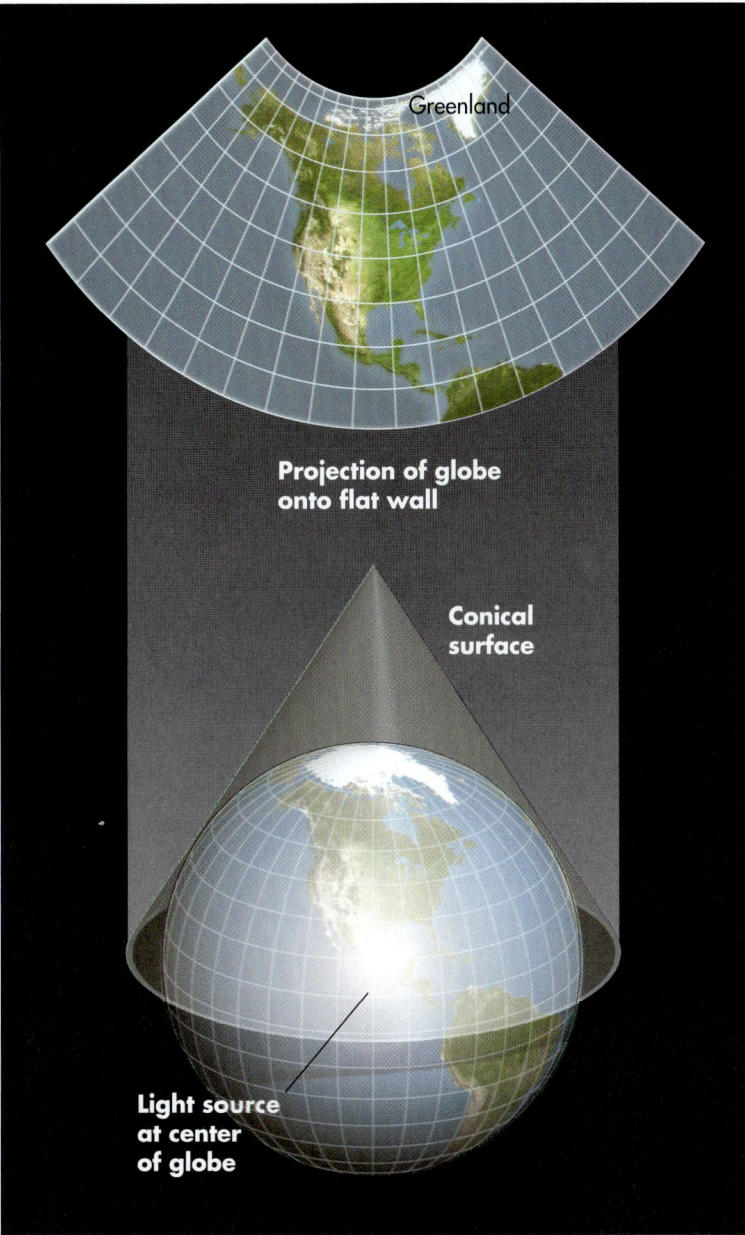

Figure 2.15 How to construct an equivalent projection. The resulting map in this diagram is a specific kind of equivalent projection called an Albers equal-area projection.

Labels in figure:
Greenland

Projection of globe onto flat wall

Conical surface

Light source at center of globe

nents at high latitudes is severely distorted by a conformal projection. Thus, the only way to compare accurately the length of coastline between Greenland and the U.S. would be to use an equivalent-area projection.

It is important to emphasize that many different kinds of projections exist, although it is beyond the scope of this text to describe each of them thoroughly. Depending upon the kind of information you want to see, a projection exists for that purpose. Some projections intensively exaggerate the distortion of one property in order to represent another more accurately. Other projections attempt to show all spatial information as accurately as possible, given the limitations of the technique, and therefore moderate the distortion of both shape and size. A good example of such a projection is the Robinson projection (Figure 2.16), which was developed in 1963 by the American cartographer Arthur H. Robinson. Notice how this particular projection seems to be evenly balanced relative to the Mercator (Figure 2.14) and Albers (Figure 2.15) projections. Greenland and Antarctica are still somewhat enlarged relative to their accurate size, but some semblance of the Earth's curvature exists.

KEY CONCEPTS TO REMEMBER ABOUT MAP PROJECTIONS

1. Map projection is a critical part of cartography because it allows the three-dimensional Earth to be represented in two dimensions.

2. A map projection is designed to preserve either the shape or size of geographic features.

3. Conformal projections maintain the angular relationship between geographic features, but distort the relative size.

4. Equivalent projections maintain the relative size of geographic features at the expense of shape.

5. There are many different kinds of projections, each of which is designed to portray size and shape in a specific way.

---

the Albers projection, the shape of places at high latitudes is distorted so that the relative size of places can be more accurately observed.

You might ask, why does it matter whether or not the sizes of land masses on Earth on a map are accurate? Remember that the Mercator projection, because it preserves shape, allows you to compare regional features such as the configuration of coastlines. However, would a Mercator projection provide an accurate comparison of the *length of* coastline between the U.S. and Greenland? The answer to this question is "no" because the size of conti-

Map Scale

Aside from understanding how the concept of map projection works, another important consideration involving maps is the scale. Map scale is a critical part of cartography because it relates the size of features on a map to their actual size in the real world. Similarly, the scale relates distance between features on the map to the corresponding actual distances on the Earth. In other words, **map scale**

Map scale *The distance ratio that exists between features on a map and the real world.*

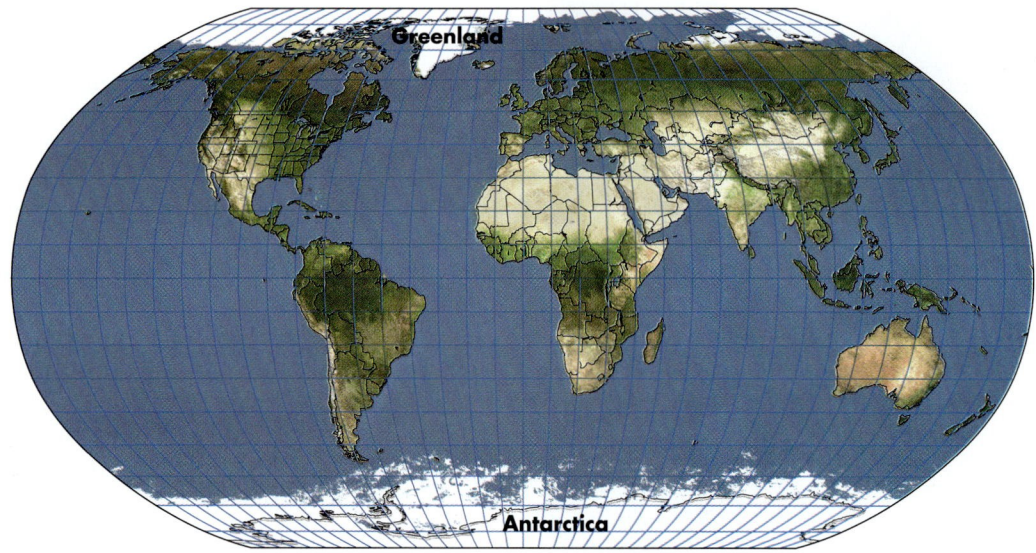

Figure 2.16 The Robinson projection. In an effort to balance the errors associated with projection, this map contains distortion in both shape and size.

represents the ratio of the size/distance on the map to the size/distance on the ground.

There are several ways to depict scale on a map (Figure 2.17).

1. *Written (Verbal) Scale.* Often written as: One Inch = 2000 Feet, which means that one inch measured on the map represents 2000 feet on Earth.

2. *Representative Fraction.* An example of a representative fraction is the ratio: 1:24000 or 1/24000. This scale means that one unit of measurement on the map represents 24,000 of the same unit on the ground. It does not matter what unit of measurement is used—it could be inches, centimeters, etc.—but the units must be the same within the ratio (inches to inches, centimeters to centimeters, etc.).

3. *Graphic Scale* or *Bar Scale.* These scales show the actual size of units on the map. The advantage of the graphic, or bar, scale is that the scale remains

accurate if the map is enlarged or reduced when photocopying.

Maps are made at different scales depending upon what geographic qualities the maps are designed to display, including the area they cover and spatial detail. In general, geographers think of maps in terms of being large scale or small scale. Although these designations appear easy to differentiate, the fact is that many nongeographers find them confusing because an inverse relationship exists between map scale and actual Earth features. You might think, for example, that a large scale map is a map that shows a large part of the Earth. Conversely, a small scale map would be a map that focuses on an area of limited geographic extent, such as a city or park. The fact is that **small scale maps** are maps that show large areas with relatively limited detail, whereas **large scale maps** represent much smaller geographic areas with a greater level of detail. In order to obtain a better feel for map scale, you should examine the information in Figure 2.17 and Table 2.2.

TABLE 2.2 — Map Scale	
Large Scale Maps	**Small Scale Maps**
Used for maps of small geographic areas.	Used for maps of large geographic areas.
Used to illustrate great detail, such as road networks, locations of parks, etc.	Used to illustrate limited geographic detail, but shows spatial relationships of large areas.
Representative fraction is a large number, such as: 1:1000.	Representative fraction is a small number, such as 1:25,000.

Small scale map   *A map that shows a relatively large geographic area with a relatively low level of detail.*

Large scale map   *A map that shows a relatively small geographic area with a relatively high level of detail.*

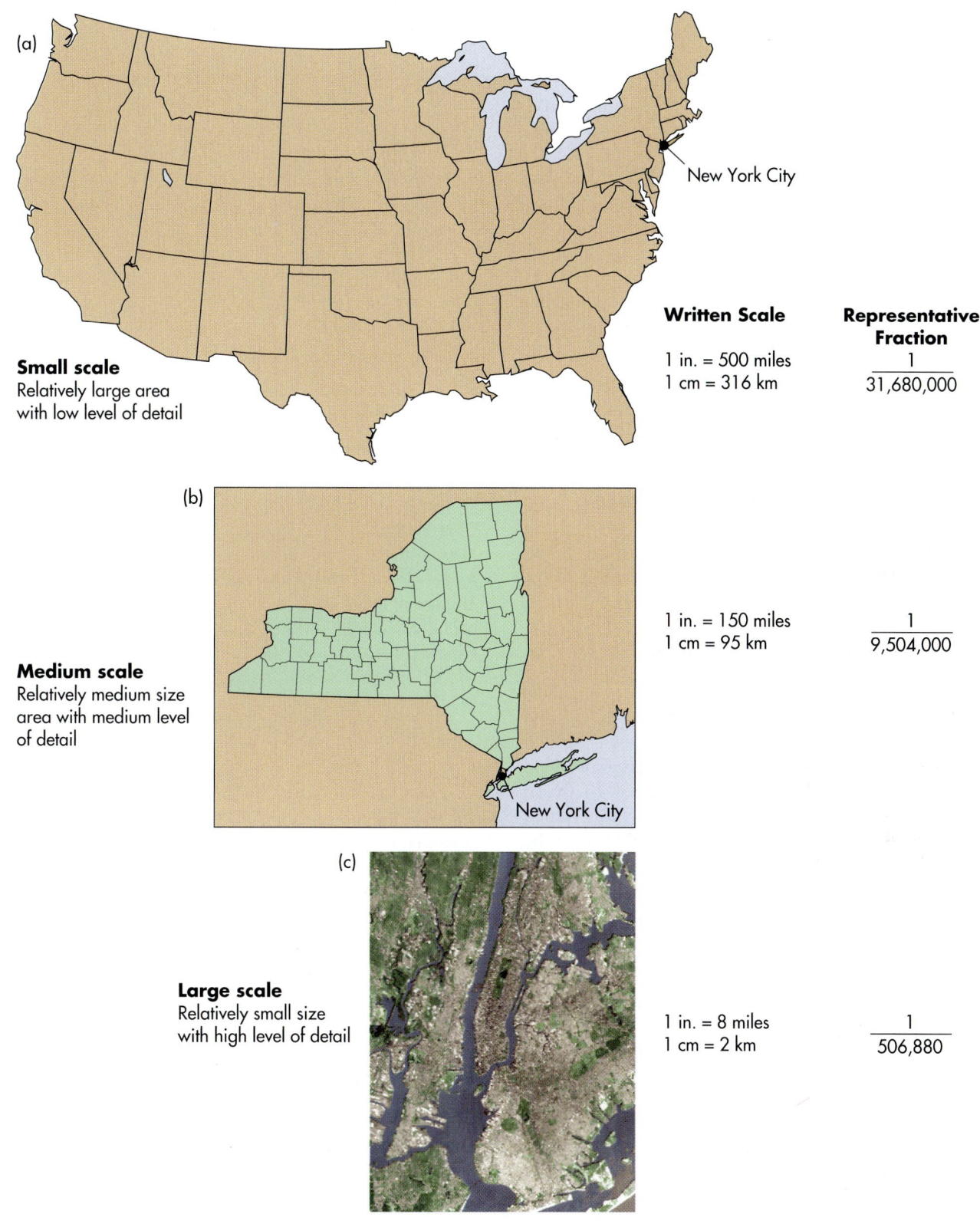

(a)

New York City

Written Scale

1 in. = 500 miles
1 cm = 316 km

Representative Fraction

$$\frac{1}{31,680,000}$$

Small scale
Relatively large area
with low level of detail

(b)

New York City

1 in. = 150 miles
1 cm = 95 km

$$\frac{1}{9,504,000}$$

Medium scale
Relatively medium size
area with medium level
of detail

(c)

1 in. = 8 miles
1 cm = 2 km

$$\frac{1}{506,880}$$

Large scale
Relatively small size
with high level of detail

Figure 2.17 Examples of map scale. Note that as scale increases, the amount of detail about a place increases. (a) Map of the U.S. that shows the location of New York City. (b) Map of the state of New York showing the location of New York City. Note that this map shows more detail (county boundaries) of the state of New York than (a). (c) Satellite image of New York City. Note how much more detail of New York City can be seen in this image than in the other maps.

Isolines

Isolines are another important map component because they are lines that connect points of equal value. These lines are particularly useful, for example, when showing regional temperature and precipitation patterns, elevation, or atmospheric pressure. These data can be derived in a variety of ways and then used to illustrate basic geographic trends. For example, imagine that you want to create a map showing the pattern of high temperatures throughout the U.S. on any given day. The first thing you would do would be to access the temperature measurements taken at all of the National Weather Service sites in the country and show them on a map. This map would contain an outline of the country, state boundaries, the various locations where temperature measurements were taken, and the temperature at each place (Figure 2.18a).

Although this map would contain the information you desire, it would be difficult to discern any geographic pattern because the map contains nothing but a series of numbers. By looking at the numbers closely you could perhaps determine that it was generally warmer in the southern part of the U.S. than in the northern part of the country, but the overall geographic patterns would be vague. Geographers typically clarify these patterns by using isolines, which are lines that connect points of equal data at predetermined intervals. In the context of our temperature map, you might decide that you wanted to show the temperature pattern in the country at 5° C (9° F) intervals, or you may choose another interval. Using a 5° C (9° F) interval as an example, you would draw an isoline connecting all points of equal value at, for example, 5° C (41° F), 10° C (50° F), 15° C (59° F), 20° C (68° F), etc. In many regions these distinct temperatures might not exist; that is, the temperature at one weather station might be 7° C (44.6° F), whereas at the adjacent station it could be 12° C (53.6° F). Where this kind of pattern exists, you would draw the line between the two weather stations in a process called *interpolation*, because you could assume that the 10° C line occurs somewhere midway between 7° C (44.6° F) and 12° C (53.6° F). Figure 2.18 shows this process more closely with some hypothetical temperature data in the upper Midwest.

In the context of isolines, several different kinds are used when mapping specific types of geographic phenomena. The various kinds of isolines are:

1. **Isobars**—connect points of equal atmospheric pressure.
2. **Isotherms**—connect points of equal air temperature.
3. **Isohyets**—connect points of equal amounts of precipitation.
4. **Isopachs**—connect points of equal sedimentary thickness.
5. **Contours**—connect points of equal elevation.

Contour lines are particularly significant because they illustrate the configuration of the three-dimensional landscape, or the **topography**, on a **topographic map**. Topographic maps contain a great deal of information, especially at a 1:24,000 scale (Figure 2.19). Here are some simple rules to follow when reading them:

1. The closer the spacing of contour lines, the steeper the slope. Conversely, contours that have wide spacing represent terrain that is relatively flat.
2. Contours that form closed circles indicate hills. Closed circles with hatch marks indicate closed depressions.
3. Where contours cross a stream, they form V's pointing upstream (which is also uphill).

Isolines *Lines on a map that connect data points of equal value.*

Isobars *Isolines that connect points of equal atmospheric pressure.*

Isotherms *Isolines that connect points of equal temperature.*

Isohyets *Isolines that connect points of equal precipitation.*

Isopachs *Isolines that connect points of equal sediment or rock thickness.*

Contours *Isolines that connect points of equal elevation.*

Topography *The shape and configuration of the Earth's surface.*

Topographic map *A map that displays elevation data regarding the Earth's surface.*

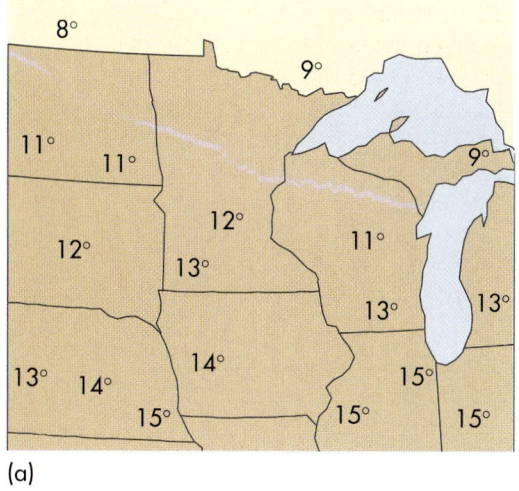

(a)

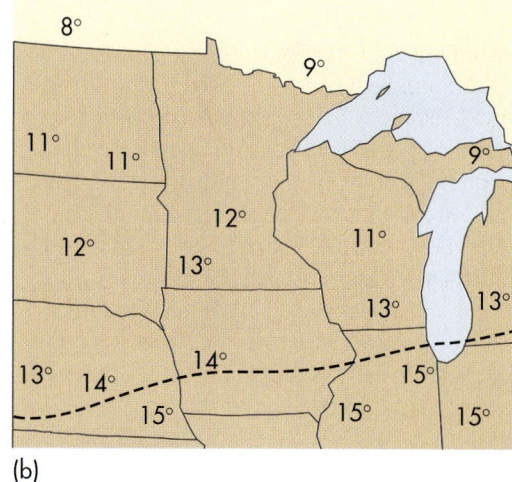

(b)

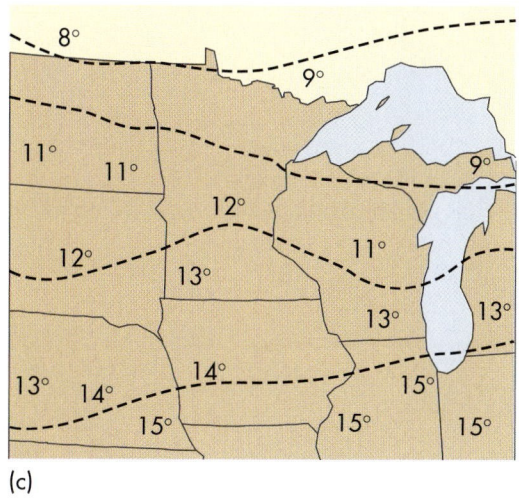

(c)

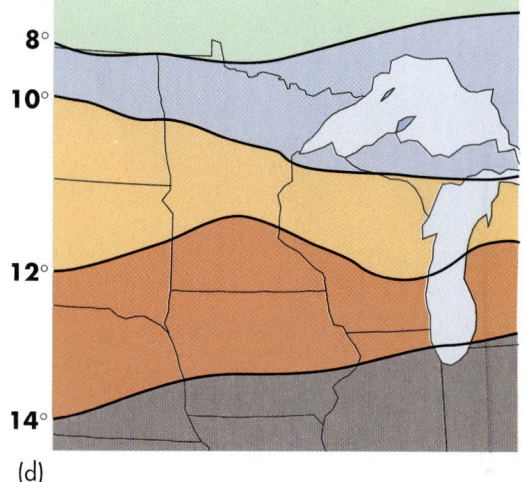

(d)

Figure 2.18 Drawing isolines. (a) Spatial distribution of data points, in this case temperature in degrees centigrade. (b) Interpolating the 14° isotherm. A line is drawn connecting points where 14° measurements were obtained. The position of the 14° isoline is esti-mated between locations where 13° and 15° temperatures were measured. (c) Isolines drawn at intervals of 2° C. (d) The resulting map shows the spatial distribution of temperature in the upper Midwest.

GEODISCOVERIES WWW.WILEY.COM/COLLEGE/ARBOGAST

USING MAPS

Now that you've learned something about basic cartography, go to the *GeoDiscoveries* website and select the module *Using Maps*. The goal of this module is to increase your understanding of maps by interacting with them online. In this particular module, you will learn more about how maps are produced, issues relating to scale, and the various kinds of data that can be presented in a map. Once you finish interacting with the various components of the exercise, be sure to answer the questions at the end of the module to test your understanding about maps.

(a)

(b)

Figure 2.19 A typical topographic map. Topographic maps show elevation patterns of a given geographic area. (a) This topographic map focuses on the Maroon Bells area near Aspen, Colorado. (b) Photograph of the Maroon Bells from Maroon Lake. The arrow in the accompanying topographic map shows the approximate line of sight from Maroon Lake toward the mountain peaks.

a) The process of map projection is the method through which three-dimensional features on the Earth are presented as a two-dimensional image. Two basic projections, conformal and equivalent, are presented here. Which of these two maps is a conformal projection? Which one is the equivalent projection? How can you tell the difference between the two?

b) The scale of a map depends upon the amount and kind of information that a cartographer wishes to present. In this case, two maps are presented that contain information about Seattle, Washington. Which one of the maps is a relatively small scale map? Which one is the large scale map? How can you tell the difference between the two?

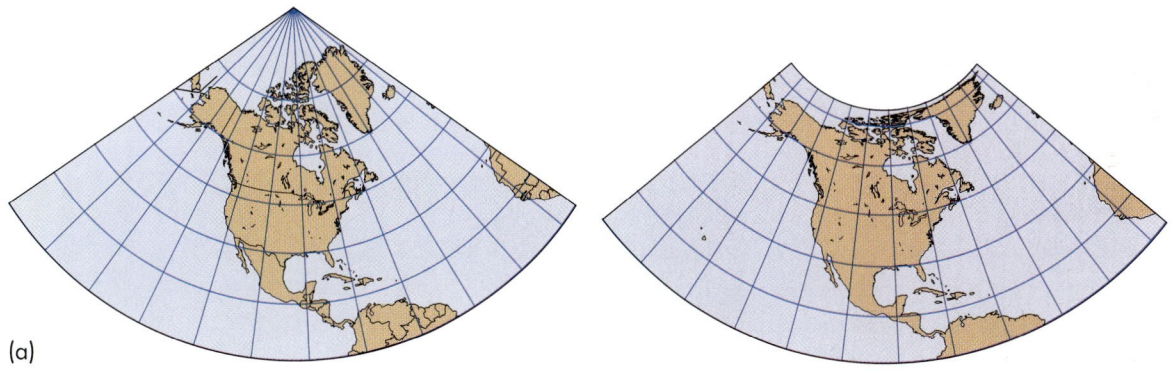

(a)

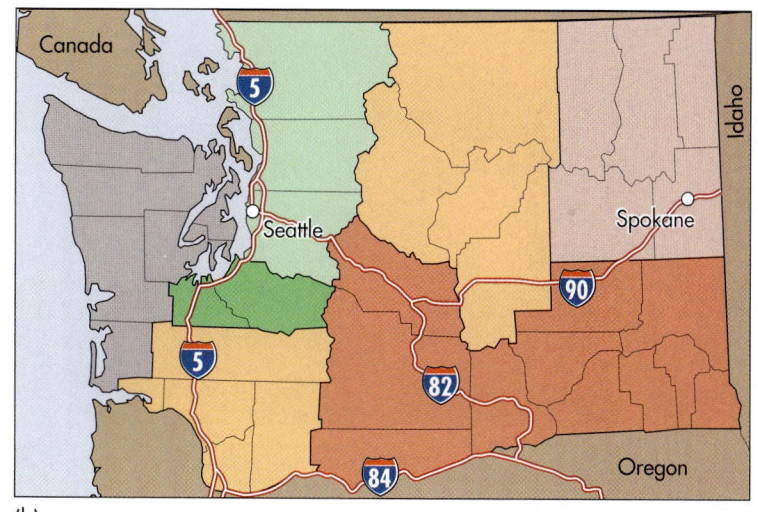

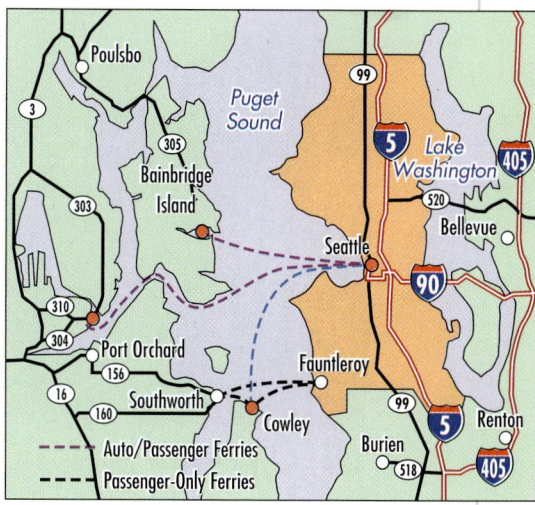

(b)

Digital Technology in Geography

As you're no doubt aware, digital technology is an essential part of our life in the industrialized world today, with personal computers, cell phones, the Internet, and satellite television common features in many households. These technological advances have also been very important in the recent evolution of geography as a scientific discipline, resulting in expanded work in the field that has increased the need for skilled people at all professional levels at excellent pay. In general, much current geographic research is focusing on the development and refinement of three digital techniques: global positioning systems (GPS), remote sensing, and geographic information systems (GIS). These techniques, used independently as well as together, enable geographers to use maps and other images in a digital format in order to increase the speed and efficiency of geographic research. In addition, the ability to display geographic information has improved remarkably. The remainder of this chapter focuses on these new and exciting tools that are revolutionizing how people are viewing the planet from a geographic perspective.

Remote Sensing

Until the early 20th century, the only way that a geographer could consistently learn anything about new places first hand was to personally conduct field work and collect samples or make observations. This limitation began to change somewhat in the early 1900s with the development of airplanes, which people could use to see places without actually setting foot on the ground. With the advent of this new technology, geographers began to use a new approach to spatial analysis called **remote sensing,** which means measuring properties of the environment without direct contact. In this manner, geographers could learn much general information about the vegetation patterns of the landscape, the routes that rivers take across the countryside, or even the configuration and distribution of agricultural fields. Remote sensing began to be used in a widespread, systematic fashion in the 1930s when on-board cameras were developed that could take **aerial photographs** (Figure 2.20) at prescribed intervals of time. With these photographs, which could be taken in a succession of linear flight lines, entire counties or even states could be viewed in a systematic way.

Figure 2.20 A typical aerial photograph. With an aerial photograph you can see a variety of landscape patterns that can't be viewed from the ground. This aerial photo, for example, shows the pattern of crop fields, trees, roads, and dwellings at Rajasthan, India.

Remote sensing *The method through which information is gathered about the Earth from a distance.*

Aerial photographs *Photographs taken of the Earth's surface from the air.*

Although aerial photography was a dramatic leap in the geographer's ability to study the landscape, this technology proved limiting because a complete set of photographs of large regions (like a state) was only taken once every decade, or at best, at irregular intervals of time. Although this temporal pattern allowed for some studies of landscape change, perhaps by taking photographs of the same place two or more times over, say, 25 years, geographers were still limited because they couldn't see the change as it was occurring. The fact is that major changes can occur on a landscape in a very short period of time and geographers usually want to see them quickly.

With the development of satellite technology in the 1950s and 1960s, the geographer's ability to remotely sense the landscape increased dramatically. As you know, satellites are platforms that continuously orbit the Earth. These systems are designed for either Sun-synchronous or geostationary orbits (Figure 2.21). A satellite with a **Sun-synchronous orbit** has an orbit that keeps pace with the Sun's westward progress as the Earth rotates. Such a position is maintained because the satellite's orbit is basically north–south between the poles with a slight (8°) angular inclination. As a result, a particular satellite may cross the Equator 12 times during the course of a day, with the local time perhaps being 3:00 P.M. at the time of each pass. A Sun-synchronous orbit is usually at an altitude of about 700 to 800 km (430 to 500 mi).

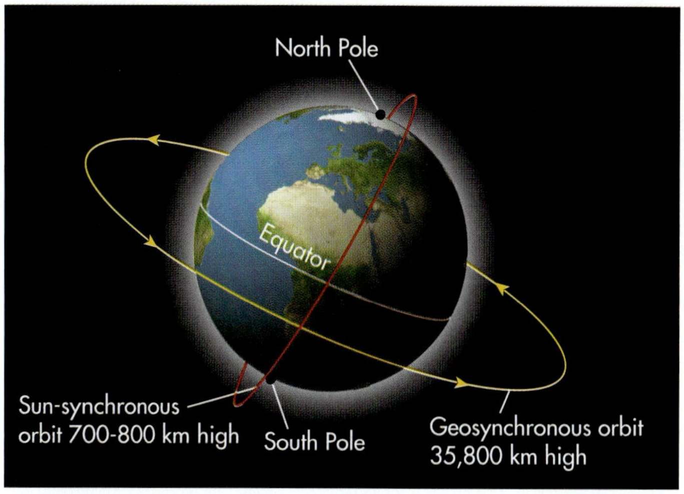

The orbit of Sun-synchronous satellites results in large gaps in image coverage between successive orbits on any given day. Given the slight inclination of the orbit, however, the satellite progresses slightly westward with each new day. In this way the satellite slightly overshoots the orbital path from the previous day, yielding imagery that overlaps a little so that no information is lost. Because satellites with Sun-synchronous orbits pass over the same part of the Earth at about the same time each day, they are particularly good for monitoring landscape change over time.

Several well-known satellite systems have Sun-synchronous orbits. Perhaps the best known of these satellites is the Landsat system, which is jointly administered by the National Aeronautics and Space Administration (NASA), the National Oceanic and Atmospheric Administration (NOAA), and the U.S. Geological Survey (USGS). This satellite system is the longest continuous Earth-observing project in history, beginning with Landsat 1 in 1972 and continuing through the launch of Landsat 7 in 1999. Landsat 7 orbits the Earth at an altitude of about 705 km (438 mi), has a swath width (it can "see") of 185 km (115 mi), and returns to the point of origin every 16 days

Figure 2.21 Sun-synchronous and geosynchronous satellite orbits. Sun-synchronous orbits are designed to keep pace with the Sun's westward progress as the Earth rotates. The orbit essentially extends between the poles with a small inclination so that the satellite moves slightly westward with each successive day. Geosynchronous orbits hover over the same point on Earth, but to do so they must stay in a very high orbit over the Equator.

(233 orbits). Landsat 7 carries two sensors: Thematic Mapper (TM) and Enhanced Multispectral Scanner Plus (EMSP+). Landsat TM has seven spectral bands, which means it can discriminate emitted energy in several parts of the electromagnetic spectrum, whereas the EMSP+ sensor has four spectral bands. The **spatial resolution** of the various Landsat sensors ranges from 15 m (49.2 ft) to 60 m (197 ft). Spatial resolution refers to the size of the ground area that the sensor can "see" in detail. A 15-m (49.2 ft) spatial resolution, for example, means that the satellite can discern objects that are at least 15 m × 15 m in size. Each of these 15 m × 15 m (49.2 ft × 49.2 ft) blocks is referred to as a **pixel.**

In addition to the Landsat platform, another satellite with a Sun-synchronous orbit is the Advanced Very High Resolution Radiometer (AVHRR), which is managed by NOAA. This satellite orbits the Earth 14 times per day from an altitude of 833 km (517 mi) and has a spatial resolution that ranges between 1.1 km (0.7 mi) and 4 km (2.5 mi), depending on whether the place viewed is directly below the satellite or on the fringe of the area seen.

Sun-synchronous orbit   *A slightly inclined polar orbit that keeps pace with the Sun's westward progress as the Earth rotates, resulting in regular return intervals over every location on Earth.*

Spatial resolution   *The area on the ground that can be viewed with detail from the air or space.*

Pixel   *The smallest definable area of detail on an image; short for* pixel element.

Figure 2.22 A typical GOES image. This example is from the GOES East satellite, which provides daily imagery of the eastern United States. Clouds are bright white, whereas darker areas are zones of clear sky. This image shows that clear skies dominated the eastern part of the country on this particular day.

A **geostationary orbit,** in contrast to a Sun-synchronous orbit, is designed to permanently remain in one place above the Earth. This consistent position is accomplished by placing the satellite in an easterly orbit, directly over the Equator, and at a very high altitude. The satellite orbits in the same direction and at the same speed that the Earth rotates. With such an orbit, the satellite can view nearly half of the Earth at any given time. This kind of orbit is mostly used for observing weather and to facilitate communications. An excellent example of satellites with a geostationary orbit is NOAA's Geostationary Operational Environmental Satellites (GOES), which are used to monitor meteorological conditions (Figure 2.22). This system consists of an array of satellites that are positioned about 35,800 km (22,300 mi) above the Earth, high enough to allow very large regions of the Earth to be viewed at any given time. The GOES-12 (or GOES East) satellite, for example, is positioned above the Equator at 75° W longitude and views most of North America, all of South America, and most of the Atlantic Ocean. The GOES-10 (or GOES West) satellite, in contrast, is located above the Equator at 135° W longitude and views the western portion of North America and South America, as well as the vast majority of the Pacific Ocean. As with other forms of satellite remote sensing, the GOES-10 and GOES-12 images slightly overlap with one another.

Remote sensing operates on the simple principle that objects on the Earth emit electromagnetic energy that can be measured. Although we'll discuss the electromagnetic spectrum in much greater detail in Chapter 4, it is sufficient for now to say that objects on Earth emit energy that is both visible and invisible to our unaided eyes. This energy can be measured by either satellites or airplane platforms and then stored digitally in a computer for detailed analysis.

Figure 2.23 shows the typical steps through which satellite imagery is processed, with this example focusing on a false-color image of Cape Cod in Massachusetts. The sequence begins with a measurement of **emissivity** by the satellite. These data are then sent to a receiving station on Earth, such as at NASA, where it is archived until it is distributed to the appropriate users in the scientific community as a black and white image. The image is then modified by the users through the application of colorizing filters to produce a false-color image that highlights particular geographic features of interest.

Although much can be learned by viewing the Earth as we normally see it, a great deal more information can be acquired by investigating the landscape in the invisible parts of the spectrum. This analysis often occurs by examining the thermal infrared energy emitted by the Earth. All objects emit this kind of energy, but the relative amount varies depending upon the temperature—warm objects emit more energy than cold ones—and the overall emissivity of the object. With a thorough understanding of these properties, geographers can see very subtle differences on the landscape and thus determine the spatial distribution of specific features such as vegetation, rock, and water. Areas such as these can be studied frequently in this fashion with infrared aerial photography, while larger regions can be investigated on an annual basis with Sun-synchronous satellites. Figure 2.24 shows a pair of infrared satellite images. This kind of analysis is especially useful, for example, as a means to show urban patterns (Figure 2.24a) and to monitor deforestation in tropical rainforests (Figure 2.24b).

The methods we've described so far are passive systems in that they depend upon energy emitted by objects on Earth to obtain measurements. Another form of remote sensing relies on active systems that send a beam of wave

Geostationary orbit *An orbit where satellites remain over the same place on Earth every day. This orbit is achieved because the satellite is placed very high above the Earth and travels at the same speed as the Earth's rotation.*

Emissivity *The amount of electromagnetic energy released by some aspect of the Earth's surface.*

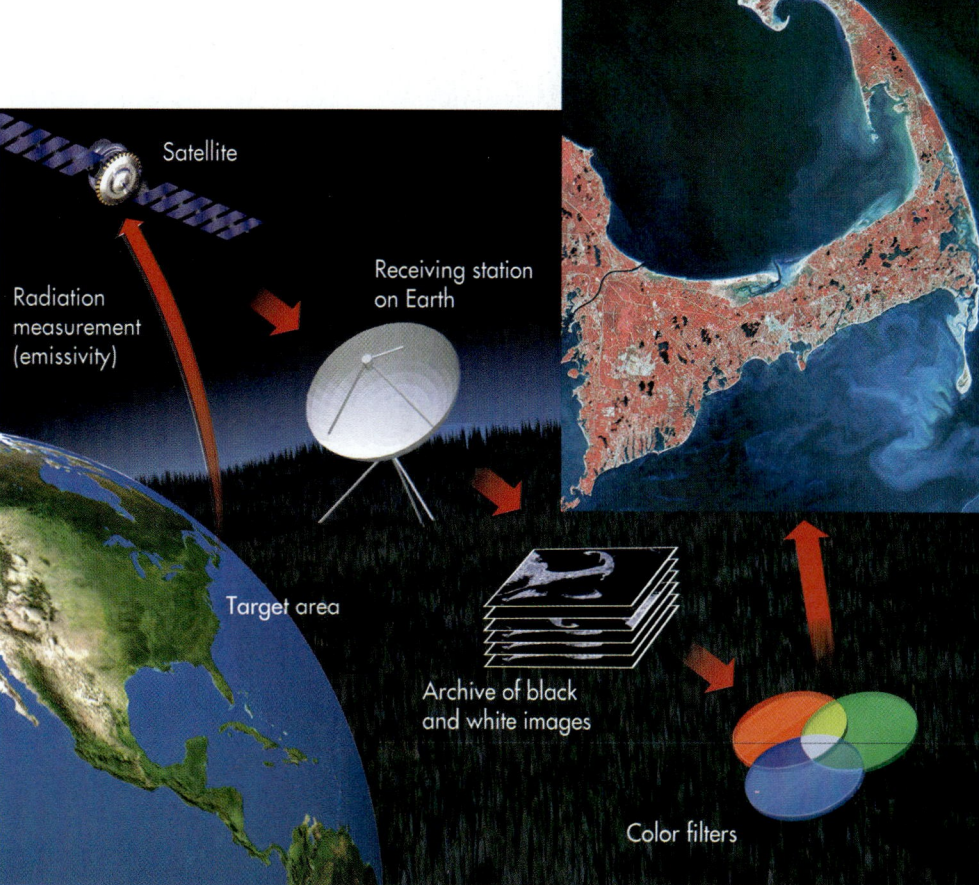

Figure 2.23 Processing of remotely sensed images. This sequence of events produces a false-color image of Cape Cod, Massachusetts, that can be geographically analyzed.

Satellite

Radiation measurement (emissivity)

Receiving station on Earth

Target area

Archive of black and white images

Color filters

(a)

(b)

Figure 2.24 Infrared satellite images. (a) This image of the region around Baltimore, Maryland, shows areas of vegetation in bright red and urban areas as bluish gray. Areas of dark shading consist of water, specifically part of Chesapeake Bay. (b) Deforestation in the Amazon rainforest as seen from NASA's Terra Satellite. Dark red represents uncut forest, whereas the lighter shades of red indicate the pattern of forest clearing. The dark wavy line in the image is the Ji-paraná River. Note how the image path appears to be tilted slightly. This shape reflects the slight inclination of the satellite's Sun-synchronous orbit.

INFRARED IMAGES

Infrared satellite images can be used to obtain many different kinds of information, depending on what range of radiation the satellite cameras are set to detect. For example, this image is an infrared satellite image of the Imperial Valley in southern California. The many red squares and rectangles are farm fields that contain maturing crops. The distinct horizontal line at the south end of the reddish zone is the border between the U.S. and Mexico. Although the agricultural area extends into Mexico, the reduced infrared output from those fields suggests that the Mexicans are using a different crop rotation than the Americans to the north. The areas of bluish gray indicate the presence of two cities, with El Centro, California, north of the border and Mexicali, Mexico, immediately to the south.

energy toward the Earth. This beam is then partially reflected back to the sensor, which can make measurements. A good example of this kind of system is *radar,* which uses energy in the microwave part of the electromagnetic spectrum. Radar systems emit short pulses of energy toward the Earth and then measure the time it takes for the pulse to return, plus its strength, to create a picture of the landscape. Such imagery can be obtained either from an airplane *(*Figure 2.25) or from space. In the latter instance, the most extensive radar imagery was acquired in the Shuttle Radar Topography Mission (SRTM), during which virtually all of the surfaces on Earth were mapped with radar. In contrast to passive systems, which can be limited when cloud cover is extensive, radar can

"see" through the clouds and can therefore be used all the time. As a result, radar has been extensively used to map the surface of Venus, which is masked by continuous cloud cover. On Earth, radar is especially effective at mapping geological formations in mountainous landscapes such as the Appalachians (Figure 2.26).

Global Positioning Systems

Another way in which digital technology has benefited geographers is with respect to determining location within the geographic grid. Until recently, the most common way that people determined their position was by looking at the grid coordinates in a large scale map. Although this method provided a general location, usu-

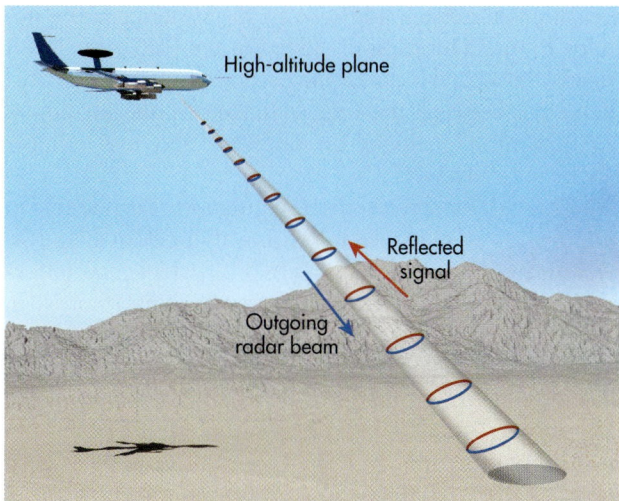

Figure 2.25 Acquisition of Synthetic Aperture Radar (SAR) imagery from an airplane. In this system, a radar unit at the base of the airplane sends radar waves to the ground in a sweeping motion across the landscape. Subsequently, these waves bounce back to the airplane where their timing (from and to the airplane) and strength are calculated and converted into an image of the landscape.

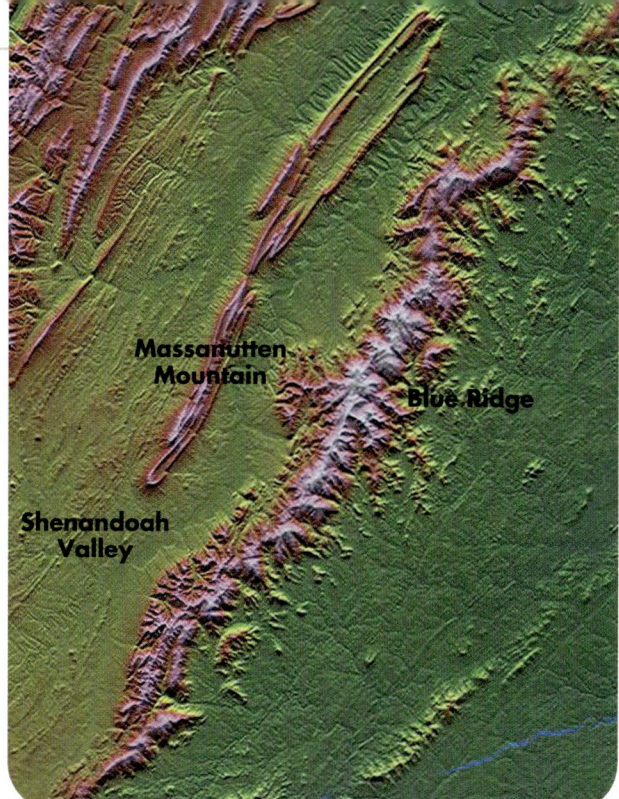

Figure 2.26 Radar image of Shenandoah National Park in the Appalachian Mountains. This image was acquired with C-Band Interferometric Radar during the Shuttle Radar Topography Mission (SRTM) and subsequently color coded to show elevation. Greens represent topographically lower areas, rising through yellow, red, and magenta, to bluish-white at the highest elevations.

ally within a degree or two, it was nevertheless a rough approximation of position.

With the proliferation of satellites in the late 20th century, the U.S. military developed a system that enables people to determine their geographic location quickly and with much greater precision than with the use of a map. This system is called the Global Positioning System, or GPS, and is based upon a network of 24 satellites that orbit the Earth every 12 hours (Figure 2.27). These satellites continuously transmit a radio signal known as Psuedo Random Code (PRC) that is unique to the GPS system. This code consists of a complicated sequence of on/off pulses that travel at the speed of light (299,300 km/sec; 186,000 mi/sec) in space and can be acquired by receivers on Earth. Figure 2.28 shows an example of a hand-held GPS receiver. The fundamental principle of GPS is that by knowing the positions of satellites in space, the exact time at which a signal is sent from a given satellite and received by a receiver, and the speed of the incoming PRC, you can use this artificial constellation of satellites as locational reference points through a system of triangulation.

Here's how the system works (Figure 2.29). Imagine your car contains a GPS receiver and receives a signal from a satellite indicating that it is 7000 km (4350 mi) away. This lone signal indicates that you are anywhere on the surface of a sphere that is 7000 km (4350 mi) from the satellite. If we add the signal from a second satellite, for example one that is 8000 km (4971 mi) away from the receiver, we can narrow your location to the region contained within the space where both spheres intersect.

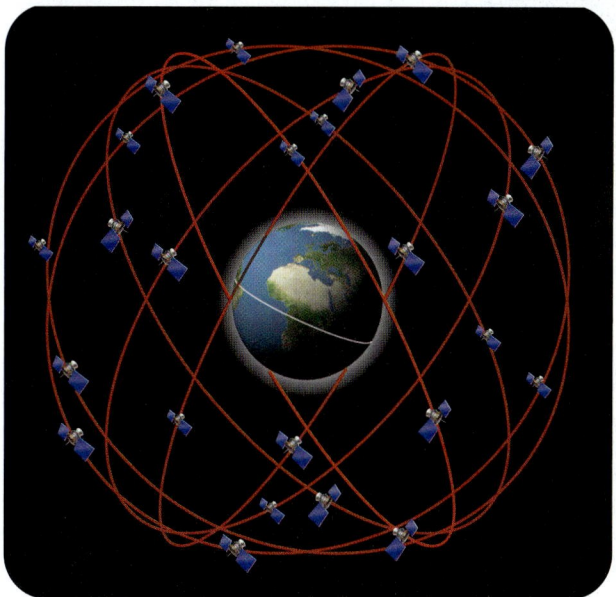

Figure 2.27 The GPS satellite system. Satellites are positioned in six orbital planes, with four satellites in each orbit (24 total) and are about 20,200 km above Earth's surface. This array provides complete coverage of the Earth.

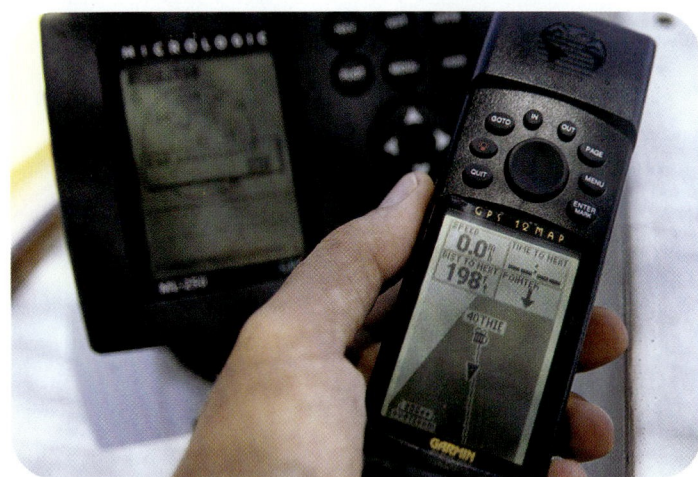

Figure 2.28 A hand-held GPS receiver. This unit receives the PRC signal from satellites in the GPS constellation.

Unfortunately, the size of this overlapping region is still considerable and might tell you that your location, for example, is someplace within the eastern United States. However, by adding the distance to a third satellite, say one that is 9000 km (5592 mi) away, we further narrow the location to a very small area where all three spheres intersect. To further increase the confidence of the measured location, the distance to a fourth satellite can be measured as a check. In this way, the system is more immune to fluctuations in the speed of the PRC signal as it travels through the atmosphere and inconsistencies that may exist between the time indicated by the incredibly accurate atomic clocks on satellites and less reliable clocks in receivers.

Through this process of triangulation, it is possible to pinpoint geographic location within about 20 m (about 90 ft) with inexpensive (less than $100) systems and to less than 1 m (3.3 ft) with more expensive systems (more than $10,000). This degree of accuracy is especially useful in the military because it has facilitated the development of weapons that can be precisely delivered, as you may have seen in television footage from recent conflicts. Within the geographic community, high-resolution GPS has been used in a wide variety of circumstances, including coastal navigation, determining the precise location of study sites, and the monitoring of plate movement along faults in earthquake zones (Figure 2.30).

Although GPS is an exciting tool that has great potential for geographic analysis, it has potential sources of error with which you should be familiar. One potential source of error is that the altitude of individual satellite orbits may vary from time to time. This variation is a problem because the receiver bases its distance calculations on a predetermined satellite altitude. Another source of error is the effect of the atmosphere on the satellite signal as it beams toward the Earth. This effect is especially a problem when dust, water vapor, and ionized particles in the atmosphere interfere with the speed of the GPS signal, which leads to errors in distance calculations. A final transmission problem is **multipath error,** which results when obstructions such as buildings and trees cause the incoming GPS signal to be deflected before it reaches the receiver. This deflection confuses the receiver because two slightly different signals arrive at the same time.

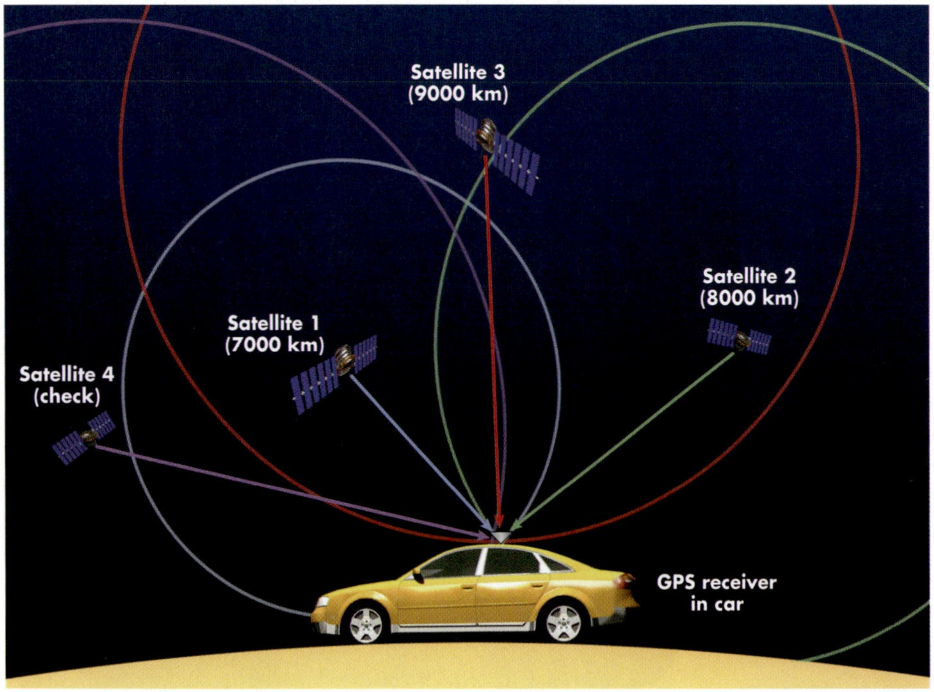

Geographic Information Systems

Another tool that has recently become very important in the way that geographers do their work is Geographic Information Systems, or GIS. A GIS is a system for storing, analyzing, and manipulating spatially referenced data, usually in dig-

Figure 2.29 The process of GPS triangulation. By determining the distance from three different satellites, the GPS receiver narrows its location to the region where three different spheres intersect. The distance to a fourth satellite can be determined as a check. Locations can be determined to within one meter with expensive receivers.

Multipath error *Disruption of the GPS signal from satellites due to obstructions such as trees and buildings.*

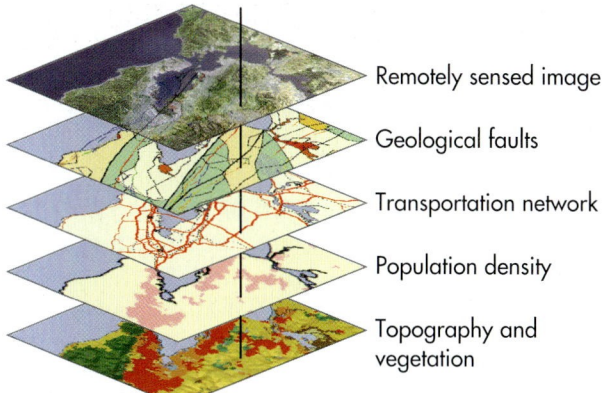

Remotely sensed image

Geological faults

Transportation network

Population density

Topography and vegetation

Figure 2.31 GIS layers. Each layer represents a distinctive part of the landscape, such as **topography, vegetation, population density,** or the local road network. When these layers are integrated, they form a detailed image of the landscape.

Figure 2.30 Using GPS to measure landscape change. Given the high accuracy of some GPS systems, it is possible to monitor change in ground elevation, such as those that may occur near the Augustine Volcano in Alaska due to renewed volcanic activity.

ital form in a computer. Most GIS databases consist of a series of individual data layers that are considered to be relevant for the study being conducted. A data layer contains measurements obtained with respect to a specific geographic variable, such as vegetation, soils, road networks, municipal boundaries, and the distribution of surface water (hydrology) in lakes and rivers.

Let's choose vegetation as an example. A geographer may choose to map the spatial distribution of trees and grassland in a region, which can be determined through infrared remote sensing or by digitally tracing old maps. In so doing, a GIS layer is created that can be identified as "vegetation" in the overall database. A variety of attributes can be subsequently assigned to this layer, such as species composition, height, the slope of the land, and condition. Subsequently, similar maps can be created for other desired variables, resulting in layers for each of them. Now look at Figure 2.31, which shows a potential

series of GIS layers from a theoretical study area. Notice how these layers exist as distinct units that can stand alone. In addition, the layers can be combined to illustrate the overall geographic character of the region. It's beyond the scope of this book to describe GIS in detail, but you should understand that it is an extremely powerful tool that is used in a wide variety of ways, including:

- **Environmental management** — GIS can be used to manage spatial information associated with soils, wetlands, vegetation species, topography, and the location of data collection sites.

- **Municipal planning** — GIS is used in virtually all cities, large towns, and counties to manage spatial information such as road networks, location of sewer and utility lines, and emergency traffic routes.

- **Business needs** — A growing number of individual companies are incorporating GIS into their operations. GIS can be used, for example, to preselect the most efficient delivery routes that will save fuel costs. In addition, GIS can be used to identify the best location for new shopping centers based on surrounding demographic characteristics.

USING A GEOGRAPHIC INFORMATION SYSTEM

To get a basic feel for GIS, go to the *GeoDiscoveries* website and select the module ***Using a Geographic Information System***. This simulation will allow you to interact with a geographic database that includes several layers of information.

You will be able to construct a series of GIS data layers that include data related to soils, vegetation, and various human features within a regional framework. After you complete the tasks, be sure to answer the questions at the end of the module to test your understanding of Geographic Information Systems.

A variety of geographic information can be acquired, stored, and analyzed using digital technology. This image is an infrared satellite image of a portion of Washington, D.C. What do the red, black, and bluish-gray areas represent in this image? What are at least four data layers that could be constructed in a GIS?

The Big Picture

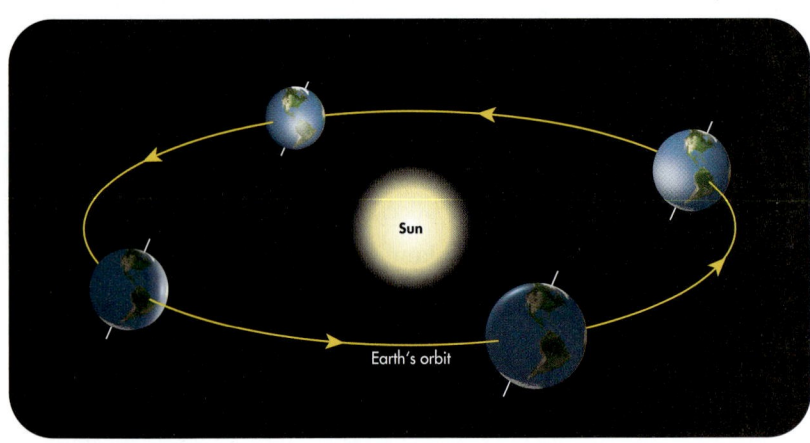

Now that we have introduced the kinds of fundamental tools that geographers use, and how they can be applied to spatial analysis, we can begin the investigation of natural processes and patterns associated with physical geography. The logical place to begin is by examining the relationship that the Earth has with the Sun. You will discover that this relationship is not only critical to life on Earth, but drives many of the processes on the planet, such as atmospheric circulation, climate, and even the character and distribution of soils. The next chapter begins this discussion by looking at the geometric relationship of the Earth and Sun and how that influences our seasons. This is a very important chapter because it provides the foundation for subsequent discussions throughout the book.

Summary of Key Concepts

1. Lines of latitude are called parallels and determine location north and south relative to the Equator. Latitude designations are calculated using simple geometric principles and extend to points 90° N and S, which are the North and South Poles, respectively. Lines of longitude, in contrast, are called meridians and determine location east and west of the Prime Meridian in Greenwich, England. These designations are also calculated using the geometry of a circle and extend to 180° E and W. Latitude and longitude overlay to form a geographic grid.

2. Maps are two-dimensional representations of three-dimensional aspects of the Earth. They are created through the process of map projection, which creates distortion in some aspect of the map. A conformal projection, for example, main-

tains the correct shape of features at the expense of their relative size.

3. Map scale refers to the ratio of the size of features on a map relative to the real world. Small scale maps show relatively large areas, but with less detail. A large scale map, in contrast, shows a relatively small area but with greater detail.

4. "Remote sensing" refers to the ability to observe and monitor aspects of the Earth from a distance. The most systematic way to accomplish this task is with satellite platforms such as the Landsat system. These systems measure emitted energy that can be stored as digital data that can be ultimately accessed and manipulated by scientific users.

5. A Geographic Information System (GIS) is a digital database that contains relevant geographic information such as the density of roads, location of human structures, and vegetation type, to name but a few. These various data sets are stored as individual layers that can be individually or collectively manipulated and analyzed. In contrast to GIS, GPS stands for Global Positioning Systems and is the method through which precise locations can be determined through triangulation with a constellation of satellites.

Check Your Understanding

1. What are the two components of the geographic grid and how are they measured? How are they similar? How are they different?

2. What does the term "low latitudes" refer to?

3. What kind of map projection would provide the most accurate comparison of the shape of national boundaries between countries in the low and high latitudes?

4. Which would have the larger scale: a map showing the detail of New York City, or one that showed the position of New York City within the state of New York?

5. What is the difference between remote sensing and personally collecting samples from a landscape?

6. What is the difference between active and passive remote sensing?

7. Compare and contrast a Sun-synchronous orbit with a geostationary orbit.

8. Why can vegetation be considered to be a data layer in a GIS?

9. How can remote sensing be used to construct a GIS?

10. What are the three potential sources of error associated with GPS?

ANSWERS TO VISUAL CONCEPT CHECKS

Visual Concept Check 2.1

a) Mont Blanc is located in the Alps in France.
b) Lake Chad is located in central Africa.
c) Valle de Luna (Valley of the Moon) is located in eastern Chile.

Visual Concept Check 2.2

a) The map on the left-hand side of the figure is a conformal projection, whereas the one on the right is an equivalent projection. You can tell the difference because the conformal projection maintains the shape of features on the Earth, but distorts their relative size, especially at the high latitudes. The equivalent projection, in contrast, maintains relative shape, with size being distorted. To see the difference between the two, note the relative shape and size of Greenland compared to the rest of North America.

b) The map of the state of Washington has the relatively small scale, whereas the map of just the Seattle area has the larger scale. You can tell the difference because the map of Washington shows a relatively large geographic area that contains very little detail about Seattle. The map of Seattle, in contrast, shows much more detail about the city, including the road pattern and shape of water bodies.

Visual Concept Check 2.3

This infrared satellite image contains a variety of geographic information. The red, black, and bluish-gray areas represent vegetation, water (mostly the Potomac River), and urban development, respectively. Potential GIS layers include (1) the configuration of water bodies, (2) distribution of vegetation, (3) the road network, and (4) the pattern of urban development.

CHAPTER THREE
EARTH–SUN GEOMETRY AND THE SEASONS

Of all the spatial relationships that geographers study, the Earth–Sun

relationship has the biggest impact on the way the Earth functions and the

existence and character of life on the planet. In the context of physical

geography, this relationship is especially important because most of the

physical processes on Earth are powered by incoming radiation received

from the Sun. Although the Earth receives only about one-half of one-

billionth the amount of energy that the Sun radiates, this amount is

nonetheless sufficient to power the amazing things that occur on Earth.

This time-lapse photograph shows the change in Sun angle that occurs over the course of a day above the Arctic Circle in summer. The Sun is at its lowest point in the sky at midnight and is at its highest point at solar noon.

At first glance it might appear to you that the Earth–Sun relationship is a simple association that's easy to comprehend. Isn't it sufficient to know that we receive our light and energy from the Sun; what else is there to know? In fact, this relationship is very complex because it depends upon many different geographic variables such as latitude, the position of the Sun in the sky, time of year, and the nature of the Earth's rotation and axial tilt. It's very important that you understand the information in this chapter because it provides the foundation for many geographic patterns you'll see in the

Our Place in Space

A good way to begin our understanding of Earth–Sun relationships is with a brief discussion of the Universe. The origin of the Universe seems to be best explained by the **Big Bang theory,** which is based on the concept that the Universe emerged from a singular, enormously dense and hot state about 14 billion years ago. The primary evidence for this theory comes in two forms, specifically that (1) the most distant star clusters from us are moving away from us at faster speeds than those that are relatively close, and (2) the amount of cosmic microwave background radiation is remarkably uniform throughout the Universe, which implies that it represents the leftover energy from an early period of rapid expansion.

According to present estimates, the Universe is approximately 20 billion light years across, with a light year being the distance that light travels over the course of a year at the speed of 1,079,252,848.8 km/h (670,616,629.4 mi/h). With respect to the ultimate fate of the Universe, it will either continue to expand for all of eternity or will one day collapse upon itself, depending on the average density of matter and energy within the entire Universe. Given that the speed of distant star clusters seems to be accelerating, it appears that the Universe may indeed expand forever.

The largest clearly definable unit within the Universe is a galaxy, which is a collection of billions of stars. According to astronomers, the Universe has approximately 50 billion galaxies, with each galaxy containing billions

of stars. We happen to live in the Milky Way Galaxy (Figure 3.1), which includes approximately 400 billion stars. The Milky Way is a typical spiral galaxy, consisting of a bright central region with a high density of stars and a flat circular region containing most of the other stars. Younger, brighter stars form in long spiral arms that extend out from the galactic center. Our solar system lies in one of these spiral arms.

The center of our particular solar system is the star that we refer to as the Sun (see *Amazing Places: The Sun*). Nine planets orbit the Sun (Figure 3.2), although astronomers debate whether Pluto is a planet or a large asteroid. Regardless of the specific number of recognized planets, it's safe to say that the solar system is a very big place. To give you an idea of its size, consider that Pluto is approximately 5.9 billion km (3.6 billion mi) from the Sun. A probe launched in January 2006 to study Pluto is currently zooming through space at about 14,500 km/h

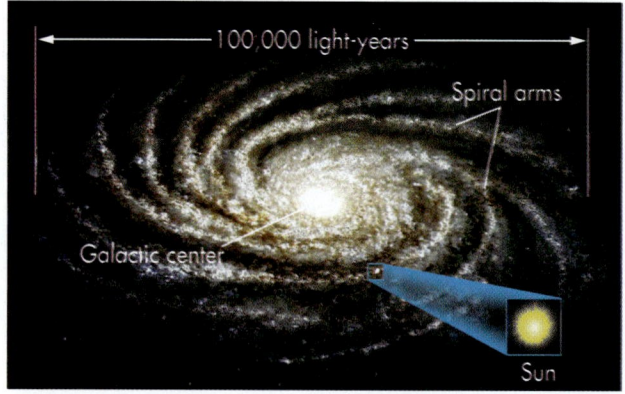

Figure 3.1 The Milky Way Galaxy. Our Sun is but one of the approximately 400 billion stars in this galaxy. Note our Sun's location in one of the spiral arms of the galaxy.

Big Bang theory *The theory that the Universe originated about 14 billion years ago when all matter and energy erupted from a singular mass of extremely high density and temperature.*

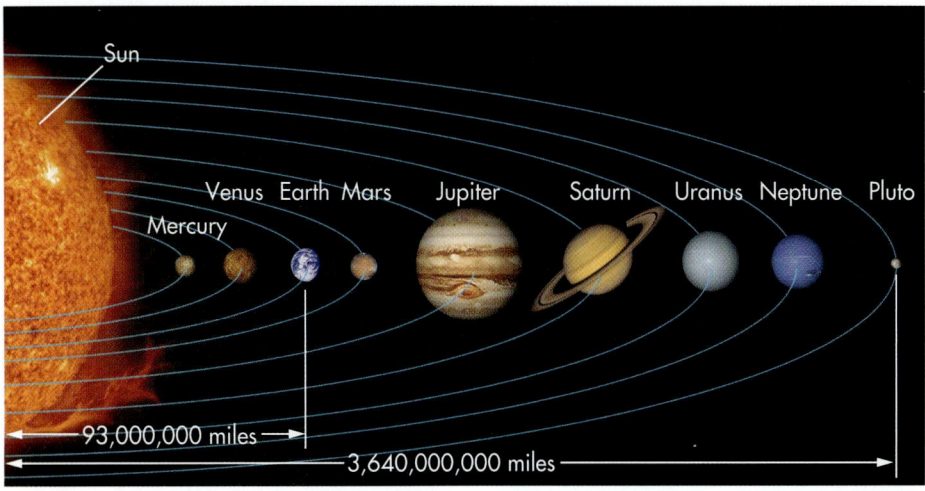

Figure 3.2 Our solar system. The solar system consists of nine planets that orbit the Sun. The Earth is the third planet from the Sun. (Not drawn to scale.)

(9000 mi/h). Even at that remarkable speed, the probe is not scheduled to arrive at Pluto until July 2015! Given this great distance, Pluto is much too far away from the Sun to receive sufficient energy to support life. Of the remaining planets, most are either too far away from the Sun (and thus are too cold) or too close (and thus are too hot) for life as we know it to exist. The lone exception is Earth, the third planet from the Sun, whose orbit happens to be a perfect distance from the Sun for supporting life.

The Shape of the Earth

Let's now turn our attention to the shape of the Earth. Although the Earth appears to be a perfect sphere, it bulges slightly at the Equator and is flattened somewhat at the poles to form a shape known as an *oblate spheroid*. The circumference of the Earth measured at the Equator is 40,075 km (24,902 mi), while the circumference measured through the poles is slightly less: 40,008 km (24,860 mi). The bulge at the Equator, which gives the Earth a slightly thicker middle, is caused mainly by the centrifugal force of the Earth's rotation and by differences in density of the Earth's crust and gravitational field. This centrifugal force is similar to the sideways push you feel when rounding a curve quickly in your car.

The roughly spherical shape of the Earth is an important factor to consider regarding the Earth–Sun relationship. Because the Earth presents a curved surface to the arriving rays of the Sun, the intensity of solar radiation re-

ceived varies by latitude. This difference is a reflection of the **Sun angle,** which is the angle at which the Sun's rays strike the Earth's surface. In general, the Sun angle is relatively high at low latitudes and becomes progressively less toward the poles because the Earth's surface is curved. The net effect of this variation is that the solar energy received at higher latitudes is spread over a relatively large surface area compared to the same amount of energy received at lower latitudes. As a result, the amount of incoming energy per unit area is relatively low in areas where the Sun angle is low.

Figure 3.3 illustrates this difference. Lower latitudes generally receive more intense solar radiation because the Sun angle is high (between 75° and 90°). The point on Earth where the Sun's rays are perpendicular to the surface (Sun angle = 90°) at any given point in time, and therefore most intense, is known as the **subsolar point.** Higher latitudes, in contrast, receive less intense radiation because the Sun's rays hit the Earth's surface at angles much less than 90° (a more oblique angle) due to the Earth's curvature. You can explore this relationship yourself by shining a desk lamp onto a surface, as in Figure 3.3a, both directly and at an angle. Note that the light is spread more diffusely when shone at a lower angle than when the angle is 90°. The Earth's curvature has an analogous effect on incoming solar radiation, which, in turn, contributes greatly to many geographic patterns that exist on Earth, including atmospheric circulation and the global distribution of climate, vegetation, and even soils.

Sun angle *The angle at which the Sun's rays strike the Earth's surface at any given point and time. This angle is high at low latitudes and is progressively less at higher latitudes.*

Subsolar point *The point on Earth where the Sun angle is 90° and solar radiation strikes the surface most directly at any given point in time.*

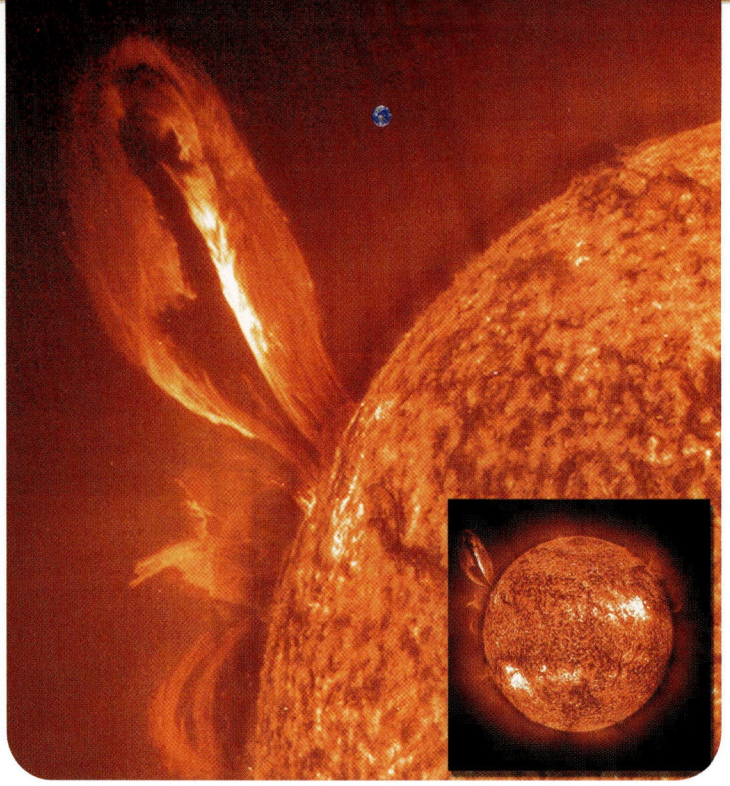

This image illustrates two interesting things about the Sun. First, notice how much bigger the Sun is than the Earth, which is the blue dot provided for scale. Second, look at the immense size of the solar flare compared to the Sun. Such flares are loops of gas that follow the pattern of the Sun's magnetic field.

THE SUN

The Sun is by far the most important natural feature regarding life on Earth. The Sun is generally believed to be about 4.6 billion years old. When compared to other stars, the Sun is not particularly special. It is quite small and mediocre compared to the millions of giant stars, multiple stars, pulsating stars, and black holes scattered throughout the Universe. Nevertheless, the diameter of the Sun is about 100 times greater than the Earth's, making the Sun an enormous object by that comparison.

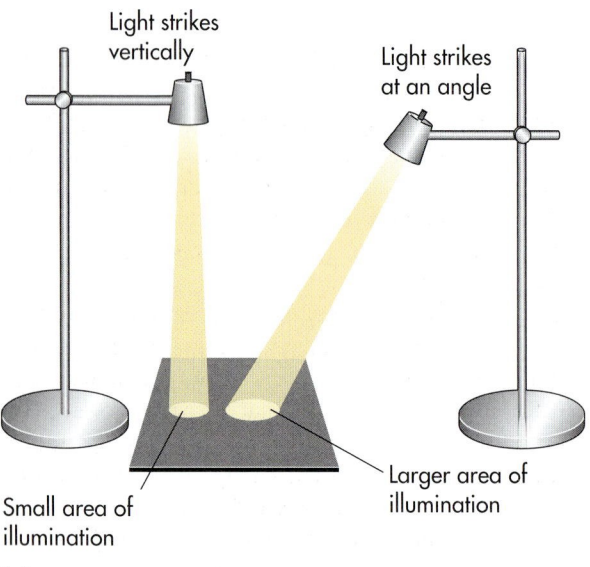

(a)

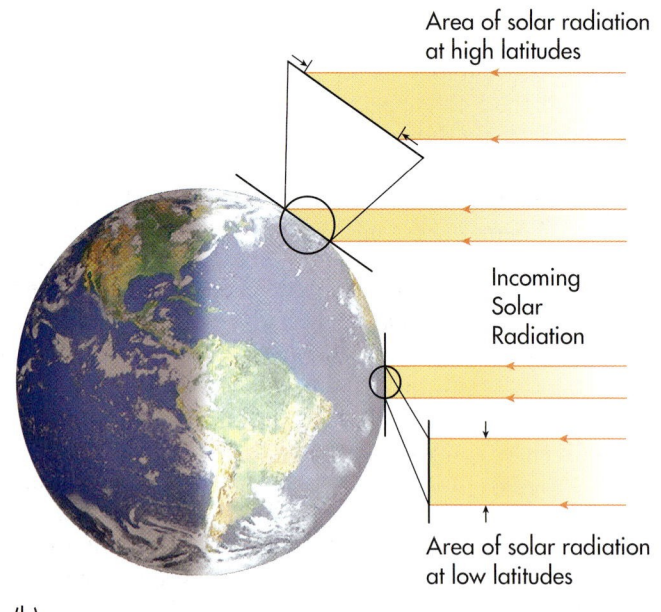

(b)

Figure 3.3 Interaction of solar radiation and the curved Earth. (a) You can use a desk lamp to show that light striking a surface at an angle illuminates a larger area (and is more diffuse) than light striking the surface vertically. (b) Solar radiation strikes the Earth most directly at low latitudes because the Sun angle is high, resulting in concentrated solar radiation in these regions. In contrast, solar radiation is more diffuse at high latitudes because the surface curves away from the radiation and the Sun angle is relatively low.

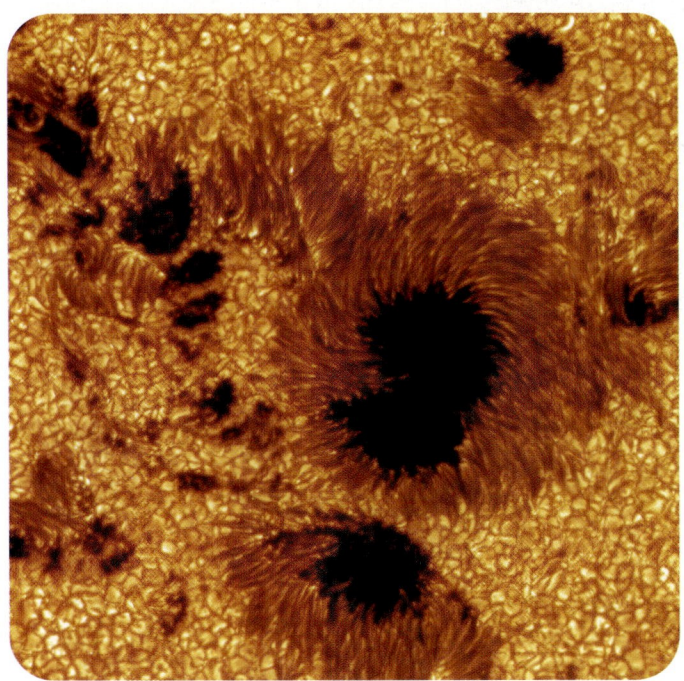

actor that converts hydrogen gas to helium. Given this process, the temperature of the Sun is incredibly high, ranging from about 15,000,000° C (27,000,000° F) at the center to about 6000° C (10,800° F) at the surface). The resulting heat energy produced by these incredible temperatures is the engine that powers the Earth. Traveling at the speed of light, it takes about 8.3 minutes for the Sun's energy to reach the Earth.

Another fascinating thing about the Sun is that it contains strong magnetic fields. These fields cause huge flares of hot gas to erupt from the surface in the shape of loops, as shown here. The Sun also has cooler areas scattered across its surface that appear dark in comparison with the hotter and brighter background; these areas are known as sunspots. The number of sunspots that appear at any one time rises and falls in a regular cycle that is closely connected to the Sun's magnetic field.

Although we take it for granted, the Sun is incredibly significant for us because it provides the power that drives most of the natural processes on Earth. This power is generated because the Sun basically functions as a gargantuan nuclear re-

Earth's Orbit Around the Sun

Let's now turn to the character of the Earth's orbit. As the Earth revolves around the Sun, it moves in a counterclockwise direction relative to a view above the North Pole on a flat (imaginary) plane referred to as the **plane of the ecliptic** (Figure 3.4). Although it may be difficult to visualize this plane, imagine that the Sun is in the middle of a large, round cake pan. If the planets were marbles on the cake pan, they would roll around the Sun only on the surface of the pan, not above or below it. In a very similar way, the Earth orbits the Sun on a flat plane, except that the plane of the ecliptic passes through the center of the Earth rather than beneath it, as marbles do on a cake pan.

It takes about 365 days (365.24 days to be exact) for the Earth to make one full revolution around the Sun. In comparison, Pluto requires 248 of our years to complete one orbit. Early astronomers who calculated Earth's or-

bital time length configured our calendar year to reflect this period of time. You may wonder why February has a variable number of days, depending on the year. Normally, February contains 28 days. Every fourth year (known as *Leap Year*), however, February contains 29 days. This day is added in order to correct for the uneven length of time (365.24 vs. 365 days) it takes for the Earth to complete its orbit.

As the Earth orbits the Sun, the Earth does not follow a perfectly circular path, but rather an elliptical one (see Figure 3.4). Note also that the Sun is not located in the exact center of our orbit and that the distance between the Earth and Sun varies over the course of the year, with an average distance of approximately 150 million km (93 million mi). The point where the Earth is closest to the Sun, approximately 147 million km (91.5 million mi) away, is called **perihelion.** In contrast, the term

Plane of the ecliptic   *The flat plane on which the Earth travels as it revolves around the Sun.*

Perihelion   *The point of the Earth's orbit where the distance between the Earth and Sun is least (~147 million km or 91.5 million mi).*

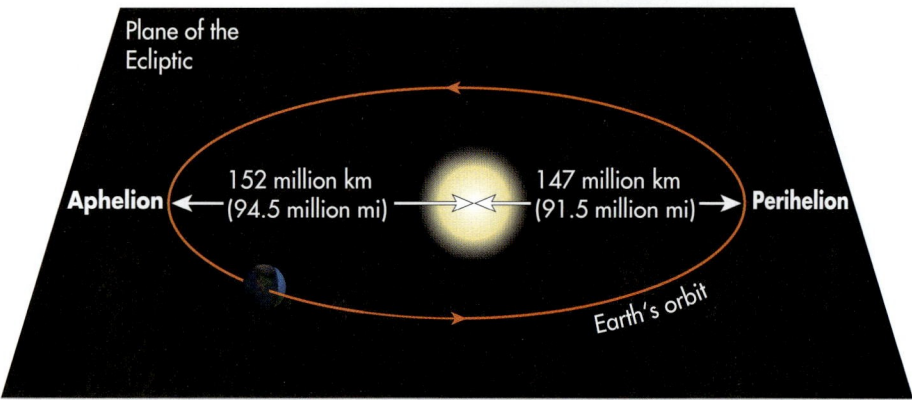

Figure 3.4 **Elliptical Earth orbit.** As the Earth travels on the plane of the ecliptic, it is closest to the Sun at perihelion on January 3. The Earth is farthest away from the Sun at aphelion on July 4.

aphelion refers to the point where the Earth is farthest from the Sun, about 152 million km (94.5 million mi) away. At the present time, perihelion and aphelion occur in early January and July, respectively.

Given the basic geometric relationship shown in Figure 3.4, what seems odd about the timing of our seasons in the Northern Hemisphere? The answer is that winter in the Northern Hemisphere occurs when we are closest to the Sun during orbital perihelion. Intuitively, you might think that we would experience our summer when the Earth is closest to the Sun. Since that is not the case, we can conclude that the distance of the Earth from the Sun does not cause our seasons. In other words, a different variable must be responsible for the seasonality we experience. That variable is the Earth's axial tilt.

The Earth's Rotation and Axial Tilt

As the Earth revolves around the Sun, it also rotates (spins) on its **axis,** which is an imaginary line that extends through the center of the Earth from pole to pole. When viewed from above the North Pole, the Earth rotates in a counterclockwise direction on its axis (Figure 3.5). It

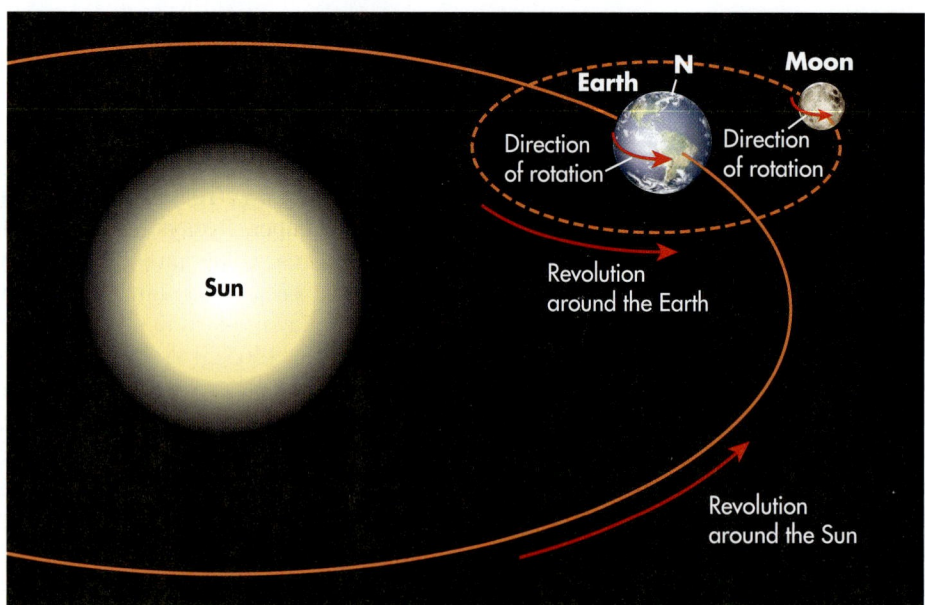

Figure 3.5 **Earth's revolution and rotation.** As the Earth revolves around the Sun, it rotates in a counterclockwise direction on its axis, when viewed from above the North Pole. Also note the geometrical relationship of the Earth and Moon.

Aphelion   *The point of the Earth's orbit where the distance between the Earth and Sun is greatest (~152 million km or 94.5 million mi).*

Axis   *The line around which the Earth rotates, extending through the poles.*

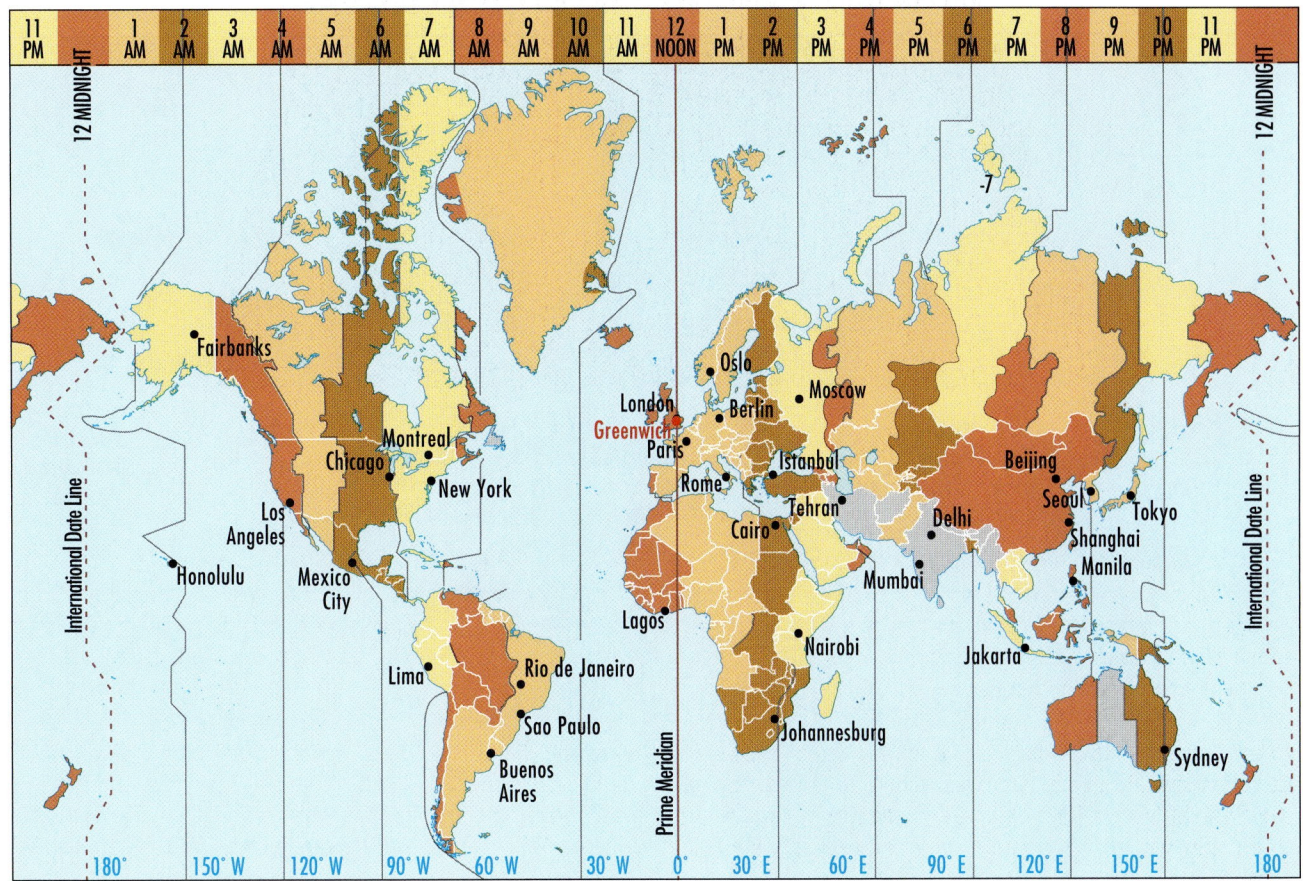

Figure 3.6 World time zones. The Earth is subdivided into 24 time zones. The international dateline, which lies at approximately 180° E and W longitude, is the arbitrary designation used to mark the beginning of each new day.

takes 24 hours for the Earth to make one complete rotation, which is the origin of our day length. During this interval, one-half of the Earth is always illuminated, whereas the other is in shadow. The boundary between day and night is known as the **circle of illumination** and is constantly migrating across the surface as the Earth rotates. You can see a good example of the circle of illumination by looking at a half Moon. Notice that one-half of the Moon is brightly illuminated, whereas the other half is completely dark. The dividing line between those two portions of the Moon is the circle of illumination.

In association with the Earth's rotational cycle, humans established 24 time zones that span specific portions of the planet (Figure 3.6). The contiguous U.S. contains four such areas, including (from east to west) the Eastern, Central, Mountain, and Pacific time zones (Figure 3.7). Each of these time zones covers about 15° of longitude because it takes 1 hour for the Earth to rotate that distance (360 degrees/24 h = 15°/h). Given the irregular nature of political boundaries, time zone boundaries are usually asymmetrical because people in individual states or countries want to be entirely within a particular time zone. Notice in Figure 3.7, for example, that the western boundary of the Eastern Time Zone jogs west to encompass most of Michigan's Upper Peninsula even though the far western part of the state nearly reaches the same longitude as St. Louis, which is in the Central Time Zone. This very irregular boundary was drawn because most people in Upper Michigan identify with Lower Michigan as far as time is concerned. A sliver of western Upper Michigan lies in the Central Time Zone because it borders Wisconsin. In this border area, it makes good economic sense that people on both sides of the state line be on the same time schedule.

The Prime Meridian in Greenwich, England, was chosen as the standard for the entire time zone system at the 1884 International Prime Meridian Conference in Washington, D.C. This system, referred to as Universal Time Coordinated (UTC), is used by the vast majority of countries on Earth, although some (such as Saudi Arabia) adhere to their own time systems. In the UTC system, time at any given place is calculated relative to how many hours ahead or behind that particular location is relative to the Greenwich meridian. The beginning and end of

Circle of illumination   *The great circle on Earth that is the border between night and day.*

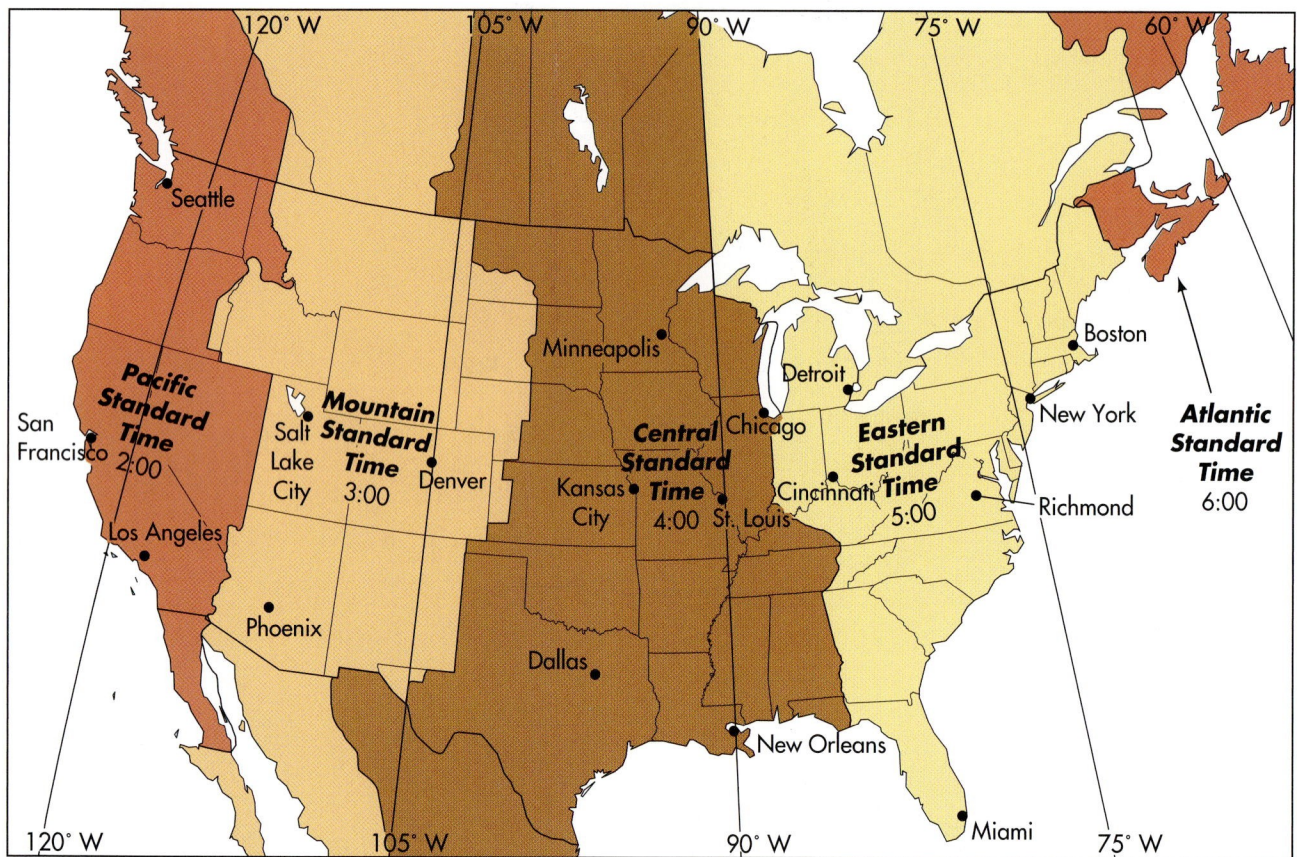

Figure 3.7 Time zones in the contiguous U.S. and Canada. The most eastern time zone in the contiguous U.S. is Eastern Standard Time, which is 3 hours ahead of the most western time zone, Pacific Standard Time.

each calendar day is the **International Date Line,** which is located at 180° E and W longitude. If you are traveling west, for example, on a flight from San Francisco to Sydney, Australia, you will cross the dateline and will immediately move into the next day (say, from Monday to Tuesday). Similarly, when you cross the dateline while traveling east (from Sydney to San Francisco), you will return to the previous day (such as from Tuesday back to Monday).

The axis of the Earth is the anchor around which the daily rotation occurs. However, it is a very important point of reference for another reason as well. On first consideration, you might think that the axis is perpendicular to the plane of the ecliptic; that is, at a 90° angle relative to the plane on which the Earth orbits the Sun. In fact, the Earth's axis is not perpendicular to the plane of the ecliptic, but is instead tilted 23.5° from a line perpendicular to the plane of the ecliptic as it revolves around the Sun (Figure 3.8). It is important to remember that the Earth

International Date Line *This line generally occurs at 180° longitude, with some variations due to political boundaries, and marks the transition from one day to another on Earth.*

maintains this degree of tilt *and* orientation with respect to the Sun throughout the year. A common misconception is that the tilt of the Earth somehow changes over the course of the year; *it does not*! This geometric relationship is important for you to understand because it is the reason for the seasons that we experience, as we explain in the next section.

> ### KEY CONCEPTS TO REMEMBER ABOUT OUR PLACE IN SPACE
>
> 1. Earth is the third planet from the Sun.
>
> 2. Although the Earth is essentially spherical, it is not perfectly round; in fact, it is an oblate spheroid.
>
> 3. Earth orbits the Sun on the plane of the ecliptic in an elliptical orbit.
>
> 4. Earth rotates on its axis, which results in our day/night cycle.
>
> 5. The axis of the Earth is tilted at 23.5° from a line perpendicular to the plane of the ecliptic.

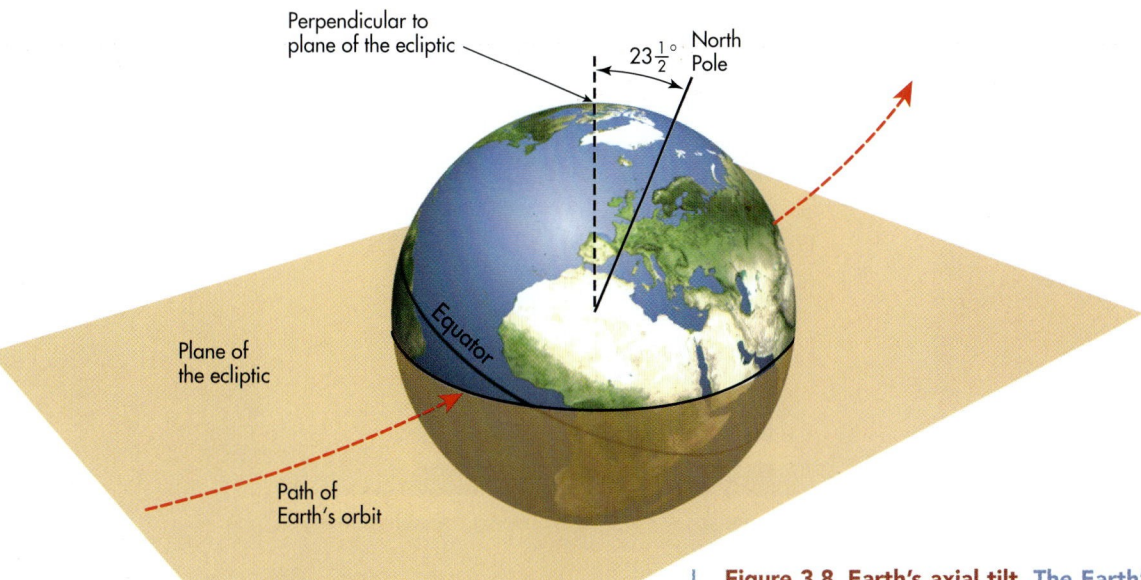

Perpendicular to plane of the ecliptic

$23\frac{1}{2}°$ North Pole

Plane of the ecliptic

Equator

Path of Earth's orbit

Figure 3.8 Earth's axial tilt. The Earth's axis is not perpendicular to the plane of the ecliptic. Instead, the axis is tilted 23.5° from a line perpendicular to the plane of ecliptic.

The Seasons

To understand how the Earth's axial tilt causes the seasons, first imagine what would happen if the Earth's axis were *not* tilted (Figure 3.9). Under these circumstances, as the Earth traveled around the Sun on the plane of the ecliptic, the Sun's rays would always strike the Earth most directly at the Equator. As a result, the subsolar point would perpetually be at the Equator and the Sun's rays would always strike the Earth at progressively lower angles with increasing higher latitudes (greater than 0°). In this scenario, as the Earth revolves around the Sun, all places on Earth would receive a consistent intensity of solar radiation (although this amount would differ with latitude) over the course of the year and there would be no distinct seasons at any one place on the planet.

Now picture the Earth's actual axial tilt, which is 23.5° from a line perpendicular to the plane of ecliptic. Keep in mind that the axis maintains a constant tilt angle *and* orientation as the Earth travels around the Sun (Figure 3.10). These geometric relationships—angle and orientation—are the fundamental cause of our seasons because one of the hemispheres is always in the gradual process of tilting toward the Sun, while the other one is slowly tilting away from the Sun. You might be thinking, "wait a minute, you just said that the axial tilt and orientation never change; isn't that contradictory?" To see why it's not, examine Figure 3.10 and focus on the Northern Hemisphere. Notice that during June, the Northern Hemisphere is tilted *toward* the Sun, whereas in December, it is titled *away*. The opposite is true for the Southern Hemisphere. In both cases the tilt did not change, nor did the orientation of the tilt. Instead, *the position of the Earth relative to the Sun changed as its orbit progressed*, which

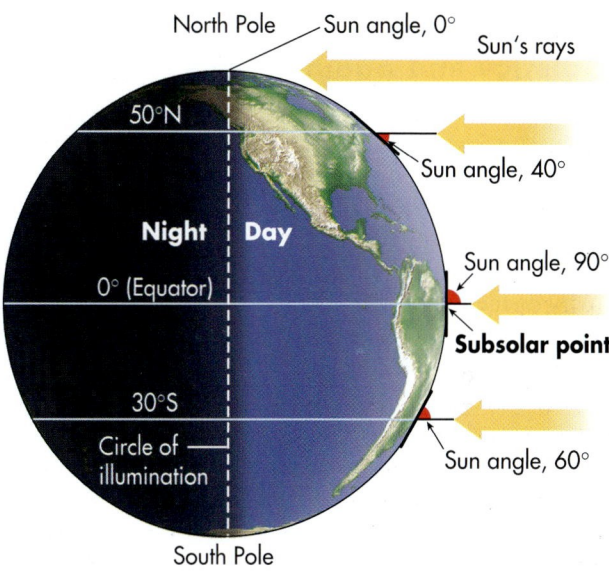

Figure 3.9 Geometric relationship if the Earth's axis were perpendicular to the plane of the ecliptic. If these conditions existed, the subsolar point would always be located at the Equator and the circle of illumination would extend from the North to South Poles. At 50° N the Sun angle would be 40°, whereas it would be 0° at 90° N (the North Pole).

has the effect of moving each hemisphere either toward or away from the Sun's rays. This apparent movement of the hemispheres toward or away from the Sun is the fundamental cause of the seasons because it results in the migration of the subsolar point 23.5° north or south of the Equator over the course of the year.

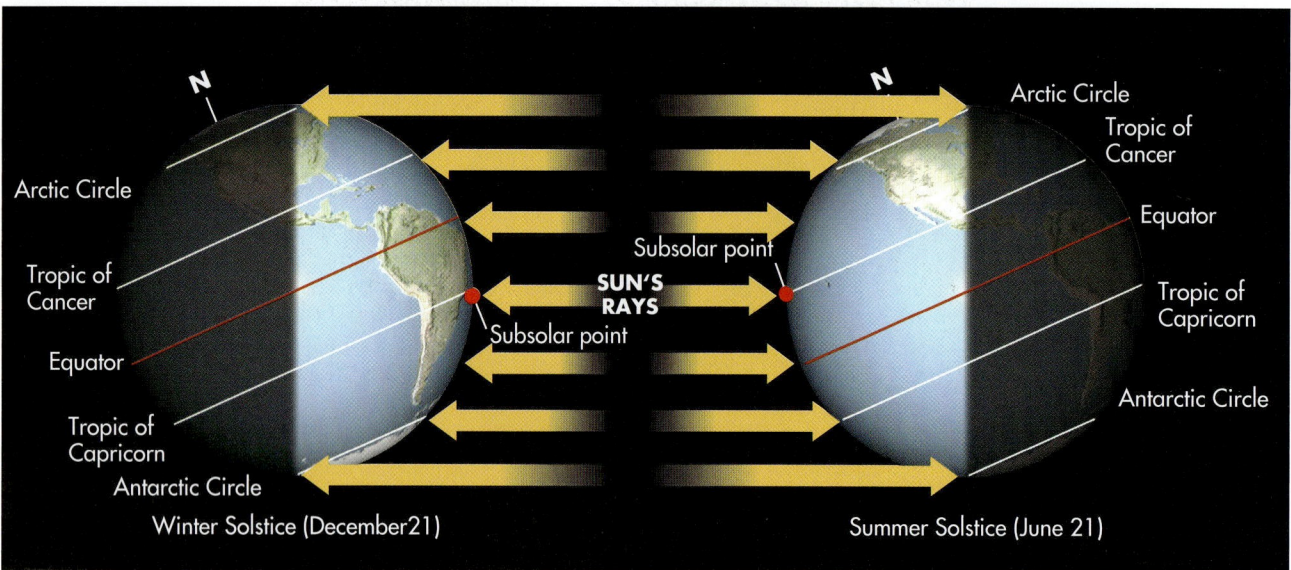

| Winter Solstice (December 21) | Summer Solstice (June 21) |

Figure 3.10 Axial geometry of the Earth during June and December. Note that the axis is tilted at 23.5°, which causes the orientation of the hemispheres to change with respect to the location of the subsolar point. On December 21, the subsolar point is at the Tropic of Capricorn, whereas it is at the Tropic of Cancer on June 21.

Solstice and Equinox

If you live in the Northern Hemisphere, you probably know that the recognized first day of summer is some time toward the end of June. Conversely, in the Southern Hemisphere, the first day of summer is toward the end of December. You also probably know that certain dates mark the first days of autumn, spring, or winter on the calendar. Why do these dates exist and what is their significance?

These dates are approximately the same every year because they indicate key periods within the Earth–Sun geometrical relationship. Remember that the Earth's axial tilt, in combination with its orbit, causes the subsolar point to move between 23.5° N and 23.5° S over the course of a year (Figure 3.11). In other words, the Sun angle is 90° at some point on the Earth's surface between these latitudes every day. Although this migration is a seamless process that we barely notice from day to day, early astronomers found it useful to mark the passage of the seasons by noting the date when the Sun's rays are perpendicular to the Equator and to the Tropic of Cancer (23.5° N) and Tropic of Capricorn (23.5° S). These two latitudes represent the highest latitude that the subsolar point reaches in each hemisphere during its seasonal cycle. The term *equinox* refers to the time when the subsolar point is at the Equator and all localities on Earth experience equal hours of daylight and darkness (12 of each). In contrast, the term *solstice* is used to denote when the Sun angle is 90° at either of the tropical boundaries.

In an effort to better understand when these dates occur and how they are related to Earth–Sun geometry, let's examine the passage of the seasons by beginning in March and moving forward in time. It will help if you refer to

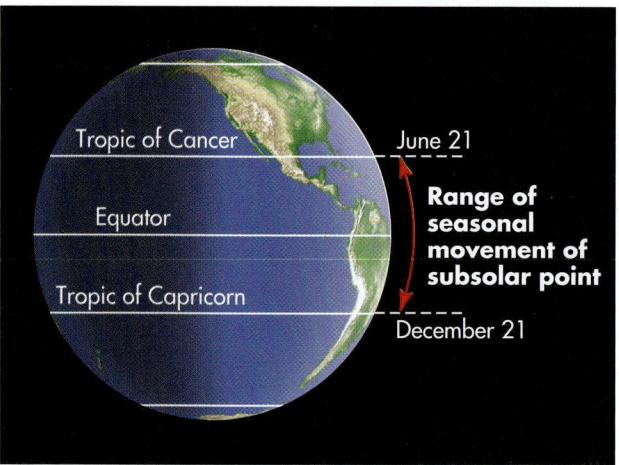

Figure 3.11 Range of seasonal subsolar point movement. The subsolar point migrates between the Tropic of Cancer and Tropic of Capricorn over the course of the year.

Figure 3.12 to view a graphical representation of the process and assume that your viewpoint of time is from the perspective of living in the Northern Hemisphere.

March 20–21: This date is the **Spring Equinox** (also called the *vernal equinox*) and represents the official

Spring Equinox *Assuming a Northern Hemisphere seasonal reference, the Spring (or vernal) Equinox occurs on March 20 or 21, when the subsolar point is located at the Equator (0°).*

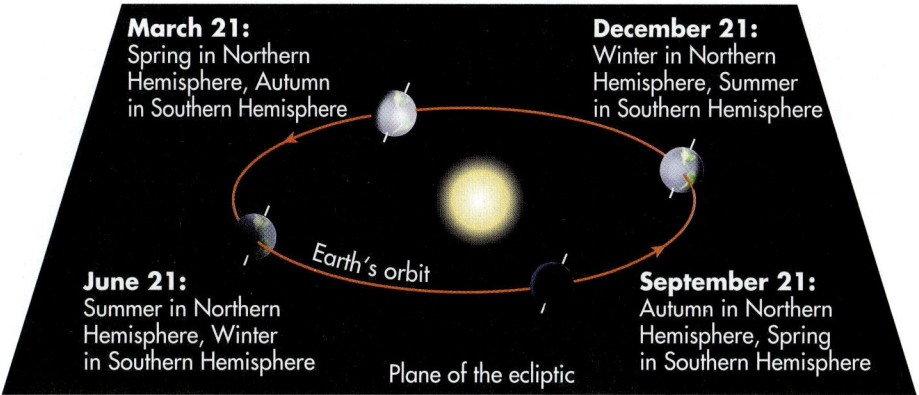

Figure 3.12 The orbital positions of the Earth during the Solstices and Equinoxes.
Note that the tilt of the Earth does not change; instead, the position of the Earth relative to the Sun changes. Due to this relationship, the intensity of the solar radiation that strikes any particular place on Earth varies over the course of the year.

beginning of the spring season in the Northern Hemisphere and the fall season in the Southern Hemisphere. This noteworthy date occurs because the subsolar point is located at the Equator and neither the Northern nor the Southern Hemisphere is tilted toward the Sun on this date. In Figure 3.12, notice that the circle of illumination extends through each pole.

June 20–21: This date is the **Summer Solstice** and represents the first official day of summer in the Northern Hemisphere and winter in the Southern Hemisphere. The Summer Solstice occurs because, following the Spring Equinox, the Earth continues to orbit the Sun in a counterclockwise direction when viewed from above. As the orbit progresses, the Northern Hemisphere slowly reaches its position of greatest tilt toward the Sun. Meanwhile, the Southern Hemisphere is tilted away from the Sun. During this solstice the subsolar point is located at the **Tropic of Cancer** (23.5° N). In Figure 3.12, notice that all latitudes above 66.5° N (the Arctic Circle) receive 24 hours of daylight during this period of time; this is why these high latitudes are called the *land of the midnight Sun*. At the same time the northern arctic regions are in perpetual sunlight, all locations above 66.5° S (the Antarctic Circle) experience 24 hours of darkness. This

continual darkness occurs because the Southern Hemisphere is tilted away from the Sun during this time and those latitudes above the Antarctic Circle never rotate into the illuminated part of the Earth.

September 22–23: This date is the **Fall Equinox** (also called the *autumnal equinox*) and represents the first official day of fall in the Northern Hemisphere and spring in the Southern Hemisphere. At this time the Earth's orbit has progressed such that the subsolar point is now once again at the Equator (Figure 3.12). As at the Spring Equinox, all locations on Earth get equal hours of day and night (12 hours daylight, 12 hours darkness) because neither the Northern nor the Southern Hemisphere is tilted toward the Sun and the great circle of illumination extends from pole to pole.

December 21–22: This date is the **Winter Solstice** and represents the first official day of winter in the Northern Hemisphere and summer in the Southern Hemisphere. Given the progression of the orbit since the Fall Equinox, the Earth is now positioned such that the Northern Hemisphere is at its greatest tilt away from the Sun, while the Southern Hemisphere is tilted toward it. At this time the subsolar point is located at 23.5° S, which is also known as the **Tropic of Capricorn.** As a result of this

---

Summer Solstice *Assuming a Northern Hemisphere seasonal reference, the Summer Solstice occurs on June 20 or 21, when the subsolar point is located at the Tropic of Cancer (23.5° N).*

Fall Equinox *Assuming a Northern Hemisphere seasonal reference, the fall (or autumnal) equinox occurs on September 22 or 23, when the subsolar point is located at the Equator (0°).*

Winter Solstice *Assuming a Northern Hemisphere seasonal reference, the Winter Solstice occurs on December 21 or 22, when the subsolar point is at the Tropic of Capricorn (23.5° S)*

Suppose the subsolar point is at 10° S latitude. In other words, the Sun angle at that latitude is 90° at solar noon. Which of the following months are logical times of the year when the subsolar point would be at this latitude? Explain.

a) October
b) December
c) February
d) June

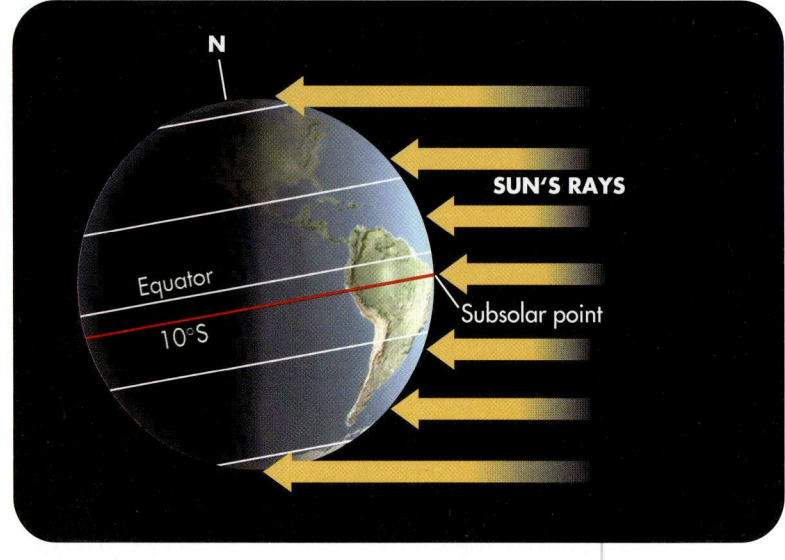

subsolar position, all latitudes above the Arctic Circle experience 24 hours of darkness, whereas those higher than the Antarctic Circle experience 24 hours of daylight.

1. If the axis of the Earth were not tilted, there would be no seasons.

2. The subsolar point is the place where the Sun's rays strike the Earth most directly. This occurs because the Sun angle is 90° at that point and the rays are perpendicular to the surface.

3. Due to the axial tilt, the subsolar point migrates between the Tropic of Cancer and Tropic of Capricorn over the course of the year.

4. A solstice occurs whenever the subsolar point is at either 23.5° N or S latitude. An equinox, in contrast, occurs when the Sun is directly over the Equator.

5. The position of the Earth relative to the Sun changes as the orbit progresses.

GEODISCOVERIES   WWW.WILEY.COM/COLLEGE/ARBOGAST

ORBITAL VIEW AND EARTH AS VIEWED FROM THE SUN

Let's reinforce the important concept of axial tilt by interacting with some simulations. First, go to the *GeoDiscoveries* website and select the module **Orbital View**. This simulation shows the Earth orbiting the Sun, with the Earth's axis indicated. This simulation has close-up views of Earth at the solstices and equinoxes, showing the Earth rotating (so lengths of daylight periods are indicated), the Earth's axis (indicating axial tilt), and

the portion of Earth that is illuminated at each of the four times. Click on the "Orbit," "Fall Equinox," "Winter Solstice," "Spring Equinox," and "Summer Solstice" buttons to see different perspectives. Pay particularly close attention to the tilt and orientation of the Earth, and the fact that these variables do not change during the orbit. In addition, be sure to notice where the subsolar point is at any given point in time and compare this migration to Figure 3.12 in the text. Lastly, change the axial tilt and explore how the migration of the subsolar point changes. Remember that the subsolar point represents the place where solar radiation is striking the Earth most intensely. Notice how, with increased tilt, the subsolar point migrates a great deal. With less tilt, the subsolar point migrates less.

After you interact with the **Orbital View** simulation, select the module **Earth as Viewed from the Sun**. This part of the module provides a different view of the Earth–Sun relationship by simulating the Earth as it would appear if viewed from the Sun. In this simulation, imagine that you're the Sun and that radiation is streaming (like a laser beam) from your eyes directly toward Earth. As the months progress in time, notice how the *apparent* tilt of the Earth changes, relative to your view (as the Sun). Remember that

the *actual* tilt does not change; instead, the position of the Earth in its orbit around the Sun progresses, which gives the impression of a fluctuating tilt when seen from the Sun. In this simulation, you will also be able to adjust the amount of axial tilt on the Earth and see how the migration of the subsolar point changes as viewed by you (the Sun). Again, notice that the subsolar point migrates more with increased tilt and less with decreased tilt. Once you complete this simulation, examine Figure 3.13 to review where the subsolar point strikes the Earth, with the tilt at 23.5° on each of the solstices and equinoxes. Be sure to answer the questions at the end of the module to make sure you understand this concept.

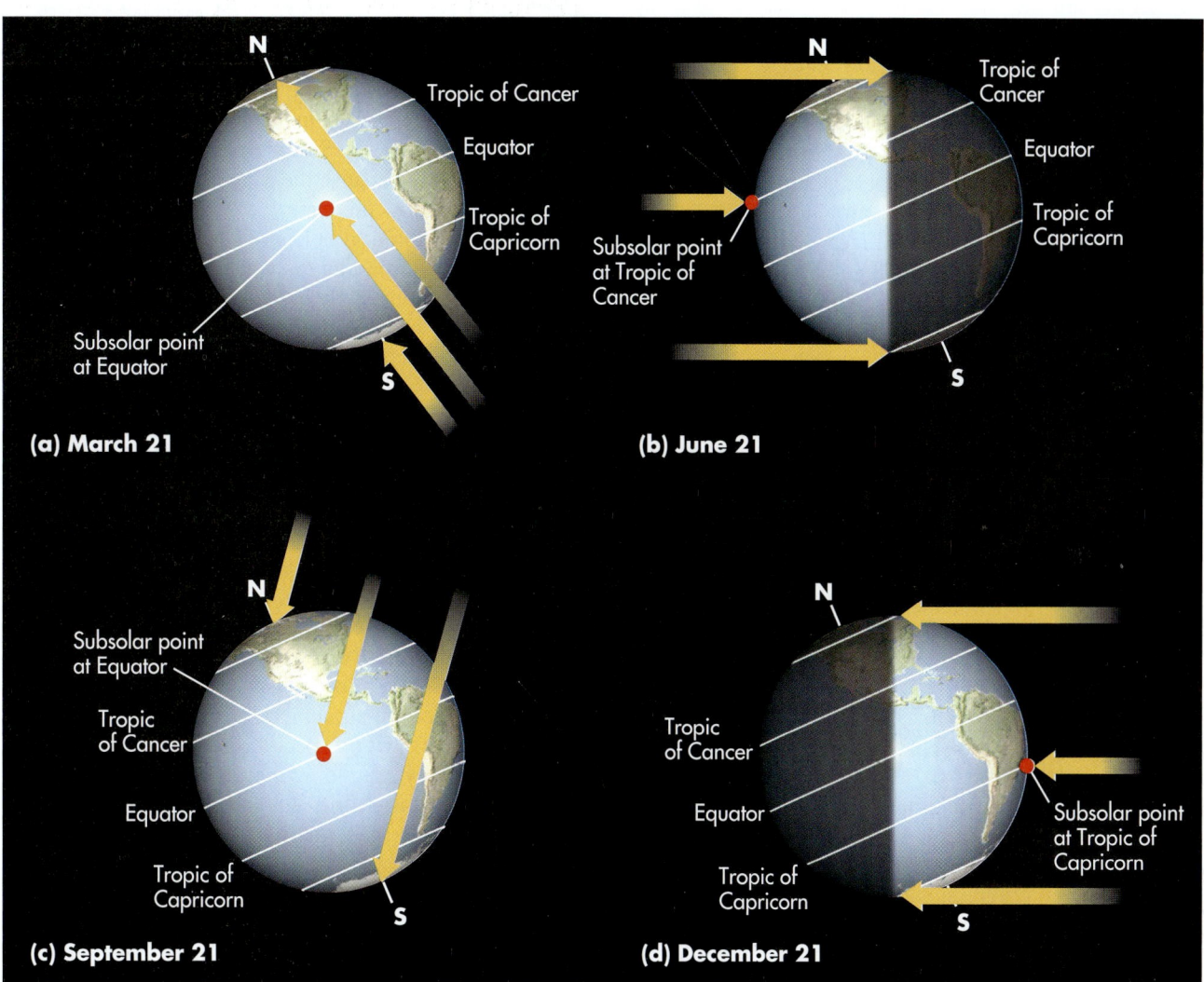

Figure 3.13 Seasonal migration of the subsolar point. Due to the tilt of the Earth's axis, the subsolar point migrates between the tropics over the course of the year. For perspective, imagine that the Sun is in front of the Earth in (a) and behind the Earth in (c). Further, imagine that it's to the left of the Earth in (b) and right of the Earth in (d).

How We See Earth–Sun Geometry on Earth

We've examined Earth–Sun geometry as it looks when viewed from space. Now let's explore how this geometry looks from the surface of the Earth; in other words, how *we* see and experience it. Let's begin with the basic principle of day and night, which is related to the Earth's rotation on its axis.

Day and Night

Over the course of the day, you can "see" the Earth's rotation by noting the position of the Sun in the sky. Everyone knows that the Sun "rises" in the east and then arcs across the sky toward the west. As the arc progresses over the course of the day, the Sun reaches its highest position at solar noon and then gradually lowers as it "sets" in the west. Solar noon at any given place is not necessarily equivalent to standard clock time because the time of sunrise and sunset varies across a time zone due to the curvature of the Earth. Although it appears that the Sun is actually rising in the morning and moving across the sky during the course of the day, the fact is that the apparent motion of the Sun is caused by the rotation of the Earth on its axis. In short, your day begins, wherever you may live, because your part of the Earth has rotated to a position where the Sun begins to illuminate your home. If you ever have the chance to watch the Sun rise, watch very closely and you can actually see the Earth rotate. (Never look directly at the Sun, however, because it can cause serious eye damage!)

At the same time that your day is beginning on one side of the circle of illumination, night has begun on the opposite side of the Earth because that place has rotated such that the Sun is no longer illuminating it. Remember, the Sun is always illuminating half of the Earth at any given time, whereas the other half is in shadow.

Seasonal Changes in Sun Position (Angle) and Length of Day

In addition to the daily or diurnal cycle of day and night, which is related to the Earth's rotation, we can also see a seasonal cycle in the Sun's position in the sky, which depends on orbital progression. Whereas the day–night cycle is obvious and can be easily seen because of the apparent east–west arc of the Sun, the seasonal migration of the Sun's position is a slow progression in north–south directions relative to the Equator. This migration results in a seasonal variation in the angle of the Sun's rays with respect to any location on Earth. To follow this migration, the best point of reference is the daily position of the Sun at solar noon at any given place because that is when the Sun is at its highest point in the arc at that locality. As we examine the seasonal migration of the noontime Sun, imagine that you're living at 45° N (but remember that what you see is opposite for those who reside in the Southern Hemisphere). Also, keep in mind that the overall seasonal north–south migration of the Sun is occurring in combination with the Sun's daily arc.

If you live in the middle latitudes of the Northern Hemisphere, you will notice that the Sun arcs across the southern part of the sky over the course of the day. At solar noon, it is directly to the south rather than overhead because the subsolar point never reaches a point higher than 23.5° N at any time in the year. Given that the Sun is in the southern sky at solar noon, shadows from buildings and trees project toward the north. In the context of the seasons, the noon Sun is highest in the sky during summer. In other words, the Sun angle is greatest during summer. This occurs because the Sun is perpendicular to the Tropic of Cancer at this time.

As the season moves from summer to fall, the position of the noontime Sun (as you see it) migrates farther south in the sky on a daily basis. Why? This southerly migration occurs because the subsolar point moves toward the Equator (which is to the south) as fall approaches and Sun angle decreases. With the gradual approach of winter, the noontime Sun moves still farther to the south because the subsolar point is migrating toward the Tropic of Capricorn. At this time, the Earth is in a position within its orbit where the Northern Hemisphere is tilted away from the Sun. If you happen to live in the Northern Hemisphere, the noontime Sun is lowest in the southern sky (in other words, the Sun angle is the least) on December 21, or the Winter Solstice. From this point on, the noontime Sun slowly migrates back to the north, as Sun angle increases, until the Summer Solstice.

You can see the combined daily and seasonal migrations of the Sun, with respect to the Earth's surface, in a graphical way by examining Figure 3.14. This diagram is called a celestial dome and shows the Sun's daily arc and seasonal migration relative to the Earth's surface. In this particular diagram, the dome is oriented with respect to a person standing at 45° N latitude. The surface of the Earth is presented as a two-dimensional feature that extends from pole to pole, with the arc and position of the Sun shown at three periods: Summer Solstice, Winter Solstice, and Equinox. In other words, this diagram illus-

---

Solar noon *The time of day when the Sun angle reaches its highest point as the Sun arcs across the sky.*

Diurnal cycle *A 24-hour cycle.*

Celestial dome *A sphere that shows the Sun's arc, relative to the Earth, in the sky.*

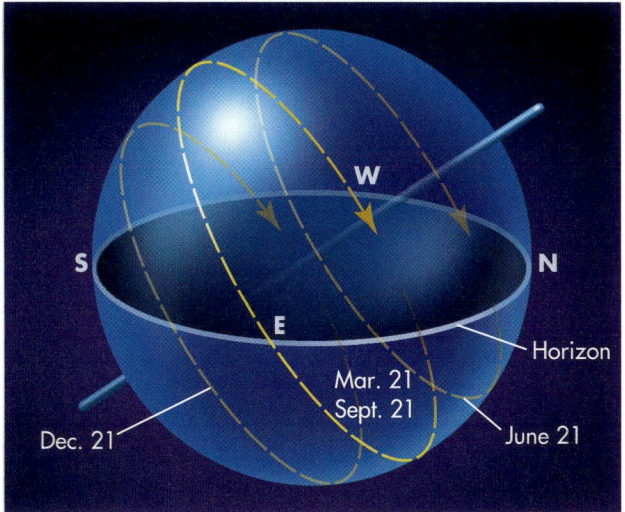

(a)

trates how you would see the Sun as the day progresses at three different times of year.

One of the most important ways in which we notice the seasonal change in Sun angle is with respect to the number of daylight hours, or the one length of day. Look again at the celestial dome in Figure 3.14 or the one in the *Celestial Dome* module and notice how the curvature of the Earth influences when the Sun emerges from the horizon in the east or disappears beneath it in the west. You should be able to see that the Sun emerges from the horizon earlier and disappears later in the summer months than during winter. It should make sense, too, that this seasonal variation is more pronounced at higher latitudes than at lower ones. Table 3.1 illustrates this concept by showing the day length at five different latitudes in the Northern Hemisphere at the equinoxes and solstices. Notice how the day length varies among the five sites.

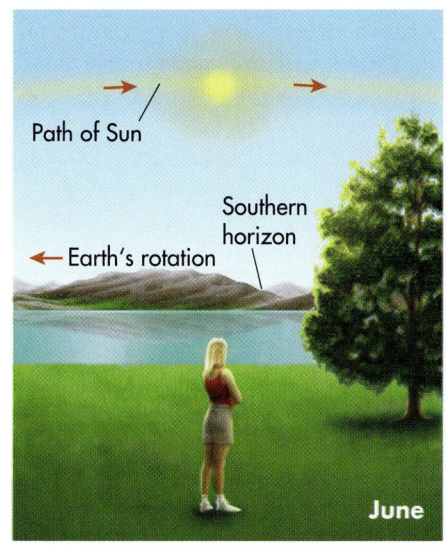
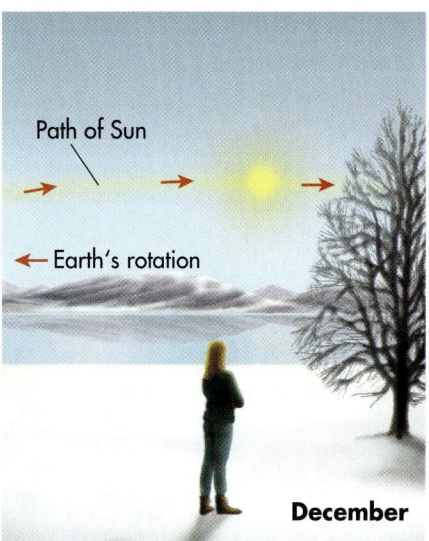

(b)

Figure 3.14 The celestial dome. (a) The Sun's arc at the Winter Solstice, Equinoxes, and Summer Solstice when viewed from 45° N latitude. (b) The position of the Sun at solar noon varies dramatically between the winter and summer seasons. It is much higher in the sky in summer and lower in the sky in winter. This difference occurs due to the Earth's axial tilt.

TABLE 3.1	Variations in Day Length on Equinoxes and Solstices Across Latitude				
		Day Length on Equinox or Solstice			
Location	Latitude (N)	March 20	June 21	September 22	December 21
Bogota, Colombia	4°32′	12 h, 06 min	12 h, 24 min	12 h, 07 min	11 h, 51 min
Miami, Florida	25°46′	12 h, 08 min	13 h, 45 min	12 h, 09 min	10 h, 032 min
Topeka, Kansas	39°07°	12 h, 07 min	14 h, 55 min	12 h, 13 min	9 h, 26 min
Calgary, Alberta	45°01′	12 h, 12 min	16 h, 33 min	12 h, 15 min	7 h, 56 min
Barrow, Alaska	71°03′	12 h, 23 min	24 h, 00 min	12 h, 22 min	00 h, 00 min

GEODISCOVERIES WWW.WILEY.COM/COLLEGE/ARBOGAST

CELESTIAL DOME

To enhance your understanding of Sun angle and seasonality, go to the *GeoDiscoveries* website and select the module, **Celestial Dome**. This module provides you with an opportunity to interact with an animated celestial dome. Change the latitude from several options to examine seasonal change in the arc of the Sun, and its height above the horizon at equinox and solstice. Be sure to notice how the Sun arc and angle vary between low- and high-latitude locations, as well as between the hemispheres. Once you complete the module, be sure to answer the questions at the end to make sure you understand this concept.

GEODISCOVERIES WWW.WILEY.COM/COLLEGE/ARBOGAST

SUN ANGLE AND LENGTH OF DAY

Let's now take a comprehensive look at Earth–Sun geometric relationships and how they influence the length of day. Go once again to the *GeoDiscoveries* website and select the module **Sun Angle and Length of Day**. As you watch this animation, notice how and why day length at three different latitudes varies over the course of the year. Relate your observations with the **Celestial Dome** animation and Table 3.1. After you complete the animation, be sure to answer the questions at the end of the module to test your understanding of this concept.

VISUAL CONCEPT CHECK 3.2

This cross section of the Eastern Time Zone in the United States ranges from Boston, Massachusetts, on the east to Holland, Michigan, on the west. Although both cities occur at the same line of latitude (42° N), the timing of sunrise and sunset varies between the two locations. The Sun rises about an hour earlier in Boston than it does in Holland. Conversely, the Sun sets about an hour later in Holland than it does in Boston. Considering the rotation of the Earth and basic Earth–Sun geometry, why does this variation occur?

SUNRISE IN THE SOUTHERN HEMISPHERE

This beautiful sunrise is seen from the top of Te Mata Peak, which lies on the east coast of New Zealand, about one hour south of the city of Napier. A geographer knows that because New Zealand is in the Southern Hemisphere, south of the Tropic of Capricorn, the Sun will arc across the northern part of the sky, rather than the southern sky as it does in the Northern Hemisphere. Thus, in the Southern Hemisphere,

as you face the sunrise, you're looking northeast. In contrast, if you watched the sunrise from the top of Mt. Rainier, in the northern United States, you would be facing southeast.

Te Mata Peak has an interesting geographic distinction. Because it is just to the west of the International Date Line, it is one of the first places on Earth where the Sun can be seen on any given day.

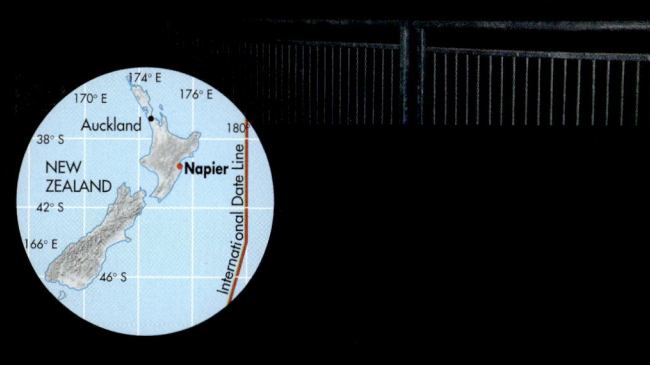

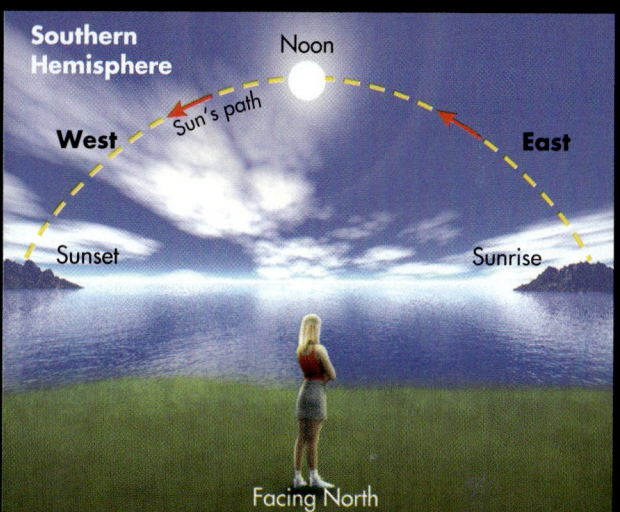

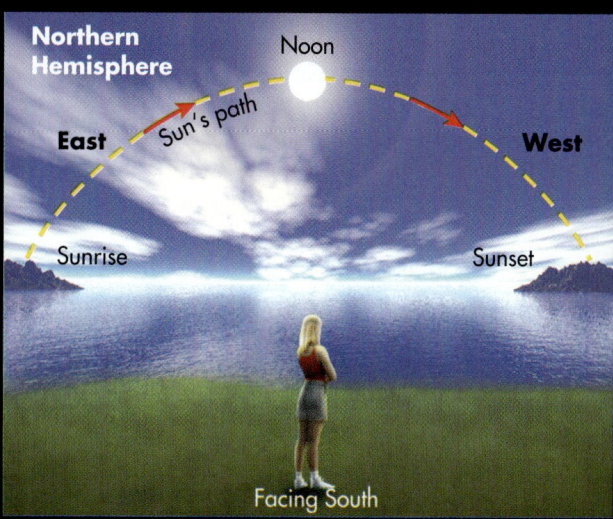

The Big Picture

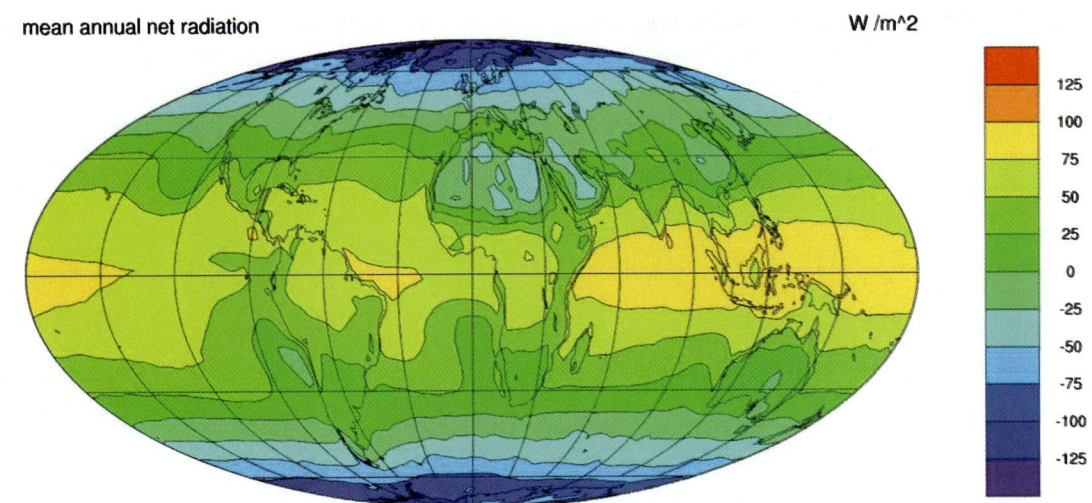

The Earth–Sun relationship—especially the tilt of the Earth's rotation axis as it travels around the Sun—causes variations in the spatial distribution of solar radiation on Earth. This variability has a profound impact on a variety of geographic patterns on Earth, including atmospheric circulation, the distribution of global climates, and even vegetation. In the next chapter, we explore how this solar radiation interacts with the atmosphere around the planet and how this interaction affects surface temperatures. To see a preview of the material covered in the next chapter, examine the accompanying figure, which shows the geography of net radiation on Earth. Net radiation is the difference between incoming and outgoing solar radiation. Notice that the highest amounts of net radiation are generally at low latitudes, whereas the lowest amounts are at high latitudes. Given what was covered in this chapter, you should have a basic idea of why this pattern occurs. In the next chapter we will investigate these relationships, as well as others, more closely.

Summary of Key Concepts

1. The Earth revolves around the Sun within an elliptical orbit in the plane of the ecliptic. As the Earth revolves, it rotates on its axis, which is tilted at 23.5° from a line perpendicular to the plane of the ecliptic. It takes approximately 365 days for one complete revolution to occur.

2. The subsolar point is the band of latitude on Earth where the Sun angle is 90° and thus, where solar radiation strikes the Earth most directly. Due to the axial tilt, the subsolar point migrates between the Tropic of Cancer and Tropic of Capricorn between June 21–22 (Northern Hemisphere Summer Solstice) and December 21–22 (Northern Hemisphere Winter Solstice), respectively. The Sun is directly over the Equator at solar noon on both the Spring and Fall Equinoxes (March 21–22 and September 21–22, respectively).

3. Day and night on Earth occur because the Earth rotates on its axis. It takes 24 hours for a complete rotation to occur.

Check Your Understanding

1. Chicago, Illinois, and Omaha, Nebraska, are both located in the Central Time Zone, with Chicago being on the east side of the zone and Omaha on the west. With respect to time, which of these two cities is the first to "see" the sunrise? In your answer, remember that the Earth is essentially round.

2. If the axis of the Earth were perpendicular to the plane of the ecliptic, where would the subsolar point be on December 21?

3. Imagine that the axis of the Earth were tilted at 45°. In this case, where would the Tropic of Cancer be located?

4. If the axis of the Earth were tilted at 55°, where would the subsolar point be located on June 20–21? Do you know why?

5. If the axis of the Earth were tilted at 40°, what would the Sun angle be at 40° S latitude on December 21–22?

6. Which direction, north or south, do shadows project during solar noon on December 21–22 at 10° S latitude?

7. Which direction, north or south, do shadows project during solar noon on December 21–22 at 40° S latitude?

8. Where does the greatest range in daylight hours occur, at high or low latitudes?

9. If the axis of the Earth were perpendicular to the plane of the ecliptic, how many hours of daylight would there be at 45° S latitude on December 21–22?

10. Assume that you are standing on the Equator and it's June 21–22. Given your perspective, is the noontime Sun in the northern or southern part of the sky?

ANSWERS TO VISUAL CONCEPT CHECKS

Visual Concept Check 3.1

Both October and February are logical times of the year when the subsolar point would be at 10° S latitude. In October, the Sun is in the process of moving from the Equator (0°) to the Tropic of Capricorn (23.5° S). After the subsolar point reaches 23.5° S in late December, it begins to move back to the Equator, which it will reach on or about March 21. To do so, however, it must pass by 10° S, which it does in February.

Visual Concept Check 3.2

The Sun rises and sets earlier in Boston, Massachusetts, than in Holland, Michigan, because the surface of the Earth is curved. When the Sun rises in Boston, Holland lies on the part of the Earth that is curved away from the Sun's light. As the Earth rotates, the Sun ultimately "rises" in Holland, but about an hour later than in Boston. The Sun sets later in Holland than in Boston because Boston rotates out of sunlight about an hour before it does in Holland.

CHAPTER FOUR

THE GLOBAL ENERGY SYSTEM

We've examined the Earth's geometric relationship with the Sun and seen how it generally

influences the distribution of solar radiation seasonally and by latitude. Now we'll look more

closely at the way this radiation is intercepted by the Earth, how it flows through the

atmosphere, how it interacts with the land and ocean surfaces, and how, on a global scale, it

maintains a balance with outgoing radiation. Understanding these relationships is critical to

Sunrise over the Pacific Ocean. The Sun sends a stream of electromagnetic energy to Earth. This energy interacts with the Earth in a wide variety of ways that are discussed in this chapter.

comprehending many Earth processes that we'll discuss in later chapters, such as wind,

weather, ocean currents, and rivers, because the Sun provides the energy required to power

them, either directly or indirectly. We'll begin by investigating the concept of radiation energy

and how it's measured.

The Electromagnetic Spectrum and Solar Energy

The Electromagnetic Spectrum

Radiation is electromagnetic energy that is transmitted in the form of waves. These waves can be measured in terms of their length and amplitude, with **wavelength** being the distance between wave crests, and the term **wave amplitude** referring to half the height between the wave crest and wave trough (Figure 4.1). The entire wavelength range of electromagnetic energy is called the **electromagnetic spectrum.**

Within the electromagnetic spectrum, wavelengths range from *gamma rays* that are less than a nanometer long (one billionth or 10^{-9} of a meter) to *radio waves* that are tens of meters in length (Figure 4.2). Humans can see radiation directly only in the *visible light* part of the spectrum, which ranges from violet with the shortest wavelength (375 nm) to red with the longest wavelength (740 nm).

You should remember two very important principles about electromagnetic radiation. The first principle is that an *inverse* relationship exists between the temperature of an object and the wavelength of the electromagnetic radiation it emits. In other words, hotter objects emit radiation with shorter wavelengths than do cooler objects. For example, the Sun has a surface temperature of about 6000° C (11,000° F) and emits energy as **shortwave radiation,** which includes gamma rays, X-rays, ultraviolet (UV) radiation, visible light, and near-infrared radiation. The Earth, in contrast, is a much cooler object (~16° C, 61° F) and thus emits **longwave radiation** in the thermal infrared part of the spectrum (Figure 4.3). Keep in mind, though, that the vast majority of energy released by the Earth initially originated at the Sun.

The second principle of electromagnetic radiation is that there is a *direct* relationship between the absolute temperature of the object and the amount of radiation it emits. This relationship is described by the Stefan-Boltzmann Law and basically means that hotter objects emit more radiation than cooler ones. As a result, the Earth emits much less radiation than the Sun. This temperature/emitted radiation relationship is exponential, meaning that a small temperature increase in an object can result in very large

Wavelength *The distance between adjacent wave crests or wave troughs.*

Wave amplitude *The overall height of any given wave as measured from the wave trough to the wave crest.*

Electromagnetic spectrum *The radiant energy produced by the Sun that is measured in progressive wavelengths.*

Shortwave radiation *The portion of the electromagnetic spectrum that includes gamma rays, X-rays, ultraviolet radiation, visible light, and near-infrared radiation.*

Longwave radiation *The portion of the electromagnetic spectrum that includes thermal infrared radiation.*

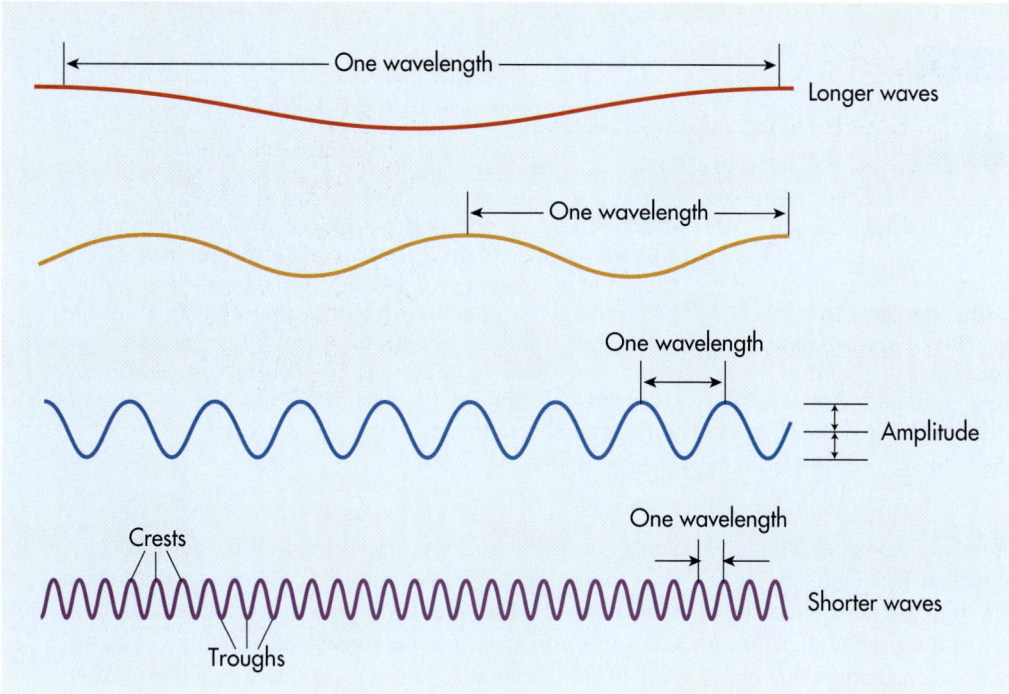

Figure 4.1 **Relative sizes of electromagnetic waves.** The distance from one crest to the other, or from one trough to another, is called the wavelength. Wave amplitude refers to half the distance from the wave crest to the wave trough.

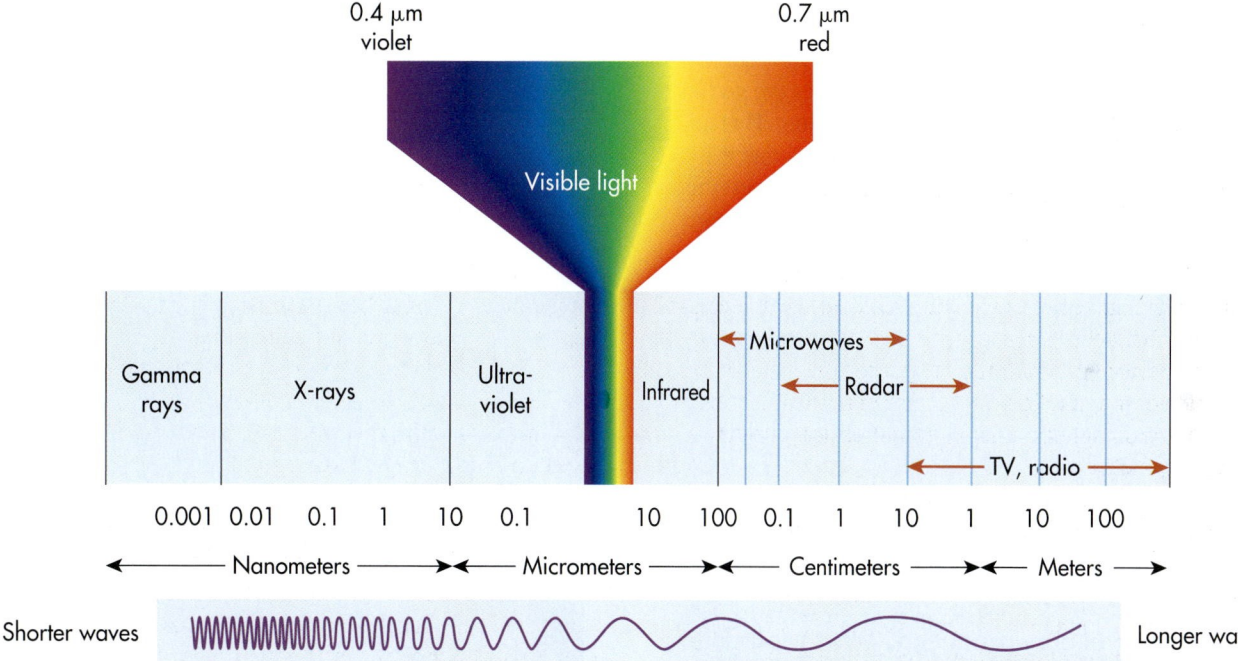

Figure 4.2 **The electromagnetic spectrum.** Wavelengths range from gamma rays, which are the shortest, to long radio waves.

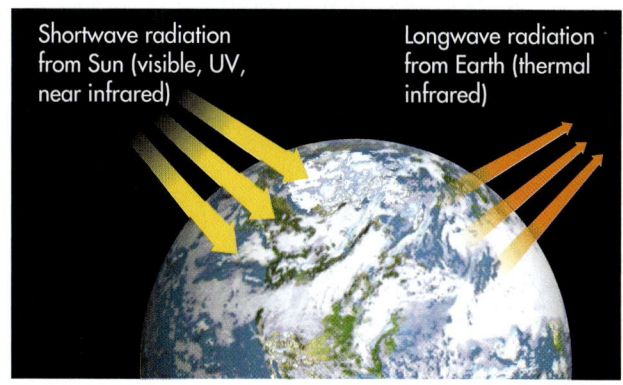

Figure 4.3 **Radiation to and from Earth.** The Earth receives shortwave radiation from the Sun. Some of this radiation reaches the surface of the Earth where it is absorbed and then emitted as longwave energy.

THE ELECTROMAGNETIC SPECTRUM

In an effort to better visualize the various components of the electromagnetic spectrum, go to the *GeoDiscoveries* website and select the module **The Electromagnetic Spectrum**. You will be able to interact with elements of the electromagnetic spectrum by selecting different wavelengths, such as those in the infrared and X-ray parts of the spectrum. When you select a specific wavelength, an image will appear that is representative of that part of the spectrum. In this way you will develop a better feel for the electromagnetic spectrum, how it is a pervasive part of the Earth, and how geographers use it to study the environment. Once you complete the interaction, be sure to answer the questions at the end of the module to test your understanding of this concept.

increases in emitted radiation. For example, you've probably experienced how a tar or concrete surface emits much more radiation, and is therefore much hotter, than a plowed field or grassy park that it might surround. This higher release of radiation occurs because the highway or sidewalk is warmer than the field or park.

Solar Energy and the Solar Constant

As we discussed in Chapter 3, solar energy is created in vast quantities by nuclear fusion within the Sun. This energy works its way to the surface of the Sun, where it's emitted as electromagnetic radiation. From this point this energy travels along straight lines (or rays) through space at the speed of light.

Given that the Sun produces energy at a nearly constant rate, the output of solar radiation is also nearly constant. Although some variability in solar output occurs due to the waxing and waning of energy production within an 11-year solar cycle, the overall output of solar energy is considered to be consistent over time. As a result of this generally constant production and emission, the amount of solar radiation received in an area of fixed size in space (outside the Earth's atmosphere), and at right angles to the Sun, is also constant. This amount of received energy, which is referred to as the **solar constant,** has a value of about 1370 watts per square meter (W/m^2) at the top of the atmosphere.

Solar constant   *The average amount of solar radiation ($\sim$1370 W/m²) received at the top of the atmosphere.*

VISUAL CONCEPT CHECK 4.1

Although we take the Sun for granted, it's by far the most important variable with respect to the presence of life on Earth. How is solar energy produced? What form does solar energy take as it travels through space? How long does this energy take to reach the Earth?

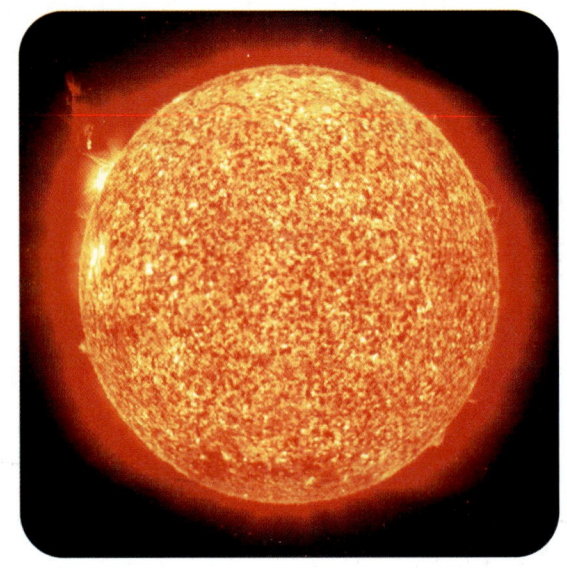

Figure 4.4 The Earth's atmosphere as viewed from space. This image nicely demonstrates the overall thinness of the atmosphere, which is the faint blue haze that appears on the horizon. Note, in contrast, the infinite blackness of deep space.

Composition of the Atmosphere

Before we can examine how solar radiation interacts with the atmosphere, we first need to know the basic composition of the atmosphere. Earth's atmosphere is the medium through which radiation flows on its way to the surface. The atmosphere is unique in our solar system because it supports life by providing oxygen and carbon dioxide for animals and photosynthesis, respectively. In addition, the atmosphere serves as a buffer that shields the Earth from the potentially deadly effects of UV radiation from the Sun, allowing mostly visible and infrared wavelengths to reach the Earth. The atmosphere behaves as a fluid in much the same manner as water, with flowing currents and eddies. Fluctuations in these currents and eddies shape the course of environmental conditions on the Earth's surface at every moment. Knowing how solar radiation flows in the atmosphere is key to understanding the Earth's temperature, atmospheric circulation, and precipitation patterns, which are discussed in later chapters.

In general terms, the atmosphere consists of air, an invisible medium that surrounds the Earth (Figure 4.4). The basic components of the atmosphere can be divided into three categories: (1) constant gases; (2) variable gases; and (3) particulates. Each of these elements is critical to the way the atmosphere functions because it performs a unique role that is essential to life on Earth.

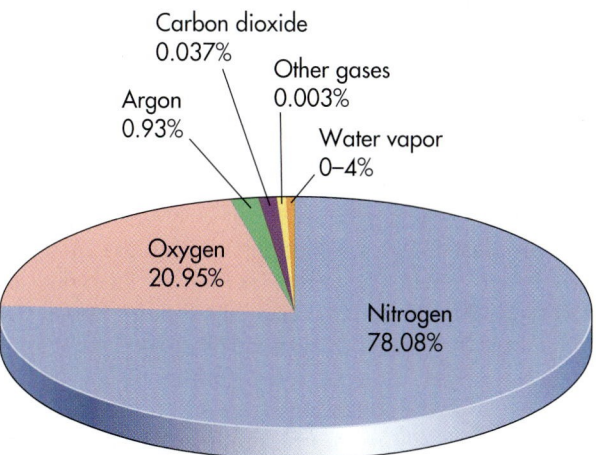

Figure 4.5 Proportion of various gases in the atmosphere. Most of the atmosphere consists of nitrogen and oxygen, which are constant gases. Other gases, such as carbon dioxide and water vapor, are variable gases.

up 99% of the atmosphere (Figure 4.5). Argon is the other constant gas, composing approximately 1% of the atmosphere. Argon is inert and of little importance in natural processes.

Constant Gases

Constant gases maintain more or less the same proportion in the atmosphere. In our atmosphere, nitrogen and oxygen are the primary constant gases and together make

Constant gases *Atmospheric gases such as nitrogen, oxygen, and argon that maintain relatively consistent levels in space and time.*

Nitrogen exists in molecular form as two nitrogen atoms bonded together (N_2). This gas makes up 78% of the atmosphere and is derived from the decay and burning of organic material, volcanic eruptions, and the chemical breakdown of specific kinds of rocks. Although nitrogen is largely inert in the atmosphere, it is critical to plant life because it can be transformed, or fixed, into chemical compounds (ammonia or nitrates) in the soil. These compounds are absorbed by plants and incorporated in their tissues as proteins. Nitrogen maintains a constant proportion of the total atmosphere because what is added is balanced by what is removed through precipitation and various biological processes.

Oxygen makes up 21% of the atmosphere and is a by-product of photosynthesis. In contrast to the inert (chemically inactive) nature of nitrogen, oxygen gas (O_2) is very active and can combine with a variety of other elements through the process of oxidation. Oxygen is absolutely essential to animal respiration because it is required to convert foods into energy. Oxygen is a constant gas because the amount produced by plants balances the amount absorbed by various organisms through respiration.

Variable Gases

Variable gases differ in their proportion of the atmosphere over time and space, depending on environmental conditions. Although some of these gases are highly significant to life, they make up only a tiny portion (less than 1%) of the atmosphere. The most important variable gases are carbon dioxide, water vapor, and ozone.

Water Vapor   An important variable gas in the atmosphere is water vapor. Water is found in three physical states: liquid, solid (ice), and gas (water vapor). The amount of water vapor in the atmosphere near the Earth's surface is about 2% in most parts of the planet, but can range from just less than 1% over deserts and polar regions to about 4% in tropical areas. Atmospheric water vapor is especially vital because it absorbs and stores heat energy from the Sun and is thus an important component, along with carbon dioxide, of the greenhouse effect. As airflow within the atmosphere moves vapor around, the effect is to moderate temperature and transport energy over the entire Earth.

The amount of water vapor at any given place in the atmosphere depends upon many variables, such as the proximity to a large body of water or the air temperature. In general, there is a direct relationship between the temperature of the air and the amount of water vapor it can hold, with warmer air capable of holding more vapor than

cooler air. You've probably experienced this relationship yourself. Try to remember what it feels like on a hot, muggy day in the summer. If you haven't experienced this kind of air, it is best described as being very "sticky" and uncomfortable. It feels that way in large part because the air holds a lot of water vapor, which is possible because the air is so warm. This vapor/temperature relationship is an important principle in physical geography because it directly influences the process of precipitation. As we will discuss more thoroughly in Chapter 7, in basic terms, precipitation occurs when bodies of air cool and water changes from vapor to liquid.

Figure 4.6 shows atmospheric water vapor over the North American continent in October 2001. Notice that on this particular day, the highest concentration of water vapor in North America was in the north-central Great Plains in states such as Minnesota, South Dakota, and Nebraska. If you happened to be in that region, you would have likely noticed numerous clouds in the sky. The water vapor in those clouds flowed into the region in association with a body of warm air originating over the tropical Pacific Ocean. Areas of relatively dry air appeared over the Gulf of Mexico and northern Florida.

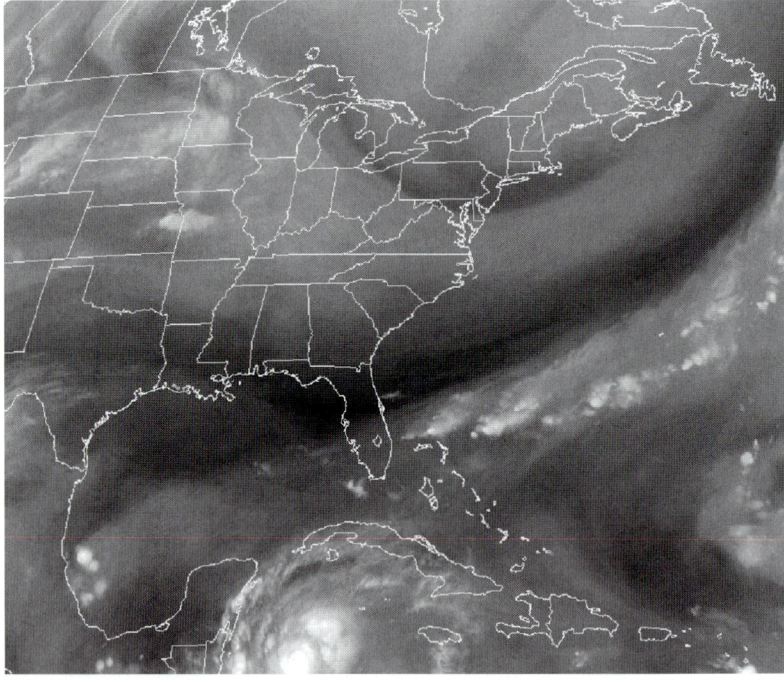

Figure 4.6 Water vapor over the North American continent. This image was taken by the Geostationary Operational Environmental Satellite (GOES) on October 8, 2001. Dark zones represent regions in the atmosphere about 6 to 10 km (~3.7 to 6.2 mi) above the Earth's surface containing very little water vapor, whereas bright areas represent relatively high concentrations of water vapor. In areas other than the tropics and poles, the average water vapor content is about 2% of the air.

Variable gases *Atmospheric gases such as carbon dioxide, water vapor, and ozone that vary in concentration in space and time.*

Carbon Dioxide Another important variable gas is carbon dioxide (CO_2), which currently comprises about 0.037% of the atmosphere (or 368 parts CO_2 for every 1 million parts of the atmosphere). Despite this very small percentage, CO_2 is a critical part of the atmosphere for two very important reasons. First, plants absorb CO_2 and release oxygen as a by-product. Second, atmospheric CO_2 contributes significantly to the **greenhouse effect,** which is the process through which the atmosphere traps longwave radiation. This process warms the atmosphere, which, in turn, warms the Earth.

You may associate the greenhouse effect most closely with the concept of global warming, an environmental issue that concerns many scientists today. In fact, a common misconception among many people is that the greenhouse effect is *entirely* a problem because it is associated with the important issue of global warming. Although the greenhouse effect is indeed linked to this issue, it has historically been a positive part of the Earth's system because it has helped maintain a generally constant atmospheric temperature that is hospitable to life. The problem now is that CO_2 levels are rapidly rising beyond recent historical norms due to human industrial activity, specifically the consumption of fossil fuels. As a result, the Earth appears to be in the midst of a rapid warming trend that may have significant environmental consequences in the future.

You can examine the processes associated with the greenhouse effect more closely in Figure 4.7. Although we'll discuss the flow of solar radiation more thoroughly later in the chapter, this diagram shows the significance of CO_2 in the atmosphere and is relevant here. As illustrated in Figure 4.7, the Earth receives shortwave radiation from the Sun, which passes through the atmosphere and is absorbed by the surface. Subsequently, the Earth releases energy as longwave radiation. Some of this radiation flows directly back into space, but most is absorbed by the atmosphere, where it is held in large part by CO_2 (water vapor is another greenhouse gas). The atmosphere then redirects some longwave energy

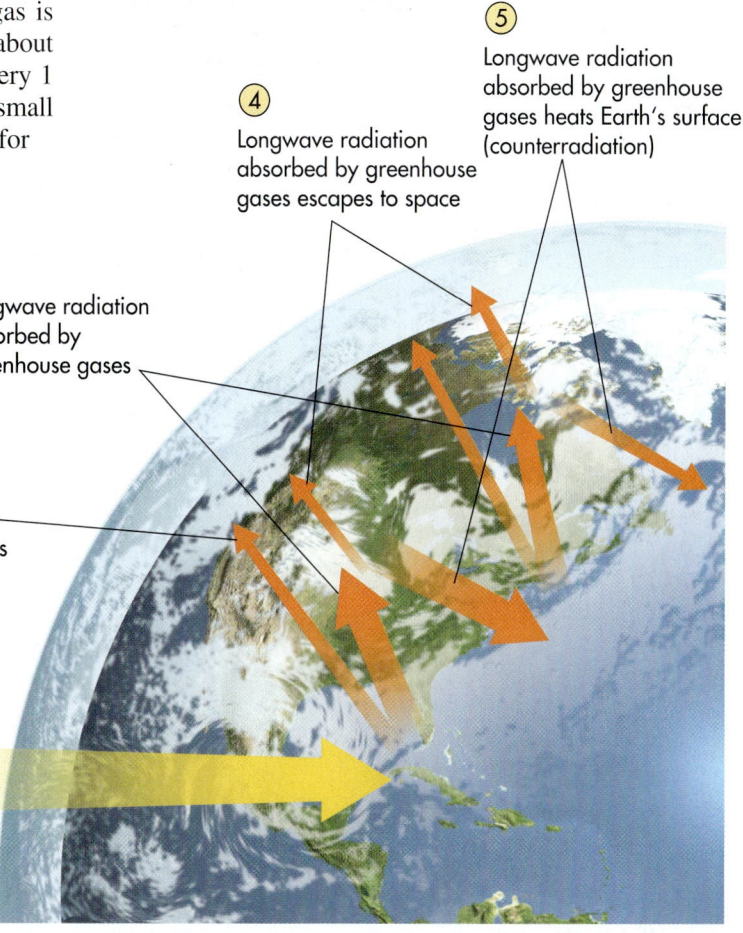

⑤
Longwave radiation absorbed by greenhouse gases heats Earth's surface (counterradiation)

④
Longwave radiation absorbed by greenhouse gases escapes to space

③
Longwave radiation absorbed by greenhouse gases

② Some longwave radiation escapes directly to space

① Shortwave radiation from the Sun absorbed at the surface

Figure 4.7 The greenhouse effect. The Earth absorbs shortwave radiation from the Sun and emits it as longwave energy. Some of this emitted longwave radiation flows directly to space, but most is absorbed by carbon dioxide (CO_2) and water vapor in the atmosphere. Subsequently, this longwave radiation either escapes into space or reflects back to the Earth's surface through the process of counterradiation.

back to the surface as **counter-radiation**. Some atmospheric longwave energy escapes to space.

The greenhouse effect is the process through which counterradiation is returned to the surface. The really fascinating aspect of the greenhouse effect, as far as life on Earth is concerned, is that the balance is just right for the planet to be neither too hot nor too cold. To see the significance of this balance, consider the planets Mars and Venus. In both places, CO_2 is the dominant component of the atmosphere, comprising about 96% of the volume on each planet. On Mars, the overall density (measured by

Greenhouse effect *The process through which the lower part of the atmosphere is warmed because longwave radiation from the Earth is trapped by carbon dioxide (CO_2) and other greenhouse gases.*

Counter-radiation *Longwave radiation that is redirected towards the Earth's surface from the atmosphere.*

A rainbow at Banff National Park in Canada.

THE FORMATION OF RAINBOWS

In addition to producing clouds, the presence of water vapor in the air can have a very beautiful side effect: it causes rainbows. A rainbow occurs when white light from the Sun peeks through the clouds and strikes water droplets in the air from a rainstorm. The water

pressure) of the atmosphere at the surface is about 1000 times less than that of the Earth; thus, there's comparatively much less CO_2 on Mars than there is on Earth. As a result, Mars is generally much colder than Earth, with an average temperature of $-63°$ C ($-81°$ F). With respect to Venus, the atmosphere there is about 90 times more dense at the surface than Earth's atmosphere; thus, the concentration of CO_2 on Venus is significantly much higher than on Earth. Consequently, Venus is considered to have a runaway greenhouse effect, with an average temperature of $500°$ C ($932°$ F)!

Although Venus provides an extreme example of the greenhouse effect, it is a context through which we can view the concerns that many scientists have regarding the ongoing human impact on the greenhouse process through the combustion of fossil fuels for energy. Since the middle of the 19th century, levels of atmospheric CO_2 have steadily increased largely through our consumption of fossil fuels. To see how these levels have changed, examine Figure 4.8, which shows CO_2 measurements obtained from Mauna Loa in Hawaii. These data indicate that atmospheric CO_2 increased approximately 50 parts per million between 1958 and the late 1990s. This increase is thought to be a primary cause of global climate change and is the negative side of the greenhouse effect that many people associate with the process. This important issue will be covered in greater detail in Chapter 9.

Ozone   Ozone is a form of oxygen that has three oxygen atoms (O_3), rather than the two found in normal oxygen gas (O_2). Atmospheric ozone forms when gaseous chemicals react in the upper atmosphere (Figure 4.9). In the initial stage of this reaction, O_2 absorbs UV energy, which causes the oxygen molecule to split into two oxygen atoms ($O + O$).

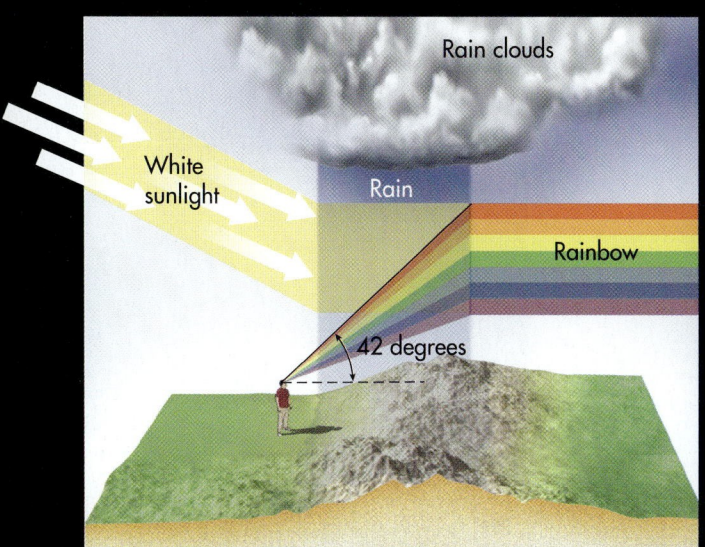

Water droplets in the air bend sunlight by an angle that depends on the wavelength (color) of the light. We see the result as a rainbow. Notice that red appears at the top of the rainbow and blue at the bottom because red light is bent more than blue light.

droplets bend the light rays, but by a different amount for each color of light. The result is that the white light is spread out into bands of colors. These colors are reflected back down toward Earth by the water droplets, at an average angle of about 42°. If you're standing in the right place at the right time, you see these colors as a rainbow.

A geographer knows that a rainbow requires bright sunlight, so you can't see one during the height of a rainstorm. On the other hand, a rainbow requires the presence of water droplets in the air, so you can't see one after the rain has gone. In other words, the sight of a rainbow usually means that the rain is clearing out and fair weather is ahead.

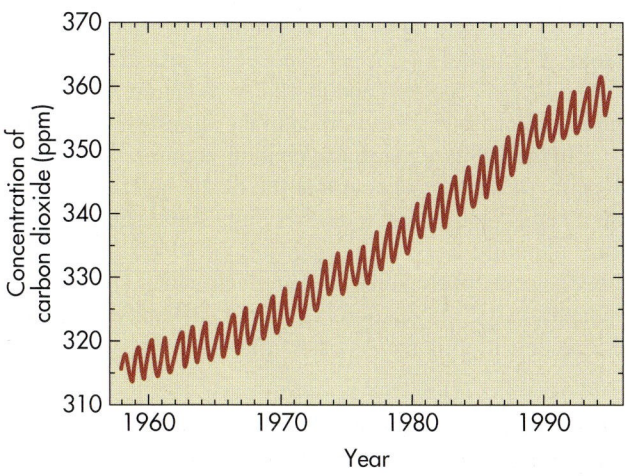

Figure 4.8 Changes in atmospheric carbon dioxide from 1958 to 1995 as measured at Mauna Loa, Hawaii. The vertical axis is the concentration of CO_2 in parts per million (ppm). Note the steady increase during this period of time, which has been attributed to human consumption of fossil fuels. Peaks and valleys in the curve represent seasonal fluctuations, with low points occurring during the Northern Hemisphere summer when plants take up CO_2. C. D. (Source: Keeling and T. P. Whorf. Atmospheric CO2 records from sites in the SIO Air Sampling Network. In Trends: A compedium of data on global change. U.S. Department of Energy.)

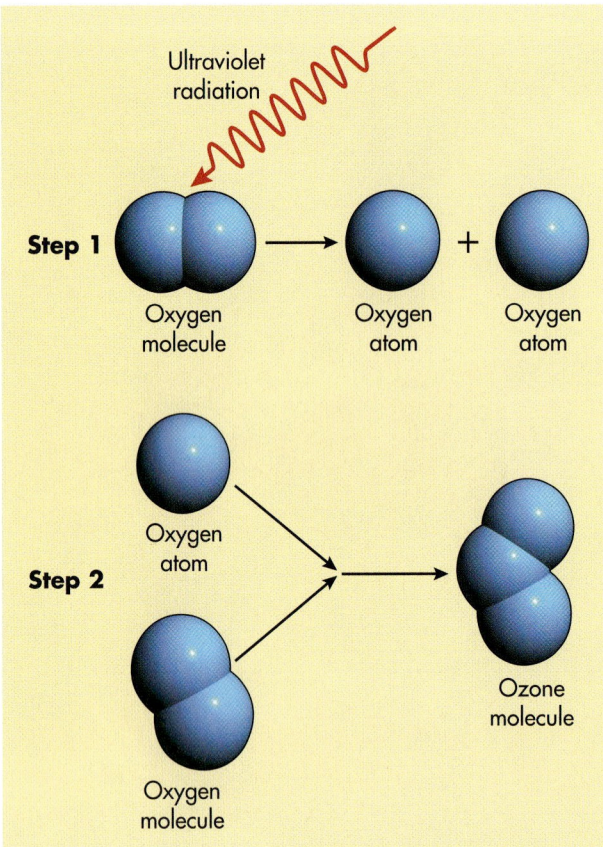

Figure 4.9 Formation of atmospheric ozone. In the first step, UV radiation is absorbed by oxygen molecules, creating free oxygen atoms. In the second step, these free oxygen atoms combine with oxygen molecules to form ozone.

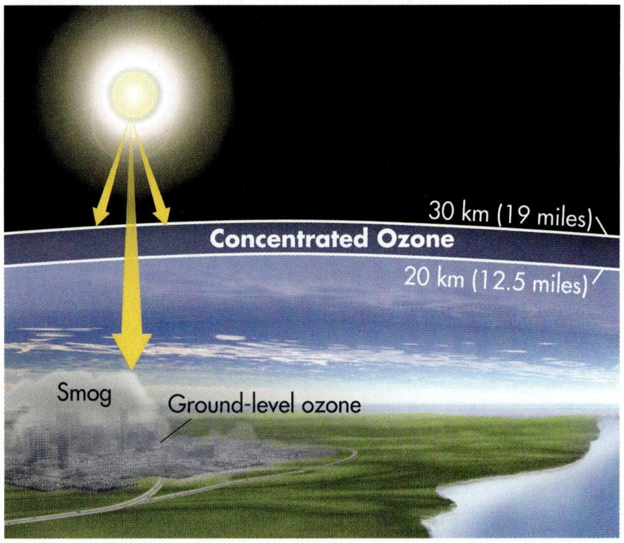

Figure 4.10 The two layers of the atmosphere where high concentrations of ozone occur. One layer is at high altitudes and absorbs UV radiation from the Sun. Most of this ozone is concentrated in the ozone layer. The second layer occurs in the lower part of the atmosphere and is associated with pollution, particularly chemical smog in cities.

Subsequently, one of the O atoms combines with an O_2 molecule to form ozone. Ozone is naturally destroyed when it absorbs UV radiation and splits from O_3 into O_2 + O. The single oxygen atom can then recombine with another O_2 molecule to form ozone. Ultimately, O_3, O_2, and oxygen atoms (O) are repeatedly formed, destroyed, and reformed in the ozone layer in a way that absorbs UV radiation each time a transformation occurs.

Ozone primarily occurs in two layers in the atmosphere: at ground level and within the stratosphere. Although diffuse amounts of ozone exist as high as 50 km (32 mi) above the Earth, it is most concentrated in the ozone layer between about 20 and 30 km (12.5 to 19 mi; see Figure 4.10). Ground-level ozone is a form of pollution (Figure 4.11) created when nitrogen oxides and organic gases emitted by automobiles and industrial sources react. This form of ozone has been linked to cardiorespiratory illness and may also cause crop damage.

The greatest concentrations of ozone are found in the **ozone layer.** This layer is critical to life because it absorbs harmful UV radiation from the Sun that would otherwise reach the Earth's surface. A major environmental issue that developed in the last half of the 20th century is the depletion of the ozone layer. This depletion occurred on a global basis due, in part, to the release of chlorofluorocarbons (CFCs) associated with air conditioners and other cooling systems. CFC molecules are very stable in the atmosphere and gradually diffuse upward from the surface to the ozone layer. Once they reach this altitude, CFCs absorb UV radiation and break down into chlorine oxide (ClO) molecules that, in turn, attack ozone molecules and convert them into oxygen atoms (Figure 4.12). The net effect of this process is that the ozone layer is depleted and more UV radiation reaches the ground because less is absorbed in the atmosphere.

In the 1980s it was discovered through satellite remote sensing of the atmosphere that the ozone layer was being depleted at an alarming rate. Although some depletion occurs naturally by volcanic aerosols periodically injected into the atmosphere, most of the reduction was attributed to industrial CFC production. The most striking example of ozone depletion is the Antarctic **ozone hole,** which was

Ozone layer *The layer of the atmosphere that contains high concentrations of ozone, which protect the Earth from ultraviolet (UV) radiation.*

Ozone hole *The decrease in stratospheric ozone observed on a seasonal basis over Antarctica, and to a lesser extent over the Arctic.*

Figure 4.11
Smog in Singapore.
A large component of this haze is ozone created by the interaction of nitrogen oxides and organic gases emitted by automobiles and factories.

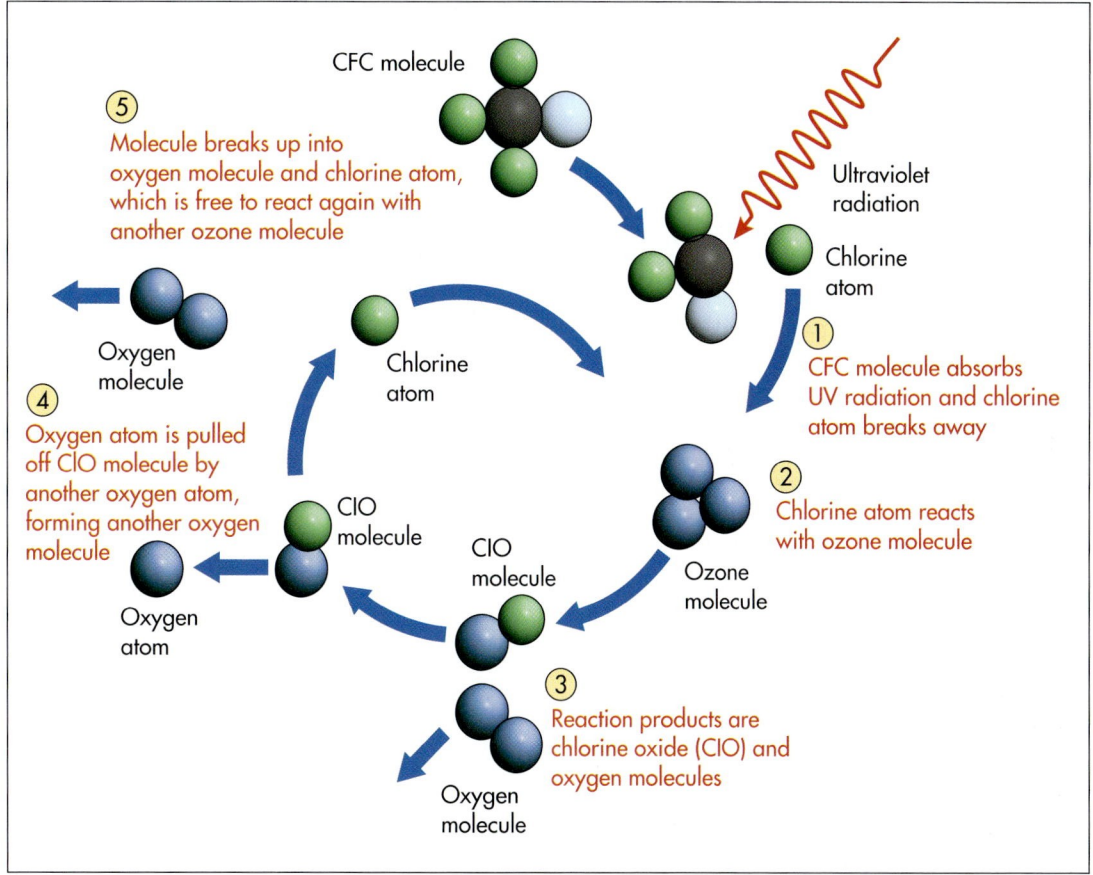

CFC molecule

⑤ Molecule breaks up into oxygen molecule and chlorine atom, which is free to react again with another ozone molecule

Ultraviolet radiation

Chlorine atom

① CFC molecule absorbs UV radiation and chlorine atom breaks away

Oxygen molecule

Chlorine atom

④ Oxygen atom is pulled off ClO molecule by another oxygen atom, forming another oxygen molecule

② Chlorine atom reacts with ozone molecule

ClO molecule

ClO molecule

Ozone molecule

Oxygen atom

③ Reaction products are chlorine oxide (ClO) and oxygen molecules

Oxygen molecule

Figure 4.12 Destruction of the ozone layer. Industrial chlorofluorocarbons (CFCs) interact with UV radiation in the ozone layer and release chlorine atoms. Follow the numbered steps to see how this process works. It begins with the breaking away of a chlorine atom from a CFC molecule due to the absorption of UV radiation. The resultant free chlorine atom then begins to interact with an ozone molecule. The chlorine atoms transform ozone molecules into ordinary oxygen molecules without being used up themselves. Thus, one CFC molecule can destroy many ozone molecules.

Sep 24 2002

(a)

Sep 22 2004

(b)

Figure 4.13 Antarctic ozone hole. (a) Ozone hole, September 2002. During this year, the ozone hole split in two, the first time such a pattern has been observed. Data for both images were acquired by the NASA Total Ozone Mapping Spectrometer (TOMS).

(b) Ozone hole during September 2004. The area of the hole was 24,200,000 km² (9,400,000 mi²), or larger than the combined area of the U.S., Canada, and Mexico. Dark blue represents levels of ozone that are about 20% less than normal.

discovered in the mid-1980s. This feature forms every spring in the Southern Hemisphere (Figure 4.13), because a polar vortex develops during the previous winter that traps the air within it. It's useful to think of this vortex as being analogous to a whirlpool in water that doesn't allow a floating object out as it spins.

Over the course of the polar vortex's seasonal existence the air becomes very cold and thin clouds of ice form within the atmospheric whirlpool. The floating ice crystals in turn facilitate the transformation of chlorine to chlorine oxide that destroys ozone, with the peak loss occurring in September and October. As a result of this decrease in ozone, the amount of surface exposure to UV radiation at 55° S has increased about 10% per decade since the late 1970s. In addition to the Antarctic ozone hole, a similar, but weaker hole develops over the Arctic region in the Northern Hemisphere winter. Significant depletions of ozone have also been measured in the midlatitudes of the Northern Hemisphere. Due to these Northern Hemisphere decreases in atmospheric ozone, the average amount of annual exposure to UV radiation at 55° N has increased about 7% per decade since the late 1970s. These increases in UV exposure are of great concern for several reasons including damage to some forms of aquatic life, reduced crop yields, and increases in the incidence of skin cancer in humans.

Although there are legitimate concerns about ozone depletion and its impact on life and human health, the threat may diminish in our lifetime. Shortly after the ozone issue emerged, the world community began working diligently to solve the problem of CFC emission. These efforts culminated in the 1987 Montreal Protocol,

an environmental treaty sponsored by the United Nations and signed by 23 nations. This treaty called for developed countries such as the U.S. and the nations of the European Union to cut CFC emissions by 50% by 1999 and for developing countries such as the various African nations and India to halt CFC use by 2010. The effect of this treaty has been positive, with stratospheric CFC concentrations peaking in 1997 and falling since that time. If the current rate of decline continues, the ozone layer should be completely restored by the middle of the 21st century.

Particulates

The last important variable component of the atmosphere is particulates, which are microscopic bodies carried in the air, existing in both liquid and solid form. The liquid variety comes in the form of clouds and rain, which occur when water vapor changes its physical state due to temperature changes in the atmosphere. Solid particulates come in an assortment of forms, including snow, hail, pollutants, wind-blown soil (dust), smoke from wildfires, volcanic ash, pollen grains from plants, and salt spray from breaking waves. Most of the time, particulate concentrations are densest near their place of origin, such as during an intense dust storm (Figure 4.14) or near a volcano. Overall, however, they comprise less than 1% of the Earth's atmosphere.

Despite their small proportion of the atmosphere, particulates nevertheless play an important role in weather and climate. Precipitation would not occur if dust particles were not present in the atmosphere because they provide a nucleus around which water condenses in the

Figure 4.14 Dust storm in New South Wales, Australia. The dust in this cloud was derived from poorly vegetated soils that were severely eroded by the wind.

1. Although the atmosphere is technically about 10,000 km (6000 mi) thick, the vast majority of the air occurs in the lowermost 30 km (10 mi).

2. The atmosphere contains constant gases (primarily nitrogen and oxygen), variable gases (mainly carbon dioxide and water vapor), and particulates.

3. The atmosphere consists largely of nitrogen (78%) and, to a lesser extent, oxygen (21%).

4. Although variable gases and particulates compose less than 1% of the atmosphere, they nevertheless significantly affect atmospheric processes and climate. A good example of the impact of these gases is the greenhouse effect and the way it moderates temperature on Earth.

first step of cloud formation. You can see this relationship the next time it rains right after you wash your car, because your car will be coated with fine dust. Other particulates, such as smoke and volcanic ash, are important because they either absorb or reflect solar energy. These combined processes influence local weather and regional climate by moderating the temperature of the atmosphere.

Although most particulates are a positive component of the atmosphere, some can cause negative environmental and health effects, especially the toxic air pollutants associated with human activities. According to the U.S. Environmental Protection Agency, over 100 pollutants are of particular concern. Benzene and perchlorethlyene, for examples, are pollutants derived from gasoline and dry cleaning facilities, respectively. Other toxic pollutants include dioxin, asbestos, toluene, and metals such as cadmium, mercury, chromium, and lead compounds. These pollutants can originate from stationary locations, such as factories and power plants, or mobile sources, such as automobiles. When ingested in large concentrations over a long period of time, these pollutants can accumulate in human tissue and ultimately cause cancer, immune disorders, and a variety of neurological, reproductive (for example, reduced fertility), developmental, respiratory, and other health problems.

The Flow of Solar Radiation on Earth

Once solar radiation reaches the Earth's atmosphere, it flows along several pathways within the atmosphere and to the surface. Some radiation flows straight from the Sun to the surface of the Earth as **insolation,** which is most accurately defined as the amount of solar radiation measured in watts per meter (W/m^2) that strikes a surface perpendicular to the Sun's incoming rays. Most solar radiation, however, follows a very indirect path. Some of it bounces around like a ping-pong ball, while some of it is directly absorbed by various components of the atmosphere. These various pathways directly influence climate on Earth and will be discussed thoroughly later in the chapter.

Heat Transfer

Before we explore the various pathways that solar radiation follows on Earth, it's important to understand how heat is transferred to the Earth's atmosphere, land, and oceans. Heat is a form of energy, related to the motion of atoms and molecules, that is lost or transferred in several different ways that are easy to imagine (Figure 4.15). The next time that you are around a campfire, consider that heat is transferred by **radiation.** This process involves

Insolation *Amount of solar radiation measured in watts per meter (W/m^2) that strikes a surface perpendicular to the Sun's incoming rays.*

Radiation *Energy that is transmitted in the form of rays or waves.*

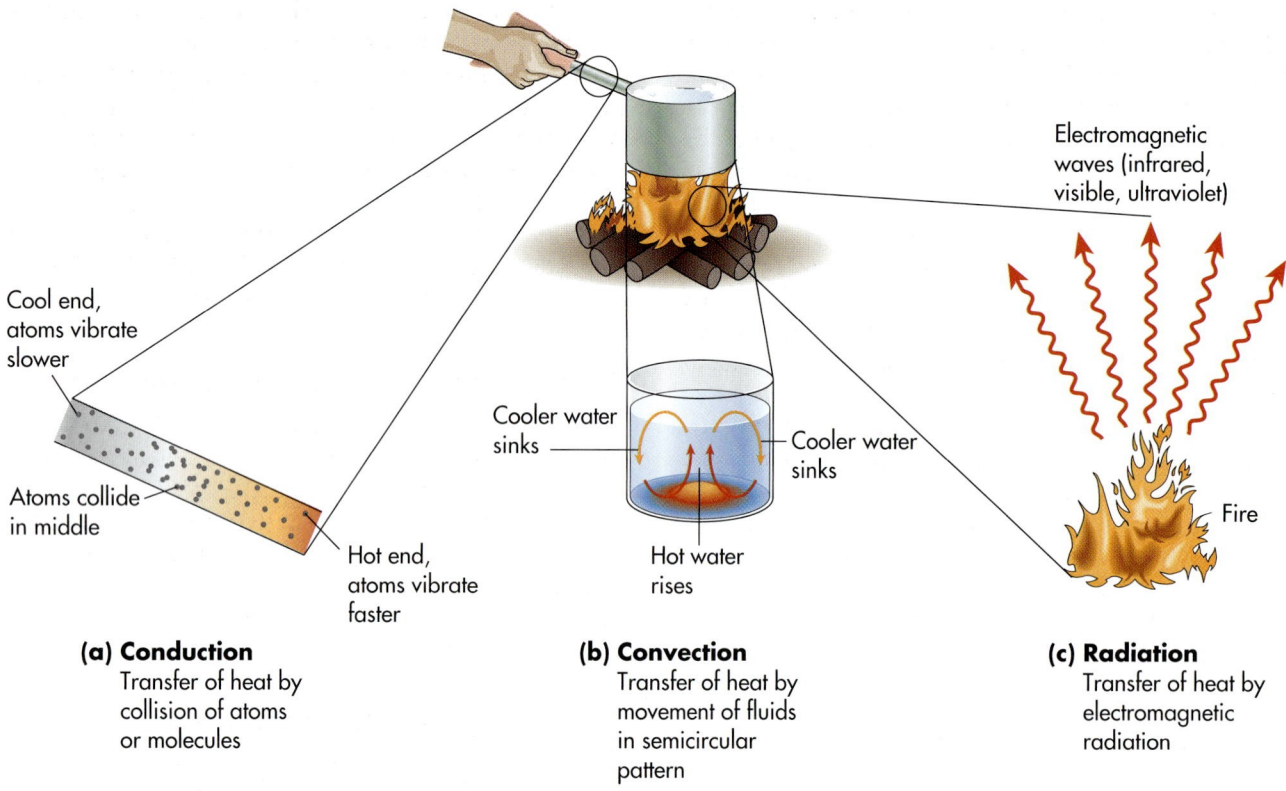

Cool end, atoms vibrate slower

Atoms collide in middle

Hot end, atoms vibrate faster

Cooler water sinks

Cooler water sinks

Hot water rises

Electromagnetic waves (infrared, visible, ultraviolet)

Fire

(a) Conduction
Transfer of heat by collision of atoms or molecules

(b) Convection
Transfer of heat by movement of fluids in semicircular pattern

(c) Radiation
Transfer of heat by electromagnetic radiation

Figure 4.15 Mechanisms of heat transfer. (a) Conduction transfers heat by collisions between fast-vibrating atoms or molecules (hot) and slower-vibrating atoms or molecules (cool). (b) Convection is the transfer of heat by the large-scale movement of matter. (c) Radiation is the transfer of heat by electromagnetic radiation, usually in the form of infrared, visible, or UV waves.

VISUAL CONCEPT CHECK 4.2

The atmosphere contains a wide variety of natural constant gases, variable gases, and particulates. People also contribute to the overall character of the atmosphere. How do you suppose that this factory complex changes the general composition of the atmosphere?

the creation of electromagnetic waves, either as visible light or invisible to the naked eye. Because this radiation exists in wave form, it carries energy and thus can move from one place to another without requiring an intervening medium. When this radiation reaches you, part of the energy of the wave gets converted back into heat, which is why you feel warm sitting beside a campfire. Some of the radiation can be in the form of visible light that we can see, but a great deal of the radiation emitted is infrared waves, whose longer wavelength is usually detectable only with special infrared detectors. Sometimes you can see these waves as the shimmer on a warm parking lot in summer.

Another form of heat transfer is **conduction,** which involves the diffusion of energy through molecules that are in contact with one another. This diffusion occurs because, as the temperature of molecules increases, they begin to vibrate more rapidly, causing collisions that produce similar motions in adjacent molecules. In this fashion, sensible heat always moves from areas of relatively high temperature to zones of relatively low temper-

ature. You can experience conduction yourself if you pick up a heated pot from a stove or other fire.

The third form of heat transfer is **convection,** which involves the upward movement of heat. The next time you boil water on the stove, notice how a small circular current of water arises within the pot where hot water moves up and is replaced at the bottom by relatively cooler water. As you will see in later chapters, convection is a very important mechanism of heat transfer on Earth and is associated with atmospheric circulation and precipitation.

Flow of Solar Radiation in the Atmosphere

With the mechanisms of heat transfer in mind, let's now explore the various ways that solar radiation interacts with the Earth. As you work your way through this discussion refer to Figure 4.16, which illustrates the flow of solar radiation on the planet. Of all the incoming radiation that reaches the Earth, about 25% flows uninterrupted to the surface as **direct radiation.** However, this amount can vary greatly depending upon local geographic variables

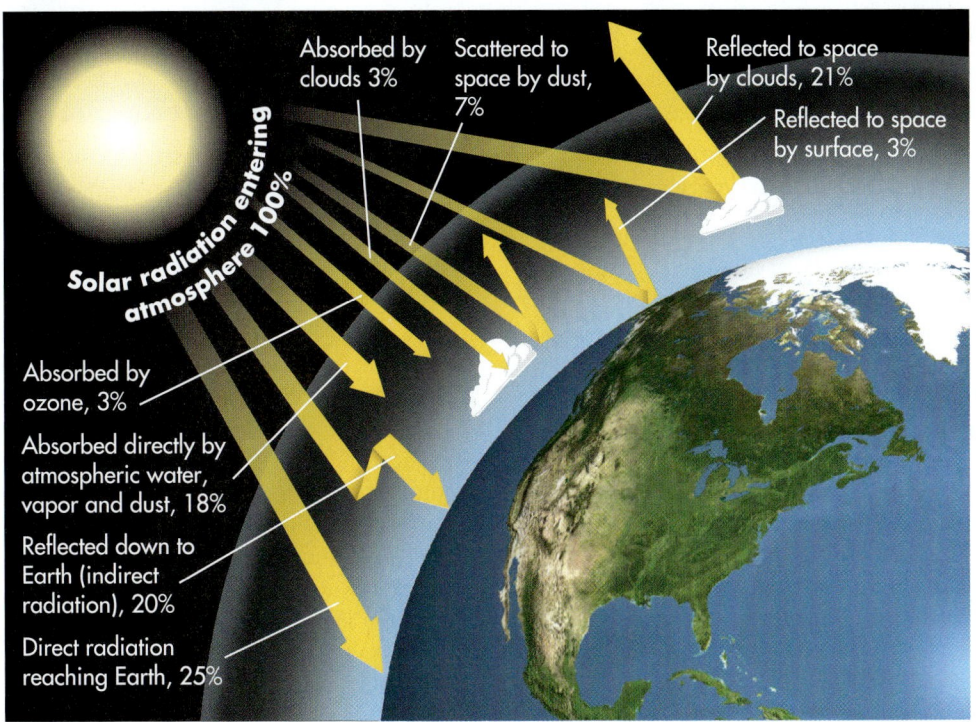

Figure 4.16 Interaction of solar radiation with various components of the atmosphere. Note that solar radiation is absorbed, stored, and reflected in many ways, and that about half of the overall amount actually reaches the Earth's surface.

Conduction *The transfer of heat energy from one substance to another by direct physical contact.*

Convection *A circular cell of moving matter that contains warm material moving up and cooler matter moving down.*

Direct radiation *Solar radiation that flows directly to the surface of the Earth and is absorbed.*

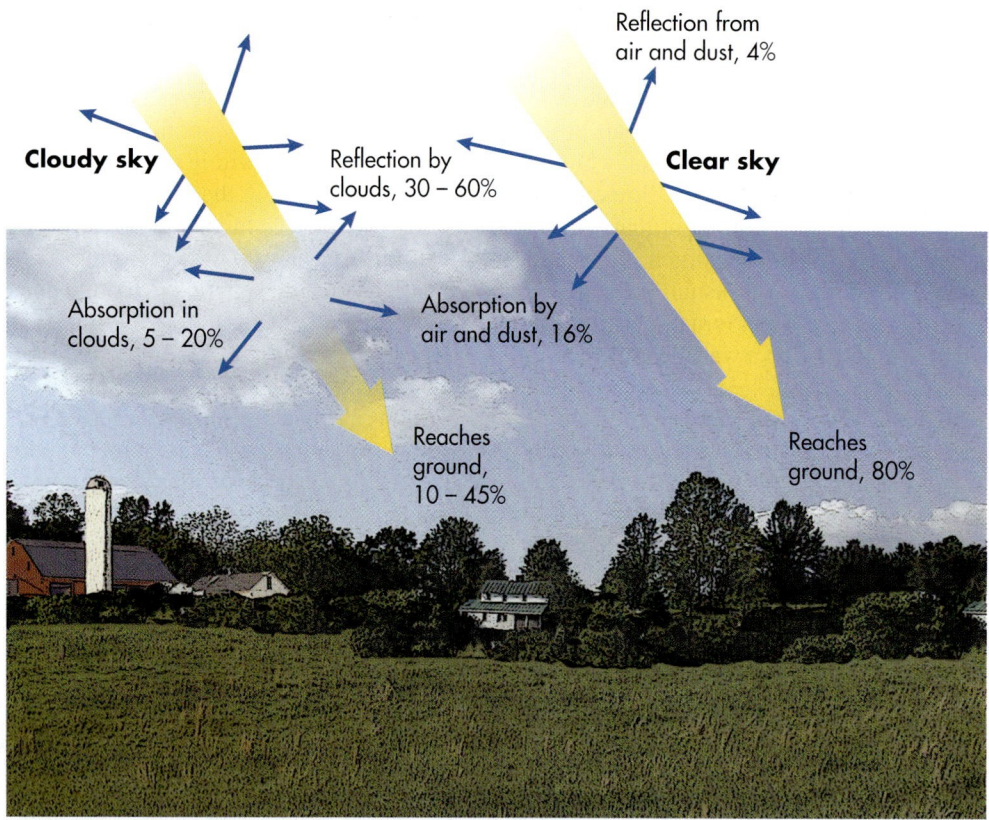

Figure 4.17 Variation in the receipt of solar radiation based on cloud cover. Areas of dense cloud reflect more radiation than areas of clear sky.

such as cloud cover or density of atmospheric dust. Figure 4.17 shows, for example, the impact that cloud cover can have on direct solar radiation.

The remaining 75% of incoming insolation is either absorbed or otherwise redirected in the atmosphere (see Figure 4.16). **Absorption** occurs when variable gases and particulates in the atmosphere interrupt the flow of solar radiation by absorbing specific wavelengths. For example, almost all UV wavelengths (those less than 0.3 μm* in length) are absorbed by oxygen and ozone. Similarly, radiation at the 1.3 μm and 1.9 μm wavelengths is absorbed very strongly by water vapor and CO_2. Overall, approximately 24% of incoming solar radiation is absorbed, with 18% absorbed by atmospheric water vapor

* Micrometer

and dust, 3% by ozone, and 3% by clouds. Remember, this absorption of solar radiation is important because it helps to moderate temperature in the atmosphere.

Some incoming solar radiation in the atmosphere is also redirected by **reflection** or **scattering.** The amount of reflection that occurs depends upon the **albedo** of a surface and occurs in the atmosphere when insolation bounces off bright cloud tops. (Albedo refers to the amount of reflection a given surface can cause; a high albedo is typical of bright surfaces such as a mirror. We'll come back to this factor shortly.) Approximately 21% of radiation is reflected back to space in this manner. Scattering occurs when dust particles in the atmosphere cause solar radiation to bounce around in the air. Approximately 7% of this scattered radiation is redirected back into space. Reflected and scattered radiation may also, in turn,

Absorption *The assimilation and conversion of solar radiation into another form of energy by a medium such as water vapor. In this process, the temperature of the absorbing medium is raised.*

Reflection *The process through which solar radiation is returned directly to space without being absorbed by the Earth.*

Scattering *The redirection and deflection of solar radiation by atmospheric gases or particulates.*

Albedo *The reflectivity of features on the Earth's surface or in the atmosphere.*

(a)

(b)

Figure 4.18 Colors in the sky. (a) The sky is blue because particulates scatter light in the blue wavelength. (b) Geometric relationship of insolation, atmospheric thickness, and sky color. (c) Colors at sunset along the Atlantic coast at Rhode Island. Oranges and reds appear because the blue wavelength is completely scattered before it reaches your eye.

Labels in figure: White light from Sun · Observer sees white Sun, blue sky at noon · North Pole · Direction of rotation · White light from Sun · Observer sees red sunset · Blue light scattered

be redirected downward toward the surface of the Earth as **indirect radiation.** Approximately 20% of solar radiation that reaches the Earth does so in this indirect fashion.

Although scattering is a process that can't directly be seen, you can see the indirect effects very clearly in the apparent colors of the sky (Figure 4.18). Atmospheric dust particles happen to be just the right size to reflect radiation in the blue wavelength. Thus, the sky appears to be blue on a clear day when the Sun is high in the sky and solar radiation is streaming more or less directly toward you. However, when the Sun is near the horizon at dawn or dusk, the sky becomes more colorful in the red and orange regions of the spectrum. This change occurs because solar radiation is streaming toward your eye from a much lower angle than when the Sun is high in the sky. As a result, the incoming radiation is passing through a thicker slice (when viewed horizontally) of the atmosphere to reach your line of sight. This thickened atmosphere causes the blue wavelengths to be scattered out before they reach you, leaving the longer wavelengths (oranges and red) for you to see.

Indirect radiation *Radiation that reaches the Earth after it has been scattered or reflected.*

Interaction of Solar Radiation and the Earth's Surface

Let's now examine the approximately 45% of all solar energy (direct and indirect) that strikes the Earth's surface. What happens to this portion of incoming radiation is important because it influences variables such as temperature, atmospheric circulation, the density and kind of vegetation in a region, soils, and even where glaciers occur. Relationships such as these will be discussed in later chapters. In general, either of two things happens when solar radiation strikes the ground: (1) it is absorbed, or (2) it is reflected.

Absorbed Radiation   Of the amount of solar energy that reaches the ground, 96% is absorbed by the various land and water bodies on the surface, thus heating the Earth. One way this heat energy can be stored is in the form of **sensible heat,** which is heat that can be sensed by touching or feeling and can be measured by a thermometer. A second way it can be stored is as latent heat, when water from land and ocean/lake surfaces is transformed into water vapor within the atmosphere through evaporation. **Latent heat** is a form of heat that is hidden and cannot be measured with a thermometer because it goes into breaking bonds between molecules when a substance changes physical state, such as from a liquid to a gas.

Stored energy can be lost from the Earth in several ways. Sensible heat can be transferred from the Earth's surface to the atmosphere through the process of convection. In this case, warm air rises upward from the Earth and cooler air descends to replace it. Heat can also be removed through the process of **evaporation,** which is the change of liquid water to water vapor by absorption of heat. In addition, absorbed radiation can simply be re-radiated back into space as longwave radiation. A good example of this release of longwave radiation is the shimmer of waves that you see ahead of you when you're driving down a dark asphalt highway on a particularly hot day. When longwave radiation is released, it can either escape directly back into space, be absorbed by atmospheric CO_2 and water vapor, or be backscattered by dust particles.

Reflected Radiation   Reflected radiation is energy that bounces off the Earth's surface in such a way that it does not provide heat. Overall, approximately 3% of incoming radiation is reflected in this way. The proportion of incoming radiation reflected by the surface is, just as for the atmosphere, dependent upon the albedo of an object. Although you may not be directly aware of surface albedo, you are probably indirectly aware of its effects. Almost everyone knows how uncomfortable it feels to wear a dark-colored shirt on a very hot day with bright sunshine. The reason for this is that dark objects absorb radiation and therefore increase in temperature. A dark asphalt highway, for example, reflects only about 5% to 10% of incoming solar energy. A snow-capped mountain peak, in contrast, reflects 80% to 95% of incoming solar radiation and thus absorbs little heat. Figures 4.19 and 4.20 show

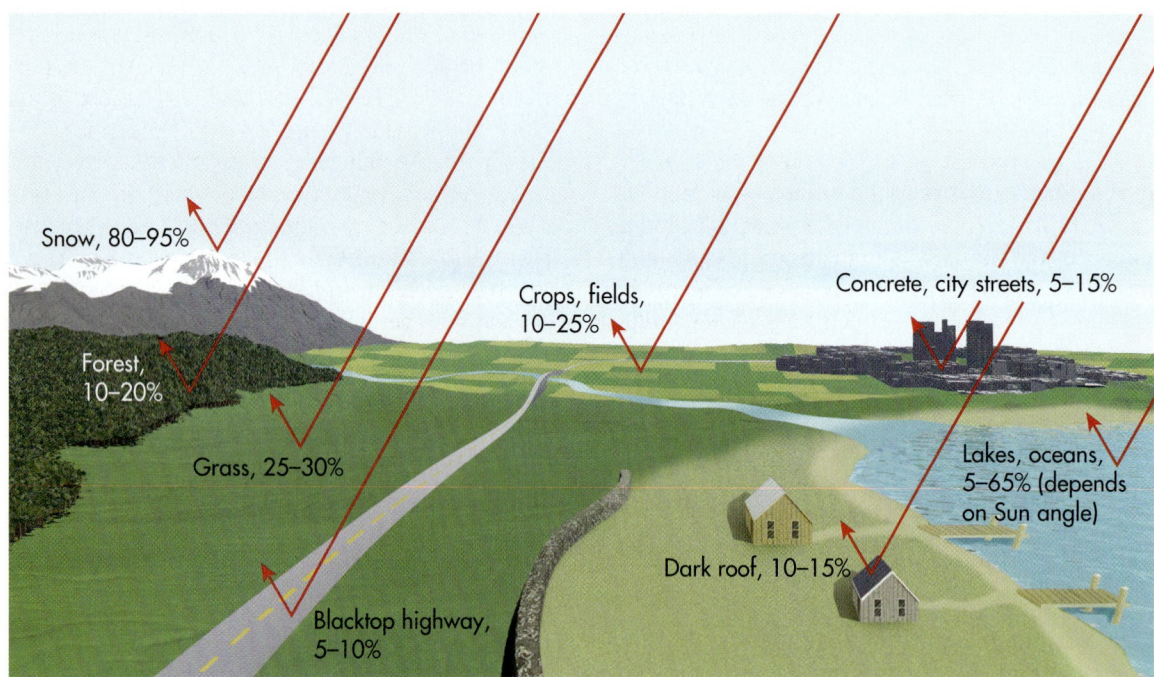

Figure 4.19 Variation of surface albedo. In general, albedo values are higher on brighter surfaces than on darker ones.

Sensible heat   *Heat that can be felt and measured with a thermometer.*

Latent heat   *Heat stored in molecular bonds that cannot be measured.*

Evaporation   *The process by which atoms and molecules of liquid water gain sufficient energy to enter the gaseous phase.*

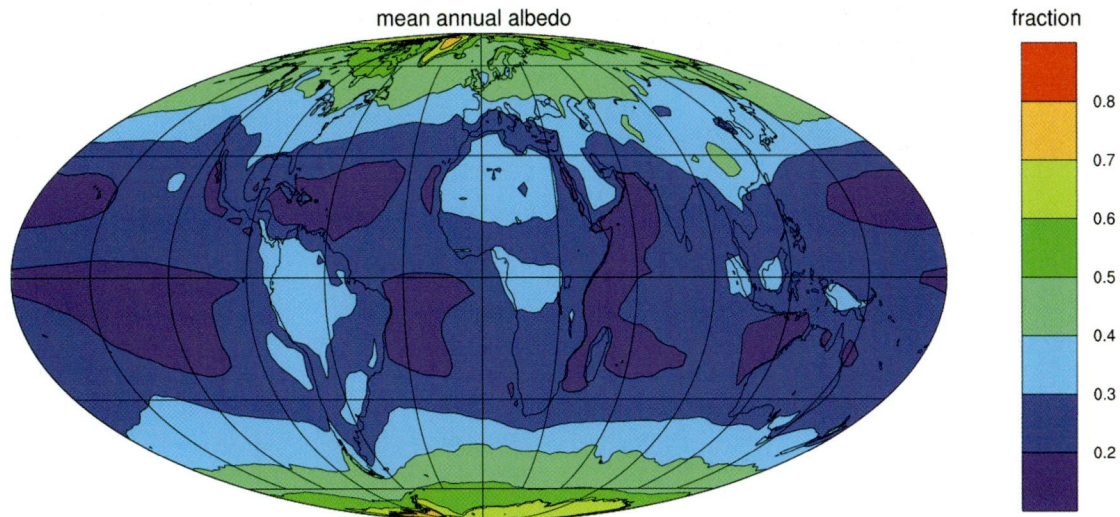

mean annual albedo

fraction

0.8
0.7
0.6
0.5
0.4
0.3
0.2

Figure 4.20 Mean annual albedo across the Earth's surface. The bar code at the right represents percentage albedo. Blue areas indicate zones of relatively low albedo and are concentrated in the oceans where radiation is absorbed. Polar ice caps, in contrast, have relatively high albedo, almost as much as 70%.

how albedo can influence the amount of energy reflection on different surfaces.

The amount of radiation reflected depends not only upon albedo, but also on the Sun angle in the sky. Remember from Chapter 3 that the Sun is always at some angle in the sky relative to any location on the Earth's surface. This concept was illustrated in Figure 3.3, if you wish to review. With respect to incoming solar radiation, this angle is called the **angle of incidence.** Think of the angle of incidence as being similar to the angle at which you might throw a rock into the water. If you drop the rock straight down into the water, it's immediately absorbed by the water and sinks to the bottom. However, if you throw the rock at a shallower angle (relative to the

water), it might skip across the water. The angle of incidence works in much the same fashion in that solar radiation from a high-angle Sun penetrates the Earth directly; in addition, the Sun's rays are confined to a relatively small area and are thus intense at that point. When the Sun is lower in the sky, however, more radiation is reflected when it strikes the Earth; also, the radiation is spread over a larger area. When this occurs, the radiation is not as intense compared to higher-angle Sun locations.

When the Sun is directly overhead, water has an albedo value of around 5%, but when the Sun is near the horizon, the albedo value is around 65% because radiation reflects off of the surface more readily. You can see this difference clearly in Figure 4.20, which shows that oceans at higher latitudes have a higher albedo than those at lower latitudes. This variability occurs because the angle of incidence is less at higher latitudes and more radiation is reflected off the oceans in those regions.

Angle of incidence *The angle at which the Sun strikes the Earth at any given place and time.*

GEODISCOVERIES   WWW.WILEY.COM/COLLEGE/ARBOGAST

THE ANGLE OF INCIDENCE

Now that we have discussed the basic components associated with solar radiation and how it interacts with the Earth, let's explore this extremely important concept by interacting with a simulation on the *GeoDiscoveries* website. Go to the website and select the module **The Angle of Incidence**. This module allows you to interact with the angle of incidence as it relates to latitude. The foundation of this simulation is Figure 3.3, which shows how the angle of incidence influences how much solar radiation is received at any given location.

You will be able to choose between several latitude locations in the Northern and Southern Hemispheres to see how the angle of incidence varies between locations. As you work through the various scenarios, notice not only how the flow of solar radiation changes, but also the direction from which it comes. You should notice a big difference between the Northern and Southern Hemispheres. Pay attention to these differences and relate them back to what you have learned about Earth–Sun geometry and seasons.

1. Once solar radiation reaches the atmosphere, it begins to flow along several different pathways. Some of it flows directly to the surface, whereas other parts are reflected off clouds or scattered by particulates.

2. Insolation that reaches the surface is either absorbed or reflected (a function of albedo).

3. Radiation that is absorbed is re-radiated as longwave radiation.

4. The amount of energy reflected or absorbed depends in large part upon the angle of incidence, which, in turn, varies by latitude and season.

The Global Radiation Budget

To fit our radiation discussion into a coherent model, think of the Earth's energy flow as being analogous to money that flows in and out of your bank account. You have money that flows from your job into an account, which in turn is stored temporarily until you spend it in various ways. When you take in more money than you spend, you have a net surplus of money. When you spend more than you earn, you have a net deficit of money.

Like your bank account, the Earth's **radiation budget** refers to the balance between incoming (shortwave) radiation and outgoing radiation, which is either re-radiated through reflecting and scattering processes described earlier or released from the Earth in longwave form. The difference between incoming and outgoing values is the **net radiation.** For the Earth as a whole, the long-term radiation budget must be balanced; otherwise the Earth would become progressively warmer or cooler. This balance is achieved because incoming radiation flows are matched by outgoing flows.

Figure 4.21 illustrates the basic nature of the Earth's radiation budget by incorporating the data from Figure 4.16. To simplify the concept, remember that solar energy is in the form of shortwave radiation and that 69% of it is actually absorbed by some aspect of the Earth, either by clouds, water, or the surface. The remaining 31% of shortwave energy is reflected directly back into space due to

Radiation budget *The overall balance between incoming and outgoing radiation on Earth.*

Net radiation *The difference between incoming and outgoing flows of radiation.*

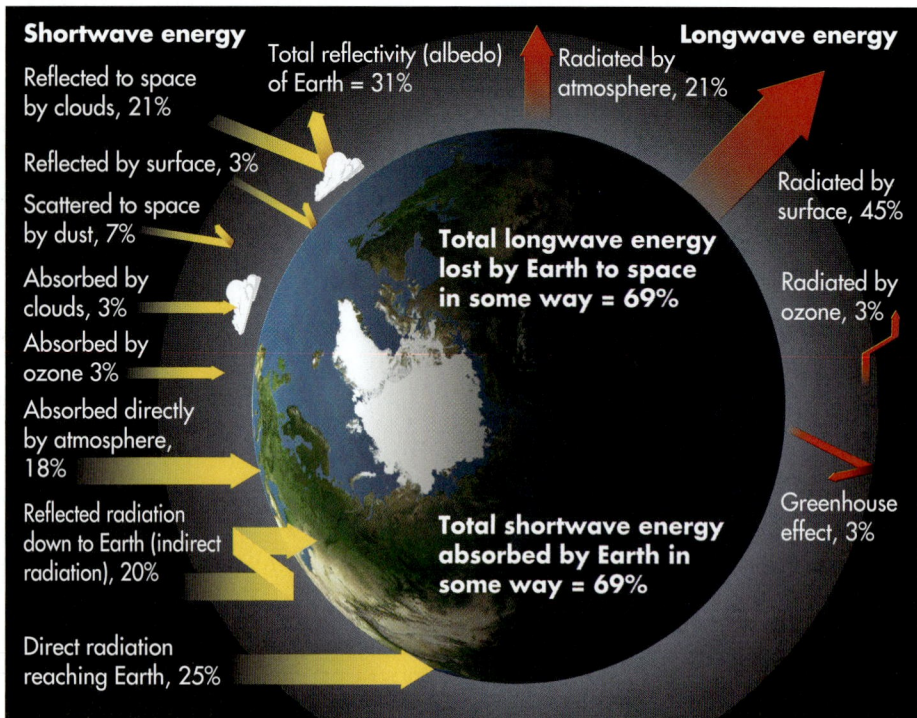

Figure 4.21 The global radiation budget. The difference between incoming (shortwave) radiation and outgoing (longwave) radiation must be in balance, otherwise the temperature of the Earth would either cool or warm dramatically.

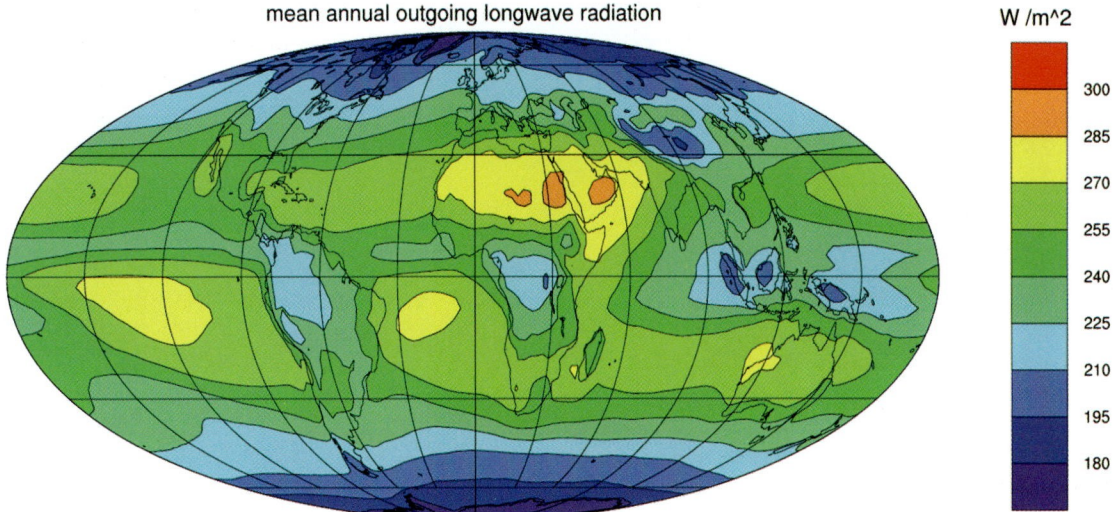

mean annual outgoing longwave radiation

W /m^2

300
285
270
255
240
225
210
195
180

Figure 4.22 Mean annual outgoing longwave radiation on Earth in watts per cubic meter. The original source of this radiation was the Sun. The radiation was then absorbed and stored by the Earth until its release as longwave energy. Note the geography of this process, with most longwave radiation being released in the tropical latitudes.

albedo or scattering. For the global radiation budget to remain in balance, therefore, the Earth must emit all of the absorbed energy back to space, which it does in longwave form. This release, plus the 31% of reflected shortwave radiation, keeps the system balanced. The Earth's average temperature is moderated because greenhouse gases such as CO_2 and water vapor trap a small amount of longwave radiation.

Aside from the amount of longwave energy that is trapped by the atmosphere, the bulk of it is lost in a way that balances the global radiation budget. This balance is accomplished in a variety of ways (see Figure 4.21). The majority (45%) of longwave energy is lost to space by direct radiation from the surface. This form of loss does not occur evenly across the Earth's surface, but rather has a distinct geographic pattern of its own, with lower latitudes emitting more energy than higher latitudes (Figure 4.22). This pattern makes sense, of course, because low latitudes receive more shortwave radiation than do high latitudes. In addition to direct energy loss from the surface, 21% of

longwave energy is lost directly by radiation from the atmosphere and another 3% is emitted by ozone.

Although the long-term energy budget for the Earth balances on a global scale, it can vary markedly over the short term in different regions, due to the interaction of various environmental factors. Because of this variability certain locations have a net surplus of radiation, whereas others have a net deficit. In addition, a particular place may have a net surplus of radiation during one time of year but a net deficit at another. These net differences are very important because they are the driving mechanism behind climate and many physical processes, such as atmospheric circulation and evaporation.

Four primary factors influence net radiation around the globe: (1) the Sun's angle of incidence, (2) latitude, (3) seasonality, and (4) length of day. Other, secondary factors include the varying output of the Sun due to sunspots and solar flares, the elliptical nature of the Earth's orbit, changes in the thickness and properties of the atmosphere, and variability in the length of day. The

GEODISCOVERIES WWW.WILEY.COM/COLLEGE/ARBOGAST

THE GLOBAL ENERGY BUDGET

After you have examined Figure 4.21, go to the *GeoDiscoveries* website and select the module **The Global Energy Budget.** This module allows you to better visualize the flow of solar radiation from the Sun to the Earth. As you watch the animation, notice the various pathways that radiation takes

as it interacts with the atmosphere. In addition, try to integrate all of the concepts so far in this chapter to solidify your understanding about the global energy budget. Once you complete the animation, be sure to answer the questions at the end of the module to test your understanding of this concept.

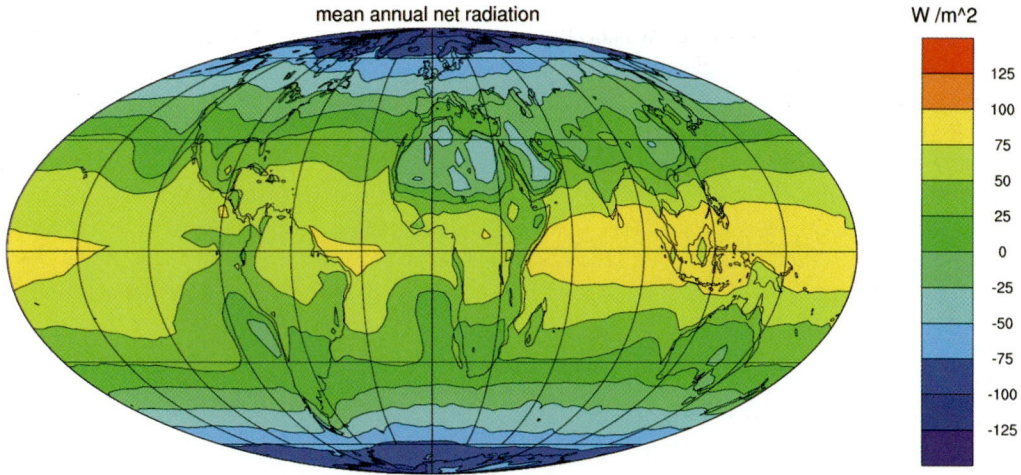

mean annual net radiation

W /m^2

125
100
75
50
25
0
-25
-50
-75
-100
-125

Figure 4.23 Mean annual net radiation at the top of the atmosphere on Earth. Note the relationship between color and value (W/m²), which is indicated in the key to the right of the image. Low to middle latitudes have a net surplus of radiation, whereas high latitudes have a net deficit.

best way to understand the changes in net radiation across the globe is by focusing on the four primary factors because they work in an interactive way. We'll begin with the angle of incidence.

Remember that the angle at which solar radiation strikes the Earth directly affects absorption and reflection. This angle is primarily a function of latitude, with low latitudes having high Sun angles and high latitudes having low Sun angles. Therefore, low latitudes generally receive high amounts of radiation and thus have a net annual sur-

plus, whereas the high latitudes have a net deficit because they receive less direct radiation (Figure 4.23).

How does this variation in net radiation between latitudes influence natural processes on Earth? We will cover the significance of this spatial relationship in greater detail later in the text, but for now it's sufficient to say that the difference in net radiation is a primary driving force behind many atmospheric circulatory processes, especially large-scale wind patterns. Figure 4.24 shows how this transfer generally works.

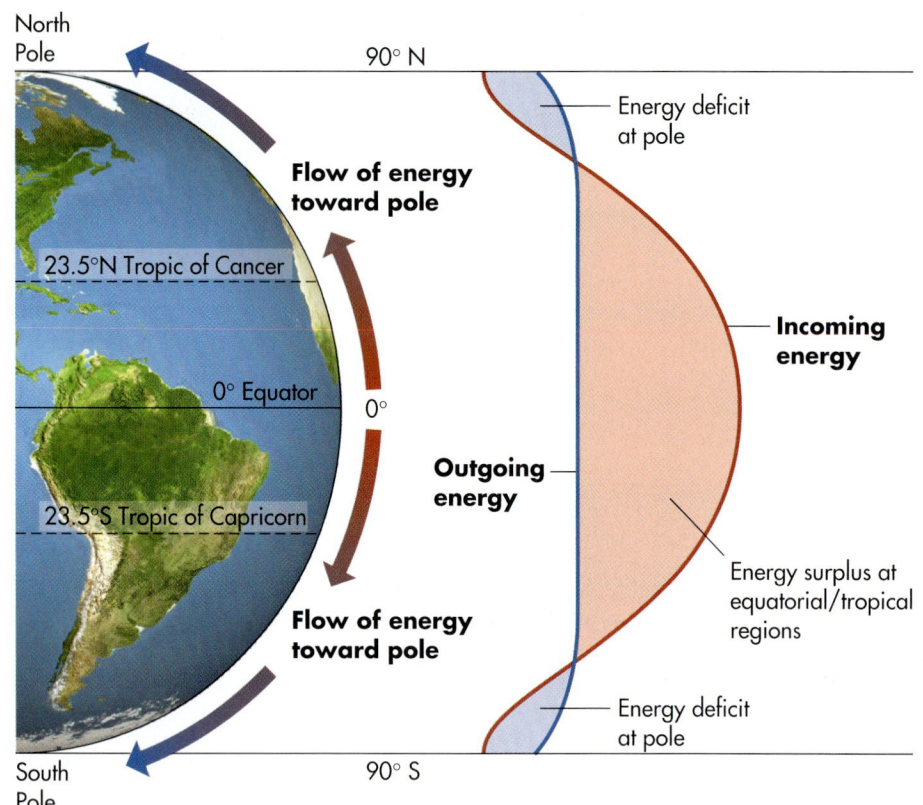

North Pole · 90° N

Energy deficit at pole

Flow of energy toward pole

23.5°N Tropic of Cancer

Incoming energy

0° Equator · 0°

Outgoing energy

23.5°S Tropic of Capricorn

Flow of energy toward pole

Energy surplus at equatorial/tropical regions

Energy deficit at pole

South Pole · 90° S

Figure 4.24 Net radiation and the transfer of heat energy on Earth. Compare this diagram to Figure 4.23 and note where areas of net surpluses and deficits occur and how energy flows on the planet.

Let's see how net radiation changes when we add the effect of the seasons to the mix of variables. Recall that this seasonal effect occurs because of the geometric relationship between the Earth and the Sun, which results in changes in the angle of incidence over the course of the year. Remember that at any given point on Earth, the position of the Sun at solar noon migrates over the course of the year because of axial tilt and orbital position. This seasonal migration of the Sun, in turn, causes the subsolar point (on Earth) to change position with respect to latitude. This shift of the subsolar point north and south of the Equator has a profound impact on the seasonal distribution of net radiation, as you can see in Figure 4.25.

Figure 4.26 compares two curves of annual insolation by latitude—one with the Earth's actual axial tilt (23.5°) and the other if the axis were not tilted. With a tilted Earth, you can see that annual radiation at low latitudes is about 40% greater than at high latitudes, but that high latitudes still receive a considerable amount. However, if the Earth's axis were perpendicular to the plane of the ecliptic, annual radiation would range from a high value at the Equator, where the Sun angle would consistently be 90°, to *zero* at the poles, where the Sun would always be at the horizon. You can see that axial tilt has a profound impact on daily insolation at various latitudes and that as a result the poles receive nearly half the amount of radiation as the Equator, even though they're in darkness for 6 months of the year.

A good way to conclude this discussion about the global radiation balance is to examine the effects when all

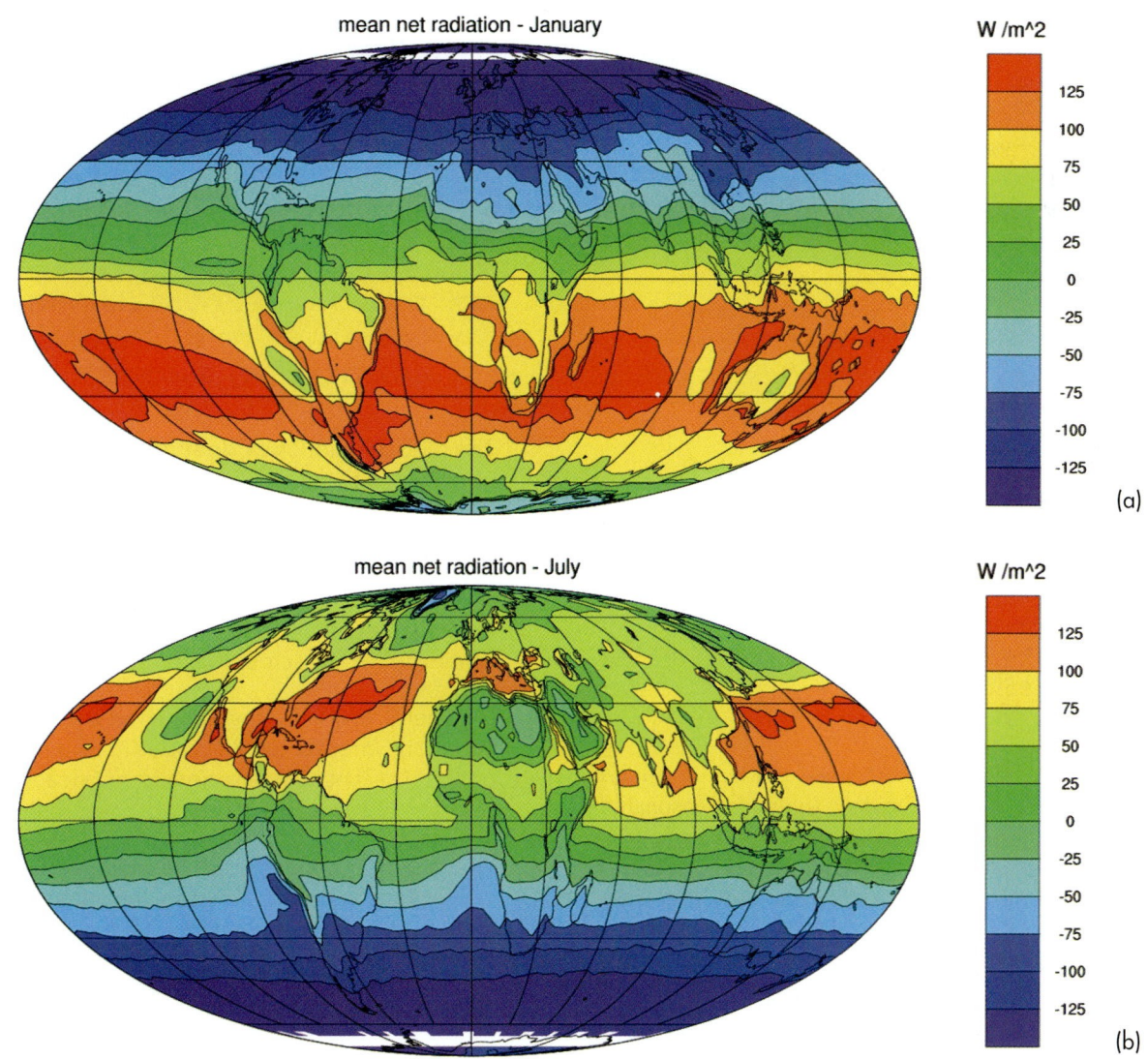

Figure 4.25 Average seasonal changes in net radiation on Earth. (a) Mean January net radiation. (b) Mean July net radiation. In both figures, reds indicate high values of net radiation, measured in Watts/m², whereas blues equal low or deficit amounts.

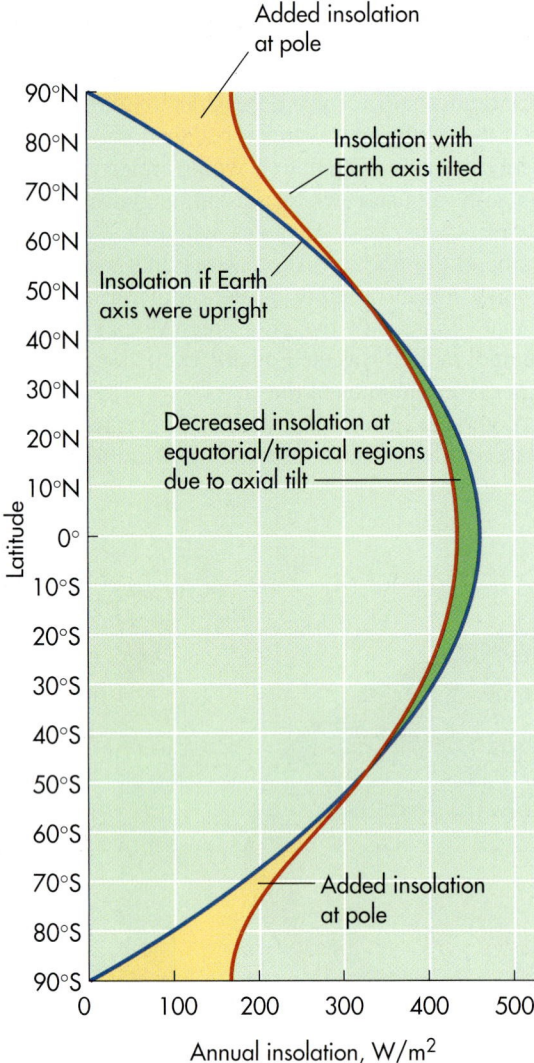

Added insolation at pole

Insolation with Earth axis tilted

Insolation if Earth axis were upright

Decreased insolation at equatorial/tropical regions due to axial tilt

Added insolation at pole

Figure 4.26 Variation in annual insolation (by latitude) on the Earth. The red line shows the variability that exists as a function of the actual axial tilt. The blue line indicates what insolation would be if the axis were not tilted. Note the added insolation at high latitudes that occurs due to the tilt of the Earth.

of the variables—angle of incidence, latitude, seasonality and length of day—are considered together. Figure 4.27 graphically illustrates these combined effects on seasonal insolation at the top of the atmosphere at three places: the Equator (0°), 45° N, and the North Pole (90° N). Notice in this graph that the North Pole has the greatest annual

range of daily insolation, with a distinct peak at the Summer Solstice and no radiation received in winter. The Equator, in contrast, receives consistently high amounts of radiation all year. Given its location midway between the Equator and the North Pole, 45° N has a moderate range of daily insolation, with a peak during the summer months and lower values during winter.

In addition to these basic patterns, Figure 4.27 reveals two other interesting patterns. The first is that the Equator shows two peaks in daily insolation, rather than one, over the course of the year. The reason for these dual peaks is that the Sun is directly over the Equator two times during the year— specifically, at each Equinox. The second interesting pattern is that the North Pole (90° N) receives more insolation around the Summer Solstice than does the Equator, even though the angle of incidence at very high latitudes is still relatively low. Why does this occur? The answer is that latitudes above the Arctic Circle receive radiation for 24 hours because the Sun never sets. At the Equator, in contrast, day length is always about 12 hours. This pattern shows the significance of day length on the amount of daily insolation received, which, in turn, depends upon axial tilt and orbital position.

KEY CONCEPTS TO REMEMBER ABOUT THE GLOBAL RADIATION BUDGET

1. The global radiation budget refers to the balance between incoming and outgoing radiation on Earth.

2. Over the long term, the global radiation budget is balanced. However, even though the long-term radiation budget is balanced, a great deal of variability occurs across the Earth as a whole.

3. In general, low latitudes have a net surplus of radiation, whereas high latitudes have a net deficit. This imbalance is significant because it drives atmospheric circulatory processes.

4. Net radiation depends upon a complex interaction of several variables, including angle of incidence, latitude, season, day length, and albedo.

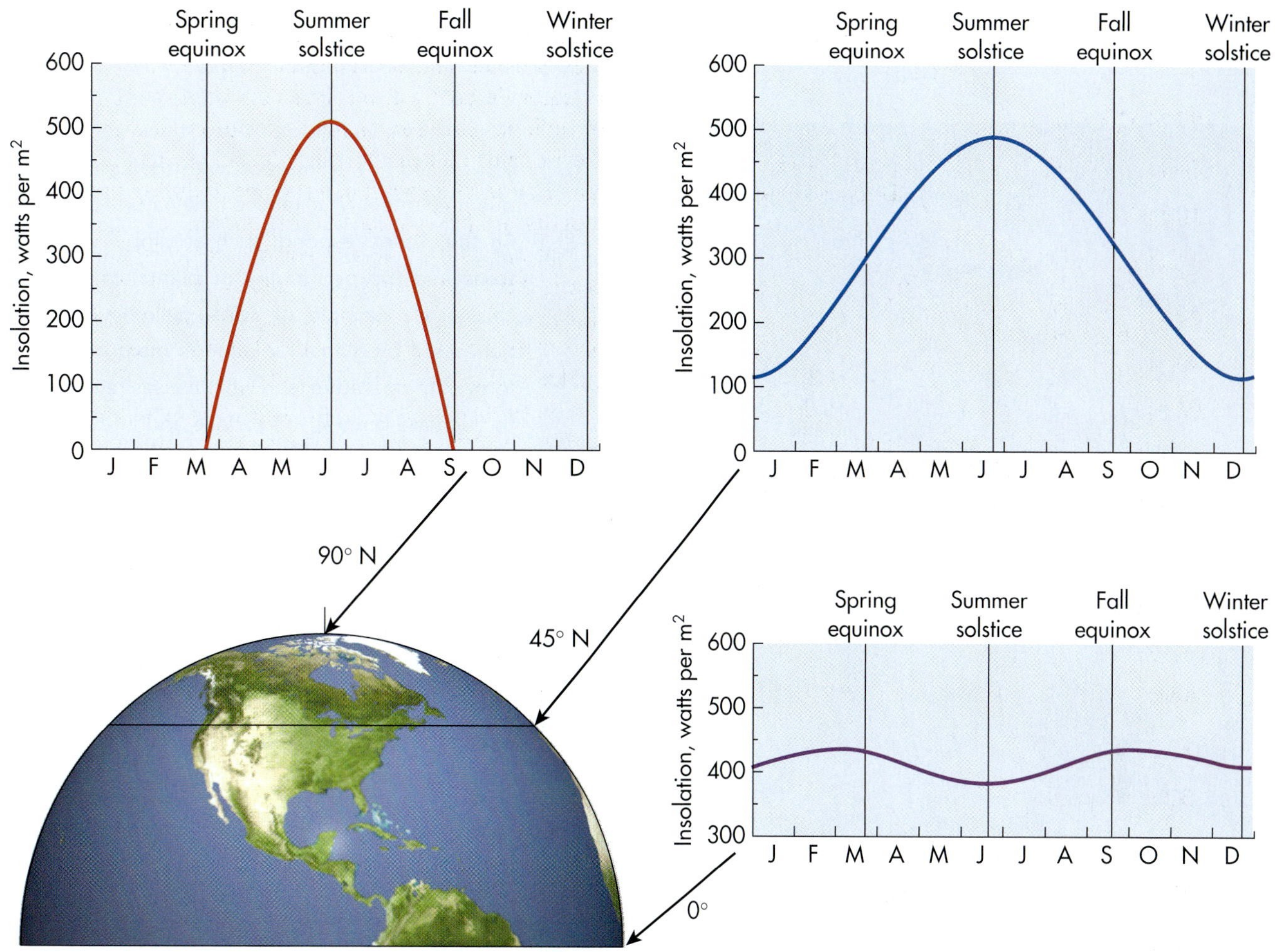

Spring equinox
Summer solstice
Fall equinox
Winter solstice
Insolation, watts per m²
600
500
400
300
200
100
0
J F M A M J J A S O N D
90° N
45° N
0°
Spring equinox
Summer solstice
Fall equinox
Winter solstice
Insolation, watts per m²
600
500
400
300
200
100
0
J F M A M J J A S O N D
Spring equinox
Summer solstice
Fall equinox
Winter solstice
Insolation, watts per m²
600
500
400
300
J F M A M J J A S O N D

Figure 4.27 Daily insolation through the year at the top of the atmosphere at select latitudes in the Northern Hemisphere. Note how insolation changes at each of these locations on a seasonal basis and that the annual range increases at progressively higher latitudes.

VISUAL CONCEPT CHECK 4.3

-150
-100
0
50
100
150

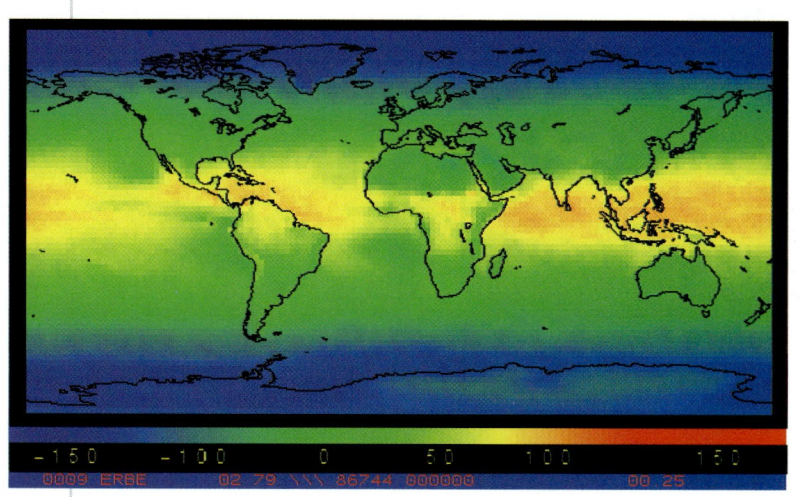

Net radiation refers to the total amount of solar radiation received by the Earth at any given place and time. This image shows the geography of mean net radiation during a particular time of year, with yellows and reddish hues representing high values in W/m^2. Darker colors, in contrast, represent low values of net radiation. By looking at this diagram, can you determine what time of year it must be? What must the Earth–Sun geometric relationship be like during this time of year?

The Big Picture

Now that we've followed the pathways of solar radiation and how it interacts with the Earth, subsequent chapters will examine how natural processes on the planet respond, such as global wind patterns and the geography of plants. In the next chapter for example, we will explain how the heat generated by global radiation is measured in the form of temperature, and what causes temperature to fluctuate both at the surface and in the various layers of the atmosphere. For example, the coldest place on Earth is in Antarctica, whereas the hottest is the Sahara Desert. The next chapter will help explain why these geographic patterns occur.

Summary of Key Concepts

1. The atmosphere serves as a protective shield that filters the potentially deadly effects of ultraviolet radiation from the Sun, allowing mostly visible and infrared wavelengths to reach the Earth.

2. The lower part of the atmosphere is warmed primarily due to the greenhouse effect. This warming occurs because variable gases in the atmosphere, such as carbon dioxide and water vapor (as well as others), trap longwave radiation that is emitted by the Earth.

3. Insolation refers to the amount of solar radiation received at the top of the atmosphere. From this point, radiation follows several paths. Approximately 50% of all radiation reaches the Earth, with some arriving directly and some indirectly. The remaining radiation is either absorbed in the atmosphere, scattered by dust, or reflected directly back into space.

4. Albedo refers to the reflectivity of surfaces on Earth. Snowy surfaces have the highest albedo, whereas darker surfaces such as roads and oceans have the lowest albedo. Overall, the Earth reflects about 31% of all incoming solar radiation through albedo.

5. The global radiation budget refers to the overall balance of incoming versus outgoing radiation. The Earth receives shortwave radiation from the Sun. This radiation is either absorbed by the atmosphere, the various Earth surfaces, or reflected and scattered back to the surface or back into space. The radiation that reaches the Earth's surface is absorbed and then re-radiated as longwave radiation. The overall radiation budget must be in balance, otherwise the Earth would become progressively hotter or colder.

Check Your Understanding

1. Are wavelengths in the visible part of the electro-magnetic spectrum short or long?

2. Why does the Sun emit shortwave radiation, whereas the Earth emits longwave radiation?

3. Why does a constant amount of radiation reach the Earth?

4. List the variable gases and describe why they change with respect to their proportion of the atmosphere.

5. Why is the sky blue during the day, but red and orange at dawn and dusk?

6. Describe some of the variables that influence the amount of radiation that is absorbed on Earth.

7. How does the angle of incidence vary by latitude?

8. If the axis of the Earth were perpendicular to the plane of ecliptic, how would net radiation change at high latitudes. Would these regions have a greater net surplus or net deficit of radiation?

9. Why does the zone of net radiation surplus migrate (on Earth) over the course of the year?

10. Describe the global radiation budget. What are some of the things that happen to solar radiation when it reaches the Earth? Why is there a global balance in the radiation budget?

ANSWERS TO VISUAL CONCEPT CHECKS

Visual Concept Check 4.1

Solar energy is produced by the process of nuclear fusion within the Sun. In this fashion, hydrogen is converted to helium at very high temperatures and pressures. Solar radiation travels in the form of electromagnetic energy at 300,000 km (186,000 mi) per second. It reaches the Earth in about 8 minutes.

Visual Concept Check 4.2

This factory is producing a variety of particulates and variable gases that are streaming into the atmosphere. Some of these industrial by-products will cause solar radiation to be reflected, whereas others will cause absorption or scattering of electromagnetic radiation.

Visual Concept Check 4.3

This map of net radiation reflects patterns observed at either Equinox. Note that the highest values of net radiation occur in the equatorial regions, which reflect the location of the subsolar point and a maximum (90°) Sun angle. Such relationships can only occur along the Equator during the Spring and Fall Equinoxes.

CHAPTER FIVE

GLOBAL TEMPERATURE PATTERNS

In the previous chapter we examined the various ways that energy flows in the atmosphere. At this point we turn to see how we measure its primary effects upon atmospheric temperature patterns. The word *atmosphere* is derived from the Greek phrase meaning *sphere of air*, which implies that the air is uniform in its consistency and character. In fact, the atmosphere consists of several layers, each having slightly different characteristics and

This sunset image from the space shuttle shows the blue atmosphere against the darkness of space. In this chapter, we'll discuss the layers of the atmosphere and how temperature varies within it and across the Earth's surface.

interactions with solar radiation. We'll examine these layers in this chapter

and how the measure of warmth or coolness—that is, temperature—varies

between and within them. Most of the discussion will focus on the

interaction of the environmental variables that influence temperature in the

lower part of the atmosphere and how these variables interact to form

distinct geographic temperature patterns on Earth.

Layered Structure of the Atmosphere

The atmosphere consists of several distinct layers, which differ in density and composition. Within the context of air temperature, the configuration of the atmosphere is important because of the way that solar energy moves through and interacts with the various layers. These interactions affect air temperature in both the horizontal and vertical dimensions of the atmosphere.

The atmosphere extends from a shallow depth within the Earth (because soil contains air and water) to a height of about 480 km (300 mi). Most of the atmosphere's mass, however, lies below an altitude of 30 km (18.6 mi). Although we generally think of the atmosphere as a uniform gaseous medium that envelops the Earth, it actually contains four distinct layers: the troposphere, stratosphere, mesosphere, and thermosphere. These layers are distinguished by temperature and also by what elements they contain. Figure 5.1 illustrates these major layers and the temperature trends within them.

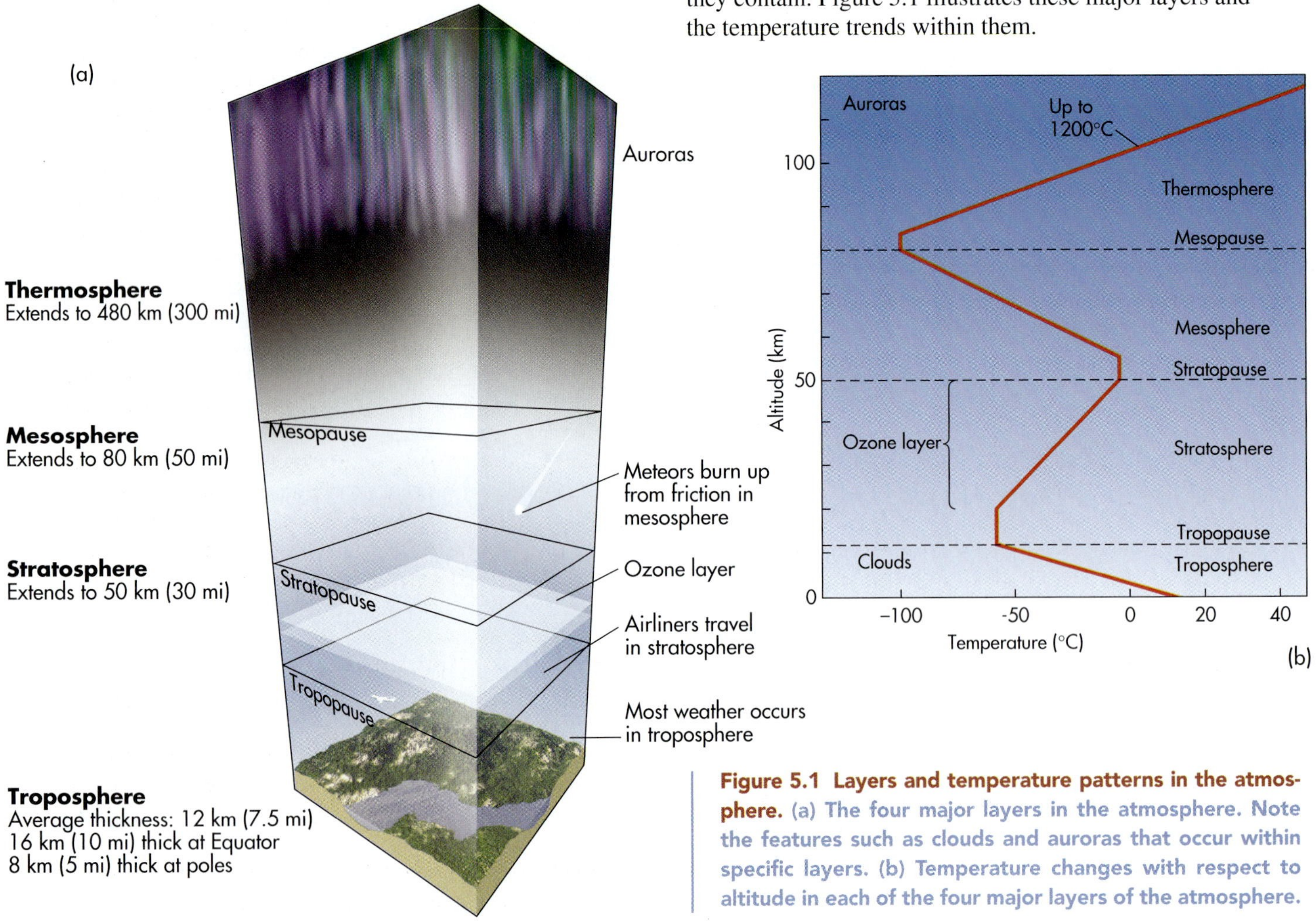

(a)

Thermosphere
Extends to 480 km (300 mi)

Mesosphere
Extends to 80 km (50 mi)

Mesopause

Stratosphere
Extends to 50 km (30 mi)

Stratopause

Tropopause

Troposphere
Average thickness: 12 km (7.5 mi)
16 km (10 mi) thick at Equator
8 km (5 mi) thick at poles

Auroras

Meteors burn up from friction in mesosphere

Ozone layer

Airliners travel in stratosphere

Most weather occurs in troposphere

(b)

Figure 5.1 Layers and temperature patterns in the atmosphere. (a) The four major layers in the atmosphere. Note the features such as clouds and auroras that occur within specific layers. (b) Temperature changes with respect to altitude in each of the four major layers of the atmosphere.

The Troposphere

The **troposphere** is the lowest layer of the atmosphere, extending from the surface to an average altitude of about 12 km (~7.5 mi). The name of this layer is derived from the Greek word *tropos* for *turning* or *mixing* because it is the most active zone of the atmosphere, with vigorously moving currents of air. Physical geographers and meteorologists are most interested in this atmospheric layer because it contains the vast majority of nonmarine living organisms and is the zone where most weather occurs, such as rain, wind, and snow. Much of this weather occurs because the troposphere contains most of the atmospheric water vapor and particulates.

The troposphere is generally warmed by longwave radiation emitted from Earth as part of the global energy balance discussed in Chapter 4. Because the source of heat in the troposphere is the Earth, temperature decreases with increasing altitude (Figure 5.2). This change in temperature is known as the **environmental lapse rate.** As altitude (from the Earth) increases in the troposphere, the average temperature drops 6.4° C per 1000 m or 3.5° F per 1000 ft (a negative lapse rate). At the upper limit of the troposphere, which is called the **tropopause,** the temperature stops decreasing and begins to increase into the next layer of the atmosphere. The tropopause occurs at the altitude where the air temperature is −57° C (−70° F), which again varies depending upon latitude.

The thickness of the troposphere depends largely upon temperatures at the Earth's surface and so it varies with season and latitude, ranging in thickness from about 16 km (10 mi) at the Equator to approximately 8 km (5 mi) at the poles. Why is this so? Recall that the low latitudes have a net surplus of radiation because these regions receive more intense solar radiation. Thus, near the Equator, more radiation is transferred back to the atmosphere as longwave radiation, which heats the lower atmosphere. The higher temperature causes the atmospheric gases to expand, making the troposphere extend higher from the Earth's surface at the Equator than at the poles.

The Stratosphere

The layer of the atmosphere that lies immediately above the troposphere is the **stratosphere** (see Figure 5.1),

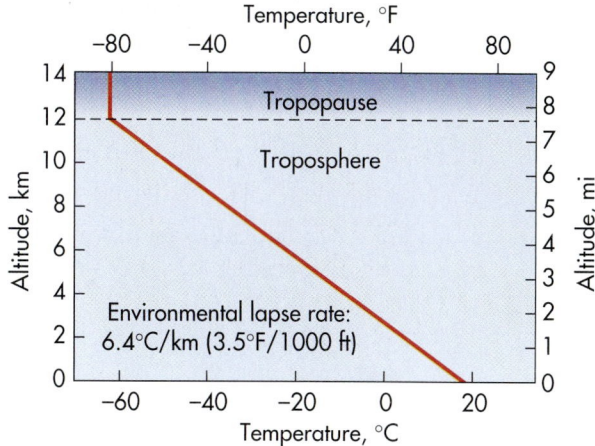

Figure 5.2 Environmental lapse rate in the troposphere. The average temperature decreases with increasing altitude in the troposphere until it reaches −57° C (−70° F).

which ranges in altitude between about 12 km and 50 km (between ~7.5 mi and 30 mi). This portion of the atmosphere derives its name from the Greek word *stratos*, which means *layer*. The stratosphere is critically important to life on Earth because it contains the ozone layer, which occurs at an altitude between 20 and 50 km (12 to 31 mi). The concentration of ozone in this portion of the atmosphere is about 10 parts per million by volume (ppmv), whereas it is only 0.04 ppmv in the troposphere.

As discussed in Chapter 4, the ozone layer filters ultraviolet radiation from the Sun and re-radiates it as infrared energy. Stratospheric temperature trends reflect this filtering process and the overall thickness of the ozone layer. From the top of the tropopause to the base of the ozone layer, temperatures are consistently about −57° C (−63° F) (see Figure 5.1b). Above that altitude, however, temperatures increase with altitude. The top of the stratosphere, or **stratopause,** is marked by the altitude where temperature stops increasing. At this altitude, average temperature is about −5° C (−23° F).

In addition to its filtering effects, the stratosphere is also important in the context of human travel because it is the portion of the atmosphere in which commercial jets fly. The stratosphere is a perfect medium through which to fly jet aircraft because it contains very little water

---

Troposphere *The lowermost layer of the atmosphere, which lies between the Earth's surface and an altitude of about 12 km (~7.5 mi).*

Environmental lapse rate *The decrease in temperature that generally occurs with respect to altitude in the troposphere. This rate is 6.4° C per km or 3.5° F per 1000 feet (a negative lapse rate).*

Tropopause *The top part of the troposphere, which is identified by where the air temperature is −57° C (−70° F).*

Stratosphere *The layer of the atmosphere, between the troposphere and mesosphere, that ranges between about 12 km and 50 km (between ~7.5 mi and 31 mi) in altitude.*

Stratopause *The upper boundary of the stratosphere where temperature reaches its highest point.*

1. An airliner in the midlatitudes takes off and climbs through the troposphere to its cruising altitude. As it rises through the troposphere, does the outside temperature (a) increase, (b) decrease, (c) remain the same?
2. The pilot informs you that the plane's cruising altitude is 36,000 feet. Is this in the (a) troposphere, (b) stratosphere, (c) mesosphere?
3. The pilot decides to avoid some turbulence by having the plane climb to 40,000 feet. At this new altitude, is the outside temperature (a) higher, (b) lower, (c) the same?

vapor and impurities; thus, pilots find relatively few clouds and good visibility. In addition, the air is relatively calm compared to the turbulent troposphere because air in the stratosphere generally flows parallel to the surface of the Earth. The next time you fly somewhere look outside the plane while you're cruising and try to remember why you're in the stratosphere.

The Mesosphere

The **mesosphere** is a layer of decreasing temperature that occurs from about 50 to 80 km (30 to 50 mi) in altitude (see Figure 5.1); it is the coldest of the atmospheric layers. Its name comes from the Greek word *mesos*, which means *middle*. Vertical temperature trends in the mesosphere have a negative lapse rate because temperature decreases with increasing distance from the ozone layer located in the stratosphere below. The altitude at which temperature stops decreasing is known as the **mesopause** and is the upper boundary of the mesosphere. At this altitude, air temperature is about −100° C (−148° F). In the mesosphere, solar radiation reduces gas molecules to individual electrically charged particles called ions. This deep layer of charged particles can disrupt communications between astronauts and ground control and can interfere with various satellite communications, such as transmission of television signals.

You can see the mesosphere indirectly, particularly at night, because most meteors burn up in this layer as they fall through the atmosphere. Most of these *shooting stars* are about the size of one sand grain and are destroyed because they collide with billions of ions and gas particles as they fall through the mesosphere. These collisions create sufficient heat to burn the tiny rock fragments, creating a short-lived streaking path, long before they strike the ground. Occasionally, the largest rock fragments reach the Earth's surface. If they do, they're called *meteorites*.

The Thermosphere

The **thermosphere** is the upper layer of the atmosphere and occurs between about 80 and 480 km (50 to 300 mi) in altitude. In this portion of the atmosphere, atmospheric gases are sorted into a variety of sublayers based on their molecular mass. Oxygen molecules are few and literally many kilometers (miles) apart from one another. In fact, they are so widely spaced that the boundary with space is very diffuse and thus difficult to precisely determine.

The name of this atmospheric layer is derived from the Greek *thermo*, which implies *heat*, and makes perfect sense if you look at the temperature patterns of the thermosphere in Figure 5.1b. Notice that they increase drastically about 10 km (6 mi) above the mesopause, ultimately reaching 1200° C (2200° F) and higher. These high tem-

Mesosphere *A layer of decreasing temperature in the atmosphere that occurs from about 50 to 80 km (~30 to 50 mi) in altitude.*

Mesopause *The upper boundary of the mesosphere where temperature reaches its lowest point.*

Thermosphere *The upper layer of the atmosphere, which occurs between about 80 and 480 km (~50 and 300 mi) in altitude.*

peratures occur because intense solar radiation interacts with the upper part of the atmosphere, causing the few oxygen molecules present to vibrate at tremendously high speeds; this creates **kinetic energy** that is measured as a high temperature for any specific molecule. This vibrating layer of the atmosphere is important for human communications because it enables radiowaves from one location at the surface to bounce off and be received at locations beyond the horizon.

Despite the very high temperatures of the thermosphere, this atmospheric layer would not feel "hot" to you in the same way you'd feel heat on the Earth's surface. The reason for this apparent disconnect is that individual oxygen molecules in the thermosphere are so far apart from one another. Because of these great distances the molecules hardly ever come in contact with each other, which means that very little heat is transferred from one to another. At lower levels of the atmosphere, in contrast, temperatures are felt to be genuinely higher because trillions of molecules are constantly colliding and a great deal of heat is exchanged.

> ## KEY CONCEPTS TO REMEMBER ABOUT THE STRUCTURE OF THE ATMOSPHERE
>
> 1. The atmosphere consists of four major layers, each with distinct temperature characteristics.
>
> 2. The troposphere lies closest to the surface of the Earth and has a negative environmental lapse rate of 6.4° C per 1000 m or 3.5° F per 1000 ft.
>
> 3. The next lowest layer is the stratosphere, which contains the ozone layer and thus warms with increasing altitude.
>
> 4. Immediately above the stratosphere is the mesosphere, which cools with increasing altitude and is the part of the atmosphere where solar radiation reduces individual molecules to ions.
>
> 5. The uppermost layer of the atmosphere is the thermosphere. Although the temperature in this part of the atmosphere is very high, it does not feel hot because the individual air molecules are very far apart.

Kinetic energy *The energy of motion in a body, measured as temperature, that is derived from movement of molecules within the body.*

Surface and Air Temperatures

Most of the atmospheric behavior of interest to geographers occurs near the Earth's surface. Of particular significance are global surface and air temperatures, which measure the amount of sensible heat at the surface or in the atmosphere. In general, atmospheric temperature is a measure of the kinetic energy contained within a unit of geographical space within the air. Surface temperature in contrast, is a measure of the kinetic energy contained in a region very close to the Earth's surface. Thus, changes in surface temperature measure the ebb and flow of energy at ground level. As noted in Chapter 4, this variation in energy on Earth depends largely upon net radiation. When a net surplus of radiation occurs, temperature increases because the surface absorbs more radiant energy than it is emitting in longwave form. Conversely, temperature decreases with a net deficit of radiation because the surface emits more energy than it absorbs. Recall from Chapter 4 that surface temperature can also change through the processes of conduction, evaporation, and convection.

In contrast to surface temperature, the term *air temperature* refers to the degree of warmth or coolness of a portion of the atmosphere. The standard altitude at which air temperature is measured is usually about 1.2 m (4 ft) above the ground surface. Although air temperature usually differs from surface temperature, the amount of heat energy in the ground influences the temperature of the air above it. As you know, air temperature is measured with a thermometer, which is traditionally a hollow glass tube containing a liquid, often mercury, that expands or contracts depending upon the amount of energy present. More recently, digital thermometers have become common, relying on small electrical devices that sense temperature.

Three temperature scales are used around the world (Figure 5.3). You are probably most familiar with the Fahrenheit scale, which was named after the German physicist Gabriel Daniel Fahrenheit, who devised the scale in the 18th century. This scale is used by the United States National Weather Service and American news media to report temperature. The common reference temperatures used in the Fahrenheit scale are the freezing and boiling points of water, defined as 32° F and 212° F, respectively.

Another temperature scale that you may be somewhat familiar with is the Celsius scale, named after Anders Celsius, the 18th-century Swedish astronomer who devised it. The Celsius scale is part of the International System of Measurement (SI) because it is a decimal scale, with 0° C and 100° C being the freezing and boiling points of water, respectively. The vast majority of the world uses the Celsius scale, including scientists in the United States. Because most of the world uses the Celsius scale, the stated goal of the U.S. government is to one day fully convert to this form of measurement rather than the Fahrenheit scale.

| Colorful Aurora Borealis near Fairbanks, Alaska.

AURORAS

Some of the most beautiful sights on Earth can usually be seen only at the high latitudes, around 60° and higher. These sights are called auroras, specifically the *Aurora Borealis* (Northern Lights) and *Aurora Australis* (Southern Lights). They appear as sheets of color shimmering high in the atmosphere. What are they?

The Sun emits more than just electromagnetic radiation; it also emits streams of high-energy particles. These particles are often referred to as the "solar wind" and extend to

The conversions required to convert from Fahrenheit to Celsius and from Celsius to Fahrenheit are (respectively):

$$F = 9/5C + 32°$$

$$C = 5/9 (F - 32°)$$

Thus, 1° C equals 1.8° F and 1° F equals 0.56° C.

The third temperature scale in wide use is the Kelvin scale, named for the 19th-century British physicist William Thomson, Lord Kelvin. This scale is used in a great deal of scientific research because it measures absolute temperature, with absolute zero theoretically being the point where an object has no measurable temperature. On the Celsius scale, absolute zero is −273° C. The Kelvin scale has no negative values as the other scales do, but it is similar to the Celsius scale because it maintains a

100° temperature range between the boiling and freezing points of water. Given that the Kelvin scale is used by climatologists and meteorologists only for more advanced work, it will not be used further in this book. For reference purposes, however, the conversions between Kelvin and Celsius scales are:

$$C° = K - 273$$

$$K = C° + 273$$

Calculating the Heat Index and Wind Chill

Have you ever noticed that the air occasionally seems warmer or colder than the temperature that is given? This kind of variation occurs when additional environmental factors, such as wind speed and the amount of atmospheric

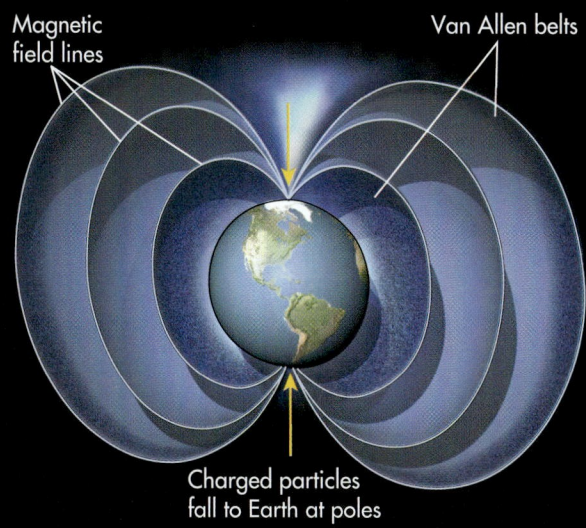

Charged particles
fall to Earth at poles

Magnetic field lines

Van Allen belts

The Earth's magnetic field extends outward to several hundred thousand kilometers. However, it bends into the upper atmosphere toward the North and South Poles.

Aurora Australis, seen from the space shuttle in May 1991, follows a line of the Earth's magnetic field.

the farthest reaches of the solar system. When these particles reach Earth, they encounter the Earth's magnetic field, which surrounds the planet high above the atmosphere. The charged particles are trapped within the magnetic field, especially within the two regions called the Van Allen belts, and do not reach Earth. However, the Van

Allen belts bend down close to the Earth at its poles, and the Sun's charged particles interact with atmospheric gases high in the thermosphere. The result of the interaction is the colorful glow of gases we see as auroras.

Auroras have been photographed from the space shuttle, where they can be seen following the Earth's magnetic field lines. In fact, space probes to Jupiter and Saturn have shown that auroras occur on those planets as well, again following the magnetic field lines.

water vapor, come into play. The combined impact of these factors with air temperature has significant implications for human comfort because they can make the air feel much warmer or colder than it really is.

The measures of human comfort that we use are the *wind chill index* in winter and the *heat index* in summer. The wind chill index is calculated using a variety of parameters, such as actual air temperature, wind speed, average face height, and components of modern heat loss theory. The wind chill index chart in Table 5.1 presents temperature data in the Fahrenheit scale and wind speed in miles per hour. In contrast to wind chill, the heat index measures the apparent temperature based on the combined variables of actual air temperature and relative humidity. (As we'll discuss in Chapter 7, relative humidity is the ratio of the specific amount of vapor relative to the

amount the air could hold at a given temperature.) The resulting heat index chart is shown in Table 5.2 and also uses the Fahrenheit temperature scale.

Large-Scale Geographic Factors That Influence Air Temperature

A common theme discussed so far in this text is that geography is a discipline that requires you to examine how variables are interrelated. These interrelationships are especially important for understanding how temperature varies across the Earth, which incorporates some of the concepts covered in Chapters 3 and 4. Some of these factors may be intuitive for you at this stage, but it's nevertheless a good time to review them and their relationships. Later in this section we'll look at some factors that can in-

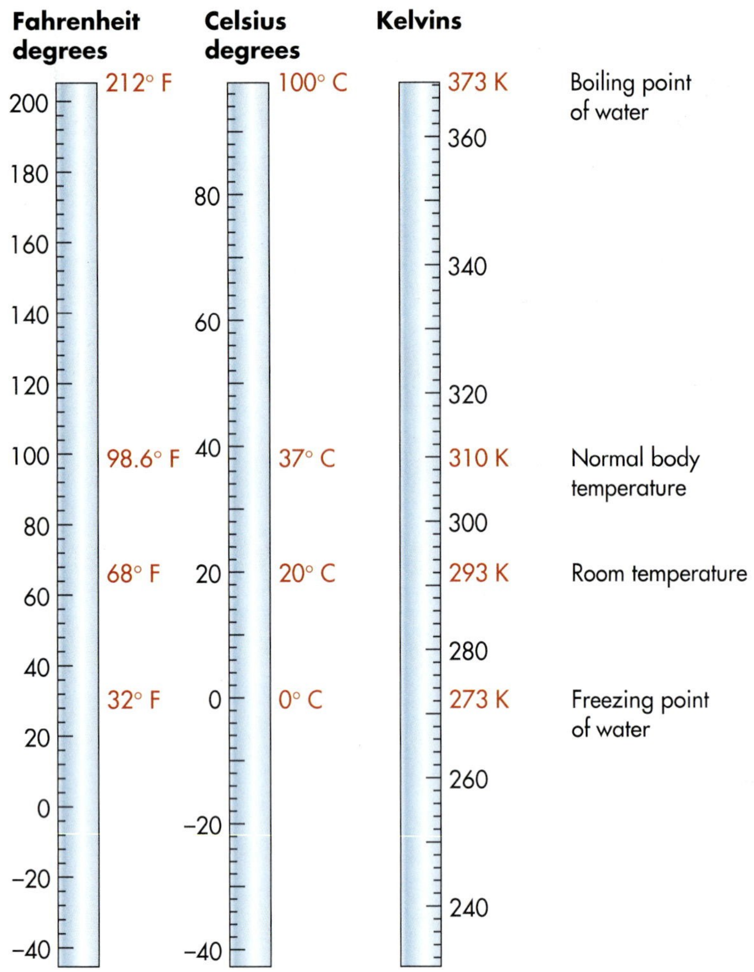

Fahrenheit degrees | **Celsius degrees** | **Kelvins**

212° F | 100° C | 373 K — Boiling point of water

98.6° F | 37° C | 310 K — Normal body temperature

68° F | 20° C | 293 K — Room temperature

32° F | 0° C | 273 K — Freezing point of water

(a)

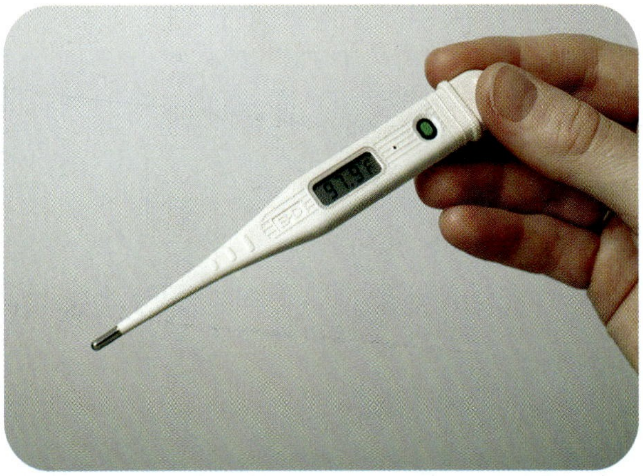

(b)

(c)

(d)

(e)

Figure 5.3 Measures of temperature. (a) The Fahrenheit, Celsius, and Kelvin temperature scales. The size of one degree on the Celsius scale is the same as one Kelvin; a degree on the Fahrenheit scale is smaller. (Note that units on the Kelvin scale are not called degrees, but are called Kelvins.) (b) Temperature is measured with a thermometer. Until recently, most thermometers were filled with alcohol or mercury, which rose or fell within a glass tube if warmed or cooled, respectively. Today, most thermometers contain temperature-sensitive electrodes and display information digitally. (c) For most people a comfortable room temperature is about 21° to 23° C (69° to 73° F; 294 to 296 Kelvins). (d) Water boils at a temperature of 100° C (212° F; 373 Kelvins). (e) The freezing/melting point of water is 0° C (32° F; 273 Kelvins).

TABLE 5.1   The Wind Chill Index (Source: NOAA)

Temperature (°F)

Wind speed (mph)	40	35	30	25	20	15	10	5	0	−5	−10	−15	−20	−25
5	36	31	25	19	13	7	1	−5	−11	−16	−22	−28	−34	−40
10	34	27	21	15	9	3	−4	−10	−16	−22	−28	−35	−41	−47
15	32	25	19	13	6	0	−7	−13	−19	−26	−32	−39	−45	−53
20	30	24	17	11	4	−2	−9	−15	−22	−29	−35	−42	−48	−55
25	29	23	16	9	3	−4	−11	−17	−24	−31	−37	−44	−51	−58
30	28	22	15	8	1	−5	−12	−19	−26	−33	−39	−46	−53	−60
35	28	21	14	7	0	−7	−14	−21	−27	−34	−41	−48	−55	−62
40	27	20	13	6	−1	−8	−15	−22	−29	−36	−43	−50	−57	−64
45	26	19	12	5	−2	−9	−16	−23	−30	−37	−44	−51	−56	−65
50	26	19	12	4	−3	−10	−17	−24	−31	−38	−45	−52	−60	−67
55	25	18	11	4	−3	−11	−18	−25	−32	−39	−46	−54	−61	−68

Frostbite occurs in 10 minutes or less

TABLE 5.2   The Heat Index (Source: NOAA)

Relative Humidity (%)

Air Temperature (°F)	20	25	30	35	40	45	50	55	60	65	70	75	80	85	90	95	100
125	141																
120	130	139	148														
115	120	127	135	143	151												
110	112	117	123	130	137	143	150										
105	105	109	113	118	123	129	135	142	149								
100	99	101	104	107	110	115	120	126	132	138	144						
95	93	94	96	98	101	104	107	110	114	119	124	130	136				
90	87	88	90	91	93	95	96	98	100	102	106	109	113	117	122		
85	82	83	84	85	86	87	88	89	90	91	93	95	97	99	102	105	106
80	77	77	78	79	79	80	81	81	82	83	85	86	86	87	88	89	91
75	72	72	73	73	74	74	75	75	76	76	77	77	78	78	79	79	80
70	66	66	67	67	68	68	69	69	70	70	70	70	71	71	71	71	72

Celsius	Fahrenheit	Impact on Humans
27–32° C	80–90° F	Caution: Fatigue is possible with continued exposure
32–41° C	90–105° F	Extreme Caution: Possible sunstroke, heat cramps, and exhaustion
41–54° C	105–125° F	Danger: Likely occurrence of sunstroke, heat cramps, and exhaustion
54° C	above 130° F	Extreme Danger: Heat stroke or sunstroke likely

AMAZING PLACES

TEMPERATURE EXTREMES ON EARTH

The Sahara Desert is the largest desert in the world. It is about the same size as the U.S. and is consistently the hottest place on Earth.

The warmest recorded temperature on Earth was measured in the Sahara Desert of Africa. Using the Celsius scale the temperature was 57.8° C, which probably doesn't sound that warm in your (Fahrenheit) frame of reference. When you consider that the temperature on the Fahrenheit scale was 136° F, then it sounds incredibly hot!

In contrast, the coldest recorded temperature on Earth was measured on the continent of Antarctica, which is the ice-covered landmass sur-

fluence temperature at a local scale. For now, however, we'll focus on three large-scale factors that influence temperature across the Earth, no matter the location.

Latitude Recall from Chapter 4 that the spherical shape of the Earth causes rays hitting the surface to differ in the area they cover, according to the latitude at which they strike. In other words, differences in the angle of incidence cause the same amount of energy to be directed at a smaller or larger area on the Earth's surface. When a larger area is covered at a lower incidence angle, it results in less energy per unit area on that surface compared to a smaller area struck at a higher incidence angle. This difference is most pronounced when comparing high and low latitudes and results in distinct temperature differences between two such regions (see Figure 4.24).

Seasons and Length of Day As was covered in earlier chapters, the Earth's axial tilt causes seasonal mi

gration of the subsolar point because the hemispheres are tilted either toward or away from the Sun, depending on the time of year. This migration of the subsolar point occurs only in the tropics, but results in dramatic changes in the angle of incidence at all latitudes, which, in turn, influences net radiation. If you need to review this concept again, examine Figures 4.24 and 4.26. Although these illustrations show only global net radiation in winter and summer, this seasonal relationship strongly influences temperature all year round. A secondary impact of axial tilt and seasons is that length of day can fluctuate a great deal, depending on latitude. Longer days mean that more radiation is received, which, in turn, also influences temperature. Think of your own experience regarding the temperatures where you live during winter and summer.

Time of Day As the day progresses from the morning sunrise, the Sun arcs westward across the sky. Recall from Chapter 3 that the Sun is at its highest point, resulting in

Antarctica is the coldest place on Earth; some areas have not been above the freezing point in more than 2 million years.

rounding the South Pole. Using the Celsius scale the temperature was −89.4° C. That sounds pretty cold, doesn't it? If you consider that this temperature on the Fahrenheit scale is −129° F, then it sounds even colder!

the day's most intense radiation, at solar noon. Temperature usually follows the Sun in that it increases as the Sun rises and decreases when it sets. The relationship isn't quite that simple, however, because a **temporal lag** occurs between the highest Sun angle and the warmest temperature of the day. In other words, the warmest part of the day occurs later in the day than the peak period of insolation and net radiation.

A good way to see how these variables affect one another is in graphical form. For example, Figure 5.4 shows how season and time of day are interrelated with respect to temperature. These diagrams show three sets of data from a typical observing station at 40° to 45° N latitude in the interior of North America: (a) insolation, (b) net radiation, and (c) air temperature. Parameter values, such as the

Temporal lag   *The difference in time between two events, such as when peak insolation and temperature occur.*

amount of net radiation, are presented on the vertical (or "y") axis of the respective diagrams, whereas time of day ranges across the horizontal (or "x") axis.

Notice in Figure 5.4 that each of the measured variables at the observation station varies both by season and time of day. With respect to insolation (Figure 5.4a), the amount of insolation on the June solstice is greater than during the December solstice. This seasonal variation in insolation results in a greater surplus (above the bold horizontal line) of net radiation during the summer months than in the winter (Figure 5.4b). Notice also that the number of hours of daily insolation and net radiation also vary per day. During the summer, for example, insolation is received for 16 hours of the day, whereas during the winter it's only 8 hours. This decrease occurs because the Earth's orbital position has changed. As a result of this seasonal insolation variation, the time of a net deficit of radiation (below the bold horizontal line) increases from about 8 hours in summer to 16 hours in winter.

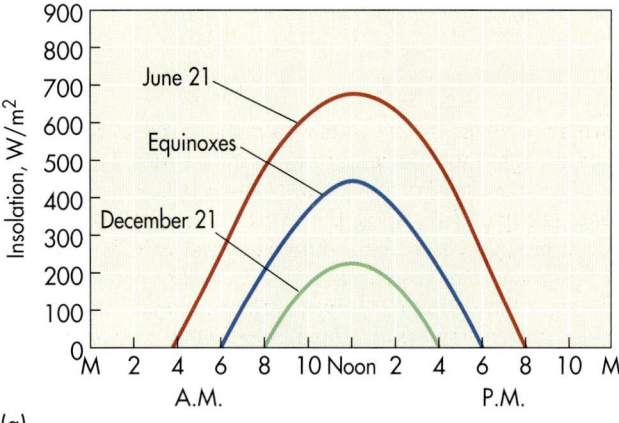

(a)

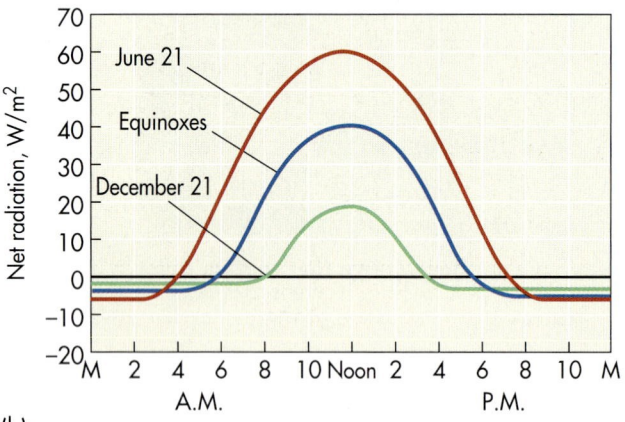

(b)

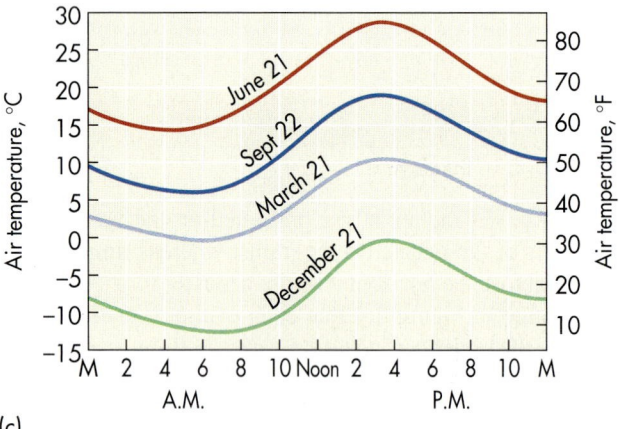

(c)

Figure 5.4 Interaction of (a) insolation, (b) net radiation, and (c) air temperature. These graphs were made from data taken at a midlatitude location in the Northern Hemisphere. Notice the close relationship among these three variables. Also note that a distinct lag occurs with respect to the temperature response.

To see how annual insolation/net radiation patterns influence temperature, study Figure 5.4c. As you can imagine, temperature is related to insolation and net radiation in that maxima and minima of each factor occur at about the same time. The pattern is predictable from a seasonal perspective in that temperatures are warmer during the summer and colder during the winter. Examination of Figures 5.4a and 5.4b shows a clear correlation between temperature, insolation, and net radiation. Notice, however, that the maximum and minimum temperatures occur at different times than insolation/net radiation fluctuations on a daily basis. In other words, a temporal lag occurs. With respect to maximum temperature, this lag exists because radiant energy is stored in the atmosphere for a brief period (following the insolation peak) before it is released as heat. Similarly, the minimum temperature occurs only after all of the stored heat from the day is released.

Local Factors That Influence Air Temperature

In the previous discussion we examined the important factors that influence air temperature on a large scale across the Earth. Besides these basic factors, additional, smaller-scale factors can influence air temperature in specific places. This section of the chapter focuses on some of these kinds of geographic patterns.

Maritime vs. Continental Locations Although the radiation budget is the largest influence upon air and surface temperatures, other factors that have yet to be discussed can play an important role. A good example of a local or regional variable that influences temperature is the **maritime vs. continental effect**. By definition, **maritime** places are located within or near a very large body of water, such as San Francisco, California, which is on the coast of the Pacific Ocean. In contrast, **continental** localities are surrounded by large landmasses, such as Topeka, Kansas, which is on the eastern edge of the Great Plains. This maritime vs. continental effect can be a critical factor with respect to explaining temperature patterns in specific places.

For example, if you compare Topeka and San Francisco, you can see that the annual temperature curves vary

Maritime vs. continental effect *The difference in annual and daily temperature that exists between coastal locations and those that are surrounded by large bodies of land.*

Maritime *A place that is close to a large body of water that moderates temperature.*

Continental *A place that is surrounded by a large body of land and that experiences a large annual range of temperature.*

SURFACE TEMPERATURE

At this time let's examine the factors that affect temperature more closely. Go to the *GeoDiscoveries* website and select the module **Surface Temperature**. This module will allow you to observe how the variables insolation, net radiation, and time of day influence temperature. The focus of this animation is a hypothetical location at 45° N latitude, one that lies deep within the interior of a continent. As you watch this animation, observe how air temperature near the Earth's surface changes over the course of the day on June 21, which is the Summer Solstice in the Northern Hemisphere and the longest day of the year. In particular, notice the lag that exists with respect to increases and decreases in insolation and temperature over the course of the day. After you complete the animation, be sure to answer the questions at the end of the module to test your understanding of this concept.

significantly (Figure 5.5). Notice that the San Francisco curve doesn't have a steep peak, but rather changes gradually over the course of the year. In addition, the range between the high and low monthly temperatures is narrow. At Topeka, in contrast, the curve has a much higher peak with a more distinct change between seasons and a greater overall range. This variability exists even though both places are located at about the same latitude (38° N), which would (correctly) lead you to believe that each locality receives about the same amount of insolation.

Why does this temperature variation occur? The main reason is the maritime vs. continental effect (Figure 5.6). Large bodies of water, such as the Pacific Ocean, store tremendous amounts of thermal energy, primarily because water has a high specific heat; that is, water absorbs a relatively large amount of energy before its temperature rises. In addition, solar radiation can penetrate to great depths in the ocean, heating water below the surface. The

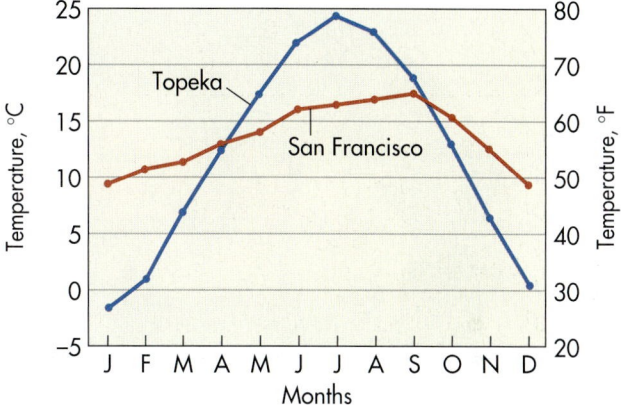

Figure 5.5 Annual temperature data from Topeka, Kansas, and San Francisco, California. A distinct peak in temperature occurs in Topeka during the summer, whereas San Francisco is comparatively moderate all year round.

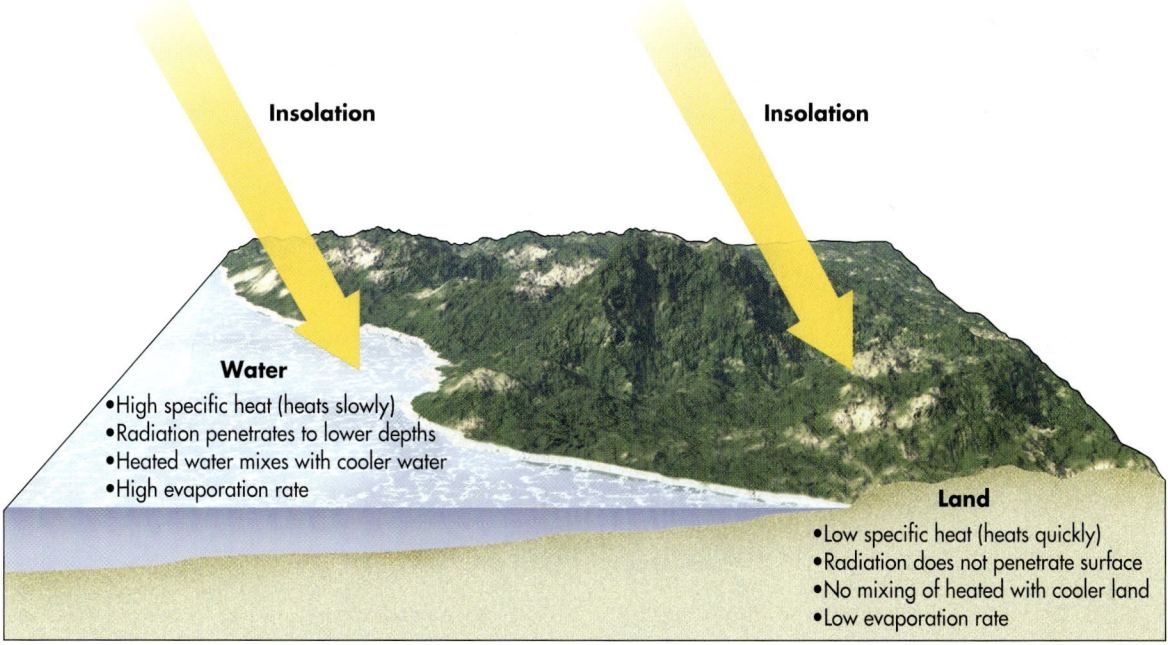

Figure 5.6 Maritime–continental contrasts. Several factors explain why temperature differences occur between coastal places and locations deep within continents. Note the different ways in which insolation interacts with land and water.

ocean currents constantly mix this warm water with the cooler water not exposed to sunlight. As a result, large water bodies maintain a more or less constant temperature for most of the year (Figure 5.7). This effect can be additionally enhanced if the ocean current that flows along the coast is a cold or warm current. Such currents will be described further in Chapter 6.

As water evaporates from the ocean, energy is transferred to the atmosphere in the form of latent heat that, in turn, moderates the air temperature of coastal places such as San Francisco. This moderating effect is magnified on the west coast of North America because the prevailing winds are westerly; that is, they blow from west to east. This circulatory pattern will be discussed in more detail in Chapter 6, but for now it's sufficient for you to know that these westerly winds blow relatively warm, moist air off the ocean some distance into the interior. In this way, San Francisco maintains a moderate air temperature all year round.

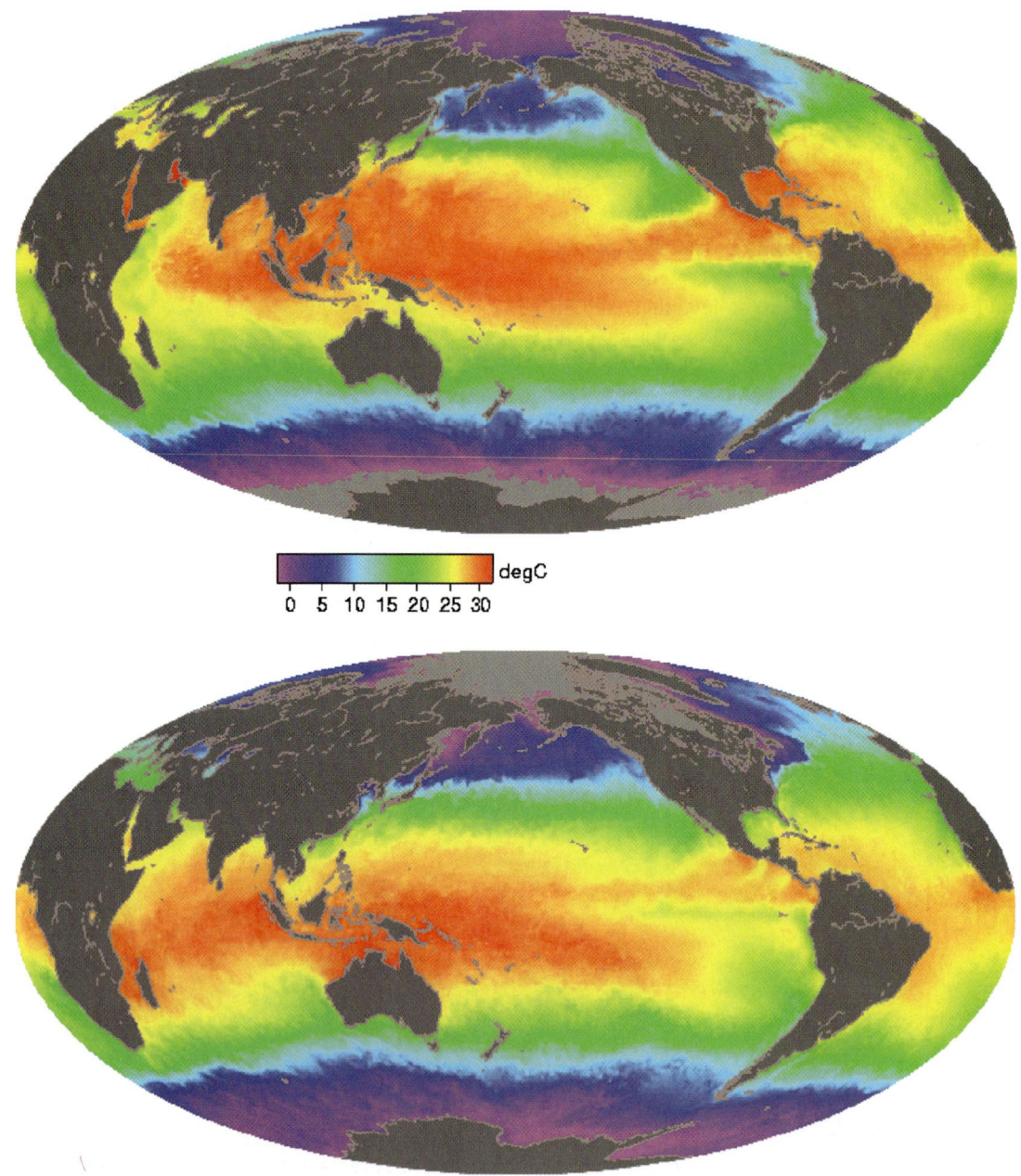

Figure 5.7 Global sea surface temperatures during June and December, 2000. Although the seasons differ dramatically between the images, notice the similarity in the basic geographic pattern. For example, the sea surface temperature in the eastern Pacific shows a small change between the months. Temperatures are in the Celsius scale and are coded by color as indicated by the horizontal bar between the maps.

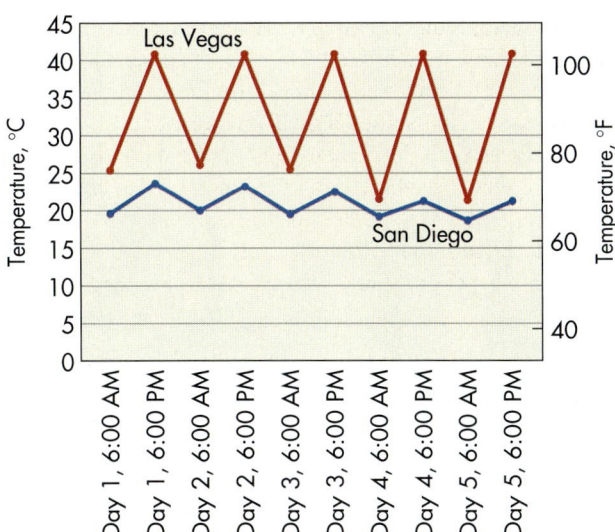

Figure 5.8 Daily cycle of temperature at San Diego, California, and Las Vegas, Nevada, from July 1–July 5, 2003. San Diego is a maritime city, whereas Las Vegas is located deep within the desert. Notice how pronounced the daily cycle of temperature is in Las Vegas compared to San Diego.

At continental localities like Topeka, however, the surrounding landmass does not store as much thermal energy as oceans do, largely because of their low specific heat; that is, the land temperature starts to rise after absorbing relatively small amounts of energy. In addition, radiation doesn't penetrate to great depths in landmasses and mixing of warm land and cool land doesn't occur. As a result, continental localities exhibit dramatic annual temperature variations compared to maritime places. This pattern not only exists over the course of the year, but can be seen in the daily cycle of temperature as well (Figure 5.8).

The Urban Effect on Temperature   In addition to the maritime vs. continental effect, another factor that influences temperature on a local scale is the presence of large cities. In general, cities are growing all around the world in response to human population increases. As cities grow, more of the natural landscape is being developed, and this results in new buildings, homes, and parking lots (Figure 5.9). In this process, vegetation is removed and soils are covered by pavement or other structures. This human-induced change in land cover can cause a distinct temperature increase in and around a city through a variety of ways, resulting in a distinct geographic region known as an **urban heat island** where temperatures may be as much as 3° to 4° C (6° to 8° F) warmer (or more) than the surrounding countryside.

Urban environments tend to be warmer than rural areas for several reasons. One of the most important reasons for this effect is that urban surfaces generally consist of metal, glass, asphalt, and concrete, whereas rural areas are covered with soil in which forest and grass grow. As a result, urban surfaces tend to be darker than rural surfaces, which leads to greater net radiation in cities because albedo is lower. In the context of this higher net radiation, up to 70% of net radiation in cities is converted to sensible heat, much more so than in the countryside. Urban surfaces also conduct more energy than rural soils and are thus warmer. This effect is magnified by the fact that forest canopies in rural areas shade the ground beneath them, keeping them relatively cool. In urban environments, on the other hand, solar radiation strikes the surface directly and is absorbed, later to be released as warming longwave radiation. This re-radiation tends to be trapped for longer periods of time in cities, further warming them, because the irregular geometry of city buildings limits wind speed by about 20% to 30% over the course of the year. This decreased flow of air reduces the loss of heat that would otherwise occur.

Urban heat island   *The relatively warm temperatures associated with cities that occur because paved surfaces and urban structures absorb and release radiation differently than the surrounding countryside.*

GEODISCOVERIES   WWW.WILEY.COM/COLLEGE/ARBOGAST

MARITIME VS. CONTINENTAL EFFECT

To see how the martime vs. continental effect occurs, go to the *GeoDiscoveries* website and select the module **Maritime vs. Continental Effect**. In this animation, you will be able to observe how temperature varies over the course of the day at Yuma, Arizona and San Diego, California. Although San Diego and Yuma are both located essentially at the same line of latitude (~33° N), and thus receive about the same amount of daily insolation, the range in daily temperature differs dramatically between the two places. Watch how air temperature varies between these two places over the course of the day and year and relate what you see with the previous discussion in this text. After you complete the animation, be sure to answer the questions at the end of the module to test your understanding of this concept.

(a)

(b)

Figure 5.9 Urban vs. rural landscapes. The different characteristics of city and rural surfaces can significantly influence temperature. (a) In general, rural surfaces such as this Scottish landscape are cooler because less radiation strikes the ground due to tree cover, and there is more water in soil and on vegetation surfaces. (b) Urban surfaces, in contrast, receive more direct radiation and absorb more radiation. This image shows a portion of the cityscape in Tokyo, Japan.

VISUAL CONCEPT CHECK 5.2

One of the most beautiful cities in North America is Vancouver, British Columbia. This Canadian city is particularly scenic because it lies between the Haro Strait, which is connected to the Pacific Ocean to the west, and the Cascade Mountains to the east. Although Vancouver lies at a fairly high latitude (49° N), it has a very moderate climate with average high temperatures that range from 22° C (71° F) in July to 7° C (45° F) in December. Which one of the following reasons account for this temperature pattern?

a) Vancouver is a continental location.

b) Vancouver lies next to a large body of water that has a generally consistent annual temperature.

c) The wind in Vancouver generally flows from east to west.

d) The Pacific Ocean has a wide range of annual temperature.

Another reason why cities are warmer than rural places is related to the way water is stored and moves in the two areas. In rural areas where soils cover the landscape, rainfall and snowmelt gradually soak into the ground. As described earlier, water has a modifying effect on temperature, which results in soils generally being cooler than urban surfaces. The presence of water in rural soils also reduces the conductivity of those surfaces, keeping them relatively cool. In the context of water absorption, it's useful to think of urban surfaces as being sealed because they're paved or covered with buildings. In the core of cities, for example, as much as 75% of surfaces are covered in this way. The effect of this change in land cover on the way water is stored and moves is significant because water doesn't absorb into the ground in cities like it does in the countryside. Instead, precipitation rapidly runs across these surfaces and into storm drains that carry it away. Thus, the modifying effect that water has on the temperature of soils in rural areas doesn't exist in cities, causing them to be warmer.

Yet another reason why cities are warmer than rural areas is the impact that human activities have on urban environments. During the summer months, for example, electricity production and fossil fuel consumption release a great deal of energy, perhaps as much as 25% to 50% of insolation. During the winter months, in contrast, a great deal of sensible heat is generated through artificial heating activities. Humans also alter the heating characteristics of cities through the production of pollutants such as ground-level ozone and other aerosols. In general, about 10 times more human-produced particulates are present in urban environments than in rural areas. Although these pollutants increase the reflectivity of the atmosphere above cities, thus reducing insolation, they cause an increase in the amount of infrared energy re-radiated downward to the surface.

An excellent example of an urban heat island is Atlanta, Georgia. This southeastern city has grown rapidly since the middle of the 20th century, becoming the leading commercial, industrial, and transportation area of the region. During this time, Atlanta has been one of the fastest growing metropolitan areas in the U.S., with population increasing approximately 30% between 1970 and 1990. In association with this explosive growth has been a dramatic expansion of the urban environment at the expense of agricultural land and forest. As a result, the air quality of the Atlanta region has decreased significantly, with increased amounts of ozone and volatile organic compounds polluting the air. Another impact has been the well-defined urban heat island that has evolved, as you can see in Figure 5.10. As you examine this image, note that May temperatures in the central business district reached 45° C (113° F) in some areas, whereas they were as low as 22° C (71° F) in the surrounding countryside. This wide disparity is the primary reason that the city is known as "Hotlanta" by many people who live there.

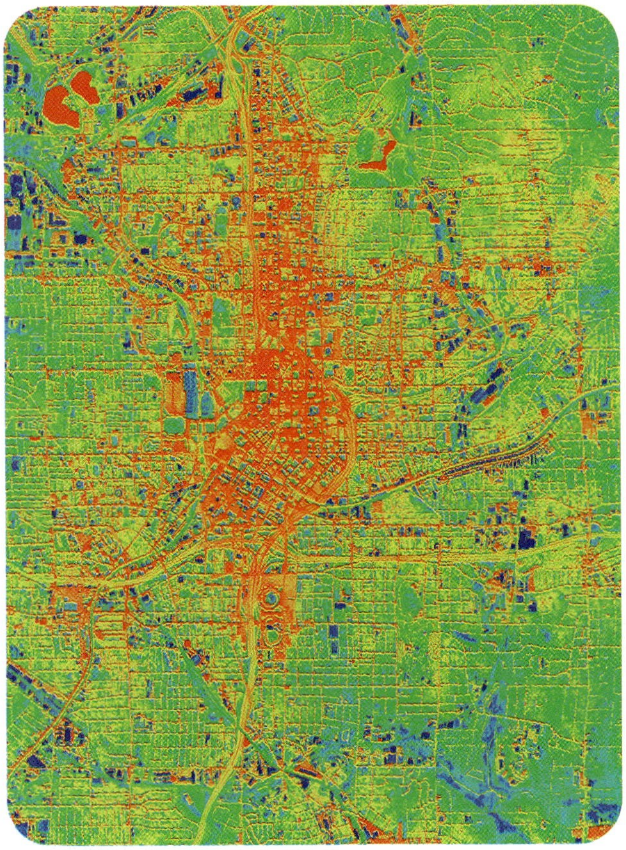

Figure 5.10 Urban heat island in Atlanta, Georgia, May 1997. This thermal infrared image was acquired during Project ATLANTA (ATlanta Land-use ANalysis: Temperature and Air-quality) with an airborne sensor. The inner city temperatures (orange) are significantly warmer than the surrounding countryside, which are represented in greens and yellows. (Credit: NASA)

Other Local Factors That Influence Temperature In addition to the maritime vs. continental effect and urban impact on temperature, a variety of other variables can influence temperature. These variables include altitude, the position of topographic barriers, and wind-flow patterns. Altitude influences temperature because, as we saw in Figure 5.1b, temperature tends to decrease with increased elevation. As you will see in Chapter 7, topographic barriers such as mountain ranges can have a significant impact on temperature that goes beyond the simple variable of increased altitude. In areas where wind descends a mountain range, temperature can warm considerably due to increased molecular friction as air is compacted.

The Annual Range of Surface Temperature (Putting It All Together)

Let's now bring together some of the variables that influence temperature around the world in an effort to see some basic global geographic patterns. A good place to begin is by considering Figure 5.11, which shows a hypothetical continent bordered on the east and west by oceans. This theoretical continent straddles the Equator and contains, as a point of reference, the hypothetical position of the 15° C (59° F) isotherm in January and July. This diagram shows the combined effects of the maritime/continental relation-

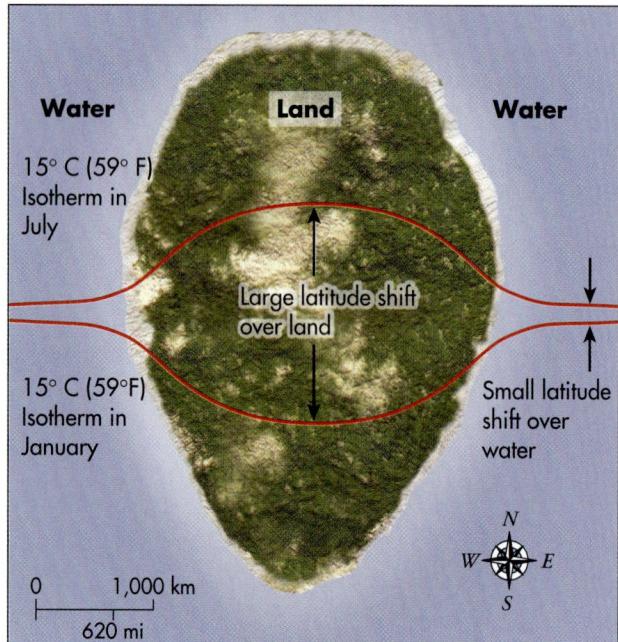

Figure 5.11 Theoretical seasonal migration of the 15° C (59° F) isotherm on a hypothetical continent. This hypothetical continent straddles the Equator, with 15° C-isotherms in each hemisphere. Notice how far the isotherms migrate on continents compared to their movements on the oceans.

ship and seasonality. Due to seasonality, the isotherm shifts into the Southern Hemisphere during that hemisphere's summer season (January) and back into the Northern Hemisphere during July. This migration occurs because a net surplus of radiation occurs in each hemisphere during the respective summer months due to high Sun angles and increased insolation. The maritime/continental effect is evident in the range of latitude that the isotherm shifts in each hemisphere. Notice that in the oceans, the isotherm doesn't move much compared to how it migrates on the landmass. This difference is due to the fact that large water bodies heat and cool much more slowly than do continents.

Given the basic seasonal and geographic pattern observed in Figure 5.11, let's consider the broad-scale range and cycle of temperature on Earth. Once again, a good place to begin is by examining some illustrations that show the basic patterns. With this in mind, take a look at Figure 5.12, which shows mean air temperatures on Earth during January and July.

Beginning with the January image, you can see that the landmasses in the Northern Hemisphere are quite cold, with temperatures in northeastern Asia (Siberia) of −50° C (−58° F) and −35° C (−31° F) in northern Canada. An interesting pattern exists in North America, where the 0° C (32° F) isotherm extends across the center of the U.S. in the interior of the continent, but crosses into the Pacific Ocean much farther to the north, virtually in Alaska. In other words, January temperatures are much warmer at higher latitudes along the west coast of North America than in the continental interior.

As expected for the Southern Hemisphere summer, mid- to high-latitude temperatures in that part of the world are also relatively warm, ranging from 10° C (50° F) on the southern tip of South America to 30° C (86° F) in the core of Australia. Predictably, temperatures in most of the tropical regions are warm, with an average of about 25° C (77° F) over a large part of the globe. The only significant deviation from this overall tropical pattern exists in western South America, where cool temperatures penetrate into very low latitudes. This pattern exists in part because the Andes Mountains follow much of the west coast of South America and are over 6000 m (19,680 ft) high. In addition, the cold Humboldt Current flows northward along the coast, which contributes to cooler temperatures in this area.

Now turn your attention to the July image in Figure 5.12. What different patterns emerge in this diagram? For one, you can see that the Northern Hemisphere landmasses are much warmer than they were in January, with temperatures reaching 10° C (50° F) in northeastern Asia and northern Canada. In other words, a 60° C (108° F) annual range of temperature exists at high latitudes in the Northern Hemisphere. Similarly, the interior of the U.S. is much warmer as well. Interestingly, the high latitudes in the Southern Hemisphere do not experience such a

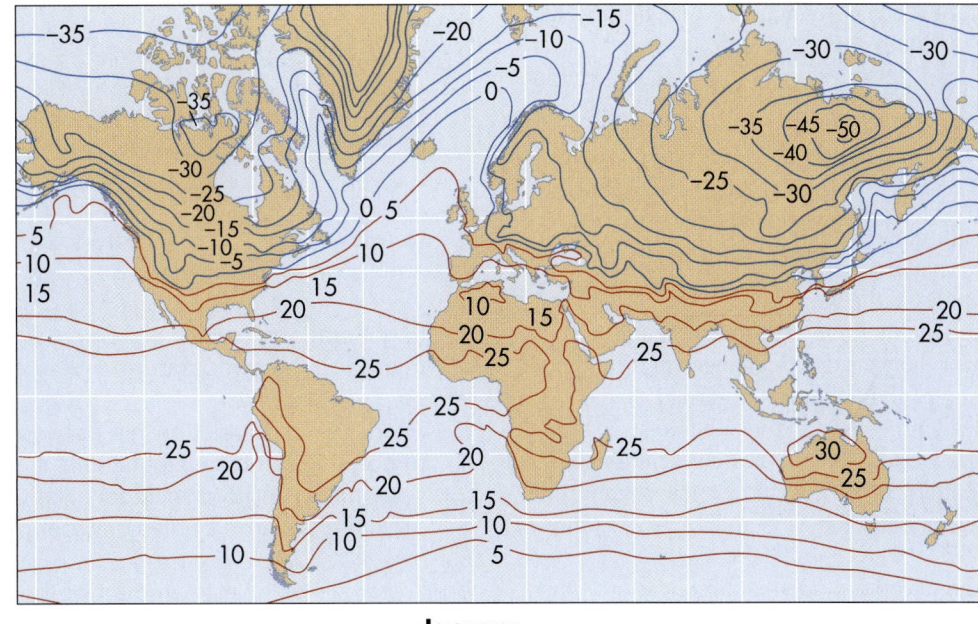

January

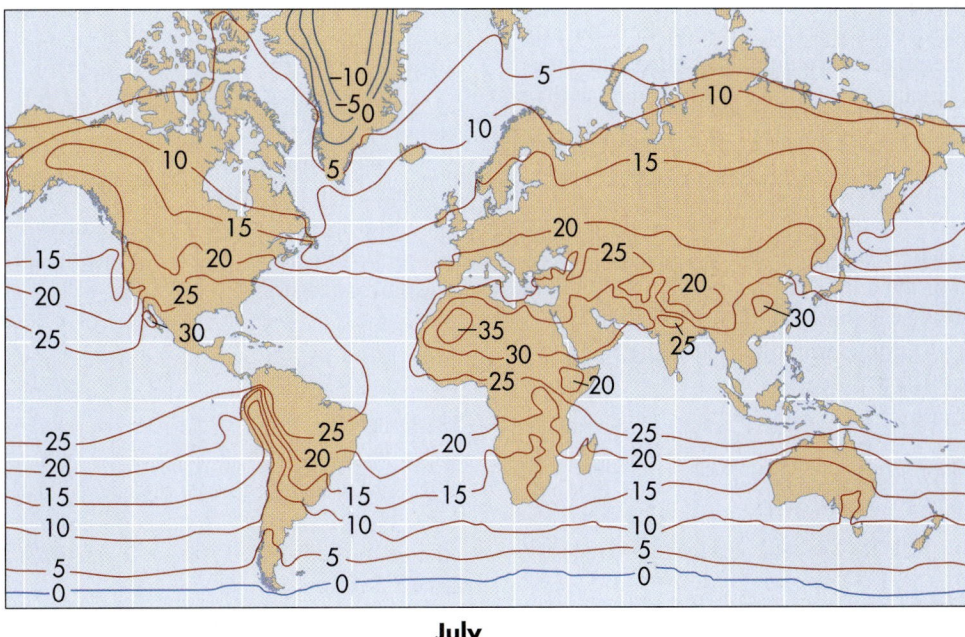

July

Figure 5.12 January and July distribution of surface air temperatures on Earth. Notice the basic geographic patterns and variations that occur between months. Blue isotherms indicate temperatures below 0° C (32° F), whereas red isotherms represent temperatures above 0° C (32° F).

range of temperature. Winter temperatures on the southern tip of South America, for example, cool to only about 0° C (32° F), which means the annual range there is approximately 10° C (18° F). Predictably, the tropical regions are warm, with a slight northerly migration of the 25° C (77° F) isotherm. Again, the only variation from this overall tropical pattern is the relatively cool temperatures that follow the Andes Mountains and Humboldt Current in South America.

Why do these annual patterns exist? The answer lies in a thorough interrelationship of the factors discussed so far, including seasonality, insolation, net radiation, latitude, the maritime vs. continental effect, and the environmental lapse rate in the troposphere. Some patterns are easy to explain. For example, the consistently warm temperatures that occur in the tropics are naturally a function of high Sun angles and associated intense radiation all year long. The exception to this tropical pattern exists along the axis of the Andes Mountains, which borders the western margin of South America. Here, as noted previously, relatively cool temperatures can be found within the equatorial zone. This pattern is easily explained because the Andes are over 6000 m (19,600 ft) high and are influenced from the perspective of temperature by the negative lapse rate in the troposphere. In other words, it is cooler at higher altitudes than at lower altitudes.

Another clear pattern is the extreme range of temperature at high latitudes in the Northern Hemisphere, which is easily explained by the highly variable amount of radiation received over the year due to changes in orbital position and the angle of incidence. It should also make sense to you that temperatures along the west coast of North America are warmer than in the continental interior in winter. This pattern is represented by the variation in temperature between San Francisco and Topeka, as illustrated in Figure 5.5.

Although the Northern Hemisphere pattern may make sense to you, you might also ask yourself the following (very good) question: Why don't the high latitudes in the Southern Hemisphere experience the same extreme temperature range? To answer this question, note the size of the Southern Hemisphere continents compared to those in the northern Hemisphere; they're much smaller than their Northern Hemisphere counterparts. In other words, there is more water in the Southern Hemisphere than in the Northern Hemisphere. Thus, more of a maritime effect occurs in the Southern Hemisphere than in the Northern Hemisphere, where the continental effect dominates. A good example of the continental effect in the Southern Hemisphere, however, is the core of Australia, where the annual range shows more variability due to the relatively large landmass of that country.

GEODISCOVERIES WWW.WILEY.COM/COLLEGE/ARBOGAST

TEMPERATURE AND LOCATION

To explore these patterns and relationships in a more interactive way, go to the *GeoDiscoveries* website and select the module **Temperature and Location**. In this module you will be able to "visit" five cities that experience the kinds of seasonal temperature patterns discussed in this chapter. These cities are Cordoba, Argentina; Yakutsk, Russia; Manaus, Brazil; St. Louis, Missouri; and San Francisco, California. You will visit each of these cities during each of the four seasons and will be able to access insolation and temperature data from each location. As you work through this simulation, note how the environmental factors we've discussed influence the seasonal range of temperature at these locations. Once you complete the simulation, be sure to answer the questions at the end of the module to test your understanding of this concept.

This image shows global temperature at a particular time of year, with the key showing the range of temperature in °C. Given your understanding of the factors that influence temperature, which one of the following choices best explains the pattern you see?

a) It must be summer in the Northern Hemisphere.

b) The Northern Hemisphere must be tilted away from the Sun.

c) The interior of North America and Asia are influenced significantly by the maritime effect on temperature.

d) The Southern Hemisphere is receiving the most insolation.

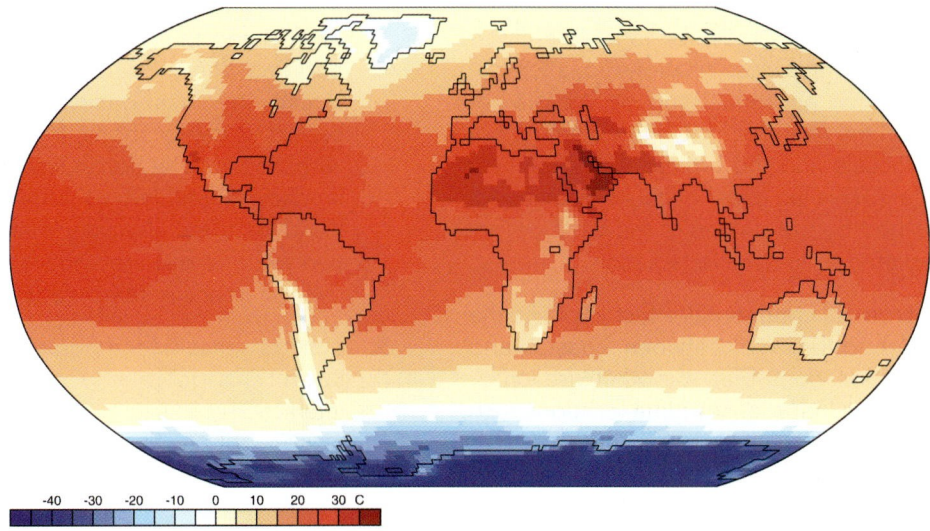

The Big Picture

Now that we've discussed how the factors of insolation and net radiation influence air temperature, we can examine how various atmospheric processes operate, such as wind patterns and the development of storm systems. A good place to begin this discussion is Chapter 6, which focuses on atmospheric air pressure and circulation and the way these concepts relate to Earth/Sun relationships and global temperature patterns. This image is a nice example of this relationship. Notice the stream of clouds across the image. This line of clouds is associated with an atmospheric feature called the jet stream, which is a band of strong winds that exists in the lower stratosphere. These winds can be particularly strong in the midlatitudes when large differences exist with respect to temperature at higher and lower latitudes. This pattern, as well as a variety of others related to the flow of air in the atmosphere, will be discussed in Chapter 6. As in previous chapters, be sure to consider how all of these concepts are interrelated and how they relate to previous discussions.

Summary of Key Concepts

1. The atmosphere contains four major layers: the troposphere, stratosphere, mesosphere, and thermosphere. Distinct temperature trends appear in each of these layers. One important change in temperature occurs in the troposphere, which cools with increasing altitude at a rate of 6.4° C per km or 3.5° F per 1000 feet. This rate is called the environmental lapse rate.

2. On a large scale, there is a very close relationship among insolation, net radiation, and air temperature. In spite of this close relationship, a temporal lag exists on an annual and daily basis between when the receipt of the highest amount of radiation occurs and the timing of the warmest temperature. This lag exists because insolation is first absorbed by the surface of the Earth before it is re-radiated as longwave radiation.

3. The maritime vs. continental effect explains the temperature difference between coastal and land-locked locations that are otherwise on or about the same latitude. Maritime locations such as San Francisco, California have a relatively narrow temperature range due to the thermal characteristics of large bodies of water like the Pacific Ocean. In contrast, continental locations such as Topeka, Kansas, have a much broader temperature range because landmasses do not absorb nor release heat as consistently as water bodies.

4. Measurable differences occur with respect to temperature between urban and rural localities. Cities are typically warmer than the countryside due to the urban heat island effect, because paved surfaces and buildings have relatively low albedo and thus absorb more solar radiation. In addition, the moderating effect of water is reduced in cities because urban surfaces are sealed. Rural locations, in contrast, are covered with soil that absorbs water, which cools the landscape relative to an urban environment. Wind patterns also vary markedly between the two regions and are another cause for the relative difference in temperature.

5. Distinct seasonal temperature patterns occur on Earth. This geography is dependent upon the interaction of several factors, including net radiation, season, latitude, the maritime vs. continental effect, and altitude.

Check Your Understanding

1. Why is latitude an important consideration when it comes to determining air temperature?

2. Where does the greatest range in annual temperature occur, at high or low latitudes? Why does this pattern exist?

3. Which location, Los Angeles, California, or Phoenix, Arizona, would experience the warmest peak temperature? Why does this pattern occur?

4. Central Canada and the southern tip of South America are both located at high latitudes. Despite this similarity, central Canada experiences a much greater range in temperature than the southern tip of South America. Why does this occur?

5. Which place would have a greater range in temperature, central Australia or Hawaii? Why?

6. Why does it make sense that the tropopause is at a higher altitude at the Equator than in northern Canada?

7. List three reasons why cities are typically warmer than rural landscapes.

8. Which place would be colder, the top or bottom of a high mountain? Why?

9. Why are temperatures in the lower stratosphere warmer than the upper part of the troposphere?

ANSWERS TO VISUAL CONCEPT CHECKS

Visual Concept Check 5.1

1. The answer is *b*; temperature decreases as you rise through the troposphere. This decrease occurs because there is a negative environmental lapse rate from the surface of the Earth to the tropopause.
2. The correct choice is *a*. This altitude occurs in the upper part of the troposphere.
3. The answer is *a*. At this altitude the temperature will begin to increase because the airplane is climbing closer to the ozone layer, which is found in the lower part of the stratosphere.

Visual Concept Check 5.2

The answer is *b*. Vancouver lies next to a large body of water (the Pacific Ocean) that has a generally consistent temperature over the course of the year. As a result, the air temperature at Vancouver remains cool in the summer and warm in the winter relative to continental locations.

Visual Concept Check 5.3

The answer is *a*. This image represents summer conditions in the Northern Hemisphere, because the high latitudes in Canada, Europe, and Asia are warm. Given that these are continental locations, which have a large annual temperature range, it must be summer and the Northern Hemisphere is tilted toward the Sun.

CHAPTER SIX

ATMOSPHERIC PRESSURE, WIND, AND GLOBAL CIRCULATION

Have you ever gone outside to discover that the wind was blowing very hard

and wondered why it was happening? In this chapter, we'll investigate the

way that air flows through the process of atmospheric circulation. The

circulatory processes described in this chapter have dramatic effects that

will be seen in later chapters, including daily weather (Chapter 8) and global

climate patterns (Chapter 9). Flowing air also has significant impact on the

Flowing air is usually invisible to the naked eye. However, in some cases, such as this sandstorm in the Kalahari Desert, strong winds blow dust in a way that allows us to visualize air movement. Flowing air is the focus of this chapter.

circulation of ocean currents, coastal erosion by waves (Chapter 19), and the

formation of sand dunes (Chapter 18). Many factors influence the direction

and speed of air flow, including pressure and temperature differences, the

Earth's rotation, and surface friction. After we discuss these variables, we'll

consider the geographic patterns of air flow around the globe. Toward the

end of the chapter we'll turn to how winds affect oceanic circulation

Atmospheric Pressure

Atmospheric circulation is an incredibly important process on Earth for a variety of reasons. The primary impact of air flow is to move heat energy around the globe in a way that moderates temperature on Earth. Air flow also affects global and local air quality and pollution levels. For example, winds can carry ash and gases from volcanic eruptions many kilometers from the volcano (Figure 6.1), or they can clear smog from large cities. Similarly, dust from eroded agricultural fields can be transported great distances before it settles back to the Earth. Before we can begin to describe the nature of air flow, however, we must first introduce the concept of air pressure because it directly influences the character of large- and small-scale wind patterns.

As we discussed in Chapter 4, the atmosphere is made up of a variety of gases that we collectively call "air." Like anything else, air is held to the Earth by gravity and thus has weight. The weight of the air exerts pressure on the surface of the Earth, which is measured as **air pressure** (also called *atmospheric pressure* or *barometric pressure*).

Factors That Influence Air Pressure

Atmospheric pressure is most closely associated with the temperature and density of air. The concept of air density is easy to understand if you consider that air expands or contracts depending upon the environmental setting. Imagine, for example, that you have a specific number of air molecules that are contained within an average shoebox. If you can somehow move all of the molecules in

Figure 6.1 Ash plume from Mt. Etna in Sicily. Northwesterly winds carry volcanic ash from the erupting volcano toward the southeast, in an image taken on July 22, 2001, from Space Station Alpha.

Air pressure *The force that air molecules exert on a surface due to their weight.*

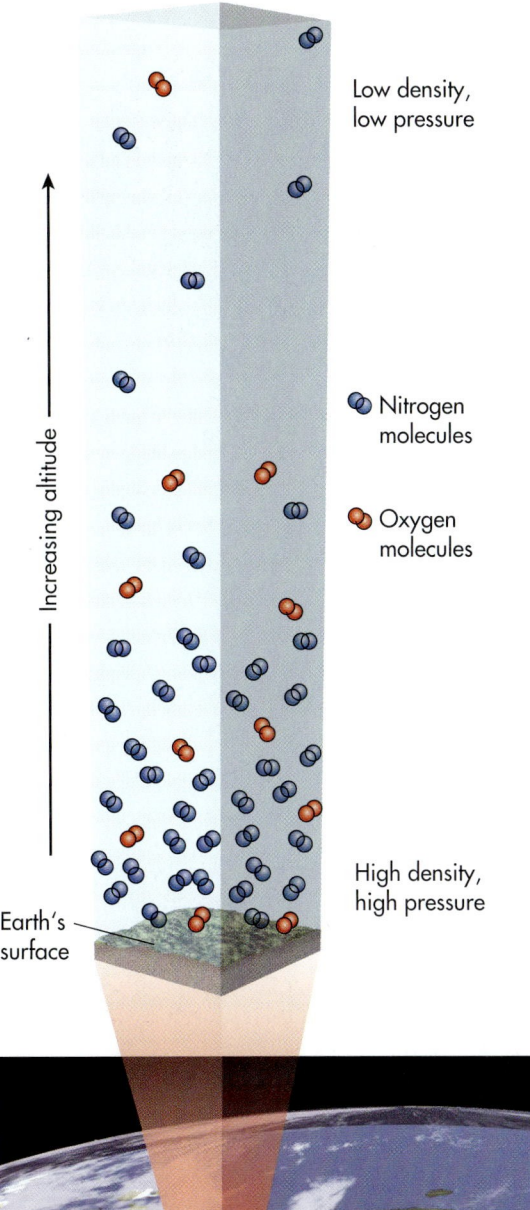

Low density,
low pressure

Nitrogen
molecules

Oxygen
molecules

High density,
high pressure

Increasing altitude

Earth's
surface

Figure 6.2 Air density, altitude, and atmospheric pressure. At low altitudes air molecules are held close to the Earth by gravity and thus are more dense, resulting in high atmospheric pressure. In contrast, the density of air molecules is low at high altitudes and air pressure is thus relatively low.

the shoebox to a larger box, the gas will expand to fit comfortably within the new container. In this example, the air molecules are closer together in the shoebox, and thus the density is greater, than when they are in the larger container.

You can see the basic effect of density on air pressure by noting how pressure changes with altitude (Figure 6.2). In general, air pressure decreases with increasing altitude because most air molecules are held close to the surface by gravity. As a result, the density of air molecules is greater closer to the Earth's surface, which means that air pressure is relatively high in that part of the atmosphere. With increased altitude the density of air molecules becomes progressively less, resulting in progressively lower air pressure.

In addition to the impact of gravity on the molecular density of air, atmospheric pressure is also strongly influenced by air temperature. The most obvious way that air temperature influences atmospheric pressure occurs when air close the Earth's surface is warmed a great deal. Such warming causes air molecules to scatter and density thus decreases, resulting in relatively low atmospheric pressure. Consider the analogy of a hot air balloon. When heat is added to the air within the balloon, it causes the air to expand and lift within the relatively cooler (more dense) air that surrounds it. Low atmospheric pressure also results when air is forced to rise vigorously. Very cold surface air is usually associated with high atmospheric pressure because cold air sinks and is thus very dense. In some instances, air from the upper atmosphere descends vigorously toward the Earth's surface. This process also results in high pressure.

Measuring and Mapping Air Pressure

Air pressure is often measured in units called millibars (mb) with an instrument called a barometer (Figure 6.3). A common type of barometer consists of a long glass tube, closed at one end, which is filled with a liquid (usually mercury) and inverted into a dish containing the same liquid. The liquid in the tube drops down slightly, leaving a vacuum at the closed end of the tube. When the liquid in the tube comes to rest, the force due to atmospheric pressure, pressing down on the liquid in the dish, exactly balances the weight of the column of liquid. The tube can then be calibrated to measure atmospheric pressure in terms of inches or millimeters of mercury, which can be converted to millibars or any other pressure unit.

You can see how air pressure changes with altitude by looking at three separate locations at different elevations (Figure 6.4). At sea level, for example, the average pressure of the air is 1013.25 mb. In contrast, at Denver, Colorado (5280 ft—"the Mile High City"), the air pressure is 840 mb. At a still higher elevation, such as at the top of Mt. Everest, the Earth's tallest mountain at 8850 m (29,035 ft), the average air pressure is only 320 mb.

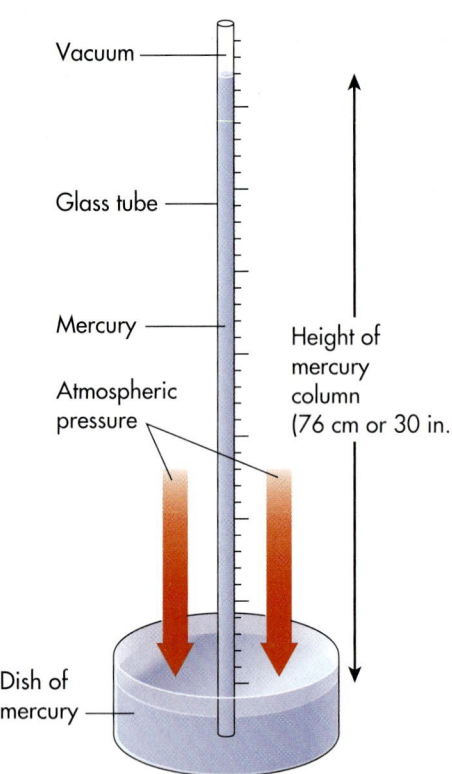

Vacuum

Glass tube

Mercury

Atmospheric
pressure

Height of
mercury
column
(76 cm or 30 in.)

Dish of
mercury

Figure 6.3 Measurement of atmospheric pressure. The pressure of the atmosphere is measured by the height of a column of mercury that can be supported by that pressure.

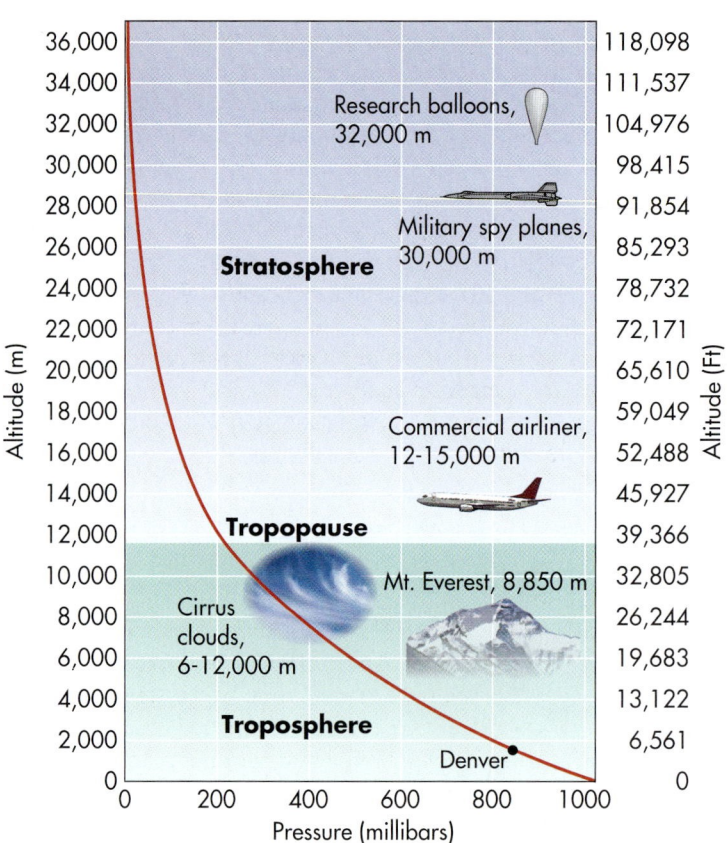

Altitude (m)	Altitude (Ft)
36,000	118,098
34,000	111,537
32,000	104,976
30,000	98,415
28,000	91,854
26,000	85,293
24,000	78,732
22,000	72,171
20,000	65,610
18,000	59,049
16,000	52,488
14,000	45,927
12,000	39,366
10,000	32,805
8,000	26,244
6,000	19,683
4,000	13,122
2,000	6,561
0	0

Research balloons, 32,000 m

Military spy planes, 30,000 m

Stratosphere

Commercial airliner, 12-15,000 m

Tropopause

Cirrus clouds, 6-12,000 m

Mt. Everest, 8,850 m

Troposphere

Denver

Pressure (millibars)

Figure 6.4 Atmospheric pressure and altitude. Average atmospheric pressure decreases with increasing elevation and altitude above the Earth's surface.

VISUAL CONCEPT CHECK 6.1

Mt. McKinley is the tallest mountain in North America, rising 6194 m (20,320 ft) in Alaska's Denali National Park. How do you suppose air pressure changes from the base of the mountain to the top—does it increase or decrease? Why does this change occur?

Atmospheric Pressure Systems

In addition to the pressure changes that occur with respect to altitude, air pressure also varies horizontally across the Earth's surface. A **high-pressure system** is a circulating body of air that exerts relatively high pressure on the surface of the Earth because air descends (toward the surface) in the center of the system. In contrast, a **low-pressure system** is a circulatory body of air where relatively less pressure exists on the Earth's surface because the air is rising (away from the surface) in the system's core. This portion of the chapter focuses on the nature of these large-scale pressure systems and how air flows within and between them.

Low-Pressure Systems

Low-pressure systems are often referred to as **cyclones.** If you look at a low-pressure system from the side (Figure 6.5), notice that the vertical flow of air consists of rising air, with the most vigorous upward flow in the center of the system. As a result, the central part of the system has the lowest pressure and is designated with the letter "L" on a weather map. In the Northern Hemisphere, the horizontal flow of air around the center of a low is counterclockwise (looking at it from above) as the air flows into (or converges at) the core of the system. In the Southern Hemisphere, the horizontal flow of air is clockwise. This inward flow occurs because the rising air at the center of the system creates a void in which air must flow locally to balance the atmospheric pressure. Low pressure centers are generally associated with cloudy or stormy weather because rising air cools with altitude and can thus hold less moisture than warmer air. We'll discuss this relationship in more detail in Chapter 7.

High-Pressure Systems

In contrast to low-pressure systems, large areas of relatively high air pressure are referred to as **anticyclones.**

High-pressure system *A rotating column of air that descends toward the surface of the Earth where it diverges. These systems spin clockwise in the Northern Hemisphere and counterclockwise in the Southern Hemisphere. Also called an anticyclone.*

Low-pressure system *A rotating column of air where air converges at the surface and subsequently lifts. These systems spin counterclockwise in the Northern Hemisphere and clockwise in the Southern Hemisphere. Also called cyclones.*

Cyclones *Low-pressure systems.*

Anticyclones *High-pressure systems.*

In these systems the vertical flow of air consists of descending air, which diverges (spreads apart) at the surface (Figure 6.5). Because air sinks most vigorously in the center of a high, the highest pressure is in the central part of the system, which is designated with the letter "H." The horizontal air flow around the center of a high is clockwise in the Northern Hemisphere and counterclockwise in the Southern Hemisphere, just opposite to the direction

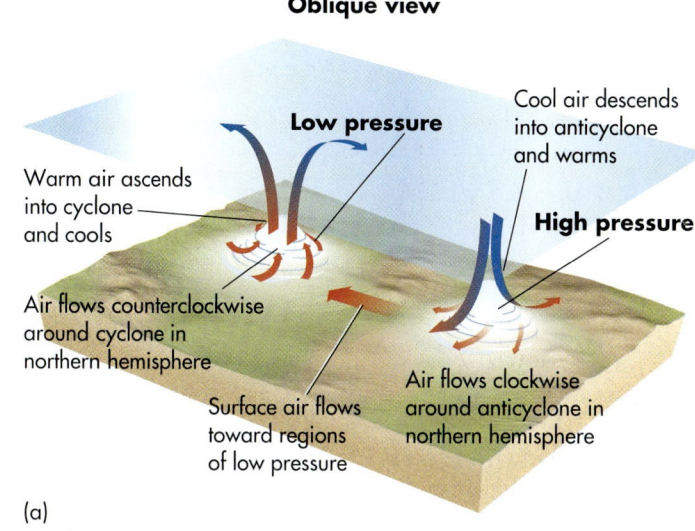

(a)

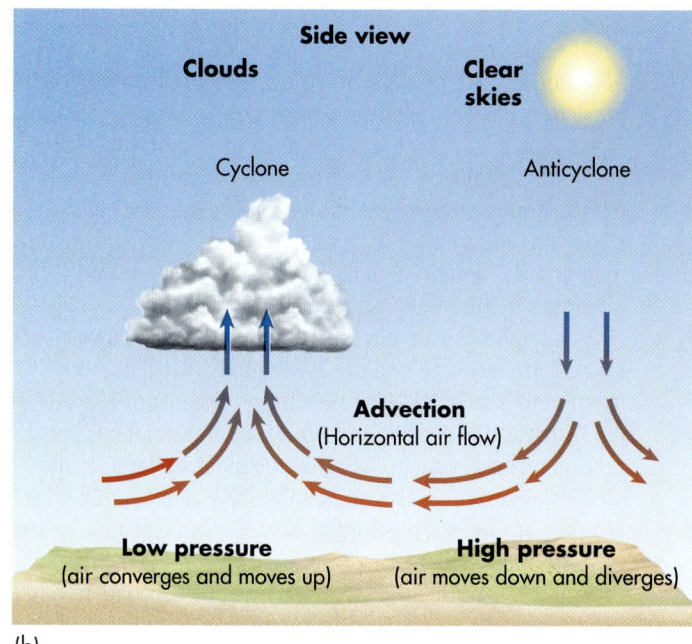

(b)

Figure 6.5 Atmospheric pressure systems. Oblique view (a) and side view (b) of typical low and high pressure systems. In a low-pressure system, air converging at the surface rises and forms clouds. In a high-pressure system, air descends and diverges at the surface; these systems are usually associated with clear skies.

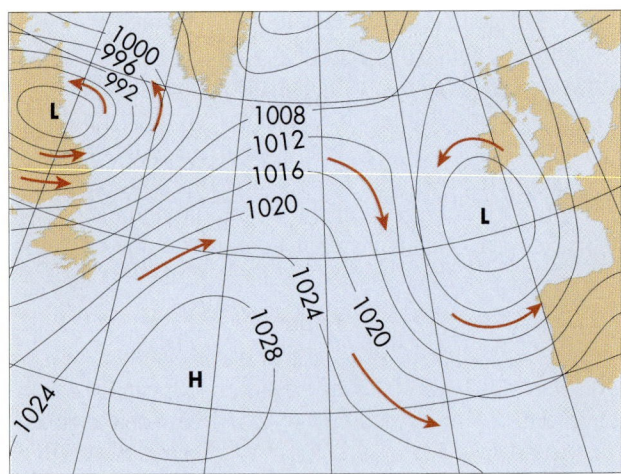

Figure 6.6 Atmospheric pressure map of the North Atlantic. The red arrows represent the direction of the winds. Notice the pressure variability across the ocean and the way air flows.

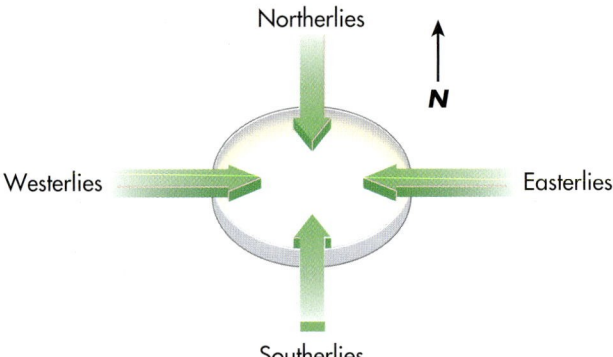

Figure 6.7 Compass headings and wind directions. Winds are named for the direction in which they originate. Westerly winds, for example, originate in the west and flow toward the east.

of flow around a low pressure system. High pressure centers are generally associated with fair weather because descending air warms as it approaches the surface. As mentioned earlier, warm air can hold more moisture than cold air can and this makes precipitation less likely. Again, we will say more about this relationship in Chapter 7.

In general, high and low pressure systems (anticyclones and cyclones) occur next to one another and form large-scale circulatory systems that are interconnected by air flow. Look again at Figure 6.5 and see how the air flows horizontally, in a process called **advection,** from the high-pressure system to the low-pressure system at the surface. You can see the geographic pattern of low and high pressure systems clearly on a barometric pressure map. Figure 6.6 illustrates pressure variations in the upper part of the atmosphere across the North Atlantic. The (gray) lines on the map are the isobars. The red arrows illustrate the way that air is moving relative to the pressure systems.

Notice in Figure 6.6 that the isobars in the eastern Atlantic form a rough oval or egg-shaped area. This region is a center of a low-pressure system (or the area of lowest air pressure) at this particular point in time. From the wind direction arrows, you can see that the air is moving counterclockwise around this low. Another low exists in the western part of the North Atlantic, over eastern Canada, where the winds are also circulating counterclockwise. In the central Atlantic, between the two lows, a broad area of high pressure exists that is rotating clockwise. On the western side of this system the air is flowing southwest to northeast, whereas on the eastern side of the system the air is moving northwest to southeast. Another way to describe these patterns is that the winds on the western side of the high are southwesterly and that they are northwesterly on

Advection *The horizontal transfer of air.*

the system's eastern side. You will learn more about the direction of wind flow later in this chapter, but a simple rule to remember is that wind direction is noted by the direction from where the wind is flowing.

Many people mistakenly believe that the named direction reflects the direction in which the air is moving. Instead, the name of the wind direction, such as "westerlies," reflects the direction from which the air flows (Figure 6.7). Westerly winds, for example, originate in the west and flow toward the east. Similarly, a north wind in the Northern Hemisphere brings in colder air south from the higher latitudes.

Although atmospheric pressure maps such as Figure 6.6 are very useful to illustrate detailed information, you can also see basic barometric patterns by viewing satellite images. Figure 6.8, which focuses on western Europe, is an example. Notice that France, Spain, and Germany are cloud-free, whereas Great Britain and Ireland are shrouded in a swirling band of clouds. Zones of clear sky, such as the large region in western Europe, are places dominated by high pressure. Remember that air pressure is low in the center of a cyclone and high in the middle of an anticyclone. Because air rises in the center of a low, a void is created above the surface of the Earth that must be filled. This void is ultimately filled by air that flows down and outward from the center of the high (remember that air descends in a high pressure system) and into the low. We feel this exchange of air as wind. Look at Figure 6.5 again to visualize how this process works and be sure you understand it thoroughly.

> **KEY CONCEPTS TO REMEMBER ABOUT ATMOSPHERIC PRESSURE SYSTEMS**
>
> 1. Air pressure refers to the weight of air distributed on the surface of the Earth. Air pressure generally decreases with increasing altitude.

The Direction of Air Flow

Now that we've discussed the fundamentals of air pressure, let's now look more closely at the concept of air flow and the factors that govern its movement. We will examine several factors, including unequal heating of land surfaces, the pressure gradient force, the Coriolis force, and various frictional forces.

Unequal Heating of Land Surfaces

The ultimate cause for all wind patterns on Earth is the unequal heating of land surfaces that result from variations in the amount of solar radiation received between regions. This spatial variation in surface air temperature means that air density, and thus pressure, differs from place to place. At a fundamental level, surface air flows from areas of high pressure to low pressure because the atmosphere works to balance the difference between the two areas.

The best example of unequal heating on Earth is the difference that exists between the tropics and the poles. Recall from Chapters 4 and 5 that the tropics are much warmer than the poles because the equatorial regions receive the most direct insolation throughout the year. If there were no mechanism to balance this difference, specifically through atmospheric circulation, then the tropical and polar regions would become excessively hot and cold, respectively.

Instead, this process of unequal heating causes motion of the atmosphere through the process of convection. Recall from Chapter 4 that convection is the vertical mixing of fluid material (in this case air) due to differences in temperature. In contrast, remember that the term *advection* refers to the horizontal movement of air or water. Figure 6.9 shows how the processes of convection and

Figure 6.8 Atmospheric pressure systems in Europe. The low is indicated by the clouds that cover Great Britain and Ireland, whereas the high is the clear sky to the east over France, Spain, and Germany.

2. Low pressure systems are called cyclones and consist of rotating air masses that lift air from the surface. In the Northern Hemisphere, these systems rotate counterclockwise, whereas they rotate clockwise in the Southern Hemisphere.

3. Low pressure systems are usually associated with clouds and precipitation.

4. High pressure systems are called anticyclones and consist of rotating air masses that descend toward the surface. In the Northern Hemisphere, these systems rotate clockwise, whereas they rotate counterclockwise in the Southern Hemisphere.

5. High pressure systems are usually associated with clear skies.

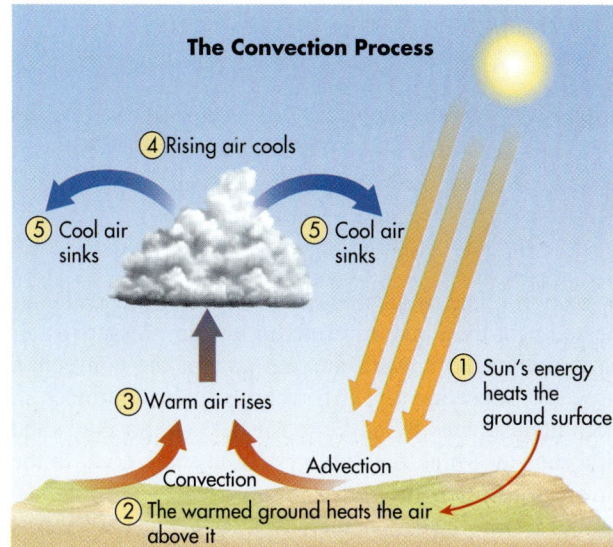

The Convection Process

④ Rising air cools
⑤ Cool air sinks
⑤ Cool air sinks
③ Warm air rises
Convection
Advection
① Sun's energy heats the ground surface
② The warmed ground heats the air above it

Figure 6.9 Atmospheric convection. Convection occurs when a portion of the Earth's surface is heated relative to another. When this happens, a large "bubble" of air lifts from the surface.

Atmospheric pressure systems are often easy to see on satellite images. This particular image focuses on western Europe in October 2001, with Spain in the lower left part of the image and clouds showing up in bright white. Given your understanding of **pressure systems, where is the high pressure system in this image? Where is the approximate center of the high? In what direction is this system spinning and where are the southerly and northerly winds in the system?**

advection relate to air movement. On a global scale, air heated near the Equator results in low air pressure as air rises within the upward-moving part of the convection process. Subsequently, it travels to higher latitudes in both hemispheres by advection, where the air cools and descends as a high pressure system at some point in the downward part of the convection process. The combined processes of convection and advection represent the first stage of atmospheric circulation on Earth and are the overall method through which heat is distributed around the planet. Once the air is set in motion in this manner, the other forces—pressure gradient, Coriolis, frictional—then directly influence the movement of the air.

Pressure Gradient Force

As we discussed earlier, air generally flows from areas of high pressure to regions of low pressure. The variable that drives the movement of air between two areas at different pressures is referred to as the **pressure gradient force.** It's useful to think of the concept of "gradient" as being analogous to the slope between two places on the surface of the Earth. If the elevation of one place is significantly

Pressure gradient force *The difference in barometric pressure that exists between adjacent zones of low and high pressure that result in air flow.*

A GOES visible satellite image of the contiguous U.S. in June 2004.

MIGRATING PRESSURE SYSTEMS

Atmospheric pressure systems in the midlatitudes generally migrate from west to east. A geographer looking at this GOES satellite image sees that clouds and perhaps rain, associated with a low pressure system, occurred in the southeastern part of the U.S. in early June 2004. At the same time, a strong high pressure system and sunny skies dominated the western part of the country. Given the migration of these pressure systems, a geographer knows that, in all probability, the western high will move slowly to the east, ultimately bringing clear skies over the eastern part of the country. This motion of air pressure systems is the basic idea behind satellite weather forecasting. In the midlatitudes, for example, you can see tomorrow's weather by looking at today's weather farther to the west.

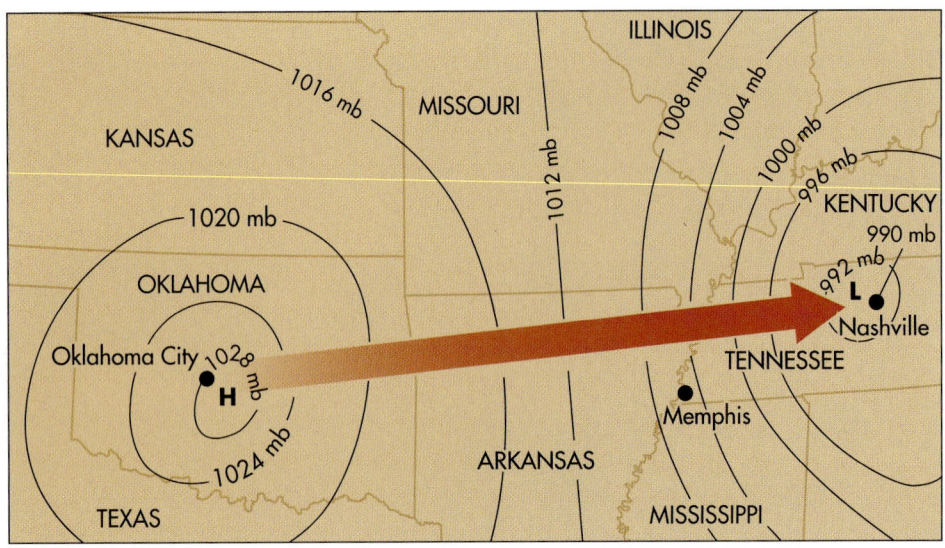

Figure 6.10 Hypothetical pressure gradient between Oklahoma City, Oklahoma and Nashville, Tennessee. Notice the direction of air flow and the spacing of isobars across the region. A steep pressure gradient exists in regions where isobars are closely spaced together, whereas the gradient is shallow in places where the isobars are far apart.

higher than the other, then the slope is said to be steeper than if the locations were located at (or near) the same elevation. The pressure gradient force operates on the same principle: the greater the difference in surface pressures between two regions, the steeper the pressure gradient. The process of air flow is significantly affected by the pressure gradient because the greater the pressure difference between two places, the faster the air flows to equalize the pressure.

Figure 6.10 is a barometric map that illustrates the pressure gradient force. The map focuses on a hypothetical difference in surface air pressure between Oklahoma City, Oklahoma, and Nashville, Tennessee. In this image, the center of the high pressure (indicated by the "H") is located in Oklahoma City, whereas the center of the low (indicated by the "L") is located in Nashville. Air pressure in this figure ranges from 1028 mb at the center of the high to 993 mb at the center of the low. It should make sense to you that as a result of this variation, the air is generally flowing from west to east (left to right on the map)—that is, from the Oklahoma City area toward Nashville.

Close to the center of the high—say, from Oklahoma City to central Arkansas—the isobars are widely spaced. This means that very limited pressure change occurs across this portion of the Earth's surface at this particular point in time. Thus, the pressure gradient in this part of the circulatory system is shallow. If you were in this area on this particular day, you would find the winds to be light, because no immediate void needed to be filled by inflowing air into that region. As you approach the center of the low, however, the isobars become closer together.

Given that isobars represent lines of equal atmospheric pressure, this close spacing can only mean a rapid change in surface pressure occurs over a relatively small geographical space—for example, from Memphis to Nashville. In other words, this part of the circulatory system has a steep pressure gradient and the air flows faster as it moves toward the center of the low, to fill the relative void created by the less dense air at the surface. If you happened to be in this area on this particular day, you would notice that the wind is strong.

Coriolis Force

In addition to the pressure gradient force, another factor strongly influences the process of air flow in the atmosphere: the **Coriolis force.** The Coriolis force is a very simple effect that is often difficult to explain and comprehend. Whereas the pressure gradient force arises because of differences in atmospheric pressure between regions, the Coriolis force is related to the rotation of the Earth on its axis. Given this rotation, objects in the atmosphere, including the air, appear to be deflected or pulled sideways as the Earth rotates under them. In the Northern Hemisphere, the direction of deflection for an object moving toward the Equator is to the right when viewed from above the North Pole. Objects moving from the Equator

Coriolis force *The force created by the Earth's rotation that causes winds to be deflected to the right in the Northern Hemisphere and to the left in the Southern Hemisphere.*

FLUCTUATIONS IN THE PRESSURE GRADIENT

In order to gain a better understanding of how the pressure gradient force influences the process of air flow, go to the *GeoDiscoveries* website and select the module **Fluctuations in the Pressure Gradient**. This simulation is based on Figure 6.10, which shows a hypothetical pressure gradient between Oklahoma City, Oklahoma, and Nashville, Tennessee. You will be able to adjust the atmospheric pressure between the two places to see how these changes impact the flow of air between them. As you make these adjustments, be sure to notice the impact that they have on wind speed and the overall direction of the winds. After you complete the simulation, answer the questions at the end of the module to test your understanding of this concept.

to the North Pole, in contrast, appear to veer left. In the Southern Hemisphere, the deflection of an object moving toward the Equator is to the left when viewed from below the South Pole. For objects moving from the Equator southward, the apparent deflection is to the right.

A good way for you to visualize how the Coriolis force works is to study the apparent path that a hypothetical rocket would take between the North Pole (90° N) and New York City (about 40° N). Look at Figure 6.11 as you work through this discussion. During the initial period of flight, the rocket follows the 74° W meridian as it flies south. The course appears to change, however, as the rocket continues southward. Why does the course change? It's not because the rocket's direction actually varies from its original destination; instead, the change occurs because the Earth rotates *under* (or eastward of) the rocket while it is in the air. In this fashion the rocket would land somewhere to the *west* of New York City unless the course is corrected in some way to account for the Earth's rotation. Figure 6.11 also shows the apparent path of a rocket fired from the South Pole (90° S) toward the Equator. Note that this rocket is deflected to the west as well, but this time from right to left relative to the path of initial motion.

We can see how a rocket's path might be altered by the Earth's rotation, but how does the Coriolis force influence the wind? Remember that at ground level the direction of air flow is influenced by the pressure gradient force and results in air flow that is perpendicular to the isobars (as in Figure 6.10). Once the air rises into the upper troposphere, however, its speed increases because it flows freely without obstruction. Once this altitude is reached, the air is influenced most directly by the Coriolis force, which causes winds to spiral to the right in the Northern Hemisphere and to the left in the Southern Hemisphere.

In this fashion, convection loops consist of spiraling masses of descending or rising air that are linked horizontally by advection (Figure 6.12). This process is most pronounced at higher latitudes and generally results in westerly air flow from the subtropics to the poles. Once again, you might be inclined to think that the air is flowing toward the west; instead, westerly winds are those that

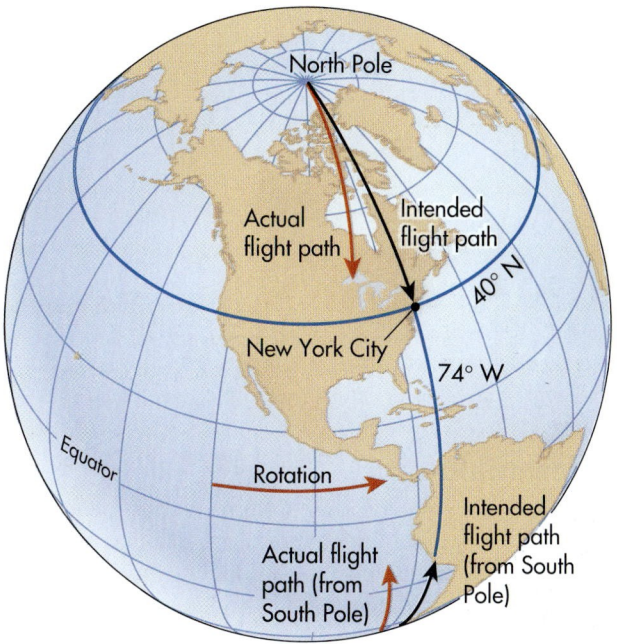

Figure 6.11 The Coriolis force. The Coriolis force influences the path of a rocket traveling from the North Pole to New York City, deflecting it to the west. Notice the direction of the Earth's rotation and how the rocket path is diverted more as it approaches the Equator. Also note that a rocket traveling from the South Pole to the Equator is deflected west as well.

flow from west to east. At this point the Coriolis force and the pressure gradient force effectively balance each other, resulting in upper air flow that is *parallel* to the isobars rather than perpendicular as seen at the surface. The net result of this flow pattern is that air moves *around* pressure systems in the upper atmosphere. Such winds are called **geostrophic winds.**

Geostrophic winds *Air flow that moves parallel to isobars because of the combined effect of the pressure gradient force and Coriolis force.*

What you see on a weather map

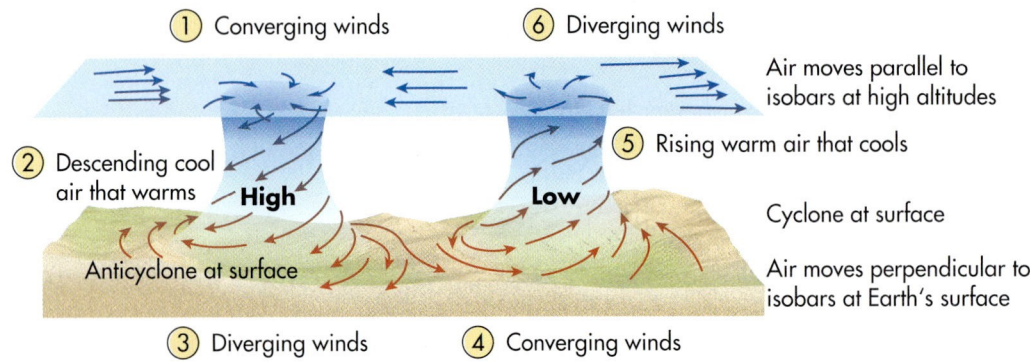

What is happening in the atmosphere

① Converging winds ⑥ Diverging winds

Air moves parallel to isobars at high altitudes

② Descending cool air that warms **High**

⑤ Rising warm air that cools

Low

Cyclone at surface

Anticyclone at surface

Air moves perpendicular to isobars at Earth's surface

③ Diverging winds ④ Converging winds

Figure 6.12 A dynamic convection loop. Cyclones and anticyclones are linked together in a convection loop consisting of air masses that spiral due to the Coriolis force. Note how the air masses move vertically within the high and low pressure systems and horizontally between them.

Frictional Forces

As briefly mentioned earlier, a third force influences the process of air flow in the atmosphere, one that occurs at ground level and that operates in direct opposition to the winds. This force is the force of friction and occurs because of the drag and impediments created by features on the surface of the Earth, such as mountains, trees, and even buildings. As you can imagine, these features cause winds to slow down and move in irregular ways (Figure 6.13). Remember from the discussion about the urban heat island in Chapter 5, for example, that winds typically flow less strongly in cities than they do in the surrounding countryside. The force of friction results in air flow that is somewhere between that driven by the pressure gradient (that

is, perpendicular to isobars) and the Coriolis force (which is parallel to isobars). As a general rule, the effect of friction is strongest at the surface and diminishes progressively to an altitude of about 1500 m (about 5000 ft). In response to this variability, the wind flows at an angle relative to isobars at ground level. At higher altitudes, however, winds follow a geostrophic course that is parallel to isobars.

In an effort to integrate all of the major factors that influence atmospheric circulatory processes, let's now turn to Figure 6.14 to review. Recall that the pressure gradient force causes winds to flow at right angles to isobars in the direction of the lower pressure. The effect of this force can be seen in Figure 6.14a. When the influence of the

THE CORIOLIS FORCE

To see how the Coriolis force looks in animation, go to the *GeoDiscoveries* website and select the module **The Coriolis Force**. This module examines how the Coriolis force influences atmospheric circulation. Be sure to watch how the rotation of the Earth on its axis impacts the way that air flows on the planet. After you finish watching the animation, be sure to answer the questions at the end of the module to test your understanding of this concept.

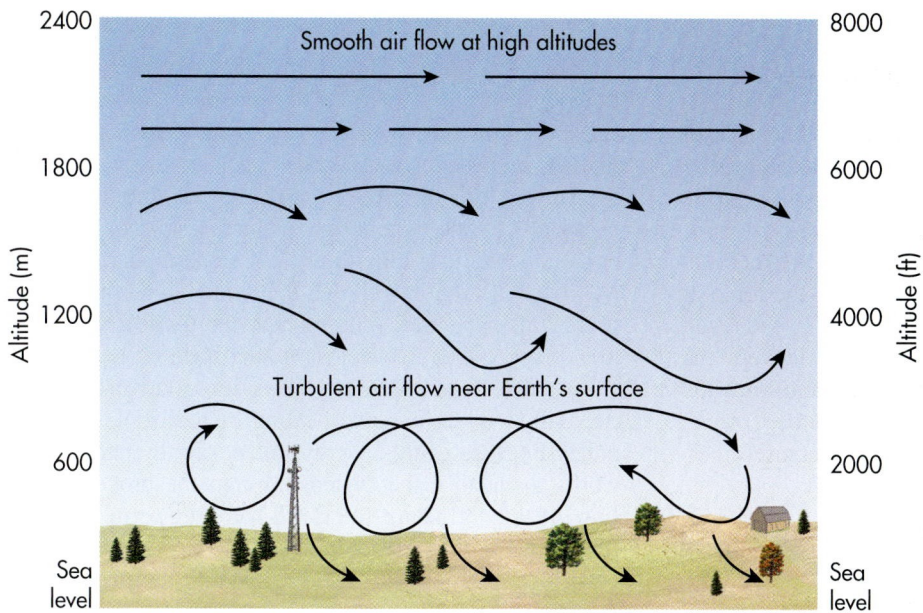

Figure 6.13 The effect of friction on wind flow near the Earth's surface. Compared to winds at higher altitudes, the flow of surface air is significantly modified by features on the Earth's surface.

Northern Hemisphere

(a) Pressure gradient force:

Out 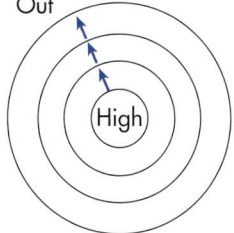 In

(b) Coriolis force:

Clockwise 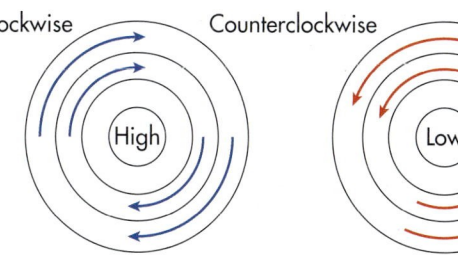 Counterclockwise

(c) Combined with frictional force:

Spiral out 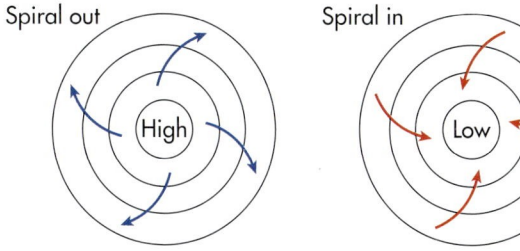 Spiral in

Southern Hemisphere

(a) Pressure gradient force:

Out 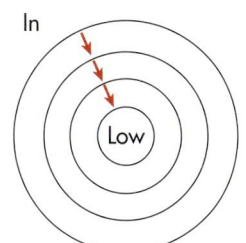 In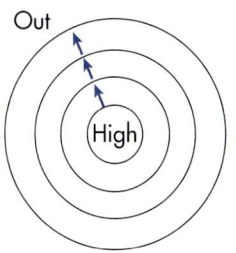

(b) Coriolis force:

Counterclockwise 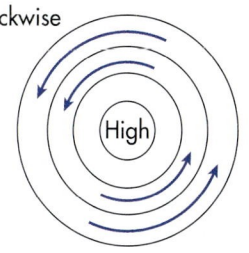 Clockwise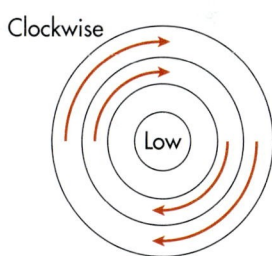

(c) Combined with frictional force:

Spiral out 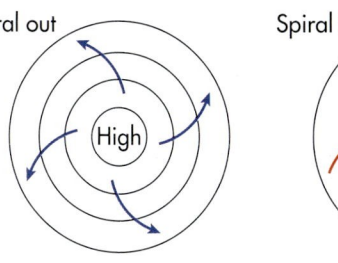 Spiral in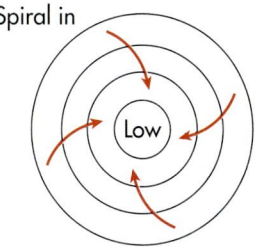

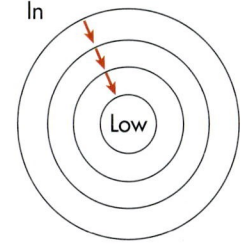

Figure 6.14 The process of large-scale atmospheric circulation is influenced by three integrated factors. (a) The pressure gradient force causes air to flow perpendicular to isobars. (b) The Coriolis force causes movement of air that is parallel to isobars. (c) In combination with the pressure gradient and Coriolis forces, frictional forces result in winds that flow somewhere intermediately between 0° and 90° of isobars.

Coriolis force is combined with the pressure gradient force (Figure 6.14b), winds then flow parallel to the isobars as geostrophic winds. This process occurs because of the balancing effect that the Coriolis force and pressure gradient force have on one another. In other words, the Coriolis force keeps wind from flowing across isobars, whereas the pressure gradient force stops winds from curving up the pressure slope. Finally, when the force of friction is taken into account (Figure 6.14c), the end result is winds that follow an intermediate course relative to the isobars, somewhere between perpendicular (due to the pressure gradient force) and parallel (due to the Coriolis force) to those lines of equal atmospheric pressure.

Global Pressure and Atmospheric Circulation

In the preceding sections, we looked at the fundamentals associated with air pressure systems and the variables that influence the process of air flow in the atmosphere. Within that context, let's now examine the general circulation of air around the globe. This discussion will refer both to the flow of air in the upper part of the atmosphere and at the surface. Implied by these distinctions is that air flows differently in these respective parts of the atmosphere. If you want to see this difference yourself sometime, look for a partly cloudy day where two distinct layers of clouds occur, one low and another high. When these conditions exist, you will often notice that the clouds in the upper part of the atmosphere are moving in a slightly different direction or speed than those closer to the surface.

As discussed previously, the primary driver of global circulation is the unequal heating of the tropics and the poles. Because of this energy imbalance, the atmosphere works to balance the system through the process of air flow. If the surface of the Earth had a uniform character (that is, no distinction between continents and oceans), did not rotate, and was not tilted relative to the plane of the ecliptic, the basic circulatory system would be very easy to understand. In this simplistic scenario, which is illustrated in Figure 6.15, low pressure would occur at the Equator because the air is very warm and high pressure would occur at the poles because the air is very cold. Very simply, air would rise away from the surface at the Equator, within a low pressure system, and would then flow toward the poles in the upper atmosphere by advection. Once it reached the poles, it would descend toward the surface in a rotating high pressure system where it would then diverge and flow back toward the Equator by advection.

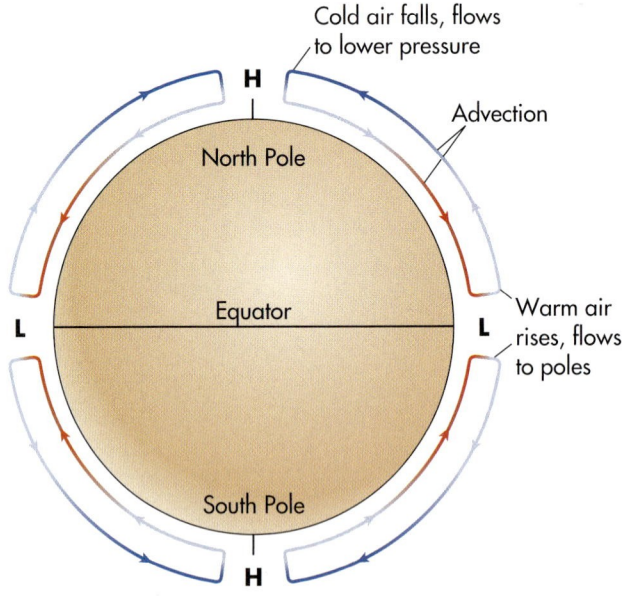

Figure 6.15 Global circulation on a nonrotating, untilted Earth with a consistent surface composition. In this model, a few simple convection loops would dominate the system, with rising air (low pressure) at the Equator, descending air (high pressure) at the poles, and horizontal flow of air by advection in between.

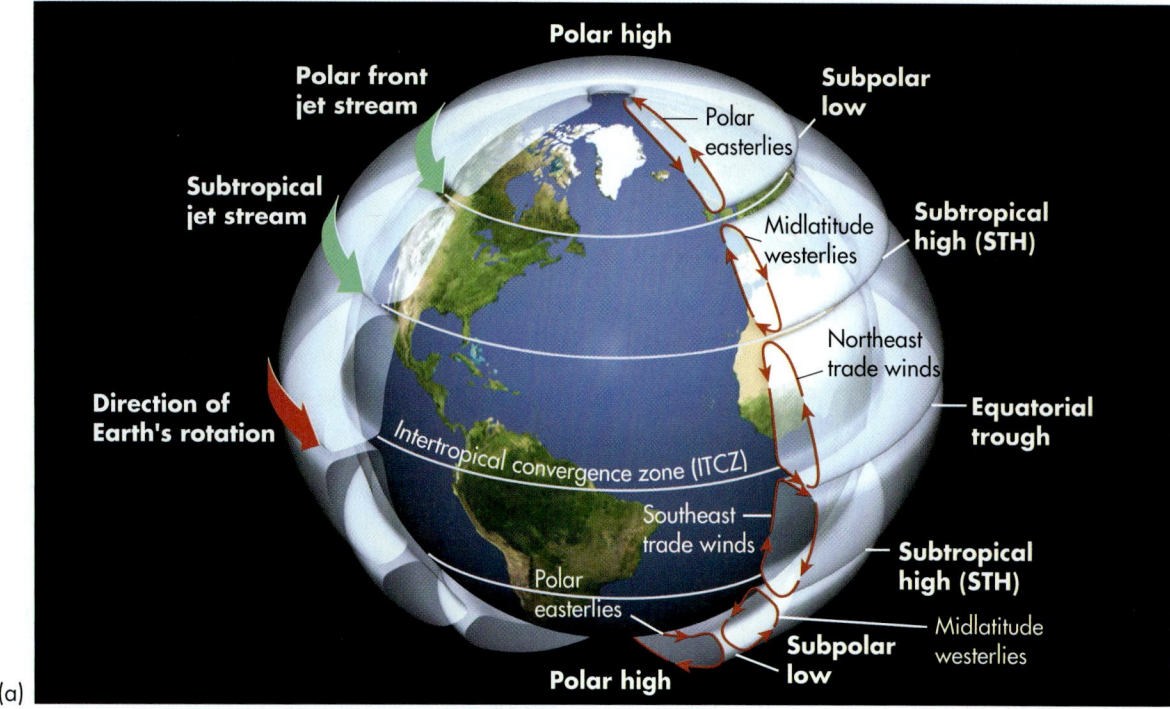

(a)

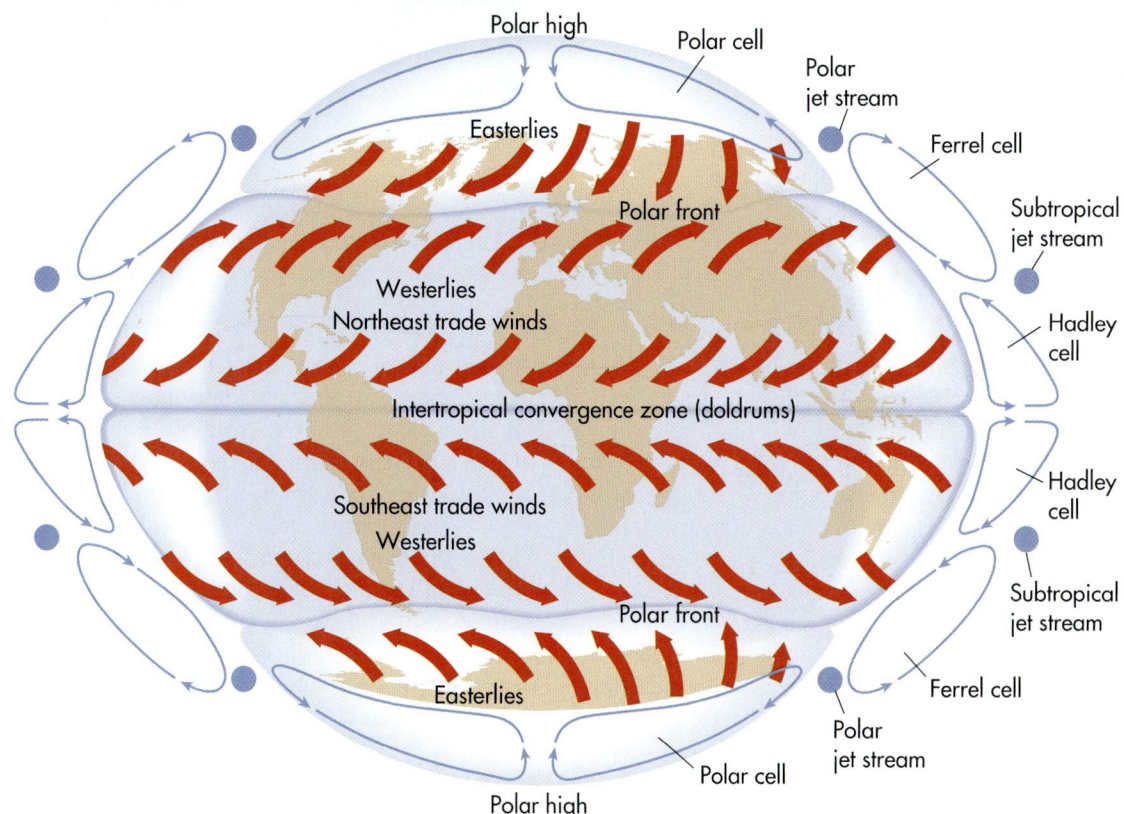

(b)

Figure 6.16 The global circulation model. (a) This three-dimensional diagram illustrates the basic flow of air in cross section as well as top views. Notice the various convection loops that exist around the surface of the Earth. **(b) A two-dimensional map of the world shows the directions of winds across various bodies of land and water.**

As you know, however, the Earth does not have a uniform surface, it does rotate, and it is tilted with respect to the plane of the ecliptic. Given these factors, the global circulatory system is more complex than the simplified model in Figure 6.15. Figure 6.16 illustrates the general global circulation model as it truly functions, showing the major wind systems on Earth. Let's look at the model in more detail in order to explain how these wind patterns occur. We'll begin at the Equator and proceed systematically toward higher latitudes, focusing on the directions of winds at both the surface and the upper part of the troposphere.

Tropical Circulation

The logical place to begin discussion of global atmospheric circulation is the tropics. As noted previously in this chapter, atmospheric circulatory processes are set in motion in the tropics because of the extremely warm temperatures that occur there. Although tropical air eventually finds its way into the midlatitudes, it's useful at this point for you to think of the tropical circulatory system as being a convection loop, with air flowing between the Equator and about 30° N and S latitude. In that context, the discussion of tropical circulation focuses on the Intertropical Convergence Zone and the Subtropical High pressure system.

Intertropical Convergence Zone   Tropical circulatory processes begin at the Equator, where air is warmed due to year-round receipt of direct sunlight. This warming helps create a zone of low pressure, called the **equatorial trough,** because warm air is less dense and more buoyant. As a result of the rising air mass, air from higher tropical latitudes in both hemispheres flows toward the equatorial trough along the surface by advection, converging in a narrow band known as the **Intertropical Convergence Zone** or **ITCZ** (Figure 6.16). The converging winds are known as **trade winds** because they were systematically used to power sailing vessels during the Age of Exploration between the 1400s and 1800s. Notice in Figure 6.16b that the trade winds north of the ITCZ are northeasterly, whereas trade winds south of the region are southeasterly. Together, they form the **tropical easterlies.** At the point where the northeasterly and southeasterly trade winds converge at the ITCZ, the winds can become relatively calm and highly variable because the pressure gradient is very weak.

An important characteristic of the ITCZ is that it is a region of cloudiness and frequent rains. This occurs for two primary reasons. One is that the trade winds flow over warm oceans across much of their length; therefore,

Figure 6.17 The Intertropical Convergence Zone. The ITCZ is a zone of low pressure, as indicated by the band of dense clouds that extends from Central America west into the Pacific Ocean in this satellite image.

a great deal of evaporation from these surfaces takes place and the air contains abundant moisture. In this context, high levels of atmospheric moisture can be reached because the air is warm and expanding. A second reason is that, due to the warmth of the air, the ITCZ is a place where the air rises vertically from the Earth's surface. As the air rises, it cools and can subsequently hold less moisture than it did at lower (warmer) altitudes. This cooling effect results in clouds and rainfall as water condenses. Due to these combined processes, the ITCZ is usually easy to find on a satellite image because it consists of a band of clouds (Figure 6.17).

Equatorial trough   *Core of low pressure zone associated with the Intertropical Convergence Zone.*

Intertropical Convergence Zone (ITCZ)   *Band of low pressure, calm winds, and clouds in tropical latitudes where air converges from the Southern and Northern Hemispheres.*

Trade winds   *The primary wind system in the tropics that flows toward the Intertropical Convergence Zone on*

the equatorial side of the Subtropical High pressure system. These winds flow to the southwest in the Northern Hemisphere and to the northwest in the Southern Hemisphere.

Tropical easterlies   *Band of easterly winds that exist where northern and southern trade winds converge.*

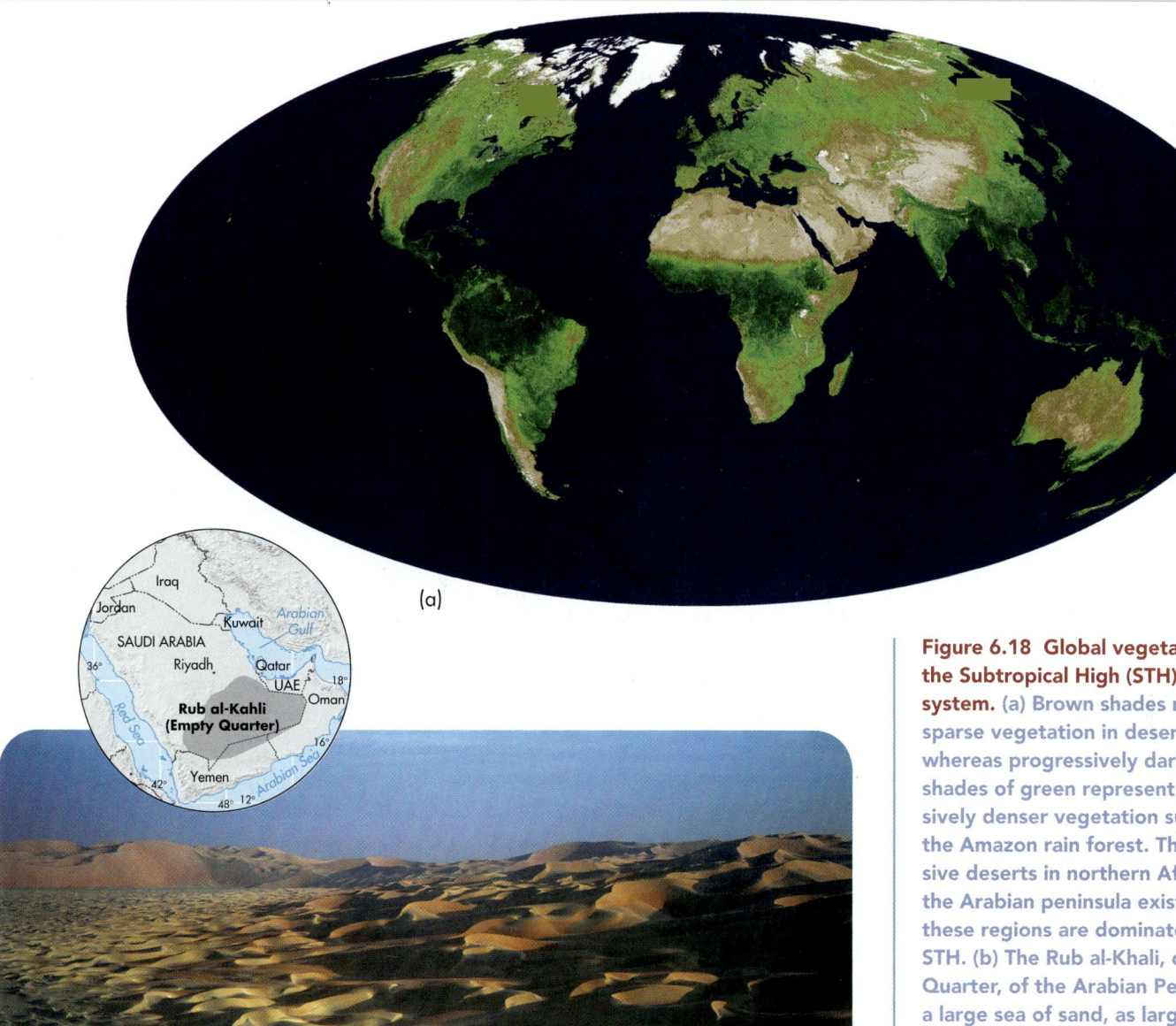

(a)

(b)

Figure 6.18 Global vegetation and the Subtropical High (STH) pressure system. (a) Brown shades represent sparse vegetation in deserts, whereas progressively darker shades of green represent progressively denser vegetation such as the Amazon rain forest. The extensive deserts in northern Africa and the Arabian peninsula exist because these regions are dominated by the STH. (b) The Rub al-Khali, or Empty Quarter, of the Arabian Peninsula is a large sea of sand, as large as all of France, with no roads and hardly any rainfall.

Subtropical High Pressure System   We have just seen that the ITCZ is an area of low pressure that exists only at very low latitudes. As we move our way through the tropical circulatory loop, let's see what happens at slightly higher latitudes.

As air in the equatorial trough rises from the Earth's surface, it spirals upward to the upper part of the troposphere. During this process, the temperature of the air drops as it rises. Although we'll discuss this cooling process more thoroughly in Chapter 7, for now it's sufficient that you know that it occurs because the air is expanding as it rises.

High-altitude air on the northern side of the ITCZ flows northward by advection, while air on the southern side of the ITCZ flows to the south. Given that the air has cooled, it must sink, and does so at approximately 25° to 30° N or S latitude. These zones of sinking air are referred to as the **Subtropical High pressure system**, or **STH** (see Figure 6.16a). The descending air is dry because much of its moisture was lost as precipitation over the Equator. The air is also compressed as it descends, making it denser and warmer as it approaches the surface of the Earth. This compression creates a high pressure zone of hot, dry air. On the Earth's surface, this zone is characterized by extensive deserts, such as the Sahara Desert in Africa and the Arabian Desert in Saudi Arabia (Figure 6.18).

Subtropical High (STH) pressure system   *Band of high air pressure, calm winds, and clear skies that exists at about 25° to 30° N and S latitude.*

The STH functions in the same way as the high pressure systems discussed earlier in this chapter. That is, as the air descends it rotates in a clockwise direction in the Northern Hemisphere and counterclockwise in the Southern Hemisphere. Once the downward spiraling airstreams reach the surface, they diverge. As they diverge on the southern side of the STH in the Northern Hemisphere, they flow back to the ITCZ, forming the northeasterly trade winds. On the northern side of the STH in the Southern Hemisphere, the airstreams form the southeasterly trade winds. At low latitudes, this flow to and from the ITCZ and STH forms a convection loop known as a **Hadley cell** (see Figure 6.16 again).

Midlatitude Circulation

With a basic understanding of tropical circulation, we can proceed to the way the atmosphere circulates in the midlatitudes. In general, the primary purpose of midlatitude circulation is to mix the cool polar air that originates at high latitudes and the warm tropical air that exists at lower latitudes. With this mixing in mind, the midlatitudes are the regions where these contrasting air masses converge. The following discussion describes the circulatory processes in the midlatitudes and how they function in the context of balancing temperature differences.

The focal point of midlatitude circulation is the **polar front,** which on average, occurs at about 60° N and S latitudes (Figure 6.19). This atmospheric feature is the contact between cold air that originates at very high latitudes and the relatively warm air that streams northward from the tropical latitudes. You can see how tropical air is pumped into the midlatitudes by looking again at Figure 6.16. For the purposes of this discussion, simply focus your attention on the Northern Hemisphere and notice that the midlatitudes are north of the STH. Remember that this pressure system is centered at about 30° N and consists of warm, dry air that has spiraled in a clockwise fashion down toward the surface of the Earth. As this descending air reaches the surface, it diverges, with air on the southern side of the system flowing toward the ITCZ in the form of northeasterly winds. Air on the north side of the STH, in contrast, flows northward in the form of southwesterly winds that are part of a circulatory loop known as a *Ferrel Cell*. It is this southwesterly air that flows toward the polar front, where it converges with air flowing southward from the highest latitudes. This southwesterly flow of air contributes to the westerly winds, or **westerlies,** that prevail in the midlatitudes.

The midlatitude westerlies are an especially significant feature because they often flow at very high speeds, reaching velocities of between 350 to 450 km/h (about 200 to 250 mi/h) in the upper troposphere. The winds

---

Hadley cell   *Large-scale convection loop in the tropical latitudes that connects the Intertropical Convergence Zone (ITCZ) and the Subtropical High (STH).*

Polar front   *The contact in the midlatitudes between warm, tropical air and colder polar air.*

Westerlies   *Midlatitude winds that generally flow from west to east.*

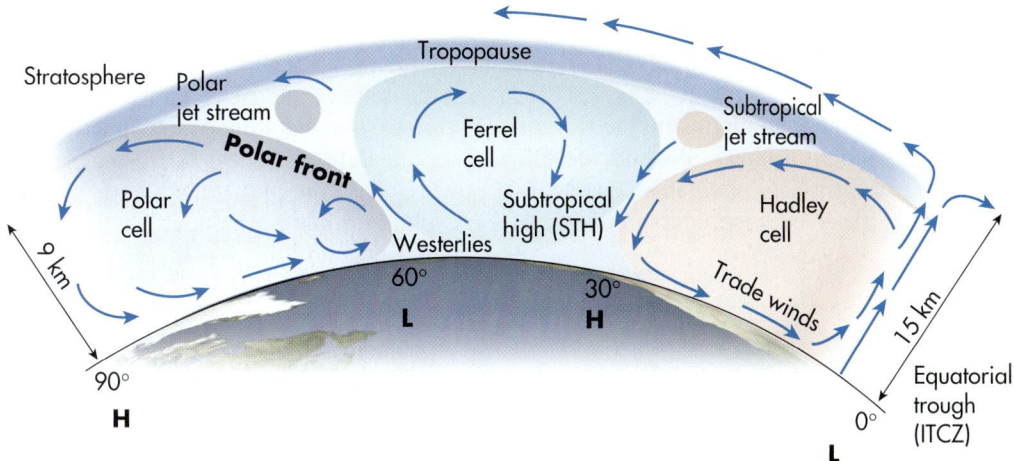

Figure 6.19 **The polar front in the Northern Hemisphere.** The polar front is the boundary between warm air (to the south) and cold polar air (to the north). Note how air from the Subtropical High (STH) pressure system flows toward the polar front.

reach these extreme speeds due to a distinct temperature gradient along the polar front, which results in a very steep pressure gradient as well. Given the seasonal effect created by the Earth–Sun geometric relationship, this temperature/pressure gradient is very weak during the summer months but strengthens with the progression of winter. The resulting river of rapidly moving air is known as the **polar front jet stream** and occurs at altitudes of 10 to 12 km (about 30,000 to 40,000 ft; see Figures 6.19 and 6.20). Like the ITCZ, the polar jet stream is often

easy to see on satellite imagery because a band of clouds forms along this line (such as in Figure 6.20b).

Although the term "westerly winds" implies that midlatitude winds are flowing straight from west to east, this is not always the case. In fact, the smooth westward flow in the upper-air circulatory system frequently develops distinct undulations, called **Rossby waves.** Rossby waves form along the polar front and are the mechanism through which significant temperature differences on either side of the front are moderated,

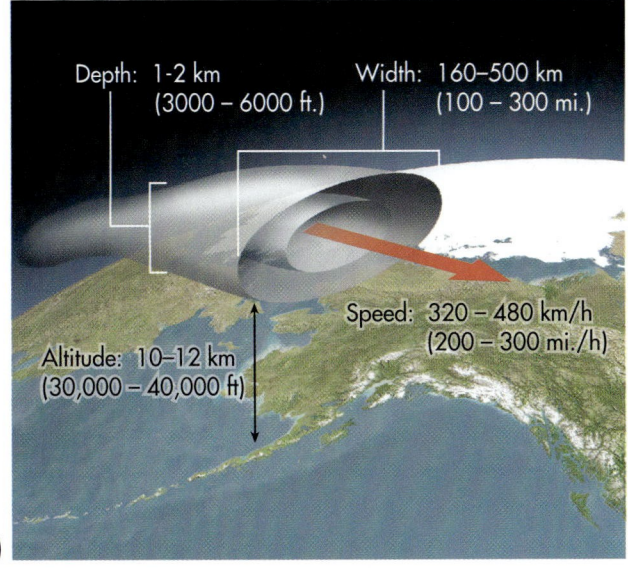

(a)

(b)

Figure 6.20 **Jet streams.** (a) Jet streams consist of rivers of air that flow at very high speeds. (b) Several jet streams flow in the upper atmosphere and can frequently be identified by a band of clouds, such as in this image of the Nile/Red Sea area in the Middle East.

Polar front jet stream *River of high-speed air in the upper atmosphere that flows along the polar front.*

Rossby waves *Undulations that develop in the polar front jet stream when significant temperature differences exist between tropical and polar air masses.*

especially during the winter months. Figure 6.21 shows how the process develops.

At any given location in the midlatitudes, the flow of the polar jet stream essentially follows a smooth west to east path for several weeks. This circulatory pattern is typically called **zonal flow** because cold Arctic air is confined to a small zone at very high latitudes. As the temperature contrast on either side of the polar front increases, the midlatitude atmosphere responds by forming an undulation in the jet stream, beginning the Rossby wave. The core of this undulation becomes a midlatitude cyclone rotating counterclockwise. With the continued development of this Rossby wave, the westerly winds no longer flow directly west to east, but have significant northerly and southerly components as well, due to the counterclockwise circulation. In contrast to zonal flow, the variable flow associated with a developing Rossby wave is called **meridional flow,** because the air frequently flows parallel to the meridians.

With the onset of meridional flow, warm air on the east side of the wave pushes poleward because the winds are southerly. At the same time that this northward push of warm air occurs, cold air from the north plunges southward on

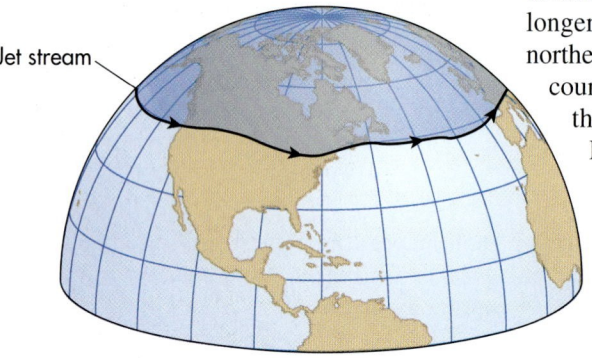

(a) Polar jet stream with small undulations

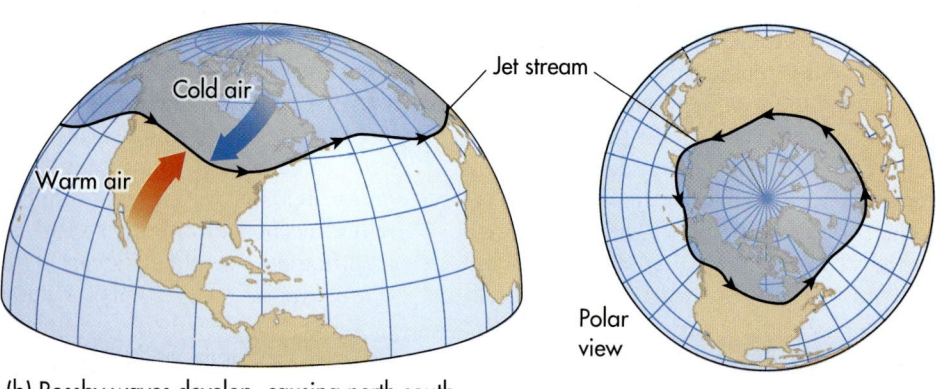

(b) Rossby waves develop, causing north-south motion of large masses of warm and cold air

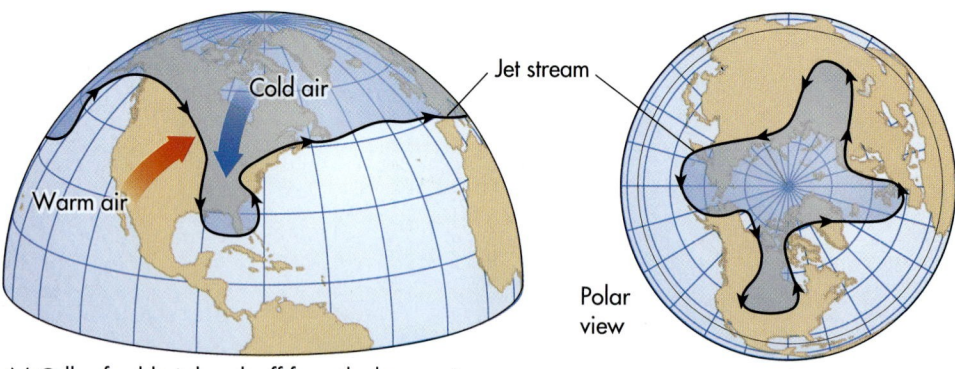

(c) Cells of cold air break off from the larger air mass, forming isolated cyclones of cold air

Figure 6.21 Development of Rossby waves in the midlatitudes of the Northern Hemisphere. The progressive development of jet stream undulations along the polar front ultimately results in pinching off of cold air pools and reestablishment of zonal flow.

Zonal flow *Jet stream pattern that is tightly confined to the high latitudes and is thus circular to semicircular in polar view.*

Meridional flow *Jet stream pattern that develops when distinct Rossby waves exist and the polar front jet stream flows parallel to the meridians in many places.*

the west side of the system, where the winds are northerly. Such an influx of cold air into a region is called an *Arctic outbreak* and can result in extremely cold air reaching latitudes far south of its origin. If you have spent any winters in the northern part of the U.S., you probably know that an Arctic outbreak can bring extremely cold temperatures into the region for a long time. An extreme Arctic outbreak can even bring freezing temperatures at night as far south as Florida. As time progresses, the cold pool of air can literally be pinched off from the main body of Arctic air, resulting in the reestablishment of zonal flow conditions. Such a pool of cold air can persist in a region for several weeks, gradually warming because of the higher Sun angles at these somewhat lower latitudes.

Polar Circulation

In contrast to the complex patterns associated with tropical and midlatitude circulation, atmospheric circulation in the polar regions is associated with a simple circulatory loop known as a *Polar Cell*. Air that flows northward at the polar front cools considerably and subsequently descends at very high latitudes, producing a weak high pressure system (see Figure 6.16 again). This system is called the **Polar High** and consists of a mass of descending air that rotates in a clockwise fashion in the Northern Hemisphere and counterclockwise in the Southern Hemisphere. These rotating systems direct cold, dry air toward the polar front in the form of **polar easterlies** and are strongest in the Northern Hemisphere, where large landmasses exist. If you want to review why this geographic pattern exists, return to Figure 5.12 to view the annual range of temperature on Earth and note the variation that

exists between the hemispheres. As a result of this variability, the Polar High located over northern continental regions of the Earth makes northernmost Canada and Siberia bitterly cold and dry during the winter.

Seasonal Migration of Pressure Systems

Up to this point in the discussion, the focus has been on the basic distribution of air pressure systems and how they distribute heat energy in the atmosphere. Given your understanding of the Earth–Sun geometric relationship, however, you might also suspect that a distinct seasonal component occurs in association with global atmospheric circulation. In fact, a powerful seasonal component does come into play. Figure 6.22 shows how the seasonal migration occurs.

A good reference point for studying the seasonal pressure migration is the subsolar point and its relationship with the ITCZ. Remember that the ITCZ migrates with the subsolar point. For example, during the Spring and Fall Equinoxes, the Sun is directly overhead at the Equator. This geographic position results in the Equator receiving the most intense radiation. Because the Equator is generally the warmest zone on Earth during these months, the ITCZ is generally located there as well (Figure 6.22a). During winter in the Northern Hemisphere, the subsolar point is located at the Tropic of Capricorn because the Sun is directly overhead at that latitude. Because this zone receives the most intense radiation, this is where the ITCZ is generally located (Figure 6.22b).

Similarly, the pattern is reversed with the approach of the Northern Hemisphere spring and summer. With the coming of the Northern Hemisphere Spring Equinox, the

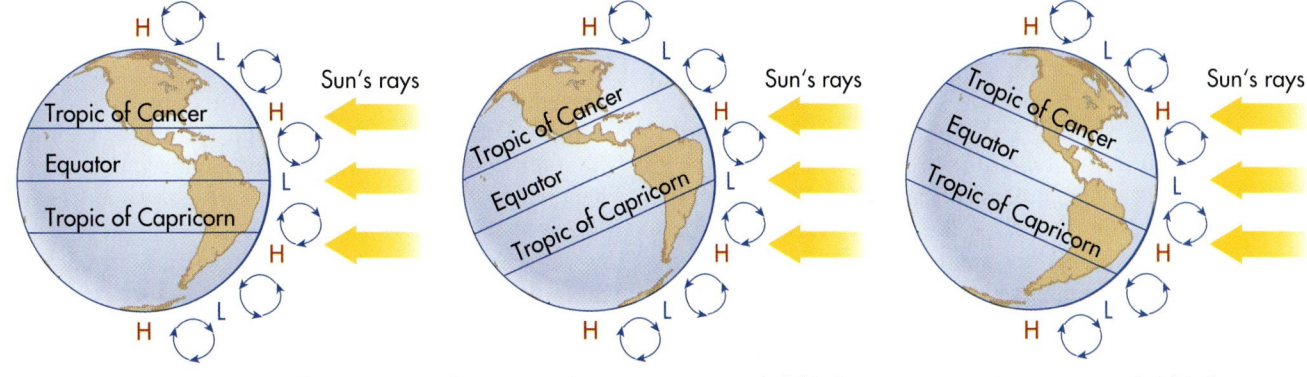

(a) Equinoxes, equatorial low, ITCZ around Equator

(b) December 21, equatorial high, ITCZ around Tropic of Capricorn

(c) June 21, equatorial high, ITCZ around Tropic of Cancer

Figure 6.22 Seasonal migration of atmospheric pressure. The positions of the major high and low pressure systems during (a) Equinoxes, (b) Northern Hemi-sphere Winter Solstice, and (c) Northern Hemisphere Summer Solstice.

Polar High *Zone of high atmospheric pressure at high latitudes.*

Polar easterlies *Band of easterly winds at high latitudes.*

(a)

MT. WASHINGTON

Where do you think the worst weather in the world would be? The North Pole? The South Pole? Many meteorologists would say that neither pole has the worst weather; instead, it occurs at Mt. Washington, in the White Mountains of New Hampshire. Here the polar jet stream slams unimpeded into the highest point of land along the eastern U.S. (6288 ft), with results that rival the weather at observatories in Antarctica.

The highest wind speed ever recorded anywhere on Earth was measured in 1934 at the weather station on Mt. Washington's summit: 231 mi/h. Annual average temperature at the summit is only 26.5° F, with annual average snow-

ITCZ follows the subsolar point back to the Equator. Subsequently, the ITCZ migrates into the Northern Hemisphere during April and May, reaching the Tropic of Cancer in the latter part of June. As the days shorten in July and August, the ITCZ migrates back to the south.

Remember, however, that the ITCZ is not the only large pressure system that migrates seasonally. *All of them do.* This migration occurs because the pressure systems maintain a more or less consistent distance from one another across the surface of the Earth. Thus, when the ITCZ migrates, all of the systems move (see Figure 6.22). You will discover in Chapter 9 that this seasonal migration of pressure systems influences the distribution of global climates in a major way.

Monsoonal Winds   One of the most significant features that results from the seasonal migration of air pressure systems is the **monsoon.** The monsoon is a cyclical

---

Monsoon   *The seasonal change in wind direction that occurs in subtropical locations due to the migration of the Intertropical Convergence Zone (ITCZ) and Subtropical High (STH) pressure system.*

shift of the prevailing wind direction that occurs at a subcontinental scale over the course of a year. This process is a good example of how closely geographic variables are integrated because it is related both to the seasonal migration of the ITCZ and the maritime/continental effect described in Chapter 5. Although a form of monsoon occurs in many subtropical places around the Earth, including the southwestern U.S., it is most pronounced in Southeast Asia because this is where the greatest seasonal fluctuation of the ITCZ occurs.

In order to understand how the monsoon functions, let's compare the geographic variability that exists with respect to the seasonal migration of the ITCZ (Figure 6.23). Beginning first with the Americas, you can see that the ITCZ is located in South America throughout the year, and migrates from the central part of the continent in January to the northern part of the continent in July. In other words, the ITCZ migrates approximately 20° latitude in South America over the course of the year. How does this migration range compare with Asia? Figure 6.23 shows that the position of the ITCZ varies dramatically in Asia, ranging from northern Australia in January to northern India in July. This is a migration of about 45° latitude, or twice that which occurs in South America. The primary reason for

(a) Tuckerman's Ravine is famous for spring skiing.
(b) Rime ice on a building. Note the view of clouds below the summit.

(b)

fall of 256 inches. (One year the snowfall reached 566 inches, more than 47 feet!) Visitors often marvel at the pictures showing the summit buildings covered in thick snow, but it isn't snow at all. It's rime ice, which forms when fog droplets crash into objects at subfreezing temperatures and freeze on contact. The rime ice grows out into the wind, which is the direction from which the particles are coming, giving it a feathery appearance. Since the summit is covered with dense fog some 60% of the time, the rime ice builds up into thick layers.

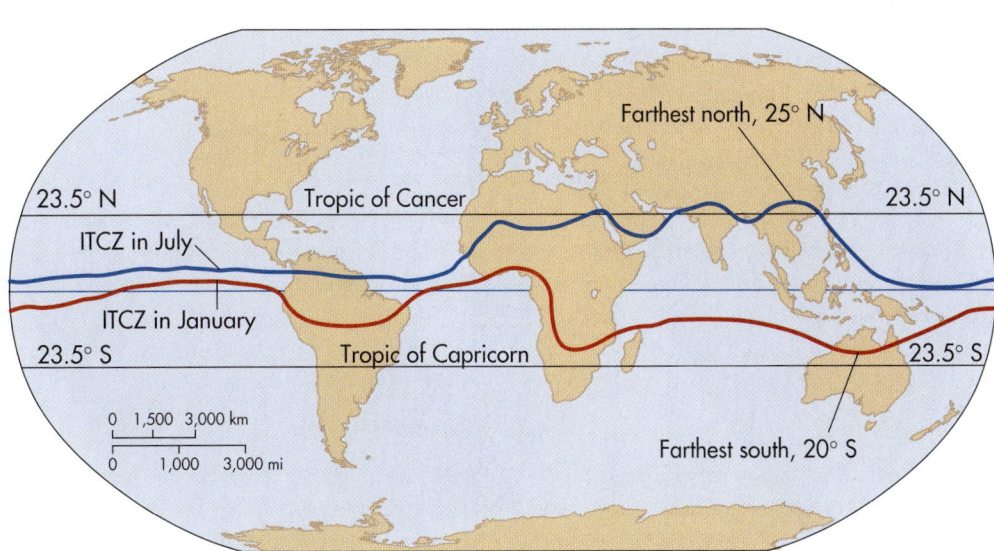

Figure 6.23 Position of ITCZ during the Solstices. The red line marks the position of the ITCZ in January (bottom line) and the blue line indicates its position in July (top line). Note the extent of seasonal migration in Asia as compared to South America.

GLOBAL ATMOSPHERIC CIRCULATION

Now that you have read about the fundamental processes associated with atmospheric circulation and the seasonal migration of pressure systems, it would be helpful for you to review the concept in animated form. To facilitate this review, go to the *GeoDiscoveries* website and select the mod- ule *Global Atmospheric Circulation*. As you watch this animation, be sure to remember the specific components of the global circulatory system and observe how they fit into the integrated network. Once you view the animation, be sure to answer the questions at the end of the module to test your understanding of this process.

this tremendous migration differential is the huge size of the Asian continent (as compared to the Americas), which results in the extreme range of temperature you saw in Figure 5.12.

How does this large temperature range and seasonal ITCZ migration create the monsoon in Asia? A good place to examine the monsoon is in India (Figure 6.24). In the winter a strong outflow of very dry, continental air moves from central Asia south across India. This *winter monsoon* occurs because the pressure gradient slopes steeply from the very strong Siberian High pressure system (in central Asia) toward the ITCZ, which is far to the south over the Indian Ocean (Figure 6.24a).

During the summer, the direction of surface winds reverses, causing the *summer monsoon*. This reversal occurs because temperatures over the Asian landmass increase significantly due to the high Sun angle in summer. As a result, the ITCZ shifts northward over Asia (see Figure 6.23 again), bringing low pressure into the region. Given the seasonal shift in the large pressure systems, a subtropical high is now located over the Indian Ocean and the pressure gradient slopes steeply toward the continent (Figure 6.24b). The warm winds from this high pressure anticyclone pick up moisture as they pass over the Indian Ocean. As the air moves north along the pressure gradient, it is lifted initially by the warmth of the Indian landmass and then intensively by the high Himalaya Mountains. As the air rises, it begins the process of cooling and precipitation occurs. The high precipitation of the monsoonal rainy season lasts from June to September.

KEY CONCEPTS TO REMEMBER ABOUT GLOBAL CIRCULATION

1. The fundamental purpose of global air movement is to equalize the differences in temperature and pressure between the tropics and the poles.

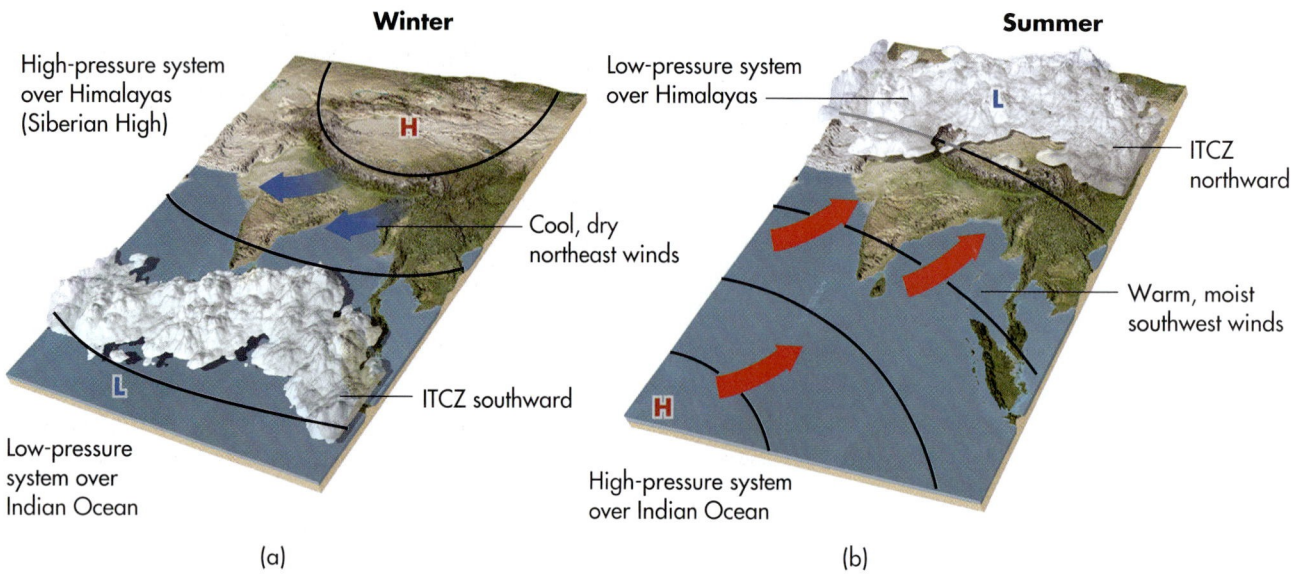

Figure 6.24 Monsoon flow in India during (a) winter and (b) summer. The flow of air is distinctly related to the seasonal position of the ITCZ and the relative temperature of the Asian landmass to the north.

2. The global circulatory system consists of a series of separate, but connected, pressure systems that distribute air around the Earth.

3. Tropical circulation occurs in association with convection processes in Hadley cells, forming the Intertropical Convergence Zone (ITCZ) and the Subtropical High (STH) pressure system. The ITCZ is a zone of low pressure where the winds are easterly. In contrast, the STH is a high-pressure system that rotates clockwise in the Northern Hemisphere and counterclockwise in the Southern Hemisphere.

4. Midlatitude circulation is driven by strongly contrasting air temperatures on either side of the polar front. A major feature of midlatitude circulation is the jet stream, which consists of high-speed westerly winds that flow along the polar front. During the summer the jet stream exhibits zonal flow that is tightly compacted at high latitude. As fall approaches, the jet stream begins to develop distinct undulations that are associated with Rossby waves.

5. The monsoon is a distinct circulatory feature that exists in tropical areas where large land bodies border oceans. Landmasses warm relative to water in summer, resulting in low pressure over land and onshore air flow. During winter, water bodies warm relative to land and the winds reverse their flow direction.

6. All pressure systems migrate on a seasonal basis, according to Earth–Sun geometry.

Local Wind Systems

So far, we have examined circulatory systems and associated wind patterns that occur over regional and global scales. On a much smaller scale, local wind systems result from temperature differences created by topography, proximity to large bodies of water, or other geographic features.

Land–Sea Breezes

Land–sea wind systems are localized along the shores of major water bodies, such as oceans or large lakes. These winds form due to the different heating and cooling characteristics of continents and water, as we discussed in Chapter 5 (the maritime vs. continental effect). During the day, air over land heats up more rapidly than air over water, resulting in a small zone of low pressure that forms over the land near the shore. Over water, in contrast, the air pressure is relatively high because the air is cooler and denser as it sinks. This difference sets in motion a **sea breeze:** the air flows from the sea toward the land; that is, from high to low pressure (Figure 6.25a).

This local circulatory pattern reverses at night because air over the land cools more rapidly than air over the sea. As a result, the pressure systems switch locations, with relatively higher pressure over land and low pressure over the water. When this adjustment occurs, the air

---

Sea breeze *Daytime circulatory system along coasts where winds flow from a zone of high pressure over water to a zone of relatively low pressure over land.*

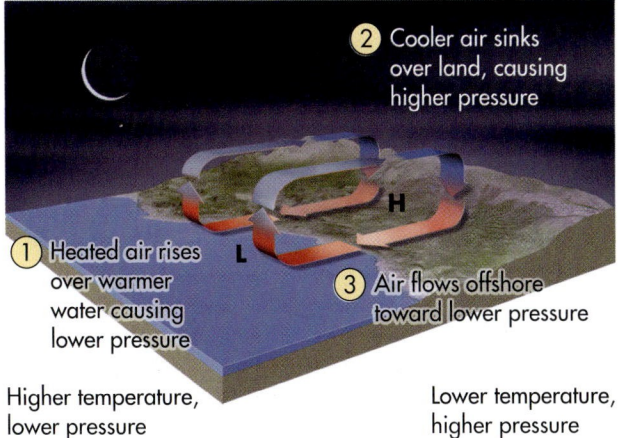

(a) Daytime sea breeze

(b) Nighttime land breeze

Figure 6.25 Coastline wind systems that develop due to the maritime vs. continental effect. (a) A sea breeze develops when air flows inland from the water during the daytime. (b) A land breeze occurs at night when air flows toward the water.

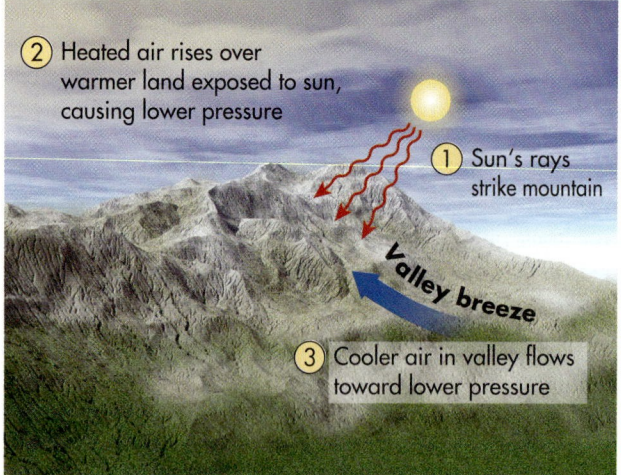

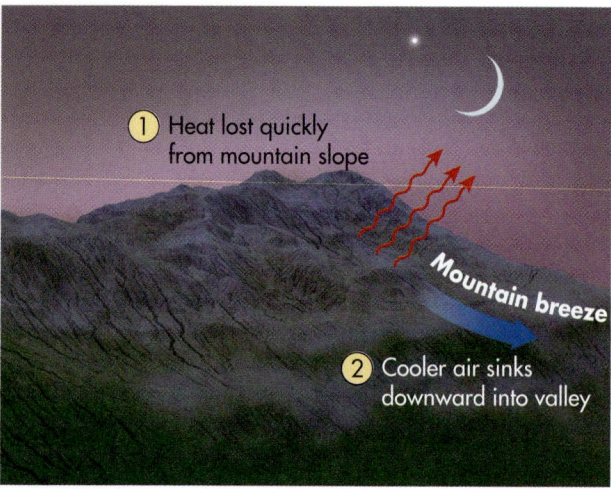

(a) Daytime valley breeze

(b) Nighttime mountain breeze

Figure 6.26 Local wind flow caused by differences in elevation. (a) During the daytime, air flows upslope in a valley breeze. (b) At night, the flow reverses in a mountain breeze.

reverses direction and moves from land toward sea in a **land breeze** (Figure 6.25b).

Topographic Winds

In addition to the winds that develop along coastlines, several kinds of winds form due to variations in topography (Figure 6.26). These winds generally form because large temperature differences can occur with elevation in hilly or mountainous landscapes. For example, a **valley breeze** is a wind system that develops when mountain slopes heat up due to re-radiation and conduction over the course of the day. When this occurs, a zone of relatively low pressure develops on the mountain slopes, whereas high pressure is found in the lowlands below. As a result, air can flow upslope to fill the pressure void created by the upward movement of air over the heated slopes. Like the coastline breezes, the situation in mountainous regions reverses at night, resulting in a downslope **mountain breeze.**

A more extreme example of mountain breezes are called **katabatic winds,** which form in particularly cold places

such as Greenland and Antarctica. Named from the Greek *katabatik,* which means "descending," this process is most common during the winter months, when extremely cold air accumulates over higher-altitude regions covered by ice sheets. This extremely cold, dense air then flows downhill under the force of gravity. These winds sometimes flow at great speeds that can be quite destructive, especially where the air is funneled down a narrow valley. Although the air associated with katabatic winds typically warms as it descends, it usually remains colder than the air which it replaced.

The last kind of topographically related wind process is the **Chinook winds** (also called *foehn winds*). These winds also form in mountainous regions, but occur only when a steep pressure gradient develops along the range (Figure 6.27). For this gradient to develop, a high-pressure system must be on the side of the range that faces the direction of oncoming winds; this side is called the **windward side.** On the other side of the range, or **leeward side,** a low-pressure system must be in place. When this combination of systems occurs, air flows along the pressure gradient—that is, from the windward to leeward side of the mountain

Land breeze *Nighttime circulatory system along coasts where winds from a zone of high pressure over land flow to a zone of relatively low pressure over water.*

Valley breeze *Upslope air flow that develops when mountain slopes heat up due to re-radiation and conduction over the course of the day.*

Mountain breeze *Downslope air flow that develops when mountain slopes cool off at night and relatively low pressure exists in valleys.*

Katabatic winds *Downslope air flow that evolves when pools of cold air develop over ice caps and subsequently descend into valleys.*

Chinook winds *Downslope air flow that results when a zone of high air pressure exists on one side of a mountain range and a zone of low pressure exists on the other.*

Windward side *The side of a mountain range that faces oncoming winds.*

Leeward side *The side of a mountain range that faces away from prevailing winds.*

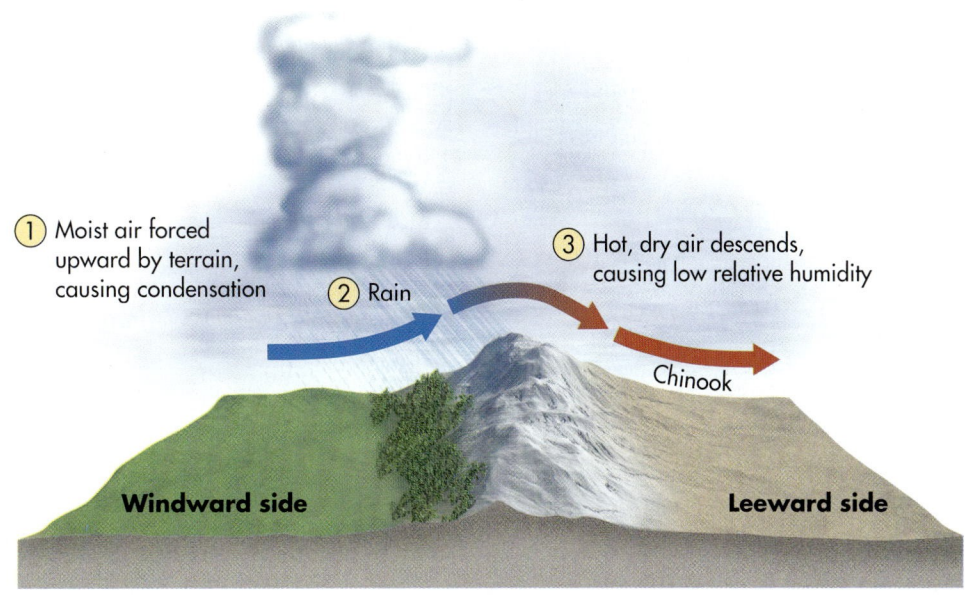

① Moist air forced upward by terrain, causing condensation

② Rain

③ Hot, dry air descends, causing low relative humidity

Chinook

Windward side

Leeward side

Figure 6.27 Formation required for the development of Chinook (foehn) winds. For these winds to develop, the pressure gradient must slope from the windward to leeward side of a mountain range.

range. The resulting air flow moves downslope on the leeward side, resulting in dry air that warms as it descends.

An excellent example of a valley wind is the *mistral wind*, which occurs along the eastern Mediterranean coast of France during the winter months. This wind develops when cold air sweeps across France from the English Channel, hits the Alps, and spills down the Rhône valley into the coastal region. On days when this process is especially active, winds can gust to over 120 km (75 mi) per hour. Another great example of chinook winds is the *Santa Ana* winds in Southern California. These winds occur in winter when high-pressure systems dominate over Nevada and Utah. On the southern side of the high air flows west toward California, rising up and over the Sierra Nevada Mountains. When the air descends on the western side of the Sierra it picks up speed as it rushes into the Los Angeles area. These winds are often beneficial because they clear smog from the city. On the other hand, they can fan wildfires and pose a danger in that regard.

Oceanic Circulation

Gyres and Thermohaline Circulation

So far in this chapter we have seen how atmospheric circulatory systems operate and how they distribute heat energy around the Earth. Another way that heat energy is moved is through oceanic circulation (Figure 6.28), which is strongly related to atmospheric circulation in many ways. In fact, surface currents in the ocean are driven by winds through the transfer of energy from the air

to the water by friction. As a result, the direction of surface ocean currents is related to the same pressure factors that influence atmospheric circulation. For example, the tropical easterly winds produce easterly oceanic currents in the low latitudes.

The Coriolis force also plays a major role in the movement of oceanic currents. Notice in Figure 6.28, for example, that ocean currents in the Northern Hemisphere generally move in a clockwise direction, moving warm tropical water into higher latitudes. A great example of such a current is the Gulf Stream, which transports warm water north from the tropical latitudes along the east coast of the U.S. (Figure 6.29). Currents in the Southern Hemisphere perform the same function, but circulate counterclockwise. A great example of such a current is the Humboldt Current, which flows northward along the west coast of South America. This current delivers cold water from the Antarctic latitudes to the equatorial region where it warms and then flows west as the South Equatorial Current.

The primary difference between oceanic and atmospheric circulatory processes is that the continents block the movement of water. This blockage results in distinct circulatory systems, called **gyres,** such as the one in the Atlantic basin, whereas air generally flows more freely because no large obstructions of this magnitude exist.

In addition to the oceanic circulation across the surface, a slow mix of water also occurs vertically between

Gyres *Large oceanic circulatory systems that form because currents are deflected by landmasses.*

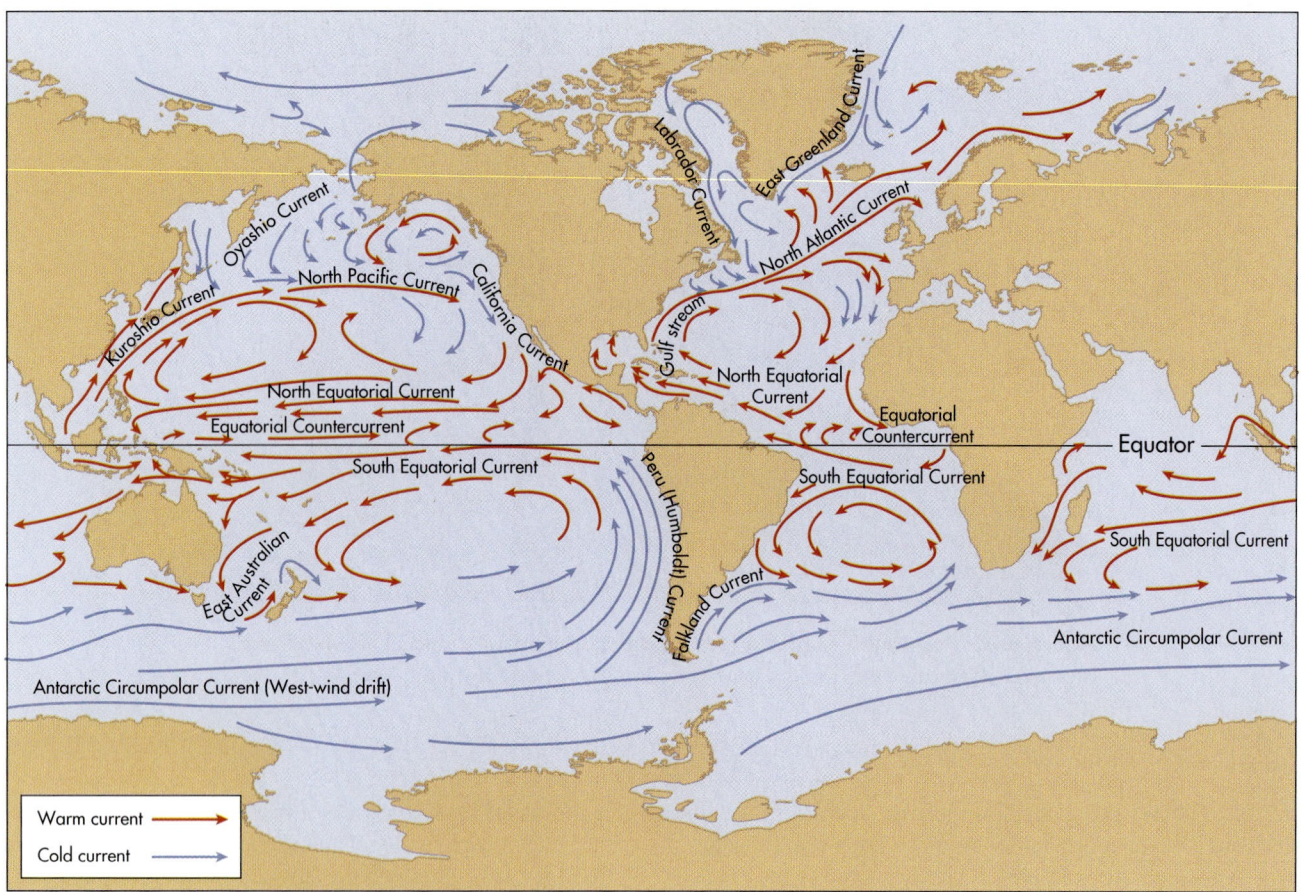

Figure 6.28 Global oceanic circulation. Ocean currents fundamentally move in the same direction as the winds. Note, for example, the easterly current along the Equator and how it relates to the easterly winds.

The primary difference between ocean and air currents is that the continents interrupt the ocean currents, forming distinct gyres.

layers of the ocean. These currents, which together make up the **thermohaline circulation** that links all of the world's oceans, are generated because of regional differences in water density that evolve through variations in temperature and salinity. In general, water at the surface of the ocean is typically warmer and less salty than deeper water. A good place to watch the beginning of this process is in the western part of the tropical Atlantic basin in Figure 6.30. Here, as part of the North Atlantic gyre, warm water flows northward in the Gulf Stream (as shown in Figure 6.29). As this warm water travels to higher latitudes, extensive evaporation occurs, which increases the relative salinity of the current. Along with this increased salinity, the surface waters gradually cool as they continue northward.

Due to decreasing temperature and high salinity, the water density increases and the water sinks in the northern Atlantic in Figure 6.30. In other words, it becomes a **downwelling current** that subsequently flows at great depths to the southern part of the Atlantic basin. From there, it flows to the east between Antarctica and Australia to the Pacific Ocean, where it then flows northward toward Alaska. As the water moves into the tropical regions of the Pacific, it warms and becomes an **upwelling current** that moves back to the surface. This pattern is repeated in several places around the world's oceans. To give you an idea of the time involved for ocean currents like this to move, consider that recent research demon-

Thermohaline circulation *The global oceanic circulatory system that is driven by differences in salinity.*

Downwelling current *A current that sinks to great depths within the ocean because water temperature drops and salinity increases.*

Upwelling current *A current that ascends to the surface of the ocean because water temperature warms and salinity decreases.*

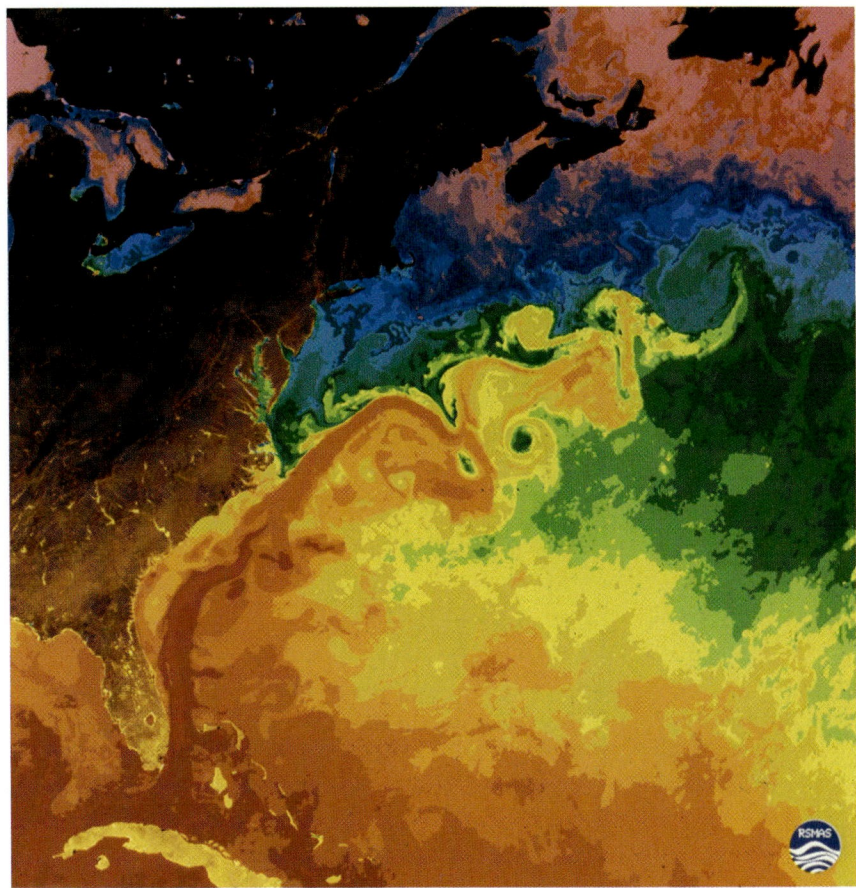

Figure 6.29 Sea-surface temperatures in the western Atlantic. NOAA-7 orbiting satellite image showing sea-surface temperatures for a week in April. Red represents the warmest water, which is flowing northeastward along the east coast of North America in the Gulf Stream. Greens, blues, and purples are progressively cooler sea-surface temperatures.

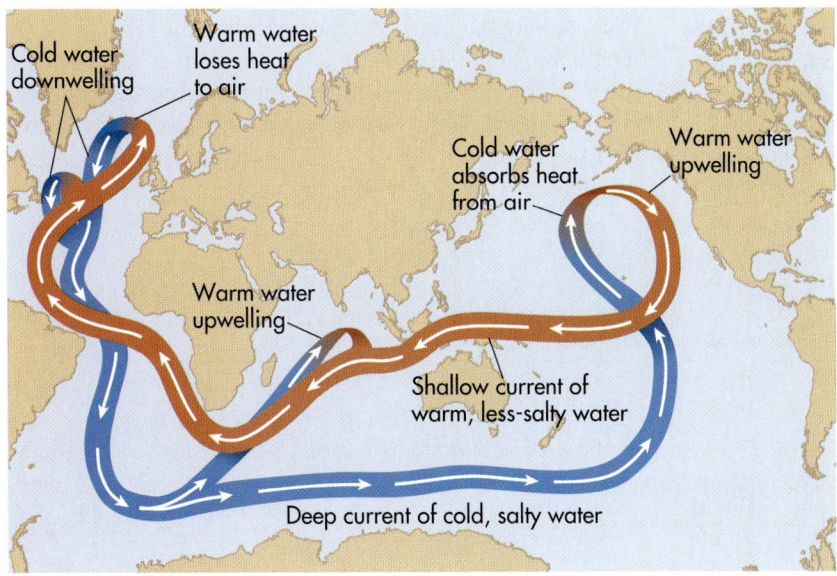

Figure 6.30 Surface and deep water circulation loop in the world oceans. Note the locations where water upwells and downwells.

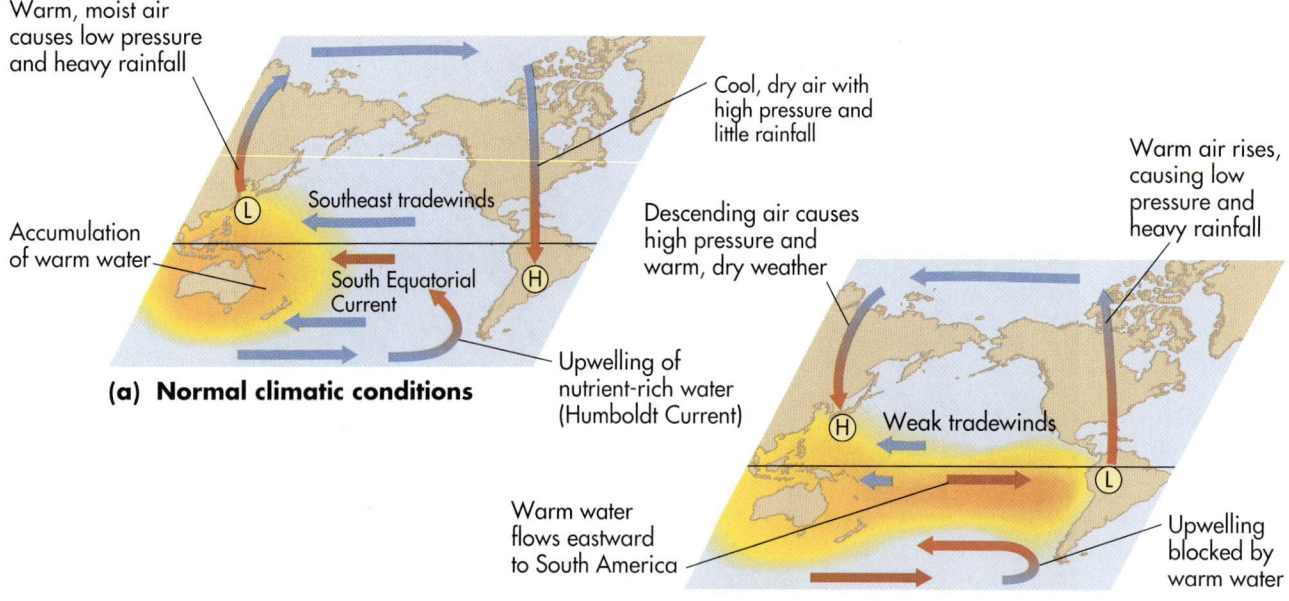

(a) Normal climatic conditions

Warm, moist air causes low pressure and heavy rainfall

Accumulation of warm water

Southeast tradewinds

South Equatorial Current

Cool, dry air with high pressure and little rainfall

Descending air causes high pressure and warm, dry weather

Upwelling of nutrient-rich water (Humboldt Current)

(b) El Niño conditions

Warm air rises, causing low pressure and heavy rainfall

Weak tradewinds

Warm water flows eastward to South America

Upwelling blocked by warm water

Figure 6.31 Relationship of atmospheric circulation to surface water flow in the Equatorial Pacific. (a) Normal conditions occur when strong easterly flow pushes warm water into the western Pacific. **(b)** An El Niño occurs when easterly flow weakens, allowing warm water to collect along the South American coast. Note the relationship between precipitation and the location of pressure systems.

strates that it takes about 2000 years for water to flow from Great Britain in the northern Atlantic Ocean to the southern Pacific Ocean!

El Niño

As with many other processes in physical geography that frequently depart from the normal pattern, the oceanic circulatory system sometimes functions in unusual ways. One of the most significant changes that occurs in the way the oceans circulate is the *El Niño* phenomenon, which is a reversal of the "normal" flow in the tropical Pacific Ocean. During most years, the dominant current in the tropical Pacific is easterly along the South Equatorial Current (see Figure 6.28). As you might suspect, this easterly flow is caused by the easterly trade winds that flow from strong subtropical high pressure systems to the ITCZ (Figure 6.31).

When the easterly pattern is in place, the flow of water away from the South American coast allows cold water from the deep-flowing Humboldt Current (which flows northward from Antarctica along the coast of Chile) to upwell (rise to the surface) in the eastern Pacific near the Equator. Due to this upwelling, the overlying air is cool and descending, which, in turn, results in little evaporation of moisture from the ocean, and, consequently, little precipitation. You can see the effect of this low precipitation in the lack of vegetation along the west coast of South America in Figure 6.18. As the water flows westward along the Equator, however, it warms because of the high Sun angle. By the time the current reaches the western Pacific the water is quite warm, which results in overlying (atmospheric) low pressure and heavy rains in that part of the basin.

GEODISCOVERIES   WWW.WILEY.COM/COLLEGE/ARBOGAST

EL NIÑO

To see an animated version of how the El Niño process functions, go to the *GeoDiscoveries* website and select the module **El Niño**. This animation will show you how sea-surface temperature and the height of the ocean changes during an El Niño year. Once you complete the animation, be sure to answer the questions at the end of the module to test your understanding of this concept.

For some as yet unexplained reason, the "normal" circulatory pattern in the Pacific Ocean reverses every 3 to 8 years. The apparent cause of this reversal is that the normally strong tropical easterlies weaken, allowing a distinct westerly flow to develop in surface ocean waters. This opposing pattern is called El Niño, derived from the expression *Corriente del Niño*, or the "Current of the Christ Child," used by Peruvian fishermen to describe the influx of warm waters into the eastern Pacific Ocean. This reversal of surface flow usually occurs around Christmastime (hence the name) and is particularly noticeable to fishermen because the warmer waters are relatively sterile when compared to the nutrient-rich upwelling (Humboldt) current that normally supports superb fishing. In addition to the depleted fish industry, an El Niño brings intense storms to the normally dry parts of South America (and even the southwestern U.S.) and drought to the western Pacific. This weather reversal occurs because the atmospheric pressure in the eastern Pacific is relatively low, whereas it is relatively high in the western Pacific.

The Big Picture

In this chapter you learned about how the atmosphere flows and the various factors that cause distinct circulatory patterns and their geographic distributions. Not only is this understanding critical to why winds fundamentally occur, but it also helps explain other physical processes on Earth, such as the distribution of vegetation, the evolution of soils, and the location of major rivers.

One of the most important processes associated with atmospheric circulation is condensation and precipitation. You have undoubtedly witnessed a rain-shower many times and probably have experienced heavy rainfall or snowfall at other times in your life. Weren't you amazed by these events, which may have resulted in flash flooding, impassable roads, and school and business shutdowns? Even so, did you really understand why the rain or snow was falling? With these ideas in mind, in Chapter 7 we will discuss the various ways that water flows in the atmosphere, including why clouds form and precipitation occurs.

Summary of Key Concepts

1. Air pressure refers to the weight of air distributed on the surface of the Earth. The two basic kinds of pressure systems are high (anticyclones) and low (cyclones). High-pressure systems in the Northern Hemisphere are columns of air that rotate clockwise and descend toward the surface of the Earth within their core. In contrast, low-pressure systems in the Northern Hemisphere are columns of air that lift from the Earth's surface within their core as they rotate counterclockwise. The rotation of these respective systems is opposite in the Southern Hemisphere.

2. The speed and direction of air flow depends upon many variables. The pressure gradient force refers to the difference in air pressure that exists between anticyclones and cyclones. The Coriolis force refers to the deflection that occurs in winds due to the rotation of the Earth on its axis. Features on the Earth's surface, such as mountains, forests, and buildings, create a frictional force that acts opposite to the wind's direction. The combined effect of the pressure gra-

dient force, Coriolis force, and frictional force causes a spiral motion of air in both low- and high-pressure systems.

3. The global circulatory system consists of a series of separate, but connected, pressure systems that distribute air around the Earth.

4. In addition to large-scale wind systems, several regional and local-scale circulation systems occur, including monsoon winds, land- and sea breezes, katabatic winds, and Chinook/foehn winds.

5. Oceanic circulation is driven by winds and generally flows in the same direction as atmospheric circulation. Due to the presence of the continents, however, distinct circulatory cells, or gyres, exist in the oceans.

6. An El Niño occurs when the normal (equatorial) circulatory system in the Pacific is reversed. This reversal results in drought in the western Pacific and strong storms in the eastern Pacific.

Check Your Understanding

1. Why is air pressure greater at low altitudes rather than high altitudes?

2. Which direction does air flow at the surface, from cyclone to anticyclone, or the reverse?

3. Which place is more likely to be a zone of convection, the ITCZ or STH? Why does this pattern occur?

4. Why is the polar front an important feature in the midlatitudes?

5. Why are Rossby waves more likely to form during the winter months than in the summer months?

6. How does Earth–Sun geometry influence the position of atmospheric pressure systems?

7. How is a land–sea circulatory system like a tropical Hadley cell?

8. How are land–sea breezes and mountain breezes similar? What conditions are required for these circulatory systems to develop?

9. Within the Atlantic gyre, what is the primary purpose of the Gulf Stream?

10. In the equatorial Pacific, the normal circulatory flow of surface waters is easterly. Why does this flow pattern normally occur?

Visual Concept Check 6.1

Air pressure decreases from the base of Mt. McKinley to the top of the mountain. This occurs because air molecules tend to be held closely to the Earth due to the force of gravity. Given that more air molecules occur at the base of the mountain than the top, the air pressure (or weight of the air) is greater at the base of the mountain. Conversely, air pressure decreases with altitude because progressively fewer air molecules exist.

Visual Concept Check 6.2

The high pressure system in this image is centered over central Europe. This can be determined because that is the approximate center of the zone of clear (or sunny) skies. Given that this anticyclone is in the Northern Hemisphere, it spins clockwise. Thus, northerly winds exist on the eastern side of the system, whereas southerly breezes are on the west. Spain is under the influence of a low pressure system, as indicated by the dense cloud cover.

Visual Concept Check 6.3

The answer to this question is c. The ITCZ consists of the band of clouds that extends east and west through the northern part of South America.

CHAPTER SEVEN

ATMOSPHERIC MOISTURE AND PRECIPITATION

In Chapter 6 we focused on air pressure and the processes associated with atmospheric circulation. As air flows, it carries many things that float in suspension, including dust, volcanic ash, and even small organisms. One of the most important of these constituents of the atmosphere is water, which moves within the air in vapor, liquid, and solid forms.

Water is a very important part of the global ecosystem because it is necessary for all living things. In addition, water serves an important

Have you ever wondered why clouds form? These beautiful cumulous clouds developed because large bubbles of relatively warm air rose within the atmosphere, causing water vapor to condense. Processes such as these are the focus of this chapter.

regulatory function in climate patterns and processes because it moderates temperature. Water vapor in the atmosphere is especially vital because it absorbs heat from the Sun. As water vapor moves around the planet, driven by atmospheric currents, the effect is to moderate surface temperature, which is beneficial to life.

In this chapter we'll first examine the chemical and physical composition of water and how it changes phases. Then we'll discuss how clouds form and

Physical Properties of Water

Water is found almost everywhere on Earth and is absolutely crucial to life as we know it. Water covers 71% of our planet's surface and even comprises about 70% of the human body by weight. It is a very important component of the atmosphere because it stores energy that contributes to global wind patterns. In addition, it also plays a major role in the regulation of climate. In order to understand how water behaves in the physical environment, let's first review its basic properties.

Hydrogen Bonding

A drop of water is made up of billions of molecules, with each consisting of two hydrogen atoms combined with one oxygen atom (Figure 7.1a). This combination is the basis for water's well-known chemical formula: H_2O. Within a water molecule, oxygen attracts the bonding electrons more strongly than do the hydrogen atoms. Because of this unequal attraction the oxygen end of the molecule has a partial negative charge, whereas the hydrogen side has a partial positive charge.

Water molecules are attracted to each other because of these contrasting positive and negative charges. The negative end of one molecule is attracted toward the positive end of an adjacent molecule. This is why water molecules stick together in a drop of water and why water occurs in liquid form at normal surface temperatures. (Most chemical compounds are solids or gases at normal surface temperatures.) The attraction between molecules of water is known as *hydrogen bonding* because hydrogen atoms form particularly good bonds between molecules. These bonds are strongest, of course, when water is frozen as ice. In these circumstances, the molecules firmly bond to each other in distinct hexagonal forms (Figure 7.1b).

Another interesting attribute of water is that it has high surface tension. Surface tension results when molecules at the surface of a liquid have a strong attachment to each other but not to the molecules of air above them. This attraction between molecules is particularly strong in water because of hydrogen bonding. Thus, the water molecules pull harder to the sides to create the smallest possible amount of surface area, forming spherical bubbles and droplets. The high surface tension also creates an elastic "skin" on the surface of water, strong enough to allow some kinds of insects (such as water striders) and even small lizards to walk on the surface.

Yet another important feature of water is its ability to move upward in thin openings (or capillaries) against the force of gravity within the soil and plants in a process called **capillary action.** This motion occurs because, through hydrogen bonding, water molecules pull other water molecules along. This process is very important be-

---

Capillary action   *The process through which water is able to move upward against the force of gravity.*

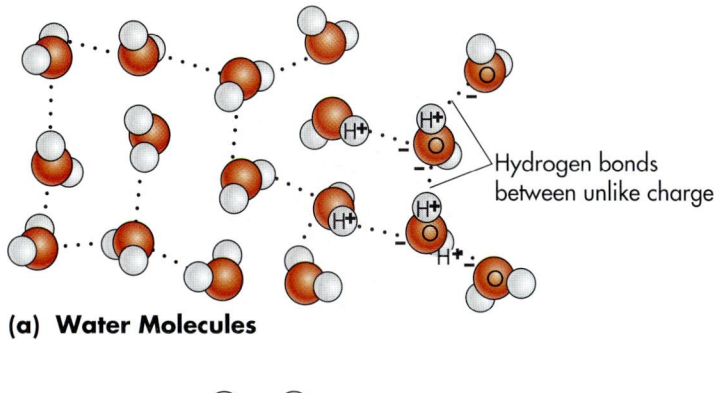

(a) Water Molecules

Hydrogen bonds
between unlike charges

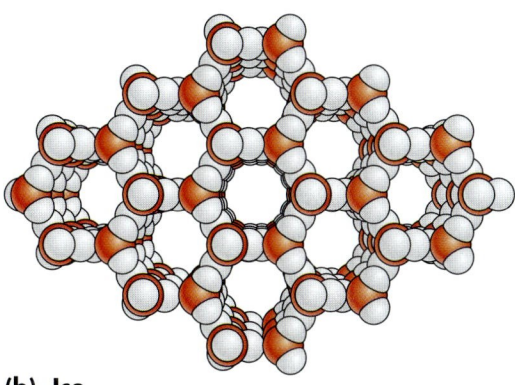

(b) Ice

Figure 7.1 Chemical composition of water. (a) Water molecules are composed of two hydrogen atoms and one oxygen atom, which form partially positive ends of molecules (the hydrogens) and partially negative ends (the oxygen). Given the attraction between positive and negative charges, relatively weak hydrogen bonds are produced between molecules. (b) Hydrogen bonds are the reason liquid water molecules form solid ice at a comparatively high temperature; it takes more energy to break these bonds than is the case for most other common liquids.

cause it enables plants to transport nutrients from their roots up into their stems and leaves. Without this process, trees and other tall vegetation could not exist.

Thermal Properties of Water and Its Physical States

One of the most important characteristics of water is that it absorbs and releases abundant amounts of latent heat, which, if you recall from Chapter 4, is hidden energy stored in molecular bonds. This capability of water is important because it contributes significantly to atmospheric circulation and helps regulate climate. Understanding this property is also important because it explains how water can exist in the three physical states, or phases, that you're familiar with: solid (as *ice*), liquid (as *water*), and gas (as *water vapor*). In short, heat energy is required for water to change phases because hydrogen bonds must be

formed, loosened, broken, or tightened. A simple rule of thumb is that heat energy must be applied to molecules if hydrogen bonds are to be loosened or broken, and extracted from molecules if they are to be formed or tightened. When bonds are loosened energy is transformed from sensible heat to latent heat, which has a net cooling effect because less sensible heat is "sensed." In contrast, when bonds are strengthened energy is converted from latent heat to sensible heat, which has a net warming effect.

The thermal properties of water are important because they strongly influence weather patterns on Earth, providing over 30% of the energy required to drive atmospheric processes. For example, much of the energy contained within a thunderstorm is produced when latent heat is released, because water changes rapidly from gas to liquid form (this will be discussed in more detail later). This section of our chapter focuses on the way in which water changes physical states and the amount of latent heat energy released or absorbed during these transformations. As you read through this discussion, refer frequently to Figure 7.2, which shows the various pathways that water follows and the amount of latent heat that's absorbed or released.

Let's arbitrarily begin the examination of water phases and latent heat transfers by considering what happens when liquid water turns to ice in the process of **freezing.** Water exists in its most dense form at a temperature of 4° C (39° F). Below that temperature the motion of water molecules further slows and more hydrogen bonds progressively form. When liquid water cools down to 0° C (32° F), the motion of the water molecules slows even further and still more hydrogen bonds develop. At this temperature water changes to ice, its crystalline phase, under normal air pressure. Water has to cool a bit more to freeze at higher air pressure because it is squeezed more and more nitrogen and oxygen molecules go into solution. During the process of freezing, 80 calories of heat energy are released as the *latent heat of freezing*.

Have you ever wondered why ice floats on liquid water, even though it's solid? The reason for this apparent oddity is that ice is actually less dense than liquid water. This decreased density occurs because as more hydrogen bonds form through freezing, the medium expands and volume increases. In fact, water can increase by 9% in volume, which is why water pipes sometimes break in homes during the winter. As you will see in Chapters 14 (Weathering and Mass Movement) and 17 (Glacial Geomorphology: Processes and Landforms), the ability of water to expand and contract has had profound impact on the shape of the Earth.

When heat energy is applied to and absorbed by ice at 0° C (32° F), the motion of the water molecules increases

Freezing   *The process through which water changes from the liquid to solid phase.*

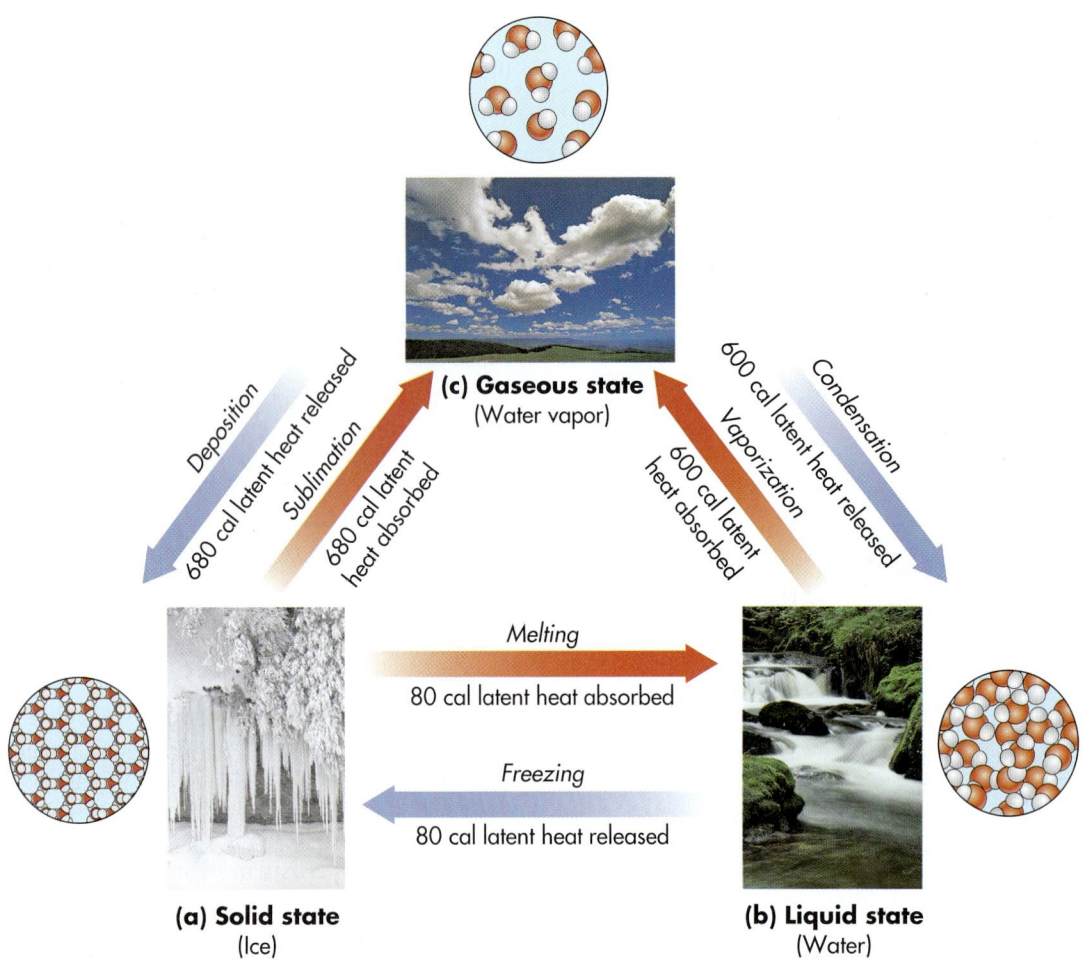

Figure 7.2 Physical states of water. Arrows indicate the pathways that water follows in changing between various states, when latent heat is either absorbed or released. Arrow colors reflect the net cooling or warming effect due to the transfer of sensible heat to latent heat or latent heat to sensible heat, respectively. (a) Water exists as ice at temperatures below 0° C (32° F) under normal air pressure because the molecules slow down enough to lock into hexagonal crystals. (b) At temperatures between 0° C (32° F) and 100° C (212° F), water molecules move freely, but slowly enough to remain attached to one another in the liquid phase. (c) When temperatures exceed 100° C (212° F), all water molecules move fast enough to become completely detached from one another in the vapor phase. (Note that clouds actually contain liquid water as well as water vapor; we see them because the liquid water reflects light.) The number of calories absorbed or released in any given phase change assumes 1 gram of water.

and some of the hydrogen bonds begin to break. As a result, the ice melts and becomes liquid water, which is a noncompressible fluid that conforms to the shape of its container (such as a glass). During this transformation 80 calories of latent heat is absorbed as the *latent heat of melting* to change 1 gram (g) of ice to 1 g of water. At this point the water molecules are still bound together, but the bonds are sufficiently weak that water flows.

Now let's examine what happens when liquid water changes to water vapor. This process begins at room temperature when added energy causes hydrogen bonds to

further loosen, resulting in the liberation of some water molecules to the vapor phase in the process of **evaporation.** The added heat is referred to as the *latent heat of vaporization* and amounts to 600 calories for every 1 g of water. If the temperature of water increases to 100° C (212° F), the tremendous amount of added energy causes all molecular bonds to break and all water molecules thus move into the atmosphere as vapor. Conversely, when water vapor changes back to liquid form through the process of **condensation,** 600 calories of energy are released as the *latent heat of condensation.*

Evaporation   *The process through which water changes from the liquid to vapor phase.*

Condensation   *The process through which water changes from the vapor to liquid phase.*

LATENT HEAT

Let's review the concept of latent heat and how it relates to the phases of water. Go to the *GeoDiscoveries* website and select the module **Latent Heat**. The animation nicely illus-trates the various phases of water and the way that latent heat is either absorbed or released. After you view the animation, be sure to answer the questions at the end of the module to test your understanding of this concept.

In rare circumstances, it's possible for water to change directly from the ice phase to vapor in a process called **sublimation.** This transformation occurs if the temperature of ice rapidly changes from 0° C (32° F) to 100° C (212° F). Under these conditions, 680 calories of energy known as the *latent heat of sublimation* are added to the medium. This amount of energy is derived from the addition of the 600 calories associated with vaporization *plus* the 80 calories absorbed when ice changes to liquid water. In other words, the energy associated with both the melting and vaporization processes must be accounted for. Conversely, water can change directly from the vapor phase to ice through the process of **deposition.** When this transformation occurs, 680 calories of latent heat are released, which reflects the addition of the energy released *both* by condensation and freezing.

The Hydrosphere and the Hydrologic Cycle

In our discussion of the physical states of water, an implied assumption is that water is stored in various places on Earth such as the atmosphere, lakes, and oceans. Over-all, the total water realm of the Earth is known as the **hydrosphere.** Within the hydrosphere, approximately 97.2% of water is stored in the oceans as salt water (Figure 7.3). The remaining 2.8% is largely fresh water, most of which is stored in massive polar ice sheets and smaller mountain glaciers. This frozen water accounts for about 2.15% of the total global water volume. Approximately 0.63% is stored within the Earth as groundwater.

What about the remaining 0.02% of hydrosphere water? Although this amount is analogous to one drop of liquid in a liter of water, it is the most important to us because it includes the water available for plants, animals, and human use. The right-hand pie chart in Figure 7.3 shows the geographic distribution of this part of the Earth's water realm. Of this small amount, most (0.009% of the total volume) is stored in fresh water lakes. Another 0.008% is contained within salty lakes and inland seas. Soil water, which is held at shallow depths within reach of plant roots, comprises 0.005%. Of the remaining 0.002% of the total water volume, the atmosphere and streams contain 0.001% and 0.0001%, respectively.

As you probably know at an intuitive level, water migrates from one place to another within the hydrosphere. Although you may not have thought about it much,

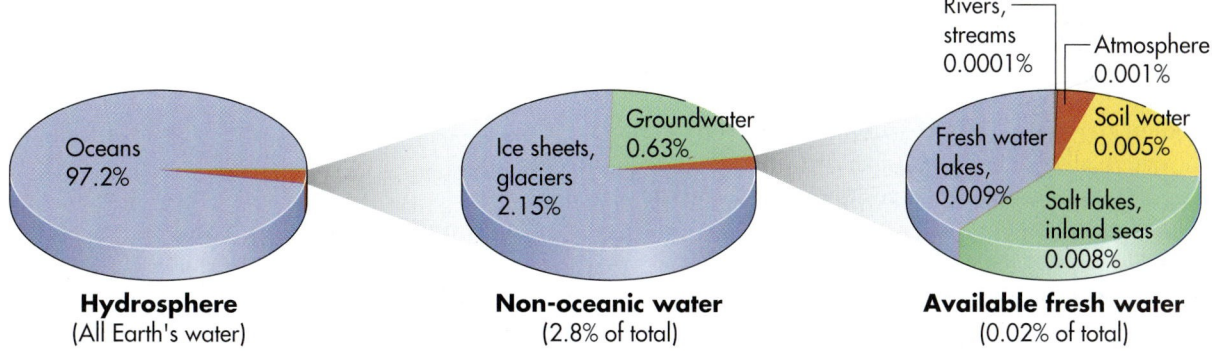

Figure 7.3 Geographic distribution of water within the hydrosphere. Most water is contained within the oceans, with proportionately smaller amounts stored in polar ice sheets and as fresh surface and soil water.

Sublimation *The process through which water changes directly from ice to the vapor phase.*

Deposition *The process by which water vapor changes directly to ice without first becoming a liquid.*

Hydrosphere *The water realm on Earth.*

The hydrologic cycle has many different components. This image shows a typical farm pond in the Midwest. Explain how this pond is part of the hydrologic cycle. What are two ways in which water could fill the pond? What happens to the water once it becomes stored within the pond? How does water return to the atmosphere from the pond?

you've no doubt seen water move from the atmosphere to the Earth as various forms of precipitation and from the Earth to the atmosphere through the gradual evaporation of street puddles over several hot days. Water not only moves back and forth from the air to the ground, but also to and from places on and within the ground. Taken together, this movement of water between storage locations (or reservoirs) constitutes the **hydrologic cycle**.

Overall, the hydrologic cycle is balanced in the sense that the Earth has a finite amount of water and the amount evaporated generally equals the amount precipitated on a global scale. However, a variety of local and regional imbalances occur. To see how the hydrologic cycle fits in the context of the global water balance, examine Figure 7.4, which shows the flow of water on Earth in thousands of cubic kilometers (km^3) per year. With the notion of a global water budget in mind, consider that water flows to the land and ocean surfaces are positive inputs, whereas those leaving are negative. As you examine these geographic patterns, note the positive balance ($+36$) on landmasses, which reflects the fact that land gains 36,000 km^3 more water through precipitation than it loses through evaporation. In contrast, oceans have a negative balance (-36). This negative balance exists because oceans lose 36,000 km^3 more water by evaporation than they gain through precipitation. Obviously, if there were no direct

connection between the land and the oceans, the landmasses would gradually become submerged by water and the oceans would empty. This reversal does not occur because 36,000 km^3 of water runs off the land per year and returns to the oceans.

Hydrologic cycle *A general model that illustrates the way that water is stored and moves on Earth from one reservoir to another.*

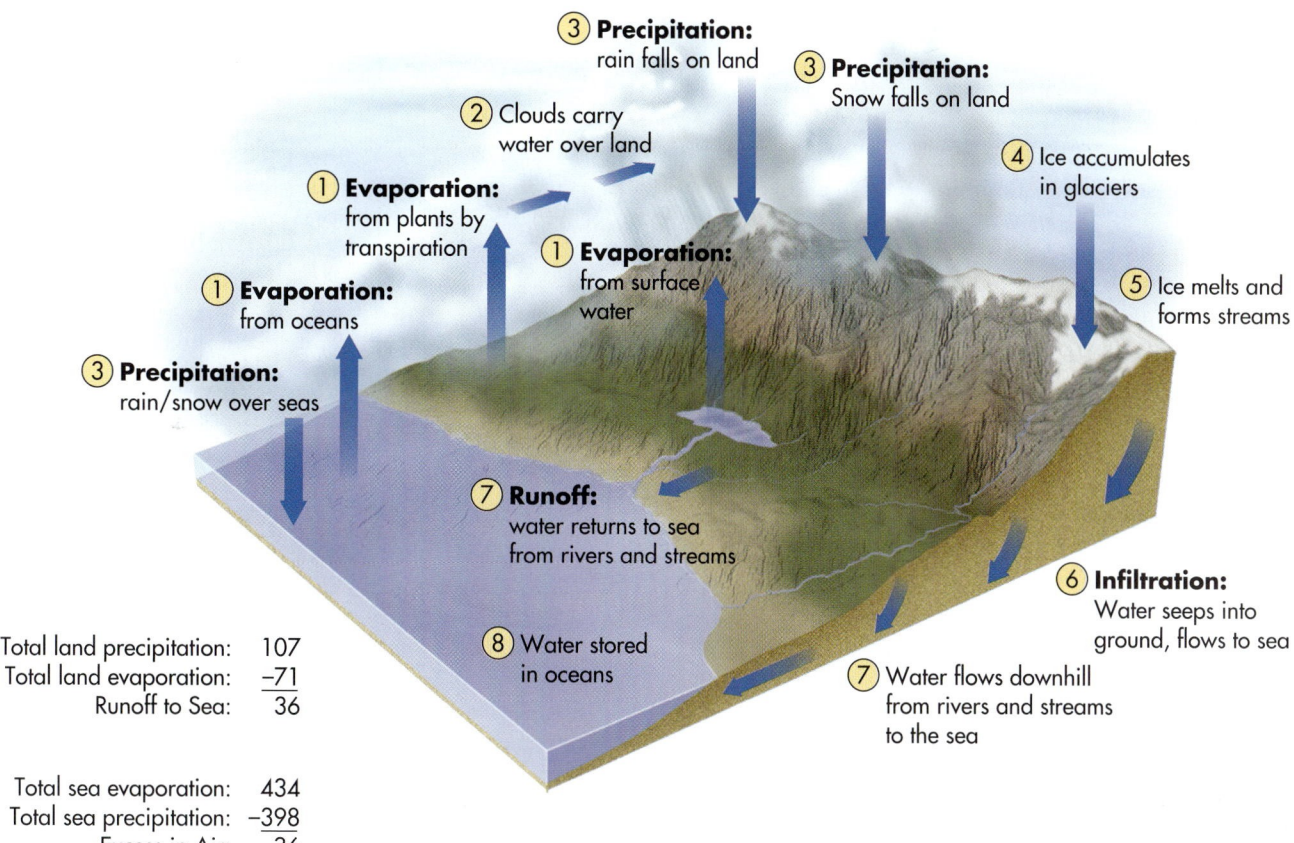

Total land precipitation: 107
Total land evaporation: −71
Runoff to Sea: 36

Total sea evaporation: 434
Total sea precipitation: −398
Excess in Air: 36

Figure 7.4 The global water balance. The global water balance refers to the way that water moves within the hydrologic cycle and the relationship between evaporation, precipitation, and runoff. Oceans are a net source of evaporation, whereas more precipitation occurs on landmasses than evaporation. The system stays balanced because water runs off the land and returns to the ocean. Values are in cubic kilometers (km³) and indicate average annual water flows to and from the Earth's land and ocean areas.

Humidity

Let's now look more closely at how water actually changes physical state from vapor to liquid and the factors that govern this process. In the atmosphere, this change of state ultimately results in precipitation.

Maximum, Specific, and Relative Humidity

We begin our examination with the concept of **humidity**, which is the concentration of water vapor within the air. This concentration is expressed in three ways: *maximum humidity*, *specific humidity*, and *relative humidity*. Although at first it may seem confusing that three different humidity types occur, their definitions and interrelationships are really quite simple.

Imagine that you're about to fill a drinking glass to some level with water. Given that the glass can hold only so much water, you probably fill it to some level below the maximum amount of water that the glass can hold. If you try to put more water in the glass than it can hold, the water will spill over the top. If you keep this very simple analogy in mind, you can understand the various kinds of humidity and how they interact.

Maximum humidity refers to the maximum amount of water vapor that a definable body of air can hold. While the amount of liquid that a drinking glass can hold depends upon the size of the glass, the maximum humidity of an air parcel is completely governed by the air temperature. *Warmer air is capable of holding much more water vapor than colder air*, because air expands in the process of warming, which leaves room for more water

Humidity *A measure of how much water vapor is in the air. The ability of air to hold water vapor is dependent on temperature.*

Maximum humidity *The maximum amount of water vapor that a definable body of air can hold at a given temperature.*

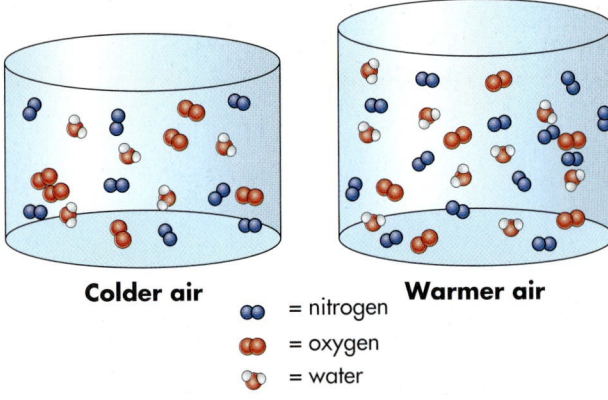

Colder air   **Warmer air**

○○ = nitrogen

○○ = oxygen

○○ = water

Figure 7.5 Maximum humidity and air temperature. A given parcel of warmer air can hold more water vapor than a similar parcel of colder air. The reason is that molecules at higher temperature are farther apart, so more water molecules can fit in a given amount of air.

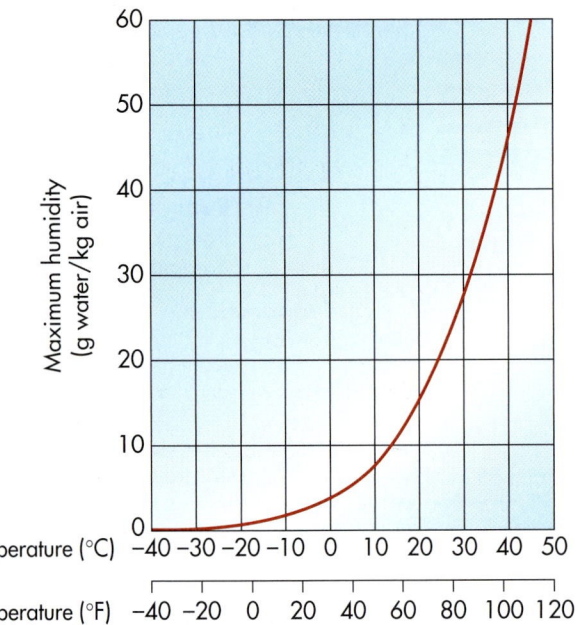

Figure 7.6 The saturation curve. Maximum humidity increases as temperature increases. In other words, warmer air can hold more water vapor than cooler air can hold.

vapor. This is a very simple, but important rule in physical geography that explains the processes of evaporation and precipitation. Figure 7.5 presents a simplified example of this concept.

The relationship between air temperature and maximum humidity is described by a *saturation curve*, such as the one presented in Figure 7.6. In the case of the atmosphere, the term *saturation* refers to the point where the air cannot hold any more water vapor based on the temperature at that time. To help make sense of this concept, simply think of what happens when a sponge absorbs water. The sponge absorbs water until it is saturated and unable to hold any more. The atmosphere works on much the same principle. The line on the saturation curve merely illustrates the respective intersections of maximum humidity and air temperature. Notice that maximum humidity is measured as grams of water per kilogram of air, or g/kg in abbreviated form.

You can see, for example, that at an air temperature of −10° C (14° F) the maximum humidity is about 2 g/kg. Subsequently, the maximum humidity increases exponentially with increasing temperature. At 20° C (68° F) the maximum humidity is about 15 g/kg and at 30° C (86° F) the maximum humidity nearly doubles again to about 26 g/kg. If you're having difficulty relating this concept to the glass of water, try to imagine what would happen if the glass could expand and contract with warmer and colder temperatures, respectively. In this scenario, the glass can hold a larger amount of *total* water when it is warm, and less as it contracts with progressively decreasing temperature. The atmosphere operates on the same principle in that it expands when warm and contracts when cold.

Whereas the term *maximum humidity* refers to the maximum amount of water vapor that a mass of air can hold, the term **specific humidity** refers to how much water vapor is *actually* in the air. Think back to the glass analogy. Recall that the glass can hold a maximum amount of water, which, in fact, rarely occurs in the course of everyday use. Instead, you usually fill the glass to some random point that's short of the container lip. The volume of this *specific* amount of liquid can be measured; for example, the glass may contain 1 cup of water, even though the glass can actually hold 2 cups.

Finally, we have the concept of **relative humidity**, which is the ratio of specific to maximum humidity—in other words, how close the air is to being saturated. In the case of the glass, you could say that the glass is half full if it contains one cup of water even though it can hold two cups. With respect to the atmosphere, this ratio is usually expressed as a percentage and is calculated as

$$\text{Relative humidity (\%)} = \left(\frac{\text{Specific humidity}}{\text{Maximum humidity}} \right) \times 100$$

$$\text{RH (\%)} = (\text{SH / MH}) \times 100$$

The following example should make these humidity concepts clearer for you. Let's say that a given mass of air at a certain temperature can potentially hold 2 units of wa-

Specific humidity   *The measurable amount of water vapor that is in a definable body of air.*

Relative humidity   *The ratio between the specific and maximum humidity of a definable body of air.*

(a) Maximum humidity = 2 units

(b) Specific humidity = 1/2 unit

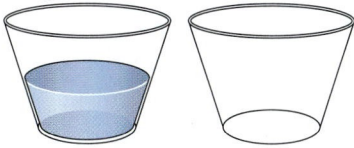

(c) Relative humidity = 25%

Figure 7.7 Hypothetical humidity values. (a) In this example, the maximum humidity is 2 units of water vapor. (b) In this example, the specific humidity is $\frac{1}{2}$ unit. (c) The mass of air can hold 2 units of water vapor, but only $\frac{1}{2}$ of a unit is present. Thus, the relative humidity is 25%.

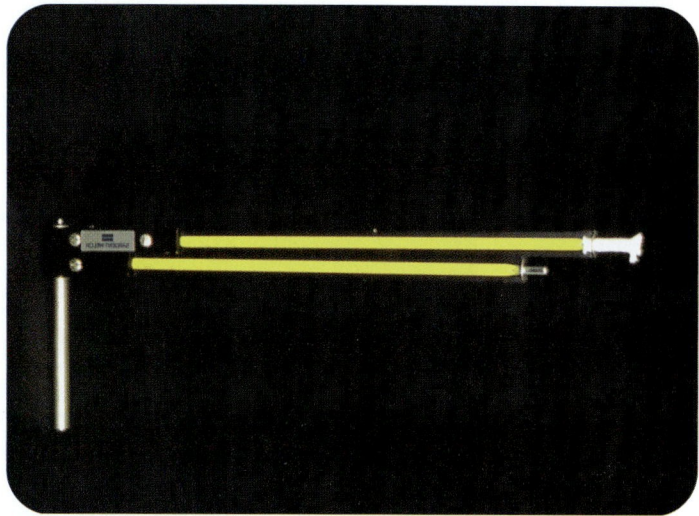

Figure 7.8 Sling psychrometer. This system uses a pair of thermometers that detect the amount of cooling due to evaporation. The dry-bulb thermometer shows the actual temperature, whereas the wet-bulb thermometer indicates the amount of cooling.

ter vapor, as seen in the simplified diagram in Figure 7.7a. In other words, its maximum humidity is 2 units.

But how much water vapor is actually there; in other words, what is the specific amount of vapor present? Well, let's say for the case of this example that the mass of air contains $\frac{1}{2}$ of a unit of water vapor, as in Figure 7.7b. Under these circumstances, you could say that the specific humidity is $\frac{1}{2}$ a unit. In this simplified example, you can conclude that the volume of air actually has $\frac{1}{2}$ a unit but has a potential of 2 units, so the air contains $\frac{1}{2}$ of the water vapor that it can *potentially* hold (see Figure 7.7c). In other words, *the relative humidity is 25%.*

Humidity is usually measured with an instrument known as a *sling psychrometer* (Figure 7.8). This measurement system uses two thermometers—a wet bulb and a dry bulb—mounted side by side on a small platform. The wet bulb is covered with a cotton sleeve and wetted, whereas the dry bulb remains dry. The two thermometers are then whirled in the air. The wet bulb cools through the process of evaporation—that is, if the air is not already saturated. If the air is saturated then no cooling takes place, because no water is evaporated from the wet cotton sleeve, and the thermometers show the same temperature. On the other hand, with progressively drier air the wet-bulb thermometer becomes progressively cooler relative to the dry-bulb thermometer. In these conditions, the wet-

bulb and dry-bulb temperatures are set on a scale and the relative humidity is calculated.

Other devices can measure relative humidity directly. One such system absorbs an amount of water vapor that depends on the relative humidity.

No matter how relative humidity is measured, when the relative humidity is low, the air is relatively dry. When relative humidity is high, the air is relatively moist and close to being saturated. Remember that warm air can hold more water vapor than cold air because warm air is rising and expanding and has room for more water molecules. Cold air, on the other hand, is descending and contracting, so it has less room for molecules of water.

You can see how specific humidity changes with air temperature by examining Figure 7.9a, which shows the geographic relationship of specific humidity to latitude in the middle atmosphere during January 2004. Specific humidity is greater in the lower latitudes where the temperatures are generally warm. In contrast, the specific humidity is generally low at the higher latitudes because air temperature is relatively cold.

A common mistake that many students make is to assume that the relationship illustrated in Figure 7.9a means that relative humidity varies in the same geographic pattern, with relative humidity being higher at lower latitudes than at high latitudes. While this pattern may sometimes follow, it isn't necessarily so. In fact, the relative humidity at higher latitudes is higher, on average, than in warmer ones. Figure 7.9b shows the nature of this geographic distribution. Why do you suppose this inverse relationship exists? The reason is that cold air can hold less vapor than warm air. So even though the high latitudes

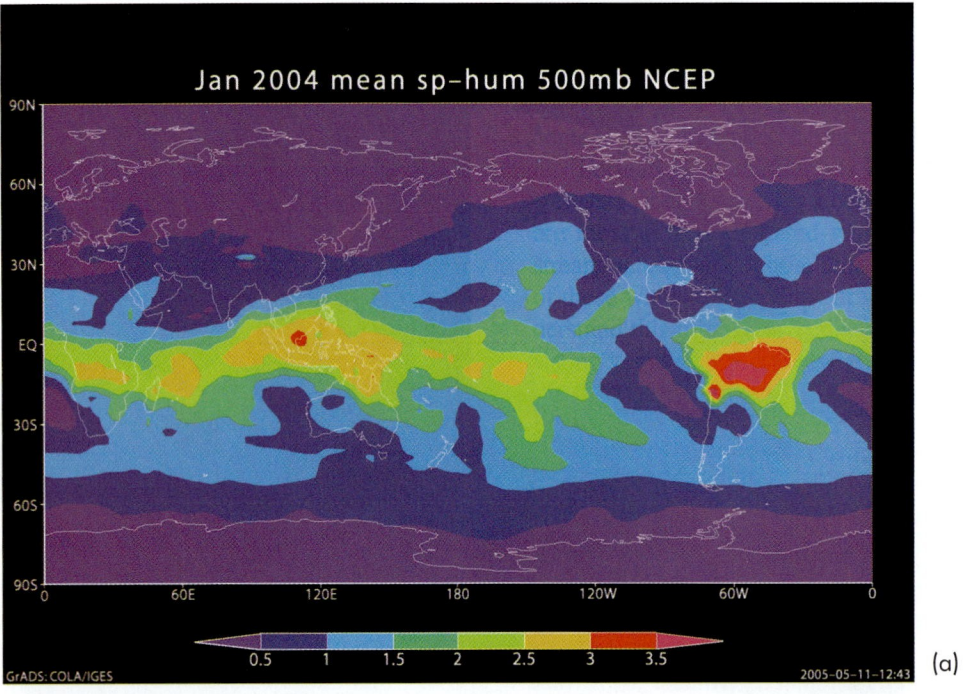

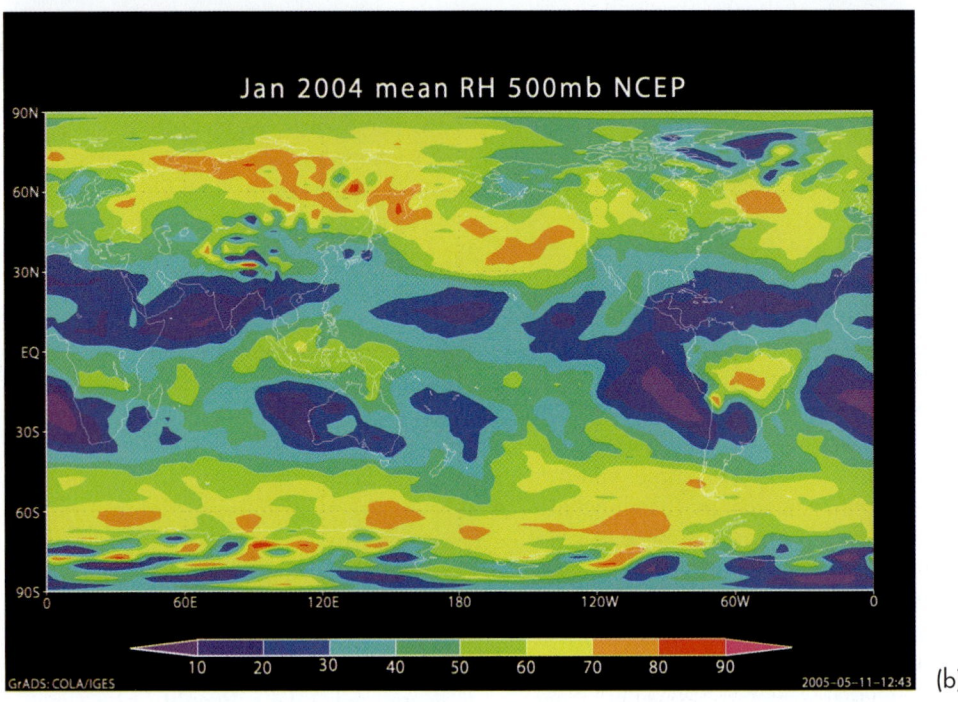

Figure 7.9 Humidity maps. (a) A direct correlation appears between specific humidity (in cm) and latitude in the lower atmosphere (as reflected at the 500-mb pressure surface). Reds and yellows indicate high values of average specific humidity, whereas greens, blues, and purples are progressively lower levels of specific humidity. (b) Higher latitudes generally have higher relative humidity, as indicated by the orange and yellow shades. Greens, blues, and purples represent areas with lower amounts of relative humidity.

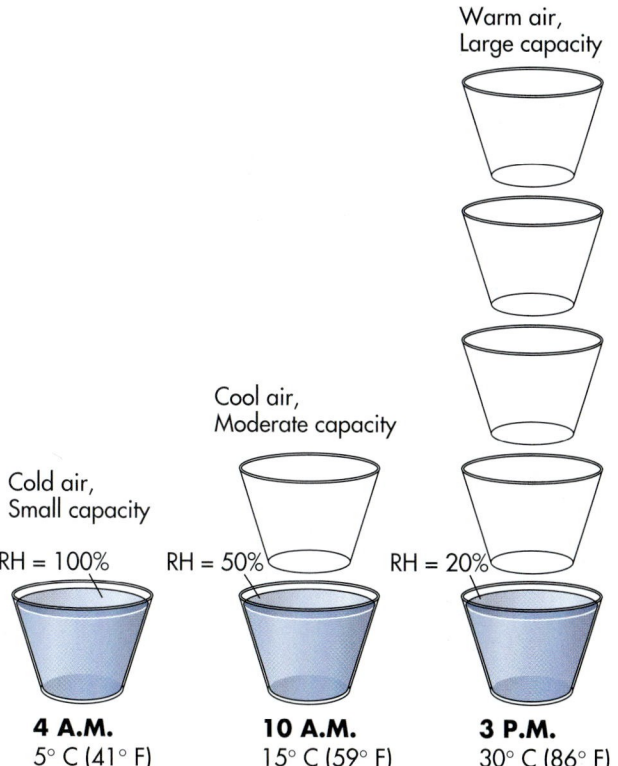

Warm air,
Large capacity

Cool air,
Moderate capacity

Cold air,
Small capacity

RH = 100%   RH = 50%   RH = 20%

4 A.M.   **10 A.M.**   **3 P.M.**
5° C (41° F)   15° C (59° F)   30° C (86° F)

Figure 7.10. Relationship between air temperature, time of day, and relative humidity. Although specific humidity does not change in this example, maximum humidity increases while relative humidity decreases.

Figure 7.11 Formation of dew. Water condenses on grass when nighttime temperatures cool to the dew-point temperature.

generally contain less vapor, the inherently colder air is proportionately closer to saturation than the warmer air at lower latitudes. Thus, the relative humidity is usually greater at high latitudes compared to low latitudes.

The same kinds of changes in relative humidity associated with latitude can also be seen over a typical day at any location that experiences strong diurnal changes in temperature; in other words, changes that occur over the course of a day and night. Figure 7.10 shows air temperature warming over the course of the day from 4:00 A.M. to 3:00 P.M. As the warming process continues, the maximum humidity increases because the capacity to hold vapor becomes greater. In contrast, the specific humidity, represented by the blue shading, is not changing between the morning and afternoon hours. What happens to relative humidity as the day progresses? It gradually decreases, as the ratio of specific humidity to maximum humidity becomes less. If you lived in a place where such a daily temperature/humidity change occurred, you would notice that the air might become slightly less sticky as it warmed over the course of the day.

Dew-Point Temperature

As shown in Figure 7.10, the relative humidity of a hypothetical air mass decreased over the course of an imaginary day because the temperature warmed from 5° C (41° F) to 30° C (86° F), creating a larger capacity to hold water vapor in the afternoon. This decrease in relative humidity occurred despite that fact that the specific humidity *did not change*. Assuming that the air mass would cool over the course of the evening and early morning back to 5° C (41° F), without any change in specific humidity, what would the relative humidity be? To calculate this value, simply work backward in the process and watch how the capacity of the air mass to hold moisture decreases until the air mass is saturated. In this particular example, if the air cooled to 15° C (59° F), the relative humidity would increase to 50%. If the air continued to cool to 5° C (41° F), then the relative humidity would rise to 100%.

Another way to look at this example is to say that the **dew-point temperature** of this air mass is 5° C (41° F). The dew-point temperature, therefore, is an important designation and is defined as the temperature at which a mass of air becomes saturated. The term *dew-point* originates from the formation of dew on grass (Figure 7.11). This occurs when air at ground level cools to the dew-point temperature at night and water condenses on blades of grass. As the year progresses, dew typically forms during the early fall and latter part of the spring, when nighttime temperatures cool sufficiently for condensation to occur at ground level, but not so much that water freezes as it does in winter. Dew is less likely in the summer months because nighttime temper-

Dew-point temperature   *The temperature at which condensation occurs in a definable body of air.*

AMAZING PLACES

DEATH VALLEY

Do you feel physically comfortable right now? Your comfort depends on several factors, including temperature, humidity, and wind speed that are associated with the Heat Index discussed in Chapter 6. The most uncomfortable location ever recorded in the U.S. is Death Valley, California. In July 1966, the temperature there was 119° F with 31% relative humidity, and in August 1970, it was 117° F and 37% relative humidity. On both of those days, the temperature felt like it was about 64° C (148° F)!

Death Valley is the hottest and driest place in North America, with temperatures reaching 57° C (134° F), only slightly below the highest temperatures ever recorded, and going without any rain at all for years at a time. The elevation at the floor of the valley is 86 m (282 ft) below sea level, making it the lowest place in North America as well. Death Valley's reputation for hot weather, however, has not stopped it from becoming a National Park and tourist attraction, noted for striking desert scenery and wildlife that has adapted to the severe conditions.

atures remain relatively high and the dew-point temperature isn't reached at ground level.

Higher in the atmosphere the dew-point temperature is significant because it marks the temperature at which water begins to condense into liquid form as clouds, fog, and ultimately various forms of precipitation. This condensation occurs because the air mass can no longer hold any more water vapor, either because the air cools to the dew-point temperature or because more water vapor is added to the air mass through evaporation of surface water. Whereas the maximum humidity of an air mass is a function of its temperature, the dew-point temperature depends upon the specific humidity, which is another way of referring to how much water vapor is actually in the air.

You can see the trend in dew-point temperature by re-examining the saturation curve in Figure 7.6. This time, consider first the humidity values on the vertical (y) axis, then the temperature data on the horizontal (x) axis. In this fashion, you are simply reversing the association you made earlier when you determined the maximum humidity at certain temperatures. For example, if the specific humidity of the air is 2 g/kg, then the dew-point temperature is $-10°$ C ($14°$ F). If the specific humidity increases to 15 g/kg or 26 g/kg, then the dew-point temperature also rises, in this case, to $20°$ C ($68°$ F) and $30°$ C ($86°$ F), respectively.

In an effort to integrate all of the humidity concepts into a coherent mathematical model that explains how precipitation occurs, let's look at some examples in Figure 7.12. Let's first consider an air mass with a temperature of $20°$ C ($68°$ F), which you can see in Figure 7.12a. At this temperature, the maximum humidity is 17.3 g/kg. That is, the air mass has the *potential* to contain 17.3 g/kg of water vapor. If the specific humidity is also 17.3 g/kg, that means that the parcel is saturated, the relative humidity is 100%, and the dew-point temperature is $20°$ C

($68°$ F). On the other hand, if the specific humidity is only 9.4 g/kg (Figure 7.12b) then the relative humidity is about 54% and the dew-point temperature is approximately $10°$ C ($50°$ F). If the air temperature were to increase without a change in specific humidity, the relative humidity would further decrease.

How could we get the relative humidity of the drier air mass to increase? This kind of increase can happen in one of two ways. One way would be to somehow increase the amount of water vapor in the air mass, perhaps through the process of evaporation at a nearby ocean source. For example, if another 4 g/kg of vapor was added to the air mass through this process (Figure 7.12c), then the specific humidity would be 13.4 g/kg and the relative humidity would increase to 78%. At the same time, the dew-point temperature would increase to about $16°$ C ($61°$ F). The second way would be to lower the temperature (Figure 7.12d), let's say to $15°$ C ($59°$ F). If this cooling process occurred at the same time that specific humidity remained 9.4 g/kg, then the relative humidity would increase to about 74% because the air could hold less moisture; in other words, the maximum humidity would be less.

An important reason why understanding atmospheric humidity is relevant to your life is that it allows meteorologists to predict when and how precipitation will occur. As far as you're concerned, it has relevance because weather conditions can directly influence your day. You may have encountered such a discussion on your local weather station, when the weatherperson talks about how much moisture is in the air at a given time, especially when thunderstorms are a possibility. When these scenarios occur, your local weatherperson may say that the dew-point temperatures in your region are "currently high." This is simply another way of saying that the specific humidity is high and the air is laden with water vapor that

ATMOSPHERIC HUMIDITY

Although the concept of humidity is fundamentally a simple one, students often have difficulty with it because the terms can seem to be interchangeable. Another problem is that it's sometimes difficult to understand these processes when reading about them in an abstract form such as a textbook. Like many concepts in physical geography, humidity is particularly well suited for simulation to enhance understanding. In this context, go to the *GeoDiscoveries* website and select the module **Atmospheric Humidity**. This module is designed so you can interact with the concept of humidity and

assess how it varies depending upon environmental circumstances. You will be presented with a variety of scenarios in which you can manipulate the amount of maximum humidity and specific humidity in a hypothetical air mass such that changes in the relative humidity and dew-point temperature occur. As you work your way through these scenarios, remember that they are representative of an almost infinite number of combinations. When you complete the simulation, be sure to answer the questions at the end of the module to test your understanding of this important concept.

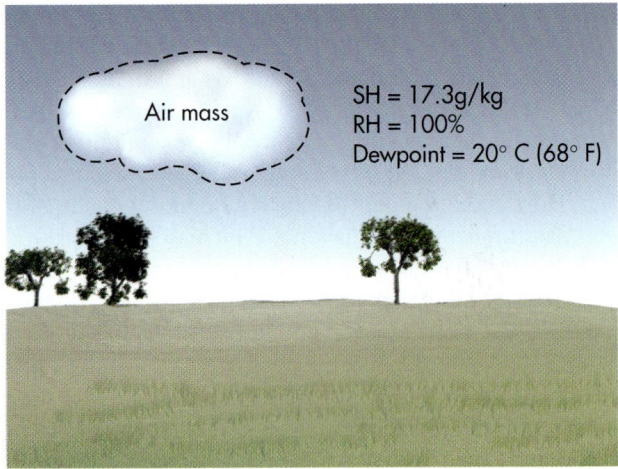

(a) Air mass temperature = 20° C (68° F); MH = 17.3 g/kg

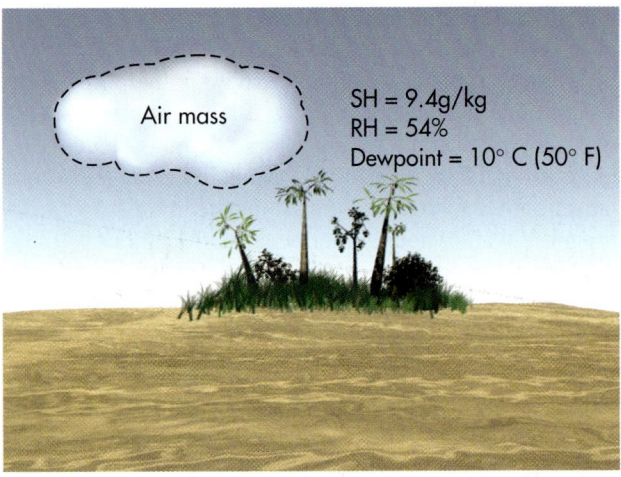

(b) Air mass temperature = 20° C (68° F); MH = 17.3 g/kg

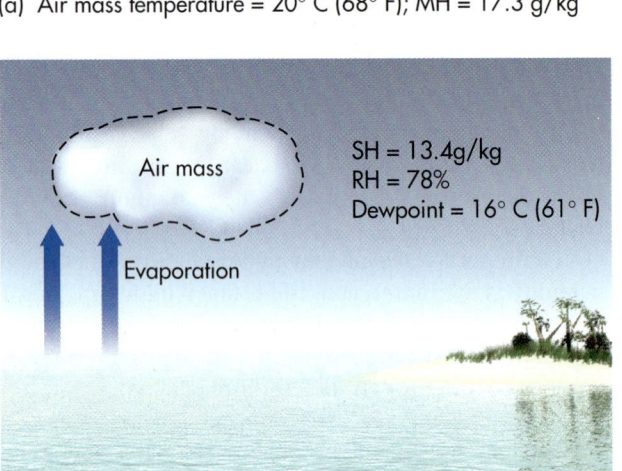

(c) Air mass temperature = 20° C (68° F); MH = 17.3 g/kg

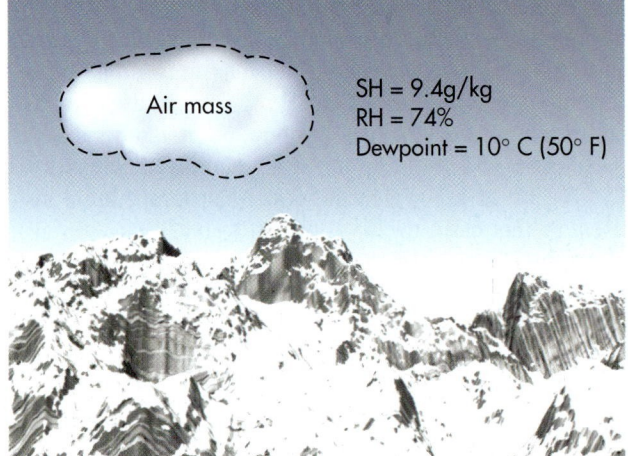

(d) Air mass temperature = 15° C (59° F); MH = 12.83 g/kg

Figure 7.12 Humidity examples. (a) Hypothetical air mass with air temperature of 20° C (68° F) and relative humidity of 100%. (b) Same air mass, now with a relative humidity of 54%. (c) Same air mass, but now with increased relative humidity due to higher specific humidity. (d) Same air mass with a cooled air temperature of 15° C (59° F) and relative humidity of 74%.

may be released as moisture in a storm should the right atmospheric conditions exist.

At another level, understanding atmospheric humidity is critical for agriculture because it is indirectly associated with crop production. Farmers are keenly aware of atmospheric moisture conditions because their crops depend on the appropriate amount of water, which, in turn, is relevant to you because the price of food is directly related to supply. If an extended period of drought occurs in a major agricultural region, production yields decrease dramatically and the cost of food increases.

Atmospheric humidity conditions also have ramifications for human comfort and health, especially during the warm, humid days of summer when dew-point tempera-tures are elevated. When the Heat Index is high, most people are very uncomfortable working outside for any length of time. The reason for this discomfort is that our bodies do not regulate internal temperature effectively when the air is warm and muggy. The primary way that we regulate internal temperature is through the production of sweat. When the air is dry, sweat evaporates and body heat is removed in a manner consistent with the earlier discussion in this chapter about latent heat transfers. On warm/humid days, however, sweat evaporates less efficiently and our core temperature increases. If the sweat-producing activity continues, body temperatures can rise to the point that heat stroke occurs and the temperature regulation system shuts down completely.

1. Three kinds of humidity exist: maximum, specific, and relative. Maximum humidity refers to how much vapor a parcel of air can hold. This variable depends upon temperature, with warm air having a higher maximum humidity than colder air. Specific humidity measures how much water vapor is in the air. Relative humidity is the ratio of specific humidity to maximum humidity.

2. When the relative humidity is 100%, the air is saturated and can hold no more water vapor.

3. Although the specific humidity may not change on any given day, the relative humidity can change dramatically due to rising and falling temperature. In this case, the warmer the temperature, the less the relative humidity.

4. The dew-point temperature is the temperature at which a mass of air becomes saturated. This is the temperature at which condensation occurs in that air mass. As specific humidity increases, so does dew-point temperature.

5. Once the dew-point temperature is reached, the specific humidity begins to decrease as water vapor condenses to liquid form. This decrease in specific humidity continues as long as the process of cooling continues.

Evaporation

Where does water vapor in the atmosphere come from? Liquid water is transformed into water vapor through two processes, evaporation and transpiration. As described earlier in the chapter, the term *evaporation* is used when water is lost from a surface such as soil, water, or pavement. Liquid water is also transformed into vapor when it flows through leaf pores in plants to the atmosphere. This process of water loss is called **transpiration** and can result in the transformation of tremendous amounts of water. In tropical areas, for example, some trees can transpire over 300 liters (~ 80 gallons) of water a day! You can indirectly see the process of transpiration by noting the rigid upright posture of plants. This posture is maintained by water flowing upward through the plant by capillary action toward the leaf surface. A plant wilts when the

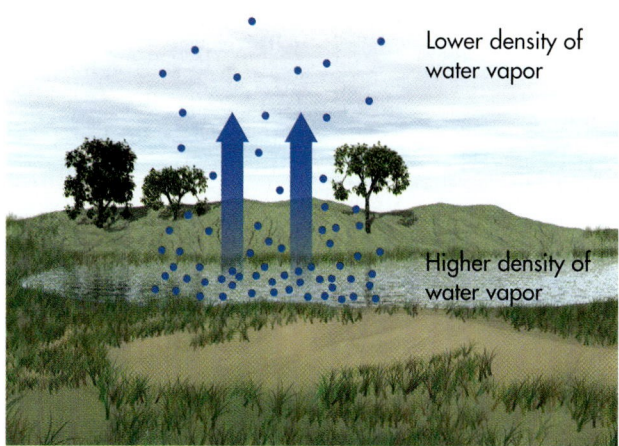

Lower density of water vapor

Higher density of water vapor

Figure 7.13 Water vapor gradient. Water vapor flows from areas of high-density toward areas of low-density vapor content. This flow creates room for additional water to be evaporated from the source area.

amount of water in the soil reduces and therefore less flows upward through the plant.

When geographers consider the humidity characteristics of a particular region, they typically combine the processes of evaporation and transpiration into the singular concept of **evapotranspiration.** Evapotranspiration rates are dependent upon many factors, including:

- **Net Radiation:** Heating of plants or ground surface. Evaporation rates are higher when a lot of energy (sunlight) exists.

- **Air Temperature:** Warm air can hold more moisture than cold air.

- **Relative Humidity:** How much moisture the air is already holding relative to its capacity to contain water vapor. Evaporation rates are higher when the air is dry because plenty of room exists for more water vapor molecules in the air.

We have seen that in the atmosphere, air flows along a pressure gradient from high pressure to low pressure. Similarly, water vapor flows along a vapor pressure gradient from areas of high vapor pressure to areas of low vapor pressure (Figure 7.13). Close to the water surface, the air contains more molecules of water vapor. As you move higher above the water, however, the air contains fewer water vapor molecules. Water vapor flows upward from the areas with more water vapor to the areas with less water vapor.

Transpiration   *The passage of water from leaf pores to the atmosphere.*

Evapotranspiration   *The combined processes of evaporation and transpiration.*

No wind

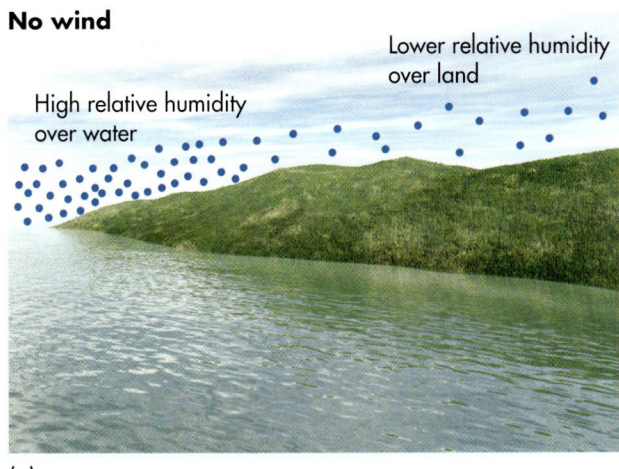

(a)

With wind

(b)

Figure 7.14 Wind and evaporation. (a) When the wind is calm, little evaporation occurs because the relative humidity of the air mass above the water is high. (b) More evaporation takes place during windy days because water vapor continually flows away from the moisture source.

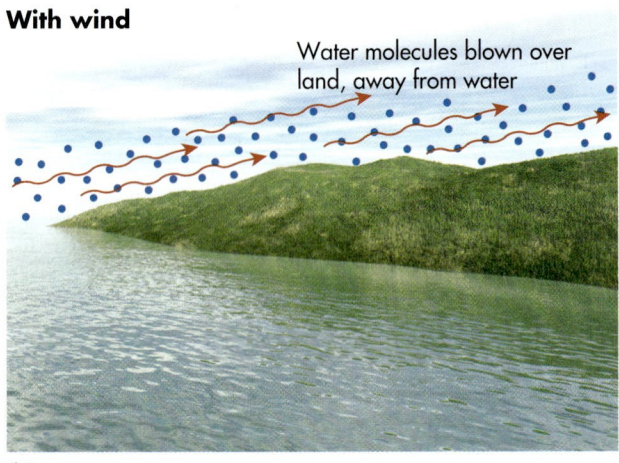

A final factor that influences evapotranspiration rates is wind speed. Wind moves moist air away from the source of the water vapor. Evaporation rates are higher in windy conditions when the air containing water vapor molecules can be moved away from the source, making room for more water molecules to move into the air. Figure 7.14 shows an example of this concept. This is also why fans are useful cooling devices; they create a wind to increase evaporation.

Adiabatic Processes

When water evaporates into vapor it moves into the atmosphere where it can continue to exist in the gas phase. It can, however, recondense into liquid water to form clouds where it is temporarily stored until it either vaporizes again or falls to the Earth as precipitation. This section deals with the processes through which condensation occurs, with the ultimate goal of providing the foundation for understanding the precipitation process. Although many of the terms in this section may seem like jargon to you at first, they are everyday terms to many geographers because they are used when we monitor atmospheric conditions and predict the weather on any given day.

From the discussion about relative humidity, you might surmise that condensation must be somehow related to the cooling of air to the dew-point temperature. This is a correct assumption, but, as you might also imagine, the process is more complex than it first appears. After all, nighttime cooling can cause condensation of near-surface water vapor if the saturation point of the air is reached immediately above the ground, resulting in dew or even frost if vapor changes phase directly to ice crystals. Neither of these processes, however, are associated with precipitation.

Precipitation occurs only when a substantial parcel of air experiences a steady drop in temperature below the dew-point temperature, which can only occur if the air parcel is somehow lifted enough for it to cool adiabatically. For purposes of discussion, think of a parcel of air as having a diameter of about 300 m (1000 ft) with uniform humidity and temperature characteristics.

Recall from Chapter 6 that the dominant way in which air rises or descends is through changes in atmospheric pressure. When this kind of movement occurs, the air either expands as it rises or compresses as it descends (Figure 7.15). Of particular relevance to this discussion is that air cools as it rises and expands and warms when it descends and compresses. These kinds of temperature changes are called **adiabatic processes** because they result solely from pressure fluctuations.

The Dry Adiabatic Lapse Rate

In most cases that involve a parcel of surface air, the air is not saturated and the relative humidity is thus less than 100% before it begins to rise. When these conditions occur, the air temperature decreases at what is called the **dry adiabatic lapse rate (DAR)** as the air starts to rise. This rate is not to be confused with the *environmental lapse*

Adiabatic processes *Changes in temperature that occur due to variations in air pressure.*

Dry adiabatic lapse rate (DAR) *The rate at which an unsaturated body of air cools while lifting or warms while descending. This rate is 10° C per 1000 m (5.5° F per 1000 ft).*

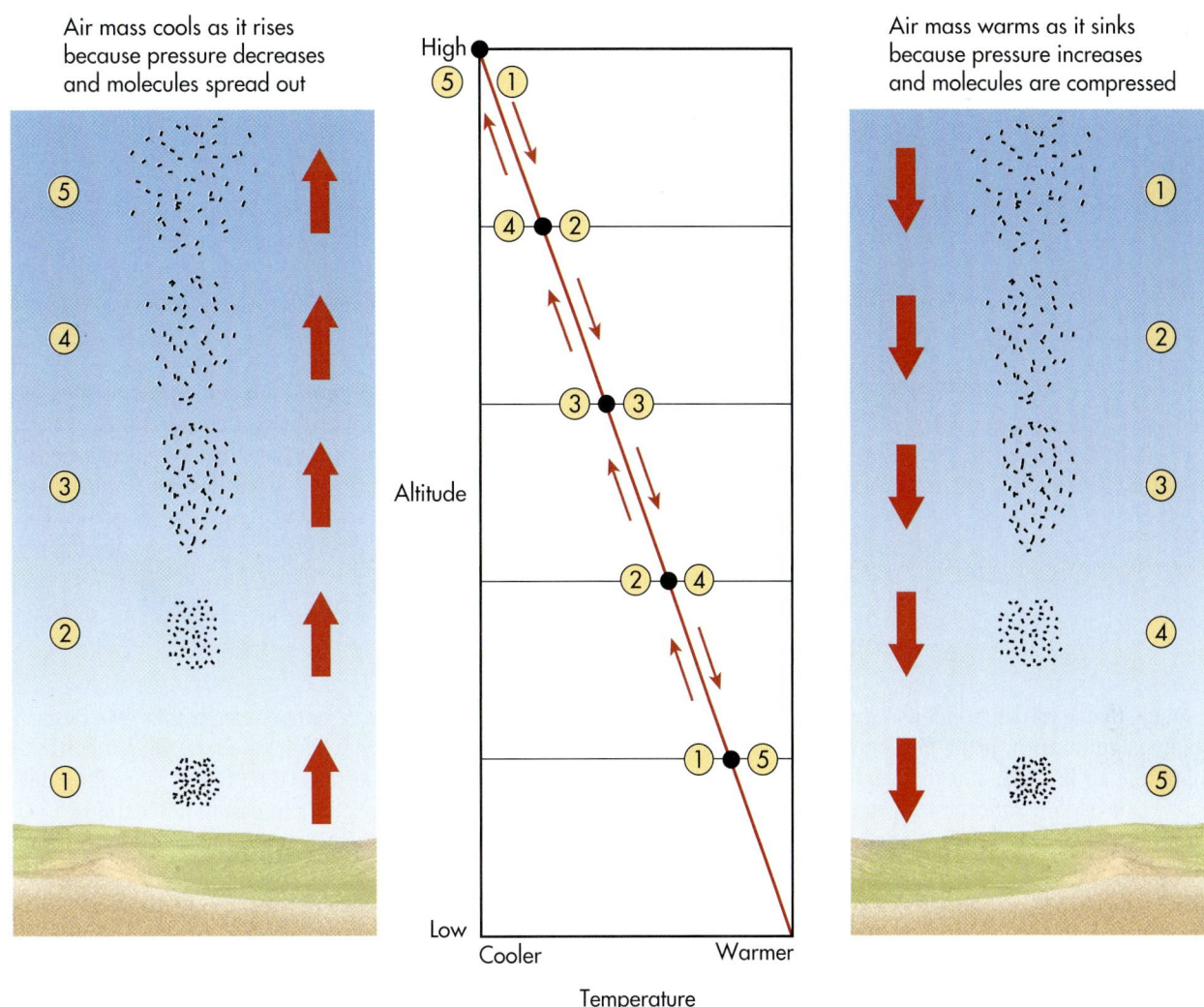

Air mass cools as it rises because pressure decreases and molecules spread out

Air mass warms as it sinks because pressure increases and molecules are compressed

High

Altitude

Low

Cooler   Warmer

Temperature

Figure 7.15 Adiabatic heating and cooling. Air cools as it rises because the pressure drops and air molecules disperse. When air descends, it warms adiabatically because the air molecules are compressed by higher pressure.

rate that we discussed in Chapter 5. The environmental lapse rate refers to the average change in temperature of *still* air with altitude, which can change with time or place. In contrast, the DAR is always constant and is applied to a parcel of rising air according to physical laws. The DAR has a value of about 10° C per 1000 m (5.5° F per 1000 ft) of vertical lift and is applied *only* when the air mass is not saturated. *This is a very important rule to remember.* You can see this rate of change if you examine Figure 7.16 and focus on the area at the bottom of the graph, which indicates altitudes of less than 1000 meters.

The DAR is relevant not only to lifting air masses that are not saturated, but descending ones as well. In the case of a descending air mass, the air *warms* at the DAR. That is, for every 1000 m that the air descends, it warms 10° C (5.5° F per 1000 ft) due to compression of the air under pressure.

The Wet Adiabatic Lapse Rate

The DAR applies when air is not saturated and the relative humidity is less than 100%. How does temperature change once the dew-point temperature is reached, saturation occurs, and the relative humidity is 100%? The altitude at which the saturation point is reached is known as the **level of condensation,** which varies depending upon the particular temperature and humidity characteristics in any given air parcel. Air parcels that have relatively high humidity when they begin to lift reach the level of condensation at a lower elevation than air parcels that are comparatively dry.

Level of condensation *The altitude at which water changes from the vapor to liquid phase.*

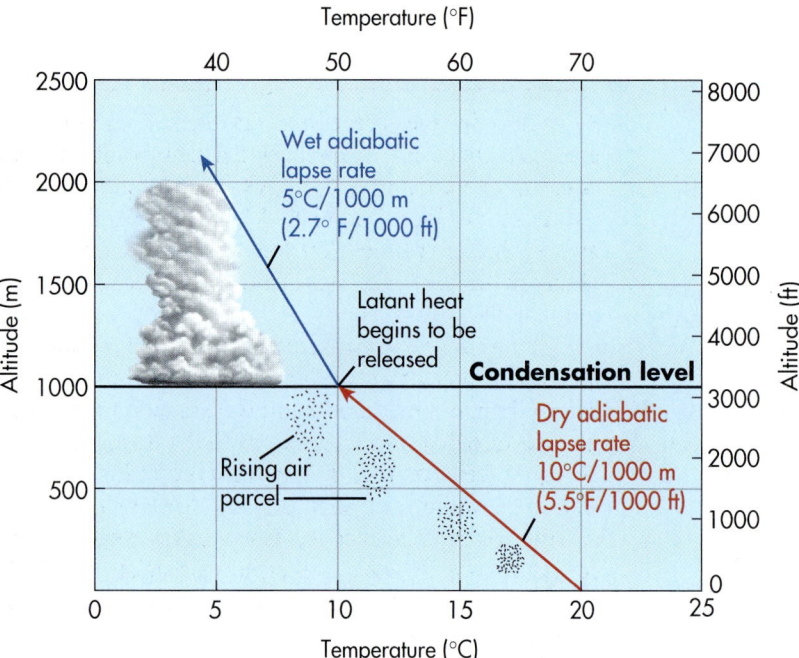

Wet adiabatic lapse rate 5°C/1000 m (2.7° F/1000 ft)

Latant heat begins to be released **Condensation level**

Dry adiabatic lapse rate 10°C/1000 m (5.5°F/1000 ft)

Rising air parcel

Figure 7.16 Adiabatic cooling in a hypothetical air parcel. The air cools at the dry adiabatic lapse rate (DAR) as it rises until it reaches the level of condensation. At that point the rate of temperature change switches to the wet adiabatic lapse rate (WAR) due to the release of latent heat.

Once the level of condensation is reached, water droplets begin to form if the air continues to cool. In this fashion, the air is analogous to a saturated sponge that releases water as you squeeze it at the same time that you lift it. Give it a try. All you have to do is take a sponge and let it absorb a small amount of water in the sink. Starting at the surface of the sink, slowly raise the sponge and squeeze it slowly as you lift it. At some point, the sponge will begin to release water. After you reach this point, start over but let the sponge absorb more water before you begin. Raise the sponge as before and try to exert the same amount of force as you did previously. You should notice that the sponge begins to release water at a lower elevation than before. This occurs because the sponge was closer to being saturated the second time you ran this simple experiment. The atmosphere operates on much the same principle as this sponge test.

If the air parcel continues to rise and cool, increasing amounts of water vapor condense into liquid water. During this process, another principle comes into effect—the release of latent heat of condensation that was described earlier in this chapter. This heat energy has the effect of warming the air at the same time that it is cooling due to the reduction in atmospheric pressure and expansion. The cooling effect is stronger overall, however, so the overall temperature continues to decrease if the air continues to rise. Nevertheless, the rate of cooling in this saturated state is less than in the unsaturated state because of the release of latent heat.

Look at Figure 7.16 again and focus this time on the area above the condensation level to see how this change appears graphically. The cooling rate of saturated air is called the **wet adiabatic lapse rate (WAR).** Although the WAR varies with moisture content and temperature, the average rate is about 5° C per 1000 m (2.7° F per 1000 ft) and this value will be used hereafter in this text.

Wet adiabatic lapse rate (WAR) *The rate at which a saturated body of air cools as it lifts. The average rate is about 5° C per 1000 m (2.7° F per 1000 ft).*

GEODISCOVERIES   WWW.WILEY.COM/COLLEGE/ARBOGAST

ADIABATIC PROCESSES

In an effort to integrate these concepts into an animated format, go to the *GeoDiscoveries* website and select the module **Adiabatic Processes**. This module reviews all of the concepts that have been discussed in this section of the text. As you watch the animation, pay particularly close attention to when the DAR and WAR are used. Also, watch what happens to the air parcel as it descends and warms at the DAR. When this process occurs, the effect is to produce a warmer air mass at the surface than the original parcel of air. After you complete the animation, be sure to answer the questions at the end of the module to test your understanding of this concept.

KEY CONCEPTS TO REMEMBER ABOUT ADIABATIC PROCESSES

1. Adiabatic processes refer to the temperature changes that occur in an air parcel due solely to changes in air pressure. When air pressure increases, a parcel of air is compressed and warms adiabatically. In contrast, when air pressure decreases, a parcel of air expands and thus cools internally because air molecules are spaced farther apart.

2. The level of condensation is the elevation at which condensation occurs; that is, when the relative humidity is 100%.

3. Air cools at the dry adiabatic lapse rate (DAR) until condensation occurs. This rate is 10° C per 1000 m (5.5° F per 1000 ft). When air warms adiabatically, it always does so at the DAR because condensation is not occurring.

4. Once the level of condensation is reached, the air temperature begins to decrease at the wet adiabatic lapse rate (WAR), which is 5° C per 1000 m (2.7° F per 1000 ft). This lesser rate of cooling occurs because latent heat of condensation is released when water changes from vapor to liquid.

Cloud Formation and Classification

As noted in the preceding discussion on adiabatic processes, once the air reaches the dew-point temperature, through cooling at the DAR, water vapor begins to condense into liquid droplets. We see the results of this process through the development of clouds, with the base of any particular cloud representing the level of condensation in the atmosphere.

Clouds are visible masses of suspended, minute water droplets or ice crystals. Figure 7.17 shows the close relationship between atmospheric water vapor and the geographic distribution of clouds in the U.S. on April 24, 2005. Notice in part (a) that high concentrations of vapor (shown in white) are present in the western part of the country, as well as over the Great Lakes region. These concentrations of vapor are closely associated with the band of clouds in part (b) that generally cover the same geographic spaces.

Two conditions are necessary in order for clouds to form:

1. The air must be saturated. Air can become saturated two different ways: When air cools below the dew-point temperature or when more water vapor is added to the air.

2. There must be a substantial quantity of small particles, such as dust or pollutants, about which water

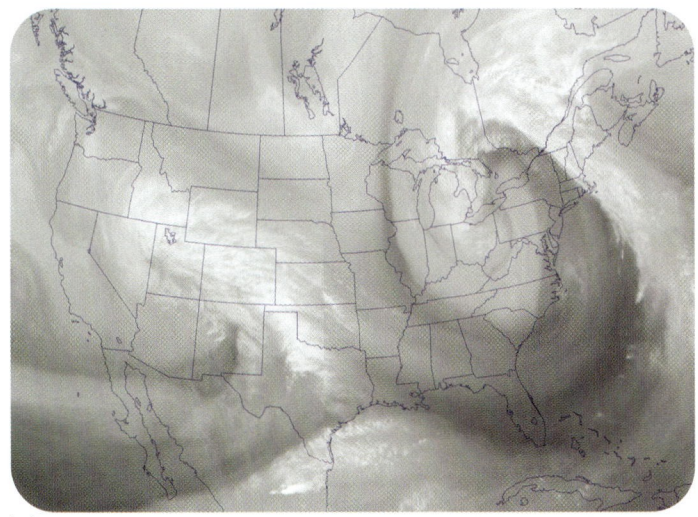

(a)

(b)

Figure 7.17 Satellite images showing atmospheric moisture on April 24, 2005. (a) GOES water vapor image. Notice the particularly high concentrations of water vapor (in white) in the west and northeast.

(b) GOES visible image. Note how the concentrations of clouds are closely associated with the vapor patterns in (a).

Figure 7.18 Types of fog. (a) An excellent example of radiation fog in the Blue Mountains of Australia. This fog developed when nighttime cold air collected in the valley bottom.

It was entirely gone by late morning because the air warmed above the dew-point temperature. (b) Sea fog at San Francisco, California. This type of fog develops when moisture-laden air flows over cold water, causing the air temperature to drop.

(a)

(b)

vapor can collect or condense. These particles are called **condensation nuclei.** Have you ever noticed that, if it rains after you washed your car, little circles of dust exist on your vehicle that resulted from the rain? The tiny particles of dust were suspended in the atmosphere prior to the rainfall and served as the nuclei for condensation to occur.

Fog

A direct correlation exists between atmospheric water vapor content and the density of cloud cover. As shown in Figure 7.17, for example, higher water vapor content generally results in more clouds. Although most clouds form and remain at some elevation above the Earth, they can come into direct contact with the surface in the form of fog. Fog forms when the air at low elevations is saturated with moisture.

Several different kinds of fog result from distinct processes. **Radiation fog** forms at night when a temperature inversion exists in the lower troposphere. Recall from Chapter 5 that temperature generally cools at a consistent rate (the environmental lapse rate) from the surface up into the higher levels of the troposphere. Occasionally, a **temperature inversion** develops whereby a body of cooler air lies beneath warmer air. If this cooler air reaches the dew-point temperature, radiation fog develops. This kind of scenario sometimes develops in deep valleys when cool air collects at the bottom of the valley (Figure 7.18a).

Another type of fog is **advection fog.** This type of fog develops when warm air flows over a cooler surface, such as snow or a body of water. When this process occurs, the warmer air cools and can reach the dew-point temperature.

A third type of fog is **sea fog,** which develops when cool marine air comes in direct contact with the colder ocean water. In the U.S., a place where sea fog frequently develops is along the coast of California, along which the cool California current flows (Figure 7.18b).

Condensation nuclei *Microscopic dust particles around which atmospheric water coalesces to form raindrops.*

Radiation fog *Fog that develops at night when a temperature inversion exists.*

Temperature inversion *A layer of the atmosphere in which the air temperature increases, rather than cools, with altitude.*

Advection fog *Fog that develops when warm air flows over cooler air.*

Sea fog *Fog that develops when cool, marine air comes in direct contact with colder ocean water.*

Water vapor is produced in the exhaust of jet engines, such as those on large airliners. As the vapor cools in the air behind the plane, it often forms condensation trails, or contrails. What conditions must be present for the contrail to form?

a) Air must be close to saturation.

b) Condensation nuclei (dust) must be present.

c) The plane must be above the condensation level of the air.

d) All of the above.

Regardless of the type of fog, you've probably noticed that fog typically dissipates or *burns off* during the day. This process occurs because the cool air in which the fog forms will warm above the dew-point temperature as the day progresses. Next time you experience morning fog, notice how this process works.

Cloud Classification

Like many other things in nature, it is possible to classify clouds based on particular characteristics, specifically with respect to altitude and form (Figure 7.19). The international cloud classification scheme recognizes three categories based on their form. **Cirrus clouds** are thin and wispy clouds composed of ice crystals rather than water droplets. **Cumulus clouds** consist of individual, puffy clouds with a flat, horizontal base. **Stratus clouds** consist of layer-like, grayish sheets that cover most or all of the sky. As you can see in Figure 7.19, these clouds occur in various combinations. For example, stratocumulus clouds have characteristics of both stratus and cumulus clouds because they are individual clouds that spread out more than regular cumulus clouds.

Clouds can be further classified based on their altitude, with three categories present that you can see in Figure 7.19. *High clouds* are generally found at altitudes greater than 6 km (20,000 ft) and generally consist of cirrus clouds because very little water vapor is present and the temperatures are cold. *Middle clouds* range between approximately 2 to 6 km (6500 to 20,000 ft) and include cumulus and stratus clouds that form during intervals of stable and changing weather, respectively. *Low clouds* typically occur below 2 km (6500 ft), usually in association with stratus and cumulus clouds. These low clouds are the source of most precipitation, with *nimbostratus clouds* affiliated with long-term rain or snow events and *cumulonimbus clouds* growing to great heights during short-term, severe storms.

With a basic understanding of clouds and their formation, it's possible to determine basic atmospheric conditions and to even predict weather a few days in advance. For example, if the sky is filled with individual puffy, cumulous clouds, it means that the atmosphere is fairly stable and fair weather will probably last for a little while. If you happen to notice that high-level cirrus clouds are becoming more numerous and dense, it probably means that a storm system is approaching. These clouds often give way to altocumulous clouds, which indicate that the storm system is closing in and that moisture is flowing into the region from hundreds of kilometers away. If the clouds begin to thicken and expand into nimbostratus clouds, it means that some form of precipitation will most likely occur soon.

Cirrus clouds *Thin, wispy clouds that develop high in the troposphere.*

Cumulus clouds *Individual puffy clouds that develop due to convection.*

Stratus clouds *Layered sheets of clouds that have a thick and dark appearance.*

Cirrocumulus: appears as white patches; usually formed from cirrus or cirrostratus torn by winds; indicates approaching surface winds

Cirrus: thin and whispy high-level clouds; known as mares' tails

Altocumulus: puffy with dark undersides; sometimes known as a mackeral sky (resembling fish scales); usually a sign of fair weather

Altostratus: thick sheets of pale gray; enough to obscure Sun or Moon; usually a sign of rain or snow

Cumulus: puffy, fleecy, fair-weather clouds, most common in summer, with blue sky visible between clouds; often evaporate at night

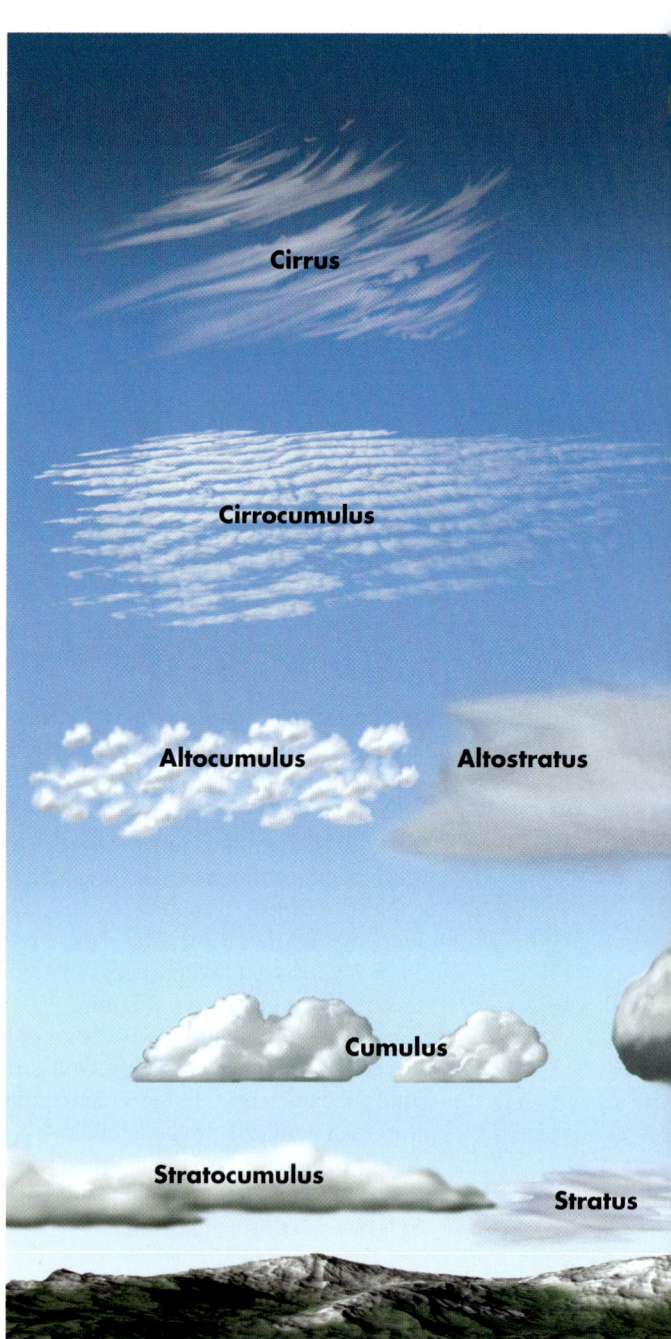

Cirrus

Cirrocumulus

Altocumulus

Altostratus

Cumulus

Stratocumulus

Stratus

Cirrostratus: thin, gauze-like sheets; usually gray and featureless but produce spectacular colors at sunsets; usually a sign of approaching rain

Cumulonimbus: thunderheads, extending to the tropopause or beyond, with typical anvil-shaped top; most storm clouds are cumulonimbus

Tropopause

Cirrostratus

High level
above 6 km
(above 20,000 feet)

Cumulonimbus

Middle level
2-6 km
(6,500 to 20,000 ft)

Low level
below 2 km
(0 to 6,500 ft)

Nimbostratus

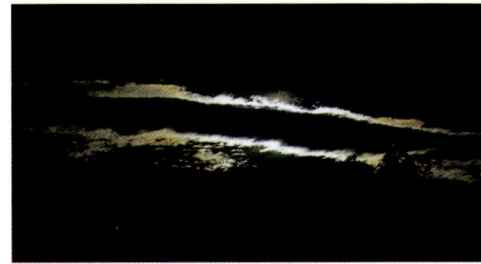

Stratus: thick, dull gray, low-lying layers; occurs as fog at ground level, often produces mist or drizzle

Nimbostratus: dark rain clouds; form near the ground but usually extend upward

Stratocumulus: forms in patches, sheets, or layers of white or gray; can produce overcast winter skies but clears quickly in summer

Figure 7.19 Cloud classification. Clouds are classified according to their height and form. Meteorologists recognize ten cloud types, which are variations on the three basic forms: cirrus, stratus, and cumulus.

UNUSUAL CLOUDS

Some cloud forms are typical of certain conditions, but are very rarely seen. An example is a lenticular cloud, so-called because of its lens-like appearance. Lenticular clouds are actually a variation of altocumulus clouds that form when an air parcel rises very rapidly to

pass over a mountain peak. If you happen to live east of the Rocky Mountains, you will rarely see this kind of cloud.

Another unusual type of cloud is called a noctilucent, or nacreous, cloud. These clouds form at high altitudes—so high, in fact, that they are in the mesosphere, not the tropo-sphere. They become visible when they reflect the light of the Sun, which has usually set below the horizon at ground level. The result is a bright cloud seen at night. Noctilucent clouds are most often seen at high latitudes during the summer, when the Sun is often not far below the horizon.

Precipitation

As we just discussed, condensation of water vapor leads to the formation of clouds. Cloud formation does not always lead to precipitation, as clouds usually appear in the sky in one form or another and it certainly doesn't rain or snow every time they do. This logically leads to the question, *why does precipitation occur?*

You probably already know that precipitation must involve some kind of a phase change from vapor to liquid or from liquid or vapor to solid. On the other hand, the nature of the process might be a mystery to you, as it is with most people. The bottom line is that precipitation does not occur until droplets of water are heavy enough for them to fall under the influence of gravity. Until then they remain suspended in the atmosphere.

Types of Precipitation

Droplets grow in size and become heavier by two processes: ice-crystal formation and coalescence of water droplets. Ice crystallization is the dominant process in most places outside the tropical regions and occurs be-cause most clouds or portions of clouds extend to altitudes where air temperatures are below the freezing point of liquid water. In these places, ice crystals occur in a matrix of water vapor and super-cooled water droplets that form around condensation nuclei (Figure 7.20a). The ice crystals "feed" off of the water droplets in that they cause droplets to rapidly evaporate when they are in close proximity to one another. Subsequently, the ice crystals absorb the water vapor that was released through the evaporation of the droplets, causing them to grow. Assuming the crystals grow sufficiently large in order to fall, they precipitate either as snowflakes or melt on their way to the surface and become raindrops that can grow further through the coalescence process.

In contrast to ice crystallization, which is a process of vapor attraction, raindrop coalescence is a process that merges small water droplets into large ones (Figure 7.20b). This process occurs mostly in tropical regions where the cloud tops do not reach sufficiently high for liquid water to freeze. Raindrop coalescence is the simple growth of water droplets through the persistent collision of small droplets in the upper atmosphere. One way in which

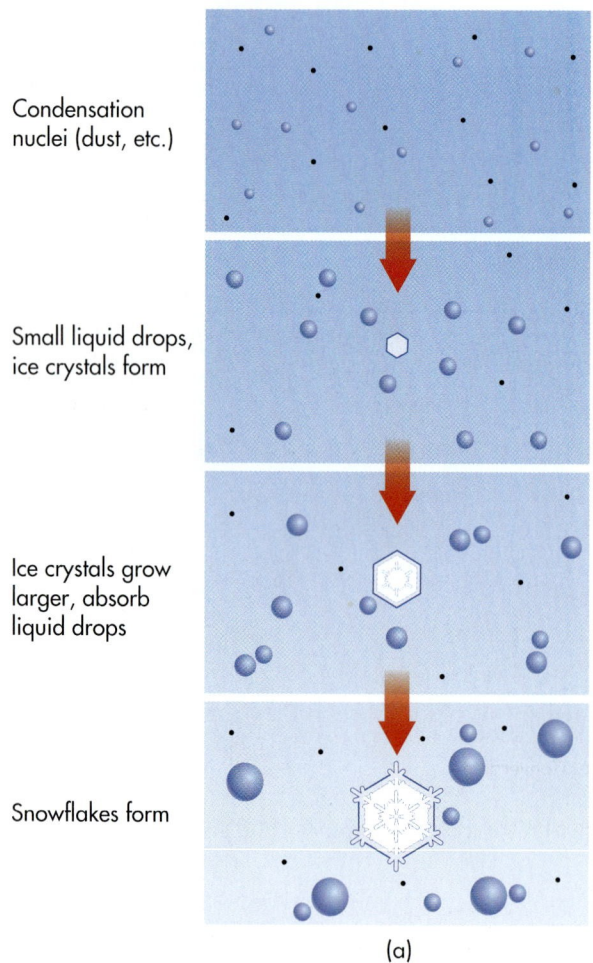

Condensation nuclei (dust, etc.)

Small liquid drops, ice crystals form

Ice crystals grow larger, absorb liquid drops

Snowflakes form

(a)

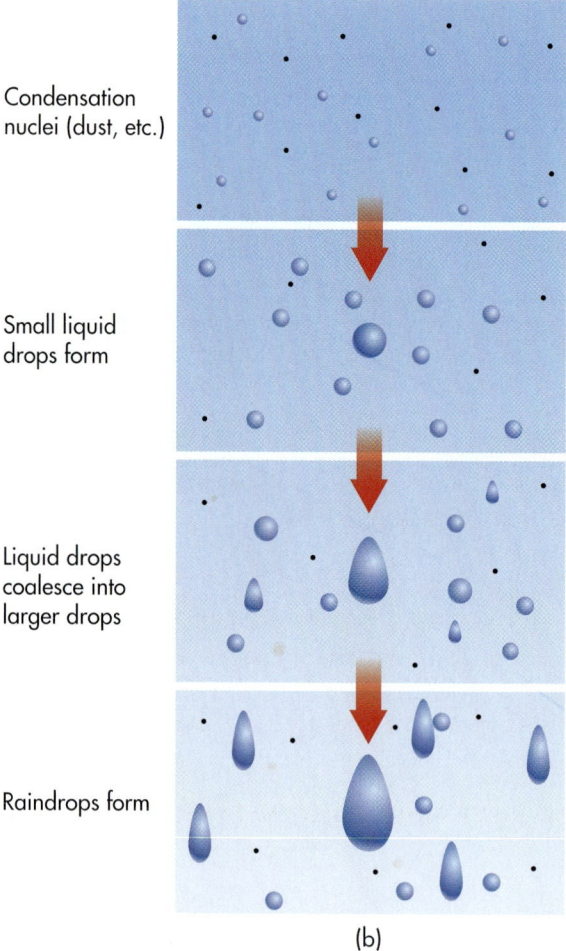

Condensation nuclei (dust, etc.)

Small liquid drops form

Liquid drops coalesce into larger drops

Raindrops form

(b)

Figure 7.20 Types of precipitation. (a) Snowflakes form by the crystallization of ice particles. (b) Raindrops form through coalescence of small water droplets.

this occurs is that large raindrops fall faster than smaller ones, allowing the larger raindrops to absorb the smaller droplets as they overtake them on their fall to Earth.

Precipitation can occur in several forms:

- Rain—large, unfrozen water droplets.
- Snow—ice crystals that do not melt before reaching the surface.
- Sleet—rain that freezes before hitting the ground.
- Freezing rain—rain that freezes on impact with ground that is below the freezing point of water.
- Hail—ice crystals that are continually pulled back up and grow in size during violent thunderstorms. We'll discuss this process more in Chapter 8.

Precipitation Processes

As just described, precipitation occurs when water drops or ice crystals become sufficiently large to fall under the force of gravity. The only way that this can happen is for an air mass to rise sufficiently high to condense large quantities of water. After all, the simple formation of clouds does not necessarily mean that precipitation will occur. How, then, does an air parcel rise to the point where precipitation does occur?

This kind of air lifting takes place in one of four ways (Figure 7.21). One way that air rises is through the process of **convectional uplift,** which results in distinct bubbles of air that rise through denser surrounding air. Another way is called **orographic uplift,** which occurs

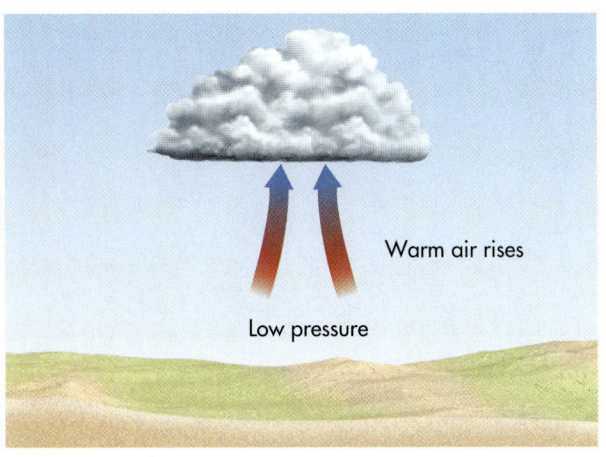

(a) Convectional

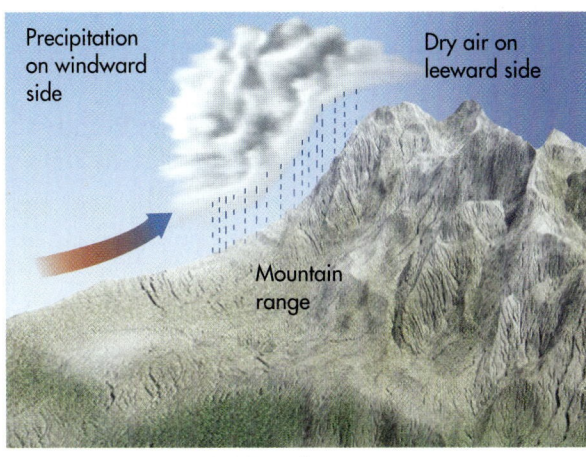

(b) Orographic

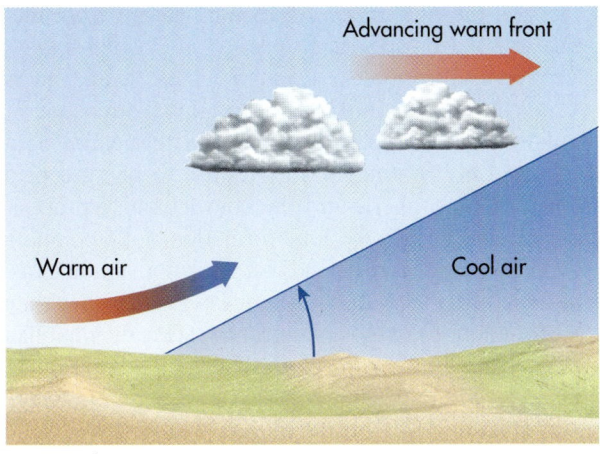

(c) Frontal

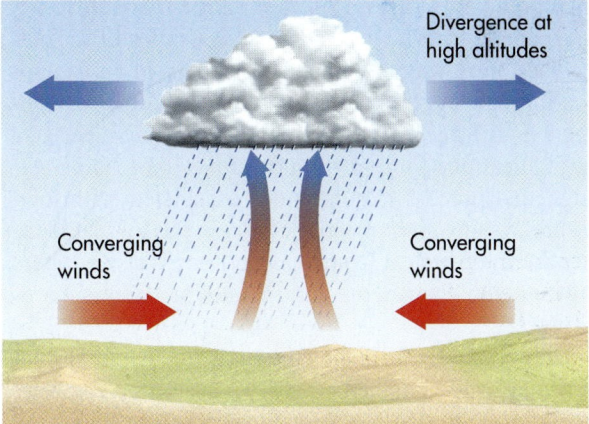

(d) Convergent

Figure 7.21 Four types of atmospheric uplift can cause precipitation. (a) convectional, (b) orographic (c) frontal, and (d) convergent.

Convectional uplift *Uplift of air that occurs when bubbles of warm air rise within an unstable body of air.*

Orographic uplift *Uplift that occurs when a flowing body of air encounters a mountain range.*

This satellite image shows precipitation over the Florida peninsula in January, 2002. East of Florida is the Atlantic Ocean; west is the Gulf of Mexico. Notice that a distinct pattern of rain is specifically associated with the Florida peninsula. Given the pattern that you see, which one of the following statements makes the most sense?

a) No convection is occurring over the Florida landmass.

b) Lots of condensation is occurring over the Gulf of Mexico.

c) Strong convection is occurring over the Atlantic Ocean.

d) The temperature of the Florida landmass is warmer than the surrounding oceans.

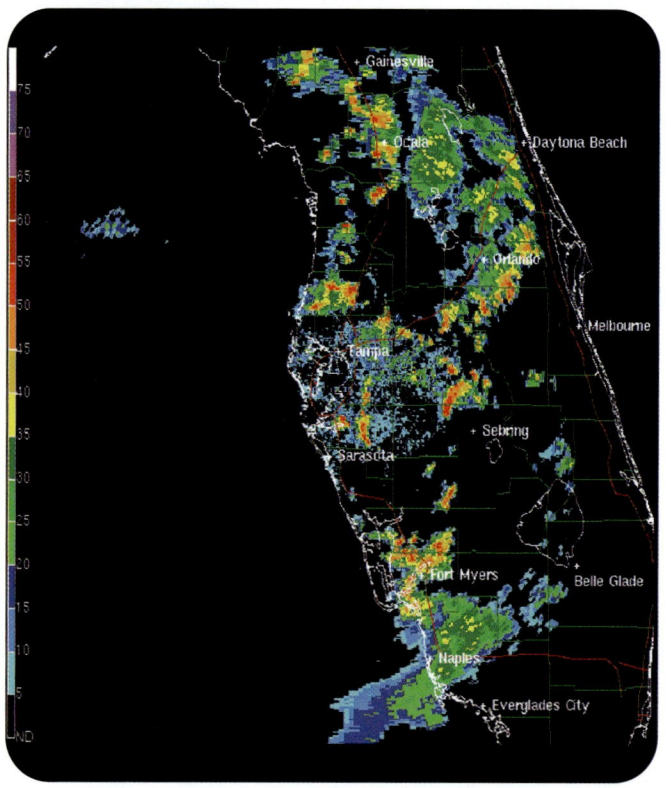

when air is forced to flow up and over mountains. A third way is through the collision of large air masses along frontal boundaries. This process is called **frontal uplift,** and typically occurs when contrasting bodies of air collide. The fourth way in which large parcels of air can uplift is through the process of **convergent uplift.** Convergent uplift occurs whenever bodies of air meet at a central location, forcing air upward at that point. Although this process is sometimes associated with low-pressure systems, it is most common in the low latitudes along the Intertropical Convergence Zone (ITCZ), where air from the Northern and Southern Hemispheres converges.

The remainder of this chapter will focus on convectional and orographic processes because they are relatively limited in a geographic extent and beautifully exemplify adiabatic processes in a way that's easy to understand. Frontal and cyclonic precipitation processes are more complex because they are associated with major weather systems that influence entire regions of the country in many different ways. We'll discuss these forms of precipitation in Chapter 8.

Convectional Uplift   As just described, one way in which the uplift of air can result in precipitation is through the process of convection. Recall from the discussion of atmospheric circulation in Chapter 6 that convection is the process through which heated air rises. Convection is the initial driver of global atmospheric circulation due to the unequal heating that exists between low and high latitudes. In addition to being a primary cause of atmospheric air flow, convection is a process that contributes greatly to cloud formation and precipitation.

Have you ever noticed that, on a hot summer day, the surface of an asphalt parking lot is much hotter than the surrounding grass-covered ground? This unequal heating creates a bubble of warm air over the parking lot. Warm air rises within the larger body of air, which is relatively still (Figure 7.22a). If the air cools sufficiently, the dew-point temperature will be reached, clouds will form, and precipitation may occur. A good indicator that convection is occurring on any given day is the presence of cumulus clouds, which are the individual puffy clouds pictured in Figure 7.22b. When you see these kinds of clouds, think of each as representing a bubble of air that lifted from the

Frontal uplift   *Uplift of air that occurs along the boundary of contrasting bodies of air.*

Convergent uplift   *Uplift of air that occurs when large bodies of air meet in a central location.*

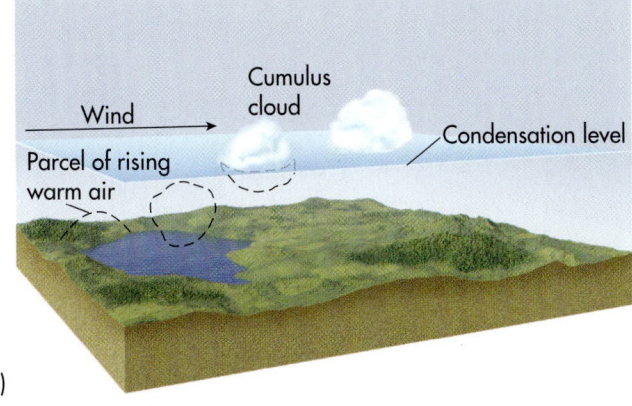

(a)

(b)

Figure 7.22 Convection. (a) When a warm bubble of air forms due to unequal heating at the surface, it rises within a body of relatively still air, causing the formation of cumulus clouds. (b) Cumulus clouds formed by convection in Brazil.

surface into the atmosphere, with the base of the clouds being the level of condensation. In most circumstances a cumulus cloud will float along for some distance downwind and then evaporate.

Although most cumulus clouds eventually dissipate before rain falls, some grow into cumulonimbus clouds that produce rain because convection continues until the water drops become sufficiently large to fall. If convection is very strong, a thunderstorm may develop. Whether or not this evolution occurs depends upon the overall stability of the air in which the convection bubble rises. At a fundamental level, air stability refers to the potential for convection within a body of air, with **stable air** being air in which strong convection cannot occur. In contrast, **unstable air** is air in which strong convection of air bubbles can occur. As we work through this part of the discussion refer to Figure 7.23, which illustrates air stability in graphical form.

Let's first consider an unstable air mass in which convection can occur to the point that clouds develop (Figure 7.23a). In other words, a bubble of warm air rises within a main body of relatively cool air that's generally stationary. This larger body of air cools with increasing altitude at the environmental lapse rate, which, on average, is 6.4° C/1000 m (3.5° F/1000 ft). Let's imagine, however, that the environmental lapse rate is 12° C/1000 m (6.6° F/1000 ft) in this particular body of air. If we further imagine that the temperature of this body of air is 32° C (89.6° F) at ground level, this lapse rate means that the temperature is 20° C (68° F) at 1000 m and 8° C (46.4° F) at 2000 m. In Figure 7.23a you can see this temperature change with elevation in the green graph line on the left as well as in the vertical list of blue temperature readings on the right.

Now, let's imagine that a bubble of relatively warm air develops over a large dark surface. This bubble has a ground temperature of 33° C (91.4° F), as you can see in the black temperature readings. Because it is warmer at ground level than the surrounding air, the bubble begins to convect, which means that it cools at the dry adiabatic lapse rate of 10° C/1000 m (5.5° F/1000 ft) due to its decreasing air pressure as it lifts. Given that this rate of cooling is less than the rate of cooling in the primary body of air that surrounds the bubble, the bubble continues to lift. Note, for example, that the convecting bubble (represented by the arrows in Figure 7.23a) has a temperature of 23° C (73.4° F) at 1000 m. If we assume that this bubble of air had a specific humidity of 20.6 g/kg at ground level, it would reach the level of condensation at 1000 m and a cloud would begin to form at that elevation.

Once condensation begins in our convection bubble, the air would then start to cool at the wet adiabatic lapse rate of 5° C/1000 m (2.7° F/1000 ft), which is still less of a rate than that observed in the surrounding body of air (that is cooling at a high environmental lapse rate). As a result, the discrepancy between the temperature of the two air bodies increases further and convection continues. This is why the air temperature of the convective bubble is 18° C (64.4° F) at 2000 m, whereas it is only 8° C (46.4° F) in the surrounding air.

Now, let's look at what happens in a stable body of air by examining Figure 7.23b. Let's imagine that in this scenario the environmental lapse rate of the main body of air is 5° C/1000 m (3.3° F/1000 ft) and that the temperature at ground level is 32° C (89.6° F), which is the same as in the previous example. In this case, the temperature of the main body of air would be 27° C (80.6° F) at 1000 m and 22° C (71.6° F) at 2000 m.

Consider what happens over the warm parking lot described in the example of unstable air. Let's imagine that the air immediately above this parking lot warms to

Stable air *A body of air that has a relatively low environmental lapse rate compared to potential uplifting air; thus, strong convection cannot occur.*

Unstable air *A body of air that has a relatively high environmental lapse rate compared to uplifting air within it; thus strong convection can occur.*

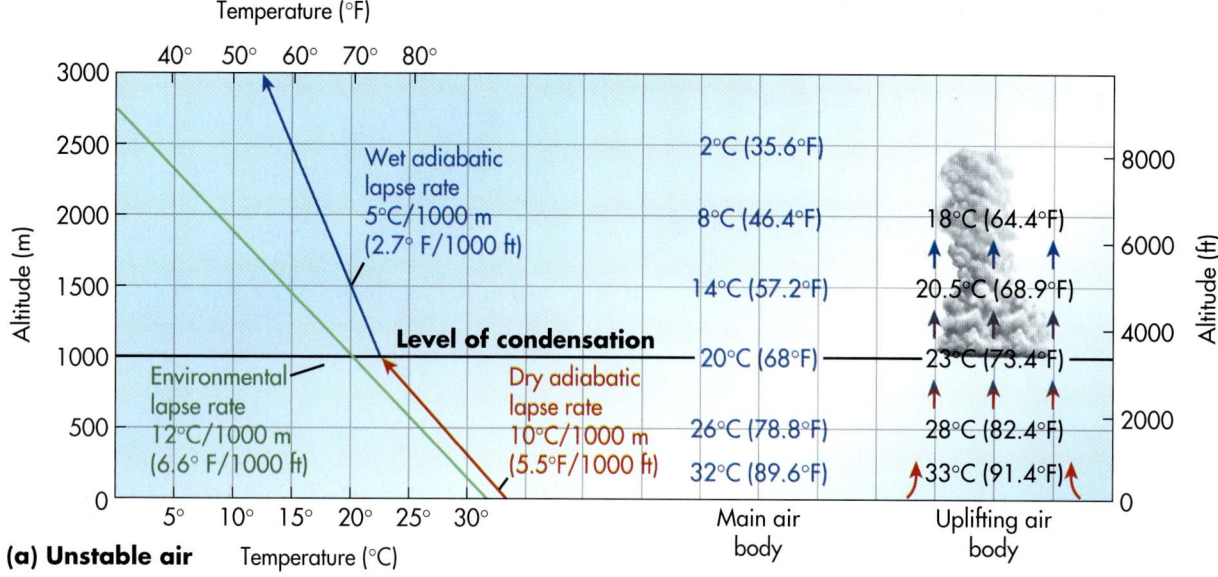

(a) Unstable air

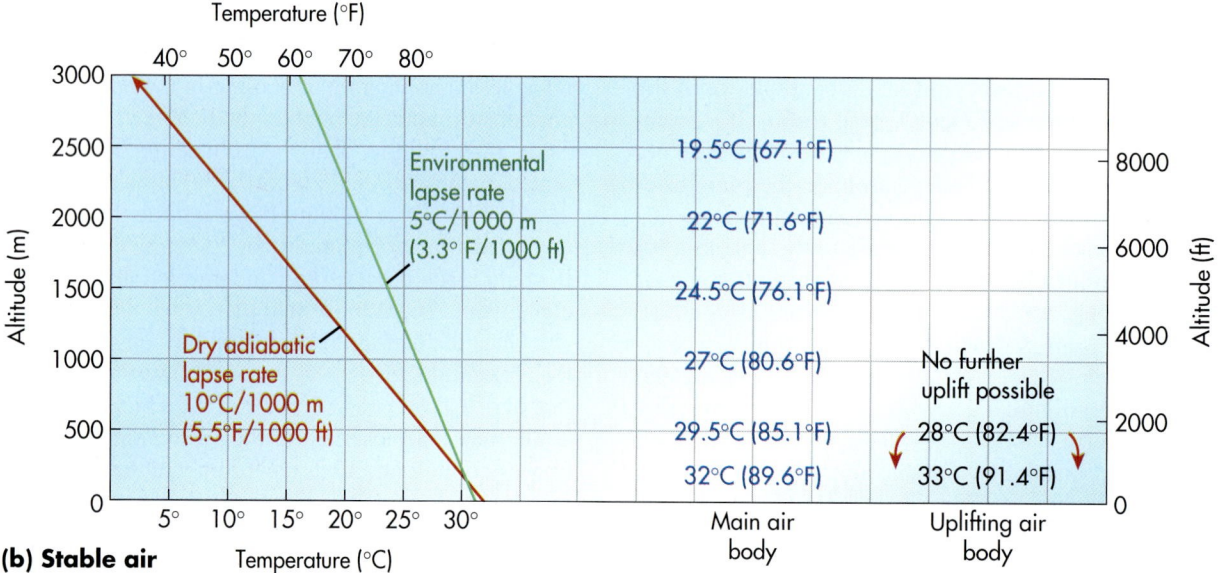

(b) Stable air

Figure 7.23. Graphical representations of unstable and stable bodies of air. (a) In this hypothetical body of unstable air, strong convection occurs when the environmental lapse rate is high and a bubble of relatively warm air forms at the surface. The graphical representation of temperature on the left shows changes with respect to altitude in the bubble and surrounding air mass; these changes can also be seen in numerical form to the right within the convecting air bubble and main air body. (b) In this hypothetical body of stable air, the environmental lapse rate is relatively low and the convecting air cannot lift because it cools more quickly than the surrounding air.

the same temperature it did previously; that is, 33° C (91.4° F). Given that this bubble of air is warmer than the surrounding air at the surface, it begins to convect. As it does, it cools at the dry adiabatic lapse rate 10° C/1000 m (5.5° F/1000 ft) because the air pressure decreases as the air lifts. In this example, however, this rate of cooling is *greater* than that observed in the surround-

ing body of air. Thus, at 500 m the temperature of the convecting air bubble is 28° C (82.4° F), whereas it is 29.5° C (89.1° F) in the surrounding body of air. Under these conditions, convection cannot continue because the air bubble is cooler than the main body of air. As a result, the air sinks toward the surface and cloud formation cannot occur.

CONVECTIONAL PRECIPITATION

Now that you've completed the discussion of convectional uplift, let's see what the effects of this process are in the real world. Go to the *GeoDiscoveries* website and select the module **Convectional Precipitation**. This module consists of a video that shows how convection results in the formation of clouds and precipitation. After you complete the video, be sure to answer the questions at the end of the module to test your comprehension of this concept.

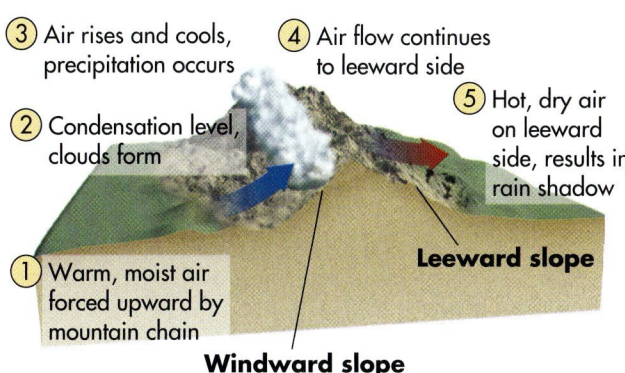

(a)

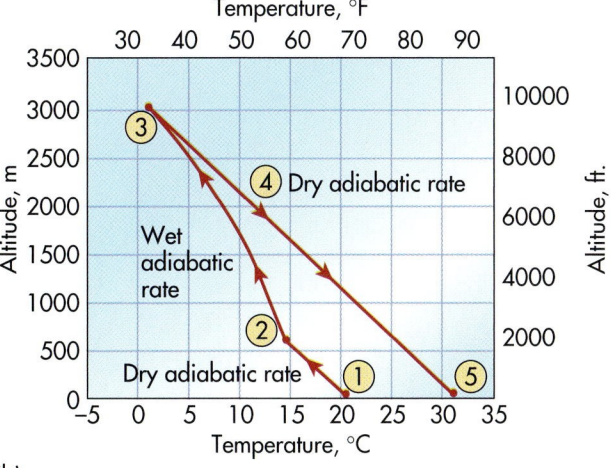

(b)

Figure 7.24 Orographic uplift. (a) When flowing air encounters a mountain range, it must flow over the top. As it does, the air adiabatically cools on the windward side and warms on the leeward side of the range, respectively. (b) Graph of the associated changes in temperature with altitude.

Orographic Uplift   Orographic uplift is probably the easiest precipitation process to understand because it makes so much visual sense. Simply put, when airflow is interrupted by a mountain range, the air must flow up and over the barrier. For point of reference, remember that the side of the mountain range that faces the direction of the oncoming wind is called the windward side, whereas the opposite, or downwind side of the range, is referred to as the leeward side. If the range is sufficiently high for the air parcel to reach its dew-point temperature through adiabatic cooling on the windward side of the range, clouds form and precipitation can potentially occur.

Figure 7.24 shows a hypothetical flow of air over a mountain and a graph of the associated temperature changes, which occur with altitude. The numbers in both images represent specific points for comparisons in the process. In this example, wind is flowing off the sea; let's say the Pacific Ocean, on the left side of the diagram (point 1). As the air encounters the mountain range, it begins to rise up the windward slope, cooling at the dry adiabatic lapse rate (DAR) until it reaches the level of condensation at point 2.

From point 2, the air begins to cool at the wet adiabatic lapse rate (WAR) because condensation, cloud formation, and precipitation are occurring, resulting in a landscape that may look very much like those pictured in Figure 7.25. The air continues to cool at this lesser rate, due to the release of latent heat of condensation, until the mountain crest is reached at point 3 (Figure 7.24). Here, the specific humidity of the air mass is less than it was at the level of condensation because moisture was lost through precipitation. This moisture loss is somewhat analogous to the water that is squeezed out of a sponge. Subsequently, the parcel of air continues over the mountain and descends on the leeward side (point 4).

As the air descends down the leeward side of the mountain range, a very interesting thing happens—the air dries rapidly. Recall that when air descends, it warms adiabatically at the DAR, which is 10° C per 1000 m (5.5° F per 1000 ft). This rate of warming takes place because as soon as the air begins its descent from the mountain crest, its maximum humidity increases while the specific humidity remains constant. In other words, the air is no

(a)

(b)

Figure 7.25 Orographic clouds. (a) These clouds formed on the windward side of a small mountain range along the northeastern coast of Australia. (b) Orographic precipitation along the Inyo Mountains in eastern California.

longer saturated and thus warms at the DAR. The air warms at this rate from the crest of the mountain range at point 3, to an elevation of 0 m (0 ft) in the leeward valley at point 5 (Figure 7.24). Given the high temperature of the air mass at this point, coupled with the very low relative humidity, the leeward side of the mountain range is dry and therefore called the **rain shadow.**

Rain shadow *The body of land on the leeward side of a mountain range that is relatively dry and hot (compared to the windward side) due to adiabatic warming and drying.*

The orographic pattern just described can result in a fascinating range of landscape variability in a relatively small geographic area, from windward slopes that receive abundant precipitation to leeward slopes that are very dry. A good place to see the results of these processes over a short distance is in California. Figure 7.26a shows the basic assemblage of landform regions of the state, including (from west to east) the low Coastal Ranges, broad Great Valley, and high Sierra Nevada Mountains. The prevailing winds are westerly in the region, bringing moisture-laden air with high dew-points in off the Pacific Ocean to the west. Notice that air flows over the Coastal

GEODISCOVERIES   WWW.WILEY.COM/COLLEGE/ARBOGAST

OROGRAPHIC PROCESSES

Now that you have a basic understanding of orographic processes, we can integrate all of the concepts that have been discussed to this point in the text in a simulated format. At this time, go to the *GeoDiscoveries* website and select the module **Orographic Processes**. This module is an opportunity for you to interact with the flow of air across a mountain range and the various humidity changes that occur on its ascent on the windward side and descent on the leeward side. The first part of the module is an animated review of the orographic process that shows the basic patterns. The second part of the module is a simulation where you can interact with the orographic process along a fictional mountain range by adjusting variables such as temperature and

specific humidity before an air mass begins its ascent on the windward slope. You can also change the elevation of the mountains to see how this factor influences adiabatic processes. In this context, you will be presented with several scenarios that represent a minute fraction of the nearly infinite number of possibilities in nature. As you interact with this simulation, notice how the level of condensation changes as you adjust the variables. Also, pay attention to the changes that occur with respect to the various forms of humidity on both the windward and leeward mountain slopes. After you complete the animation and simulation, be sure to answer the questions at the end of the module to test your understanding of this process.

Coastal Ranges

Sierra Nevada Mtns.

San Francisco

Great Valley

Inyo mountains

Salinas Valley

Owens Valley

San Bernadino Mountains

Mojave Desert

Los Angeles

(a)

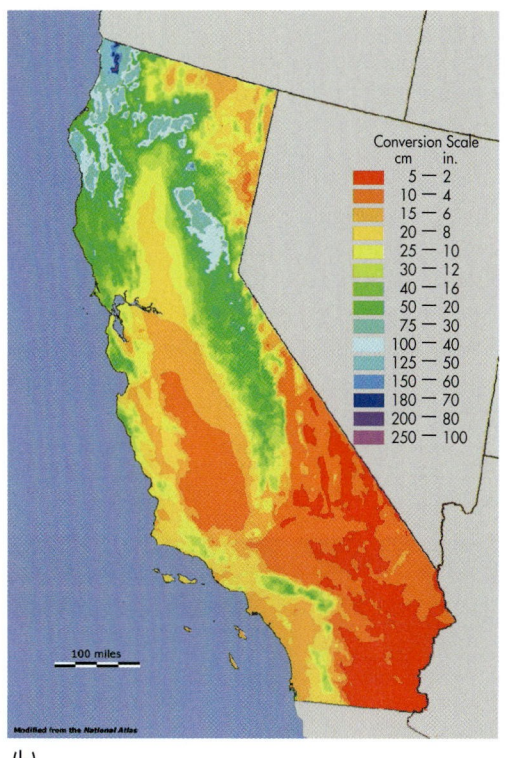

Conversion Scale
cm	in.
5	2
10	4
15	6
20	8
25	10
30	12
40	16
50	20
75	30
100	40
125	50
150	60
180	70
200	80
250	100

100 miles

Modified from the National Atlas

(b)

Figure 7.26 Air flow and precipitation patterns in California. (a) Map showing the major landform regions in California, including the prevailing wind direction. (b) Geography of mean annual precipitation. The pattern of precipitation follows the geography of the land, with greater rainfall on the windward sides of the mountain ranges and less rainfall on the leeward sides. You can see these patterns on the satellite image, with green areas representing relatively dense vegetation where high rainfall occurs, and brown zones reflecting less vegetation in areas of relatively low precipitation.

Ranges, down into the Great Valley, and then up and over the Sierra Nevada Mountains.

Figure 7.26b shows mean annual precipitation in the region. What patterns do you see? For one, you can see that annual precipitation is high along the Coastal Ranges and Sierra Nevada Mountains, with values reaching over 457 cm (180 in) in the coast ranges and about 180 cm (~ 70 in) in the northen Sierras. In contrast, the Great Valley is a distinct rain shadow, with only 40 cm (16 in.) of precipitation per year, compared to the 200 cm (80 in.) that falls in some parts of the mountains to the west. Another rain shadow exists east of the Sierras, with only 25 cm (10 in.) of precipitation per year in northeastern California.

KEY CONCEPTS TO REMEMBER ABOUT PRECIPITATION AND PRECIPITATION PROCESSES

1. In order for precipitation to occur, condensation nuclei must be present in the form of dust particles or other small solids.

2. Water droplets remain in suspension as clouds until they grow sufficiently large to fall under the force of gravity.

3. For precipitation to occur, some mechanism must be present to cause sufficient uplift of an air parcel such that high amounts of condensation take place.

4. Convectional uplift occurs when a bubble of air rises due to unequal heating of the Earth's surface. The bubble of air continues to rise as long as the air is unstable; that is, as long as the temperature of the bubble is warmer than the surrounding air. When these temperatures become equal, the air is stable.

5. Orographic uplift refers to the rising air that occurs due to a blocking mountain range. The windward side of the range faces the oncoming winds, whereas the leeward side is the downwind side. Condensation occurs on the windward side; the leeward side is typically in the rain shadow and is dry.

The Big Picture

In this chapter you learned about the nature of the hydrologic cycle and humidity. Through the course of this discussion we presented a model of adiabatic processes and described how precipitation occurs due to orographic and convectional uplift, which are easy concepts to visualize. Adiabatic processes are also associated with more complex weather systems, specifically midlatitude cyclones and hurricanes, which can influence large regions in many different ways. These systems integrate the atmospheric circulatory processes discussed in Chapter 6 with the adiabatic processes presented in Chapter 7. The interaction of these various processes can result in powerful weathermakers that release tremendous amounts of energy, as this great image of lightning demonstrates.

Summary of Key Concepts

1. Water molecules are attracted to one another through hydrogen bonding. Given the unique nature of this bond, water can exist in three physical states: liquid (water), solid (ice), and gas (water vapor).

2. Latent heat is stored in water molecules. When water changes phase, latent heat is either absorbed or released, depending upon the transformation.

3. The hydrologic cycle refers to the ways in which water moves on Earth through the combined processes of evaporation, precipitation, and runoff. Most water is stored in the oceans, which are a net source of evaporation in the context of the global water balance. On landmasses, more precipitation occurs than evaporation.

4. Three kinds of humidity occur: maximum, specific, and relative. Maximum humidity refers to how much water vapor a parcel of air can hold. This variable depends upon temperature, with warm air having a higher maximum humidity than colder air. Specific humidity measures how much water vapor is in the air. Relative humidity is the ratio of specific humidity to maximum humidity. The dew-point temperature is the temperature at which a mass of air becomes saturated. This is the temperature at which condensation occurs in that air mass. As specific humidity increases, so does dew-point temperature.

5. Adiabatic processes refer to the temperature changes that occur in an air parcel due solely to changes in air pressure. When air pressure increases, a parcel of air is compressed and warms adiabatically. In contrast, when air pressure decreases, a parcel of air expands and thus cools internally because air molecules are spaced farther

apart. Air cools at the dry adiabatic lapse rate (DAR) until condensation occurs. This rate is 10° C per 1000 m (5.5° F per 1000 ft). When air warms adiabatically, it always does so at the DAR because condensation is not occurring. Once the level of condensation is reached, the air temperature begins to decrease at the wet adiabatic lapse rate (WAR), which is 5° C per 1000 m (2.7° F per 1000 ft). This lesser rate of cooling occurs because latent heat of condensation is released when water changes from vapor to liquid.

6. Air is uplifted such that cloud formation and precipitation occur in any of four different ways. The extent of uplift and condensation depends greatly on the stability of the main body of air. Orographic lifting occurs when air flows over a mountain range. Convection occurs when bubbles of warm air develop in a body of relatively cool air that has a high environmental lapse rate. Frontal uplift occurs when two contrasting bodies of air collide. Convergence occurs when air flows into the center of a low-pressure system.

Check Your Understanding

1. How does hydrogen bonding enable water to have an unusually high surface tension?

2. In the context of the hydrologic cycle, what does the term *reservoir* mean?

3. Which body of air would have the higher maximum humidity: an air parcel that has a temperature of 25° C (77° F), or one that has an air temperature of 15° C (59° F)?

4. Of the two air masses described in the previous question, which of the two will likely have the higher specific humidity. Which one will probably have the higher relative humidity?

5. Imagine that the dew-point temperature of an air mass is 30° C (86° F). What would the specific humidity be? (*Hint:* Look at the saturation curve, Figure 7.6.) Would this be a humid air mass, or dry? Would the actual temperature of the air mass likely be more or less than the dew-point?

6. Imagine that the temperature of an air parcel is 20° C (68° F) and the specific humidity is 10 g/kg. What would the relative humidity be? What temperature would the air have to cool to reach a point of saturation?

7. Why are condensation nuclei important for precipitation to occur?

8. Why is the wet adiabatic lapse rate (WAR) less than the dry adiabatic lapse rate (DAR)? When do you use the DAR as opposed to the WAR?

9. Of the three cloud types—cirrus, cumulus, and stratus—which one is most likely associated with convection? Why?

10. Why is the specific humidity of an air parcel less on the leeward side of a mountain range than on the windward side?

CHAPTER EIGHT

AIR MASSES AND CYCLONIC WEATHER SYSTEMS

In Chapter 7 we discussed atmospheric humidity and two processes that result in precipitation—orographic uplift and convectional uplift. These processes involve distinct air parcels contained within a definable unit of geographic space, such as above a mountain range or a large bubble of air that originated over a relatively hot surface. In this chapter, we broaden the

Hurricanes are the largest storms on Earth. This satellite image shows Hurricane Katrina when it struck the Gulf Coast in August 2005. Katrina caused billions of dollars in damage and killed thousands of people, many who died when New Orleans flooded. This chapter focuses on midlatitude and tropical circulatory processes, including the evolution of hurricanes and tornadoes.

discussion to large air masses and complex atmospheric systems that

influence weather over thousands of square kilometers. We shall focus on

cyclonic weather systems that occur in the tropics and midlatitudes,

including thunderstorms, tornadoes, and hurricanes.

Air Masses and Fronts

Have you ever experienced hot and humid weather for 2 or 3 days and then, *boom*, a strong storm occurred and the temperature dropped sharply? You might have even talked with your friends about how you couldn't believe the weather changed so fast. Most likely, what you experienced was the passage of a midlatitude cyclone through your region and the associated air masses that came with it. The first air mass was warm with high dew-points that made it uncomfortable to be outside, whereas the second air mass was relatively cool (maybe even cold) and dry. Although you may not have known the reasons why this weather change occurred, you certainly noticed that it affected your life for a couple of days. In this section we'll examine the characteristics of the various air masses.

Air Masses

Before we discuss midlatitude circulatory processes and how they influence weather, we must first define the concept of an air mass. This definition is essential because the formation of midlatitude cyclones depends upon the interaction of very large bodies of air, thousands of square kilometers in size, that have distinctly contrasting physical properties.

An **air mass** is any large body of the lower atmosphere that has fairly uniform conditions of temperature and moisture. The source region of an air mass is any large body of land or water from which the air derives these characteristics. Imagine that a maritime air mass moves from

---

Air mass *A large body of air in the lower atmosphere that has distinct temperature and humidity characteristics.*

---

the northern Pacific Ocean to northern Canada, where it stagnates for a couple of weeks in January. What will happen to that air mass? How will it change character? From our previous discussions, you should realize that the temperature of the air mass will drop while it is over Canada, because the landmass is cold during winter. In addition, the air mass would lose a lot of its moisture because cold air holds less water vapor than warm air. If the air mass then moved south into the U.S., perhaps due to changes in the pressure gradient or the overall circulatory system, it would be cold and dry by the time it reached where you live.

Air masses are categorized by a combination of letter designations. The first letter is lower case and designates the moisture source of the air mass:

c = continental (dry)

m = maritime (moist)

The second letter is upper case and designates the latitude position of the source region:

A = Arctic

AA = Antarctic

P = Polar—from between 50° and 60° N or S latitude

T = Tropical—from between 20° and 35° N or S latitude

E = Equatorial

Using these designations, we can identify five principal types of air masses that most often affect North America (Figure 8.1). The most common cold air mass to affect North America is the *continental Polar* (cP), which forms over northern Canada and consists of stable, dry air. You've

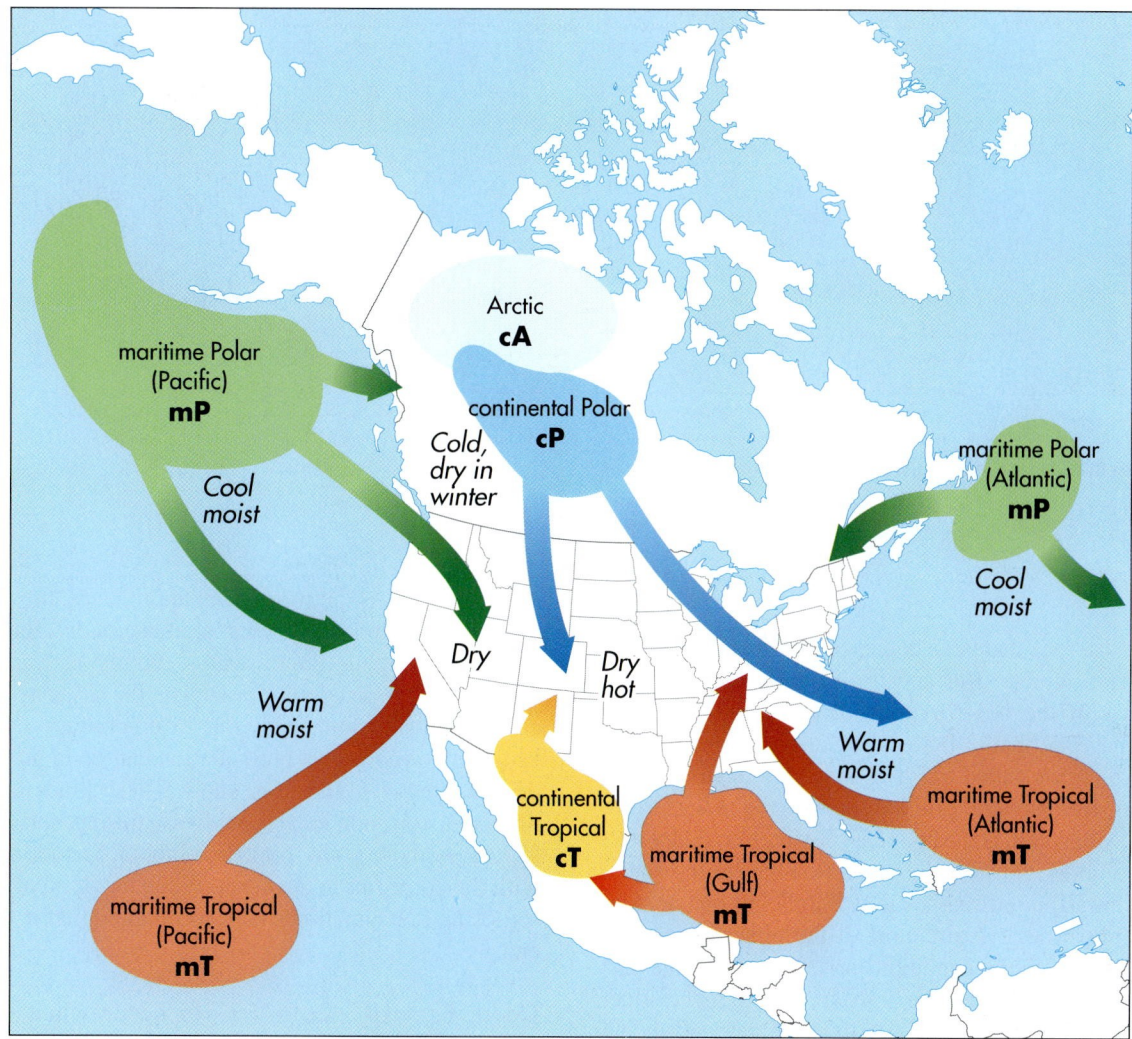

Figure 8.1 Principal air masses in North America. Five major types of air masses periodically flow into the continent.

[handwritten annotations:] cP = stable & dry air.   mP = cool & moist bodies of air.   mP: cool, damp, moist   mT: warm & moist

probably experienced this type of an air mass, though perhaps you didn't know it at the time. If you can, recall a period, maybe in the middle of January or February, when it was cold for a couple of weeks. This weather scenario likely occurred because a cP air mass moved into your region of the country. Such an air mass can be further cooled when it's infused with *continental Arctic* (cA) air, which is extremely cold. Although cP air masses are most intense in winter, they sometimes affect the weather in the summer. Such an influence is particularly noticeable when it's been hot and muggy for a few days and then it cools to perhaps 24° C (75° F) with a very comfortable northwest breeze.

Another dry, stable air mass that influences the continent is the *continental Tropical* (cT). This air mass forms over places like Mexico and Arizona and thus is hot and extremely arid. If you've ever been to the southwest deserts of the U.S. in the summer, at places such as Phoenix, Arizona, or Las Vegas, Nevada, you've no doubt experienced such an air mass. Do you remember what it felt like?

In contrast to the continental air masses, two maritime air masses periodically influence parts of the continent. One of these air masses is the *maritime Polar* (mP), which consists of cool and moist bodies of air that form over the northern Pacific Ocean and northwest Atlantic Ocean in the Northern Hemisphere. Given the influence of westerly winds in the midlatitudes, this kind of air has a particularly strong influence in the Pacific Northwest. If you happen to live there, or have visited places like Seattle, Washington, or Portland, Oregon, in the winter, you've certainly experienced the cool, damp air associated with an mP air mass. The other maritime air mass is the *maritime Tropical* (mT), which forms over places like the Gulf of Mexico and is distinctive because it's warm and moist. To associate this air mass with your experience, try to remember when the summer air was really hot and sticky and air-conditioned buildings were comfortable. This air mass was an mT air mass.

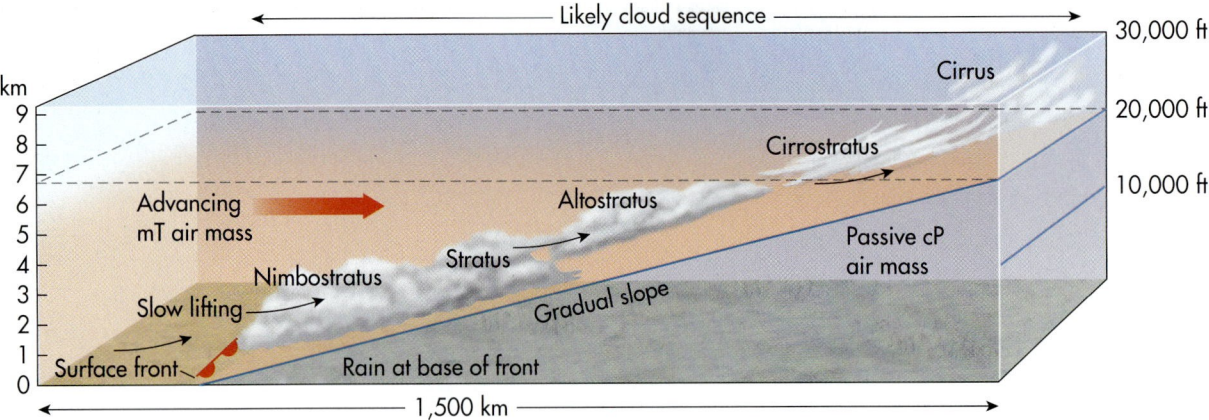

Figure 8.2 Adiabatic uplift at a warm front. At a warm front, the advancing warm air gradually slides over the top of underlying, passive cooler air, causing the formation of stratiform clouds as the warm air slowly cools adiabatically.

Fronts

We have just seen that large bodies of air with distinct characteristics originate over specific places and then move over parts of the North American continent. These air masses have definable boundaries at the surface of the Earth, which are called *fronts*. The dominant front in the Northern Hemisphere midlatitudes is the polar front, which, if you recall, is the boundary between cold, dry (cP) air to the north and warmer, moist (mT) air to the south. These contrasting air masses flow parallel to one another along a **stationary front** most of the time. Sometimes, however, atmospheric conditions arise that cause one particular air mass to advance into another along distinct frontal boundaries. It's helpful to think of these advancing air masses as being analogous to a military force that is advancing into a particular area, with the lead columns of the force being the front. This part of the chapter focuses on these atmospheric boundaries and the types of precipitation that occur along them as air is uplifted. Recall from Chapter 7 that this type of uplift is called *frontal uplift*.

Warm Fronts **Warm fronts** occur in places where warm air advances into relatively cool air. This interaction causes the warmer air to slowly slide over the top of the underlying cooler air (Figure 8.2). This process of gradual overriding of cooler air actually occurs in the upper atmosphere ahead of the surface warm front and causes the lifting mT air to cool adiabatically. As the air cools, clouds form, beginning with high-level cirrus clouds at the deepest penetration of upper level warm air

in the system. As the warm front approaches, these clouds change to progressively lower stratus clouds, culminating in rain-producing nimbostratus clouds at the surface front. When these conditions evolve, the sky may be overcast with a slow but steady rate of precipitation that may last a day or two. This precipitation can be in the form of drizzle, showers, and even snow. These kinds of conditions are common in the Midwest during spring when slow-moving mT air from the Gulf of Mexico interacts with the cooler air to the north. When this kind of weather settles in, it's sometimes hard to imagine that it will ever end.

Cold Fronts **Cold fronts** occur when cool air moves into a region that was previously dominated by warmer air. The cold air is denser and heavier than the warm air ahead of it, so the warm air is forced to rise. In this fashion, a cold front is significantly different from a warm front because the cold air hugs the surface along a cold front and vigorously drives the warm air ahead of it aloft. Notice in Figure 8.3 that the edge of a cold front is very steep when compared to the warm front in Figure 8.2.

Once the warm air begins to lift along the cold front, adiabatic cooling starts and vapor condenses, forming clouds. Because of the rapid rise of air, the air quickly cools adiabatically, which means that large amounts of latent heat energy are rapidly released as the water condenses. As a result, rainfall is intense and of short duration. If sufficient moisture is present and enough latent

---

Stationary front *A boundary where contrasting air masses are flowing parallel to one another.*

Warm front *A frontal boundary where warm air is advancing into relatively cool air. This front is typically associated with slow, steady precipitation.*

Cold front *A frontal boundary where cold air is advancing into relatively warm air. This front is typically associated with intense rain of short duration.*

Thunderstorm *A brief, but strong storm that contains strong winds, lightning, thunder, and perhaps hail.*

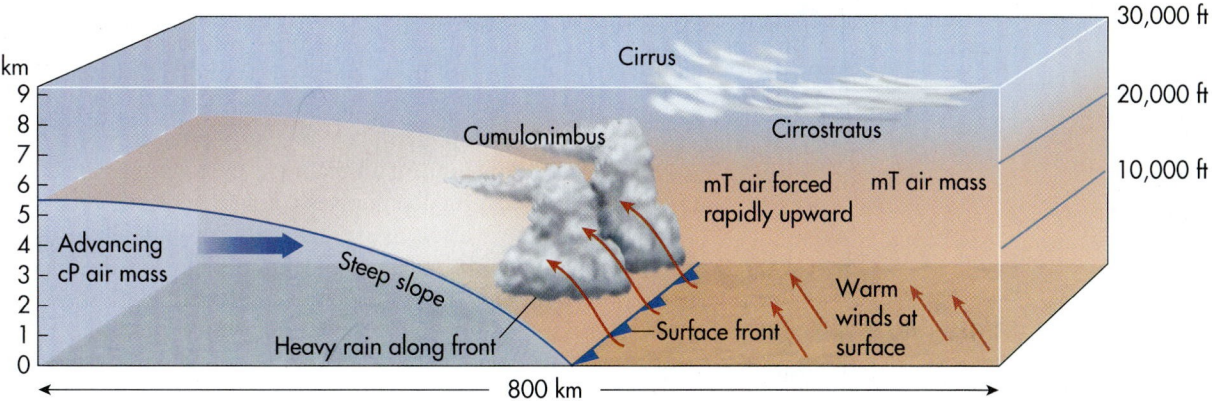

Figure 8.3 Uplift along a cold front. In contrast to a warm front, warm air ahead of a cold front is abruptly lifted and quickly cooled, causing cumulonimbus clouds to form.

heat is quickly released, a tremendous **thunderstorm** can form, with the latent heat being the fuel for its development. Although thunderstorms can develop through the simple process of convection, they are prone to be more severe and widespread along a cold front because of the rapid adiabatic cooling and latent heat release that occurs. For example, Figure 8.4 shows how widespread frontal precipitation can be. This radar image shows a cold front that stretched from Canada into Texas. A line of thunderstorms extends along this front from Michigan back into eastern Kansas. We'll describe the evolution of thunderstorms in more detail later in the chapter.

KEY CONCEPTS TO REMEMBER ABOUT AIR MASSES AND FRONTS

1. An air mass is a large body of air that has distinctive characteristics and forms in specific geographic regions.

2. Five principal types of air masses affect North America. Continental air masses include continental Polar (cP), continental Arctic (cA), and continental Tropical (cT). Maritime air masses are maritime Tropical (mT) and maritime Polar (mP).

3. Air masses have distinct boundaries that are called fronts. A stationary front is a place where contrasting air masses are flowing parallel to one another.

4. A warm front is a place where warm air is advancing into relatively cool air. Warm air slowly slides over the top of the cooler air along a warm front, causing slow and steady rainfall.

5. A cold front is a place where cold air is advancing into relatively warm air. Given the higher density of colder air, the warm air is rapidly forced aloft, where it cools quickly. As a result, rainfall is intense and of short duration along the front.

Evolution and Character of Midlatitude Cyclones

Now let's closely examine the atmospheric circulatory systems in which warm and cold fronts develop. Such a system is called a **midlatitude cyclone,** which is basically a well-organized low-pressure system that migrates across a region while it spins. In order to explain how midlatitude cyclones develop along the polar front and how they influence weather, we must first briefly review how the polar front, and the associated jet stream, migrates on a seasonal basis (you might want to review Figure 6.21).

In the summer, when high latitudes receive high amounts of solar radiation, the polar front and jet stream retreat to a location very close to the poles and exhibit a zonal-flow pattern much like that illustrated in Figure 6.21a. Although midlatitude cyclones may develop along the polar front during this time of year, they are generally weak systems because no strong temperature contrast occurs north and south of the polar front.

Midlatitude cyclone *A well-organized low-pressure system in the midlatitudes that contains warm and cold fronts.*

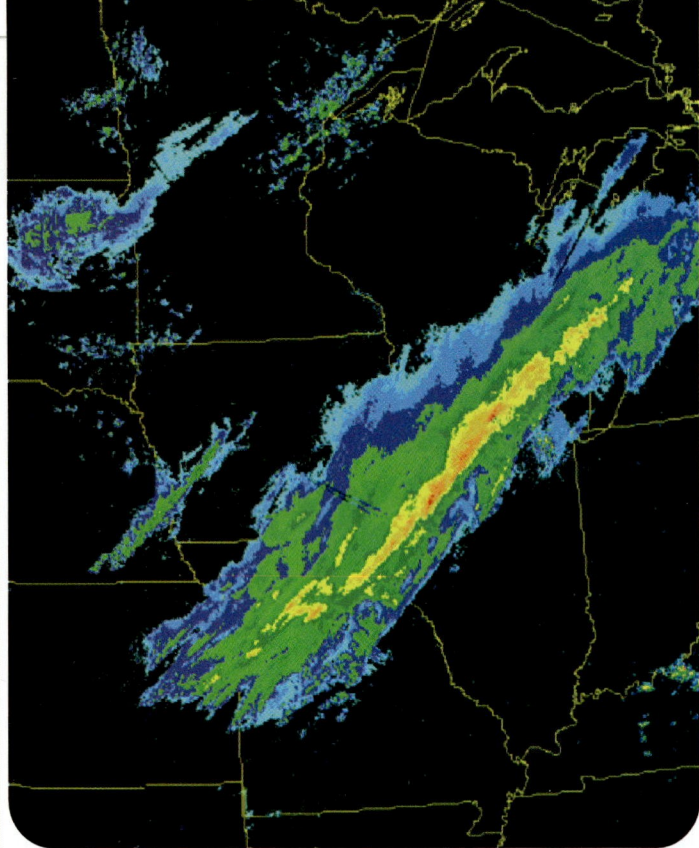

Figure 8.4 Frontal precipitation. This radar image shows a line of thunderstorms that extends from northwestern lower Michigan southwest to eastern Kansas. This line of storms developed along a strong cold front that passed through the Midwest. Orange and yellow colors represent areas of heavy rain, whereas the areas of lightest precipitation are in blue.

Once the Summer Solstice passes and daily insolation at high latitudes begins to decrease, large bodies of cP air begin to develop north of the polar front. This increase of cold, dry air north of the polar front causes the front and associated jet stream to migrate south. Essentially, the buildup of cold, dense air at high, northerly latitudes pushes the polar front south. As this migration proceeds, the path of the jet stream gradually changes from a zonal pattern to an undulating meridional pattern with well-developed Rossby waves in the upper atmosphere (see Figure 6.21 again). This development occurs because the atmosphere is working to balance the growing temperature difference on either side of the polar front. These waves allow tongues of warm air (south of the front) to penetrate northward at the same time that wedges of cold air (north of the front) expand to the south.

On a global scale, the development of Rossby waves is how the atmosphere begins the process of mixing cold and warm air. The most vigorous mixing of these air masses occurs on a more regional scale, within midlatitude cyclones that form along distinct sections of the polar front in any given Rossby wave. The term *cyclone* is sometimes confusing because tropical storms near south-ern Asia are referred to by the same name. In addition, tornadoes are also sometimes called *cyclones*. Although we'll discuss these specific kinds of storms later in this chapter, for now we'll use the term "cyclone" only in the context of midlatitude circulatory processes.

Formation of midlatitude cyclones is called **cyclogenesis.** It encompasses a complex set of processes that are most often associated with an undulating polar jet stream pattern in the upper atmosphere and the interaction of these winds with airflow at ground level. To better understand this interaction, think of the atmosphere as consisting vertically of many different pressure surfaces, with the highest air pressure exerted at ground level and progressively lower air pressure at progressively higher altitudes. Just like a particular isobar (a line of equal air pressure) can be followed on the ground, it can also be traced horizontally within different levels of the atmosphere.

Upper Air Flow and the 500-mb Pressure Surface

A good example of such an upper air-pressure surface is the 500-mb surface, which occurs at a specific but varying altitude at any given place on Earth. Recall that the average surface air pressure at sea level is 1013.2 mb and that air pressure decreases with altitude. Thus, the 500-mb pressure surface is at a relatively high altitude compared to the surface air pressure. From a meteorologic standpoint, the 500-mb pressure surface is important because it vertically divides the atmosphere in two, if we assume that the average surface air pressure is 1013 mb and 0 mb of air pressure occurs at the top of the atmosphere. In addition, the wind patterns at the 500-mb surface exert a strong steering influence on the circulatory patterns at the surface.

To understand the significance of the 500-mb pressure surface and how its height variation is important to air flow, consider Figure 8.5. Figure 8.5a shows two columns of air (Air columns 1 and 2) that contain the same mass of atmosphere. Given that each column contains the same number of molecules, the surface pressure is the same. In addition, the altitude of the 500-mb surface in both columns is equal, in this case 5460 m (17,900 ft).

What happens if we warm one of the columns of air, say Air column 2, relative to the other (Figure 8.5b)? When this warming occurs, the column of air stretches vertically—because warm air rises—and the height of the 500-mb surface in that column moves to a higher altitude; let's say, for example, 5760 m (18,900 ft). At that same altitude in Air column 1, the air pressure is 480 mb. Air column 1 has this relatively lower pressure at 5760 m (18,900 ft) because the elevation of the 500-mb surface in that column did not change from the previous example

Cyclogenesis *The sequence of atmospheric events that develops along the polar jet stream that produce midlatitude cyclones.*

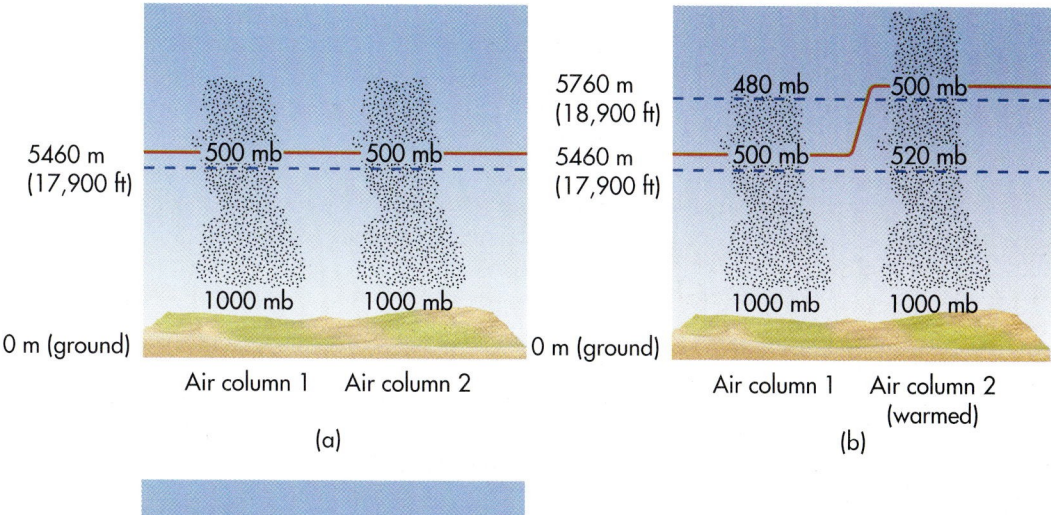

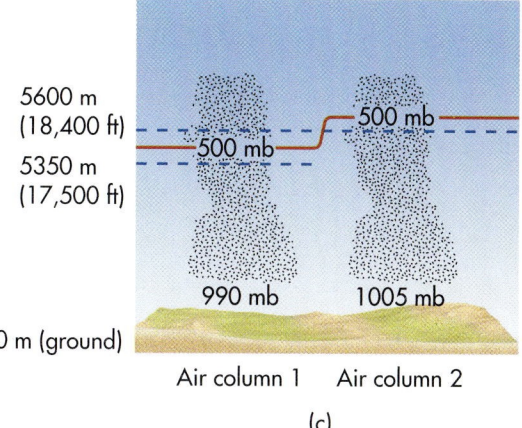

Figure 8.5 Vertical changes in air pressure. (a) The height of the 500-mb pressure surface in two columns of air with the same mass and temperature. (b) If one column of air (column 2) is warmed relative to the other, then the atmosphere stretches vertically and the height of the 500-mb surface rises. (c) The height of the 500 mb surface also fluctuates when the surface air pressure changes.

(Figure 8.5a). This pressure/altitude difference between the two air columns reflects the fact that the warmer column of air now has a lower overall density than the first (cooler) one. No change occurs in the mass of Air column 2, which is why the surface air pressure is still 1000 mb. Another way of comparing the two air masses in the second example is to say that Air column 2 has a smaller vertical pressure gradient than Air column 1.

The same kind of altitude change in the 500-mb surface also occurs when air pressure at the surface varies between two regions. Using Figure 8.5a again as a hypothetical baseline scenario, let's say that the surface air pressure in Air column 1 decreases to 990 mb at the same time it increases to 1005 mb in Air column 2 (Figure 8.5c). What happens? Notice that the altitude of the 500-mb surface drops to, let's say, 5350 m (17,500 ft) in Air column 1, whereas it rises to 5600 m (18,400 ft) in Air column 2. This change occurs because the mass of air beneath the 500-mb surface decreases in Air column 1 at the same time it increases in Air column 2.

This example shows how pressure can vary within two distinct columns of air. Now let's examine how the altitude of the 500-mb surface might vary horizontally in the continuous medium of the actual atmosphere. Figure 8.6a. presents a typical isoline map that shows the change in altitude of the 500-mb surface across the U.S. on a particular day. This reflects the fact that surface air pressure varies across the nation in a manner similar to Figure 8.5c. The map shows a strong high-pressure system (anticyclone) in the western U.S. and two areas of low pressure—one over the northeastern part of the country and another off the Pacific Northwest coast. You can see how this information translates into an actual topographic map of the 500-mb surface in Figure 8.6b. Notice that the height of the pressure surface is higher and forms a ridge-like feature where the anticyclone occurs. Such an atmospheric feature is called a **high-pressure ridge** by meteorologists. In contrast, the height of the pressure surface is lower in the areas associated with the low-pressure systems along each coast. Meteorologists call these valley-like features in the 500-mb surface **low-pressure troughs**. Regardless of whether the upper atmospheric pressure system is a ridge or a trough, each shares the common reference of an axis, which is the imaginary line that extends along their length, that is, from north to south.

High-pressure ridge   *An elongated area of elevated air pressure in the upper atmosphere that is typically associated with sunny skies and calm winds.*

Low-pressure trough   *An elongated area of depressed air pressure in the upper atmosphere that is typically associated with cloudy skies and rain.*

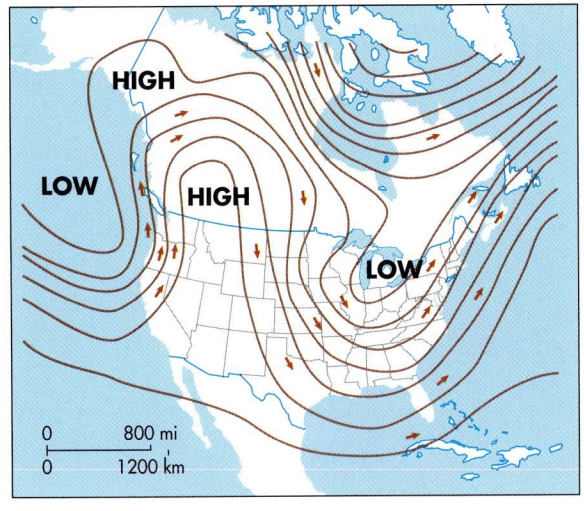

(a)

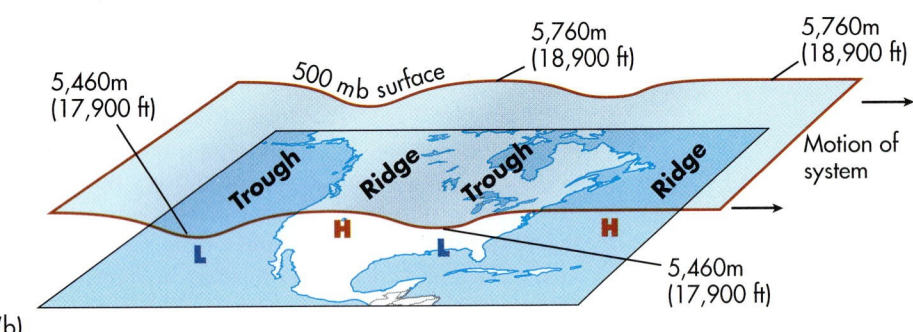

(b)

Figure 8.6 Mapping the 500-mb pressure surface. (a) Isolines of constant pressure across the U.S. on a particular day. (b) A topographic map showing the variation in the 500-mb surface.

Interaction of Upper Air Flow and Surface Air Flow

In the context of understanding how midlatitude cyclones form, it's useful to examine the interaction of air flow on the ground with air flow at the 500-mb pressure surface. As we begin this part of the discussion, follow along by studying Figure 8.7, which shows air flow both on the ground and at the 500-mb altitude. This figure is very similar to Figure 6.12, because it shows a dynamic convection loop in the atmosphere, but differs because it's viewed within the context of the 500-mb surface.

Let's imagine a scenario in which a prominent Rossby wave exists in the upper atmosphere and a low-pressure trough has developed poleward of the jet stream at the 500-mb level. When this condition develops, the upper air flow becomes disturbed, with a region of converging air flow developing behind (to the left of) the trough axis in Figure 8.7. At the same time, a zone of diverging air forms east (to the right) of the trough axis. As these air flow patterns continue to develop, the air pressure beneath them (at ground level) changes. Where the air converges aloft, more air is flowing into a given unit of space at the 500-mb level, which forces air toward the surface as a zone of high pressure. In contrast, where the

air diverges aloft on the downwind side of the trough, less air is occupying a given unit of space, which has the effect of pulling air up from beneath to fill the void. When this situation occurs, the surface and upper air flows become closely linked as columns of spinning air—in other words, a dynamic convection loop—that are set in motion by the Coriolis force, as described in Chapter 6.

To understand better how midlatitude cyclones form and how they influence the weather on the ground, let's investigate the system evolution at the surface by studying the section of the polar front outlined in Figure 8.8. Notice that prior to the development of a midlatitude cyclone, the polar front is linear, with parallel-flowing but opposite-direction winds—cold, easterly winds to the north of the front and warm westerly winds to the south.

Cyclogenesis

As previously described in the discussion about vertical air flow within Rossby waves, cyclone development begins when the upper air flow and surface air flow along a stationary front become linked through the rise of warm air and descent of cold air (Figure 8.9a). At the surface, this developing linkage causes an undulation to form in

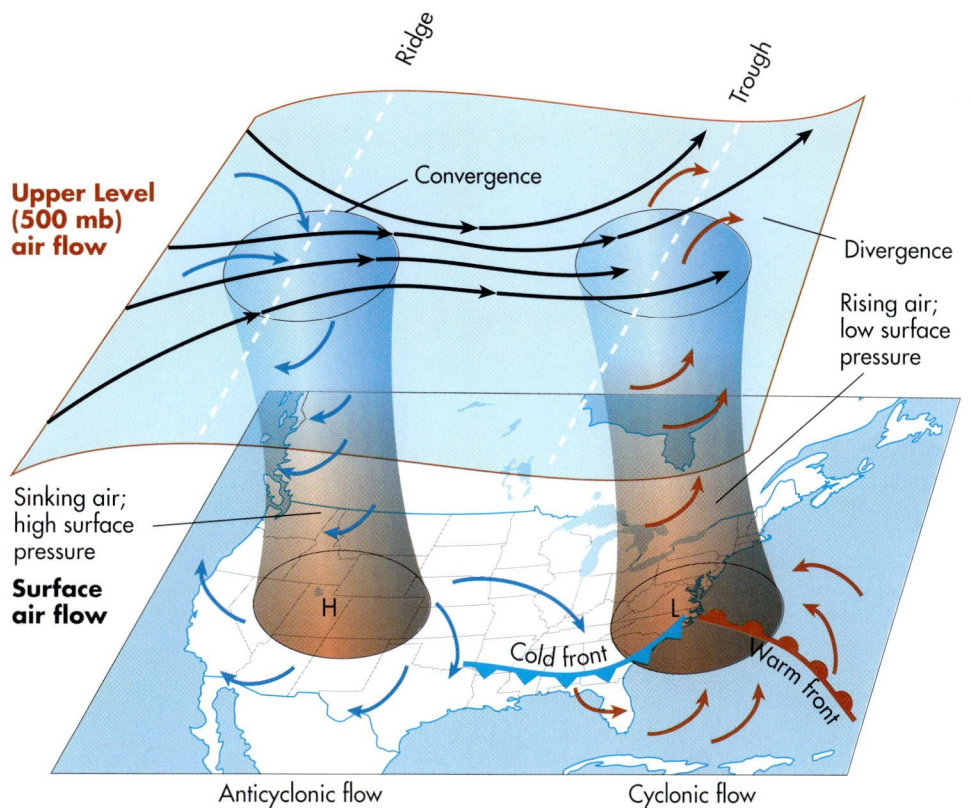

Figure 8.7 **Linkage of upper air flow and surface air flow in a midlatitude cyclone.** These systems form through a complex interaction of upper air and surface air flow. Under the high-pressure system, the weather is stable with clear sky. Where the cyclone forms, however, the atmosphere is unstable.

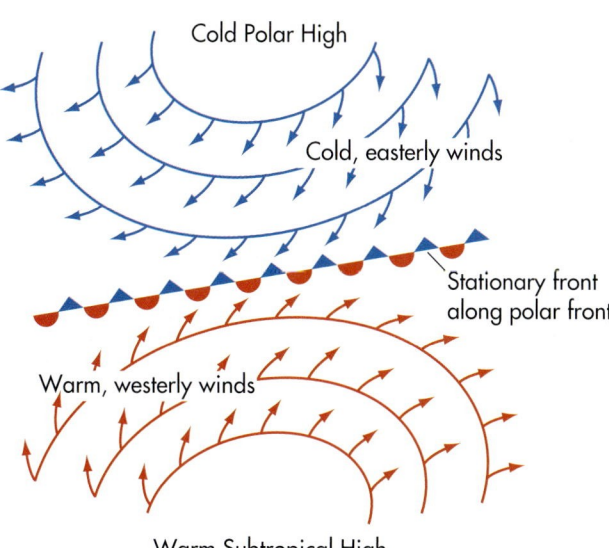

Figure 8.8 **Section of the polar front.** Prior to the development of a midlatitude cyclone, a stationary front exists along the polar front. Winds in contrasting air masses flow parallel and opposite to each other.

the polar front (Figure 8.9b). As this undulation becomes better defined, the convergence of surface winds intensifies in a counterclockwise pattern, with cold air flowing south on the backside of the system, while warm air flows northward to the east of the circulatory core. In this situation, the leading edge of the inflowing cold air mass is the cold front, whereas the leading edge of the warm air mass is the warm front. Notice in Figure 8.7 how these fronts relate to the dynamic convection loop.

The developing cyclone is said to be in the *open stage* when the warm and cold fronts are fully separated; that is, when warm air exists at the surface all the way to the system's core (Figure 8.9c). This region of warm air is referred to as the *warm sector*. Let's imagine that such a system develops over the northern Great Plains in the United States. In this scenario, the air behind the cold front (that is, to the north) would consist of cold, dry (cP) air that originated in Canada, while the air in the warm sector that lies between the cold and warm fronts is relatively warm and moist (mT) air that may have come from the Gulf of Mexico. At this stage the cyclone may very well look like the one pictured in Figure 8.10.

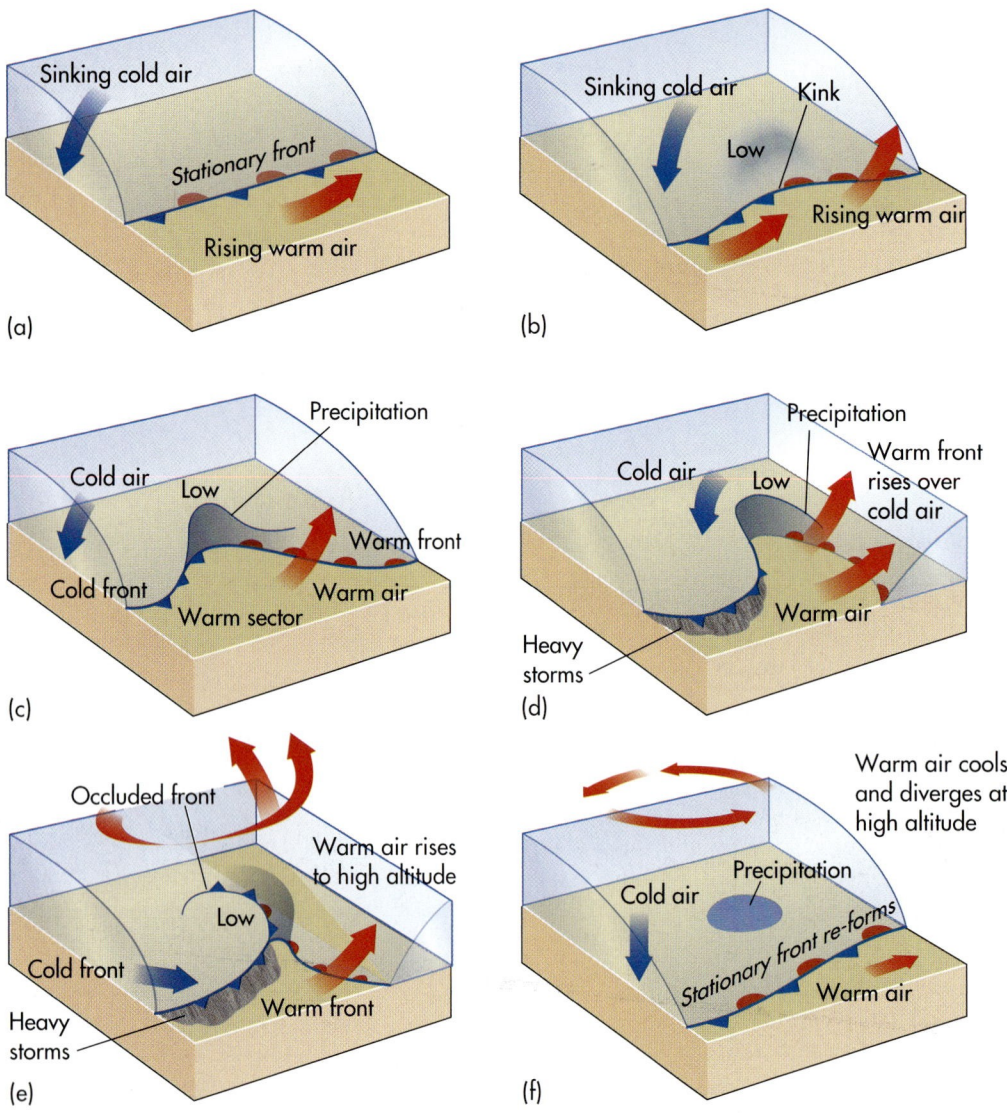

Figure 8.9 Evolution of a midlatitude cyclone. (a) Stationary front stage; (b) early stage when undulation develops along stationary front; (c) open-wave stage with prominent cold and warm fronts; (d) closing of open wave due to rapid speed of advancing cold front; (e) occluded stage when cold front begins to overtake the warm front; (f) dissolving stage when the warm sector is completely pinched off by overtaking cold front.

When the cyclone is in the open stage it is fully mature. This stage does not last very long, however, because the cold front moves faster than the warm front ahead of it. These contrasting frontal speeds occur because the air behind the cold front is denser than the warm air ahead of it and thus flows more quickly. In this fashion, the cold front drives into the warm, moist air ahead of it, whereas the warm front gradually slides over the top of the stationary cold air ahead of it (Figure 8.9d). As a result of this difference in speeds, the cold front begins to overtake the warm front in the *occluded stage* of development (Figure 8.9e). Although a distinct warm sector still exists in the southerly parts of the system, the overtaking cold air lifts warm air aloft near the center of the low. An **occluded front** develops when the cold front overtakes the warm front and begins to drive the warm air at the surface to a higher altitude. This process of occlusion continues as the system enters the *dissolving stage* (Figure 8.9f), at which time all of the warm air is pinched aloft and a stationary front is reestablished.

It's very important to remember that the evolution of midlatitude cyclones occurs within the context of the overall westerly wind pattern in the jet stream. You can think of the jet stream as the medium that steers cyclones

Occluded front   *The area where a cold front begins to overtake a warm front and thus lift warm surface air aloft.*

This image shows temperature in the United States. Where is the approximate location of the cold front?

a) Western Texas

b) Florida

c) It cuts across the states of Washington and Oregon.

d) It extends from western Michigan to Central Texas.

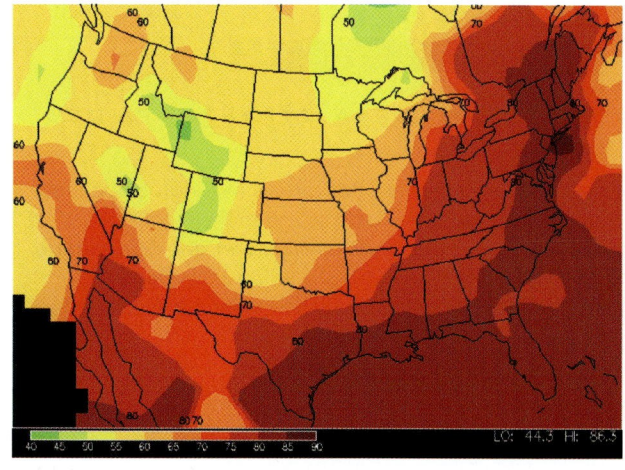

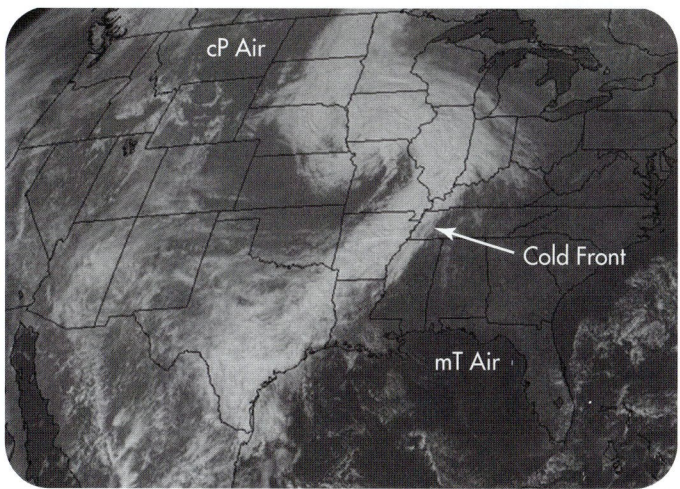

Figure 8.10 A classic midlatitude cyclone. A cyclone centered over the midwestern United States. Notice the prominent line of clouds, which developed along a strong cold front, which extends from Illinois to eastern Texas. To the east of the cold front the weather was warm and muggy, whereas west of the cold front it was relatively cold and dry, as indicated by the clear skies in western Kansas and eastern Colorado.

across the continent. Figure 8.11 shows an example of this migration, beginning with an open-stage cyclone centered in the central U.S. (Figure 8.11a). This view of the system shows the position of the warm and cold fronts and their relationship to the air mass source regions. In the open stage of development, mT air in the warm sector is clearly flowing north from the Gulf of Mexico into Louisiana and Arkansas. If you were there on that day, the weather would be warm and muggy, with southerly winds. To the north and west of the cold front, in contrast, the air is flowing in a southerly direction from Canada into places like the panhandle of Texas and Oklahoma. In these places, the temperature would probably be cold, with clear sky and northwest winds.

Now let's move on in our hypothetical sequence of cyclonic evolution to Day 2 (Figure 8.11b). Two important

GEODISCOVERIES WWW.WILEY.COM/COLLEGE/ARBOGAST

FORMATION OF A MIDLATITUDE CYCLONE

We can now examine how the development of midlatitude cyclones occurs in an animated way. Go to the *GeoDiscoveries* website and select the module **Formation of a Midlatitude Cyclone**. This module allows you to see the process of cyclogenesis in motion. Figure 8.9 is the foundation of this animation. As you watch the animation, follow how an initial

kink in the atmosphere evolves into a mature cyclone that spins counterclockwise in the Northern Hemisphere. This animation will help you better understand how these systems and associated processes cause highly variable weather when they migrate through a region. Once you have completed the animation, be sure to answer the questions at the end of the module to test your understanding of this concept.

MIGRATION OF A MIDLATITUDE CYCLONE

Let's now visually integrate the various moving components of a midlatitude cyclone. To do so, go to the *GeoDiscoveries* website and select the module ***Migration of a Midlatitude Cyclone***. This module contains a video that shows the migration of a typical midlatitude storm system in North American over the course of several days. As you watch this system evolve, pay attention to *both* its counterclockwise circulation and movement from west to east. See if you can identify the center of the system and associated cold and warm fronts. After you complete the video, be sure to answer the questions at the end of the module to test your understanding of this concept.

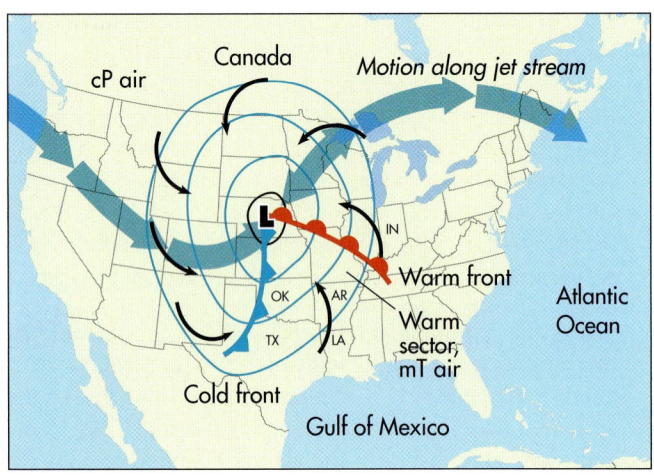

(a) Day 1, open stage

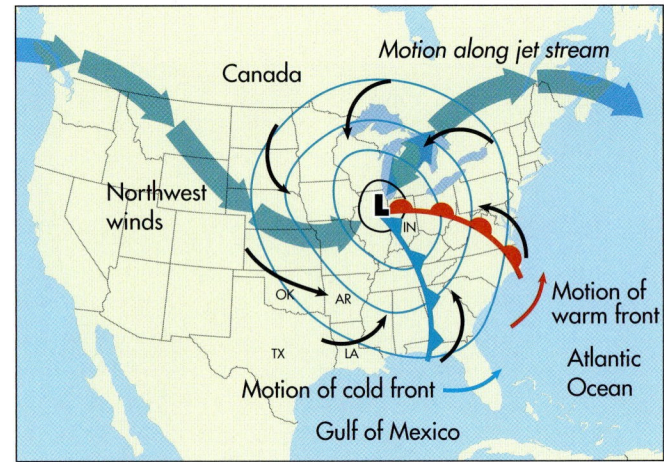

(b) Day 2, cold front overtaking warm front

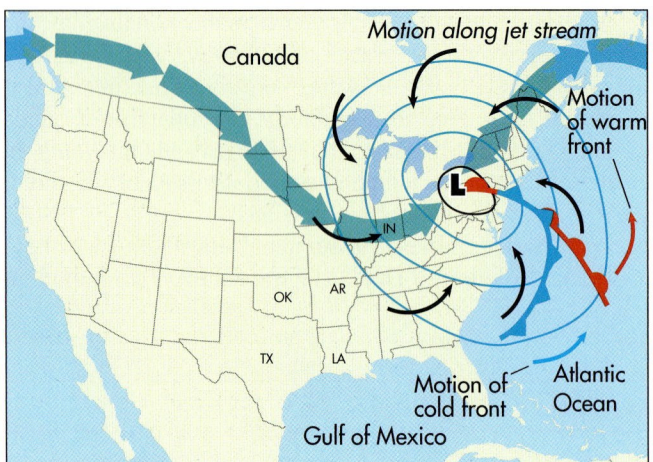

(c) Day 3, start of occluded stage

Figure 8.11 Evolution and migration of a cyclone. Midlatitude cyclones migrate from west to east as they spin counterclockwise and evolve from initial formation, through the mature stage, and into the occluded stage.

things are seen in this diagram. First, notice that the cold front is beginning to overtake the warm front, as you might have predicted. As a result, the temperature in Louisiana and Arkansas, which had been warm the previous day due to southerly air flow, is now colder because the winds became northwesterly. Second, the center of the system has migrated slightly to the northeast along with the wave in the jet stream. Eastward migration continues into Day 3 (Figure 8.11c), with the onset of the occluded stage as the system moves into the northeast part of the country.

Thunderstorms

You've probably experienced a severe thunderstorm at some point in your life. Maybe you had a sense that a storm was brewing and then suddenly it hit with heavy rain, strong winds, thunder, lightning, maybe even some hail. The storm lasted for a while and then gradually tapered off. This section focuses on the evolution and characteristics of thunderstorms. It immediately follows the discussion about midlatitude cyclones because the most severe storms in the midlatitudes are most often associated with low-pressure systems.

Evolution of Thunderstorms

Like midlatitude cyclones, thunderstorms have a distinct life cycle, illustrated in Figure 8.12. The first stage in the development of a thunderstorm is the *cumulus stage* (Figure 8.12a), which begins with convection or the advancement of a cold front into mT air. This kind of disturbance causes air to rise, which results in the formation of ice crystals and raindrops in growing cumulus clouds. Given their small size in this early stage of thunderstorm development, the ice crystals and raindrops continue to be uplifted. Condensation adds additional latent heat energy to the air during the *developing stage* and the atmosphere becomes very unstable with strong updrafts.

As the thunderstorm continues to develop, it enters the *mature stage* (Figure 8.12b). In this stage, the atmosphere is very unstable and continued convection or frontal advancement results in the massive uplift of air. As cold raindrops begin to fall to the ground, however, they pull cold air toward the surface with them at high speeds in a **downdraft** at the front of the storm. If you happen to be in this place as a storm forms, you can often feel a distinct increase in wind speed at this *gust front*. Where convection is most intense, an anvil head cloud forms at very high altitude (Figure 8.12c), which may be 6 to 12 km (20,000 to 40,000 ft). This cloud completely develops when the top of the convection bubble is pulled (or sheared) downwind by upper air flow. You can see an example of a forming anvil cloud in Figure 8.12d. If thunderstorm development occurs along an advancing cold front, a group of mature storms and associated anvils will form a feature called a *squall line* (Figure 8.12e).

Following the mature stage of storm development, thunderstorms begin to weaken. This weakening occurs during the *dissipation stage* and really begins when strong downdrafts develop when the storm has its maximum intensity. Look for these downdrafts in Figure 8.12c. These downdrafts cause the lower part of the atmosphere to cool, which serves to stabilize the air such that further convection cannot occur on a large scale. When these circumstances develop, rain can continue to fall for a short period of time, but will consist of a shower that gradually tapers off.

Severe Thunderstorms

In many cases, thunderstorms are short-lived events that are not particularly intense, consisting of brief downpours of rain, short gusts of moderately strong winds, some isolated lightning, and a few low rumbles of thunder. At other times, however, thunderstorms develop into severe storms that include very heavy rain along with extremely strong winds, intense lightning, numerous loud cracks of thunder, and hail. The most intense thunderstorms contain tornadoes that have incredibly strong winds that can cause great damage. Although these kinds of events are relatively rare, it's nonetheless important to understand how they develop because you may encounter such a storm in the future.

Severe midlatitude storms form during the mature stage of thunderstorm development. Along with very strong winds, one of the characteristics of a severe thunderstorm is abundant lightning. The production of lightning begins when collisions among ice crystals and rain droplets cause a separation to develop in the electrical charge within clouds (Figure 8.13a). The result is that the tops and bottoms of clouds have positive and negative charges, respectively, while the ground is positively charged. In stage 2 the negative charge on the bottom of

Downdraft *A rapidly moving current of cool air that flows downward in a thunderstorm.*

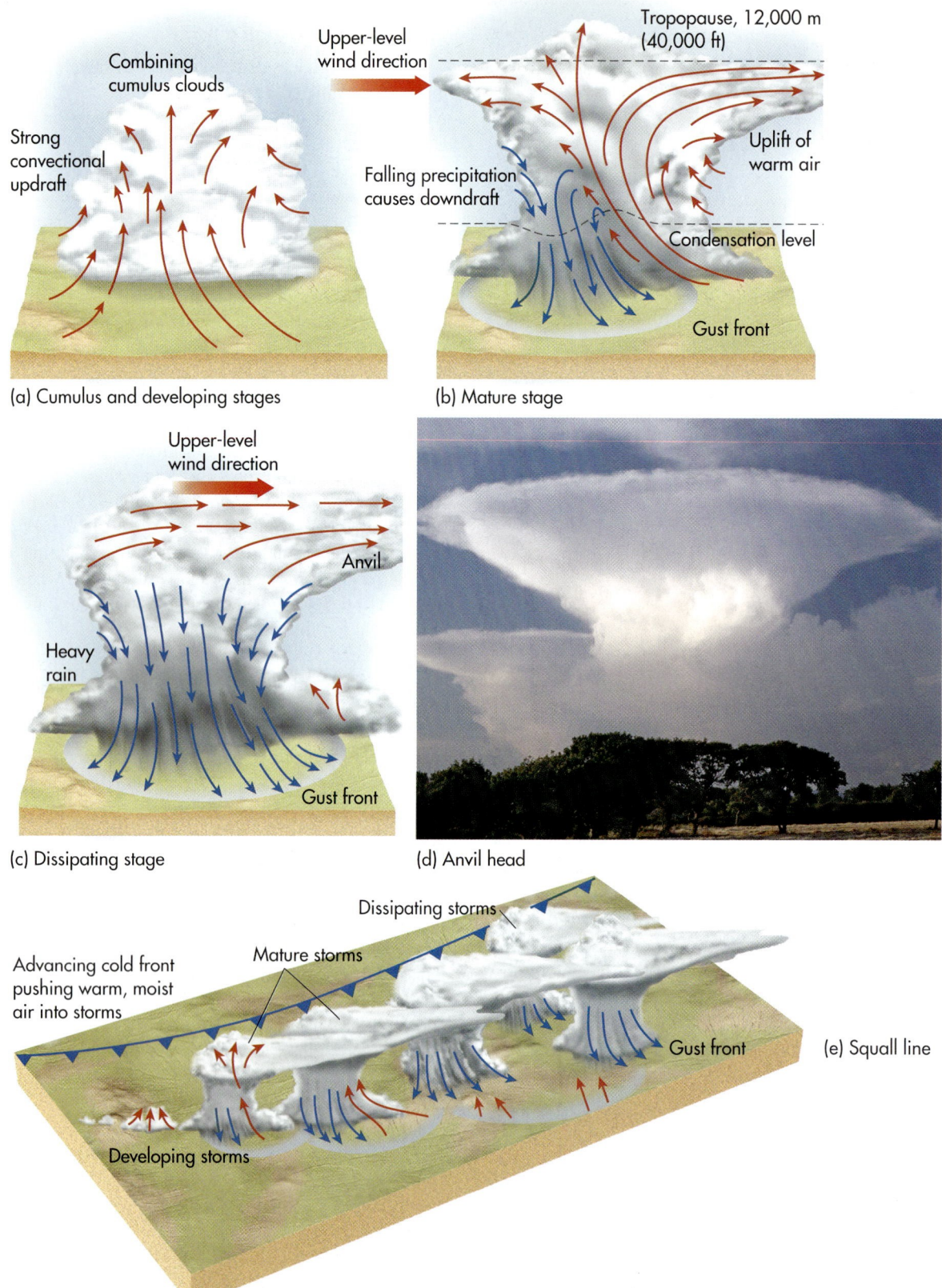

Figure 8.12 Stages of thunderstorm development. (a) Convection or advancement of a cold front forces air to rise during the cumulus stage. As the storm evolves, convection strengthens and the system moves into the developing stage. (b) The storm is most intense during the mature stage when strong updrafts occur next to strong downdrafts of cold air. (c) Development of an anvil head. This feature develops in strong storms at the altitude where convection can no longer continue, causing the upper clouds to be sheared downwind. (d) Anvil head of a thunderstorm. Note how the cloud appears to be spreading outward at the top. (e) Squall lines consist of several individual thunderstorms that form along an advancing cold front.

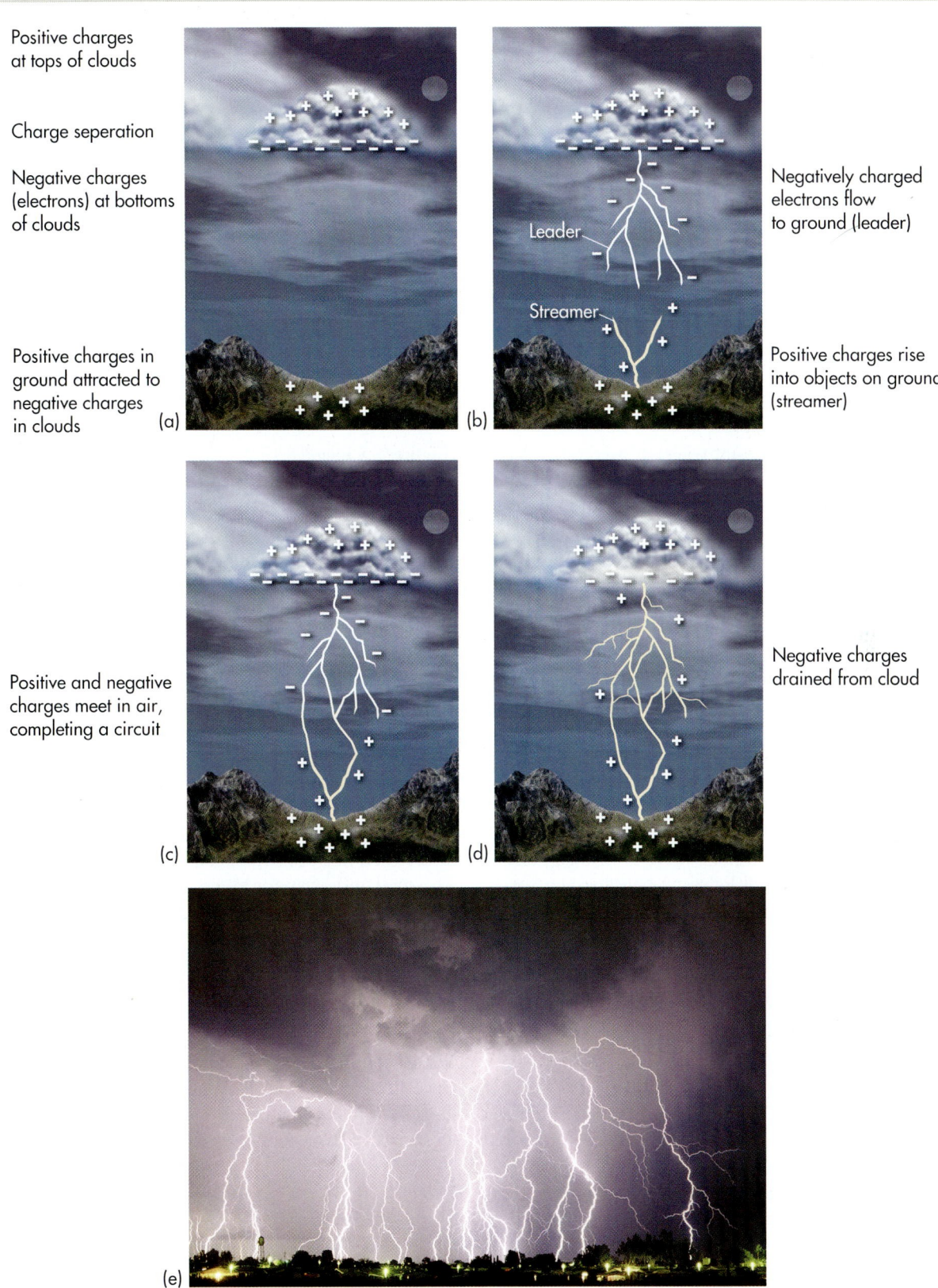

Positive charges at tops of clouds

Charge seperation

Negative charges (electrons) at bottoms of clouds

Positive charges in ground attracted to negative charges in clouds

(a)

Leader

Streamer

(b)

Negatively charged electrons flow to ground (leader)

Positive charges rise into objects on ground (streamer)

Positive and negative charges meet in air, completing a circuit

(c)

(d)

Negative charges drained from cloud

(e)

Figure 8.13 Evolution of a lightning strike. (a) Lightning begins when a separation develops in the electric charge between the bottom and top of a cloud. (b) Negatively charged electrons begin to move down toward the ground from the base of the cloud. At the same time, positively charged electrons begin to move upward through conducting material on the ground. (c) The circuit completes when the upward- and downward-moving electrons meet in the air. The process begins the visible lightning strike. (d) The lightning strike drains the cloud of excess negative electrons. (e) A massive lightning storm.

This Doppler radar image shows a line of thunderstorms along a cold front in Kentucky and Tennessee. Which one of the following choices best explains what occurred along the front?

a) The warmest air temperatures were northwest of the front.

b) The cold front caused mT air to rapidly uplift along the front.

c) The air southeast of the front consisted of cP air.

d) The most humid air was located northwest of the front.

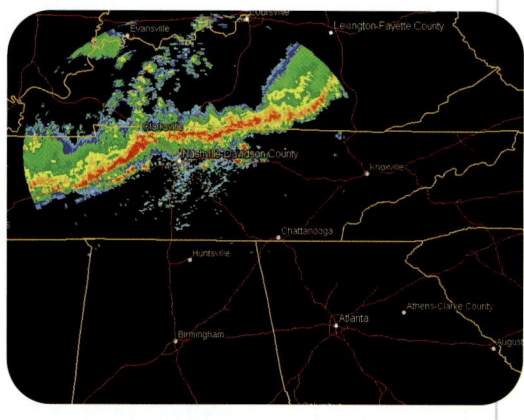

the cloud increases to the point where it overcomes the air's resistance to electrical flow. At this point, negatively charged electrons then begin flowing toward the Earth along a zigzag, forked path, called a *leader,* at about 97 km (60 mi) per second. As these electrons approach the ground, positive charge collects in the ground. These positive charges are attracted to the negative charges of the downward flowing leader and thus move upward through a channel (called a *streamer*) that begins at any conducting object, such as a tree, a house, and even people. Both leaders and streamers are invisible to the naked eye.

The electrical circuit becomes complete in stage 3 (Figure 8.13c), when the upward flowing charges meet those moving down from the clouds. This meeting typically occurs at an altitude of about 30 m (~100 ft) and is the place where the actual lightening bolt you see begins. Less than a millisecond later, millions of volts of electricity reach the ground. The actual lightning bolt that you see, however, is the return stroke from this initial ground impact (Figures 8.13d and 8.13e). This stroke moves upward at the incredible velocity of about 57,910 m/sec (190,000 ft/sec) and can reach altitudes of about 5 to 6.4 km (~3 to 4 mi). Once the return stroke reaches the cloud, it drains the cloud of its excess negative charge and the cycle begins again.

Thunder occurs in association with lightning because lightning heats the air in the path of the stroke to above 45,000° F. This incredible heat creates a shockwave, which disintegrates within a few meters (feet) of the stroke but generates soundwaves you hear as thunder. Have you ever noticed that thunder seems to rumble? This occurs because soundwaves travel from different distances within individual lightning bolts, from lightning strikes, and from sound bouncing off of clouds. If you want to approximate your distance to a lightning strike, count the number of seconds between when you see the lightning flash and hear the thunder. A good rule of thumb is that a lightning strike is about 1.6 km (1 mi) away for every 5 seconds it takes between when you see the flash and subsequently hear the thunder.

Another indicator that a thunderstorm is severe is hail, which consists of fragments of ice that plummet from the sky to the Earth. Hail indicates that strong **updrafts** are present at the rear of the storm, which repeatedly pull ice crystals back into the upper part of the storm, where they add layers of ice through condensation and grow larger. Once these developing hailstones grow sufficiently large that they cannot be pulled up anymore, they fall to Earth by the force of gravity, with large hailstones reaching speeds of 160 km (100 mi) an hour. Usually, hail ranges in size from *pea-sized* to *golf-ball-sized*. Occasionally, updrafts will be sufficiently strong to produce *softball-sized* hail. The largest hailstone ever recovered in the U.S. fell in south-central Nebraska during a severe thunderstorm in June 2003. This hailstone was 17.8 cm (7 in) wide, which is almost as large as a soccer ball! A severe hailstorm can be very destructive, causing millions of dollars in damage to agricultural crops, buildings, and automobiles (Figure 8.14).

Figure 8.14 Hail damage. Hail can cause extensive damage to homes, crops, and automobiles, as this photo demonstrates.

Updrafts *An area of rapidly flowing air that is moving upward within a thunderstorm.*

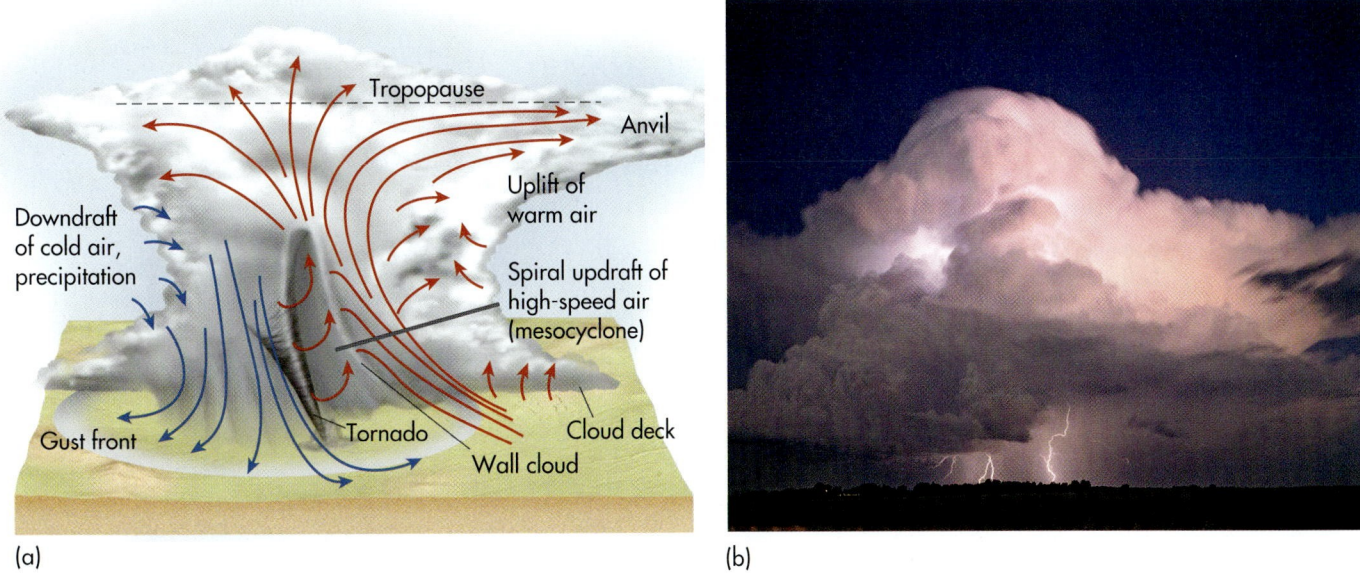

(a)　　　　　　　　　　　　　　　　　　　　**(b)**

Figure 8.15 (a) Typical development of a supercell thunderstorm. (b) Seen from the outside, a supercell thunderstorm is an ominous sight. The lower (circular) part of the cloud complex is rotating cyclonically.

Tornadoes

When certain atmospheric conditions are present, very strong **supercell thunderstorms** can develop. These extremely violent storms are unique because they contain large rotating updrafts known as **mesocyclones** (Figure 8.15) that range from 3.2 to 9.7 km (2 to 6 mi) in diameter. When these atmospheric conditions occur, they can easily evolve to include tornadoes, which are small, but deep, low-pressure cells surrounded by a violently spinning mass of air. Although generally less than 400 m (0.25 mi) in diameter, tornadoes are the most destructive of all atmospheric phenomena, with wind speeds ranging from 320 to 800 km (200 to 500 mi) per hour. These high speeds occur because tornadoes have extremely tight pressure gradients that may differ by 100 mb between the inside and outside of the funnel. Tornadoes are classified on the Fujita scale based on their strength (Table 8.1), ranging from F-0 (gale) tornado to the extremely rare F-6 (inconceivable) tornado.

Figure 8.16 shows the evolution of a tornado. Tornadoes usually form at the rear of a storm when strong updrafts occur in conjunction with wind shear at higher altitudes. Wind shear occurs when winds are moving in different directions at variable altitudes. This wind shear causes a horizontal vortex of air to form (Figure 8.16a) that is then pulled vertically in the updraft (Figure 8.16b). Assuming this process is not interrupted, a funnel devel-

ops that becomes a full-fledged tornado if it reaches the ground (Figure 8.16c). Figure 8.16d shows a supercell tornado that formed near Spearman, Texas in May 1990. Notice how similar this tornado and spawning supercell system are to the diagram in 8.16c. The air flow in a fully developed tornado can generate fierce winds of over 300 mph. Fortunately, tornadoes seldom last very long, only 15 minutes on average. The air sucked into the funnel slows down the winds—the tornado becomes clogged with the air it takes in. Eventually the funnel stretches out into a thin shape and disappears.

Although studying tornadoes is interesting purely from a meteorologic perspective, the primary goal of tornado research is to improve public safety. In this context, the National Weather Service constantly monitors atmospheric conditions and makes public announcements as weather events unfold on any given day. A *tornado watch* is issued for an area, such as the eastern part of Oklahoma or central Texas, when atmospheric conditions are favorable for the development of tornadoes. When an actual funnel or tornado has been identified, a *tornado warning* is issued for a specific place or, perhaps, county. At this time sirens are sounded and people are advised to seek shelter, either in a low-lying ditch or, preferably, in a basement. You may have experienced such an event in your life.

With respect to tornado monitoring, the most significant technological development in the past few decades has been

Supercell thunderstorms　*Large thunderstorms that contain winds moving in opposing directions and are associated with strong winds, lightning, thunder, and sometimes hail and tornadoes.*

Mesocyclones　*Strong updrafts that are rotating within a supercell thunderstorm.*

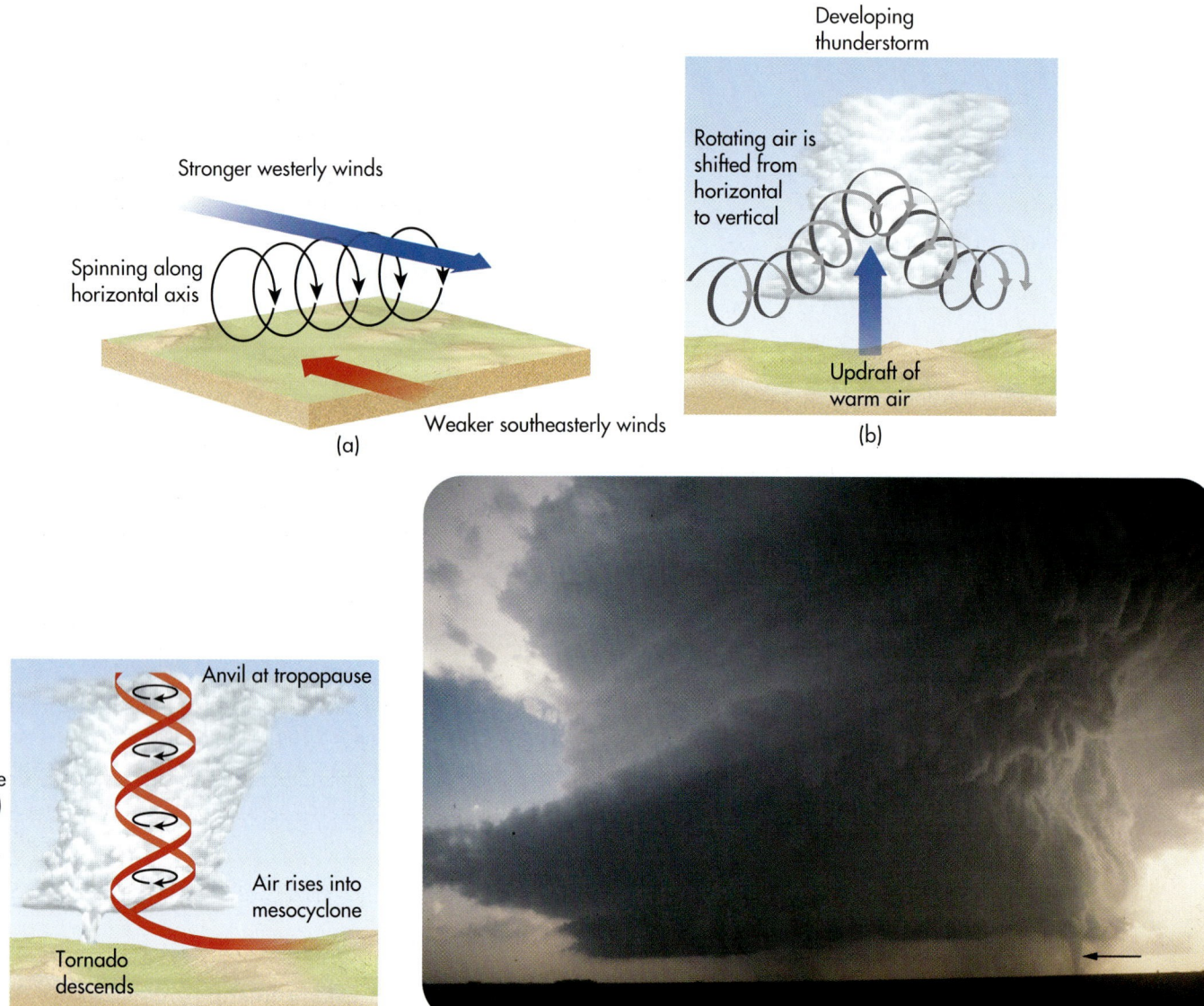

(a)

(b)

(c)

(d)

Figure 8.16 Evolution of a tornado. (a) Air begins to rotate due to wind shear; (b) updrafts pull a rotating cylinder of air upward; (c) a tornado develops when the funnel reaches the ground. (d) Supercell tornado (arrow) near Spearman, Texas in 1990.

Doppler radar, such as the Next Generation Weather Radar (NEXRAD) used by the National Weather Service (Figure 8.17). Prior to this development, tornado warnings were based largely on direct observations, which were often too late to save lives. In contrast to conventional radar, which shows the intensity of precipitation in any given storm, Doppler radar can detect *both* rainfall patterns and the actual rotation of a tornado. It does so by making use of the Doppler Effect, described first by the 19th-century physicist Christian Doppler, which states that the frequency of energy waves generated by a moving source changes relative to an observer.

A classic example of this effect is the sound a train makes as it approaches you. What happens? The pitch of the train's whistle rises with the approach of the train and then lowers after it passes. Using this effect, Doppler radar can determine that raindrops on one side of a tornado are moving toward the radar detector while those on the other side are rotating away because of the shift in energy frequency that occurs relative to each side of the twister (Figure 8.17a). This rotation is seen by the meteorologist in the weather laboratory as a distinct **hook echo** on the computer screen (Figure 8.17b). If such a feature is identified, a tornado warning is quickly issued for the area in the path of the storm and people usually have sufficient time to seek shelter before the storm strikes.

Hook echo *The diagnostic feature in Doppler radar that indicates strong rotation is occurring within a thunderstorm and tornado development is thus possible.*

TABLE 8.1	The Fujita Scale to Classify Tornadoes Based on Their Wind Speed		
Intensity Phrase	**Wind Speed**	**F-Scale Number**	**Type of Damage Done**
Gale tornado	40–72 mph	F-0	Some minor damage to chimneys, trees, and signs.
Moderate tornado	73–112 mph	F-1	Winds of this speed rip shingles from roofs and may push mobile homes off their foundations.
Significant tornado	113–157 mph	F-2	These winds can tear roofs off of frame houses, destroy mobile homes, and snap large trees in two.
Severe tornado	158–206 mph	F-3	This kind of storm tears roofs and walls off of even well-constructed houses and can uproot most trees in a forest.
Devastating tornado	207–260 mph	F-4	Winds of this magnitude destroy well-built houses and throw cars briefly through the air.
Incredible tornado	261–318 mph	F-5	These very strong winds can rip houses off of their foundations and blow them in the air for a considerable distance before they are destroyed; cars can be carried up to 30 m (100 ft).
Inconceivable tornado	319–379 mph	F-6	Although these winds are very unusual, they may briefly develop within an F4 or F5 tornado, resulting in things like cars and refrigerators being carried as missiles for considerable distance to do extensive damage.

Although tornadoes occur in every month of the year in the continental U.S., they most frequently develop between April and July. This time interval is peak tornado season because the strongest contrast in air masses exists at this time in the midlatitudes. Similarly, although tornadoes statistically occur at least once per year in every state of the U.S. (except Alaska), they are by far most common in the central part of the country (Figure 8.18). This region is prime tornado country because the relatively flat terrain provides an ideal place for Canadian cP and Gulf mT air to interact.

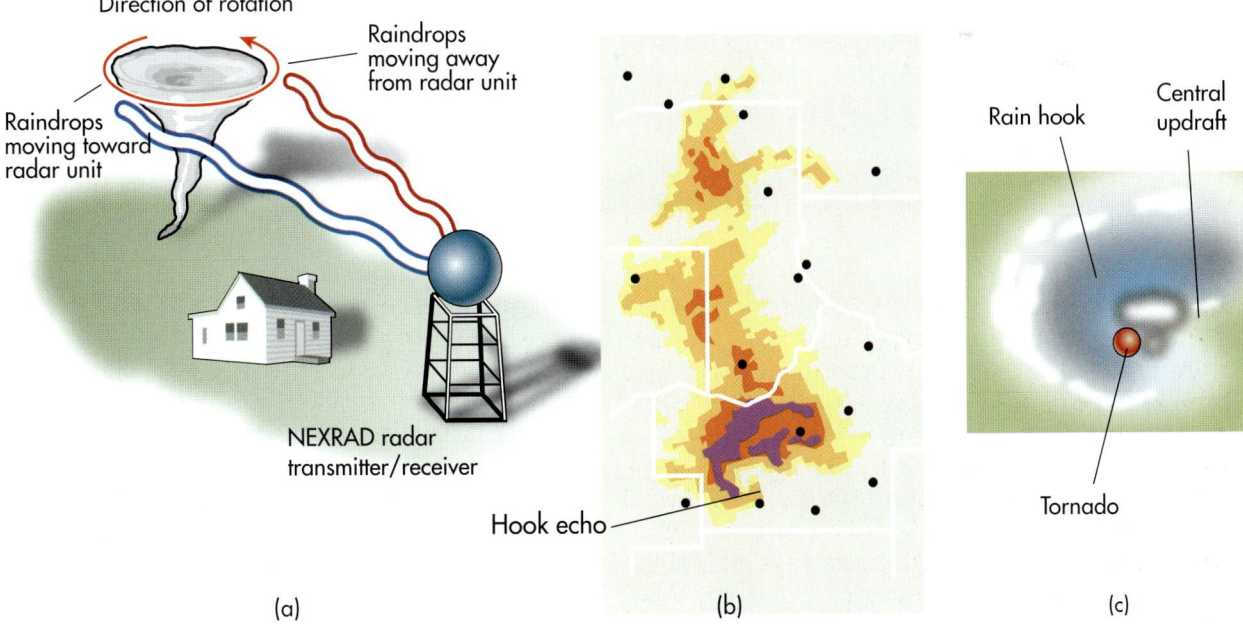

(a) (b) (c)

Figure 8.17 Doppler radar. (a) Doppler radar "sees" the rotation of a tornado because of the Doppler Effect, which, in this case, reflects the variation in returning wave frequency from either side of the system. (b) The rotation of a tornado is displayed as a hook echo on the computer screen. (c) A hook echo is indicative of aggressive cloud rotation, which usually means that a tornado is present.

TORNADO ALLEY

An area 740 km (460 mi) long and 644 km (400 mi) wide spawns more tornadoes than any other place on Earth, as many as 300 per year. Known as "Tornado Alley," the region extends from northern Texas through Oklahoma, Kansas, Missouri, and into Iowa and Illinois. Many people in this area build underground tornado shelters when they build their homes, so they have a place to go for protection when "twisters" drop down from the sky.

June 1966, Topeka tornado. (a) This storm was an F-5 tornado that traveled through much of the city of Topeka, Kansas. This tornado remains one of the five costliest tornadoes in U.S. history.

Why does this area produce so many tornadoes? This is a region where warm, moist air from the Gulf of Mexico collides with cool, dry air blowing from the west. These conditions are just right for the massive thunderstorms from which tornadoes form. Some of the most damaging tornadoes in history have struck in this region, including one that con-

TORNADOES

Tornadoes are fascinating natural phenomena that are often caught on film by storm chasers and regular citizens alike. To see some example of tornadoes, go to the *GeoDiscoveries* website and select the module **Tornadoes**. The first part of this module is a video that shows some of the various forms that tornadoes take as they move across the ground. The second part of the module is an animation that demonstrates the wind patterns of a tornado and how they cause damage to homes. After you watch this animation, be sure to answer the questions at the end of the module to test your understanding of this concept.

Moore/Oklahoma City tornado. This F-5 tornado, which was part of the May 3, 1999 tornado outbreak in central Oklahoma, had a continuous damage path which at times reached 1.5 km (0.9 mi) wide.

tributed greatly to my lifelong interest in physical geography. This storm was an F-5 tornado that struck Topeka, Kansas, where I lived as a youth, in June 1966. This tornado was about 800 m (0.5 mi) wide and was on the ground for about 30 minutes as it traveled 35 km (22 mi) through the heart of the city. The infamous "Topeka Tornado" killed 16 people, damaged 3000 homes, and caused $100 million (1966 dollars) in damage, making it still one of the costliest storms in U.S. history. Even though I was only 8 years old at the time, I remember it vividly and have been fascinated by such storms ever since.

Sometimes atmospheric conditions are ripe in Tornado Alley to produce a *tornado outbreak*, with numerous tornadoes forming in a small region over the course of a few hours. One such outbreak occurred on May 3, 1999 in central Oklahoma, when a series of eight supercell thunderstorms produced a total of 58 tornadoes. The largest of these storms was the F-5 Moore/Oklahoma City tornado, which left a 70-km (43.5-mi) long path of damage. Overall, this outbreak killed about 40 people and caused approximately $1.2 billion in damage.

Damage caused by the Moore/Oklahoma City tornado. As this photograph clearly demonstrates, tornado-force winds are extremely powerful. This portion of the Oklahoma City metro area was completely leveled.

Tropical Cyclones

So far, we've been discussing midlatitude cyclones and the storms associated with them. Now let's now turn our attention to cyclonic storms that form in the tropical regions. Although tropical systems are similar to midlatitude systems in that they consist of rotating air masses of low pressure, they differ significantly because they form in regions where there are no fronts or contrasting air masses. Instead, a tropical cyclone develops entirely within a homogeneous air mass at very low latitudes. As a result, abundant water vapor and latent heat are present to fuel these systems. About 80 tropical cyclones develop around the world every year, with the strongest becoming extremely powerful storms that, depending upon the region, are called hurricanes, typhoons, or cyclones.

You may be asking, if tropical storms don't contain contrasting air masses, how do they form? Most Atlantic hurricanes form in association with an atmospheric feature called an **easterly wave,** which is a slow-moving

Easterly wave *A slow moving trough of low pressure that develops within the tropical latitudes.*

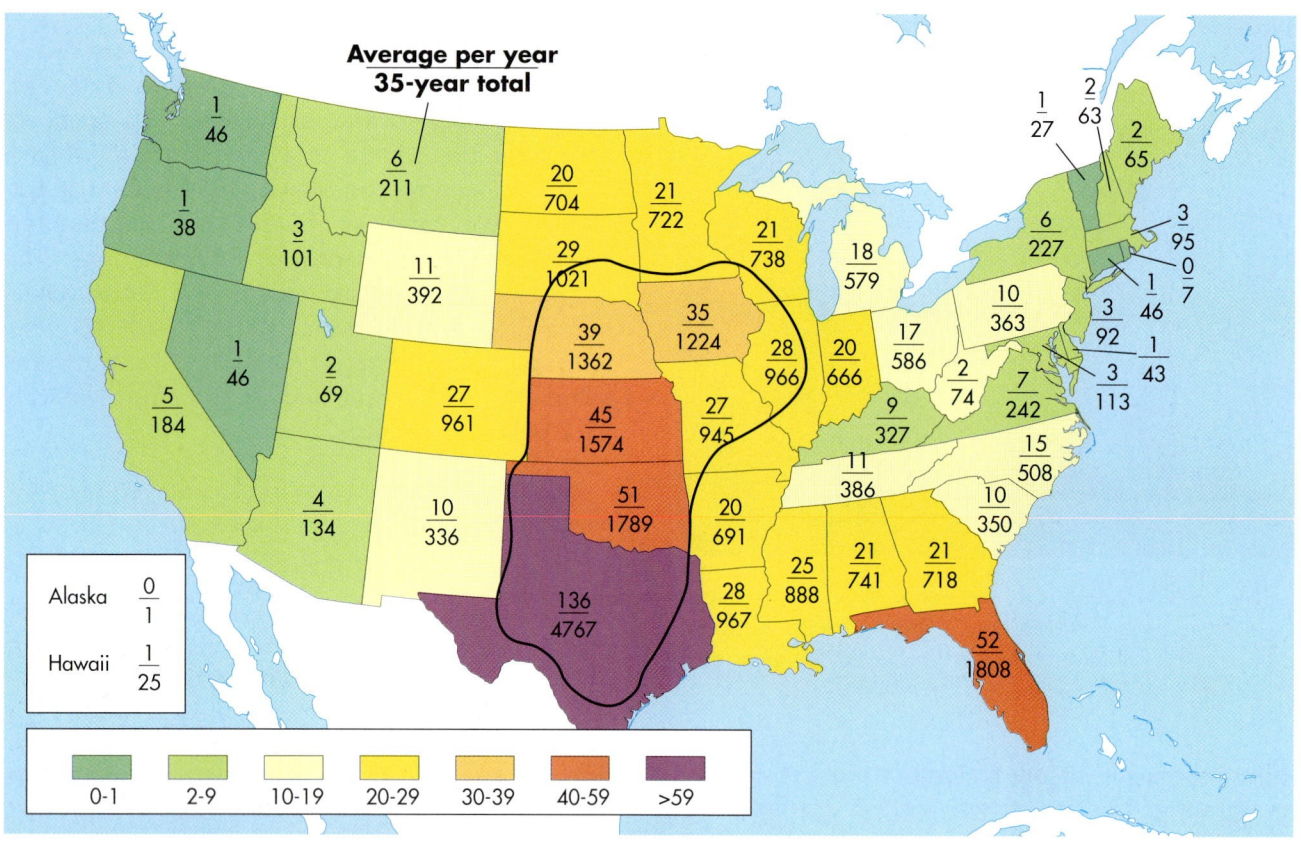

Average per year / 35-year total

Color	Range
	0-1
	2-9
	10-19
	20-29
	30-39
	40-59
	>59

Figure 8.18 Distribution of tornadoes in the mainland United States. Note the yearly average and total number observed for over 35 years per state. The approximate boundary of Tornado Alley is outlined in black.

trough of low pressure that migrates along the belt of tropical easterlies. Recall that this belt is associated with the easterly trade winds and occurs between 5° to 30° N and S latitude. This low-pressure system causes air to converge on the windward (east) side of the trough axis while air diverges in the lee (west) side of the system. The convergence of upper air on the eastern side of the wave causes further convectional uplift within the system and a zone of showers develops parallel to the trough axis. Figure 8.19 shows the basic structure of an easterly wave, including the axis, wind flow patterns, and precipitation zone.

If the storm strengthens into a **tropical depression,** it begins to rotate cyclonically around a definable center of low pressure at the same time that a complex of strong thunderstorms develops. At this time, sustained winds in the center of the system are 20 to 34 knots (23 to 39 mph). If the storm continues to strengthen, it becomes a **tropical storm** and the maximum sustained winds reach 35 to

Tropical depression *A tropical low-pressure system with central sustained winds ranging between 20 to 34 knots (23 to 39 mph).*

Tropical storm *A tropical low-pressure system with maximum sustained winds between 35 to 63 knots (39 to 73 mph).*

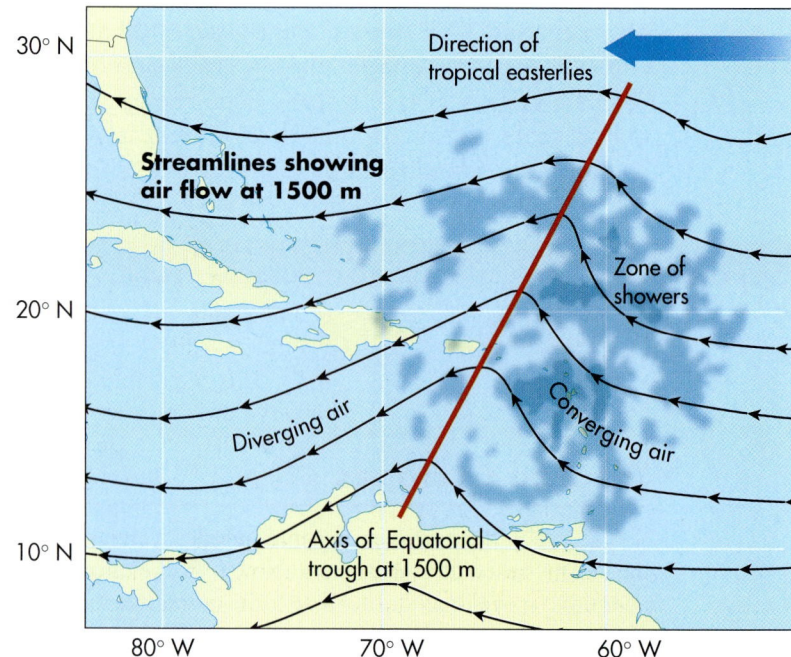

Figure 8.19 Characteristics of a tropical wave. Tropical waves consist of a center of surface low pressure that results in the convergence and divergence of air aloft. Showers develop on the eastern side of the wave axis because converging air causes convectional uplift.

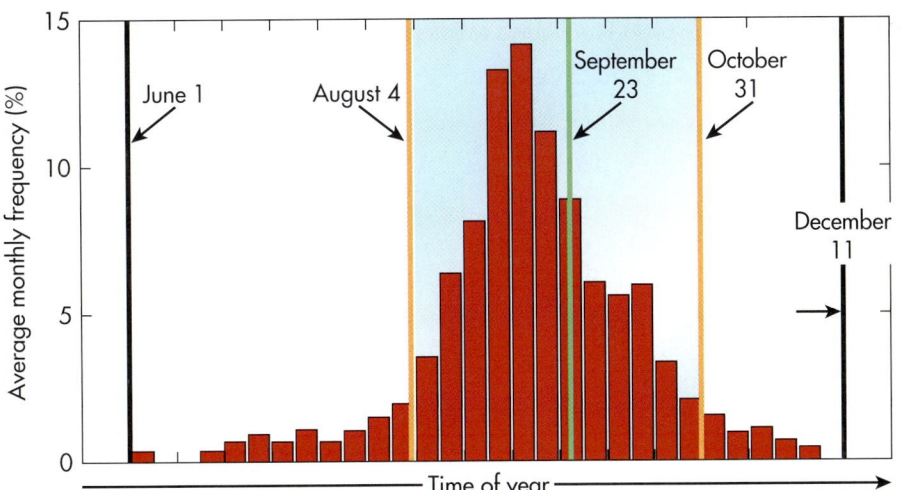

Figure 8.20 Average frequency of monthly hurricanes in the Atlantic Ocean from 1896 through 1996. Hurricanes are most likely to develop from the beginning of August to the end October because sea surface temperatures are warmest this time of year in the Northern Hemisphere. (Credit: NOAA)

63 knots (39 to 73 mph). At this time, the storm begins to become more circular in shape and is assigned a formal name (such as *Charlie*) for monitoring purposes as it proceeds on its easterly migratory path.

Hurricanes

If a tropical storm in the Atlantic Ocean or eastern Pacific Ocean strengthens to the point that the maximum sustained winds are greater than 63 knots (73 mph), then it officially becomes a **hurricane.** A storm of similar strength in the western Pacific Ocean is called a *typhoon* or a *cyclone* if it forms in the Indian Ocean. Australians call such a storm a *willy nilly* because they twist about and wreak havoc. For purposes of discussion here, we'll use the term *hurricane* and focus on Atlantic storms.

Although it's common to hear the word *hurricane*, storms of this magnitude are really quite rare because they require a particular combination of environmental variables. Most importantly, they require an abundance of latent heat energy to fuel their development. Such an abundance of heat energy can occur only when a great deal of evaporation from the underlying ocean takes place, which means that the surface waters must be warmer than about 27° C (80° F). This temperature requirement is the primary reason why hurricane season in the Atlantic Ocean largely occurs from August to October (Figure 8.20); in other words, in the period following the highest Sun and insolation.

In addition to the warm ocean waters, another requirement for hurricane development is that the upper air wind pattern must be favorable, with high air pressure aloft. This relationship may seem strange to you because hurricanes are zones of low air pressure, which might lead you to wonder how both low pressure at the

surface and high pressure in the upper atmosphere could exist at the same time. Remember that the atmosphere has many different layers and that air can flow differently in each of them. In the case of a hurricane, high pressure aloft is required because it serves to "cap" the storm in the vertical direction so it can continue to strengthen at the surface. Otherwise, if the surface air can freely convect into the upper atmosphere, or if shearing winds aloft blow the top off the storm, then the system does not fully organize.

If all the necessary ingredients are in place for hurricane development, the system can grow to immense

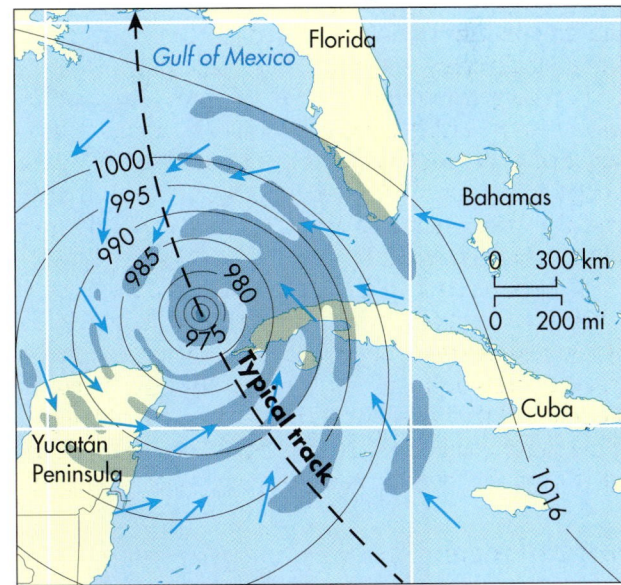

Figure 8.21 Hurricane weather map. This simplified weather map of a typical Atlantic hurricane illustrates the cyclonic rotation and pressure gradient associated with the system. Shaded areas are regions of heavy rainfall.

Hurricane *A tropical circulatory system with maximum sustained winds greater than 63 knots (73 mph).*

TABLE 8.2	Saffir-Simpson Scale of Tropical Hurricane Intensity		
Category	Central Pressure, mb	Mean Wind, m/sec (mph)	Storm Surge, m (ft)
1 Weak	>980	33–42 (74–95)	1.2–1.7 (4–5)
2 Moderate	965–979	43–49 (96–110)	1.8–2.6 (6–8)
3 Strong	945–964	50–58 (111–130)	2.7–3.8 (9–12)
4 Very strong	920–944	59–69 (131–155)	3.9–5.6 (13–18)
5 Devastating	<920	>69 (>155)	>5.6 (>18)

size and strength. Depending on the specific atmospheric conditions, hurricanes can range in diameter from 150 to 500 km (100 to 300 mi). Hurricanes acquire their strength because of the steep pressure gradient that develops from the outside of the storm to a center of extremely low pressure, which can depress to 900 mb or even lower. As a result of this steep pressure gradient, winds spiral into the core of the system at very high speed. Figure 8.21 shows an example of the pressure gradient and storm size.

Hurricanes are classified based on their wind-speed intensity on the Saffir-Simpson scale. Table 8.2 shows this scale and how it relates to the central pressure of a storm and its wind speed. Notice the inverse relationship between these two variables. The fourth column of the table shows storm surge, which is the rise in sea level that occurs because strong winds cause water to pile up ahead of the storm. Storm surge is typically highest on the right front quadrant of a hurricane, where onshore winds flow

as the storm approaches. These winds tend to push the water ahead of the storm and on to the coast. In addition to wind speed, the height of a storm surge depends on the timing of the tides and configuration of the coastline, with shallow coastlines typically more susceptible because water has a tendency to pile up at those locations.

Figure 8.22 shows the typical anatomy of a hurricane. Note the distinct bands of cumulonimbus clouds, which can drop prodigious amounts of rainfall, surrounding the core of the system. Also notice that air flows inward at the surface to a point where it rapidly flows upward at the eye wall. Remember that the entire time the air is flowing inward at the surface, and upward at the eye wall, it's spinning cyclonically at intense speed. A unique characteristic of a well-developed hurricane is the eye, which occurs in the center of the storm and is associated with high-pressure air that rapidly flows toward the surface and warms adiabatically as it descends. As a result, this region of the storm is cloud-free with no winds and can be seen in satellite images of well-developed hurricanes.

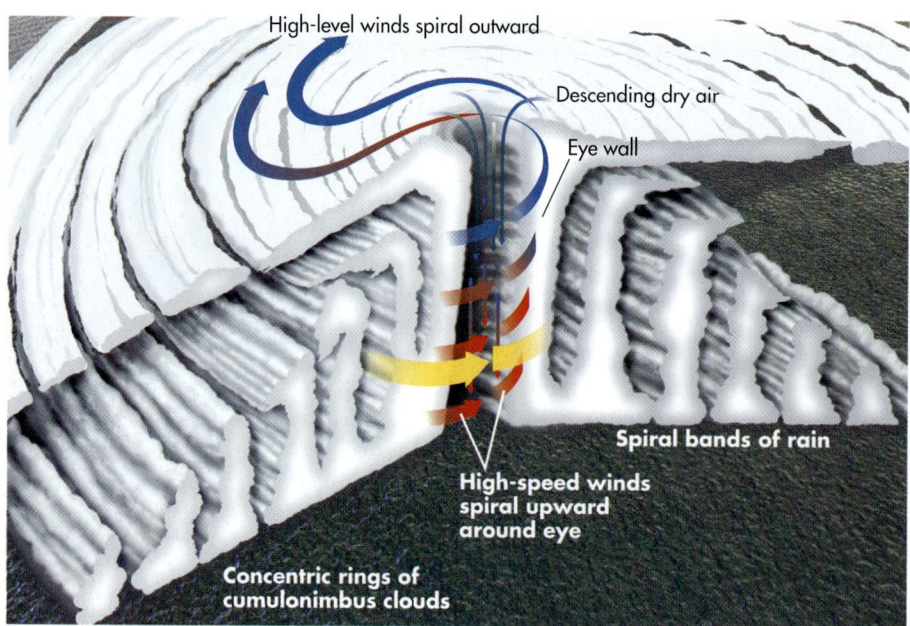

Figure 8.22 Components of a typical hurricane. Hurricanes contain numerous spiral rain bands of cumulonimbus clouds. Air spirals inward until it reaches the eye wall, where it circles upward. The eye of the storm is clear and calm, which is the result of air descending and adiabatically warming in the core of the system.

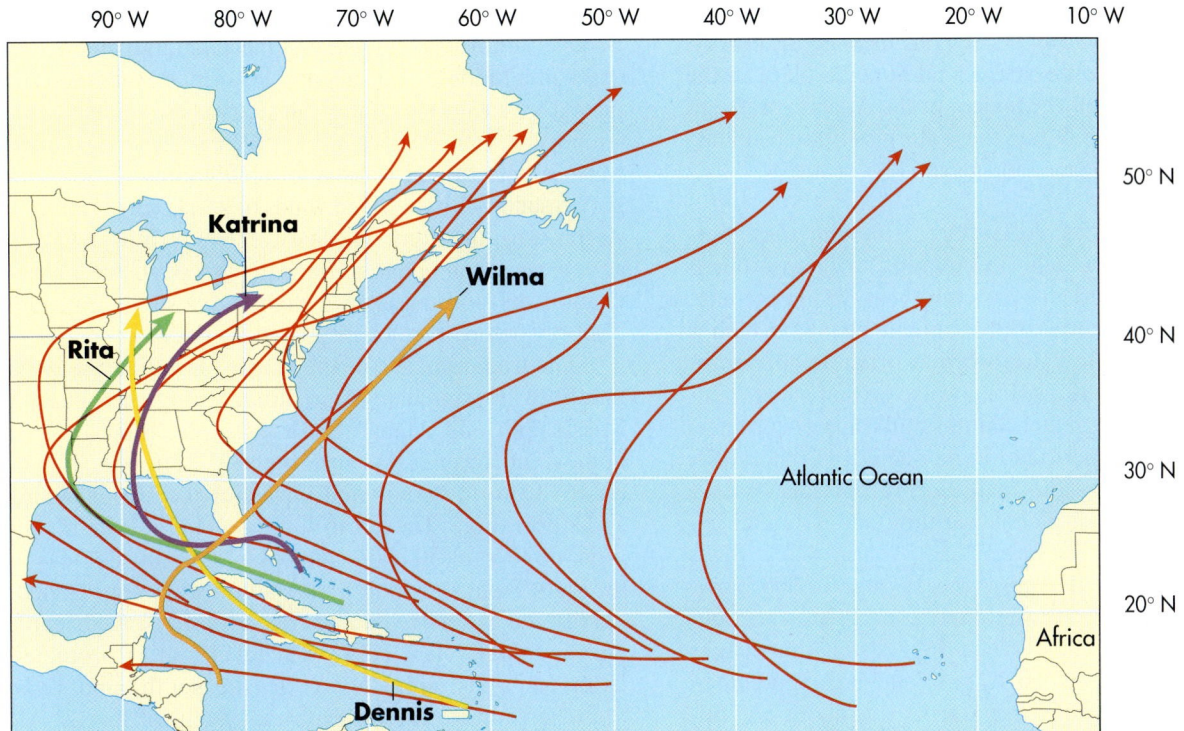

50° N
40° N
30° N
20° N

90° W 80° W 70° W 60° W 50° W 40° W 30° W 20° W 10° W

Katrina

Wilma

Rita

Atlantic Ocean

Africa

Dennis

Figure 8.23 Typical track of August hurricanes in the Atlantic Ocean. During this time of year, hurricanes begin to develop over the warm waters of the eastern Atlantic. Subsequently, they migrate with the easterly trade winds until they become influenced by the mid-latitude westerlies. This map also includes the tracks of the four major hurricanes that struck the U.S. in 2005. (Credit: NOAA)

Because hurricanes evolve in the easterly trade belt, they migrate in a fairly predictable way. Figure 8.23 shows typical tracks that hurricanes take in August. As you examine this illustration, notice that many of the storms originate in the eastern Atlantic Ocean. These storms actually begin to organize when they exit western Africa as easterly waves in the trade-wind belt. When they reach the open warm waters of the eastern Atlantic, they begin to intensify, first as tropical depressions, then as tropical storms, and finally, if conditions are right, into hurricanes. The systems continue to migrate westerly, sometimes strengthening, others times weakening. If the storm remains organized for several days, it ultimately becomes influenced by the midlatitude westerlies and is

VISUAL CONCEPT CHECK 8.3

This figure is a satellite image of Hurricane Floyd in September 1999, which at one point developed into a Category 5 storm. Note the position of Floyd in the central Atlantic Ocean in this image. Imagine that you're a hurricane forecaster who has become aware of the hurricane's presence for the first time. Describe how you would expect Floyd to migrate in the future, beginning at this position and ending when the storm would most likely dissipate.

Hurricane Floyd

deflected to the northeast. As you can see, some storms remain over open water throughout their entire history. Other storms, however, strike land at some place in the Caribbean Sea, Gulf of Mexico, or the Atlantic seaboard in the United States. When these storms strike land, they can result in significant damage and loss of life.

Recent Hurricane Activity Although some hurricanes occur every year, the number of intense hurricanes has increased the past few years. This increase may be linked in part to global climate change, which is believed by some to cause warmer ocean temperatures that fuel stronger storms. It may also be that we are simply in a period of greater natural hurricane occurrence, one similar to the period of higher activity in the 1930s and 1940s. Perhaps the increased recent activity is a combination of both global warming and a natural cycle. No one knows for sure.

Regardless of the cause, the 2004 and 2005 hurricane seasons were very intense. Florida was particularly hard hit in 2004, when four strong hurricanes struck the state between August and late September. The combined cost of these storms was about $30 billion in Florida alone and caused almost 100 deaths. In 2005, a record 27 tropical storms formed, with 15 reaching hurricane status. Seven of these hurricanes strengthened into major hurricanes ($\geq$ Category 3). Five of these major hurricanes reached Category 4 status and three (another record!) strengthened into Category 5 storms. Of these, Hurricane Wilma was the most intense storm ever recorded in the Atlantic Ocean with a central pressure of 882 mb. Four major hurricanes struck the U.S., as you can see in Figure 8.23.

Of all of the 2005 hurricanes, easily the most destructive was Hurricane Katrina, which struck the Gulf Coast near New Orleans on August 29. Katrina formed over the Bahamas on August 23 and first made landfall on the east coast of Florida as a Category 1 storm. It subsequently moved west over the Gulf of Mexico, where surface water temperatures were about 32° C (90° F). These incredibly warm waters caused Katrina to rapidly intensify to Category 5 status with sustained winds of 280 km/h (175 mph). At this time, the storm had an extremely well-defined eye, as you can see in Figure 8.24a. After this period of intensification, the storm weakened slightly to a strong Category 3 shortly before it struck the Louisiana coast.

The tragic images of the storm's aftermath may be indelibly imprinted in your mind. Along the coast of Mississippi and Alabama, for example, thousands of homes and businesses were literally flattened (Figure 8.24b) from the combined power of the strong winds and storm surge. The storm surge was highest in this region, up to 9 m (30 ft) in some places, because it was in the right front quadrant of the storm as it approached. Additional extensive damage occurred in New Orleans, which actually lies below sea level and is protected from the Mississippi River (to the south) and Lake Pontchartrain (to the north) by a system of levees. Although we'll discuss levees in more detail in Chapter 16, for now think of a levee as a wall that keeps high water out of lower areas. The storm surge in Lake Pontchartrain caused several levees to fail, resulting in severe flooding (Figure 8.24c) and the displacement of tens of thousands of people. As of January 2006, the confirmed death toll from Hurricane Katrina was over 1400 people, with almost 5000 still missing. Hundreds of thousands of people across the region were made homeless. The total cost of the storm was expected to be over $100 billion, making it by far the costliest hurricane ever in the United States.

GEODISCOVERIES : MIGRATION OF HURRICANE KATRINA

An excellent example of the need for monitoring can be seen in the video showing the path of Hurricane Katrina in 2005. To see this video, go to the *GeoDiscoveries* website and access the module **Migration of Hurricane Katrina**. The video follows the path of this catastrophic hurricane as it moved across the Gulf of Mexico. It begins when the storm was centered over the southern part of Florida, shortly before it moved over the Gulf. There are several things you should notice as you watch the video. First, look for the overall rotation of the system and how its path changed over time. Also note the rapid intensification of the system as it migrated across the Gulf. You'll be able to see this intensification because the eye of storm suddenly becomes very prominent. At this point the storm was at its most powerful, specifically a Category 5 hurricane. Next notice how after this rapid strengthening the size of the eye fluctuated; this, which occurred because the storm's strength varied some before it made landfall. Lastly, notice how the eye suddenly disappeared after the storm reached land. As you watch the video, think about why satellite images such as these are essential for good monitoring and how they save lives. Once you finish the video, be sure to answer the questions at the end of the module to test your understanding of this concept.

(a)

(b)

(c)

Figure 8.24 Hurricane Katrina. (a) Satellite image of Hurricane Katrina shortly before it struck Louisiana. Note the incredibly well-defined eye, which developed when the storm reached maximum strength. (b) Extensive damage along the Mississippi coast was caused by a combination of strong winds and storm surge, which in some places was over 9 m (30 ft) high. (c) Failure of protective levees in New Orleans caused widespread flooding.

Monitoring Hurricanes The intensity of the 2004 and 2005 hurricane seasons underscores the need for accurate hurricane forecasting and monitoring. Had forecasting not been in place, it is very possible that thousands more people would have died during Hurricane Katrina alone. Such a high death toll occurred during the infamous Galveston Hurricane that struck Galveston, Texas in 1900. Because accurate forecasting and monitoring methods did not exist at that time, the storm basically caught people unaware, resulting in somewhere between 6000 and 12,000 deaths.

Fortunately, in this day of real-time satellite imagery, it is possible to monitor the development of any individual storm as it migrates across the Atlantic or Caribbean. As a result, forecasters can predict the general path a storm will take several days in advance and provide warning of the storm's approach to residents in its path. If you happen to live in the southeastern part of the U.S., you're probably very aware of this monitoring process and how it may affect your life. Should you happen to live elsewhere, try paying attention to storm events in the Atlantic Ocean during the next hurricane season late next summer and early fall. These systems are easy to follow, can be quite dramatic, and serve as an excellent window into the way in which the atmosphere functions.

1. In general, the most severe midlatitude storms form along strong cold fronts when warm, moist (mT) air ahead of the front is rapidly forced aloft.

2. Thunderstorms evolve in predictable stages, including the cumulus stage, mature stage, and dissipating stage. These stages are related to the upward and downward flow of air.

3. Strong winds, lightning, and thunder typically accompany severe storms. Lightning occurs when opposing electronic charges develop between the base of clouds and the ground. Thunder is created by the shockwave produced when lightning superheats the atmosphere locally.

4. The strongest storms associated with midlatitude weather are tornadoes, which are localized bodies of intense low pressure that develop in association with supercell thunderstorms.

5. The strongest tropical storms are hurricanes, which develop when easterly waves strengthen beyond the depression and tropical storm phases to produce sustained winds greater than 63 knots (73 mph).

The Big Picture

So far in this book we have covered a very wide range of topics, including Earth–Sun relations, insolation, temperature, global wind patterns, adiabatic processes, and weather systems. We can now integrate all of these concepts and associated processes into a geographic framework that helps explain a variety of phenomena on Earth, including vegetation patterns, soils, the weathering of rocks, and even the location of large sand dunes. A good place to begin this integration is Chapter 9, which focuses on global climate processes and patterns.

You can see in the satellite image of Africa, an example of how this integration occurs. This image contains several distinct geographic patterns. For example, a zone of dark green at the Equator represents the tropical rain forest, which correlates to the region that receives abundant precipitation over the entire year. In contrast, the broad expanse of tan in the northern part of Africa is the Sahara Desert, which corresponds to the part of the continent that receives very little annual rainfall. These patterns, as well as numerous others in Africa and the rest of the world, are related to the wide variety of processes presented so far in this text. In the case of Africa, the geographic patterns are closely related to the seasonal migration of Hadley cells presented in Chapter 6. Can you imagine how they are related? As you work your way through this next chapter, be sure to think about how all of the factors you've learned about so far fit into the overall climate pattern.

Summary of Key Concepts

1. An air mass is a large body of air that forms in specific geographic regions and thus has distinctive characteristics. Five principal air masses affect North America. Continental air masses include continental polar (cP), continental arctic (cA), and continental tropical (cT). The maritime air masses are maritime tropical (mT) and maritime polar (mP).

2. Air masses have distinct boundaries called fronts. At a stationary front, contrasting air masses are flowing parallel to one another. A warm front is a place where warm air is advancing into relatively cool air. Given that warm air slowly slides over the top of the cooler air along a warm front, rainfall is slow and steady. A cold front is a place where cold

air is advancing into relatively warm air. Given the higher density of colder air, rainfall is intense and of short duration along the front because warm air cools quickly when it is rapidly forced aloft.

3. A midlatitude cyclone spins in a counterclockwise fashion as seen from above. These atmospheric features develop when undulations form in the polar front jet stream at the 500-mb pressure level. As a cyclone spins, it pulls warm (mT) air up from the south on its eastern side. This warm, moist air encounters cold air as it moves to the north. The cyclone also pulls cold air (cP) down from the north on its western side. This cold, dry air encounters warm air as it moves to the south.

4. In general, the most severe midlatitude storms form along strong cold fronts when warm, moist (mT) air ahead of the front is rapidly forced aloft. Thunderstorms evolve in predictable stages, including the cumulus stage, mature stage, and dissipating stage, that are related to the upward and downward flow of air. The strongest storms associated with midlatitude weather are tornadoes, which are localized bodies of intense low pressure that develop in association with supercell thunderstorms.

5. The strongest tropical storms are hurricanes, which develop when easterly waves strengthen beyond the depression and tropical storm phases to produce sustained winds greater than 64 knots (75 mph).

Check Your Understanding

1. Define the concept of an air mass.

2. What are the specific characteristics of an mT air mass and how do they differ from the characteristics of a cP air mass?

3. Which air mass is most likely to be associated with precipitation—an mT air mass or a cP air mass? Why?

4. Why are midlatitude cyclones a mechanism through which air masses of contrasting character are mixed?

5. How does the formation of an upper air trough at the 500-mb level result in the development of a midlatitude cyclone?

6. What is the basic difference between a warm front and a cold front? Why is the term *front* used in association with these concepts?

7. Precipitation along a warm front is gradual and long-lasting, whereas it is short-lived and often violent along a cold front. Why does this difference exist?

8. What is a downdraft and why is it the first step in the dissipation of a thunderstorm?

9. In what way are hurricanes a mechanism by which heat energy is transferred to higher latitudes?

10. Although the Summer Solstice occurs in June, the peak hurricane season takes place in late summer and early fall. Why does this lag exist and why is warm ocean water a necessary ingredient for hurricane development?

ANSWERS TO VISUAL CONCEPT CHECKS

Visual Concept Check 8.1

The answer is *d*. The cold front extends from western Michigan to Central Texas. You can tell the location of the cold front because of a sharp temperature difference between the east and west sides of the frontal boundary.

Visual Concept Check 8.2

The answer is *b*. The cold front caused mT air to rapidly uplift. Remember that a cold front plows aggressively into relatively warm, moist air. This interaction causes the mT air to lift rapidly.

Visual Concept Check 8.3

One would expect Hurricane Floyd to move to the west because the prevailing winds are easterly at low latitudes. At some point, the hurricane would begin to interact with the midlatitude westerlies, causing the hurricane to veer northeastward. In fact, Hurricane Floyd migrated in just this manner, moving westward toward Florida. Just before it reached Florida, however, it veered to the northeast and just grazed the mainland of the United States.

CHAPTER NINE

GLOBAL CLIMATES

Up to this point in our text we have examined a variety of features associated with

the Earth's atmosphere, including how radiation flows within it, how heat is stored

and released, how air circulates, and how weather systems produce precipitation.

Although each of these processes has stood alone in the sense that we have

discussed them in specific chapters, an underlying theme of physical geography

is that processes are interrelated, resulting in distinct causes and effects that

Climate refers to the general temperature and precipitation characteristics of an area. This image of the San Juan Mountains in Colorado shows two climate regions, one in the high mountains where cold, snowy conditions prevail, and another at a lower altitude which is sufficiently warm for trees to grow.

ripple through the atmospheric system and around the globe. A good example of this cause and effect is the development of strong midlatitude cyclones, which is related to atmospheric circulation, to precipitation, and, ultimately, to the seasonal effects caused by Earth–Sun geometry. In this chapter we tie together the discussion of atmospheric processes by focusing on another part of the big picture—the global distribution of climates.

Climate and the Factors That Affect It

When most people think of what "climate" means, they frequently think it's another word for "weather." This is a misconception. **Weather** refers to the state of the atmosphere at a specific place and time on the Earth's surface. For example, you are describing the weather when you say something like "it's cold and rainy today" where you live. Think about it for a moment—what *is* the weather as you're reading this chapter? The concept of **climate** is different from that of daily weather because it refers to the average values of weather elements, such as temperature and precipitation, over at least a 30-year period of time.

Climate is a very geographic concept because Earth exhibits distinct regional climate patterns. The study of this geographic distribution and the character of Earth's climates is a subfield of geography called *climatology*. Be careful not to confuse climatology with *meteorology*, which is the study of short-term atmospheric phenomena such as thunderstorms that constitute day-to-day weather in any particular place. In contrast to this short-term view, climatologists study long-term temperature and precipitation patterns. Their research is important to many disciplines outside the field of geography, including agricul-

Weather *Day to day changes that occur with respect to temperature and precipitation.*

Climate *Average precipitation and temperature characteristics for a region that are based on long-term records.*

ture, architecture, ecology, forestry, and economics because climate is a factor that influences human behavior and natural processes in a variety of ways.

Before studying world climates, let's review all the factors that, in some way, play a role in the long-term weather patterns for a region (Figure 9.1). Given what has been covered so far in this book, these variables should come as no surprise to you.

1. *Latitude*—The two latitudinal variables that are critical to climate are the intensity of radiation and the length of the day, which are intimately related to the basic Earth–Sun geometric relationship. These factors are first-order determinants of climate because high-latitude places are relatively cold, since they generally receive low amounts of insolation over the course of a year, whereas tropical latitudes are usually warm because insolation is relatively intense there. This insolation variability has a direct impact on ambient (overall) air temperature.

2. *Seasonality*—Seasonality is a critical variable in the character and distribution of global climates for two reasons related to Sun angle: (a) the number of hours of sunlight in a day changes in many places over the course of the year, and (b) insolation and temperature can be highly variable between the winter and summer seasons. In places like the tropical regions, however, there is very little annual difference in these variables, which is reflected in the relatively consistent temperature and precipitation characteristics of tropical climates.

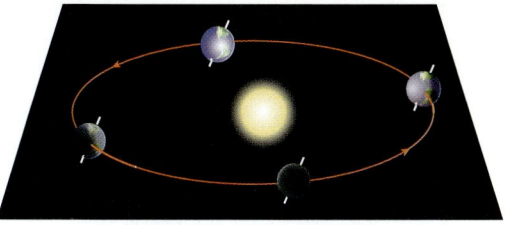

Earth–Sun Relationships
Position of Earth in orbit influences amount of radiation received

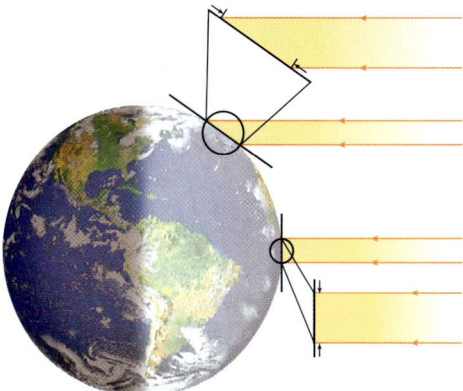

Latitude
Influences Sun angle and length of day

Air circulation
Influences flow of air and position of high and low pressure systems

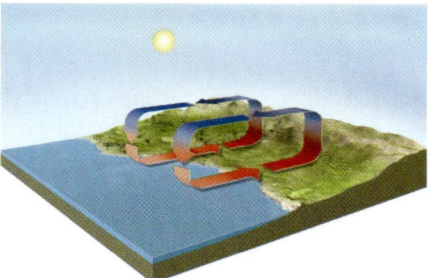

Marine/Continental
Large water bodies moderate temperature, influence air flow, supply water vapor

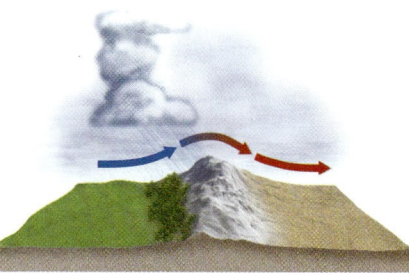

Topography
Mountains affect precipitation on windward and leeward sides, influence air flow

Figure 9.1 Variables that influence climate. The climate of any given place is determined by the interaction of a variety of factors, such as Earth–Sun relationships, latitude, marine/continental relationships, air circulation, and topography.

3. *Air mass circulation*—As we discussed in Chapter 6, the atmosphere generally flows in a predictable way, with distinct pressure systems associated with specific zones on Earth. The unique distribution of these pressure systems results in regions with heavy or persistent precipitation, such as those associated with the Intertropical Convergence Zone, while others (such as those near the Subtropical High pressure system) are relatively dry. Still other regions have distinct weather-producing systems, such as midlatitude cyclones that result in highly variable weather from day to day.

4. *Maritime/Continental relationships*—Proximity to a large body of water can have a big impact on temperature and the direction of air flow, as we saw in Chapter 7. In the interior of continents the annual temperature range can fluctuate a great deal, whereas it is moderated along ocean coasts. Large bodies of water can also be great sources of atmospheric water vapor through the process of evaporation, and air flow can transport water vapor inland where it falls as precipitation. Evaporation of large amounts of water vapor is usually associated with warm oceans, whereas cold oceans can have a drying effect on adjacent continental locations.

5. *Topographic effects*—Topography can dramatically influence atmospheric processes in a region, as shown in Chapters 6 and 7. For example, the windward side of a mountain range is often a place of heavy precipitation because air cools adiabatically there. In contrast, the leeward side of a mountain range is often a rainshadow due to the descent of air and associated adiabatic warming. Similarly, the topography of a region can also influence the flow of air, resulting in Chinook winds and cool-air drainages that otherwise would not occur.

All of these variables interact with one another to influence climate. As you study the next section on the geography of global climates, remind yourself to look for the ways in which these factors interact to form a regional climate pattern. To assist you with this process, I will frequently prompt you with questions that are intended to make you think about the interrelationship of variables. If you have trouble seeing how any particular variable fits into the picture, review parts of the preceding chapters.

Köppen Climate Classification

The purpose of climate classification is to identify certain characteristics, such as temperature and precipitation, that have definable regional patterns. Given the discussions in the previous chapters, you may already have some basic regional climate characteristics in mind. For example, tropical regions tend to be wet due to the presence of the ITCZ, whereas latitudes around 30° N and S are drier due to the influence of the STH pressure system. Nevertheless, it is difficult to classify all global climates because there are so many contributing factors. Also, the geographic distribution of climates is a spatial continuum; in other words, sharp breaks from one climate region to the next rarely happen. Despite these difficulties, a variety of climate classification systems have been devised. The most widely used classification system is the Köppen (pronounced: *Kepun*) Climate Classification System, which describes world climates based on average monthly temperature, average monthly precipitation, and total annual precipitation.

The Köppen system describes six major climate groups (*A, B, C, D, E, H*), which range from latitudes near the Equator (*A*) to latitudes near and at the poles (*E*); climate group *H* indicates a high-altitude climate, regardless of latitude. All major climate groups are distinguished initially on the basis of temperature, except for category *B*, which is based on moisture characteristics. To classify climate for smaller regions, the major Köppen groups are further subdivided on the basis of temperature and moisture by use of a second letter, and sometimes a third, lowercase letter. Table 9.1 outlines the various Köppen letter designations and their meanings. The full range of Köppen climate types and subtypes is presented in Table 9.2, with the geographic distribution of these classes shown in Figure 9.2. As you might imagine, this classification system can be cumbersome to use, especially at the introductory level; it is best not to get bogged down in its alphabetical minutia. Rather, think of

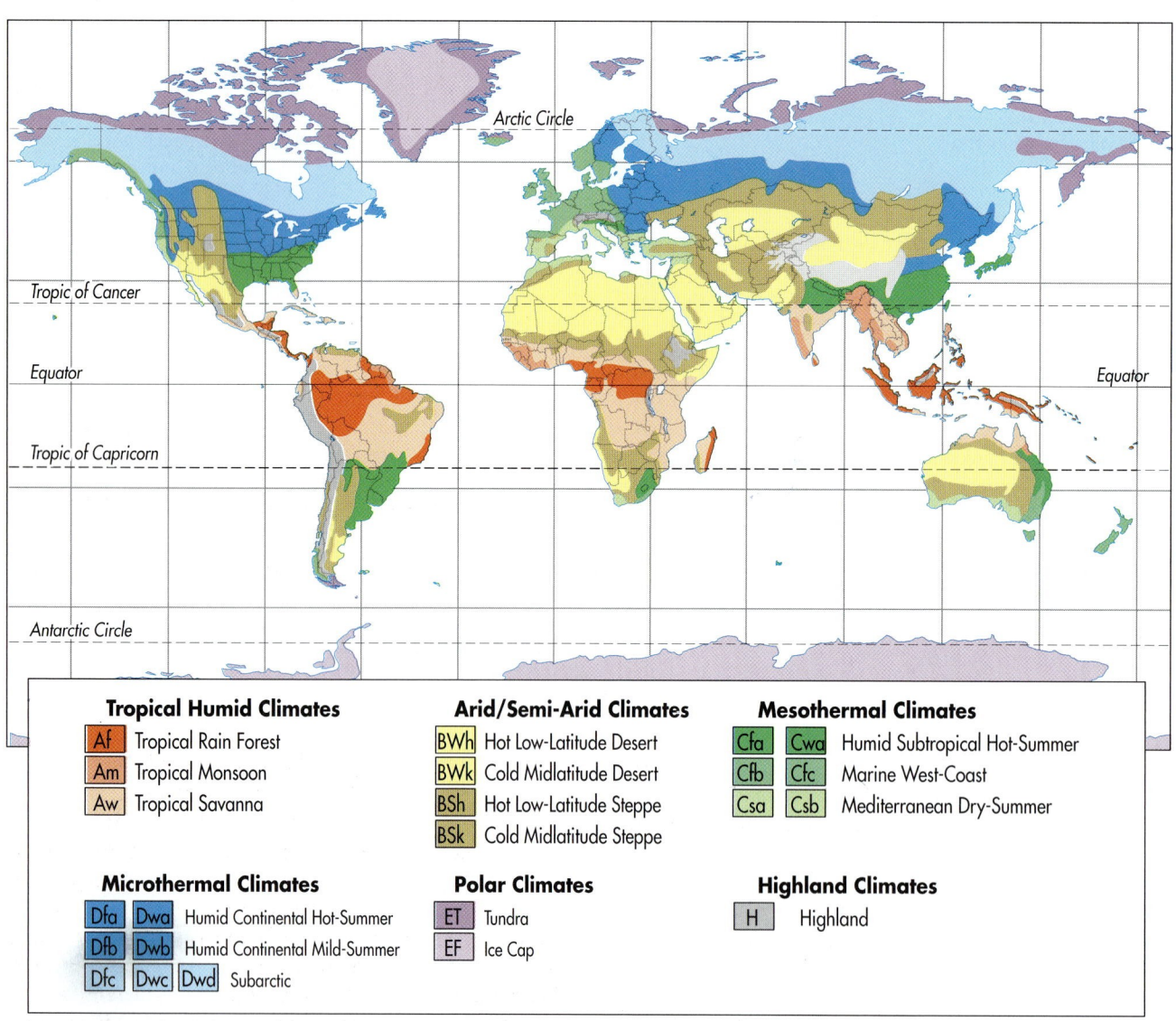

Tropical Humid Climates
- **Af** Tropical Rain Forest
- **Am** Tropical Monsoon
- **Aw** Tropical Savanna

Arid/Semi-Arid Climates
- **BWh** Hot Low-Latitude Desert
- **BWk** Cold Midlatitude Desert
- **BSh** Hot Low-Latitude Steppe
- **BSk** Cold Midlatitude Steppe

Mesothermal Climates
- **Cfa** **Cwa** Humid Subtropical Hot-Summer
- **Cfb** **Cfc** Marine West-Coast
- **Csa** **Csb** Mediterranean Dry-Summer

Microthermal Climates
- **Dfa** **Dwa** Humid Continental Hot-Summer
- **Dfb** **Dwb** Humid Continental Mild-Summer
- **Dfc** **Dwc** **Dwd** Subarctic

Polar Climates
- **ET** Tundra
- **EF** Ice Cap

Highland Climates
- **H** Highland

Figure 9.2 Global distribution of major climate groups. Six major climate groups are usually broken into as many as 23 subcategories.

TABLE 9.1	Köppen Climate Designations	
First Letters	**Derivation**	**Distinguishing Characteristic**
A	Alphabetical	Mean temperature each month > 18° C (64° F).
B	Alphabetical	Mean annual precipitation < 76 cm (30 in).
C	Alphabetical	Mean monthly warmest temperature > 10° C (50° F) in warmest month; mean monthly coldest temperature between 18° and −3° C (64°–27° F) in coldest month.
D	Alphabetical	Mean temperature > 10° C (50° F) in 4–8 months.
E	Alphabetical	Mean temperature < 10° C (50° F) in all months.
H	Highland	Significant climate changes due to altitude variations.
Second Letters		
f	German *feucht*, "moist"	≥ 6 cm (2.5 in) of mean rainfall in each month.
m	Monsoon	Only 1–3 months with mean rainfall < 6 cm (2.5 in).
s	Summer dry	Summer dry season, with driest month having $< \frac{1}{3}$ the mean precipitation of wettest winter month.
S	Steppe (semi-arid)	Mean annual precipitation in low latitudes is 38–76 cm (15–30 in), whereas it is between 25–64 cm in midlatitudes. No distinct seasonal trend in either latitude range.
T	Tundra	At least 1 month with mean temperature between 0°–10° C (32°–50° F).
w	Winter dry	Winter dry season, with 3–6 months of < 6 cm (2.5 in) of mean rainfall in A climates. In C and D climates, the driest month has $< \frac{1}{10}$ the mean precipitation of the wettest summer month.
W	German *wüste*, "desert"	Mean annual precipitation < 38 cm (15 in) in low latitudes and < 25 cm (10 in) in midlatitudes.
Third Letter:		
a	Alphabetical	Warmest month has mean temperature > 22° C (71.6° F).
b	Alphabetical	Warmest month has mean temperature < 22° C (71.6° F), but has 4 months with mean temperature > 10° C (50.0° F).
c	Alphabetical	Warmest month has mean temperature < 22° C (71.6° F); fewer than 4 months with mean temperature > 10° C (50.0° F).
d	Alphabetical	Same as c, but coldest mean monthly temperature < −38° C (−36.4° F).
h	German *heiss*, "hot"	Mean annual temperature > 18° C (64.4° F).
k	German *kalt*, "cold"	Mean annual temperature < 18° C (64.4° F).

this system as one that is meant to assist with your understanding of a very complex global climate.

As you work your way through the following descriptions of climate regions, you will find that the characteristics of a particular climate are illustrated in two different ways. The first way that climate is presented is with a **climograph**, which is really three

Climograph *A graphical representation of climate that shows average annual precipitation and temperature characteristics by month.*

graphs in one and shows the following: (1) monthly average precipitation (at the bottom of the graph); (2) mean monthly high temperature (at the top of the graph); and (3) the seasonal position of the Sun (in the middle). Figure 9.3 is an example of a climograph, one that shows the characteristics of climate at Moscow, Russia. Note that the red line in the top part of the graph indicates temperature in degrees Celsius (left side) and Fahrenheit (right side), whereas the blue bars at the base of the graph represent precipitation in centimeters (left side) and inches (right side). Months of the year are listed across the base of the diagram.

TABLE 9.2		Climate Types and Subtypes in the Köppen Climate System		
First Letter	Subcategory	Köppen Designation	Köppen Climate Name Characteristic	Basic Defining
A	Tropical humid	Af	Tropical Rainforest	No dry season
		Am	Tropical Monsoon	Short dry season; heavy monsoonal rains in other months
		Aw	Tropical Savanna	Winter dry season
B	Arid/Semi-Arid	BWh	Hot Low-Latitude Desert	Low-latitude desert
		BSh	Hot Low-Latitude Steppe	Low-latitude dry
		BSk	Cold Midlatitude Steppe	Mid-latitude dry
		BWk	Cold Midlatitude Desert	Mid-latitude desert
C	Mesothermal	Cfa	Humid Subtropical Hot-Summer	Mild with no dry season, hot summer
		Csa	Mediterranean Dry-Summer	Mild with dry, hot summer
		Csb	Mediterranean Dry-Summer	Mild with dry, warm summer
		Cwa	Humid Subtropical Hot-Summer	Mild with dry winter, hot summer
		Cfb	Marine West-Coast	Mild with no dry season, warm summer
		Cfc	Marine West-Coast	Mild with no dry season, cool summer
D	Microthermal	Dfa	Humid Continental Hot-Summer	Humid with severe winter, no dry season, hot summer
		Dfb	Humid Continental Mild-Summer	Humid with severe winter, no dry season, warm summer
		Dwa	Humid Continental Hot-Summer	Humid with severe, dry winter, hot summer
		Dwb	Humid Continental Mild-Summer	Humid with severe, dry winter, warm summer
		Dfc	Subarctic	Severe winter, no dry season, cool summer
		Dwc	Subarctic	Severe, dry winter, cool summer
		Dwd	Subarctic	Severe, very cold and dry winter, cool summer
E	Polar	ET	Tundra	Polar tundra, no true summer
		EF	Ice Cap	Perennial ice
H	Highland	Created after Köppen system was devised		High elevation with cool to cold temperature

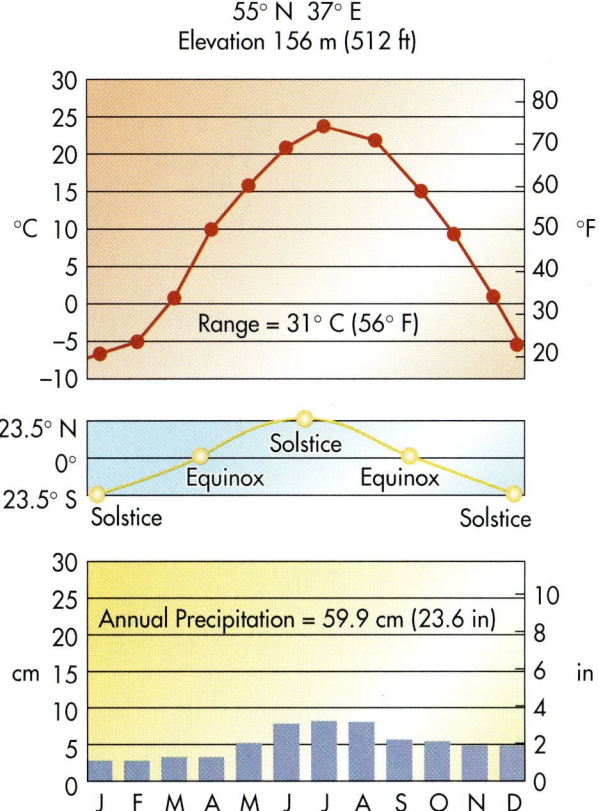

Moscow, Russia
55° N 37° E
Elevation 156 m (512 ft)

Range = 31° C (56° F)

23.5° N
0°
23.5° S

Solstice
Equinox   Equinox
Solstice   Solstice

Annual Precipitation = 59.9 cm (23.6 in)

J F M A M J J A S O N D

Figure 9.3 Climograph for Moscow, Russia. Average monthly temperature values are represented by the red line toward the top of the graph, while mean monthly precipitation is indicated by the blue bars at the bottom of the diagram. Temperature and precipitation values are in the top and bottom parts of the graph, respectively. Note the relationship of monthly patterns to Sun position, which is indicated in the center of the graph.

As you can see, annual temperature at Moscow has a distinct seasonal component, with an average high of 24° C (77° F) in July and an average high of −7° C (21° F) in January. Average precipitation ranges from 7.5 cm (about 3 in) in July to about 2.5 cm (about 1 in) in February. Winter (low) precipitation is associated with the dry Polar High pressure system, whereas the relatively wet summer season correlates with the migration of the cyclonic storm track into the region from the south.

The second way in which specific Köppen climates are represented in this chapter is with a photograph that shows the characteristic vegetation of the region. These photographs are meant to give you a sense of how a specific climate region affects the ground cover in the area. Although we will cover plant geography in more detail in Chapter 10, it is important to illustrate the association of climate and

vegetation now so you can begin to see the distribution of landscape features and understand why they exist where they do.

Geography of Köppen Climates

This section of the chapter focuses on the global geography of Köppen climates. The discussion begins with the climates found in tropical regions and progresses to those occurring in very cold regions of the world. As you work your way through this discussion, try to visualize how the various climate factors interact to produce distinctive climate regions.

Character and Geographic Distribution of Tropical (A) Climates

Tropical (*A*) climates are the climates found at low latitudes that straddle the Equator, extending to approximately 25° latitude in both Northern and Southern Hemispheres (Figure 9.4). *A* climates are warm; average monthly temperature exceeds 18° C (64° F). These climates are further subdivided on the basis of precipitation, with the tropical rainforest climate (*Af*) and tropical monsoon (*Am*) being the wettest compared to the tropical savanna climate (*Aw*). The following discussion outlines the basic characteristics of each of these climate groups.

Tropical Rainforest Climate (Af)   The tropical rainforest climate (*Af*) occurs at very low latitudes and is most closely associated with consistently high amounts of monthly solar radiation and the strong influence of the ITCZ. A good example of this climate can be found at Mbandaka, Democratic Republic of the Congo, which is located at 0° latitude. As you can see in the climograph (Figure 9.5a), monthly average high temperatures are consistently warm, about 30° C (87° F), with very little change over the course of the year at this locality. (*Do you know why? [Think about Sun position and yearly insolation.]*) In fact, the daily change in temperature is greater than the annual one. The relative humidity of this region is always high and there is a lot of rainfall every month due to the consistent presence of the ITCZ and afternoon convection. There is some variability in monthly average precipitation, however, ranging from a low of about 8 cm (3.2 in) in January to a high of about 22 cm (~8.7 in) in October. As a result of this heavy rainfall (as well as other factors to be covered in Chapters 10 and 11), vegetation in this region is dense rainforest (Figure 9.5b). (*Why does the presence of the ITCZ result in a tremendous amount of rainfall?*)

Tropical Monsoon Climate (Am)   Closely related to the tropical rainforest climate (*Af*) is the tropical monsoon climate (*Am*). Like the tropical rainforest climate, regions

THE RELATIONSHIP OF CLIMATE AND VEGETATION

In a very general sense, much can be learned about climate by looking at global vegetation patterns. This satellite image of Africa provides an excellent example of how geographers see this relationship. The green shades that you see in the tropical latitudes represent vegetation, whereas the tan and brownish colors in the northern and southern parts of Africa reflect desert landscapes. This spatial variability is directly related to climate, with the tropical latitudes receiving much more precipitation than the higher latitudes on the continent. After you work through this chapter, you will have a much better understanding of why these patterns exist around the world.

influenced by the tropical monsoon climate also receive abundant precipitation. This climate region differs, however, because it has a more prominent seasonal pattern that is associated with Sun position and the flow of air. Thus, average monthly precipitation drops below 6 cm (2.5 in)

for a month or two. Nevertheless, this slightly reduced season precipitation is not sufficient to change the nature of the vegetation, which is also rainforest in these regions.

The tropical monsoon climate (*Am*) is directly related to the onshore flow of mT air that occurs in places like

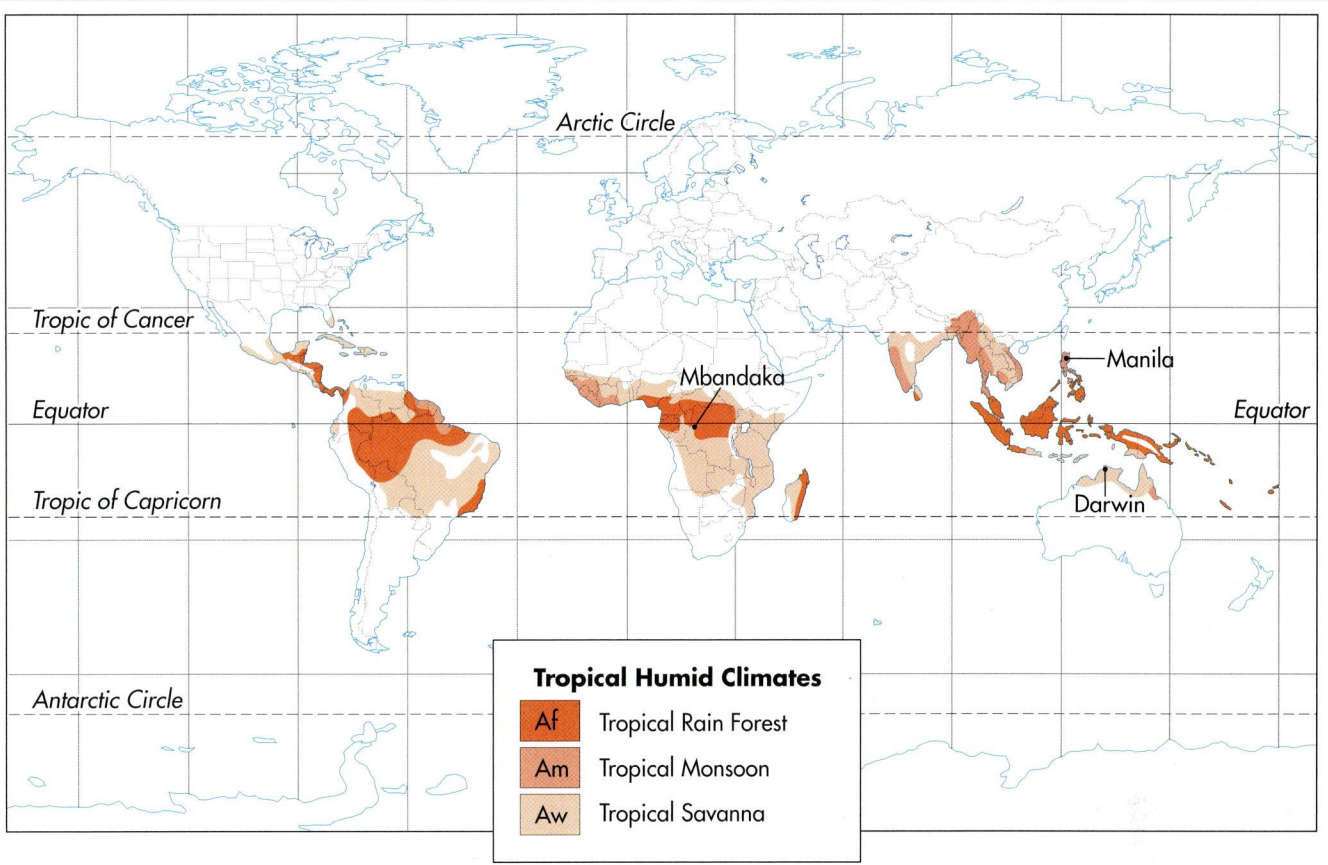

Figure 9.4 **Global distribution of tropical A climates.** The cities shown are discussed in the text as being representative of a specific climate subcategory.

Tropical Humid Climates

Af	Tropical Rain Forest
Am	Tropical Monsoon
Aw	Tropical Savanna

(a) **Mbandaka, Democratic Republic of Congo**

0° 18° E
Elevation 317 m (1040 ft)

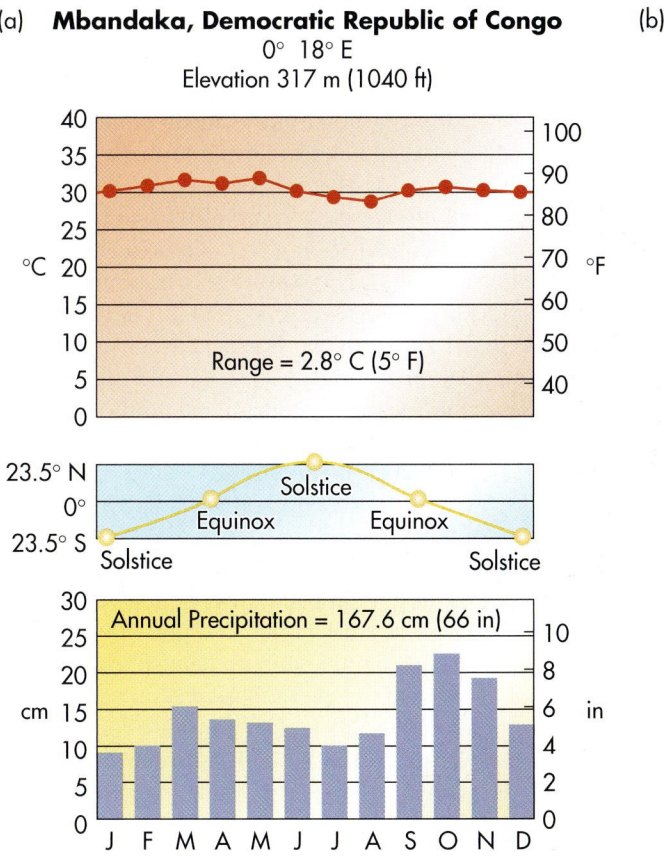

Range = 2.8° C (5° F)

Annual Precipitation = 167.6 cm (66 in)

J F M A M J J A S O N D

(b)

Figure 9.5 **The tropical rainforest climate (Af).** (a) Climograph at Mbandaka, Democratic Republic of Congo. Temperatures are consistently warm throughout the year and there is abundant rainfall. (b) A tropical rainforest in Ecuador. Due to the very warm temperatures, high humidity, and persistent rainfall in this region, the vegetation is very dense.

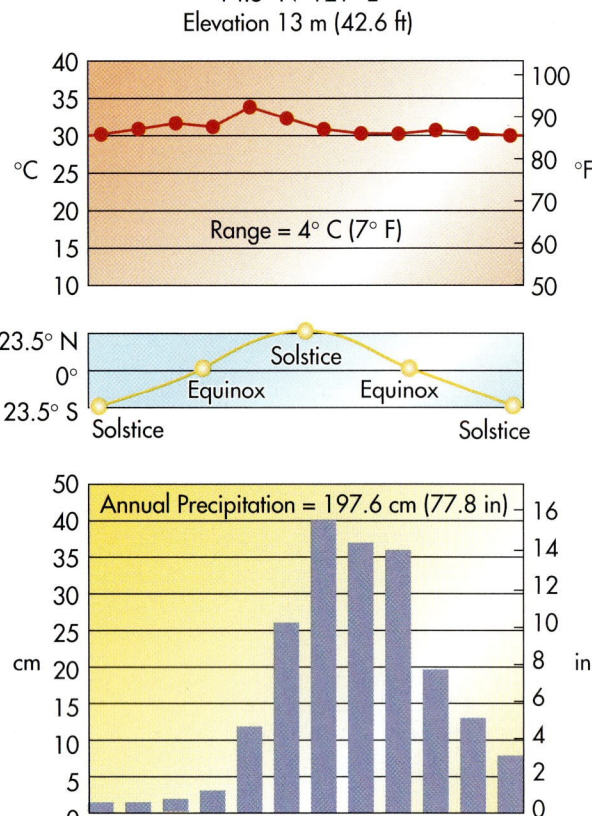

Manila, Philippines
14.5° N 121° E
Elevation 13 m (42.6 ft)

Range = 4° C (7° F)

23.5° N
0°
23.5° S
Solstice
Equinox   Equinox
Solstice   Solstice

Annual Precipitation = 197.6 cm (77.8 in)

J F M A M J J A S O N D

Figure 9.6 The tropical monsoon climate (Am). Climograph of the Tropical Monsoon climate (Am) at Manila, Philippines, about 14.5° N latitude. At this locality, precipitation has a distinct seasonal pattern, ranging from about 1 cm (0.40 in) in February to over 40 cm (15.8 in) in July. Annual temperature is warm with a very narrow range.

Southeast Asia and southwestern India (see Figure 9.4), which we discussed in detail in Chapter 6. This onshore flow occurs in summer when the ITCZ is overhead, and reverses in winter when the landmass cools. Because the onshore monsoon winds are southwesterly, the climate is largely associated with the west coasts of landmasses. Figure 9.6 is a climograph from Manila, Philippines, which illustrates the impact of this air flow on precipitation; note the distinct wet and dry seasons that occur in this region. The temperature is always warm, ranging from an average high of 30° C (86° F) in January to 34° C (93° F) in April and May. The tropical monsoon climate (Am) also occurs on the eastern side of landmasses in some places, such as the northeast coast of South America (see Figure 9.4), where northeast trade winds bring humid air into the region on a seasonal basis. Vegetation in this

climate region is somewhat similar to the rainforest (see Figure 9.5b) in the tropical rainforest climate, except that it is not quite as dense and is not called *rainforest*.

Tropical Savanna Climate (Aw)

The tropical savanna climate (Aw) is poleward of the Tropical Rainforest climate (Af) in both hemispheres and is found over large regions of the world (see Figure 9.2). A good example of this climate class occurs at Darwin, Australia, which is at 12.5° S latitude. The primary distinguishing variable of the climate at Darwin is that it has a distinct dry season in the winter (June–August) and a wet season in the summer (see climograph in Figure 9.7a), which results in a mix of grass and trees known as savanna (Figure 9.7b).

Temperatures at Darwin are consistently warm, but with a slightly greater range than in the tropical rainforest climate (Af). Peak temperature is about 33° C (91° F) from October through December and is closely associated with the presence of the STH at that time of year. The annual low temperature is about 29° C (84° F) and occurs in June and July. Precipitation ranges from about 37 cm (about 16 in) in January when the ITCZ dominates to 0 cm (0 in) during the winter months when the STH is present. As you might expect, the seasonal distribution of precipitation is intimately related to Earth–Sun geometry. (*How does Earth–Sun relationship affect the position of the ITCZ and the STH pressure system?*)

Character and Geographic Distribution of Arid and Semi-Arid (B) Climates

Arid and semi-arid (B) climates are regions with both hot and cold temperatures that are relatively dry. These climate regions are poleward of the A climates and the most widespread, with B climates occurring virtually on every continent (Figure 9.8). The most arid zones, specifically the *hot low-latitude desert climate (BWh)* and *hot low-latitude steppe climate (BSh)*, are closely associated with the subsiding air of the STH pressure belt, resulting in large regions of dry climate between 20° and 30° N and S latitude in both the Northern and Southern Hemispheres. To see this distribution, examine Figure 9.8 and note the extensive deserts in Australia, southern and northern Africa, and Asia. On the fringe of the dry tropical core are less dry, semi-arid steppe climates (BS) that are related both to the STH and rainshadow effects in continental interiors.

Hot Low-Latitude Desert Climate (BWh)

The hot low-latitude desert climate (BWh) region is found in the center and eastern sides of the STH pressure systems, spanning 15° to 30° latitude in both hemispheres. Given the subsiding air in the center of these pressure cells, very little precipitation falls and temperatures are very hot due to adiabatic warming. The ground is further heated because the skies are usually clear and solar radiation strikes the surface directly because the Sun angle is high.

(a)

Darwin, Australia
12.5° S 131° E
Elevation 27 m (88.5 ft)

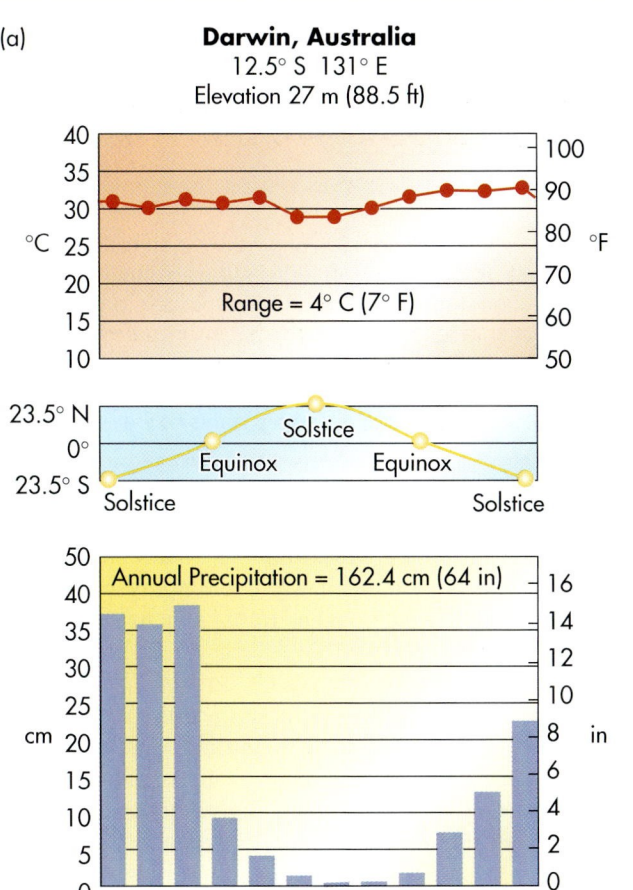

(b)

Figure 9.7 The tropical savanna climate (Aw). (a) Climograph for Darwin, Australia. Temperatures at this location are consistently warm, but with a slightly greater range than in the tropical rainforest climate (*Am*). Note the distinct seasonal pattern of average annual rainfall. (b) Tropical savanna in northern Australia. Here the vegetation consists of a mix of trees and grass due to the seasonal wet and dry cycle associated with the tropical savanna climate.

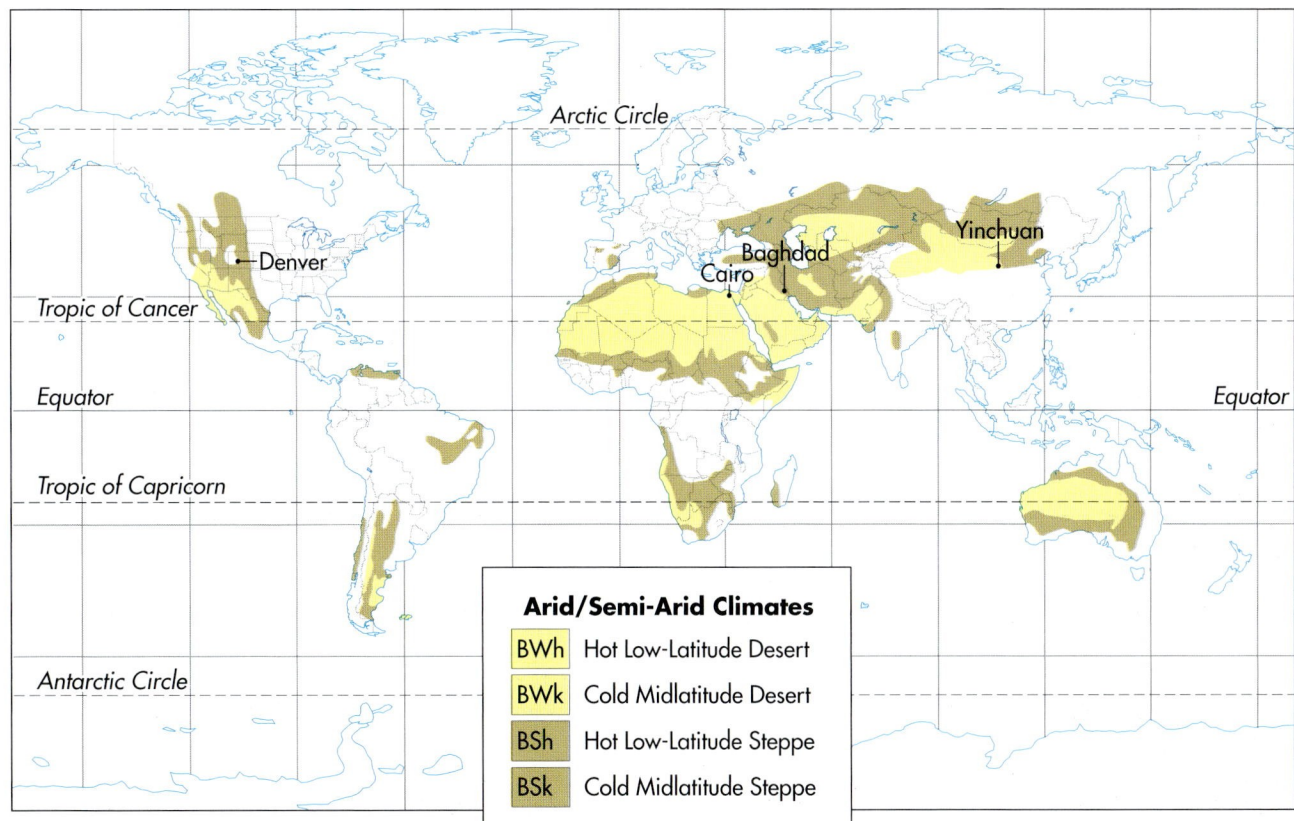

Figure 9.8 Global distribution of arid and semi-arid climates. Note the large areas covered by this climate region in Africa, Asia, and Australia. The cities shown are discussed in the text as being representative of a specific climate subcategory.

TROPICAL SAVANNA CLIMATE (Aw)

To visualize better how the tropical humid climate functions, go to the *GeoDiscoveries* website and select the module **Tropical Savanna Climate (Aw)**. This module presents an oblique view (an elevated view from the side) of Timbo, New Guinea from the Atlantic Ocean. In this fashion, you will be able to see how the migration of pressure systems influences the climate of the region. You will also be able to manipulate the migration of the ITCZ, which will allow you to better see the importance of this pressure system on precipitation in any specific geographic region. Once you complete this simulation, be sure to answer the questions at the end of the module to test your understanding of this concept.

A typical place to observe this dry tropical climate is at Cairo, Egypt, which is located at 30° N latitude. This region is extremely arid, as you can see in the regional climograph (Figure 9.9a), with virtually no precipitation over the course of the entire year. Annual temperatures are also hot, ranging from an average January maximum of 18° C (65° F) to an average high of 34° C (93° F) in July and August, but there is more annual variation than in the tropical *A* climates. (*As you examine the climograph, do you see any correlations between the Sun position and the highest temperature?*) On the ground, the hot low-latitude desert climate (*BWh*) is the classic tropical desert that you've seen pictured in many places (Figure

9.9b). The vegetation in these regions is extremely sparse and drought-tolerant, with waxy and spiny leaves designed to reduce water loss and trunks able to store water.

Hot Low-Latitude Steppe Climate (*BSh*)   The hot low-latitude steppe climate (*BSh*) zone is very closely related to the hot low-latitude desert climate (*BWh*) and results from similar air mass patterns. Because these regions are located slightly poleward of the hot low-latitude desert climate (see Figure 9.8 again), they have a greater annual temperature range and a bit more precipitation. The greater temperature range results from the occasional incursion of cP air masses, re-

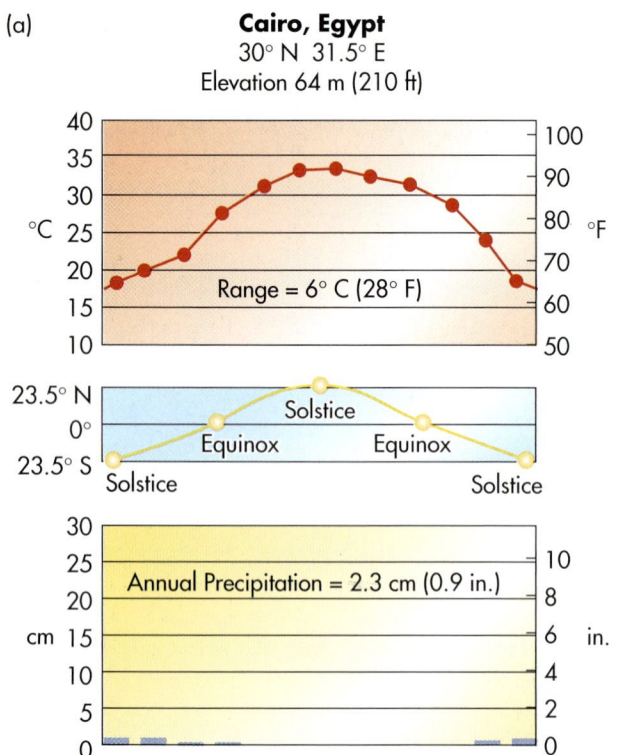

(a) **Cairo, Egypt**
30° N 31.5° E
Elevation 64 m (210 ft)

Range = 6° C (28° F)

Annual Precipitation = 2.3 cm (0.9 in.)

J F M A M J J A S O N D

Figure 9.9 The hot low-latitude desert climate (BWh). (a) Climograph of Cairo, Egypt. Note the seasonal, although hot, temperatures at this location and the extremely low amount of annual precipitation. (b) A typical tropical desert landscape, such as this one in Jordan, features very sparse vegetation due to the drying effects of STH pressure cells.

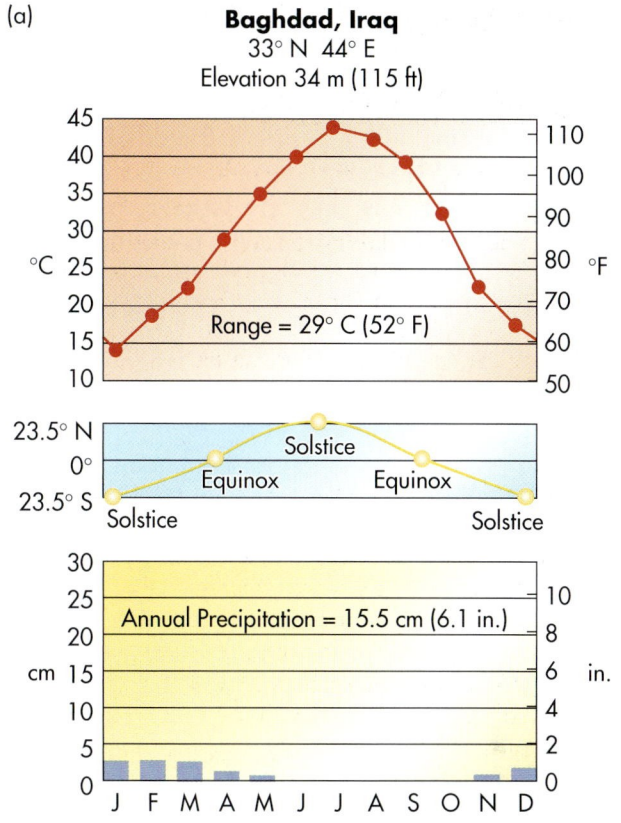

(a)

Baghdad, Iraq
33° N 44° E
Elevation 34 m (115 ft)

Range = 29° C (52° F)

Annual Precipitation = 15.5 cm (6.1 in.)

J F M A M J J A S O N D

(b)

lated to strong midlatitude cyclones, in the winter. As a
result, the periodic influence of these storm systems
causes some precipitation to fall during the winter
months.

A good place to investigate the hot low-latitude steppe
climate is at Baghdad, Iraq, which is located at 33° N latitude. Notice that Baghdad has a greater seasonal high temperature range (Figure 9.10a), from 14.4° C (58° F) in January to 43° C (110° F) in July, than Cairo (Figure 9.9).
There is also more annual precipitation (15.5 cm [6 in]) at
Baghdad (Figure 9.10a) than at Cairo, which only has 2.3
cm (0.9 in) of rainfall (Figure 9.9). Most of this precipitation in Baghdad falls during the winter months due to isolated effects of midlatitude cyclones. Given that more precipitation falls in this region than in the core of the tropical
deserts, there is sufficient moisture to support some trees.
In the U.S. the effect of this climate on vegetation can be
seen in the Mojave Desert (Figure 9.10b), which is located
in the southwestern part of the country.

Cold Midlatitude Steppe Climate (*BSk*)   The
cold midlatitude steppe climate (*BSk*) is most often
found deep within the continental interiors of North
America and Eurasia (see Figure 9.8). It is most often
associated with the rainshadow effect that occurs in the
lee of major mountain ranges, such as the Rocky and

**Figure 9.10 The hot low-latitude steppe climate
(*BSh*).** (a) Climograph from Baghdad, Iraq. Note the
seasonal range of temperature and precipitation. Although the STH pressure system dominates the region, periodic influxes of cP air associated with midlatitude cyclones lead to cooling and precipitation in
winter. (b) The landscape of the Mojave Desert in the
southwestern United States. Although the vegetation
here is generally sparse, some trees grow, such as this
Joshua tree.

This satellite image focuses on the Middle East, including northern Saudi Arabia, Iraq, Jordan, and Israel. The land bodies are tan colored because they are largely deserts with very little vegetation. Which of the following statements best explains the presence of deserts in northern Saudi Arabia and southern Iraq? (*Hint:* The latitude of Baghdad is 33° N.)

a) The region is predominantly influenced by the ITCZ.

b) The region is influenced largely by low-pressure systems.

c) The region is dominated by the STH pressure system.

d) The region lies on the windward side of a major mountain range.

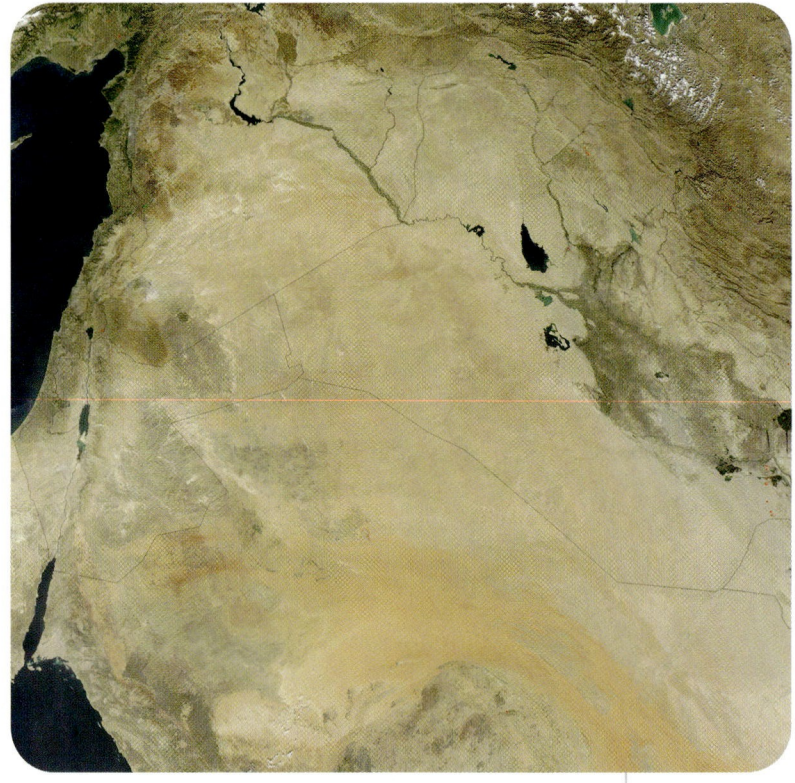

Himalayan Mountains. As discussed in Chapter 7, this effect occurs because maritime air masses are effectively blocked by the mountains, allowing relatively dry continental air masses to dominate the region. Most precipitation falls in summer and is related to convectional processes. When precipitation does fall in the winter, it is usually associated with the midlatitude storm track. Given your understanding of continental air masses, it should be no surprise to you that the annual temperature range in the dry midlatitude climate is large, ranging from hot to quite cold. (*Do you know why this pattern occurs?*)

An excellent example of the cold midlatitude steppe climate (*BSk*) is Denver, Colorado. This city lies at 40° N latitude and is just east of the Rocky Mountain front. The local climograph (Figure 9.11a) shows the large annual temperature range that extends from an average high of 6° C (41° F) in January to 31° C (88° F) in July. This range is clearly related to seasonal fluctuations in Sun angle. A seasonal distribution of precipitation also occurs, ranging from 5.8 cm (2.3 in) in May to 1.3 cm (0.5 in) in January. The dominant weather-producing system in all seasons is the midlatitude storm track. Given the relatively high amount of precipitation in this dry climate, relative to the desert-like environments examined earlier, there is a continuous cover of vegetation in the Denver region, consisting of short-grass prairie (Figure 9.11b).

Cold Midlatitude Desert Climate (*BWk*) The cold midlatitude desert climate (*BWk*) covers one of the smallest overall areas of any climate group, occurring only in central Asia, north-central China, Patagonia in Argentina, and at high elevations in the southwestern United States. The primary distinguishing characteristic of this climate is that it receives 15 cm (5.9 in) or less of rainfall per year. The annual monthly maximum temperature range can be high.

An excellent place to see the character of the cold midlatitude desert climate (*BWk*) is in Yinchuan, China (Figure 9.12a). This city is located at 38° N and lies within the Gobi Desert. The climograph clearly illustrates that average annual rainfall at Yinchuan is very low, with a little more precipitation in summer than the rest of the year because of increased convection. Average annual high temperature ranges from −1° C (30° F) in January to 29° C (85° F) in July. Given the dry character of this climate, the vegetation consists of typical desert vegetation (Figure 9.12b).

Denver, Colorado
40° N 105° W
Elevation 1609 m (5280 ft)

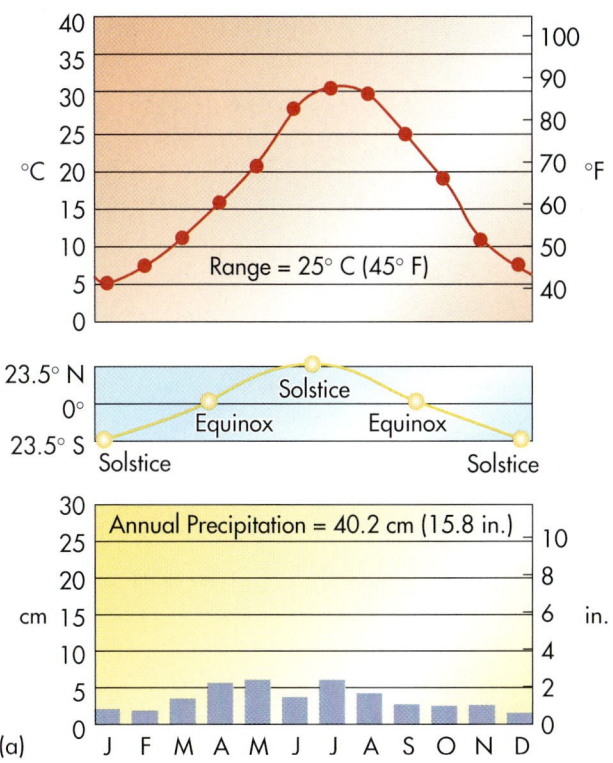

Range = 25° C (45° F)

Annual Precipitation = 40.2 cm (15.8 in.)

J F M A M J J A S O N D

(a)

(b)

Figure 9.11 The cold midlatitude steppe climate (*BSk*). (a) Climograph at Denver, Colorado. Note the seasonal temperature range and patterns associated with annual precipitation. (b) Short-grass prairie in eastern Colorado. This low-growing vegetation is in response to the low rainfall in this climate region.

Yinchuan, China
38° N 105° E
Elevation 1112 m (3,648 ft)

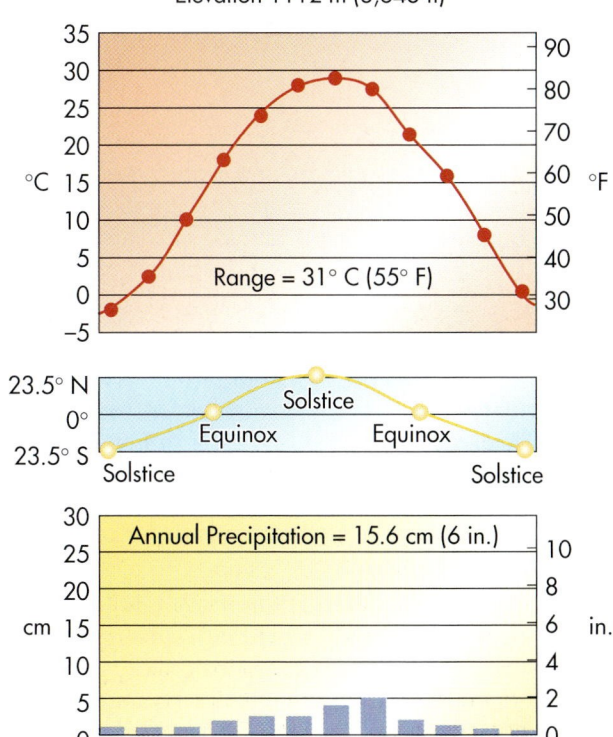

Range = 31° C (55° F)

Annual Precipitation = 15.6 cm (6 in.)

J F M A M J J A S O N D

(a)

Figure 9.12 The cold midlatitude desert climate (*BWk*). (a) Climograph from Yinchuan, China. Note the large annual temperature range and overall lack of precipitation. (b) The Gobi Desert is located in north-central China and southern Mongolia. Note the sparse vegetation and active sand dunes, which reflect this arid climate.

(b)

ATACAMA DESERT

The driest place on Earth is the Atacama Desert on the west coast of South America. This region is on the northern margin of the cold midlatitude desert climate (*BWk*) and extends about 1000 km (600 mi) from the southern border of Peru into northern Chile. The Atacama Desert is

A typical scene in the Atacama Desert. Note the very sparse vegetation in this extremely dry climate.

considered to be the driest place on Earth because there are areas within the desert that have *never* received any measurable precipitation in recorded history.

KEY CONCEPTS TO REMEMBER ABOUT THE DRY ARID AND SEMI-ARID (B) CLIMATES

1. There are four major dry arid and semi-arid climates: (a) hot low-latitude desert climate (*BWh*); (b) hot low-latitude steppe climate (*BSh*); (c) cold midlatitude steppe climate (*BSk*); and (d) cold midlatitude desert climate (*BWk*).

2. The hot low-latitude desert climate (*BWh*) occurs in places like northern Africa and is dominated largely by the STH pressure system. Vegetation is very sparse.

3. The hot low-latitude steppe climate (*BSh*) lies slightly poleward of the hot low-latitude desert climate (*BWh*), in places like Saudi Arabia and Iraq. It thus has a wider temperature range than the *BWh* climate and slightly more precipitation. This greater precipitation occurs because periodic midlatitude circulatory systems move through the region in winter. Vegetation is sparse, consisting of some grasses, cactus, and isolated trees.

4. The cold midlatitude steppe climate (*BSk*) occurs in places like Denver, Colorado, which lie in the rainshadow of major mountain ranges. Large temperature ranges and seasonal precipitation variations occur in these areas. Vegetation consists largely of short to medium grasses.

5. The cold midlatitude desert climate (*BWk*) occurs in places like the Gobi Desert in north-central China. These regions are dry because they either lie adjacent to cold ocean currents (like northern Chile) or are deep within continents. Vegetation is limited.

Several interrelated factors explain the lack of precipitation in the Atacama Desert. The region lies on the west side of the Andes Mountains, which are a significant orographic barrier to the moisture-laden easterly winds that blow across equatorial South America. Thus, the Atacama Desert lies in the rainshadow of the Andes, where descending air under high pressure compresses and warms such that precipitation is rare. In addition to being in the Andean rainshadow, the Atacama Desert lies immediately east of the cold Humboldt Current (also called the Peru Current). Given the low temperature of this current, the air above it is stable, with very little convection. In short, the Atacama Desert is so dry because it lies in the lee of a major mountain range and to the east of a cold ocean current.

This map shows the geographic relationship of the Atacama Desert, the Humboldt Current, and the Andes Mountains.

Character and Geographic Distribution of Mesothermal (C) Climates

Mesothermal (C) climates are situated in the midlatitudes, generally between about 20° and 60° N and S latitude (Figure 9.13). Although these regions are generally characterized by warm to moderate temperatures and fairly abundant precipitation, the higher latitude of these locations results in a distinct seasonal pattern relative to the tropics. As a result, C climates are usually known as "mesothermal" because they have both a warm and cool season. Most people live in this climate zone.

If you look again at the global map of climates in Figure 9.13, you can see that the distribution of C climates varies noticeably between the Northern and Southern Hemispheres. (*How can this variability be related to the hemispheric land–water differential that exists?*) In the Southern Hemisphere, C climates extend across much of the South American landmass south of 40° S latitude; this occurs because the continent is narrow. North of the Equator, in contrast, the east–west belt of C climate is interrupted by a zone of arid climate that lies in the lee of the Rocky Mountains in North America and by the Himalayas in Asia. These breaks occur in the Northern Hemisphere because the continents in that part of the world are very large, which limits the amount of oceanic-derived moisture that can reach the interior.

Humid Subtropical Hot-Summer Climates (*Cfa, Cwa*)
The humid subtropical hot-summer climate (*Cfa, Cwa*) region is found on the eastern side of continents at a latitude range of 20° to 35° N and S (Figure 9.14). The distinction between *Cfa* and *Cwa* is based on the fact that *Cwa* climate regions have a distinct winter dry period—hence, the *w* designation. Otherwise, the two regions are climatically linked because they have a hot and humid summer season.

Cfa regions are humid, with average annual precipitation ranging from 100 to 200 cm (40 to 80 in). This humid climate arises because these regions are affected by mT air masses on the western side of the adjoining

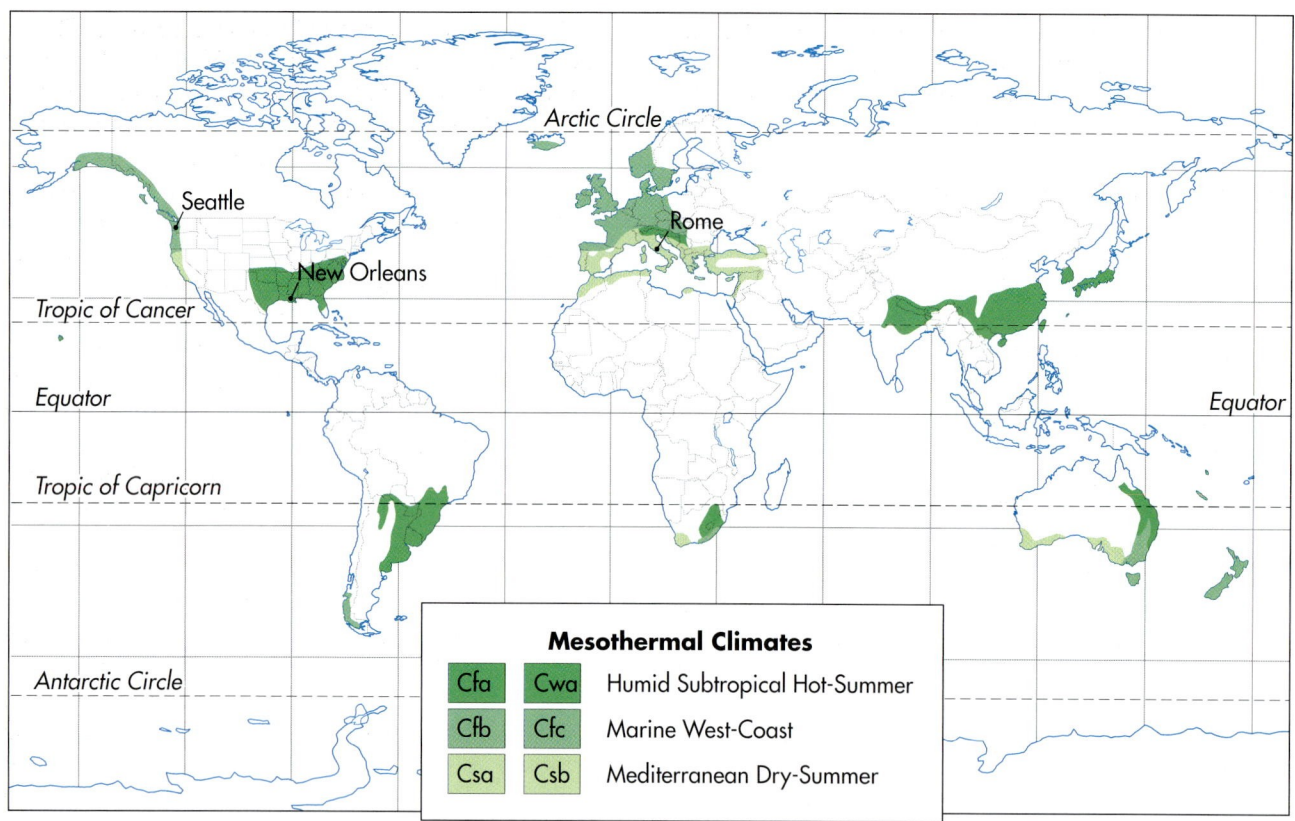

Figure 9.13 Global distribution of mesothermal (C) climates. The cities shown are discussed in the text as being representative of a specific climate subcategory.

Mesothermal Climates

Cfa	Cwa	Humid Subtropical Hot-Summer
Cfb	Cfc	Marine West-Coast
Csa	Csb	Mediterranean Dry-Summer

ocean's STH pressure zone. A good example of this geographic pattern occurs in the southeastern U.S., ranging from the Carolinas to east Texas (Figure 9.13) and represented in the climograph from New Orleans, Louisiana (Figure 9.14). During a typical summer, a strong STH pressure system known as the "Bermuda High" develops off the southeast coast of the United States. Because of the anticyclonic circulation of the system, it pumps warm, moist air from the Atlantic into the *Cfa* region, where it falls as convectional rain due to land heating. Winter precipitation is caused by frontal systems associated with strong midlatitude cyclones. (*Why is it logical that in the winter midlatitude cyclones bring precipitation to the southeastern U.S., whereas they don't in summer?*) Total annual rainfall is about 162 cm (63 in), with ample precipitation every month. Although winters are mild, there is a distinct annual high temperature cycle, ranging from 16° C (62° F) in January to 33° C (91° F) in July.

Figure 9.14 The humid subtropical hot-summer climate (*Cfa, Cwa*). Climograph from New Orleans, Louisiana, showing the character of the humid subtropical hot-summer climate (*Cfa*). Think about the relationship between the summer precipitation maximum and the STH, which is normally a drying force.

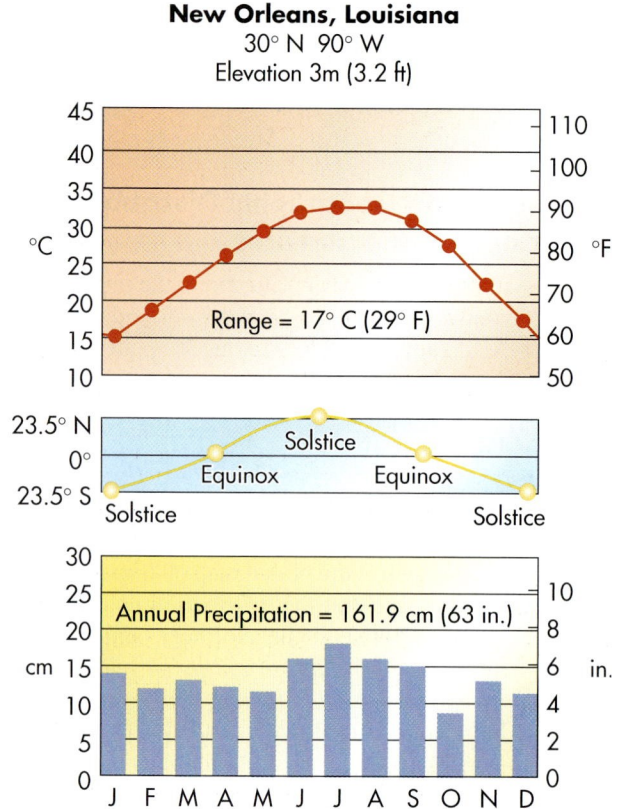

New Orleans, Louisiana
30° N 90° W
Elevation 3m (3.2 ft)

Range = 17° C (29° F)

Annual Precipitation = 161.9 cm (63 in.)

On the ground, *Cfa* regions are largely characterized by dense forests of broadleaf deciduous trees (like oak) that drop their leaves during extended cold or dry periods. In particularly warm places, like southern China, the native forest consists of broadleaf evergreen trees that remain green throughout the year.

Mediterranean Dry-Summer Climates (*Csa, Csb*)

The second climate within the *C* classification is the Mediterranean dry-summer climate (*Csa, Csb*). In contrast to the humid subtropical hot-summer climates (*Cfa, Cwa*), *Cs* climates are located along the west coast of continents, at an average latitude of about 35° N and S (see Figure 9.13). These climates are unique because the wet season occurs in the winter, as opposed to the summer when the air is dry. This pattern occurs because, in the winter, strong midlatitude cyclones bring moisture into the region from the nearby ocean to the west. During the summer months, however, the midlatitude storm track migrates north and is replaced by the STH pressure system. The amount of rainfall any particular *Cs* region receives is essentially related to latitude, with lower latitude locations generally being arid to semi-arid and those at higher latitudes more humid. (*Why do you think this pattern occurs?*) Regardless of the amount of rainfall received, the temperature in these regions is usually moderate with a very mild seasonal variation.

A great example of the *Cs* climate occurs at Rome, Italy, located in southern Europe at 42° N latitude. Notice in the climograph (Figure 9.15a) that the seasonal average monthly high temperature cycle is moderate, ranging from 13° C (55° F) in January to 28° C (83° F) in August. This moderate range is far less than what occurs at a typical midlatitude continental location. (*Do you know why? Hint: What borders Italy on three sides?*) As is typical for this climate region, maximum precipitation occurs during the winter months, with an average monthly high of 11.2 cm (4.4 in) in November. In contrast, only 1.5 cm (0.6 in) of rain falls on average in June and July. Be sure to correlate this precipitation pattern with the influencing weather system. On the ground, this region is characterized by shrubs and trees that have thick, hard leaves (Figure 9.15b) designed to retain moisture during long summer droughts.

Marine West-Coast Climates (*Cfb, Cfc*)

The third type of mesothermal climate are the marine west-coast climates (*Cfb, Cfc*), which are found dominantly on the west coast of continents between 35° to 60° N and S latitude (see Figure 9.13). In North America this climate occurs from Oregon to northern British Columbia and also occurs on the east side of the continent in the Appalachian Mountains. Elsewhere, this climate is found in the British Isles, the west coast of Europe, New Zealand, the southern tip of Australia, and southern Chile. The distinction

(a)

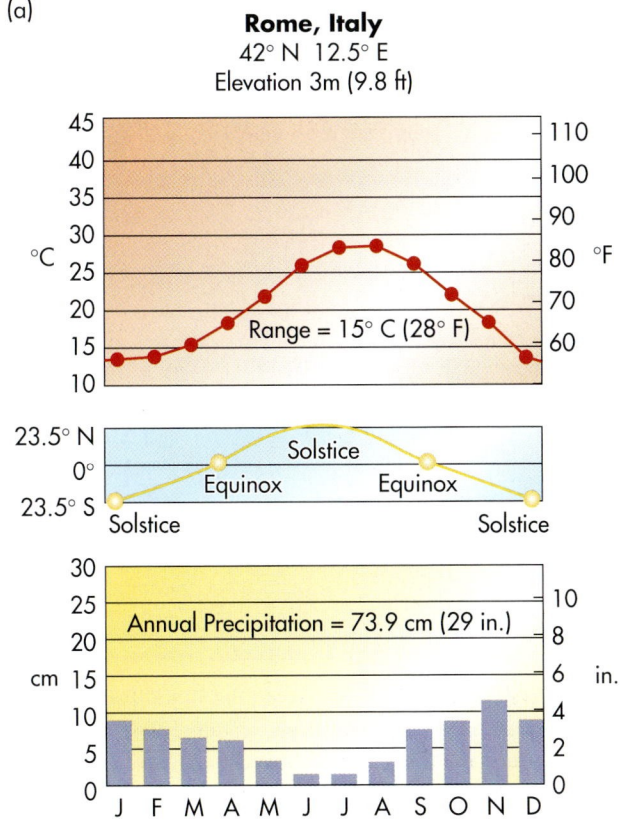

Rome, Italy
42° N 12.5° E
Elevation 3m (9.8 ft)

Range = 15° C (28° F)

Annual Precipitation = 73.9 cm (29 in.)

(b)

Figure 9.15 The Mediterranean dry-summer climates (*Csa, Csb*). (a) Climograph from *Csa* climate at Rome, Italy. Note the moderate range in temperature and the seasonal rainfall with a summer dry season. (b) The California chaparral is associated with the *Cs* climate and consists of wiry shrubs that retain moisture during summer dry periods.

HUMID SUBTROPICAL HOT-SUMMER CLIMATES (CFA, CWA)

In order to get a better feel for the variables that influence the humid subtropical hot-summer climate, go to the *Geo-Discoveries* website and select the module *Humid Subtropical Hot-Summer Climate (Cfa, Cwa)*. This module shows how the STH known as the Bermuda High pumps moisture into the southeastern U.S. during the summer. The animation will begin with a map that shows the rotating pressure

system that is migrating on a seasonal basis due to the effects of Earth–Sun geometry. As spring progresses into summer, the Bermuda High forms off of the southeastern coast. This system is rotating clockwise, and, in so doing, pumps moisture into places like Louisiana, Georgia, and North Carolina. This can be shown by clouds forming over land with precipitation. Once you complete this animation, be sure to answer the questions at the end of the module to test your understanding of this concept.

---

between the *Cfb* and *Cfc* subcategories is that the *Cfb* zones have slightly warmer summer temperatures than the areas of *Cfc* climate.

Given the higher latitude of these locations, they are more directly, and more frequently, in the path of the westerly storm track that brings cool, moist mP air masses from the large oceans to the west. Therefore, marine west-coast locations receive more precipitation than do Mediterranean

regions. This precipitation is magnified by orographic processes when mP air masses encounter large mountain ranges, such as the Cascades in the northwestern U.S., the Appalachians in the eastern U.S., or the Andes in Chile. In regions where this geographic interplay occurs, the amount of annual rainfall is prodigious, exceeding 254 cm (100 in) per year. As you might expect, marine west-coast regions have a greater annual temperature range than their Mediterranean counterparts at lower latitudes, but the winters are still very mild compared to continental locations.

A well-known city in the marine west-coast climate (*Cfb, Cfc*) is Seattle, Washington, located at 47° N latitude. Figure 9.16a is the climograph for this location.

Seattle, Washington
47.5° N 122° W
Elevation 6 m (19.7 ft)

Range = 16° C (30° F)

Annual Precipitation = 94.2 cm (37 in.)

J F M A M J J A S O N D

(a)

(b)

Figure 9.16 The marine west-coast climates (Cfb, Cfc). (a) Climograph of marine west-coast climate (Cfb, Cfc) at Seattle, Washington. Located at 47° N latitude, Seattle has a moderate temperature range and abundant precipitation, mostly due to midlatitude cyclones bringing moisture from the Pacific Ocean. (b) In the needleleaf forest at the Hoh Rainforest at Olympic National Park, Washington, abundant precipitation associated with the marine west-coast climate (Cfb, Cfc) supports a dense forest of Sitka spruce, hemlock, and various ferns and mosses.

Average annual precipitation in Seattle is about 94 cm (37 in), which is actually rather low for a marine west-coast climate (*Cfb, Cfc*), but this is because the city lies in the rainshadow of the Olympic Mountains to the west. November is the wettest month in Seattle, with 15 cm (5.9 in) of precipitation during that period of time. In contrast, only 2 cm (0.8 in) of rain falls in July. Although the annual temperature range is larger than at Rome (see Figure 9.15a), it is still relatively moderate due to the coastal location. The average high temperature ranges from 8° C (46° F) in January to 24° C (76° F) in August. On the ground, the native vegetation consists of dense needleleaf forests in the Pacific Northwest (Figure 9.16b) and deciduous forests in places such as France.

VISUAL CONCEPT CHECK 9.2

This map shows the average annual precipitation in the contiguous United States. Explain the patterns for these three regions: (1) the Southeast, including Mississippi and Tennessee; (2) the Intermountain West, including Nevada and Utah; and (3) the Pacific Northwest coast, including western Washington and Oregon.

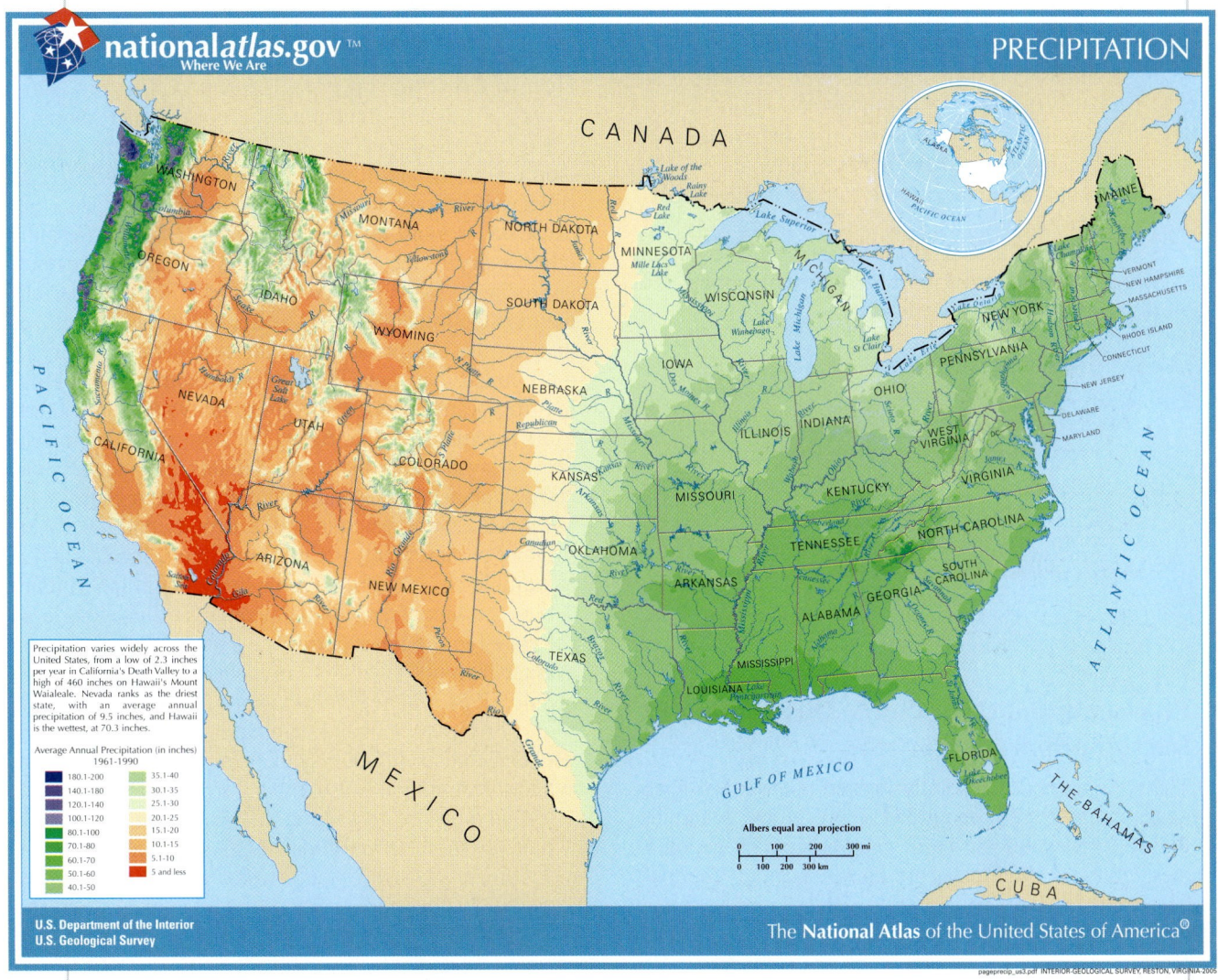

MARINE WEST-COAST CLIMATES (CFB, CFC)

In order to get a better feel for how the marine west-coast climate functions, go to the *GeoDiscoveries* website and select the module **Marine West-Coast Climates (Cfb, Cfc)**. In this module you will be able to see how the migrating midlatitude storm track brings moisture off of the eastern Pacific Ocean into the region. This animation is similar to the humid subtropical hot-summer climates (*Cwa*, *Cwb*) animation you previously viewed, in that it will show air flow and associated cloud development. As you watch the animation, notice the relationship between cyclones and moisture flow, as well as the effect that the Cascade Mountains have on precipitation intensity. Once you complete the animation, be sure to answer the questions at the end of the module to test your understanding of this concept.

---

continents in places like South Carolina and China. These regions have ample year-round precipitation, but have a distinct summer peak due to the influx of mT air by off-shore STH pressure systems. Vegetation consists dominantly of pines and deciduous trees.

3. The Mediterranean dry-summer climates (*Csa, Csb*) occur on the southwest coast of North America and in the Mediterranean region. These climates have a warm, but narrow temperature range due to their relatively low-latitude coastal location. The summer dry season is associated with dominance by the STH pressure system. Vegetation consists of a mix of trees and grass.

4. The marine west-coast climates (*Cfb, Cfc*) occur in coastal zones poleward of their Mediterranean counterparts. Thus, they have a wider temperature range that is, on average, cooler. These regions receive ample annual precipitation, but have a distinct winter wet season due to passage of midlatitude cyclones. Vegetation consists largely of needleleaf trees.

Character and Geographic Distribution of Microthermal (*D*) Climates

The *D* climates are generally located poleward of *C* climates, ranging from about 35° to 60° N and S latitude. Figure 9.17 shows the basic geographic distribution of these climates. *D* climates are known as "microthermal" because they have long winters and limited summer warmth. Thus, regions with *D* climates have a much greater temperature range than any of the other climates discussed so far.

D climates are found almost exclusively in the Northern Hemisphere because there are no large landmasses at corresponding latitudes in the Southern Hemisphere. (*Why does this distribution of D climates make sense?*) In contrast to the warmer *C* climates, which have just a warm and a cool season, *D* climates have four distinct seasons.

Humid Continental Hot-Summer Climates (*Dfa, Dwa*)

The humid continental hot-summer climates (*Dfa, Dwa*) are located in the central and eastern parts of North America and Eurasia (Figures 9.2 and 9.18). Two subclassifications are recognized in these climates because one is moist over the entire year (*Dfa*), whereas the other is dry in the winter (*Dwa*). The primary weather-producing system in humid continental hot-summer climates is the midlatitude jet stream, which steers cyclones through these regions. Given the continental locality of this climate, the annual range of temperature is high, ranging from warm, periodically hot summers to cold winters. Most precipitation falls during the summer, when mT air masses invade from the south. Winters are relatively dry because cP and cA air masses dominate the region.

An excellent example of the humid continental hot-summer climate (*Dfa*) is Chicago, Illinois, located at 42° N latitude. As you can see in the climograph (Figure 9.18a), the annual range of temperature at Chicago is large compared to the other climates examined so far, with an average high of 0° C (32° F) in January and 29° C (84° F) in July. (*Why are the coldest months in January and February?*) Average annual precipitation is 96.5 cm (38 in), with most occurring during the summer months as rain. Winter precipitation is usually in the form of snow.

A good place to see the characteristics of the dry-winter variation (*Dwa*) of the humid continental hot-summer climates is Seoul, South Korea (Figure 9.18b). Located at 37° N latitude on the Korean Peninsula, Seoul is very much like Chicago as far as seasonal temperature is concerned, ranging from an average high of 1° C (33° F) in January to 29° C (84° F) in August. Precipitation is

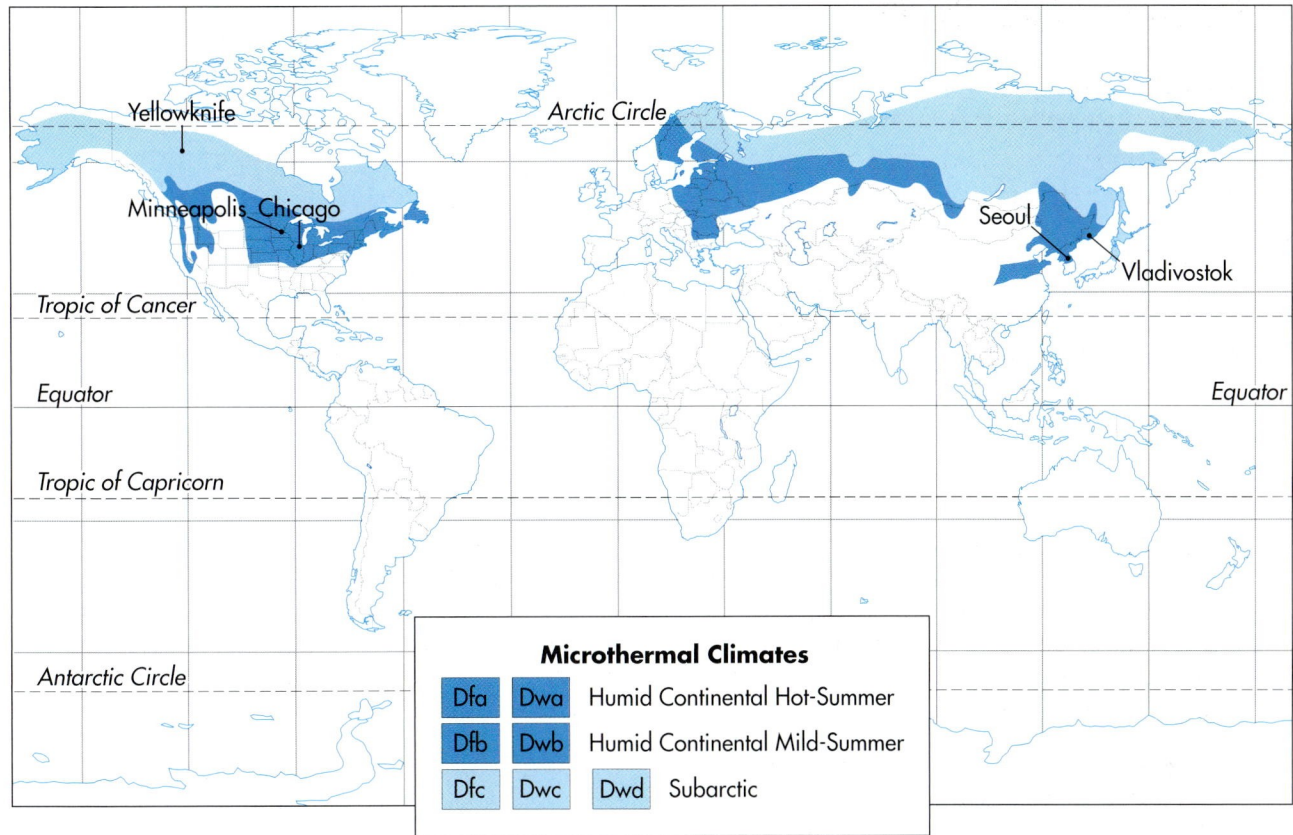

Figure 9.17 Global distribution of the microthermal *D* climates. The cities shown are discussed in the text as being representative of a specific climate subcategory.

much more seasonal in Seoul, however, than it is in Chicago. In January, about 3 cm (1 in) of precipitation falls in Seoul, whereas there is 5.5 cm (2.1 in) in Chicago. In August, there is almost 38 cm (~15 in) of rain in Seoul, compared to about 10 cm (~3.9 in) in Chicago. Some of this summer high precipitation in Seoul is due to the influence of the tropical monsoon and the occasional tropical cyclone. In places like Chicago and Seoul, where annual precipitation is relatively high, the dominant vegetation is deciduous forest (Figure 9.18c).

Humid Continental Mild-Summer Climates (*Dfb, Dwb*)

Poleward of the humid continental hot-summer climates (*Dfa, Dwa*) are the humid continental mild-summer climates (*Dfb, Dwb*). These climate regions range across central and eastern North America, western Europe, central Asia, and the far east of Asia. They occur because the summers are generally shorter and cooler at higher latitudes. This pattern should make sense by now because the overall Sun angle is less and more variable due to Earth–Sun geometry.

An excellent place to observe the moist winter aspect of the humid continental mild-summer climate (*Dfb*) is at Minneapolis, Minnesota. Located at 45° N latitude, Minneapo-

lis is known for having harsh winters and mild summers. The climograph for Minneapolis (Figure 9.19a) reflects this trend, with average high temperatures ranging from –6° C (22° F) in January to 28° C (83° F) in July. Minneapolis receives 67 cm (26.5 in) of precipitation every year on average. The driest month is February, when 2 cm (0.8 in) of precipitation occurs, mostly as snow. The wettest average month is June, when 10.9 cm (4.3 in) of rain falls. A good example of the vegetation that occurs in association with this climate is found in northern Michigan, where northern hardwood and pine forests are located (Figure 9.19b).

A good place to observe the character of the dry-winter variant (*Dwb*) of the humid continental climate is Vladivostok, Russia. Located at 43° N latitude, Vladivostok is the easternmost major city in Russia and is one of only two ice-free ports in the country. The climograph for Vladivostok (Figure 9.19c) shows a similar seasonal high temperature trend as that observed at Minneapolis (Figure 9.19a). Vladivostok is a bit colder, however, with an average high temperature in January of −11° C (12° F). The highest average monthly temperature of 24° C (75° F) occurs in August. Average annual precipitation is 79.3 cm (31.2 in), with only about 7 cm (2.8 in) falling between November and March.

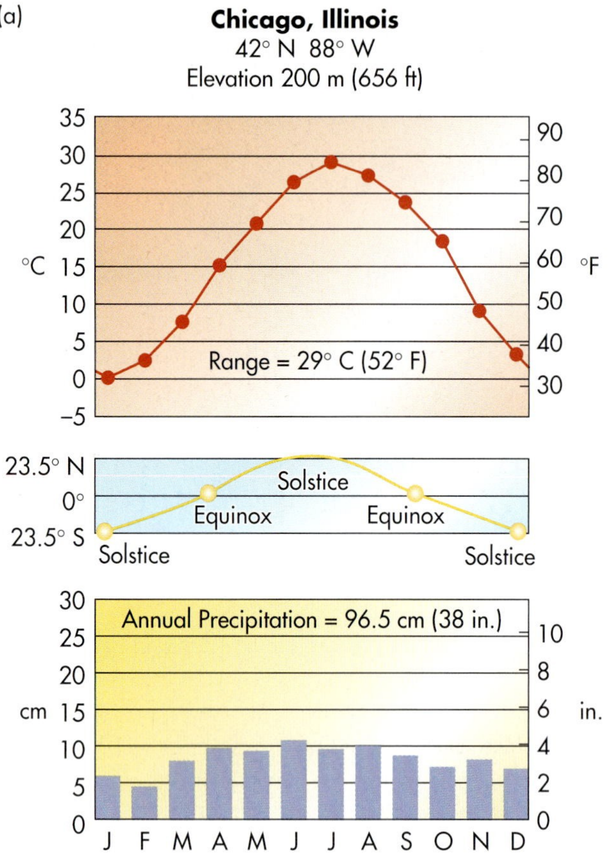

Chicago, Illinois
42° N 88° W
Elevation 200 m (656 ft)

Range = 29° C (52° F)

Solstice
Equinox · Equinox
Solstice · Solstice

Annual Precipitation = 96.5 cm (38 in.)

J F M A M J J A S O N D

(b)

Seoul, South Korea
37° N 126° E
Elevation 87 m (285.4 ft)

Range = 28° C (51° F)

Solstice
Equinox · Equinox
Solstice · Solstice

Annual Precipitation = 121 cm (47.6 in.)

J F M A M J J A S O N D

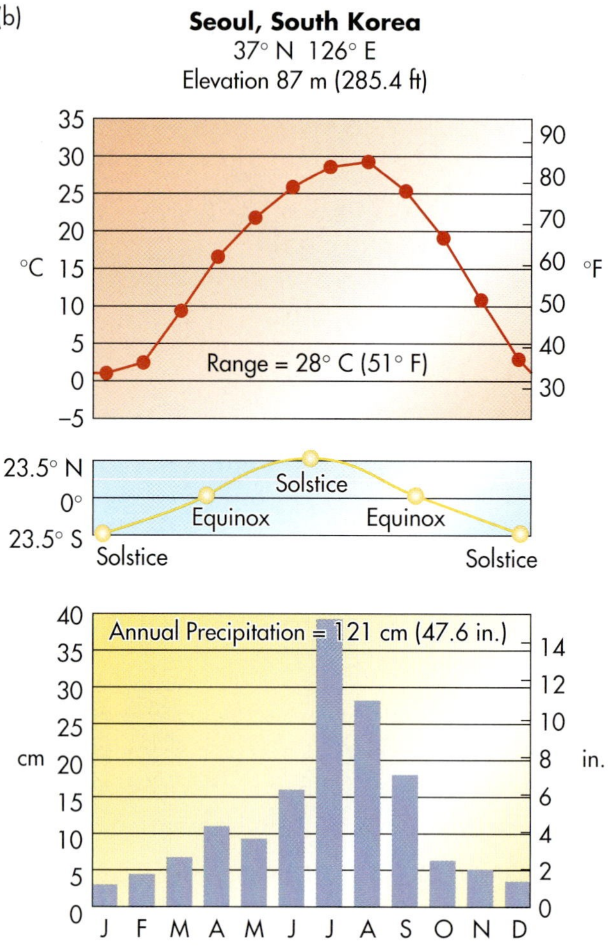

(c)

Figure 9.18 The humid continental hot-summer climates (*Dfa, Dwa*). (a) Climograph of Chicago, Illinois. This city is located at 42° N latitude and lies within the humid continental hot-summer climate belt. Note the large annual temperature range and slight seasonal variation in precipitation. (b) Climograph for Seoul, South Korea. Compare the annual temperature and precipitation with Chicago. (c) A typical deciduous forest in the humid continental hot-summer climate includes various oak and maple species. Such forests covered much of this climate region in the U.S. prior to European settlement.

Subarctic Climates (*Dfc, Dwc, Dwd*) The subarctic climates (*Dfc, Dwc, Dwd*) occur poleward of the humid continental climates (*Dfa, Dwa, Dfb, Dwb*), stretching across all of northern North America and from northern Scandinavia east across all of Asia (see Figure 9.17). It is a continental climate characterized by long, bitterly cold winters and short, cool summers. The subarctic climates (*Dfc, Dwc, Dwd*) are the source regions of

cold, dry cP air masses, and are often invaded by more frigid cA air masses that originate at still higher latitudes. Given that this region is often north of the midlatitude storm track, the annual amount of precipitation is relatively low. Most precipitation occurs during the summer months, when the belt of cyclonic storms migrates into the region from the south. Otherwise, the polar high-pressure system dominates the region, resulting in little rainfall.

(a)

Minneapolis, Minnesota
45° N 93° W
Elevation 244 m (800 ft)

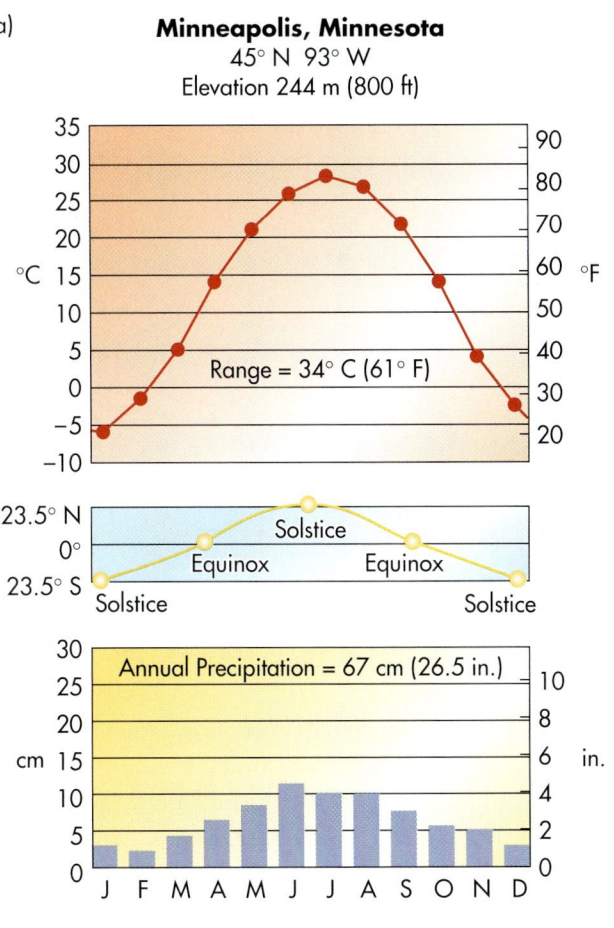

(b)

Figure 9.19 The humid continental mild-summer climates (*Dfb, Dwb*). (a) Climograph from Minneapolis, Minnesota. Note the distinct seasonal temperature trend. Although average precipitation also varies seasonally, the difference between winter and summer is not dramatic. (b) Pine forest in northern Michigan. This type of vegetation is closely associated with the wet-winter variant (*Dfb*) of this climate. (c) Climograph for Vladivostok, Russia. Note the distinct winter dry season in this area.

(c)

Vladivostok, Russia
43° N 93° E
Elevation 184 m (604 ft)

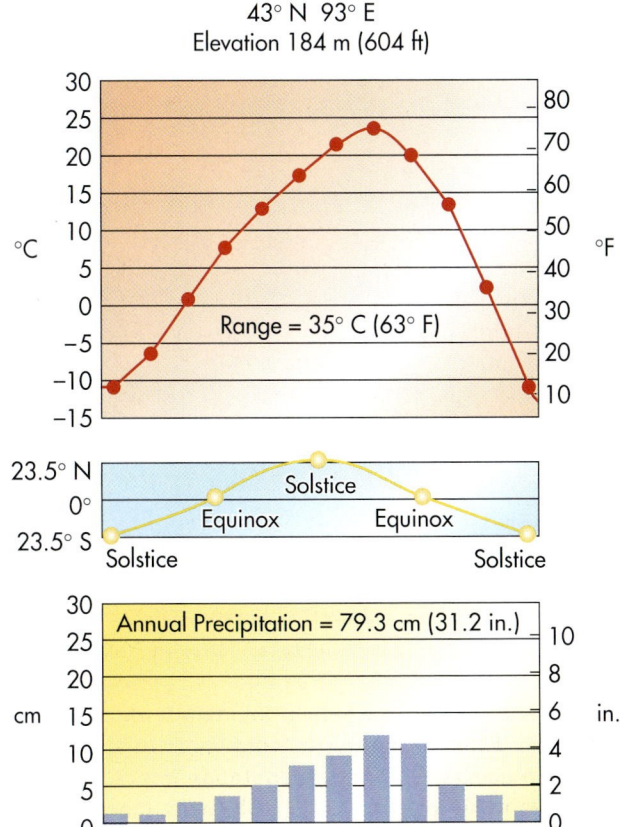

The annual range of temperature is greater within this climate than any other. (*Do you remember why this extreme range occurs? Think back to previous discussions about Earth–Sun geometry, seasonality, and insolation.*)

Let's examine the general character of the subarctic climates (*Dfc, Dwc, Dwd*) by focusing on the dominant variant within the group, the *Dfc* subcategory. This variant is found in a wide band across North America and Eurasia (see Figure 9.17) and is characterized by year-round precipitation and 1 to 4 months with temperatures above 10° C (50° F). The *Dwc* variant is found only in far eastern Russia and is distinguished from the *Dfc* type because it has a distinct winter drought. The *Dwd* variant occurs only in extreme northeastern Siberia and is distinctive because, in addition to a winter drought, its coldest month is below –38° C (–36.4° F).

A good example of the dominant subarctic climate (*Dfc*) occurs in Yellowknife, which is located at 62° N latitude in the Northwest Territories of Canada. The climograph illustrated in Figure 9.20a indicates a distinctive annual temperature cycle, one that ranges from an average high of –24° C (–11° F) in January to 20.5° C (69° F).

The amount of annual precipitation is only 26.7 cm (10.5 in), with 62% falling between June and October. Although there is relatively little winter precipitation, it consists of snow that persists for the long winter due to the bitterly cold temperatures. You can see the effects of this climate on the ground because areas like these are dominated by boreal forests consisting of needleleaf trees, such as spruce, fir, and hemlock trees (Figure 9.20b).

Character and Geographic Distribution of Polar *E* Climates

Polar *E* climates generally occur at latitudes higher than 70° N and S latitude. The tundra climate (*ET*) is one of the two polar climate subcategories and is found along the arctic coastal fringes, including such places as the island region of northern Canada, the north slope of Alaska, the Hudson Bay region, coastal Greenland, and all of northern Eurasia (Figure 9.21). Given this geographic location, a moderating climate effect occurs from the nearby oceans, resulting in winter temperatures that are less severe than in the more continental regions to the south. Nevertheless, winters are very long and cold. Very little precipitation falls in this region, largely because it is dominated by the polar high-pressure system the vast majority of the year.

A good example of the tundra climate (*ET*) is at Nome, Alaska, which is located at 64° N latitude. Figure 9.22a shows the climograph for Nome. Notice that compared to the subarctic climate (*Dfc*) in Figure 9.20, the range of average annual temperature at Nome is much less, with an average high in January of –11° C (13° F) and an average high of 15° C (59° F) in July. Also notice the correlation of temperature to hours of daylight. (*How does that compare with your understanding of Earth–Sun geometry?*) Nome receives 42.1 cm (16.6 in) of average precipitation every year, with the majority of it falling between June and October. On the ground, this climate region is characterized by short mosses and lichens that dominate the landscape (Figure 9.22b).

In contrast to the tundra climate, the ice-cap climate (*EF*) occurs in the interior of Greenland and Antarctica where enormous glaciers cover the landscape. Sun angle is consistently very low in these areas, resulting in little annual insolation. In addition, the ice-covered surfaces have high albedo, which means that most radiation is reflected and thus provides no heat. As a result, temperatures are consistently brutally cold to frigid. The climograph from McMurdo Station in Antarctica clearly illustrates the character of the ice-cap climate (Figure 9.22c). Note that temperatures range from –5° C (25° F) in December to –27° C (–17° F) in July and August. (*Why do the coldest months occur in July and August?*) Annual precipitation is very low, with a total 7.8 mm (0.3 in). The reason for this low precipitation is that the air over the Antarctic ice cap is continental Antarctic (cAA) air, which is very cold and dry. In short, this place is a polar desert.

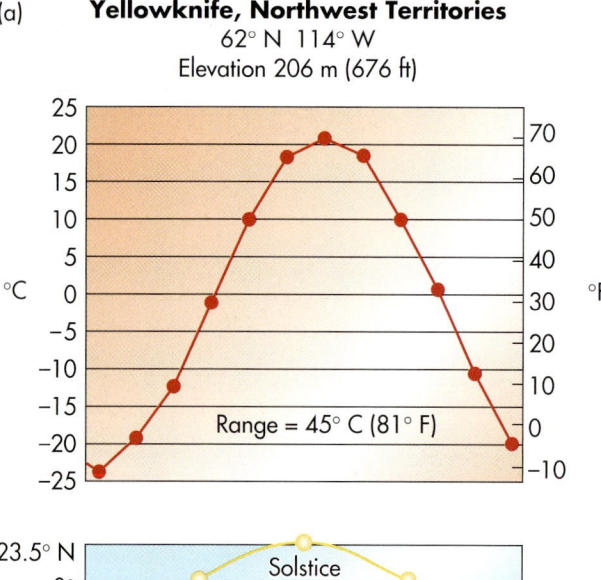

(a)
Yellowknife, Northwest Territories
62° N 114° W
Elevation 206 m (676 ft)

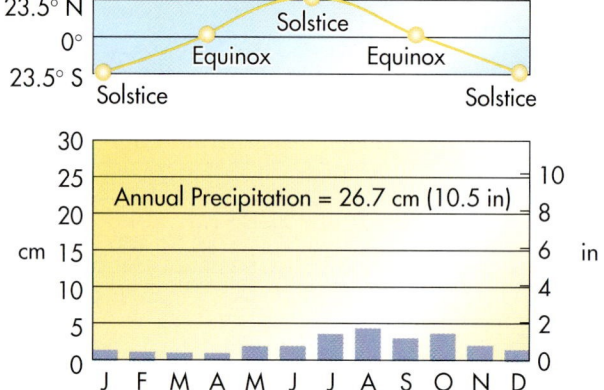

Range = 45° C (81° F)

Annual Precipitation = 26.7 cm (10.5 in)

(b)

Figure 9.20 The subarctic climate (*Dfc*). (a) Climograph of the subarctic climate (*Dfc*) at Yellowknife, Northwest Territories. This region is characterized by long winters with extremely cold temperatures. Note the short summer season in which most annual precipitation occurs. (b) Typical vegetation in the subarctic climate (*Dfc*) consists of boreal forest that contains spruce, fir, and hemlock trees.

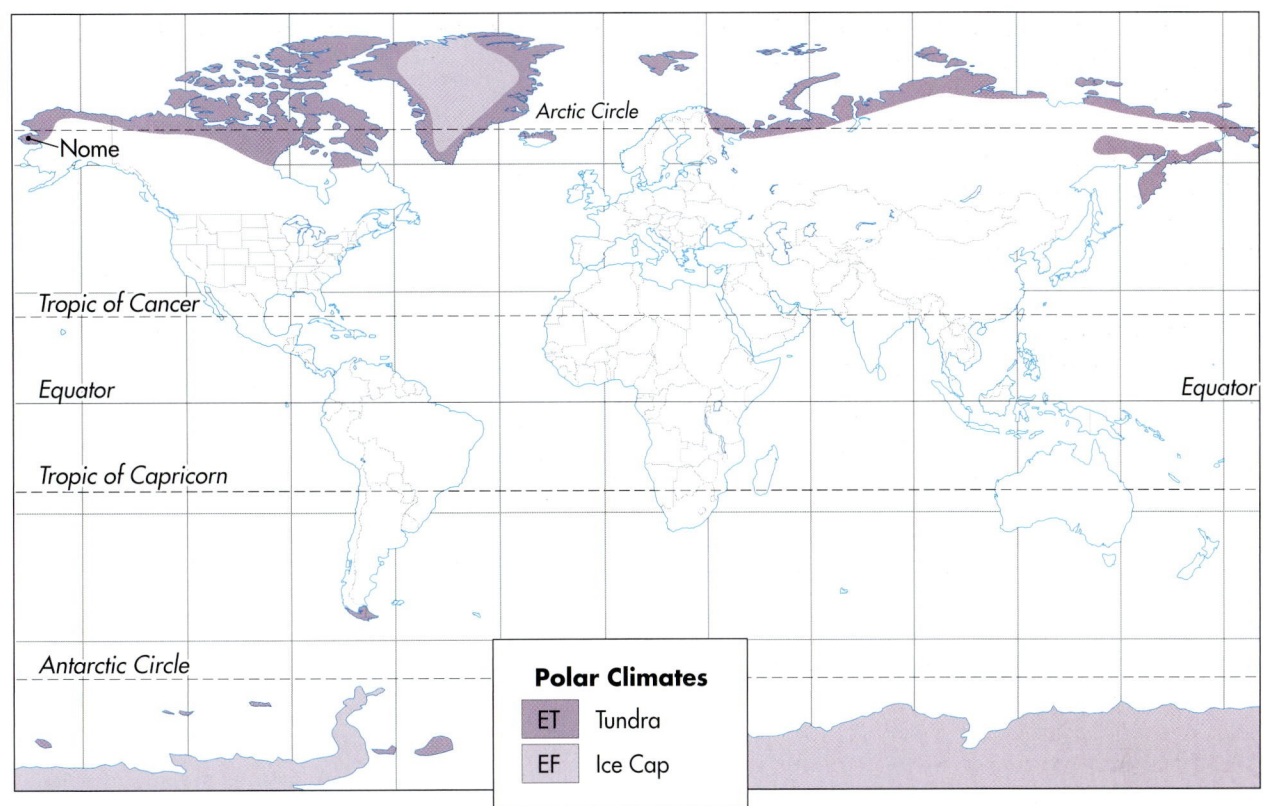

Figure 9.21 Global distribution of the polar (E) climates. This climate region is found at very high latitudes. The tundra climate (*ET*) is largely confined to the Northern Hemisphere where it occurs along the coastlines of high-latitude oceans, whereas the ice-cap climate (*EF*) occurs in both the Northern and Southern Hemispheres and is associated with the enormous glaciers on Antarctica and Greenland. An example of the tundra (*ET*) climate is Nome, Alaska, and is shown on this map. McMurdo Station, Antarctica, is an example of the ice-cap climate (*EF*), but its location is not shown on this map because of the map perspective.

In the map legend:

Polar Climates

| ET | Tundra |
| EF | Ice Cap |

KEY CONCEPTS TO REMEMBER ABOUT MICROTHERMAL (D) CLIMATES

1. There are three major subcategories of microthermal (*D*) climates: (a) humid continental hot-summer climates (*Dfa, Dwa*); (b) humid continental mild-summer climates (*Dfb, Dwb*); and (c) subarctic climates (*Dfc, Dwc, Dwd*).

2. The humid continental hot-summer climates (*Dfa, Dwa*) occur in places like the midwestern U.S. and eastern Asia. Vegetation consists largely of deciduous trees. These regions have wide annual temperature ranges and moderate yearly precipitation. Most precipitation is associated with midlatitude cyclones.

3. The humid continental mild-summer climates (*Dfb, Dwb*) occur in places like the north-central and northeastern U. S. and eastern Asia. Vegetation consists dominantly of needleleaf trees. These regions also have a wide annual temperature range, but with colder winters and milder summers than the humid continental hot-summer climates (*Dfa, Dwa*). Average annual precipitation is moderate and most closely associated with the midlatitude storm track.

4. The subarctic climates (*Dfc, Dwc, Dwd*) occur in high-latitude, continental locations that span North America and Asia. Vegetation is boreal forest. These regions have a very wide annual temperature range and have a summer wet season associated with the midlatitude storm track.

(a)

Nome, Alaska
64° N 165° W
Elevation 11.3 m (37 ft)

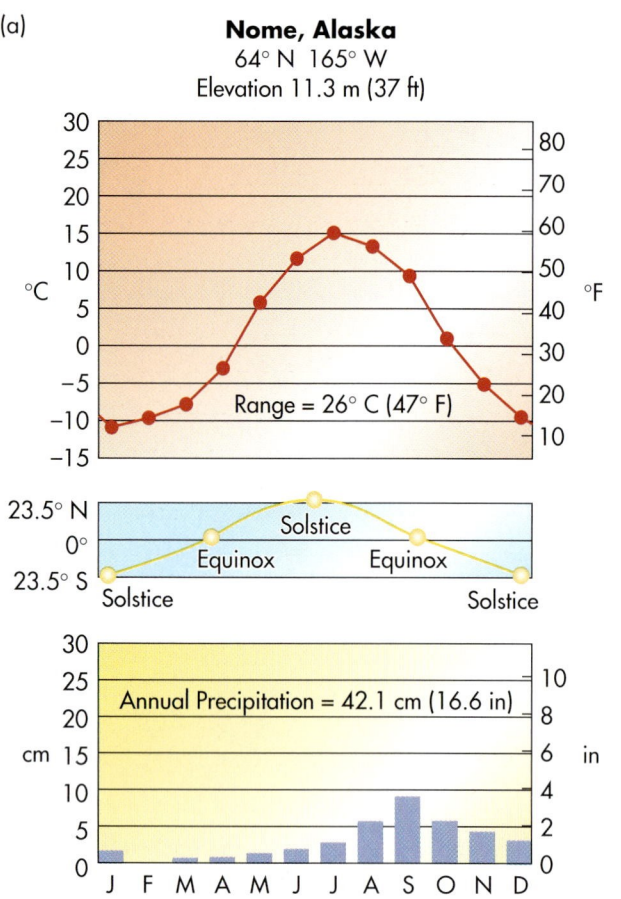

Range = 26° C (47° F)

23.5° N
0°
23.5° S

Solstice
Equinox   Equinox
Solstice   Solstice

Annual Precipitation = 42.1 cm (16.6 in)

J F M A M J J A S O N D

(b)

Figure 9.22 The polar climates (*E*) (a) Climograph of the tundra climate (*ET*) at Nome, Alaska. Notice the range of temperature and compare it to other high-latitude climates. (b) Vegetation in the tundra land-scape consists of low-lying lichens and mosses, which are the main diet of caribou. (c) Climograph of the ice-cap climate (*EF*) at McMurdo Station in Antarctica. Note the brutally cold monthly high temperatures and very low amount of precipitation.

(c)

McMurdo Station, Antarctica
77° S 166° E

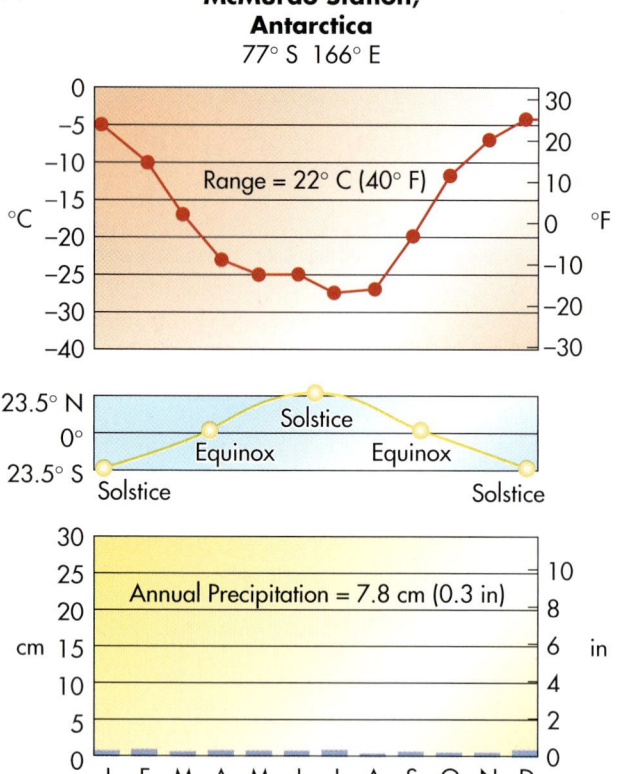

Range = 22° C (40° F)

23.5° N
0°
23.5° S

Solstice
Equinox   Equinox
Solstice   Solstice

Annual Precipitation = 7.8 cm (0.3 in)

J F M A M J J A S O N D

Character and Geographic Distribution of Highland (*H*) Climates

Highland climates are associated with the world's large mountain ranges, such as the Andes in South America, the Himalayas in Asia, the Swiss Alps in Europe, the Southern Alps in New Zealand, and the Rockies in North America. *H* climates are largely determined on the basis of temperature. Recall from Chapter 4 that temperature generally decreases with elevation in a way that mirrors the change that occurs with increasing latitude. Perhaps you've observed this yourself when hiking in the mountains, or if you've noticed that mountain peaks may be covered with snow while it is hot in the surrounding lowlands. In this fashion, a highland region can result in a zone of cool to cold climate in what is otherwise a warm to hot latitude. Figure 9.23 shows the distribution of *H* climates. Notice in this figure, for example, that a distinct band of *H* climate occurs within the wet equatorial climate region of South America. This zone of highland climate correlates with the Andes Mountains, which rise several thousand meters above the Amazon River basin.

Which one of the following choices best explains the location from which this climograph was derived?

a) It represents the climate of a city influenced by the Subtropical High (STH) pressure system.

b) It represents the climate of a mid-western city in the United States.

c) It represents the climate of a marine location in the Southern Hemisphere.

d) It represents a city located within the polar climate (*EF*).

Range = 12° C (22° F)

23.5° N — Solstice
0° — Equinox — Equinox
23.5° S — Solstice — Solstice

Annual Precipitation = 66 cm (26 in)

J F M A M J J A S O N D

Highland Climates

H Highland

Figure 9.23 Global distribution of highland (*H*) climates. These climate regions correlate with major mountain ranges. The cities shown are discussed in the text as being representative of a specific climate subcategory.

REMOTE SENSING AND CLIMATE

Global climates and the factors associated with them can be analyzed in many different ways. One way that the general character of global climates can be viewed is through satellite remote sensing. Through remote sensing, we can see the basic pattern of global climates around the Earth and some of the variables associated with them, such as atmospheric pressure systems and vegetation. To see some of these patterns, go to your *GeoDiscoveries* website and select the module **Remote Sensing and Climate**. This module will allow you to see how remote sensing can be used to observe climate patterns on Earth. You will be asked, for example, to identify a variety of atmospheric features, such as the ITCZ, the STH, and midlatitude cyclonic systems. You can also test your ability to locate hurricanes and determine the effect that cold ocean currents have on climate. As you work your way through this module, try to put together all of the concepts discussed so far in this book and visualize how they all fit together to form distinctive geographic patterns. Once you complete the module, be sure to answer the questions at the end to make sure you understand these concepts.

You can see an excellent example of the effect that highlands have on climates by examining the climographs from Munich, Germany (Figure 9.24a) and Sonnblick Mountain, Austria (Figure 9.24b). Munich is in the south-central part of Germany, lies at an elevation of 447 m (1467 ft), and is within the part of Europe classified as

(a)

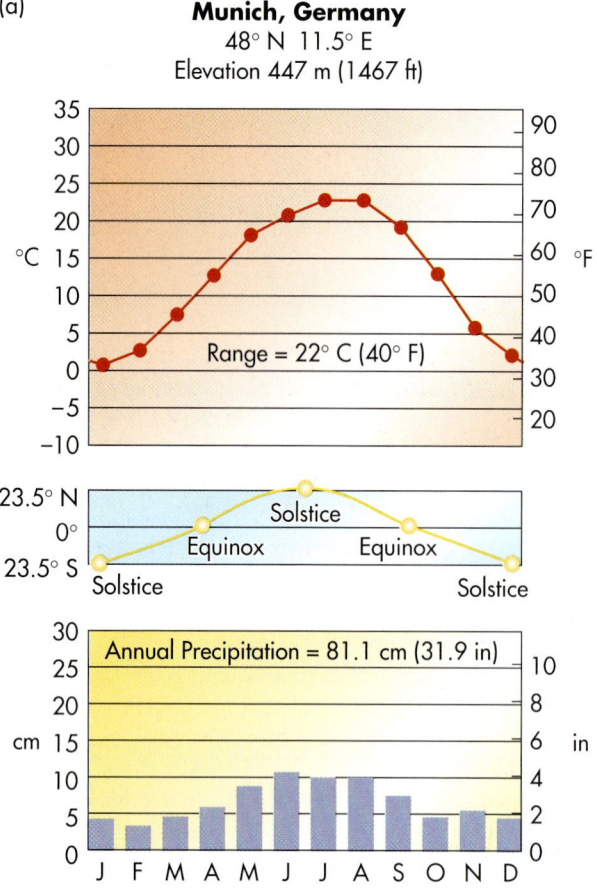

(b)

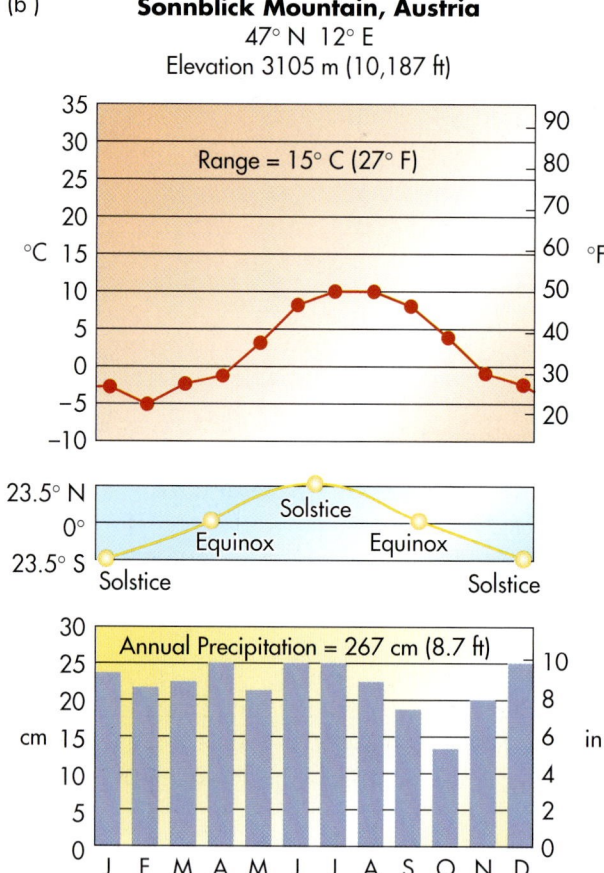

Figure 9.24 The highland climate (H). (a, b) Climographs for Munich, Germany, and Sonnblick Mountain, Austria. Both places are located in south-central Europe, but Sonnblick Mountain is at a much higher elevation. Notice the effect this change in altitude has on climate and precipitation between the two cities, which are close together.

having a marine west-coast climate (*Cfb*). Sonnblick Mountain, in contrast, is located within the Austrian Alps at an elevation of 3109 m (10,200 ft). Although these cities are less than 150 km (93 mi) apart, the difference in climate between the two places is clear. Both locations show a distinct seasonal range, but Sonnblick Mountain is much cooler than Munich. In fact, the average high hot-summer temperature at Munich is 23° C (73° F), whereas it is only 10° C (52° F) on Sonnblick Mountain. You can also see a dramatic difference in the two places with respect to precipitation. Sonnblick Mountain receives 267 cm (8.7 ft) of precipitation (most of it snow, of course), whereas Munich receives only about 81 cm (32 in) of average annual precipitation. (*Given what you know about orographic precipitation, why does this contrasting pattern make sense to you?*)

Global Climate Change

Global climates are generally stable entities, with fairly predictable conditions occurring in a region from one year to the next. For example, if you live in a continental midlatitude location, chances are that every year the summers will be very warm and the winters cold. However, climate conditions can and do fluctuate, for a variety of reasons. Short-term fluctuations can be caused by volcanic eruptions, in which resulting ash in the atmosphere blocks insolation, causing lower temperatures. Long-term fluctuations can also occur. For example, the global climate during the Jurassic period (a part of the dinosaur age), approximately 175 million years ago, was generally warmer and more humid than it is today. Similarly, the global climate cooled at the beginning of what is called the Pleistocene Epoch about 1.6 million years ago, allowing large ice sheets (glaciers) to grow and spread several times across much of the Northern Hemisphere. We'll discuss that glacial history in Chapter 17.

As you probably know, a major environmental issue currently facing the world today is "global warming," which is the notion that human industrial activity is causing the planet to warm. Global warming is the subject of daily political debates, public discussions, and news reports. This issue has become much politicized, with many interest groups involved. As a result, it can be difficult to sift through the massive amounts of information to formulate a personal opinion about whether global warming is really occurring. However, if you are to participate in future decisions regarding environmental policies associated with global warming issues, then it is essential to have some knowledge about the topic. The fact is that the vast majority of scientists agree that some kind of global climate change is occurring and will continue for the foreseeable future. In addition, the vast majority of scientists agree that the present trend of global warming is mainly due to **anthropogenic** (caused by humans) influences associated with greenhouse gas production. The remainder of this chapter is devoted to this topic.

The Carbon Cycle

In order to understand the essence of the global warming debate, it is first necessary to briefly review the carbon cycle and the "greenhouse effect." Recall from Chapter 4 that we discussed the various components of the atmosphere, including the constant and variable gases. One of the variable gases is carbon dioxide, CO_2. Although this variable gas comprises less than 1% of the atmosphere, it is an important natural regulator of temperature because it traps longwave radiation emitted by the Earth, similar to how a greenhouse functions.

As you may know, carbon is one of the basic elements of life and is a part of all living things. Simply put, carbon is stored in many different places on Earth and,

---

Anthropogenic *Environmental changes caused by humans.*

over time, moves from one *reservoir* to another. Figure 9.25 is an idealized diagram of how the carbon cycle works, with storage and transfer amounts in billions of metric tons. A simple example of a transfer is the flow of carbon that occurs when a blade of grass dies and decomposes. During its life, carbon was stored within the grass fiber. After the blade of grass dies, however, it decomposes and the carbon within it (among other things, such as minerals) is gradually incorporated into the soil, where it is stored for a period of time.

The carbon cycle progresses fairly rapidly, by geologic standards of time, and can be measured in years to centuries. Nevertheless, it is thought that less than 1% of the total amount of Earth's carbon is actively within the cycle at any moment in time. If you examine Figure 9.25 again you can see, for example, that about 120 billion metric tons of carbon transfer back and forth between plants and the atmosphere every year. (A metric ton is 1000 kg; multiply by 1.1 to get the equivalent number of English tons, each of which is 2000 lbs.) Although this sounds like a large number, it is small when you consider that 100 million billion metric tons of carbon are stored in marine sediments and rocks, mostly as calcium carbonate ($CaCO_3$). These carbonates are basically the remains of dead organisms that accumulated on ocean bottoms millions of years ago. Other large quantities of carbon are contained within coal and petroleum deposits that essentially consist of decomposed plant and animal matter.

In natural circumstances, the transfer of carbon from the inactive part of the cycle to the active zone is very gradual and usually occurs through the process of rock weathering. In this process the atmosphere is the connecting link because carbon flows to and from it in the active carbon cycle. However, since the onset of the industrial revolution in the mid-1800s, humans have dramatically accelerated the transfer of "inactive carbon" to the active part of the carbon cycle through the process of fossil fuel consumption.

Fossil fuels are carbon-containing energy sources such as coal, natural gas, and petroleum that are extracted from the ground in mines and wells. The term "fossil fuels" comes from the fact that these compounds were once living organisms that died and slowly fossilized due to intense pressure and heat, turning into other carbon compounds over thousands and millions of years. Simply put, a tremendous amount of energy is required to power the complex and interconnected web of factories, automobiles, shopping malls, home appliances, and every other conceivable modern convenience in the world today. The vast majority of the energy required in this industrial and technological age is derived from the consumption, in one

Fossil fuels *Carbon-based energy sources, such as gasoline and coal, which are derived from ancient organisms.*

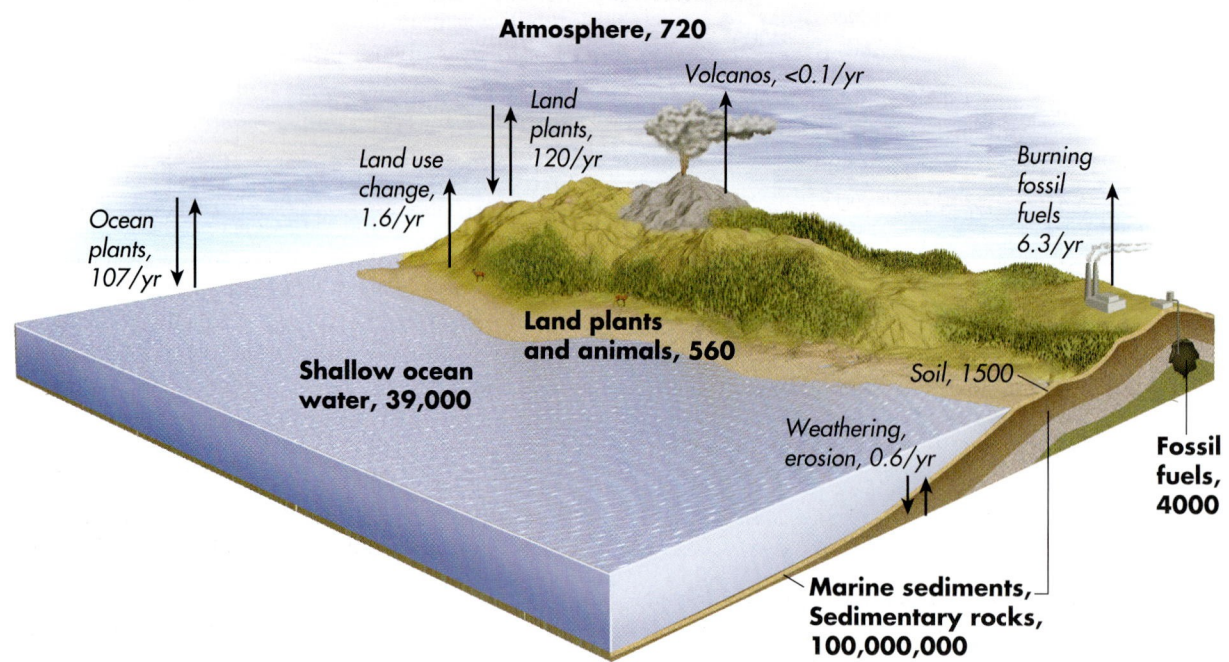

Atmosphere, 720

Volcanos, <0.1/yr

Land plants, 120/yr

Land use change, 1.6/yr

Burning fossil fuels 6.3/yr

Ocean plants, 107/yr

Land plants and animals, 560

Shallow ocean water, 39,000

Soil, 1500

Weathering, erosion, 0.6/yr

Fossil fuels, 4000

Marine sediments, Sedimentary rocks, 100,000,000

Figure 9.25 The carbon cycle. The carbon cycle represents the flow of carbon that occurs on Earth as it moves from one reservoir to another. Note the various places where carbon is stored and the amount of carbon contained within these reservoirs. (Reservoirs are noted in boldface type, transfers are in italic type.)

form or another, of fossil fuels. To put this thought in perspective, consider that 98,000 kg (98 tons/216,000 lbs) of prehistoric plant material is required to produce 3.8 liters (1 gallon) of gasoline. Looked at another way, you need to load the prehistoric equivalent of 16.2 hectares (40 acres) of wheat—stalks, roots, and all—into the tank of your car just to drive 20 miles! In this context, the U. S. currently leads the world by consuming about 25% of the annual global energy supply. This lead exists despite the fact that the U.S. contains only 5% of the world's approximately 6.5 billion people. Of the total amount of energy consumed in the U.S. in 2005, about 84% was derived from fossil fuels.

Through this global reliance on fossil fuels, humans have exponentially increased the amount of atmospheric carbon dioxide over time. In 1950, for example, approximately 1.6 billion metric tons of carbon was released into the atmosphere through industrial pathways. By 1997, this transformation of fossil fuel to atmospheric carbon dioxide had increased 400% to 6.3 billion metric tons. Approximately 50% of these total emissions have moved from the atmosphere and are stored in reservoirs such as the oceans and forests. The remaining 50%, however, has stayed in the atmosphere.

You can see the cumulative effect of this pattern in the proportional change in the amount of pre-industrial atmospheric carbon dioxide to current values. In 1850, there were about 288 parts of carbon dioxide per every million part of the overall atmosphere. By 2001 this proportion had increased to almost 370 parts per million (ppm) (Figure 9.26). This may not seem like much of an increase to you, but when you consider the importance of atmospheric carbon dioxide in the natural regulation of atmospheric temperature, it's significant. This increase is likely to continue in the foreseeable future when densely populated developing countries such as India and China, which between them contain about 2.3 billion people, fully industrialize and use the same fossil fuel power sources that have worked so well for us.

Is Anthropogenic Climate Change Really Occurring?

The "million dollar question" that currently faces the global society is whether or not human-induced climate change is really occurring. In this context, you might first ask: "Is the climate really warming?" With respect to this initial question, the evidence strongly suggests that the Earth is indeed in a warming phase. You can see the evidence of this warming by examining Figure 9.27. In this graph, the data clearly show that the observed global surface air temperature has increased about 1° C (1.8° F) since 1880. In fact, the 1990s comprised the warmest decade ever recorded. This increase in temperature correlates very closely with the overall increase in atmospheric carbon dioxide that has occurred at the same time.

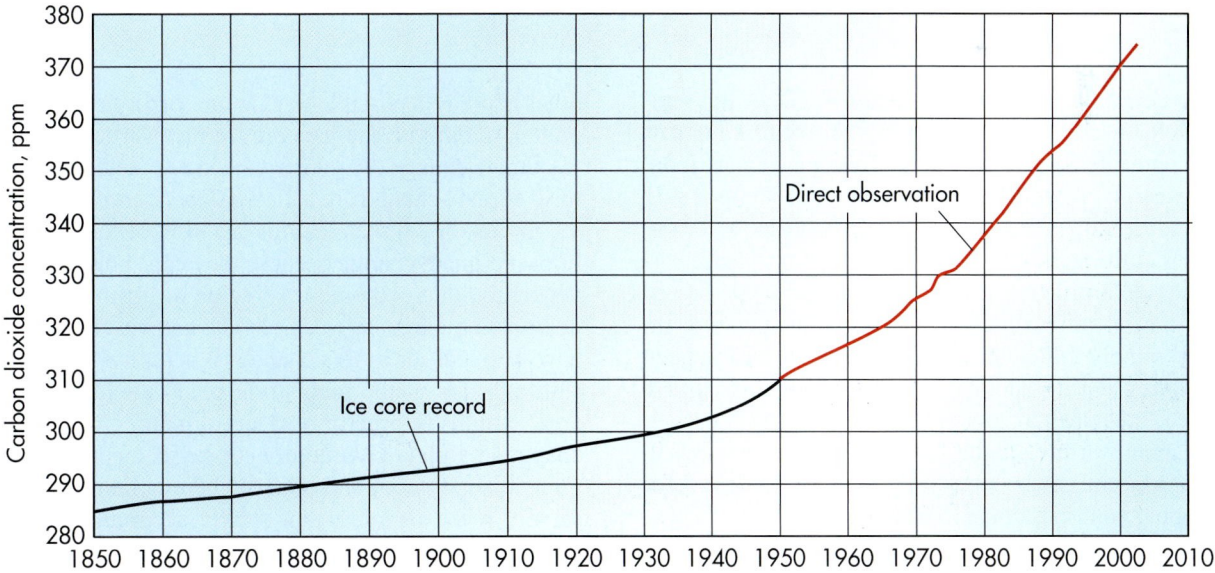

Figure 9.26 Changes in atmospheric carbon dioxide since 1850. Carbon dioxide levels have increased over 30% since 1870. The black line represents atmospheric concentrations derived from ice cores. The red line is derived from actual atmospheric measurements taken by the Scripps Institute of Oceanography. (Credits: Pre-1958: A. Neftle et al., Historical CO₂ record from the Siple Station ice core, in *Trends: A Compendium of Data on Global Change*, 1994, U.S. Department of Energy; Post-1958: C. D. Keeling and T. P. Whorf, Atmospheric CO₂ records from sites in the SIO air sampling network, in *Trends: A Compendium of Data on Global Change*, 2005, U.S. Department of Energy.)

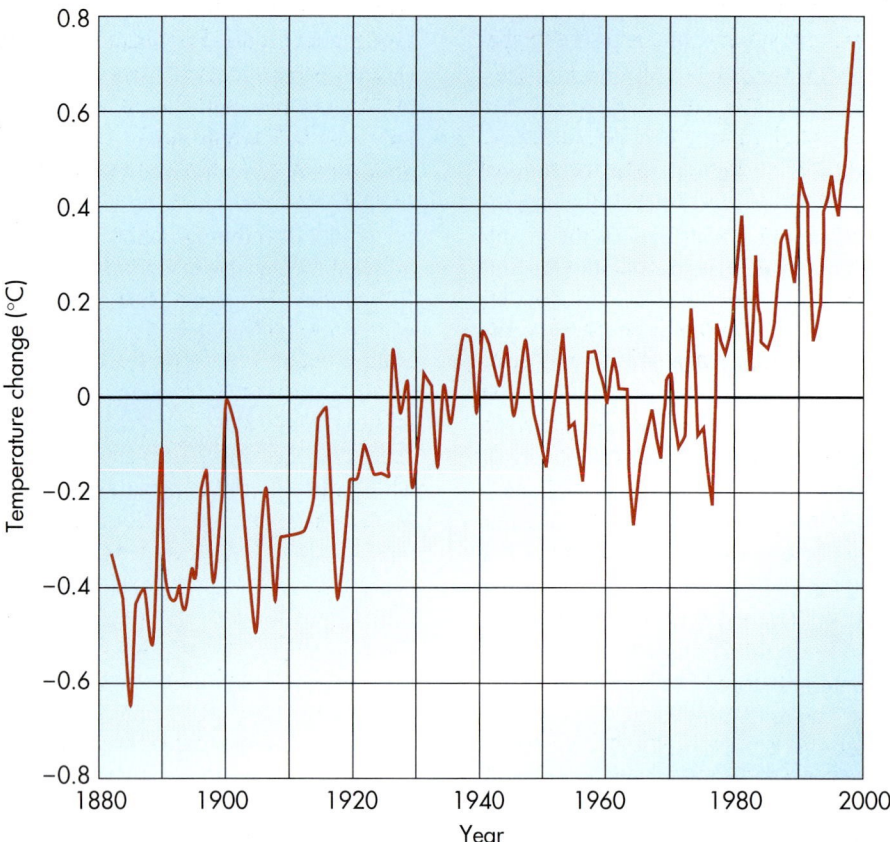

Figure 9.27 Recent changes in global temperature. This graph shows the change in average global temperature from 1880 to 1998. For comparison, note the horizontal line at 0, which represents the average global temperature from 1950 to 1980. (Credits: J. E. Hansen, Goddard Institute, NASA, and National Climate Data Center, NOAA.)

However, the next question to ask is, "Does this temperature/CO_2 correlation necessarily mean that humans are responsible for the change?" This question is more controversial. Skeptics argue that this correlation *could* be coincidental, in that temperature *happens* to be increasing at the same time that levels of atmospheric carbon dioxide are increasing. In fact, the geologic record shows that at other times in prehistory temperature increased rapidly *and* naturally, and we might just happen to be living during another such period. If this climate change is coincidental, then we do not necessarily need to worry about warming in the context of future increases in greenhouse emissions, especially if those predicted for developing countries come to pass.

But, what if the present global warming is anthropogenic? How can we know? Is there any way to test the relationship by looking at past climate changes? Unfortunately, humans have been keeping systematic climate records, including temperature and precipitation data, for only about 150 years. Although 150 years sounds like a long time in the context of the average human life span, it is really quite short from a geological perspective. On the

other hand, it is possible to examine **proxy data** such as pollen, tree rings, and ice cores to reconstruct the history of climate change. Scientists who are interested in prehistoric climate change are called *paleoclimatologists*.

Prehistoric pollen records can be excellent indicators of past climate change because they can show changes in vegetation that occurred in a region over time. Given the demonstrable link between climate and vegetation, as we have seen earlier in this chapter, it is logical that significant changes in the prehistoric plant assemblage of an area would indirectly reflect climate change. Such change in a plant community can be reconstructed in regions (usually cool and moist) where pollen is well preserved in the deposits that slowly accumulate in lakes and marshes. Scientists who study such records are called *palynologists*.

Proxy data   *Indirect evidence of an event. For example, fossil pollen is a proxy indicator of climate change because vegetation reflects climate.*

The premise of palynology is that plants produce microscopic pollen that are unique in shape and size to specific species. You can see this kind of variation in Figure 9.28. Each year, plants within a region produce pollen as part of their reproductive cycle; this pollen is then spread by wind. If you happen to be allergic to some kind of pollen, you are acutely aware of this seasonal process because of the discomfort it may cause you. Some of this pollen accumulates at the bottom of lakes and marshes in sediments that progressively thicken with each passing year. For example, given that the pollen-containing deposits at the base of a vertical sequence of lake sediments are obviously older than those at the top, because the base deposits accumulated first, it is possible to reconstruct the change in pollen composition (and therefore climate) through time. Initially, a core (vertical sample) of the lake sediments is acquired from the lake bottom and then the palynologist extracts and analyzes in the laboratory the pollen contained within each layer of sediment.

Ultimately, a pollen diagram such as the one presented from Nelson Lake, Illinois (Figure 9.29) is created that shows the change in pollen (and thus, climate) at the lake or bog over time. Here's how you interpret the diagram. Begin by looking at the vertical axis, which shows the age in years before the present time, beginning with 17,000 years ago and ending today (0). From left to right across the top of the graph are the various plant species identified in the core samples, ranging from sedge on the left to ragweed on the right. The relative percentages of each of these plant species are shown at the bottom of each plant column. For example, around 15,000 years before the present, spruce trees comprised about 60% of the overall sample, whereas the amount of oak pollen was very small. Approximately 12,000 years before the present, however, spruce pollen makes up 0 percent of the sample, whereas oak is over 40%.

Using these percentage changes through time, a palynologist can reconstruct the climate for the Nelson Lake area for the past 18,000 years. The high amount of sedges between 18,000 and 17,000 years ago reflect a cold, dry climate when tundra prevailed in the area. As you will learn in Chapter 17, this climate/vegetation combination existed because a giant glacier was nearby. By 17,000 years ago the climate became warmer and more humid, which caused tundra to be replaced by spruce, which is a needleleaf tree. Spruce dominated the local vegetation until about 12,000 years before the present when the climate apparently warmed sufficiently to allow the influx of oak (a deciduous tree) at the expense of spruce. Oak has generally dominated for the past 12,000 years at Nelson Lake, although grass has increased in the past 3000 years. This expansion of grass is probably related to the climate becoming slightly warmer and drier.

Another way that climate change is reconstructed is through the analysis of tree rings. The interior of most trees contains annual rings, such as those shown in Figure 9.30a, that reflect each year of growth in the tree's life. In trees that have annual rings, each ring consists of a dark

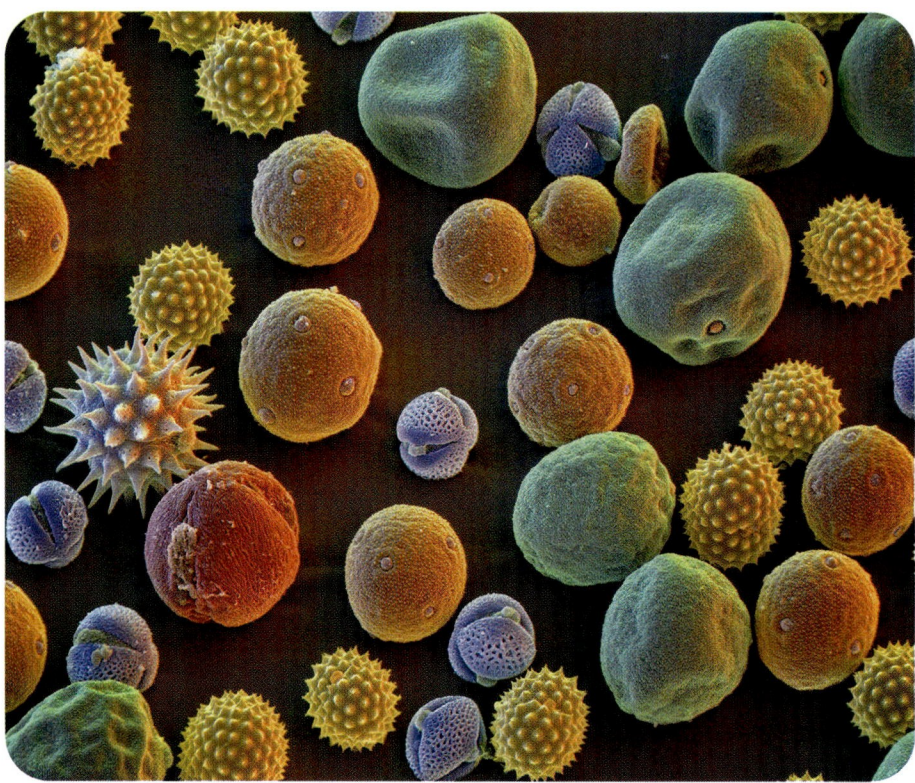

Figure 9.28 Pollen grains. Different plant species produce distinctive forms of pollen that can be seen under a microscope. Note the different shapes in this image, which is magnified 400 times.

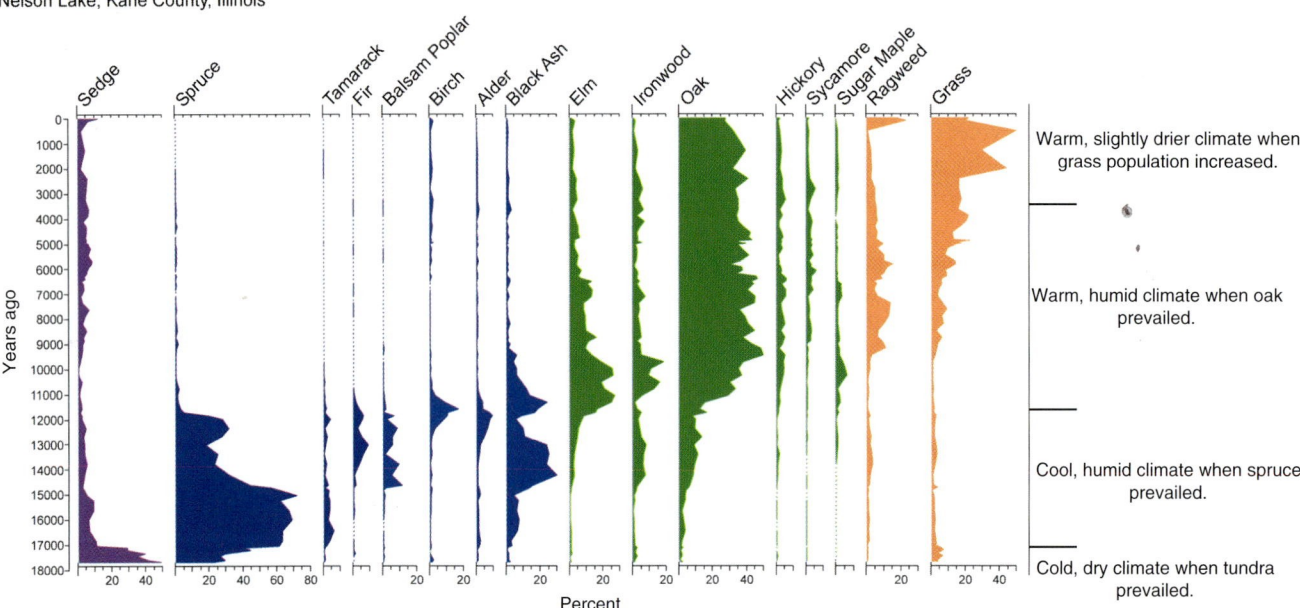

Nelson Lake, Kane County, Illinois

Warm, slightly drier climate when grass population increased.

Warm, humid climate when oak prevailed.

Cool, humid climate when spruce prevailed.

Cold, dry climate when tundra prevailed.

Figure 9.29 Pollen diagram from Nelson Lake, Illinois. Note the vegetation change through time and how it relates to climate fluctuations discussed on the right.

and light wood couplet, with the dark wood representing the winter period when growth is slow or nonexistent and the light wood reflecting rapid growth in summer. When local climate conditions are favorable the rings are wide, whereas they are narrower (reflecting less growth) when it is colder or drier, depending on the type of tree.

The study of tree rings is called **dendrochronology** and, in part, attempts to reconstruct the history of regional climate change by analyzing tree-ring patterns. These ring patterns can be accessed by coring the tree with a device called an increment borer. This method does not harm the tree. To obtain a core, the implement is first screwed horizontally into the center of the tree (Figure 9.30b) and then an approximately 6-mm (0.2-in) wide core is extracted that cuts across the tree rings, thus showing all of them. Some trees live a very long time and thus produce long, continuous tree-ring records. An excellent example of such a tree is the famous bristlecone pines in the White Mountains of California (Figure 9.30c), which are the oldest living things on Earth. Some of these trees are over 4000 years old and still growing! Thus, by looking at the rings of one tree alone it is possible to reconstruct the growth cycles, and therefore the history of climate change, for the past 4000 years in the area where the bristlecones live.

Although the living bristlecone pines have yielded an excellent record of climate change, paleoclimatologists

Dendrochronology *The dating of past events and variations in the environment and climate by studying the annual growth rates of trees.*

are constantly looking for ways to push the evidence of prehistoric climate change further and further back in time. In the context of dendrochronology, this goal is accomplished through a method called *cross dating*. The premise of cross dating is that ring patterns from similar trees that overlapped in their life spans can be matched to extend the record of ring variability back in time.

To see how this method works, look at Figure 9.30d. Imagine that a core is extracted from a living tree (*A*) that produces a distinctive ring record. Near this living tree is another tree (*B*) that is dead but still standing. Upon coring (or, in the case of a dead tree, cutting) the tree, it is discovered that the ring pattern in the latter half of its life was the same as that during the early years of the first (*A*) tree's life. It is thus logical to conclude that the trees overlapped in their life spans, and that they responded to the same climate conditions as far as their growth was concerned. Imagine, then, that another tree (*C*) is discovered that was used as a support structure in an archaeological ruin. Analysis of its ring pattern indicates that in the latter part of its life it overlapped with the first part of the second (*B*) tree's life. We now have a record of climate change that extends back to a period before the life of the intermediate tree. Using this method, paleoclimatologists have extended the history of climate change in the White Mountain area to beyond 9000 years ago!

Yet another way that prehistoric climate records have been reconstructed is by investigating samples of the atmosphere contained in ice cores recovered from glaciers in Greenland and Antarctica. These glaciers contain layers of ice that have accumulated annually for at

(a)

(b)

(c)

Figure 9.30 Using trees to reconstruct climate change. (a) The theoretical foundation of dendrochronology is that many trees produce annual growth rings, with dark and light wood representing winter and summer seasons, respectively. (b) Using an increment borer to core a tree. (c) A 4000-year-old bristlecone pine in the White Mountains of California. (d) The premise of cross-dating is that the tree-ring record can be extended back in time by overlapping ring patterns from a sequence of trees. In this simplified diagram, ring set *A* comes from a living tree, ring set *B* comes from a tree that recently died, and ring set *C* comes from wood that was used in an old house. This particular record permits climate reconstruction between the late 1930s and early 2000s.

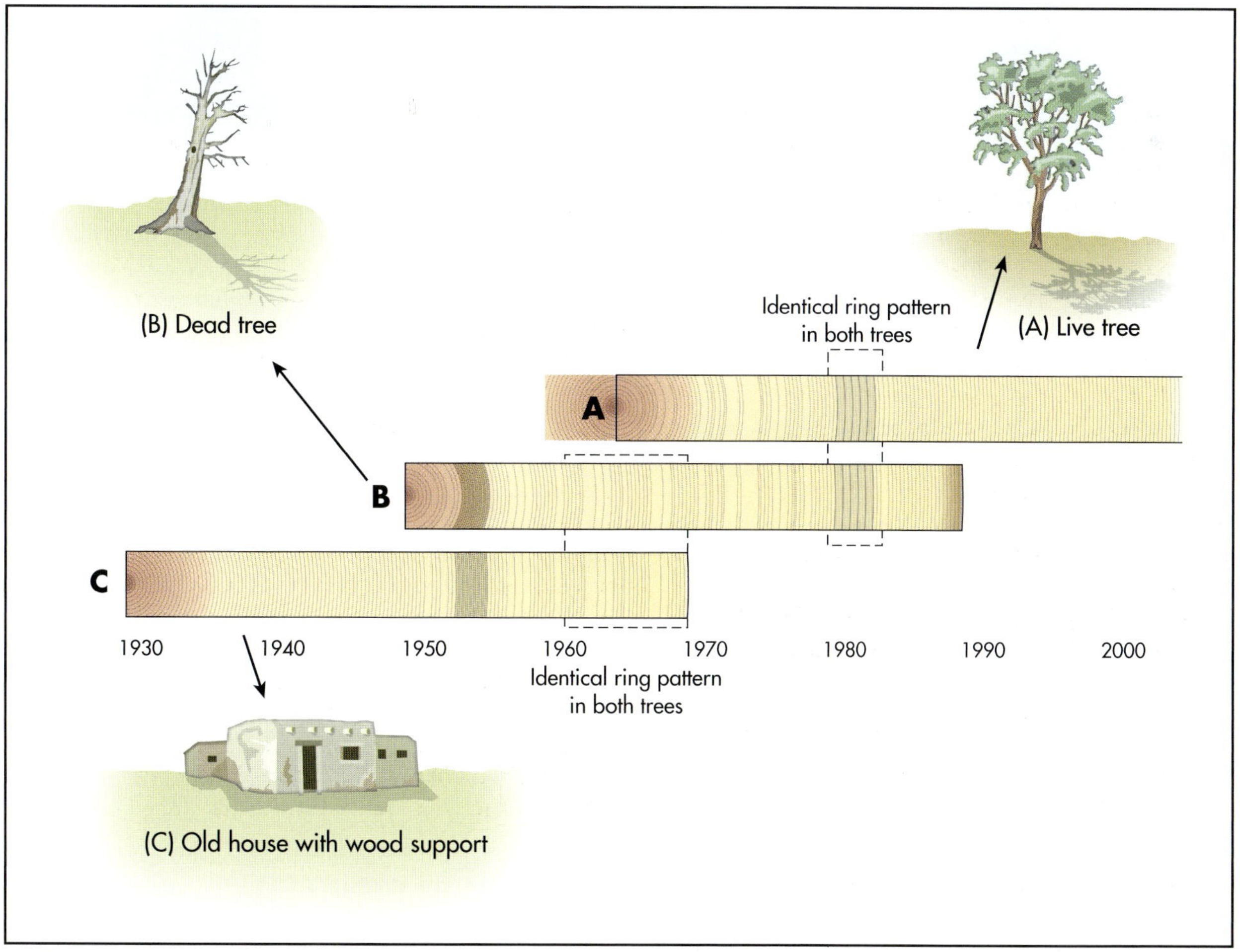

(d)

(B) Dead tree

Identical ring pattern in both trees

(A) Live tree

A

B

C

| 1930 | 1940 | 1950 | 1960 | 1970 | 1980 | 1990 | 2000 |

Identical ring pattern in both trees

(C) Old house with wood support

least the past 650,000 years in some places. Because ice is made of snow, which contains trapped atmospheric gases, it is possible to reconstruct how atmospheric carbon dioxide has varied through time by measuring the prehistoric "atmosphere" contained within each layer of ice. Very simply, the deeper the layer of ice, the older it is. Using this method, the correlation between temperature and greenhouse gases is striking, as you can see in Figure 9.31. Look at this diagram carefully and notice how well peaks and valleys in temperature (the red line) compare with atmospheric carbon dioxide (the blue line) and methane (another greenhouse gas, represented by the green line). You can see that for the past 160,000 years there has been a very close relationship between climate and greenhouse gases.

After viewing these data, you might still be skeptical about the link between atmospheric carbon dioxide and global warming. After all, 160,000 years ago, even 100,000 years ago, seems like a long time before the present and maybe environmental conditions then were just *different*. Is there any more recent information that would make you feel more confident about this link? In that context, let's focus for a moment on the past 1000 years, which encompasses a significant portion of recent human history. If you examine Figure 9.32, you can see that, beginning 1000 years ago, the temperature was about 0.2° C

(0.4° F) cooler than the average temperature between 1961 and 1990. Subsequently, between 1000 and 1900 the average temperature *further cooled* about another 0.2° C (0.4° F). This cooling was most intensive during the *Little Ice Age*, which generally occurred from the 14th to the 19th centuries.

Now, look at the change that occurred at about 1900. At this time the temperature began to warm quickly to its present state of being 0.7° C (1.3° F) warmer than the past 30-year average. In other words, the natural climate progression for the 850 years prior to the industrial revolution was one of general cooling, with a rapid warming in the past 100 years. Given the demonstrable link between temperature and greenhouse gases for the past 160,000 years (see Figure 9.31), coupled with the fact that the climate had been *cooling* between 1000 and 1850 A.D., the rapid warming of the past 100 years seems to be best explained by human-induced greenhouse gas emissions. At least, that's what the vast majority of climatologists believe.

Predicting Future Climate Change

If the one million dollar question is whether or not human-induced climate change is occurring, then the two million dollar question is "How much future warming will

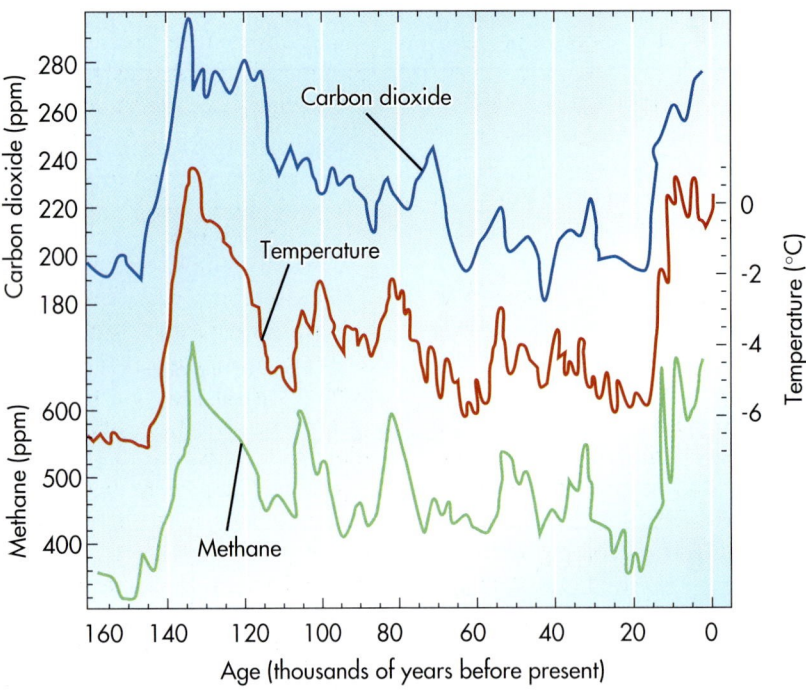

Figure 9.31 Greenhouse gases and prehistoric temperature from the Vostok ice core in Antarctica. This graph shows the close relationship between atmospheric temperature (red line), carbon dioxide (blue line), and another (lesser) greenhouse gas (methane), which is the green line. Temperature is the number of degrees above or below the average modern surface temperature of about 14° C (55° F). (Credit: J. M. Barnola, D. Raynaud, Y. S. Korotkevich, and C. Lorius. Vostok ice core provides 160,000-year record of atmospheric CO_2. *Nature* 1987; 329:408–414.)

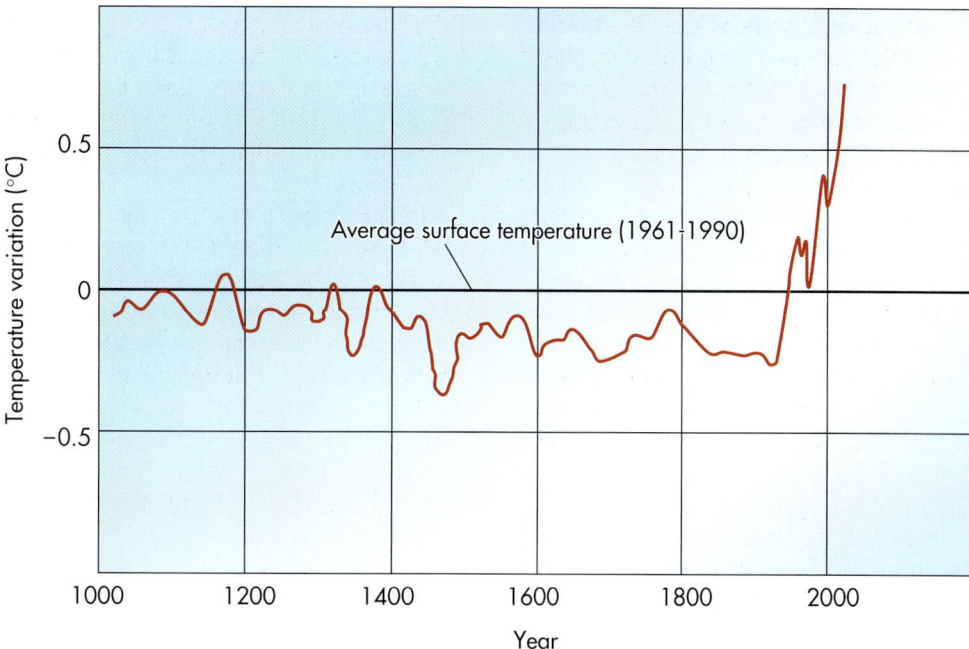

Figure 9.32 Global climate change for the past 1000 years. Temperature was generally cooling between 1000 years ago and approximately 1850. Since that time, temperature has been warming dramatically. The baseline (0) for these comparisons is the average temperature from 1961–1990. (Credit: M. E. Mann, R. S. Bradley, and M. K. Hughes, Northern Hemisphere temperatures during the past millennium: Inferences, uncertainties, and limitations. *Geophys. Res.* Lett. 1999;26:759–762.)

occur?" Remember that the current proportion of atmospheric carbon dioxide is about 370 parts per million parts of the atmosphere, and that this proportion exists with the developed world producing 73% of the excessive carbon dioxide. The big unknown is how climate will respond when the developing countries increase their carbon dioxide emissions by the predicted 25% in the next few decades.

In the context of understanding future change in an enhanced greenhouse world, the scientific challenge is to somehow predict how the climate system will respond. This complex task is best accomplished with a general circulation model (GCM), which is a mathematical model that incorporates known climate variables, such as cloud cover, insolation, latitude, and proximity to oceans, to forecast future climate events. Using these variables, large computers can run simulations involving scenarios of low or high carbon dioxide levels. Models that run moderate-case scenarios use a doubling of atmospheric carbon dioxide as a benchmark.

With this benchmark in mind, the Intergovernmental Panel on Climate Change has predicted potential amounts of warming by the year 2100 based on a variety of global economic and social scenarios. One such scenario is based in part on a rapidly growing global population (to 15 billion people) and limited international cooperation with respect to carbon emissions. Figure 9.33 illustrates the geography of warming in this scenario. Note that above average levels of warming are predicted over much of the Earth, with much greater than average warming expected at high latitudes of the Northern Hemisphere, especially during the winter months.

What would happen if the Earth really does warm to the levels indicated in Figure 9.33? Much of what occurs depends on the nature of climate feedbacks, which are indirect or secondary changes that take place within the overall climate system in response to the initial forcing mechanism—in this case, increased atmospheric carbon dioxide. For example, it is possible that increased warming will cause more water vapor to evaporate from oceans, resulting in more atmospheric vapor. This increase, in turn, can have a larger effect on the global energy balance, both as a *positive climate feedback* that further increases warming or potentially as a *negative climate feedback* that causes cooling. Increased atmospheric water vapor could cause a positive feedback because it traps solar energy, which would further warm the Earth. On the other hand, increased atmospheric vapor will increase the number of clouds, which could increase the amount of sunlight that reflects back into space, thus resulting in a net cooling effect.

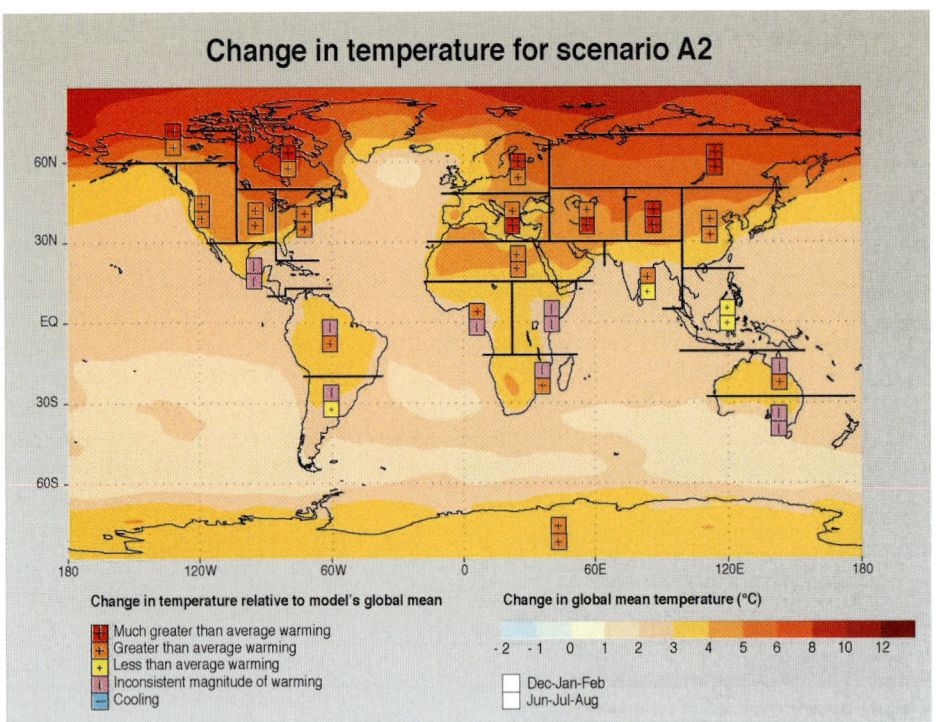

Figure 9.33 Predicted climate change by 2100. This predicted outcome is based on a specific social and economic scenario and shows the level of warming (relative to current average temperature) in December–February (upper box of box pair) and June–August (lower box of box pair).

(Credit: Intergovernmental Panel on Climate Change.)

The ultimate positive feedback is a runaway greenhouse effect where significant warming causes the polar ice caps to melt and thus less radiation is reflected to space, which in turn, results in even more warming. In contrast, the ultimate negative feedback is that melting ice caps cause warm-water currents such as the Gulf Stream to stop flowing north because of the rapid influx of cold water at high latitudes. Some models suggest that the Earth might suddenly shift into another ice age if this feedback occurs. Fortunately, neither the ultimate positive nor negative feedback is considered likely.

Although the response of the Earth's future climate to feedbacks is uncertain, the fact is that the vast majority of climatologists believe that warming will continue at least for the next century. Some scientists argue that there will be no major environmental consequence—that, in fact, a warmer world might be a good thing. In the context of this positive outlook, those who live in the northern tier of the U.S. might find some warming a welcome respite from the usually harsh winters that occur there. A potential benefit may have already occurred in the Great Lakes region, where the annual length of the growing season increased by about one week in the past 20 years.

Although some "environmental benefits" may occur in a warmer world, these data suggest that several negative results will probably occur as well. One of these negative effects is a rising sea level, which would result from increased melting of polar ice caps and thermal expansion of water. According to current predictions, sea level will rise 15 to 90 cm (5.11 to 35.4 in) by 2100, which will place 92 million people around the globe within the risk of coastal flooding. Another potential negative factor is that climate change may promote the spread of insect-borne diseases such as malaria. Climate boundaries may also shift, making some regions drier while others become wetter. This boundary shift may cause significant displacement of human populations due to changes in agricultural patterns. Still another potential change is that climate variability may be enhanced, with a higher frequency of extreme events such as snowstorms, rainstorms, and intense spells of hot and cold weather. In fact, data suggest that these kinds of extreme events have been occurring more frequently since 1980.

Because it appears that global warming really may be occurring, and many negative consequences may result, world leaders have begun efforts to control greenhouse emissions. The first truly global effort occurred at the 1992 Earth Summit in Brazil, where 150 nations signed a treaty that limited emissions of greenhouse gases. A subsequent, more comprehensive, global treaty was proposed in 1997 at Kyoto, Japan. The resulting *Kyoto Protocol* calls for the 38 leading industrial nations, including the U. S., nations of the European Union, and Japan, to reduce greenhouse emissions 5% – 8% below 1990 levels. These reductions could be accomplished either through outright curtailment of emissions or through using carbon credits as a currency between countries. In other words, each country would be allowed a certain number of carbon credits that could be bought, sold, or traded.

Although the U.S. signed the treaty, it was rejected by the Senate because it did not include guidelines for how developing countries should adjust their greenhouse gas emissions. According to many U.S. political leaders, the treaty is thus unfair because it would cause undue hardship to the national economy by forcing businesses and corporations to develop new technologies that would reduce carbon emissions. Developing these technologies would add cost to products produced in the U.S., which would theoretically put us at a competitive disadvantage compared to developing countries that have no carbon standards. In response to this argument, leaders of the developing countries argue that the increased levels of atmospheric carbon dioxide are the product of just a few countries like the U.S. and some countries in Europe, and thus we should take the lead regarding carbon standards. Given this political stalemate regarding global carbon standards, greenhouse emissions continue to rise. In all probability, these ongoing increases and the impact they have on climate will be a prominent environmental issue for the rest of your life.

KEY CONCEPTS TO REMEMBER ABOUT GLOBAL CLIMATE CHANGE

1. A major issue facing the world today is global warming, which is related to human enhancement of the natural greenhouse effect.

2. The carbon cycle refers to the way that carbon is stored in the various reservoirs (such as the atmosphere, ocean, and rocks) and how it moves between them.

3. In order to understand the extent of potential future climate change, it is necessary to understand prehistoric climate variability. This can be reconstructed by indirect methods, such as pollen analysis and dendrochronology.

4. Since the onset of the industrial revolution in the mid-1800s, atmospheric carbon dioxide has increased about 30%. This increase is almost entirely due to human activities.

5. Between the years 1000 and 1850 A.D. global climate generally cooled. Since 1850 it has been warming.

6. General circulation models predict that average global temperature will increase 2° to 4° C (4° to 7° F) this century.

The Big Picture

With this chapter you have completed the focus on atmospheric processes. As noted earlier, this chapter marked the first time in this book that you've begun to see how geographic patterns can exist on the ground, here in the form of vegetation as it responds to climate. The next chapter will build upon this transition by focusing on plant geography and the many variables that influence the distribution of plants, both at the global and local scales. You will find that there is a close correlation between global climate and vegetation patterns. Look at this map of global vegetation, for example, and see how the patterns resemble those in Figure 9.2. The next chapter will investigate these relationships, as well as more local influences.

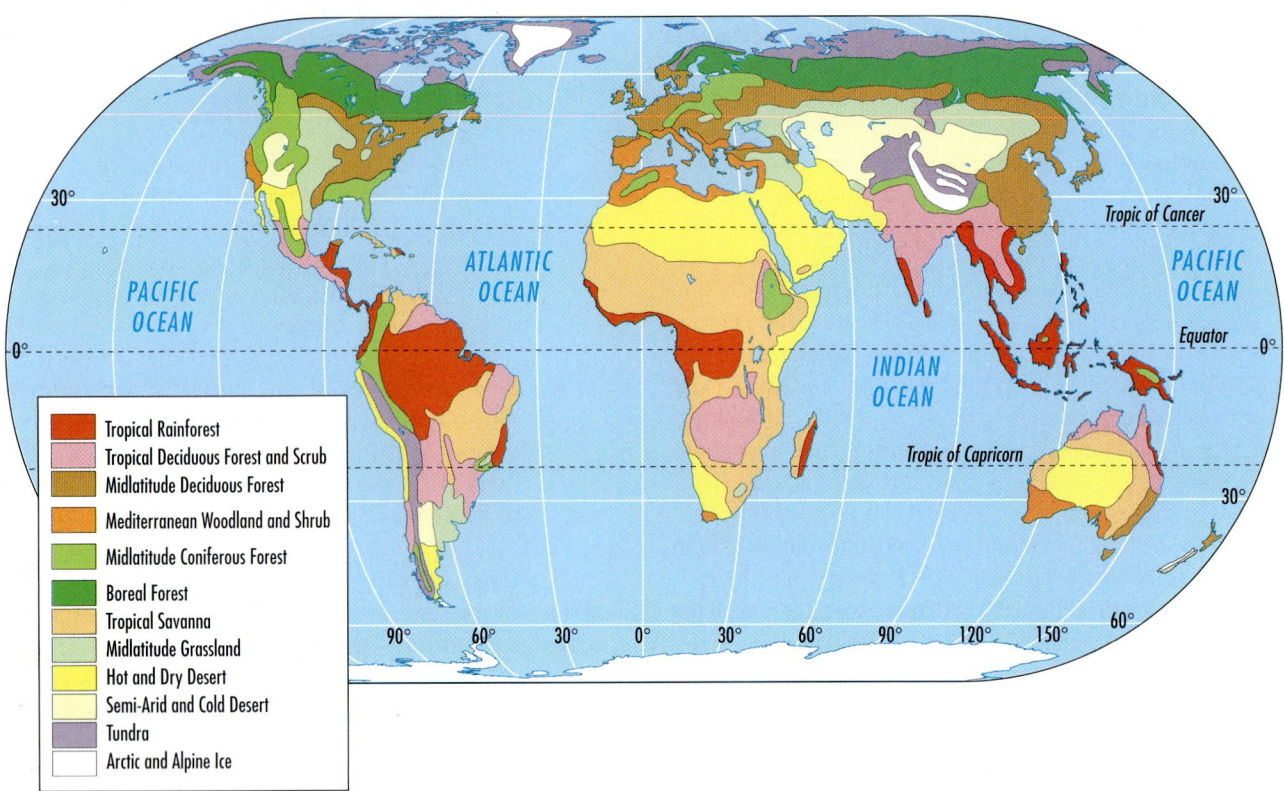

Legend:
- Tropical Rainforest
- Tropical Deciduous Forest and Scrub
- Midlatitude Deciduous Forest
- Mediterranean Woodland and Shrub
- Midlatitude Coniferous Forest
- Boreal Forest
- Tropical Savanna
- Midlatitude Grassland
- Hot and Dry Desert
- Semi-Arid and Cold Desert
- Tundra
- Arctic and Alpine Ice

Summary of Key Concepts

1. Climate refers to the average precipitation and temperature characteristics of a region. This depends on many interrelated variables, including latitude, insolation, seasonality, atmospheric circulation, and topographic effects.

2. The Köppen climate system categorizes global climates based on their average annual temperature and precipitation characteristics. There are six major categories: tropical humid climates (A), arid/semi-arid climates (B), mesothermal climates (C), microthermal climates (D), polar climates (E), and highland climates (H).

3. The humid tropical climates (A), such as the tropical rainforest climate (Af), generally have a low annual temperature range and abundant precipitation caused by the Intertropical Convergence Zone (ITCZ). Precipitation becomes distinctly seasonal with increased latitude in these regions.

4. The arid/semi-arid climates (B) are regions of very low rainfall and sparse vegetation. This low precipitation is caused by the Subtropical High (STH) in places such as the Sahara Desert, the rainshadow effect in places like Denver, Colorado, and by the continental effect in places like north-central China.

5. Mesothermal climates (C) are typically found in the midlatitudes. These climate regions have a distinct seasonal temperature and precipitation pattern. The humid subtropical hot-summer cli-

mates (*Cfa, Cwa*), for example, have hot summers and mild winters. The wet season occurs in summer due to the influx of mT air pumped in by offshore STH pressure systems.

6. Microthermal (*D*) climates are largely continental climates that occur only in the Northern Hemisphere. These climate regions have hot to mild summers and cold winters. Average annual precipitation is moderate and is usually associated with midlatitude cyclones.

7. Polar (*E*) climates occur at very high latitudes and are associated with tundra and ice caps. Although these regions have a distinct annual temperature range, they are generally cold with low to moderate precipitation. Highland climates occur in mountainous areas and have characteristics of subarctic and polar climates.

8. Since the onset of the Industrial Revolution, atmospheric carbon dioxide has increased about 20% and is expected to increase further. In that same time, average global temperature has increased about 1° C (1.8° F). In the next century, average global temperature is predicted to increase about another 2° to 4° C (4° to 7° F).

Check Your Understanding

1. How does latitude influence the distribution of global climates?

2. Why does the wet-equatorial climate receive consistent rainfall throughout the year?

3. In contrast to the wet-equatorial climate, the tropical wet-dry climate has a distinct dry season. What time of year does the dry season occur and why does it happen?

4. The Mediterranean and marine west-coast climates are similar in that both have moderate annual temperatures. However, the marine west-coast climate receives more precipitation. Why?

5. Why is the tropical desert climate dry over the course of the entire year?

6. The Subtropical High (STH) pressure system is usually associated with dry climates, but, in the case of the humid subtropical climate, it contributes to summer rainfall. Why does this pattern occur?

7. Both the moist-continental and boreal-forest climates have a large annual range of temperature. Why does this variability occur?

8. Why is there a zone of cold climate along the equatorial region of South America?

9. Why do increased levels of atmospheric carbon dioxide contribute to global warming?

10. What is the evidence that there has been a close relationship between atmospheric carbon dioxide and climate change in the prehistoric past?

11. What role has the U.S. played in the enrichment of atmospheric carbon dioxide?

ANSWERS TO VISUAL CONCEPT CHECKS

Visual Concept Check 9.1

The answer is *c*; the area is dominated by the Subtropical High pressure system. This pressure system consists of descending air that warms adiabatically. Thus, it is dry in the Middle East.

Visual Concept Check 9.2

The lower 48 states can be subdivided into three general zones of precipitation. The southeastern U.S. is a zone of high precipitation that originates from moisture from the Gulf of Mexico. This moisture is pumped into the southeastern states by the subtropical high pressure system. The western part of the U.S. is generally dry, mostly because this area lies in the rainshadow of the Rocky, Cascade, and Sierra Nevada Mountains. The West Coast of the U.S. is an area of high precipitation because orographic processes release moisture that flows inland from the Pacific Ocean.

Visual Concept Check 9.3

The correct choice is *c*; the climograph represents the climate at Christchurch, New Zealand, which is in the Southern Hemisphere (43.5° S, 173° E). Christchurch is a coastal city on the eastern shore of the South Island. As a result, it has a marine climate, which makes sense given the narrow and mild temperature range in the area.

CHAPTER TEN · PLANT GEOGRAPHY

In the preceding chapter, we examined the geographic distribution of global climates. One way in which we illustrated this distribution was through the type of vegetation that grows in a particular climate zone. For example, dense forests occur in humid regions because there is sufficient moisture to support them. In contrast, a place like the Sahara Desert is relatively devoid of vegetation because it is very dry.

Some physical geographers study the spatial distribution of plants. One element of interest to these scientists is why some plants grow in some places and not in others. This photograph is of marshland forest along the Amazon River in Brazil. In this chapter you will investigate why this dense forest grows in this kind of environment.

This chapter marks an important transition within this book because we begin to examine the geographic distribution of environmental variables *on the ground,* rather than solely within the atmosphere or oceans. The remainder of this book will focus on these kinds of geographic distributions through the study of variables such as soils, rivers, and the landscapes shaped by tectonic forces, wind, and ice. In this chapter, we'll look more closely at some of the factors that influence the distribution of vegetation around the globe.

The Process of Photosynthesis

Plant geography is an important part of physical geography because plants are absolutely critical to life on Earth. One reason for this significance is that plants convert atmospheric carbon dioxide to oxygen, which animals (including humans) breathe. In addition, plants produce and store carbohydrates, which are a primary food source for a wide variety of organisms. Plants produce these important compounds through **photosynthesis**, which is the process in which solar radiation is converted to chemical forms of energy. We discussed photosynthesis in Chapter 9, in the context of the carbon cycle. Here we present more details of the process in the context of the life cycles of plants.

Conversion of solar energy into chemical energy through photosynthesis requires the presence of water, sunlight, and carbon dioxide. Carbon dioxide is the foundation of the photosynthetic process because it is absorbed by plants through respiration and converted to carbohydrate form, specifically glucose (sugar). The process must occur within a water medium, which is contained within the plant fibers, because water provides the electrons required for the conversion of carbon dioxide to sugar. This process is represented chemically in the following equation:

$$6H_2O \ + \ 6CO_2 \ + \ \text{light energy} \rightarrow \ C_6H_{12}O_6 \ + 6O_2$$

| water | carbon dioxide | glucose (sugar) | oxygen |

Photosynthesis *The conversion of solar radiation into chemical energy. Sugars and starches are produced from carbon dioxide and water, through the interaction of light and chlorophyll in green plants. The process releases oxygen into the atmosphere.*

In other words, six molecules of water plus six molecules of carbon dioxide plus light energy produce one molecule of glucose (sugar) plus six molecules of oxygen. At this scale, the result of this chemical reaction may not seem impressive to you, but each plant produces millions of these sugar molecules each second. If you are still not impressed, consider that almost all of the oxygen in the Earth's atmosphere is created by this global process.

Figure 10.1 shows the photosynthetic and respiratory cycles, including the various pathways that oxygen, water, and carbon dioxide take within the system. As you work through this discussion, follow the numbered stages in the figure. These numbers are not meant to imply that stages occur in a specific order, but rather they help organize the flow of the presentation. For simplification, think of the cycle as broadly consisting of primary producers, such as plants, and decomposers, such as bacteria. There are also plant consumers, but they are not really relevant to the photosynthetic cycle.

A logical place to begin to trace the various pathways is in the soil, which provides water to the living plant (1). At the same time the roots of the plant are absorbing water, the green leaves of the plant absorb solar radiation (2). This enables photosynthesis to take place in the leaves. Compounds in the plant's leaves called chlorophylls and carotenoids are important ingredients in the process because they absorb sunlight of specific wavelengths. Chlorophylls absorb wavelengths in the blue and red part of the electromagnetic spectrum, whereas carotenoids absorb blue-green light (carotenoids appear orange or yellow).

Figure 10.2 shows graphically how these patterns of electromagnetic absorption occur. Notice that plants do not absorb energy in the green and orange parts of the spectrum. Instead, these wavelengths are either reflected or pass through the plants' leaves. The green wavelength is most actively reflected, which is why plants appear to

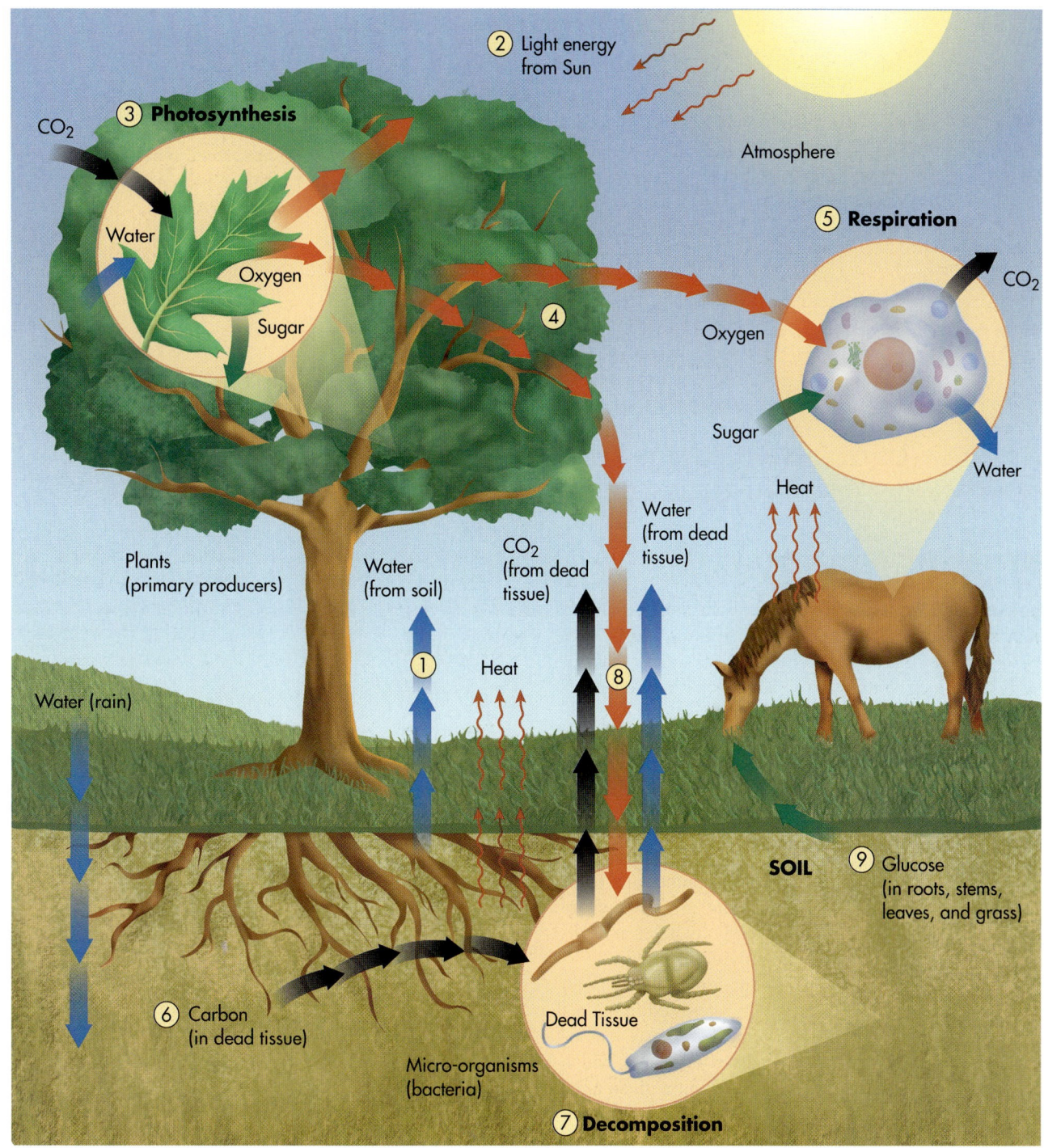

Figure 10.1 Photosynthesis, respiration, and decompositon cycles. Water, oxygen, and carbon dioxide take various pathways within the biosphere. Plants are part of these pathways through the processes of photosynthesis and decomposition, while animals (including people) are part of respiration and decomposition processes.

be green. This energy absorbed in the blue and red wavelengths is critical to photosynthesis because it excites chlorophyll and carotenoid electrons within the plant.

Turning back to Figure 10.1, notice that while photosynthesis is occurring, carbon dioxide is absorbed into the plant leaves as a part of the respiratory process (3). As this

absorption of CO_2 occurs, oxygen is released and cycles into the atmosphere (4). This oxygen is absorbed by animals, which produce CO_2 as a by-product through respiration (5). When plants and animals die, their remains fall to the ground and are consumed by decomposers (6). As the decomposers absorb oxygen through respiration from the atmosphere or

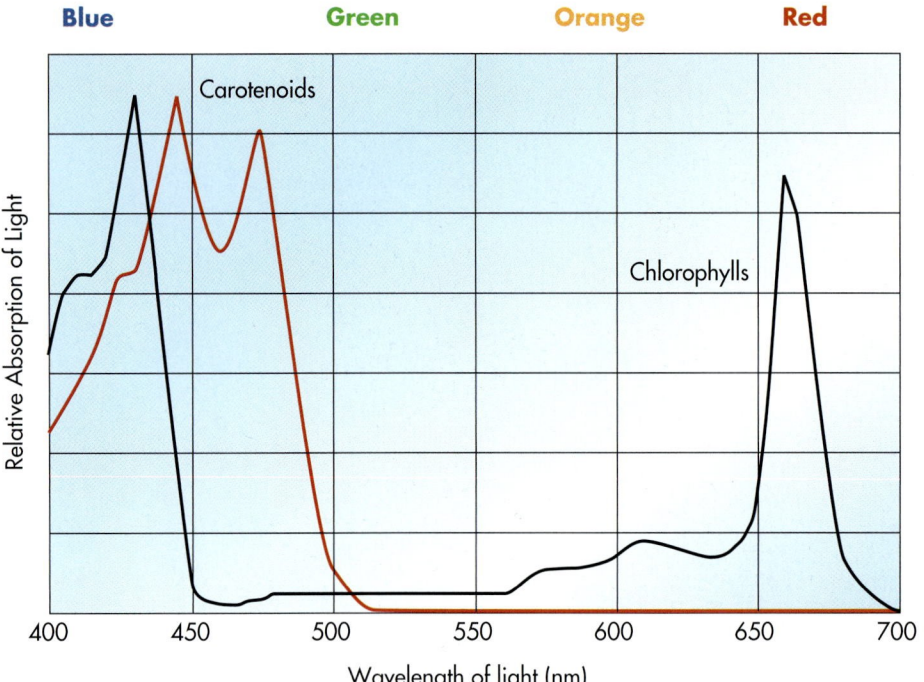

Blue **Green** **Orange** **Red**

Carotenoids

Chlorophylls

Relative Absorption of Light

400   450   500   550   600   650   700

Wavelength of light (nm)

Figure 10.2 Preferential absorption of solar wavelengths in plants. Note that blue and red wavelengths are actively absorbed, whereas the green and orange parts of the spectrum are not.

soil (7), they combine the oxygen with decomposing carbohydrates, releasing energy contained within CO_2 and water vapor that are emitted into the atmosphere (8).

Although most of the solar energy absorbed by plants is used for respiration, a great deal of it is used in photosynthesis to produce glucose, which is retained in the plant to build and sustain leaves, fruits, and seeds. In addition, plants convert glucose to cellulose, which is the structural material used to build cell walls. Although much of the produced glucose is used to sustain the life of the plant, a great deal of it is stored as starches and other carbohydrates in roots, stems, leaves, and grass (9), where it can be used later. Some of these carbohydrates are subsequently consumed by animals, providing energy.

Scientists who study plant production are called *plant ecologists*. Plant ecologists are often interested in the net amount of vegetation produced by photosynthesis. An important indicator of plant productivity is **biomass**, which is the dry weight of all living organisms in a given area. This measure is important because it reflects the amount of chemical energy that has been stored and is thus available for consumption. Biomass is typically measured in kilograms per square meter of ground or as metric tons per hectare. Given what you read in Chapter 9 regarding climate, it should be no surprise that biomass varies greatly across geographic regions. Forests have the highest biomass, whereas grasslands have less.

In the context of global biomass, examine Figure 10.3, which is a composite satellite image that shows the

geography of global biomass as constructed from data collected from 1978 to 1986. The ocean portion of the figure is a composite of more than 60,000 images collected by a sensor on the NIMBUS-7 satellite. This sensor measured the distribution and amount of phytoplankton, which are microscopic plants that grow in the upper part of the ocean where the Sun penetrates. Red and orange colors represent high concentrations of phytoplankton, whereas yellow to violet colors reflect progressively lower concentrations. The land vegetation image is a composite collected during 15,000 orbits by the Advanced Very High Resolution Radiometer (AVHRR). This sensor measured radiation from the land surface, which is then used to estimate vegetation production. The dark green areas are rainforests that have the highest growth potential, whereas the lighter greens reflect the potential vegetation growth in other tropical and subtropical forests, as well as midlatitude forests and farmland. The lightest shades of yellow represent deserts where little growth potential exists. Snow- and ice-covered regions have no vegetation potential.

The Relationship of Climate and Vegetation: The Character and Distribution of Global Biomes

Many different factors influence the geographic distribution of plants and biomass. As you have seen by now in physical geography, environmental variables affect one another to form distinct geographic patterns. The study of plant geography is no exception. Thus, it is a mistake to assume that each variable acts on its own to control the geographic distribution of plants. Nevertheless, for the

Biomass *The amount of living matter in an area, including plants, large animals, and insects.*

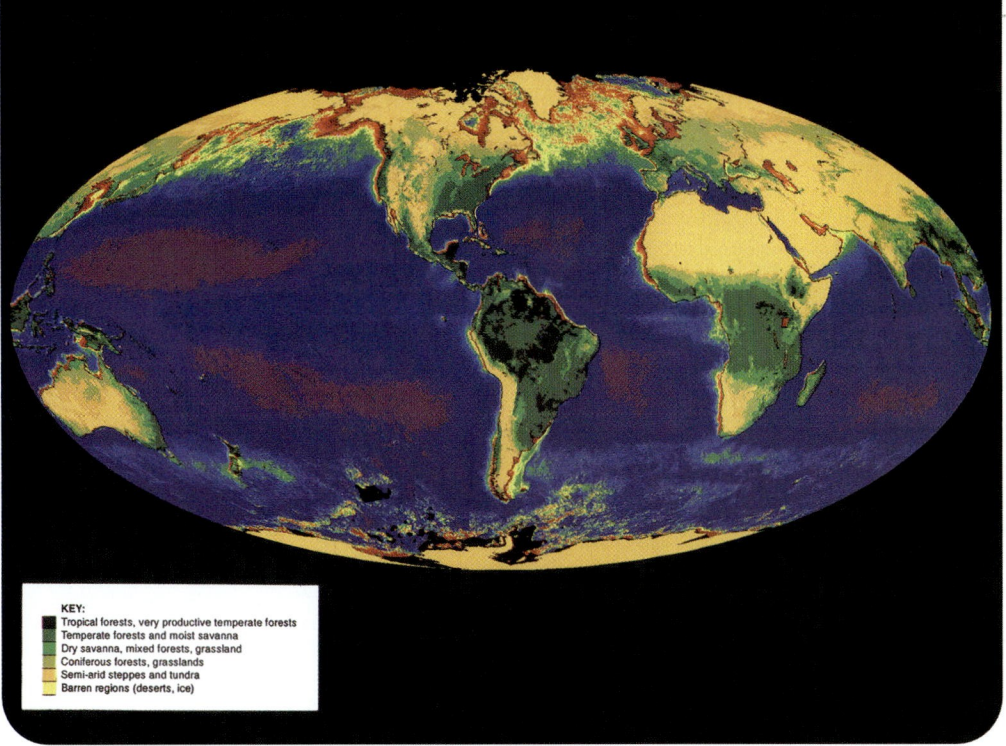

KEY:
- Tropical forests, very productive temperate forests
- Temperate forests and moist savanna
- Dry savanna, mixed forests, grassland
- Coniferous forests, grasslands
- Semi-arid steppes and tundra
- Barren regions (deserts, ice)

Figure 10.3 The global biosphere. This composite image was acquired by the NIMBUS-7 and AVHRR satellites over a period of 8 years. It shows the geographic distribution of potential global biomass.

sake of clarity, in this section of the chapter we will generalize about how certain of these variables act as independent factors to affect plant growth.

The most important factor influencing the geographic distribution of vegetation is climate. Secondary factors such as geology, soils, landscape position, and human behavior also influence the pattern of vegetation on Earth and will be discussed later in the chapter. For now, however, the focus of this part of the chapter is the relationship between climate and vegetation. You saw in Chapter 9 how vegetation reflects the geographic distribution of climates. Geographic climate patterns are related to many other primary factors, including latitude, proximity to oceans, and the global atmospheric circulatory system. In this context, to say that the geographic distribution of vegetation is related to climate is, in effect, the same as acknowledging the many different variables that influence the climate system. Be sure to remember this in the following discussion.

A good way to study the relationship between climate and vegetation is through the concept of global biomes. A **biome** is a definable geographic region that is classified and mapped according to the predominant vegetation and organisms of that particular environment. Each biome is a distinct **ecosystem**, which is a community of plants, animals, and micro-organisms that are linked by energy and nutrient flows. Ecosystems can be viewed at many different scales, ranging from local (such as a pond) to regional (such as a large desert). This part of the chapter will focus on large areas that typically cover thousands of square kilometers, with the focus being vegetation. Although the biome concept also includes information about the character of soils, we will describe the major biomes here because vegetation is the most visible component of a biome. The relationship of soils to vegetation and climate will be covered in Chapter 11.

This section of the chapter focuses on geographic units as they relate to **potential natural vegetation**—that is, the vegetation that would exist in a region if humans did not modify the landscape. Figure 10.4 is a map showing the distribution of the 12 biomes and Table 10.1 outlines their general characteristics. You should refer to this map and table frequently as you read this section.

Biome *The complex of living communities maintained by the climate of a region and characterized by a distinctive type of vegetation.*

Ecosystem *A community of plants, animals, and micro-organisms that are linked by energy and nutrient flows.*

Potential natural vegetation *The vegetation that would occur naturally within a specific area if no human influence occurred.*

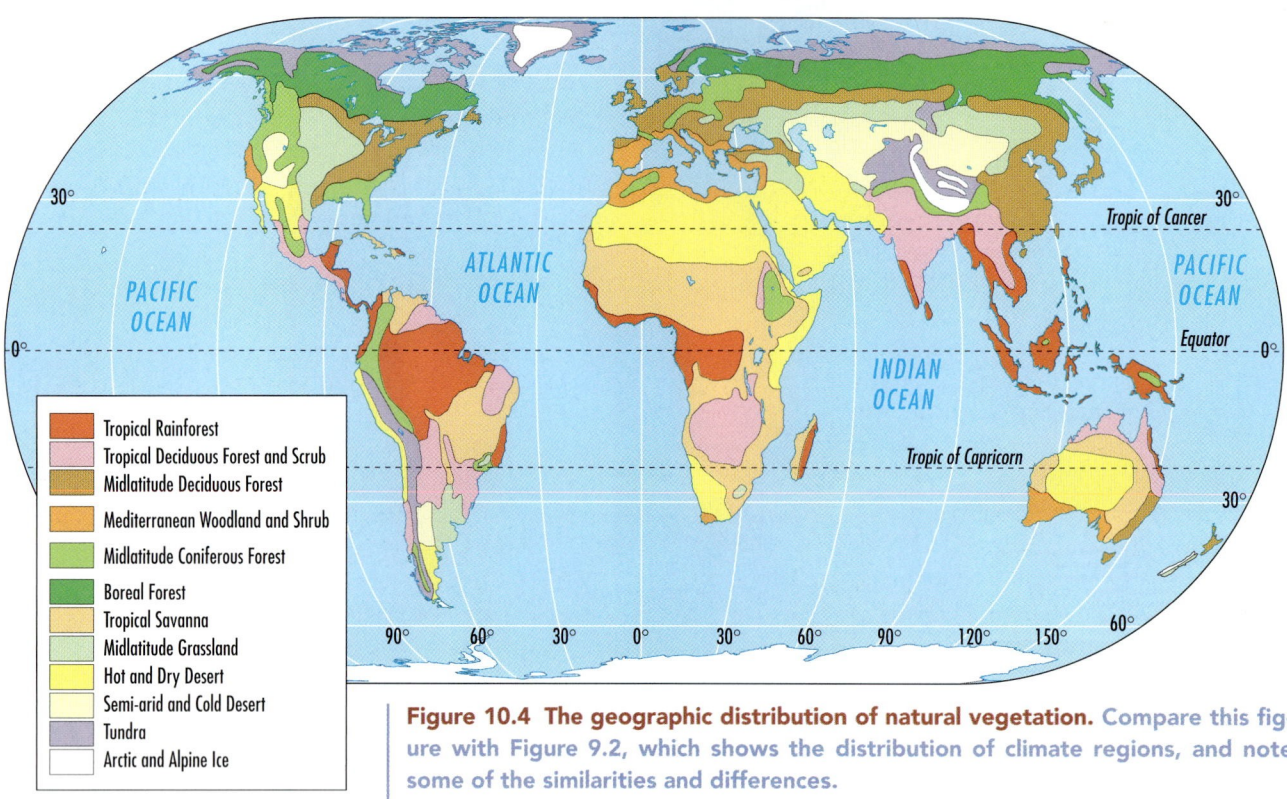

Figure 10.4 The geographic distribution of natural vegetation. Compare this figure with Figure 9.2, which shows the distribution of climate regions, and note some of the similarities and differences.

Legend:
- Tropical Rainforest
- Tropical Deciduous Forest and Scrub
- Midlatitude Deciduous Forest
- Mediterranean Woodland and Shrub
- Midlatitude Coniferous Forest
- Boreal Forest
- Tropical Savanna
- Midlatitude Grassland
- Hot and Dry Desert
- Semi-arid and Cold Desert
- Tundra
- Arctic and Alpine Ice

These biomes are organized into four categories: (1) forest biomes, (2) grassland biomes, (3) desert biomes, and (4) tundra biome. The alpine and arctic ice biome will not be discussed in detail because it simply consists of landscapes covered by bare rock and glaciers and therefore has no plants associated with it.

As you study the remaining 11 biomes, remember that the boundary between vegetation communities is rarely distinct. Instead, there is usually a transition from one vegetation type to another that may occur over several hundred kilometers. Such a vegetation transition is called an **ecotone**, because it contains elements of two distinct plant communities. These transitions occur because the boundaries between climate regions are usually difficult to precisely define, as was described in Chapter 9.

Forest Biomes

Forest biomes are geographic regions where the vegetation is dominated by trees. These vegetation assemblages occur in areas where there is a net surplus of available moisture, either because there is a high average annual precipitation or because average temperatures are sufficiently cool that little evaporation occurs. Forest biomes occur from the tropical regions to the high midlatitudes. The forest biomes include the tropical rainforest biome, midlatitude forest biomes, and the boreal forest biome.

Tropical Forest Biomes The tropical forest biomes occur in the low latitudes within South America, Africa, Asia (including Indonesia), and northern Australia. These forests are generally characterized by dense biomass and tremendous species diversity. The tropical forest biomes contain two major subdivisions, including (1) tropical rainforest, and (2) tropical deciduous forest and scrub.

Tropical Rainforest Biome The tropical rainforest biome straddles the equatorial region between 23.5° N and 23.5° S (Figure 10.4) and is very closely related to the tropical rainforest climate (*Af*) zones. It is also associated with small areas of tropical monsoon climate (*Am*) that have a very limited dry season, such as in central Mexico. Average annual precipitation in the tropical rainforest biome ranges from 180 to 400 cm (70.1 to 158 in), with monthly rainfall consistently greater than 6 cm

Ecotone *The transition area where two or more ecosystems merge.*

TABLE 10.1 **Characteristics of the Major Global Biomes**

Biome	Dominant Vegetation	Köppen Climate	Average Annual Precipitation and Moisture Patterns	Average Annual Temperature Patterns
Tropical Rainforest	Broadleaf evergreen trees. Dense canopy with open forest floor. Lianas (vines) and epiphytes.	Af Am (wet side)	180–400 cm (71–158 in)	21°–30° C (70°–86° F)
Tropical Deciduous Forest and Scrub	Ecotone between rainforest and grasslands; broadleaf trees with seasonal leaf fall.	Am, Aw	130–200 cm (51–79 in), with distinct winter dry season.	Seasonal, but always warmer than 18° C (64° F).
Midlatitude Deciduous Forest	Broadleaf trees with seasonal leaf fall.	Cfa, Cwa, Dfa, Dwa, Dwb	75–150 cm (30–60 in), with summer maximum; no seasonal water deficits.	Hot summers and cool to cold winters.
Mediterranean Woodland and Shrub	Individual short shrubs separated by grassy patches.	Csa, Csb	25–65 cm (10–26 in); winter surplus and summer deficit.	Warm summers and cool winters.
Midlatitude Coniferous Forest	Needleleaf evergreen trees, including spruce, pine, fir.	Cfa, Cfb, Cfc, Cwa, Dfb, H	75–150 cm (30–60 in) in southeastern pines; 30–100 cm (12–39 in) in northern conifers and highlands; 150–500 cm (60–197 in) in west coast rainforests.	Hot summers with cool winters in southeast; cold winters and cool summers in northern conifers and highlands. Cool winters and warm summers in west coast rainforest.
Boreal Forest	Needleleaf conifers, including spruce, pine, fir.	Dfb, Dfc, Dfd	30–100 cm (12–39 in); no seasonal water deficit; ground frozen most of year.	Long, cold winters and short, cool summers.
Tropical Savanna	Ecotone between tropical forests and deserts; grasslands with isolated trees.	Aw, BS	9–150 cm (4–59 in); short summer wet season and long dry season.	Seasonal, but always warmer than 18° C (64° F).
Midlatitude Grassland	Tallgrass prairie in humid areas to shortgrass steppes in more arid regions.	Cfa, Dfa	25–75 cm (10–30 in); summer water stress.	Seasonal, with hot summers and cool to cold winters.
Hot and Dry Desert	Xerophytes such as succulents and cacti trending to large patches of bare ground.	BWh, BSh	Less than 2 cm (0.7 in); perpetual moisture deficits.	Perpetually hot.
Semi-Arid and Cold Desert	Short grass and dry shrubs; patches of bare ground.	BSk, BWk	2–25 cm (0.7–10 in)	Average annual temperature = 18° C (64° F).
Tundra	Stunted shrubs and sedges; mosses, lichens.	ET, Dwd	15–180 cm (6–71 in); no seasonal moisture deficit; frozen ground virtually all year.	Above freezing only 2 or 3 months of year, with warmest month < 10° C (50° F).
Alpine and Arctic Ice	Some algae perhaps.	EF	Less than 10 cm (4 in).	Frigid.

Emergents
40–60 m (130–200 ft)

Canopy (densest layer)
15–40 m (50–130 ft)

Understory
5–15 m (15–50 ft)

Ground layer
0–5 m (0–15 ft)

(a)

(b)

Figure 10.5 The tropical rainforest. (a) The layered structure of the rainforest creates a continuous canopy that causes dense shading of the forest floor. (b) Do you see the emergent trees in this tropical rainforest?

(2.4 in). These areas are always warm, with an average temperature ranging from 20° to 25° C (68° to 77° F).

The tropical rainforest plant community contains a staggering assortment of trees and plants that form a very dense biomass. The trees are broadleaf and experience no seasonal leaf fall because of the consistently high temperature and abundant moisture; thus, these trees, like the needleleaf trees that you are familiar with, are *evergreen*. The rainforest also contains a variety of plants called *lianas* and *epiphytes*. Lianas are woody vines that use trees for support but are rooted in the ground. Epiphytes, in contrast, are plants that are rooted in a host plant and use the plant for support.

Classic tropical rainforest consists of four definable layers that can be seen vertically (Figure 10.5). The uppermost layer consists of randomly spaced emergent leaf crowns that exist at the tops of the tallest trees, some of which may be up to 60 m (200 ft) tall. Many of these trees are so large that they have evolved *buttress roots* (Figure 10.6), which help support them. Directly below the emergent layer is the canopy, which is the densest layer of the tropical rainforest and consists of a near-continuous web of branches and leaves. This part of the tropical rainforest is from 15 to 40 m (50 to 130 ft) above the forest floor.

Beneath the rainforest canopy is a third layer, called the *understory*, consisting of another dense network of interwoven branches and leaves that is 5 to 15 m (15 to 50 ft) high (Figure 10.5). In conjunction with the upper canopy, this dense layer allows very little sunlight to reach the forest floor where the ground layer occurs. Given the shady character of the lower forest, the vegetation in this fourth layer of the rainforest is unevenly distributed and ranges from 0 to 5 m (0 to 15 ft) above ground level. In fact, contrary to what may be your preconceived notions about the

Figure 10.6 Buttress roots at the base of an emergent tree in the rainforest. These roots give support to the tall trees that extend to heights of 40 to 60 m (130 to 200 ft).

character of the ground-level rainforest, many places are open—that is, you could walk through them more easily than you might think. Only where there are distinct openings in the rainforest canopy, such as along streams, does sufficient light reach the forest floor so that dense *jungle* vegetation can grow.

Tropical Deciduous Forest and Scrub Biome

The second major subdivision within the tropical forest biome is the tropical deciduous forest and scrub plant community. Extensive areas of tropical deciduous forest and scrub vegetation occur in Africa, South America, India, and Southeast Asia (see Figure 10.4). This plant subdivision is generally associated with the relatively dry tropical environments where annual precipitation is from 130 to 200 cm (50 to 80 in), which occur due to a distinct seasonal dry cycle. These areas exist within both the tropical monsoon climates (*Am*) and the more humid parts of the tropical savanna climates (*Aw*). Although the temperature varies more in these areas than in the tropical rainforest regions, it is always warmer than 18° C (64° F).

The tropical deciduous forest and scrub plant assemblage reflects the progressive transition to a drier climate at higher tropical latitudes. In the wetter areas, where average annual precipitation is closer to 200 cm (80 in), the vegetation is dominated by tropical deciduous forest (see Figure 10.4 again). Although these areas contain many of the same species seen in the tropical rainforest, they are not as tall and are not evergreen; that is, they drop their leaves during the dry season (hence, the name tropical *deciduous* forest). In addition, there is less species diversity and total biomass than in the rainforest. Although less overall diversity appears in these areas, a greater array of shrub-like species and lesser

plants occurs at ground level because more sunlight reaches the forest floor (Figure 10.7). This enhanced ground-level sunlight is due to the upper canopy of the forest being less dense than in the tropical rainforest.

The vegetation gradually shifts from tropical deciduous forest to tropical scrub in areas where average annual rainfall becomes closer to 130 cm (50 in). Tropical scrub consists mostly of low-growing, deciduous trees that grow to heights ranging from 3 to 9 m (10 to 30 ft; Figure 10.8). These trees tend to be ragged-looking due to the relative lack of moisture and they vary in density from being closely bunched to widely spaced. With increased distance between the trees, a lower level of tall bushes and grasses appears. In contrast to the tropical rainforest and tropical deciduous forest, the species diversity within the tropical scrub regions is relatively low, with a few species dominating many of the taller growth areas.

Midlatitude Forest Biomes The midlatitude forest biome occurs in the midlatitudes of North America, northeastern Asia, and western and central Europe (see Figure 10.4) where the climate has a distinct seasonal temperature cycle with cold winters. Several major subdivisions in this biome are based on climate variability; they include the midlatitude deciduous forest, the Mediterranean woodland and shrub forest, and the midlatitude coniferous forest.

Figure 10.7 Example of the tropical deciduous forest. Although this is a low-latitude region, note that the trees are relatively short and the canopy is more open than in the tropical rainforest biome.

Figure 10.8 Tropical scrub vegetation in the Australian Outback. Note the ragged-looking trees and the scrubby understory.

Midlatitude Deciduous Forest Biome

The dominant plant subdivision in the midlatitude forest biome is the midlatitude deciduous forest. This plant community consists of a dense network of broadleaf trees that forms a nearly continuous canopy in the summer. In the eastern U.S., the dominant species are oak, elm, and maple, to name a few that you may be familiar with. These trees have a distinct seasonal cycle in which the leaves change color and fall during the autumn months.

The midlatitude deciduous forest is associated with a variety of climate regions where average annual precipitation ranges between about 75 to 150 cm (30 to 60 in), including the humid subtropical hot-summer climates (*Cfa, Cwa*) and the humid continental hot-summer climates (*Dfa, Dwa Dwb*). Although these climates vary in some significant ways, each has sufficient moisture to support trees, either because average annual rainfall is abundant or evaporation is low because average annual temperatures are cool. In eastern North America, the midlatitude deciduous forest biome extends across a broad area from the Gulf of Mexico into southeastern Canada. Elsewhere, it occurs in an arc that extends from Europe to eastern Asia. This plant assemblage is also found in southeastern Australia and northern New Zealand (see Figure 10.4 again).

Mediterranean Woodland and Shrub Forest Biome

A second subdivision in the midlatitude forest biome is the Mediterranean woodland and shrub forest. This plant assemblage is closely associated with the Mediterranean dry-summer climate (*Csa, Csb*) and is found on the west coast of continents in the lower middle latitudes (see Figure 10.4 again). Average annual precipitation in these areas generally ranges from 25 to 65 cm (9.8 to 25.6 in). Winters are cool and wet, with a moisture surplus, whereas summers are hot and dry, with a water deficit.

The vegetation in the Mediterranean woodland and shrub biome is dominated by a dense cover of woody shrubs, known as the *chapparal* in North America (Figure 10.9). In many other places the trees are somewhat larger and more widely scattered, with some patches of grass present. Although the trees vary greatly between continents—oak is most common in California—the overall appearance is quite similar regardless of the specific region. Given the distinct dry season that occurs in summer, many of the species in this biome are adapted to fire, which, as in the case of tropical savanna, recycles nutrients in the system. In fact, some plant species in the Mediterranean woodland and shrub biome have seeds that germinate only after they are stimulated by the heat of a fire.

Midlatitude Coniferous Forest Biome

The last major subdivision in the midlatitude forest biome is the midlatitude coniferous forest. This plant community consists largely of needleleaf evergreen trees and is found in places like the west coast and mountains of North America, the southeastern U.S., and eastern Europe (see Figure 10.4). Coniferous forests in mountainous areas are associated with the highland climate (*H*) and have short growing seasons. Some of these coniferous forests occur at relatively low latitudes because temperatures are cool at high elevations. In North America, mountain forests include species such as Douglas fir and ponderosa pine (Figure 10.10). On the west coast of North America, these forests are associated with the marine west-coast climates (*Cfb, Cfc*), where annual precipitation is

This is an image of the tropical deciduous forest biome. Which one of the following statements is best associated with this biome?

a) The trees in this forest are evergreen.

b) This forest is located in the midlatitudes.

c) This biome receives less average annual rainfall than the tropical rainforest biome.

d) The canopy in this forest is closed.

e) The wet season occurs in this biome when it is influenced by the Subtropical High pressure system.

between 150 and 500 cm (59 and 197 in) and average annual temperatures are mild for a midlatitude location. Given the cool, moist environment in this region, the forest is often referred to as *temperate rainforest*. This forest also contains significant stands of deciduous trees, such as the giant redwoods in northern California and southern Oregon. (To learn more about this particular forest, see the *Amazing Places* feature.)

Figure 10.9 The Mediterranean woodland and shrub biome in southern California. This biome is adapted to dry summers and wet winters and consists of shrubby trees with intervening grass.

Figure 10.10 Midlatitude coniferous forest. Ponderosa pine in the Sierra Nevada Mountains in California.

WILDFIRES

Wildfires are a very controversial issue in the United States. Wildfires burn about 5 million acres of land every year, causing millions of dollars of damage to federal and private property, as well as the unfortunate loss of human life. With these kinds of losses in mind, it is easy to see why most people view wildfires as bad things that must be suppressed at all costs. In response to these concerns, the U.S. Forest Service monitors all wildland fire activity in the country and devotes vast resources to fighting any fires that develop. In 2003, for example,

Wildfires are one of the most impressive natural phenomena on Earth, as this photograph from a fire in Montana attests.

over 57,000 fires occurred, requiring thousands of people and associated support systems to fight them. During the winter of 2005/2006, numerous large prairie fires in Texas and Oklahoma scorched over 280,000 hectares (700,000 acres) and destroyed over 500 homes.

Although wildland fires can develop anywhere where lightning strikes and drought conditions prevail, most fires greater than 500 acres in

Another kind of midlatitude coniferous forest exists in the southeastern part of the U.S. within the humid subtropical hot-summer climate (*Cfa, Cwa*) zone. This forest is called the *southeastern coniferous forest* and stretches across southeastern Louisiana, southern Mississippi, southern Alabama, central and southern Georgia, and the Florida panhandle on sandy sites that do not hold much water. The biological diversity in this region is very high, with the dominant tree species being longleaf pine that has an understory of wiregrass.

Boreal Forest Biome The last major forest biome is the boreal forest biome. The global distribution of this biome is closely associated with the subarctic climates (*Dfc, Dwc, Dwd*), and is the broad expanse of coniferous forest that occurs across Canada, Alaska, and Russia (see Figure 10.4). This vegetation assemblage is located entirely within the higher midlatitudes in the Northern Hemisphere because the large continents have very cold winter temperatures and relatively low rainfall. If you recall, there are no large land-

Although wildfires are most commonly associated with forests, they are a critical part of the prairie ecosystem because they recycle nutrients and remove woody vegetation. This image is of a prairie fire in the Great Plains.

size develop in the western U.S. during the summer season when temperatures increase along with a corresponding drying of the landscape. Many of these large fires occur in areas where fire was previously suppressed and extensive amounts of dead wood serve as fuel. The worst of the 2003 fires occurred in the hills of Southern California, which burned 300,000 acres, destroyed at least 1000 homes (and threatened 30,000 more), and killed at least 14 people.

Plant geographers know that in spite of the negative aspects of fires, they produce significant ecological benefits. Regular fires reduce the amount of dead wood and thus lower the likelihood of potentially larger wildfires by removing their fuel. Fires often remove alien plants that compete with native species for nutrients and space. Fires also remove undergrowth in forests, which allows sunlight to reach the forest floor, thereby supporting the growth of native species. The ash produced by a fire adds nutrients to the soil to be used later by trees and other vegetation. Fires can also provide a way of controlling insect pests by killing off the older or diseased trees and leaving the younger, healthier trees. Overall, fire is a catalyst for promoting biological diversity and healthy ecosystems. It fosters new plant growth, and wildlife populations often expand as a result. Many ecologists believe that the incidence of wildfire in the world will increase if predicted levels of global warming occur.

masses at high latitudes in the Southern Hemisphere; thus, there is no significant expanse of subarctic climate and no boreal forest biome.

The boreal forest biome, or *taiga* in the Russian language, has the least diverse assemblage of plants of any biome. Most of the trees in the boreal forest biome are coniferous needleleaf evergreens such as spruce, pine, and fir, with large regions covered almost entirely by one or two of these species (Figure 10.11). These expanses of conifers are occasionally broken by small stands of broadleaf deciduous trees, specifically birch, poplar, and aspen. These trees are adapted to fire in that they are the first trees to move into a freshly burned area. Subsequently, they are succeeded by the coniferous trees, which also include the needle-dropping species tamarack and larch. The lower level of the boreal forest biome is sparsely vegetated, consisting of low shrubs, mosses, and lichens.

Although this region appears to be biologically productive, the fact is that it's not, due to the short growing

Old-growth California redwoods are indeed giant!

THE COASTAL REDWOODS

Perhaps the most majestic forest in the world is the famous redwood forest along the coast of northern California and southern Oregon. These trees are over 100 m (350 ft) tall and can be over 6 m (20 ft) in diameter. The bark alone can be as much as 75 cm (30 in) thick! These trees grow in the marine west-coast climate (*Cfb*), where the environment is particularly cool and misty along the coastal zone. This climate produces a lush forest that contains the redwoods as well as spruce, hemlock, Douglas fir, and large ferns at ground level. Heavy moisture is necessary to support the

season and the persistently wet soils in which the plants grow. The stress of this environment on plant growth is best seen in the northern parts of the boreal forest biome, where the growing season is very short and cold. Here, the trees are quite scraggly and stunted in comparison to their counterparts that live in the southern, warmer parts of the biome.

Grassland Biomes

Now let's turn our focus to the grassland biomes. These plant communities are distinct because they consist largely of grass with few trees. There are two major grassland biomes: (1) tropical savanna biome and (2) midlatitude grassland biome.

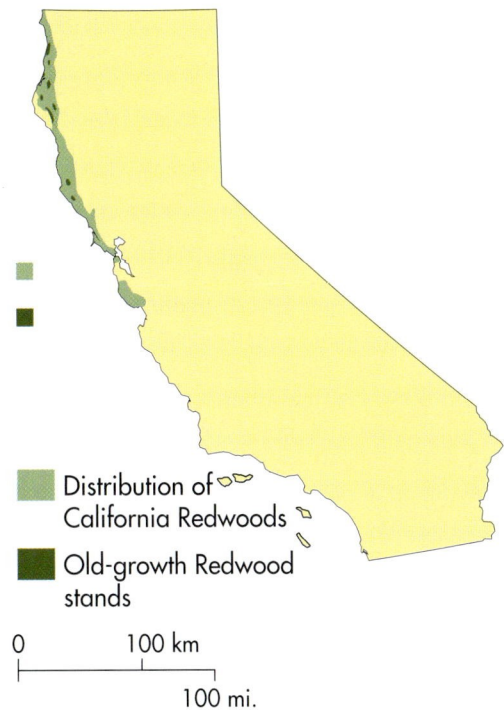

Distribution of California Redwoods

Old-growth Redwood stands

0   100 km

100 mi.

Map showing the location of the California redwoods. Most of the forest consists of second growth stands. The location of old-growth stands is shown in dark green.

the strong coastal winds that occasionally knock over a tree, the biggest enemy of redwoods is humans. In the mid-1700s, coastal redwoods covered about 2 million acres in the region. Since that time, about 96% of the forest was logged and second growth stands now cover most of the landscape. These trees were a critical natural resource in the regional economy because individual trees contained so much wood. Each of the largest redwoods, for example, contains enough wood to build 30 two-story houses of average size. Fortunately, not all of the trees were cut, with many of those remaining old-growth stands now preserved in a variety of parks such as Redwood National Park.

The old-growth redwood forest is majestic.

redwoods because the average tree uses about 1900 liters (~500 gallons) of water per day.

Although some individual redwoods may be as much as 2000 years old, their average age is about 600 years. They live such an incredibly long time because they are not susceptible to disease or fire. They don't burn easily because they contain no resin in their bark. They don't taste particularly good to insects, which bring disease, because they contain a lot of tannic acid. Other than

The Tropical Savanna Biome The tropical savanna consists of grasslands at low latitudes that contain isolated trees. This biome is most closely associated with the tropical savanna climate (*Aw*) and is present in Africa, South America, India, and Australia (see Figure 10.4). In these regions, average annual precipitation is from 9 to 150 cm (4 to 60 in) and average monthly temperature is greater than 18° C (64° F).

A good place to examine this biome is in Africa, where tropical savanna covers about 13,000,000 km^2 (5,000,000 mi^2) of the continent. Recall from Chapter 9 that this region is most directly influenced by the seasonal migration of the tropical Hadley cell and associated ITCZ and STH pressure systems, with a summer wet season and winter dry season. In response to this

Figure 10.11 The boreal forest. This scene is from Alaska. Coniferous trees dot this landscape because it is near the northern margin of this biome.

distinct wet/dry cycle, the vegetation is a mix of forest and grass Figure 10.12). Forest patches are denser where the conditions are more humid and thin considerably where it is drier. During the wet season the savanna springs to life as grasses turn green and grow tall at the same time that leaves develop on trees. When the dry season comes, however, the grass withers and browns while the trees drop their leaves. A major factor that affects the savanna biome is the occurrence of wildfires, which are common in the dry season. These wildfires are critical to the ecosystem because they reduce the dead vegetation and recycle nutrients back into the system.

Midlatitude Grassland Biome Midlatitude grasslands are broad areas where the dominant vegetation is grass. The climate of these areas is seasonal, with warm to hot summers and cool to cold winters. Average annual precipitation ranges from 25 to 75 cm (10 to 30 in). Several climate regions fall within this general description, including the dry portions of the humid continental hot-summer (*Dfa, Dwa*) climates, the humid continental mild-summer climates (*Dfb, Dwb*), and most of the cold midlatitude steppe climate (*BSk*). In many cases, the relative dryness of these climate zones is due to the continental effect in that the regions are located far from moisture sources. In other places, such as the interior of North America, distinct rainshadows exist in the lee of large mountain ranges such as the Rockies.

Extensive midlatitude grasslands occur in South Africa, eastern Europe, central Asia, and central North America (see Figure 10.4). In North America, grasslands are most closely associated with the *Great Plains*

(Figure 10.13), whereas in South America they are known as the *pampa*. In Russia, grasslands are referred to as the *steppe*. The biomass in these areas is quite dense, with a nearly continuous cover (except in the drier places) of grass and flowering plants that have an extensive root network. Most grasses are perennials that lie dormant through the winter and sprout in spring. This landscape is also adapted to fire, which usually occurs during the summer dry season. In fact, the common recurrence of fire historically contributed to the lack of trees and woody shrubs in this region, as did grazing by large animals such as bison (also known as "buffalo") in the Great Plains. Given that fire is suppressed by humans in many grasslands today, forest vegetation is actually expanding into the fringe of many grassland areas.

Desert Biomes

Desert biomes occur in areas where average annual precipitation is generally less than 25 cm (10 in). Deserts cover about one-fifth of the Earth's land area. Although deserts most commonly occur in low latitudes, they also appear in colder regions such as in Nevada and Utah in the western U.S. and in portions of western Asia. There are two major subdivisions within the desert biome: (1) the hot and dry desert biome, and (2) the semi-arid and cold desert biome.

Figure 10.12 The tropical savanna. This photograph shows savanna in Brazil. Here, the vegetation consists of patches of grass that are dotted by isolated trees and woody shrubs.

Figure 10.13 Midlatitude grassland biome. The Pampas of Argentina is one of the great grasslands on Earth.

STH pressure system. Average annual precipitation can be less than 2 cm (0.8 in) and average monthly temperature is greater than 18° C (64° F). As a result of the combined low rainfall and high temperatures there are chronic moisture deficits in these areas.

The vegetation in the hot and dry desert generally consists of bare ground that grades into **xerophytic plants**, including cacti, low-growing shrubs, and scattered bunches of grass (Figure 10.14). These plants have evolved in ways that ensure their survival in very dry regions. One survival strategy used by many plants is to be *drought-resistant*, which means that they can conserve moisture either by developing very small waxy leaves that store moisture (these plants are called *succulents*), or by having roots that penetrate deeply into the ground. Other plants are *drought-evading* in that they rapidly flower and reproduce during the rare times of relatively high precipitation.

Hot and Dry Desert Biome The hot and dry desert biome occurs in the subtropical regions and is centered on 30° latitude (see Figure 10.4). Major deserts of this kind include the Sahara in northern Africa, the Kalahari Desert in southern Africa, and the Mojave and Sonoran Deserts in the southwestern United States. Another large area of hot and dry desert occurs in central Australia. These areas are associated with the hot low-latitude desert climate (*BWh*), which is dominated by the

Semi-Arid and Cold Desert Biome The semi-arid and cold desert biome occurs extensively in western North America and central Asia (see Figure 10.4). This biome is most closely associated with the cold midlatitude desert climate (*BWk*). In North America, this climate occurs in places like Utah and Nevada where extensive

Xerophytic plants *Plants that live in very dry places that have a number of survival mechanisms in response to prolonged periods of drought.*

VISUAL CONCEPT CHECK 10.2

This image shows a typical scene in one of the major biomes on Earth. Which biome is pictured here and how do migrating atmospheric pressure systems influence the seasonal climate and thus the vegetation in this place?

Figure 10.14 Example of the hot and dry desert biome. The Sonoran Desert in the southwestern U.S. features tall, columnar saguaro cactus, as well as a variety of smaller cacti and hard-leaved shrubs.

Figure 10.15 The semi-arid and cold desert biome. Sagebrush landscape in Monument Valley, Utah.

rainshadows form in the lee of the Sierra Nevada Mountains. This climate also occurs in places like the Gobi Desert in central Asia, which lies deep within an extremely large landmass and is thus far from a moisture source. Both places feature a distinct seasonal temperature pattern, with hot, dry summers and cold winters. Average annual precipitation ranges from 2 to 4 cm (0.8 to 1.6 in), with most occurring during the winter months.

The vegetation of the semi-arid and cold desert consists of bare patches that grade to low-growing spiny and woolly plants that are randomly spaced. These uneven and sharp plant surfaces protect the plants from grazing animals and significantly reduce moisture loss. Some plants have a silver or glossy leaf, which give them a higher albedo and thus keep the plants cooler by reflecting more solar radiation. In the western U.S., plants in this vegetation community include creosote bush, sagebrush, and mesquite (Figure 10.15).

Tundra Biome

The last of the major biomes on Earth is the tundra biome, which borders the Arctic Ocean at very high latitudes across North America and Russia (see Figure 10.4). The *arctic tundra* is found largely in association with the tundra climate (*ET*), which means that winters are long with an average temperature of −34° C (−29° F). Although the summer growing season is only 60 to 80 days long, average temperature can reach 12° C (54° F) due to high Sun angle. Average annual precipitation is 15 to 25 cm (6 to 10 in). Tundra also occurs to a limited extent at very high elevations within the highland climate (*H*) as *alpine tundra*. In this context, alpine tundra is found as far south as the equatorial Andes Mountains.

Regardless of whether the tundra is alpine or arctic, the plant assemblage and thus the appearance is basically the same. The tundra biome consists of a diverse array of grasses, shrubs, sedges, mosses, and lichens (Figure 10.16) that tend to be very short because the growing season is limited and the landscapes are wind swept. Tundra landscapes are very fragile and are easily disturbed by human impact. Although people have historically avoided tundra because of the harsh conditions, it is now a growing source of interest because untapped reservoirs of fossil fuels and other minerals may exist there.

> ### KEY CONCEPTS TO REMEMBER ABOUT BIOMES
>
> 1. A biome is a definable assemblage of plants that occupy a large region of geographic space with generally recognizable boundaries.
>
> 2. The global distribution of biomes is closely related to the geography of climates.

3. Trees generally characterize humid regions, whereas dry to semi-arid regions are populated by shrubs and grasses.

4. The vegetation of deserts and the tundra biome are very similar in that they consist of low-growing plants that have adjusted to extreme environmental conditions.

(a)

(b)

Figure 10.16 The tundra biome. (a) The vegetation here consists of low-growing plants, primarily very short grasses, mosses, and lichens, which have adapted to the short, cold growing season. (b) Lichens are a mix of algae, bacteria, and fungi that favor rocky surfaces such as these pictured here. The central rock, which is about 15 cm (6 in) wide at its thickest point, is surrounded by low-growing moss.

Local and Regional Factors That Influence the Geographic Distribution of Vegetation

In addition to the basic geographic distribution of biomes, other factors can influence the types of plants that live in a region or area. It is useful to think of these variables as being local or regional in scope because they may cause a particular pattern within the overall biome. The following discussion of local factors is not meant to be a complete review of all the variables that can influence the distribution of vegetation; instead, it is meant to give you a good sense of the most important of these factors.

Slope and Aspect

The slope of an area is its degree of steepness, with high slopes being places that are steep and low slopes being relatively flat. As the slope of a landscape increases, the amount of water that runs off the surface, rather than soaking into the ground, usually increases. As a result, there is typically less available soil moisture in places of steep slopes and the associated vegetation can be different than in sites that are less steep. Soils also tend to be thinner on steep slopes than on shallow slopes, which influences the amount of nutrients available for plants.

Another factor related to the slopes within a particular landscape is the concept of aspect. *Slope aspect* refers to the orientation of the slope—that is, whether the slope faces north, south, east, or west. Aspect is an important variable in plant geography because it is related to how intensely the Sun strikes a given slope (Figure 10.17). In this context, the adret slope is the slope that faces the Sun most directly; that is, the slope at which solar radiation arrives at a high angle. Where this geographic situation occurs, the microclimate (climate over a small region) is usually warmer, there is consequently more evaporation, and the slope can be relatively dry. The opposing, or low-Sun slope, is called the ubac slope. This aspect is characterized by a relatively cooler microenvironment with less evaporation because there is less direct solar radiation.

If you live in a region that receives abundant snowfall, you can see the effect of aspect on hills that slope

Slope   *The degree of steepness of a portion of the landscape.*

Adret slope   *The slope that faces the Sun most directly.*

Microclimate   *Average temperature and precipitation characteristics within a small area within a larger climate region.*

Ubac slope   *The slope that faces away from the Sun.*

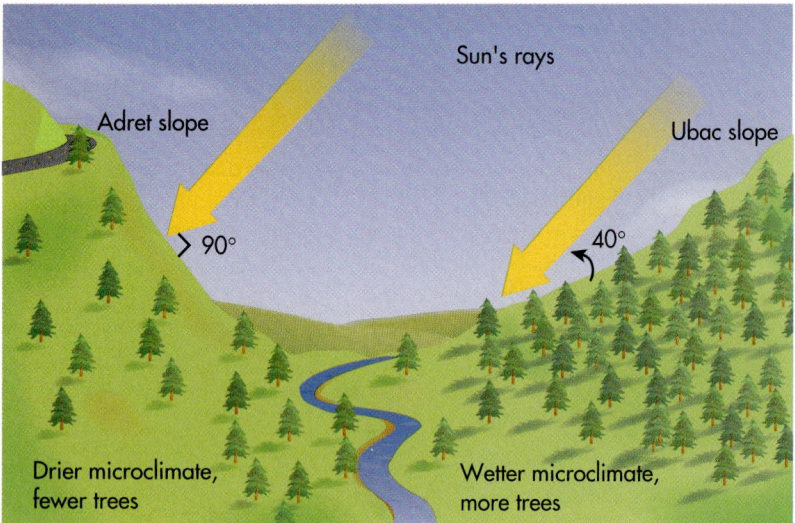

Sun's rays

Adret slope

Ubac slope

> 90°

40°

Drier microclimate, fewer trees

Wetter microclimate, more trees

Figure 10.17 Slope aspect and Sun angle. Slopes that directly face the Sun (adret slopes) have higher Sun angles and thus a warmer and drier microclimate than opposite slopes (ubac slopes).

down on the north and south sides of highways. One slope will be free of snow due to melting, whereas the other slope will have snow on it for a longer period of time.

Think about how aspect affects plant growth. In the Northern Hemisphere, where would you expect warmer temperatures and higher rates of evaporation: north- or south-facing slopes? It is warmer, with more evaporation, on the south-facing slopes because they face the Sun more directly. As a result, we see a geographic pattern where trees grow (or are more frequent) on the north-facing ubac slope while they don't occur (or are less frequent) on the south-facing slope. Figure 10.18 shows a great example of this pattern by focusing on the growth of bristlecone pine in California.

Vertical Zonation

Another way that vegetation can vary locally from what you might initially expect is through the concept of vertical zonation. **Vertical zonation** refers to the change in vegetation that occurs with respect to elevation, rather than latitude. Recall that temperature generally decreases at a consistent rate (the environmental lapse rate) as you gain altitude in the troposphere. This overall trend shows

Vertical zonation *The change in environmental characteristics that occur with respect to altitude.*

Figure 10.18 Slope aspect and growth of bristlecone pine in the White Mountains, California. (a) Growth on the south-facing (adret) slope of a mountain. Note the wide spaces between trees. (b) Growth on the north-facing (ubac) slope. Bristlecone pine favors this slope, rather than the adret slope (a) because there is more water due to less evaporation.

(a)

(b)

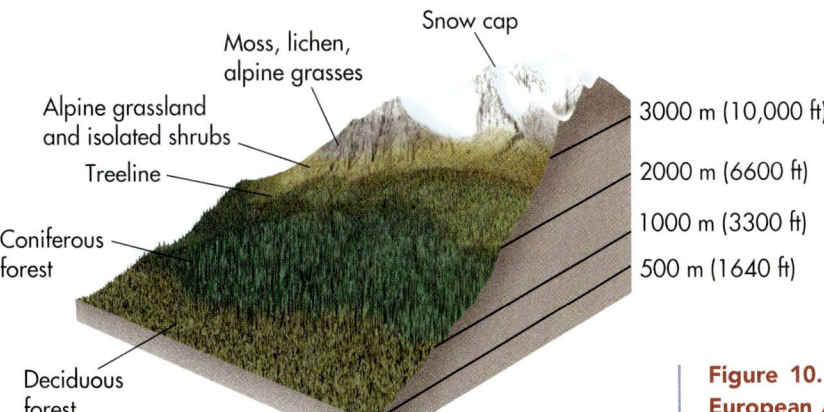

Snow cap

Moss, lichen, alpine grasses

Alpine grassland and isolated shrubs

Treeline

Coniferous forest

Deciduous forest

3000 m (10,000 ft)

2000 m (6600 ft)

1000 m (3300 ft)

500 m (1640 ft)

Figure 10.19 Vertical zonation of vegetation in the European Alps. Note the changes in vegetation that occur with elevation.

up most dramatically on the world climate map (see Figure 9.2) as highland climate (*H*) regions. Highland climates occur in very cold/dry regions associated with the Earth's large mountain chains. Highland climates even occur in tropical regions, particularly in the Andes Mountains that extend along the western margin of South America. This is a fascinating geographic relationship in the sense that near arctic conditions exist in a region that is generally thought of as being very warm and humid. This dichotomy exists because the Andes reach a very high altitude where temperatures are cold due to the environmental lapse rate.

In the context of vertical zonation, the key relationship is that environmental changes that occur with respect to elevation correlate to the same ones you see with latitude. Figure 10.19 illustrates how vertical zonation generally works in the European Alps. In the lower elevations of the mountains the vegetation often consists of deciduous forest. As you move to about 1000 m (3300 ft), the vegetation gradually shifts to coniferous trees that are adapted to cooler conditions. At about 2000 m (6600 ft), the vegetation shifts from trees to alpine grassland and scrub vegetation. The elevation where trees can no longer grow because conditions are too harsh is called the **treeline.** At still higher elevations, the vegetation becomes tundra, and, ultimately, the landscape is covered with snow and ice (Figure 10.20).

When do environmental conditions become too harsh to support the growth of trees? The critical variable for the treeline location appears to be that seasonal

mean temperature be at least 6.5° C (43.7° F) at a depth of 10 cm below the ground, where roots grow. You are probably most familiar with *alpine treeline* that occurs at high elevations in the mountains, above which are only scattered tundra, bare rocks, and even perpetual snow. Although alpine treeline exists in all major mountain ranges on Earth, a distinct pattern exists with respect to latitude and altitude, with the elevation decreasing progressively with increased latitude. Another kind of treeline is called *circumpolar treeline* and is the highest latitude at which trees can grow. Regardless of whether the treeline is alpine or circumpolar, the

Figure 10.20 Alpine vegetation in the Snowy Range in Wyoming. The scattered trees, which are similar to high latitude boreal forest, grade upward into tundra. This is an excellent example of vertical zonation.

Treeline *The line that represents the upper limit in mountains and high latitudes where environmental conditions support the growth of trees.*

appearance is basically the same, consisting of scattered trees that are stunted and deformed. They are often bent in a particular direction that reflects the prevailing winds (Figure 10.21).

Plant Succession

Plant succession refers to the natural changes that occur in a biome of a particular place over time. Generally speaking, these natural changes progress to the point where the most complex array of vegetation possible arises, given the regional climate, water, and soil variables. If the succession begins on freshly deposited sediment, then we use the term *primary succession*. If, on the other hand, the succession occurs in an area that was disturbed through fire or some other catastrophic event, then the term *secondary succession* applies.

Regardless of whether the succession is primary or secondary, the sequence of plant stages is usually predictable for a given area. The first plants that occupy the area are called *pioneer species*. These plants set the stage for future succession because they initiate the gradual change of the host soil through root penetration and the cycling of organic material and minerals as they spread across the area. In this fashion, bacteria and animals begin to inhabit the area and the site vegetation gradually becomes more complex. In time, the local micro-environment becomes hospitable for larger plants, which change the shading and associated microclimate at ground level. These adjustments in humidity and soil temperature make it possible for even more additional plants to move into the area. When this progression reaches the ultimate complexity of an area, the vegetation is said to be **climax vegetation.**

A good example of primary succession can be seen on a sequence of sand dunes. As you will learn in a later chapter, sand dunes are highly mobile geologic features that migrate due to a complex interaction of wind and sand supply. Basically, dunes are mounds of sand that grow through fresh deposition of wind-blown sand. These landforms are relatively infertile micro-environments because they are very well drained and nutrient poor, due to high amounts of quartz and feldspar and low amounts of nitrogen, calcium, and phosphorus.

In the context of plant succession, the first plants to inhabit a fresh deposit of wind-blown sand are beach grasses. These grasses are adapted to the relatively infertile conditions of a new sand dune and continue to grow even if they are slightly buried by fresh deposits of sand. As they spread across the dune, they cause it to stabilize, which means that the sand stops moving. At this time,

Figure 10.21 At the treeline. Trees at circumpolar or alpine treeline are often dwarfed and scraggly because of the harsh environmental conditions. The orientation of these trees in the Rocky Mountains indicate the direction of the prevailing winds.

new plants move in that are adapted to the relatively infertile conditions of the dune. These plants, however, cannot handle the stress of being buried, as they would have been able to if they were in place when the dune first became established. When this successional stage occurs, the soil chemistry and ground-level micro-environment continue to evolve, culminating in the establishment of trees on the older dunes.

Figure 10.22 is a scene from a coastal dune field along Lake Michigan and is an excellent example of this succession. Compare the vegetation on the low (young) dune in the foreground with that on the higher (older) dunes in the background. The younger dune is inhabited

---

Plant succession *The natural changes that occur within a landscape over a period of time.*

Climax vegetation *The vegetation within an area that has reached its ultimate complexity.*

Figure 10.22 Primary succession on Great Lakes coastal dunes. Beach grass grows on the low dune in the foreground, which is the most recent deposit of wind-blown sand. The higher and older dunes in the background are covered by forest.

Figure 10.23 Aerial photograph of riparian trees along the Arikaree River in the western Great Plains. Although the overall environment of this region only supports grass, trees can grow along the river because there is more moisture there.

by beach grass, while those in back are covered with forest. In the Great Lakes region, this succession sequence is thought to take about 150 years.

Riparian Zones

A major factor that can influence the geographic distribution of vegetation is proximity to water. This factor is especially important in semi-arid regions where water is generally scarce. An example of this kind of relationship is the **riparian zone**, which is defined as the area immediately adjacent to a stream. More water is available for plant growth near the river than away from it, especially

because elevation increases with distance away from the valley bottom. A good place to see how riparian vegetation may differ from the overall pattern can be seen in parts of the Great Plains of North America, shown in Figure 10.23. Notice in this image that trees are growing only along the river, while the surrounding landscape consists of grass and sagebrush.

Riparian zone   *The strip of land, which borders a body of water, that supports plants and animals adapted to water systems.*

GEODISCOVERIES   WWW.WILEY.COM/COLLEGE/ARBOGAST

PLANT SUCCESSION

Let's take a closer look at the concept of plant succession by watching the sequence of plants that populate a more complex landscape than the dunes just described. Go to the *GeoDiscoveries* website and select the module *Plant Succession*. This module illustrates the process of secondary plant succession in a cool, humid landscape in the higher midlatitudes. The environmental stage is set with a small stream that flows through a

forested area about 7000 years ago. Subsequently, a colony of beavers builds a dam across the stream, causing a pond to form. From this point, watch how the succession of plants proceeds in this environment, culminating in the complete disappearance of the pond and return of climax vegetation. Once you complete the animation, be sure to answer the questions at the end of the module to test your understanding of plant succession.

Human Influence on Vegetation Patterns

It would be a major omission to discuss the geographic distribution of vegetation without talking about the impact of humans. In fact, given the enormous impact that humans have had on the global distribution of vegetation, it is somewhat misleading to discuss the influence of people following a section of text that focuses on local and regional variables. Nevertheless, people are usually functioning at a local scale when it comes to decisions regarding vegetation. The combined impact of these local decisions by countless numbers of people over time often shows up as a distinctive regional pattern that occurs within the overall biome. This discussion is not meant to be comprehensive—far from it. Instead, we will try to give you a sense of how humans have modified the natural vegetation of regions all around the globe.

Deforestation and Its Consequences

A major issue that presently faces global society is **deforestation**, which is the process through which large tracts of land are cleared of trees by humans, either for commercial use or to make way for agriculture. A good place to begin the discussion of deforestation is to examine the current extent of forest cover around the world. According to the United Nations Forest Resources Assessment program,

Deforestation *The removal of trees for economic or agricultural purposes.*

approximately 3.9 billion hectares (9.6 billion acres) of the Earth were covered by forest in 2000. Of this total, 47% was in the tropics, 33% was in the boreal forest, 11% was in the middle latitudes and temperate areas like the Pacific Northwest, and 9% was in the subtropics (Figure 10.24). Overall, this extent of coverage amounts to about one-third of the Earth's total landmass, excluding Antarctica and Greenland.

Although the amount of forest cover is extensive, studies demonstrate that it is reducing at a dramatic rate due to human activities. This loss is occurring in spite of the fact that forest is expanding in some areas due to natural processes or through *afforestation*, which is the planting of trees by people in previously nonforested areas. According to the United Nations, the average rate of global deforestation per year during the 1990s was 14.6 million hectares (36.8 million acres). At the same time, the average amount of afforestation and natural expansion was 1.6 million hectares (4.0 million acres) and 3.6 million hectares (8.9 million acres), respectively. In spite of the offsetting impact of afforestation and natural expansion, the United Nations believes that approximately 94 million hectares (232 million acres) of the global forest cover was lost in the last decade of the 20th century. This amount represents about 2.4% of the estimated global forest cover in 1990 and is an area approximately two times the size of California.

As you can see in Table 10.2, the majority of deforestation on Earth between 1990 and 2000 occurred in the tropical regions of Africa and South America, with losses of 7.8% and 4.1%, respectively, by continent. These losses reflect the fact that 94% of deforestation on Earth occurs in tropical regions. Studies indicate that tropical

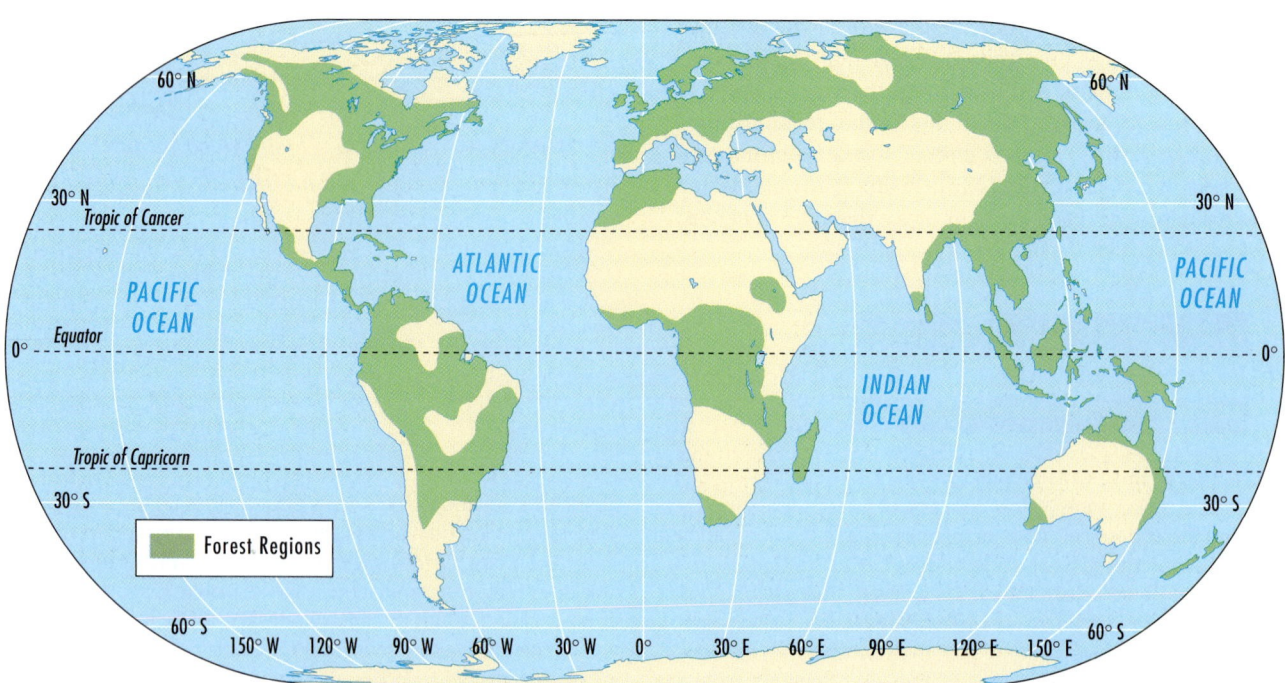

Figure 10.24 Forest regions on Earth. Forests have adapted to every zone of latitude on Earth except for the polar regions.

TABLE 10.2	Change in Forest Cover (in millions of hectares), 1990–2000		
Continent	Total Forest 1990	Total Forest 2000	Percent Change 1990–2000
Africa	702	650	−7.8
Asia	551	548	−0.7
Oceania	201	198	−1.8
Europe	1030	1039	+0.8
North and Central America	555	549	−1.0
South America	923	886	−4.1
Total World	**3963**	**3869**	**−2.4**

Source: U.N. Food and Agriculture Organization, State of the World's Forests 2001 (Rome, 2001).

Figure 10.25 Slash and burn clearing in the rainforest of Brazil. The forest in this region has been cut and burned to prepare the ground for cultivation. Organic matter from the burned forest temporarily increases soil fertility.

the consequences may be far-reaching. Of major concern is the impact that extensive deforestation has on biodiversity within the tropical forest biomes. Given the complexity of these regions, it is unknown precisely how the effect of cumulative clearing ripples through the system. In fact,

Figure 10.26 Location of wildfires in central South America in September 2001. The red dots mark the location of the fires, the largest of which produce distinctive smoke plumes. The extensive areas that are already cleared appear as tan zones in the sea of green forest.

deforestation occurs for two reasons: (1) to clear land for cattle grazing and agriculture, and (2) logging of commercially valuable wood, such as mahogany.

In the context of agriculture, poor farmers practice a technique called *slash and burn agriculture*, in which the forest is burned both to clear the land of debris and to recycle nutrients to the soil (Figure 10.25). This method of farming is very widespread, as indicated by the satellite image (Figure 10.26) acquired in September 2001 of a portion of Bolivia, Brazil, and Paraguay in South America, showing the location of numerous fires. Note the extent of forest that has already been cleared prior to the fires.

Although the amount of deforestation appears to have slowed somewhat, tropical deforestation continues and

Figure 10.27 Stump prairie in northern Michigan. These stumps are the remnants of old-growth pine trees on the Kingston Plains that were cut in the 1890s. Following this period of deforestation, an abundant amount of dried limb and branch fragments lay on the ground. This *slash* subsequently burned in a very hot fire that torched seedlings and basically cooked the soil. As a result, forest has not regenerated over much of the area.

many scientists believe that it is possible that deforestation will cause extinction of plant and animal species that have yet to be discovered. Deforestation is also significant with respect to global warming. The tropical forest regions are sometimes called the "lungs of the Earth" because they absorb carbon dioxide and release oxygen through photosynthesis. Increased atmospheric carbon dioxide is apparently related to global climate change, but its overall increase may be enhanced because the smaller tropical forests absorb less.

This current focus on tropical deforestation should not lead you to believe that forests elsewhere in the world have not been severely impacted by people: they have. In the Great Lakes region of North America, for example, virtually all of the native pine forest was cut in the late 19th and early 20th centuries (Figure 10.27) to provide timber for buildings in growing cities such as Chicago, Milwaukee, and St. Louis. The forest in this region is still heavily managed, with large jack and red pine plantations in many parts of the region that are grown for pulpwood. This practice of forest management occurs not only in the Great Lakes region, but also in the needleleaf forests of the Pacific Northwest. Most of the old growth forests are gone in this area, and the second and third growth forests continue to be extensively logged (Figure 10.28). The ongoing demand for

old growth timber may have serious consequences regarding species extinction, with the spotted owl being a high-profile example.

Agriculture in the Midlatitude Grassland Biome

In addition to the tremendous impact that people have had on global forests, another biome that has been severely modified by humans is the midlatitude grassland biome. Prior to European settlement of North America, as much as 162,000 hectares (400,000 acres) of the continental interior was grassland. You can see the potential natural distribution of grassland in Figure 10.4. Notice how much of central North America, in particular, was covered by grass. As you will learn in the next chapter, the soils in the midlatitude grassland biome are some of the most fertile in the world. This fertility is ideal for the development of large-scale agriculture, which dominates the region today in the form of extensive corn, wheat, and soybean operations. Although the benefit to people of this form of intense agriculture is obvious, the impact on the midlatitude grassland biome has been immense. In that context, current estimates are that only 1% to 4% of virgin (unplowed) prairie remains throughout the entire biome.

Overgrazing

Another way that humans have impacted the network of natural vegetation is through overgrazing of the landscape by livestock, such as cattle and sheep. This impact is most extensive in marginal landscapes that are extremely sensitive to disturbance. An excellent example of this kind of

Figure 10.28 Clear cutting in the Pacific Northwest. Deforested areas on mountain hillslopes south of Olympic National Park in Washington.

This pair of infrared satellite images is of the same region in the Amazon rainforest at different times. Remember that red on this kind of image represents vegetation. The bottom (a) image was acquired in 1975, whereas the right-hand image (b) is from 2000. The difference that you see reflects extensive deforestation that has followed road networks in the region. Which one of the following statements is accurate with respect to the impact that this kind of deforestation has on global climate change?

(b)

(a)

a) Fewer trees will result in the production of more oxygen.

b) The amount of atmospheric carbon dioxide will increase because there are fewer trees to absorb it.

c) Global temperature should decrease given that fewer trees are present in the rainforest.

d) There will be less atmospheric carbon dioxide because trees produce carbon dioxide as a by-product of respiration.

marginal landscape is the semi-arid Great Basin of the American West (Figure 10.29). Prior to European settlement, this region contained very few large herbivores that would have grazed on the vegetation there. In the 20th century, however, this region became available for ranchers because the U.S. Forest Service, Bureau of Land Management, and the National Park Service classified much of it as public lands. According to federal legislation, it is permissible to graze sheep and cattle in these areas. This practice continues to the present day and has dramatically affected this biome. Some of the potential impacts are weed invasion into disturbed areas, reduction in forest due to grazing of young saplings, and the increased likelihood of more intense and widespread fires.

GEODISCOVERIES   WWW.WILEY.COM/COLLEGE/ARBOGAST

DEFORESTATION

The topic of deforestation is of particular concern to many scientists. One way to monitor the rates and patterns of deforestation is through satellite remote sensing. To see some examples of how this kind of monitoring occurs, go to the *GeoDiscoveries* website and select the module **Deforestation**. This interactive module will allow you to see the kind of patterns that geographers see when they study deforestation around the world. As you work your way through the module, you'll be asked to identify deforested and noncut areas using both true-color and infrared satellite imagery. You'll also watch a short animation that illustrates the rate of deforestation in the Amazon rainforest. After you complete this module, be sure to answer the questions at the end of it to test your understanding of this concept.

Figure 10.29 Impact of grazing on vegetation in the American West. Overgrazing reduces the cover of vegetation on hillslopes, which can result in severe erosion when rain falls.

GEODISCOVERIES WWW.WILEY.COM/COLLEGE/ARBOGAST

REMOTE SENSING AND THE BIOSPHERE

Plant ecologists and geographers are very interested in the patterns and rates of vegetation change on Earth. In this context, one of the ways that plant geography is studied and monitored is with remote sensing. Recall from Chapter 2 that remote sensing is the method through which a landscape is viewed from afar, specifically from an airplane or a satellite in space. To give you a sense of how this form of monitoring can be done, go to the *GeoDiscoveries* website and select the module **Remote**

Sensing and the Biosphere. This module will allow you to experience what it's like to study landscape change in a variety of environments with satellite imagery. Within this interactive module you will be able to analyze (1) *Land Productivity*, (2) *Ocean Productivity*, and (3) *Fire Patterns*. In each of these sub-modules, you will be presented with a series of satellite images and prompts to correctly identify specific features on imagery. Once you complete the *Deforestation* interactivity, be sure to answer the questions at the end of the module to test your understanding of the concepts contained within it.

The Big Picture

The next chapter will focus on the processes and geographic distribution of soils on Earth. You will notice a close relationship between soil and vegetation, with a dominant influence once again being the role of climate. Chapter 11 will also be the first chapter where we begin to examine the kinds of processes and outcomes that occur beneath the surface of the Earth. For example, this photograph shows a soil profile, which is the view of a soil from the surface to some depth within the ground. A distinct layering pattern appears in the profile, with a dark zone near the surface that overlies a lighter area that contains numerous white specks. In the next chapter we will describe how such profiles develop and how they vary geographically.

Summary of Key Concepts

1. The photosynthetic, respiratory, and decomposition cycles explain how biomass is produced and consumed through the interaction of solar energy, carbon dioxide, water, and oxygen in plants, animals, and soil.

2. A biome is a definable geographic region that is classified and mapped according to the predominant vegetation and the adaptations of organisms to that particular environment. There are 12 major biomes on Earth.

3. Six major forest biomes in the world occur in places where average annual precipitation is high. They also occur in places that have moderate amounts of precipitation and average temperatures that are sufficiently cool to maintain a good water balance.

4. The two major grassland biomes in the world are (1) tropical savanna, and (2) midlatitude grassland. The tropical savanna occurs in the tropical ecotone where the ITCZ dominates in the summer (bringing the wet season) and the Subtropical High pressure system dominates in winter (bringing the dry season). Midlatitude grasslands occur deep within continents or within rainshadows.

5. There are two major desert biomes in the world: (1) hot and dry desert, and (2) semi-arid and cold desert biome. The hot and dry desert occurs in the higher tropical latitudes, dominated by the STH pressure system. The semi-arid and cold desert biome occurs deep within continents and rainshadows within the midlatitudes.

6. The tundra biome occurs at very high latitudes and elevations where environmental conditions are too harsh for trees to grow. The plants here consist of low-growing woody shrubs, lichens, and mosses.

7. The human impact on the distribution and character of natural vegetation has been dramatic, including deforestation, extensive agriculture, and overgrazing.

Check Your Understanding

1. What are the products of photosynthesis and why are they important to life on Earth?

2. Why is biomass higher in tropical regions than in desert regions?

3. Why is the floor of the tropical rainforest biome considered to be "open"?

4. Many of the trees in the tropical deciduous forest biome can be found in the tropical rainforest biome. Nevertheless, the trees in the rainforest are evergreen, whereas they drop their leaves in the tropical deciduous forest biome. Why does this difference exist?

5. The ITCZ and STH pressure system are very closely associated with the tropical savanna biome. What is this relationship?

6. In the grassland of the U.S., the grasses in the western part of the region are shorter than in the eastern part. Why does this difference exist?

7. Where would the warmest/driest slopes be in the Southern Hemisphere: on the north or south-facing slopes?

8. Where is overgrazing likely to have the biggest impact: in the Great Plains or in the eastern United States? Why does this difference exist?

9. Why is there likely to be a denser forest on the windward side of a mountain range than on the leeward side?

10. Tundra is found in the very low latitudes of South America, even though this region is generally thought to be "tropical." Why?

ANSWERS TO VISUAL CONCEPT CHECKS

Visual Concept Check 10.1

The answer to this visual concept check is *c*, which states that this biome receives less annual precipitation than the tropical rainforest biome. The reason this answer makes sense is that the relative lack of rainfall causes a more open forest to develop than in the rainforest.

Visual Concept Check 10.2

This biome is the African Serengeti. This environment exists in the part of Africa that has a distinct wet and dry season. The wet season occurs in summer when the insolation is high and the ITCZ dominates the weather. In contrast, the dry season occurs in winter when the STH pressure system moves into the region.

Visual Concept Check 10.3

The answer to this question is *b*; there will be more atmospheric carbon dioxide as a result of deforestation. This increase will occur because trees absorb carbon dioxide in the context of the global carbon cycle.

CHAPTER ELEVEN

THE GLOBAL DISTRIBUTION AND CHARACTER OF SOILS

This chapter will help you learn the difference between soil and dirt, and how they are related. We are now beginning to examine the geographic patterns associated with parts of the Earth that are within the Earth, or the *lithosphere*, rather than being on top of it, as in the case of the *biosphere*, or above it in the *atmosphere*.

Soil is one of the most important natural resources on Earth and comes in many different forms. The soil exposed in this roadcut in Brazil shows the telltale signs of having formed in a tropical environment, specifically its reddish color. This chapter focuses on the evolution of soils and how they can differ around the Earth.

We will examine how soils form, what properties affect soil fertility and development, and where the basic types of soils are found on Earth. This chapter is a logical progression from the preceding chapter on vegetation because a close relationship is usually found between plant types and soil characteristics. Most important is a distinct geographic correlation between basic soil types and climate because precipitation and temperature are important factors in soil formation. In this sense we are integrating many components of the atmosphere, hydrosphere, biosphere, and lithosphere

What Is Soil?

Remember when you were a child and you played in dirt all the time? Did you ever stop to think about where the dirt had come from, and how long it had been there? Did you ever dig holes in the ground and notice how some of the dirt was darker than other parts, and how some dirt just spilled through your fingers while some broke up into good-sized clods that you could throw? Although you generally referred to the material as "dirt," in all probability you were playing in soil. You're probably asking yourself, what difference does it make whether you call it "dirt" or "soil"; aren't they the same thing? In everyday language they are, but in the context of physical geography, they're not. The fact is that dirt is not necessarily soil, but soil *forms* in dirt, among some other related things (Figure 11.1).

   Soil is the uppermost layer of the Earth's surface and contains mineral and organic matter capable of supporting plants. Another way to think about soil is that it is the outermost rind of the Earth, similar to how an orange peel is the outermost part of the orange. Still another way to consider soil is as the transition between the atmosphere and the rocky Earth. Although all soils can be viewed in these various ways, they nevertheless vary dramatically in terms of their character and thickness, ranging from a few centimeters to several meters.

Soil *The uppermost layer of the Earth's surface that forms by the influence of parent material, climate, relief, and chemical and biological agents.*

Figure 11.1 Analyzing soil in the field. These people are scientists studying soil in a backhoe trench. Recognizing soil types is so important that most geography students will spend at least some time doing "hands-on" work like this.

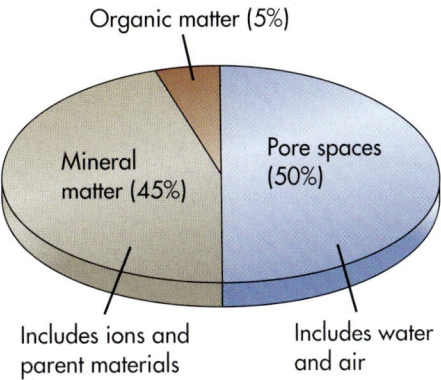

Organic matter (5%)

Mineral matter (45%)

Pore spaces (50%)

Includes ions and parent materials

Includes water and air

Figure 11.2 General composition of soil. Approximately 50% of soil is mineral and organic matter. The remaining 50% consists of pore spaces between grains; these spaces hold water and air.

You may have heard this before, but soils are one of our most important natural resources. They are fundamental to life on Earth because of their interaction with climate and vegetation. Soils provide plants with physical support as well as nutrients and water for growth. Plants, in turn, support soils by anchoring them to the Earth. Without a vegetative cover, soil is prone to erosion, which leads to reduced fertility for agriculture and can contribute to famines in the world's poorest nations.

Basic Soil Properties

The different variables that, taken together, comprise soil can be generally classified into the following four groups (Figure 11.2).

1. **Inorganic materials.** Inorganic materials are naturally occurring chemical elements or compounds that possess a crystalline structure. Each specific mineral (element or compound) owes its structure to the particular rock that it came from. Some of the most common minerals found in soil contain the elements silicon, aluminum, iron, calcium, potassium, and magnesium. The element calcium, for example, is part of the rock limestone. As limestone disintegrates through a process called *weathering* (discussed in Chapter 13), the minerals become reduced in size enough for plants to use them as nutrients. The key point is that, until the plants absorb the minerals, they are stored in the soil.

2. **Organic matter.** Organic matter forms from living and decayed organisms and accumulates in the upper part of soil. As plants and animals die and decay, bacteria and fungi decompose their remains and produce **humus,** which is partially decomposed organic matter. Humus has a sponge-like quality, which allows soil to hold water. In addition, humus increases soil fertility because (a) it stabilizes the soil and (b) microorganisms within humus give off compounds of nitrogen, phosphorus, sulfur, and other mineral substances that can be subsequently taken up by plants.

3. **Water.** Water is critical to soils for a variety of reasons. The most obvious reason is that plants require water to grow. In turn, high plant production contributes to increased biomass, which generally results in more humus production and stable, more fertile soils. Water is delivered to soils in the form of rain and melting snow (Figure 11.3). After precipitation occurs, water can take several pathways with respect to the soil. If the precipitation is heavy, or if it falls on soils that are already very wet, some

Humus *Decomposed organic matter, typically dark, that is contained within the soil.*

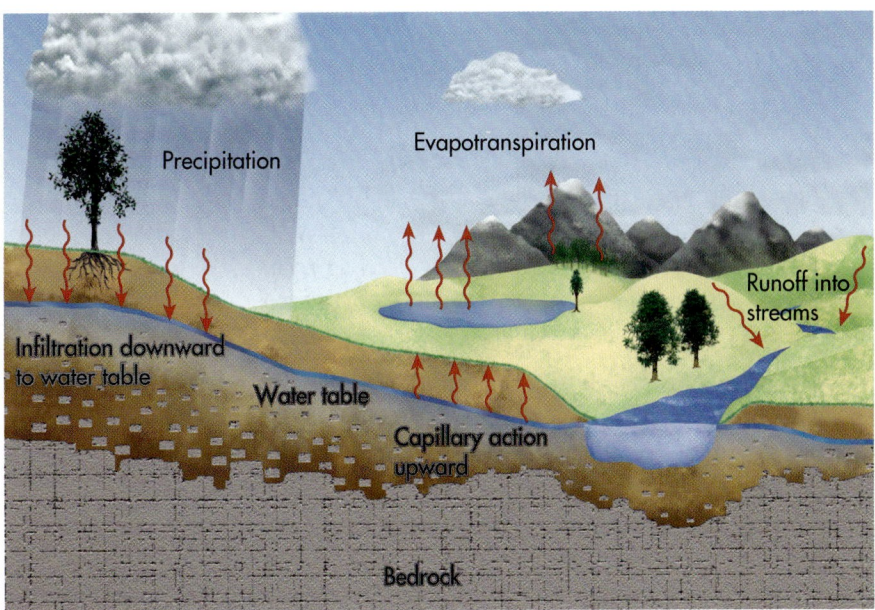

Figure 11.3 Delivery of water to soil. Soils receive most of their water from rain and melting snow. This water moves downward into the soil by infiltration. The soil also receives water from below by capillary action of water from belowground water tables.

Precipitation

Evapotranspiration

Runoff into streams

Infiltration downward to water table

Water table

Capillary action upward

Bedrock

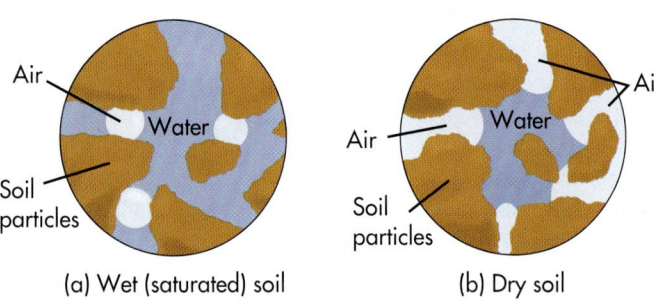

Air — Water

Soil particles

(a) Wet (saturated) soil

Air — Water — Air

Soil particles

(b) Dry soil

Figure 11.4 Wet vs. dry soils. (a) A soil is considered to be saturated when pore spaces are nearly full of water, which occurs after extended periods of precipitation. (b) Subsequently, water drains under the force of gravity, leaving water held to soil particles by surface tension. When most of the downward moving water is lost, the soil is dry.

or most of the water will flow across the surface into local rivers and streams as runoff and never be absorbed into the soil.

When snow melts slowly, rainfall is moderate, or the soil is dry, water readily infiltrates into soils. Once infiltration begins, some water will either evaporate into the atmosphere directly from the ground or when plants absorb it through their roots and carry it to their leaves through the process of transpiration. In the context of soil water loss, the combined processes of evaporation and transpiration are often lumped together in the single term *evapotranspiration*, which we described in Chapter 9.

Let's examine what happens to the soil water that is not lost immediately by evapotranspiration. A soil is *saturated* when water mostly fills all available pore spaces, except for a few small pockets of air here and there (Figure 11.4). Subsequently, within a few days, all excess water drains downward under the force of gravity toward the underlying groundwater zone, leaving larger pockets of air behind. Although most of this downward moving water is lost as far as the soil is concerned, some of it will be pulled back up into the soil at a later time through the process of **capillary action** (see Figure 11.3). This occurs when the molecular attraction at the boundary between the underlying water and soil particles is stronger than the attraction between

water molecules. To see how capillary action works, fill a transparent glass about half way with water. If you look at the side of the glass, where the water meets the inside surface, you can see that the edge of the water actually climbs a short way up the side of the glass. Water molecules have a slightly stronger attraction to the glass than they do to other water molecules. Capillary action in soils works the same way.

Although most water is flushed from the soil pores during the draining process, some remains in the soil–water belt because it is held to soil particles by the force of **surface tension.** You can see for yourself how surface tension works. Splash a few drops of water onto the surface of an upright bottle. Notice that each drop is enclosed within a film that pulls water inward into a rounded shape that withstands the force of gravity. Within soils, surface tension works in the same way, holding water onto soil particles until they are lost through evapotranspiration.

After the excess water has drained, the soil is left at field capacity. **Field capacity** refers to the maximum amount of water that a soil can hold after free gravitational drainage ceases. If additional precipitation occurs shortly after field capacity is reached, then the soil will again become saturated and the cycle begins anew. On the other hand, if drought conditions begin and little rain falls, soil water will not be replenished. Even so, soil water will continue to be lost through evapotranspiration, resulting in less and less water held by surface tension. If dry conditions persist, the soil reaches the **wilting point,** which represents the time when no soil water is available for plant use and plants literally begin to wilt.

The term **soil-water budget** reflects the balance between soil–water gains and losses. To understand how the concept of the soil–water budget works, Figure 11.5 presents a simple example that reflects changes over the course of the year in a typical midlatitude region. Let's imagine that the study location is a continental region in the midlatitudes. During most of the year, there is a surplus of soil water. This surplus occurs from October through April and, interestingly, happens to correlate with the time of year when the least precipitation falls. There are two reasons why the surplus occurs during this time of year. First, very little evaporation occurs because the Sun angle is low, temperatures are cool to cold, and it

Capillary action   *The force that causes water to rise in the small tubular conduits within the soil.*

Surface tension   *The contracting force that occurs when the water surface meets the air and acts like an elastic skin.*

Field capacity   *The maximum amount of water the soil can hold after gravitational water has moved away.*

Wilting point   *The threshold amount of soil water below which plants can no longer transpire water.*

Soil-water budget   *The balance of soil water that involves the amount of precipitation, evapotranspiration, and water storage and loss.*

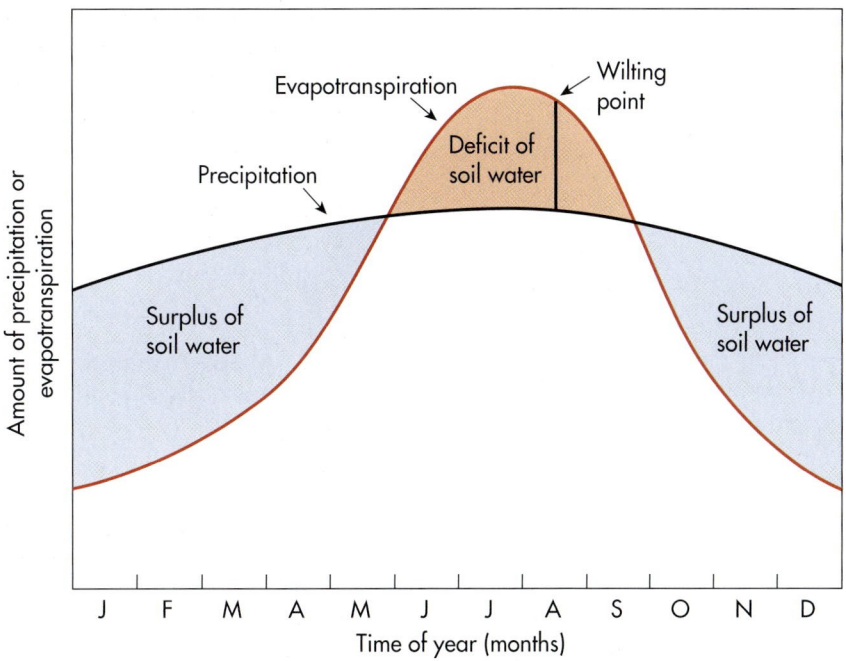

Figure 11.5 Hypothetical soil–water budget for a location in the midlatitudes of the Northern Hemisphere. Although precipitation is relatively low from October to May, it is greater than the amount of soil water lost through evapotranspiration, causing a soil–water surplus. During the summer months, however, the wilting point is sometimes reached because a great deal more soil water is lost by evapotranspiration, even though there is actually more precipitation during that time of year.

is frequently cloudy. Second, plants are not active during this time of year, so little soil water is lost due to transpiration. Taken together, more precipitation than evapotranspiration takes place during the cooler months of the year; thus, there is an excess of soil water. During the summer months, however, deficits of soil water frequently develop, even though precipitation is greater at that time of year, because there is much more evapotranspiration than rainfall.

4. **Air.** The air in soil is mostly carbon dioxide, which plants give off during respiration and then take in during photosynthesis. In fact, there is so much carbon dioxide in soil that the soil layer is sometimes considered to be part of the atmosphere and thus a transition to the solid Earth below. The amount of air in the soil fluctuates depending on how wet the soils are. As you can see in Figure 11.4, when soils are saturated the pore spaces are full of water and there is little room for air. In contrast, when soils dry out there is less water in the pore spaces and more air.

KEY CONCEPTS TO REMEMBER ABOUT SOIL PROPERTIES

1. Soils contain organic material and a variety of minerals. Soils also contain voids, called pore spaces, between grains of sediment.

2. Pore spaces contain both air and water. Following heavy rains or periods of snowmelt, soil pore spaces fill mostly with water. This water drains within a few days, leaving the soil at field capacity with water held by surface tension.

3. The soil–water budget refers to the balance between water gained by precipitation and water lost through evapotranspiration.

Pedogenic processes *The natural processes of soil formation that involve additions, translocations, transformations, and losses.*

Soil-Forming (Pedogenic) Processes

A key concept to understand at this point is that soils form and evolve through a complex sequence of interrelated **pedogenic processes.** It's useful to think of the basic processes

1) Soil Additions
- Organic matter
- Water (as precipitation, condensation, runoff)
- Oxygen, carbon dioxide (from atmosphere)
- Minerals (from precipitation)
- Sediments (from wind, water)
- Energy (from Sun)

2) Soil Translocations
- Clay, organic matter (carried by water)
- Nutrients (circulated by plants)
- Dissolved minerals (carried by water)
- Sediments (carried by animals)

3a) Upper Soil Depletions
- Water (by evapotranspiration)
- Carbon dioxide (from oxidation of organic matter)
- Nitrogen (by biological and chemical means)
- Sediments (by erosion)
- Energy (by longwave radiation)

4) Soil Transformations
- Organic matter (decomposed into humus)
- Particles made smaller (by weathering)
- Particles made into structures (by accretion)
- Minerals transformed (by weathering)

3b) Lower Soil Depletions
- Water and dissolved minerals

Figure 11.6 Basic soil-forming processes. Although all these processes usually operate to some extent in all soils, the rate at which any process operates varies from place to place.

as consisting of *additions, depletions, translocations,* and *transformations* that occur within the soil (Figure 11.6). Let's examine how these processes operate individually.

1. **Soil additions.** Soil additions consist of material and energy that is added to the soil, such as organic material, water from precipitation, various minerals, solar radiation, oxygen and carbon dioxide from the atmosphere, or sediments deposited by wind or water. A good example of a soil addition is the leaf litter that accumulates during the fall on the ground under trees. If you didn't rake up the leaves, decomposers would gradually break them down and their remains would be incorporated into the soil by simple decay or with the aid of burrowing animals and worms. A similar process occurs when grasses die at the end of the growing season. Animal remains and animal excreta that accumulate on the ground also rapidly become part of the soil.

2. **Translocations.** Translocations refer to the vertical movement of materials through the soil. Some movement occurs as nutrients are cycled through plants or

when animals move soil around. A significant amount of additional movement occurs in the presence of water, which percolates downward through the soil during wet conditions. When water moves in this fashion, it dissolves minerals and carries them deeper within the soil, where they recrystallize and collect. When water moves materials, usually in the form of minerals, clays, and organic matter, *out* of the uppermost part of the soil, the process is called **eluviation,** or leaching. When leached materials from the upper part of the soil recrystallize *in* the lower part of the soil, the process is called **illuviation**. Figure 11.7 shows the results of the combined processes of eluviation and illuviation.

Eluviation   *The dissolution and downward mobilization of minerals by water in the soil.*

Illuviation   *The recrystallization of minerals that occurs directly below the zone of eluviation.*

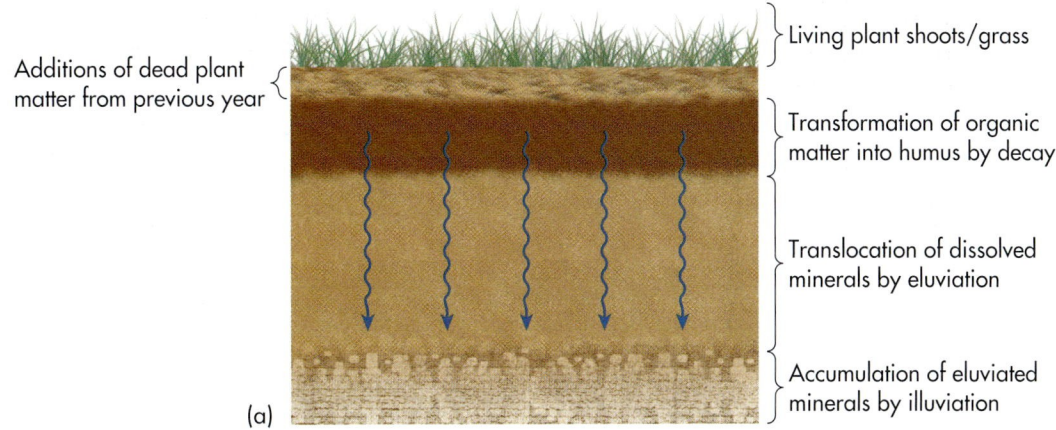

Additions of dead plant matter from previous year

Living plant shoots/grass

Transformation of organic matter into humus by decay

Translocation of dissolved minerals by eluviation

Accumulation of eluviated minerals by illuviation

(a)

Figure 11.7   How additions and translocations affect soil development. (a) One way that additions occur in soil is through the accumulation of dead plant matter, which slowly transforms into humus by decay. The processes of eluviation and illuviation move materials vertically through soil. Eluviation moves material, such as dissolved minerals, out of the upper soil. These minerals recrystallize and accumulate in the lower part of the soil by illuviation. (b) In this photo of actual soil, the dense-looking whitish layer in the lower part of the soil is calcium carbonate that was eluviated by water from the upper part of the soil.

(b)

3. **Soil depletions.** Soil depletions occur when some aspect of the soil is lost for one reason or another (see Figure 11.6). These losses can occur both in the upper and lower part of the soil. An example of a loss in the upper part of the soil is when erosion by wind, water, or glaciers removes sediment. Depletion in the upper part of the soil also occurs when plants take up nutrient minerals or water through evapotranspiration. Another form of depletion takes place when carbon is lost to the atmosphere as carbon dioxide when organic matter is oxidized. In a similar vein, nitrogen is lost to the atmosphere through biological and chemical means. The loss of longwave radiation to the atmosphere is another form of loss in the upper part of the soil. In the lower part of the soil, losses occur when water drains out of the soil and carries soluble minerals with it.

4. **Transformations.** Transformations refer to the decomposition of minerals and organic matter into other forms within the soil. An excellent example of a transformation is when organic matter, such as leaf and animal litter, decomposes into humus. Another example is when minerals are transformed into other minerals or are made smaller by weathering. In addition, soil particles begin to clump together in a recognizable structure.

Soil-Forming Factors

Now that we've seen what the components of soils are and how soils form through pedogenic processes, we can examine the factors that generally influence soil development. Five of these variables, traditionally called the **soil-forming factors,** consistently affect how soils evolve. These factors are parent material, climate, organisms, relief, and time. Although you need to consider these factors in every soil that you study, their character and relative significance vary considerably from a geographic perspective. In order to understand the discussion of specific soil

Soil-forming factors   *The variables of climate, organisms, relief, parent material, and time that collectively influence the development of soil.*

types presented later in this chapter, you must first comprehend how soil-forming factors influence soil development individually and collectively. As you examine these variables, keep in mind that all of them contribute to soil development in some way in each and every soil.

1. **Parent material.** A very important concept is that soils form in sediment. This sediment is referred to as **parent material** and is directly related to the geology of the region in which a particular soil forms. The character of the parent material at any given locality is very important in the context of soils because it sets the stage for future development and affects the way in which additions, translocations, depletions, and transformations occur. In general, there are two types of parent material: residual and transported.

Residual parent material is sediment that develops when underlying rock weathers *in place*. In this context a key concept to keep in mind is that these rocks may be millions of years old, or even hundreds of millions of years old. We will discuss more about weathering in Chapter 14, but for now, think of weathering as the process by which rock, which has been in one place for millions of years, slowly breaks up into

smaller fragments and particles. This breaking up takes place through chemical and physical changes that are largely associated with water.

As this weathering occurs—in other words, as the rock fragments become smaller and change their chemical composition—they form a material called **regolith**. Figure 11.8 shows the relationship between bedrock, regolith, and soil. Notice how the rocks in the upper part of the bedrock break up into smaller pieces. Somewhere in this part of the vertical sequence is a transition in which the bedrock is sufficiently altered, through the weathering process, to be different from the underlying (solid) rock from which it came. At the same time, it has not been modified by additions and translocations from above that occur in soil. In other words, regolith is not rock and is not soil. Rather, it lies somewhere in between the two and is the material in which the soil forms. Often, regolith production at the bedrock interface occurs at the same time that soil development is occurring in the upper part of the weathered zone.

The second kind of parent material is **transported parent material**. As the name implies, transported parent material differs from residual par-

Parent material *The mineral or organic material in which the soil forms.*

Residual parent material *Parent material that forms by the weathering of bedrock directly beneath it.*

Regolith *The fragmented and weathered rock material that overlies solid bedrock.*

Transported parent material *Parent material such as glacial or stream sediments that has recently been deposited and in which soil forms.*

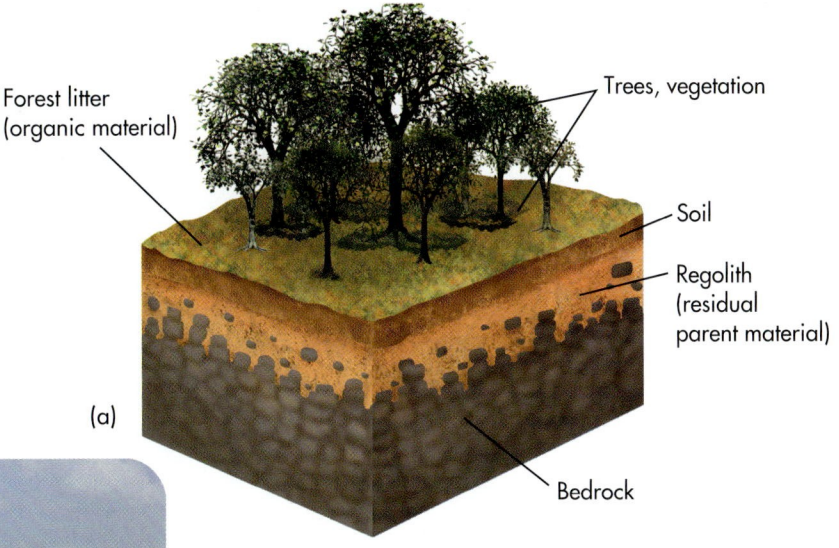

Forest litter
(organic material)

Trees, vegetation

Soil

Regolith
(residual
parent material)

(a)

Bedrock

Soil

Regolith

Bedrock

(b)

Figure 11.8 Relationship of regolith to soil. (a) Regolith consists of weathered bedrock that forms parent material in which soil forms. (b) Limestone bedrock and overlying regolith in the Great Plains. Note how the size of rock fragments becomes progressively smaller above the horizontal slab of solid rock. The soil is the dark zone at ground level.

ent material because it has been moved from one place to another. Remember that residual parent material is regolith that forms *in place* as bedrock weathers. Although the sediments that comprise this bedrock were deposited after being moved from someplace else, the deposition happened such a long time ago (millions of years ago) that, as far as regolith production is concerned, it is considered to have occurred in one place. On the other hand, transported parent material is sediment that was moved relatively recently by the wind, glaciers, or water. Imagine, for example, that a dump truck dropped a large pile of sand in your front yard. This sediment would be transported to your yard much like the wind might blow sand to form a new sand dune (Figure 11.9). If you left the pile of sand undisturbed for many years, plants and animals would begin to live in the uppermost part of the deposit and a soil would begin to form through the combined processes of additions, translocations, depletions, and transformations.

2. **Climate.** Climate is a critical variable in the context of soil development. Temperature and precipitation are the most important aspects of climate that affect

Figure 11.9 Example of transported parent material. These sand dunes along the shore of Lake Michigan consist of sand deposits that were blown there by the wind. Note the grass growing on parts of the dunes. The establishment of grass begins the process of soil formation by providing organic material to be added to the sand parent material.

AMAZING PLACES

BURIED SOILS AT THE EUSTIS ASH PIT IN NEBRASKA

Soils take a long time to form and develop best when very little erosion or deposition is taking place. Another way to view soils is that they represent a time of landscape stability when pedogenesis can operate freely within parent material. From a geologic perspective, soils are important because they can provide information about past environmental conditions within a region. They are most diagnostic in places where buried soils occur.

What is a buried soil? For simplicity, try to imagine what would happen to the soil in your front yard if you dumped a huge pile of sediment on it; let's say one that was 9 m (~30-ft) thick. The soil would become buried, right? If a geologist came to your yard years in the future, and dug a hole deep enough to reach the old (now buried) soil, he or she would conclude that more transported parent material accumulated at some point in front of your house and that this most recent sediment buried the old soil.

A great place to see how buried soils are used to reconstruct depositional histories of transported parent material is the Eustis Ash Pit in south-central Nebraska. The Eustis Ash Pit is an old quarry from which volcanic ash was once extracted. This ash is about 600,000 years old and is found at the base of the quarry wall. Exposed in the quarry wall above the ash is over 40 m (131 ft) of wind-blown silt that has accumulated sometime in the past 600,000 years. Within this silt is a variety of buried soils. These buried soils indicate that deposition of wind-blown silt occurred during distinct intervals of time. In between those periods of silt accumulation, however, the landscape was sufficiently stable to allow soils to form. These intervals of soil formation and then deposition of new parent material have been linked to regional climate change, with soils forming during cool/wet periods and wind-blown silt accumulating during colder/drier periods of time.

This map shows the location of the Eustis Ash Pit in Nebraska.

soil formation because they influence the kind and rate of biological and chemical reactions that occur in the soil. In general, more reactions occur in warmer and wetter places than in colder and drier regions. As a result, soils tend to be better developed and thicker in warm, humid regions than in cold, dry locations.

Climate also dramatically influences the process of translocation in a soil. Remember that translocation refers to the movement of organic and mineral matter from the upper part of the soil into a deeper zone. Most of this movement occurs when water percolates through the soil after it falls at the surface. To visualize this process, imagine, for example, the gradual melting of snow that occurs after a big winter storm passes through the area. As the snow melts, some of it evaporates into the atmosphere, but some of it is absorbed into the ground. When spring comes, plants rapidly take up some of this soil water, but a certain percentage of the water continues to move slowly downward within the parent material under the force of gravity. As this water comes into contact with minerals in the soil or sediment, it dissolves them—much like water dissolves sugar—and carries the dissolved particulates deeper into the soil through the process of eluviation. At some point, the water stops flowing downward and the dissolved minerals recrystallize in the zone of illuviation. Simply put, there is more eluviation in humid climates than in arid regions because there is more water percolating through the soil. Figure 11.10 shows in general terms the depth to which eluviation penetrates and the resulting depth of the illuvial zone in humid, semi-arid, and arid climate regimes.

3. **Organisms.** Organisms are the living things, both plant and animal, that reside in soil. The importance of organisms in soil formation simply cannot be overestimated. At a fundamental level, organisms give life to soil and contribute greatly to the notion of soil being more than *just dirt*. At a more complex level, organisms have a symbiotic relationship with soil in that they not only acquire their food from it, but also help regulate the soil environment in which they live.

Organisms contribute to soil health and formation in many different ways. One way is when plant roots penetrate deeply into the ground. This penetration creates passageways in the soil in which water and oxygen can flow. Many small animals, such as ants, ground squirrels, prairie dogs, and other rodents, spend the majority of their lives burrowing through the soil (Figure 11.11). Through this ongoing process, the soil is continually mixed in a process called **bioturbation**. The mixed soil is softened and aerated, which makes it easier for plants to grow.

Bioturbation   *The mixing of soil by plants or animals.*

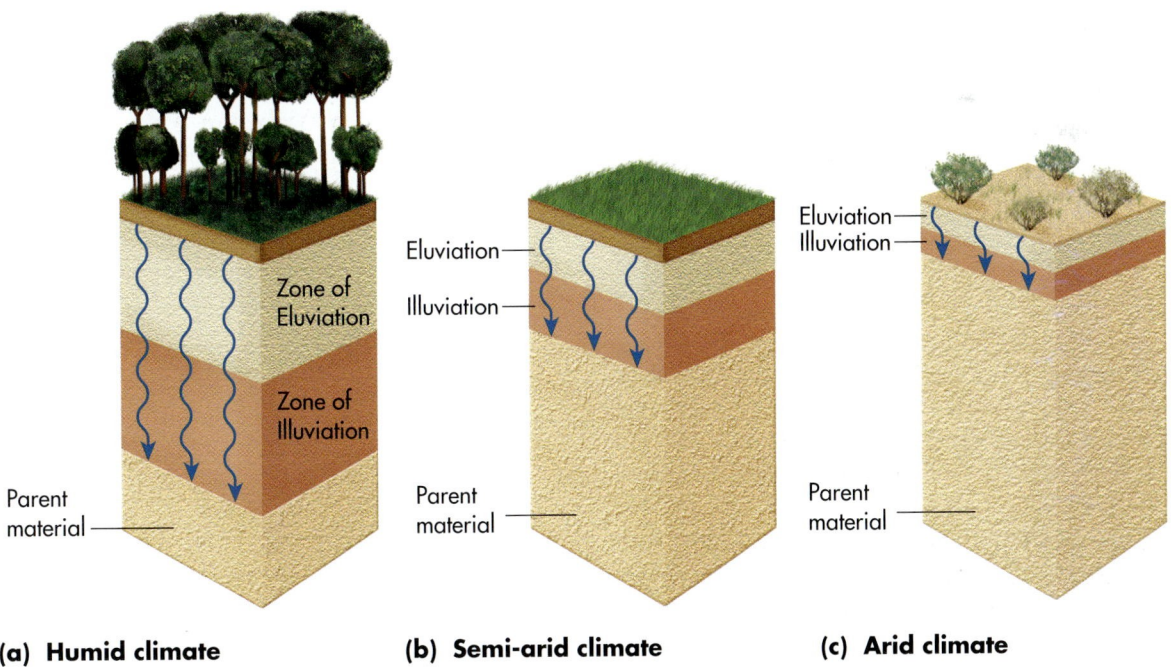

(a) Humid climate   **(b) Semi-arid climate**   **(c) Arid climate**

Figure 11.10 Eluviation and illuviation in three climate regimes. More eluviation takes place in (a) humid climates than in (b) subhumid and (c) arid climates. In addition, the zone of illuviation (where minerals precipitate from solution) is deeper and thicker in more humid climates. This zone typically is progressively thinner and less deep as the climate becomes increasingly dry.

Figure 11.11 Soil bioturbation (a) Prairie dogs in the Great Plains constantly burrow in the soil, which recycles nutrients and keeps soil loose and well aerated.

(b) Termite mounds in central Brazil are composed of earth brought to the surface by the termites. This process mixes the soil.

Earthworms are very important within the system because they recycle soil elements through consumption and waste production. Another important biological component is the huge number of micro-organisms, such as algae, fungi, and bacteria that live within the soil. The primary function of these micro-organisms is to assist with the decomposition of organic matter into humus.

4. **Relief.** Recall that topography refers to the configuration of the Earth's surface—in other words, the position, shape, and orientation of hills, valleys, and mountains. In a related vein, the term **relief** refers to the differences between the highs and lows of a landscape (Figure 11.12a). The Rocky Mountains, for example, is a region of high relief because tall peaks are separated by lower valleys. In contrast, the Great Plains is an area of relatively low relief because the landscape is relatively flat.

   The most important variable associated with relief and soil development is slope. The slope of the landscape is loosely defined as the ground surface that connects the higher areas with lower areas. Slope is significant in the context of erosion because, in general, soils are thinner and less well

Relief *The difference between the high and low elevation of an area.*

developed where slopes are steeper. Figure 11.12b shows a hypothetical landscape with a steep slope next to a relatively flat surface. Note that the soil on the steep slope is thinner; that is, the bedrock is closer to the surface. This pattern occurs because more erosion takes place on the steeper slope, predominantly due to water flowing downhill under the force of gravity. As a result, the rate of erosion on the steeper hillslope may be equal to, or even greater than, the rate of soil formation (Figure 11.12c). On the more level surface, in contrast, the soil can develop more fully because it lies essentially undisturbed from erosion. In addition, the soil is thicker because sediments eroded from the steep slope are added where the slope is more level.

5. **Time.** Time is a critical variable because soils need time to develop. If we assume that all of the other soil-forming factors (parent material, climate, organisms, relief) are equal, then the best developed soils occur on surfaces that have been stable for the longest period of time. For example, where would you expect to find the better developed soil: in a 3000-year-old sand dune or in a 100,000-year-old stream deposit? The answer is that the older stream deposit has a better developed soil because it had more time to develop. The significance of this greater amount of time is that collectively more additions, translocations, depletions, and transformations have taken place in the older soil than in the younger soil.

Area of High Relief

Area of Low Relief

(a)

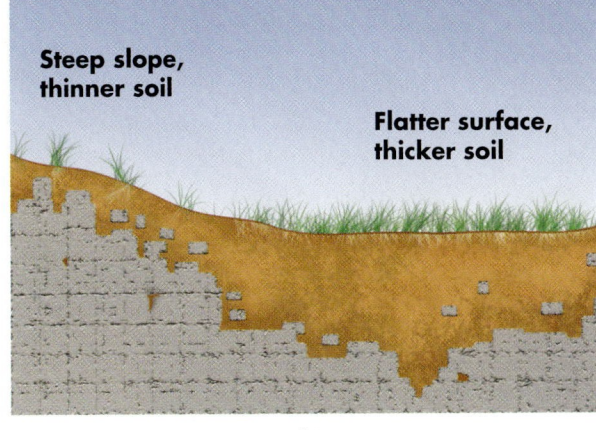

Steep slope, thinner soil

Flatter surface, thicker soil

(b)

(c)

Figure 11.12 Relief and soil formation. (a) The term *relief* refers to the differences between the highs and lows of a landscape. In this scene, the area in the foreground is a zone of low relief, whereas the mountain in the background is an area of high relief. (b) Soils are generally thinner on steeper slopes than on flatter surfaces because there is more erosion on steep slopes. (c) Relief affects soil thickness on slopes, such as the exposed bedrock (arrow) on the steep slope shown here. The soil is very thin on this slope because of erosion; soil is thicker on top of the ridge and in the valley.

Measurable Soil Characteristics

To understand soil formation thoroughly, you should recognize that the soil-forming factors we've discussed are interactive, in the sense that each contributes something to the development of any particular soil. As a soil forms under a particular set of environmental conditions, it develops properties that are unique to that particular setting. Using people as an analogy, each person that you know has different characteristics that you can readily identify, such as brown hair, blue eyes, height, etc. With respect to soils, the properties are physical characteristics that can be measured to distinguish different soils from one another. By examining the physical properties of a soil, it's possible to learn something about how the soil formed. These properties are color, texture, structure, soil chemistry, and soil pH. We will discuss these

SOIL CONSERVATION ON STEEP SLOPES

Soils on steep slopes are prone to erosion because flowing water has a lot of energy and thus moves sediment. Geographers can tell that hillslope soil erosion has been a problem for farmers when they see conservation terraces. Farmers build these features to reduce the effects of soil erosion by creating a series of stair-steps on steep hillslopes. These steps

properties individually; remember, however, that all of these variables work together to give any one soil a particular overall character.

1. **Color.** Soils vary considerably with respect to color, both within and between soils. Color can be an indicator of the composition of the soil. For example, soils with dark colors are usually high in humus or organic matter (Figure 11.13a), which might indicate that the biomass of the region is high or that the environment is cool and wet or that organic matter decomposes slowly. Soils with a high content of iron-bearing minerals are often a reddish color, whereas soils (or parts of soils) that contain abundant calcium are whitish in color. These colors might reflect the color of the local parent material or they may indicate the amount and type of eluviation that has occurred over time.

   Soil color is typically measured with a Munsell Color Chart (Figure 11.13b). This color system uses three elements of color—hue, value, and chroma—to designate a specific color. *Hue* is a measure of the chromatic composition of light that reaches the eye. The Munsell system is based on five principal hues: red (R), yellow (Y), green (G), blue (B), and purple (P). In addition, five intermediate hues represent midpoints between each pair of principal hues, resulting in ten major hue names. The intermediate hues are yellow-red (YR), green-yellow (GY), blue-green (BG), purple-blue (PB), and red-purple (RP). *Value* indicates the degree of lightness or darkness of a color in relation to a neutral gray scale under natural lighting, ranging from pure black (0/) to pure white (10/). *Chroma* is the relative strength of the spectral color. The scales of chroma for soils extend from /0 for neutral colors to a chroma of /8 for the strongest soil color. To put it all together, a soil that is pale brown, for example, is designated 10YR 6/3. In contrast, a soil that is very dark brown has a Munsell designation of 10YR 2/2. Still another example is a soil that has a yellowish red (7.5YR) hue and is strong brown in color. This soil would have a 7.5YR 5/6 color designation.

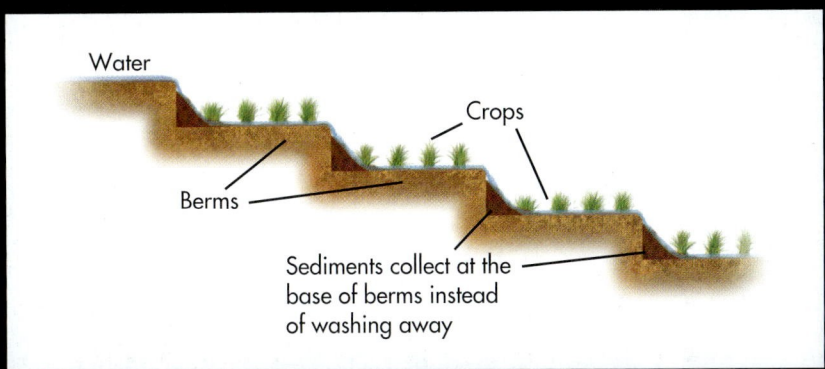

Water

Crops

Berms

Sediments collect at the
base of berms instead
of washing away

produce a series of essentially horizontal surfaces on which farmers can confidently plant
crops. Each of the terraces is separated by a
built-up ridge of dirt, called a *berm*, which
reduces the velocity of water and traps sediment. Although some soil erosion still occurs
where terraces exist, the amount is signifi-

cantly reduced. Perhaps the most impressive
conservation terraces on Earth are the rice
paddies shown here in the mountainous area
of Yunnan Province, China. Note that each of
these surfaces can hold a lot of water, which
falls as precipitation during the summer
monsoon.

2. **Texture.** As discussed earlier, soil contains a multitude of inorganic particles that are typically related to
the parent material. Most soils contain a continuum of
three distinct size categories (Figure 11.14), which are
(in order from largest to smallest) sand, silt, and clay.
Some larger (gravel) and smaller (colloid) particles
may also be present. The combined percentages of
these soil particulates are referred to as the textural
class of the soil, as shown in the soil textural triangle
in Figure 11.15.

Here's how the textural triangle works. Let's say,
for example, that a soil contains 40% sand, 30% silt,
and 30% clay. To determine the textural class, simply
follow the relative percentages of each variable to
where the lines intersect within the triangle. For percent sand, use the lines slanting from the bottom of the
triangle up to the left; for percent silt, use the lines
slanting from the right side down to the left; for percent clay, use the horizontal lines. In this particular
case, the soil would be texturally classified as *clay
loam*. Let's try another one, this time with the relative

percentages of 80% sand, 10% clay, and 10% silt.
This soil would be texturally classified as *loamy sand*.

You're probably asking yourself, what difference
does it make whether a soil is clay loam, loamy sand,
or any other textural class? These kinds of classifications may seem mind numbing rather than informa-
tive, but in the case of soils, however, individual textural classifications are important because they
provide a rough measure of how well water flows
through a soil or how appropriate the soil is for agriculture, just to name two examples. Regarding water
movement, soils that are sandy generally drain better—that is, water flows through them faster—than
clay-rich soils. Thus, soil texture has significant implications for the soil-water budget in any given locality. Although clay-rich soils hold more water than
sandy soils, they are frequently difficult to work with
because clays stick together so well. In general, the
best agricultural soils are the loamy soils because they
are nicely balanced. On the one hand they contain
some sand, which allows water to drain through. On

(a)

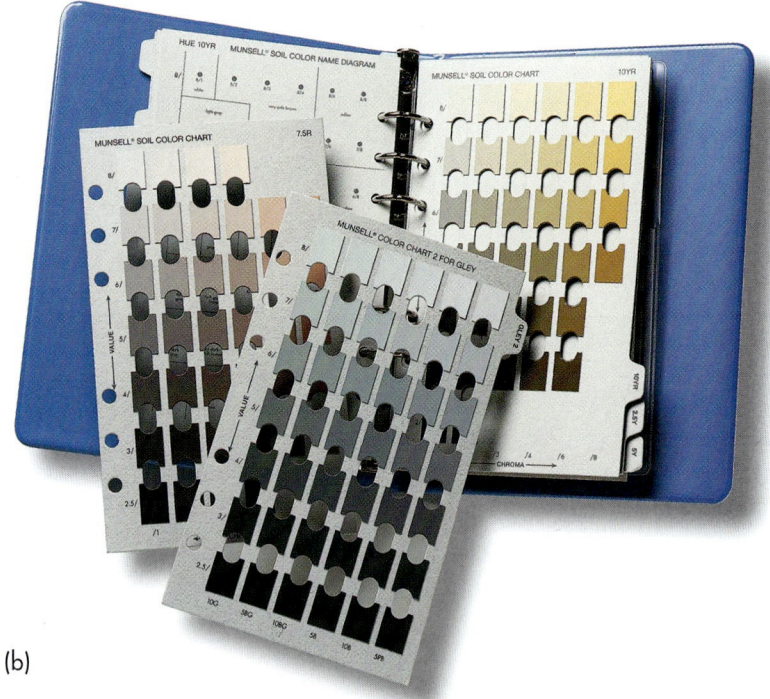

(b)

Figure 11.13 Differences in soil color. (a) The lower part of this soil in Oklahoma is red because it formed in residual parent material (regolith) derived from iron-rich bedrock. In contrast, the upper part of the soil is relatively dark because of the addition of humus to the parent material. In other words, all of the parent material was reddish at one time, but the upper part was transformed to a darker color through the addition process. (b) Soil color is measured with a Munsell Color Chart, which specifies hue, value, and chroma.

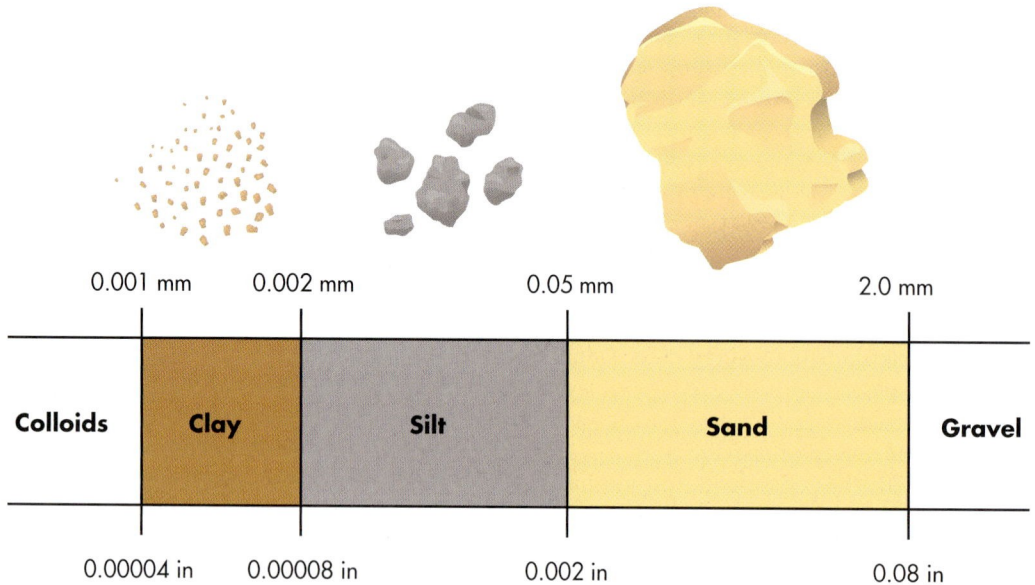

0.001 mm	0.002 mm	0.05 mm	2.0 mm

Colloids	Clay	Silt	Sand	Gravel

0.00004 in	0.00008 in	0.002 in	0.08 in

Figure 11.14 Soil texture. Soils form in parent material that contains particles ranging in size from sand to clay. Note that even a grain of sand is still much smaller than the head of a pin.

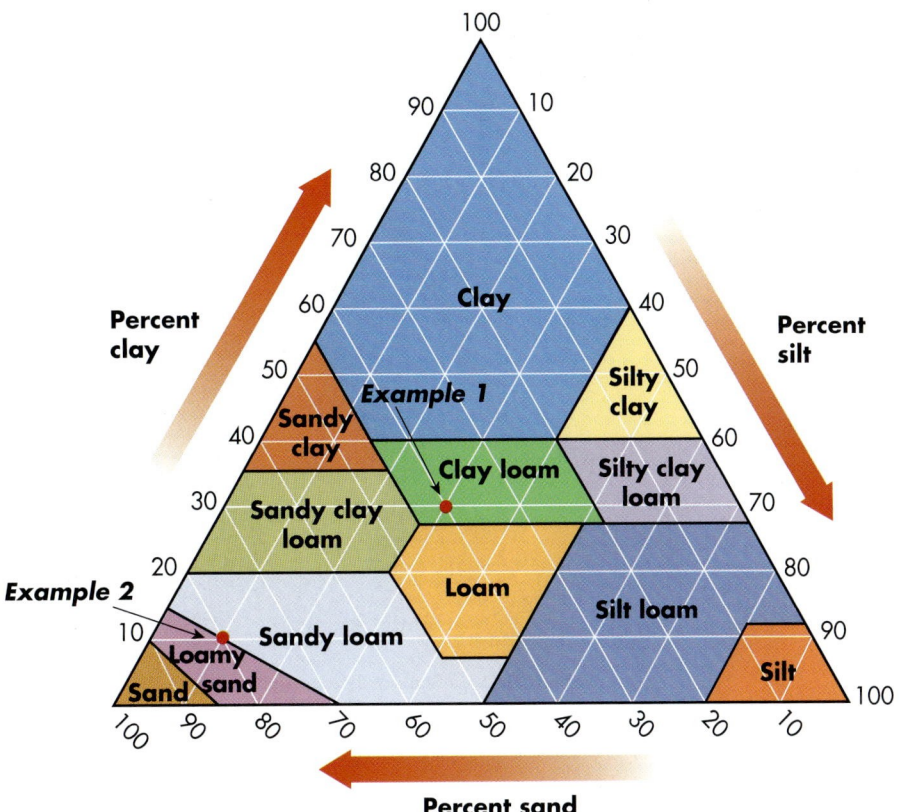

Figure 11.15 The textural triangle of soils. Soils can be texturally classified from the relative proportion of sand, silt, and clay in the parent material. For percent sand, use the lines slanting from the bottom of the triangle up to the left; for percent silt, use the lines slanting from the right side of the triangle down to the left; for percent clay, use the horizontal lines. Examples 1 and 2 described in the text are shown.

the other hand, they contain some clay, which keeps the water from draining through the soil too fast. These clays are also important with respect to overall soil fertility, as you will learn later. A high percentage of silt allows the soil to be easily mixed and aerated.

3. **Structure. Soil structure** refers to the way in which soil particles naturally clump together in peds, which are natural soil aggregates. Another way of looking at structure is that it is related to how clumps of soil break apart and their resulting shape, size, and arrangement. Soil structure is a difficult concept to illustrate; nevertheless, there are four primary types of structure (Figure 11.16).

- *Granular* structure occurs when the peds break up into very small, rounded clumps. You can recognize granular structure when the soil is slightly cohesive but still easily spills between your fingers.

Soil structure *The way soil aggregates clump to form distinct physical characteristics.*

- *Blocky* structure occurs when the clumps break up into distinct blocks that are easy to throw.

- *Platy* structure occurs when the peds are stacked in distinct plate-like forms.

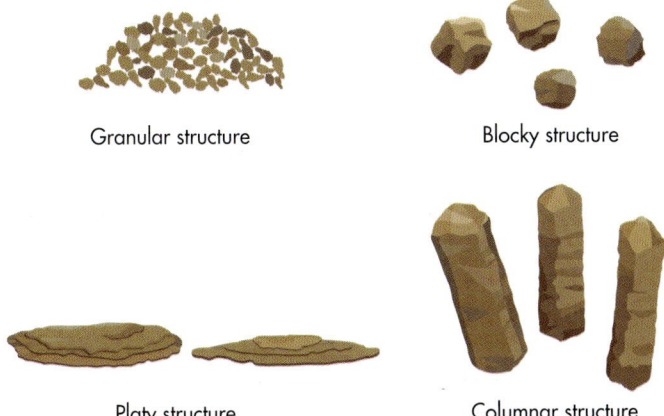

Granular structure Blocky structure

Platy structure Columnar structure

Figure 11.16 Basic soil structures. Particles within the soil clump together in various ways, forming several basic kinds of structure.

What is the texture of a soil that is 50% sand, 40% clay, and 10% silt?

   a)   loam

   b)   silt

   c)   sandy clay loam

   d)   sandy clay

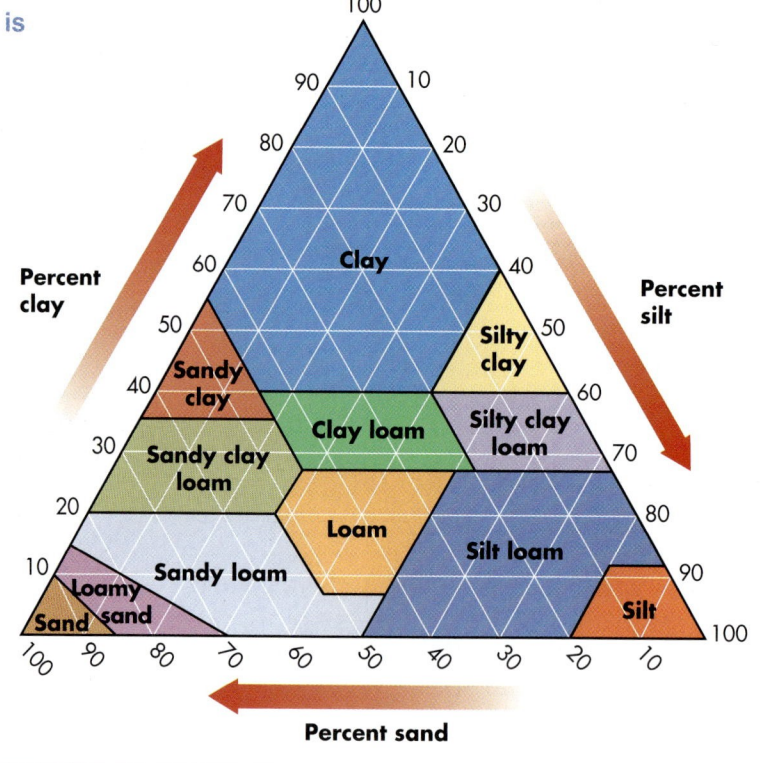

- *Columnar* structure occurs when the peds are organized vertically.

Again, you might be saying to yourself, "what difference does it make how soil particles group together?" The fact is that it's extremely important to how water percolates through the soil and how easily plant roots can penetrate it. Which soil structure presented in Figure 11.16 would be the most difficult for plants to penetrate? The answer is the platy structure because the roots can't drive straight down the soil. As far as water's infiltration is concerned, it flows best in soil with granular and blocky structure.

Soil Chemistry

Given that soils contain a variety of mineral, organic, water, and atmospheric components, each soil has a distinct chemical composition that is important regarding its formation and fertility. The chemical composition of any given soil is determined by a complex set of chemical reactions that occur within the soil solution. As noted before, when water comes into contact with minerals under the right conditions it causes many minerals to dissolve. Similarly, when water mixes with carbon dioxide and various forms of organic matter, it creates car-

bonic and organic acids, respectively, that promote further changes within the soil. A primary product of these kinds of transformations is chemical ions that are free to be absorbed by plants.

Soil pH

One of the ways to characterize the chemical composition of soil is by measuring its pH. The abbreviation **pH** refers to the relative concentration of hydrogen ions (H^+) present in solution. The pH scale ranges from 0 to 14, with lower values indicating acidic conditions and higher values indicating basic (alkaline) conditions. Neutral pH is about 7, which is the pH of pure water. To get a feel for pH and how it relates to various common products, see Figure 11.17.

In the context of soils, pH is an important indicator of soil fertility. Highly alkaline soils, for example, are not efficient with respect to the dissolution of minerals and making them available as nutrients for plants. Acidic soils, on the other hand, result in extensive leaching of minerals—so much so, in fact, that many nutrients are completely lost before they can ever be consumed by plants. In general, most plants, including agricultural crops, are adapted to soils that have a neutral or near neutral pH. It is

pH *The measure of acidity or alkalinity of a solution, ranging from 0 to 14, based on the activity of hydrogen ions (H^+).*

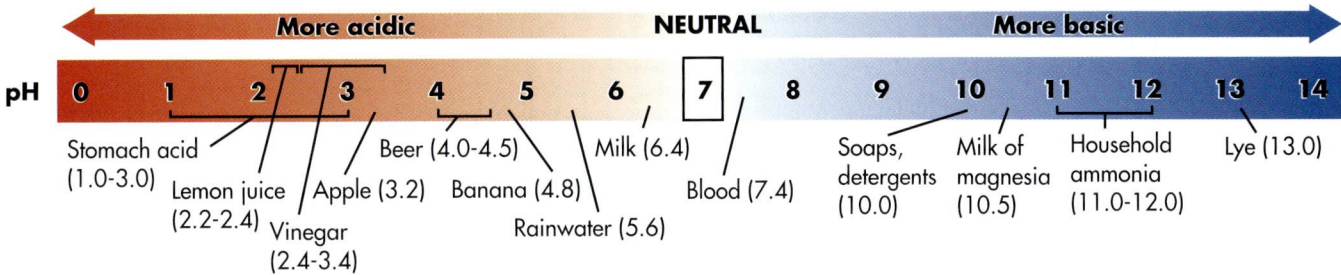

Figure 11.17 The pH scale. Note that most natural foods are slightly acidic and most common cleaning agents are basic.

possible, however, to treat slightly acidic soils with alkaline fertilizers, such as slaked lime (calcium carbonate, $CaCO_3$), to raise the pH to the appropriate level.

Colloids and Cation Exchange

An important part of soil chemistry is the concept of **colloids**, which are extremely small ($<$ 0.001 millimeter) particles that exist in inorganic and organic form. Inorganic colloids consist of crystalline clays that are thin and platy. These microscopic bodies are created when larger particles are chemically altered through weathering. Organic colloids, in contrast, consist of humus derived from organic matter. Regardless of whether colloids are inorganic or organic, they are critical to soil chemistry because they are chemically active and have a high water-holding capacity.

Colloids are also important to soil chemistry because they hold and exchange cations. **Cations** are positively charged (atomic) ions that exist in soil solution due to the dissolution of soil minerals, such as calcium, magnesium, iron, and potassium. Because soil colloids are negatively charged, they attract the positively charged cations that are suspended in the soil solution. Although colloid particles are individually small, they collectively have a very large surface area on which cations can become attached. Some ions, such as metal and hydrogen ions, are bound tightly to colloids, whereas basic ions such as calcium move relatively freely. In many instances, the loosely bound (basic) cations are replaced by the metallic and hydrogen ions through the process of cation exchange. Soils that have a high **cation exchange capacity** (CEC)

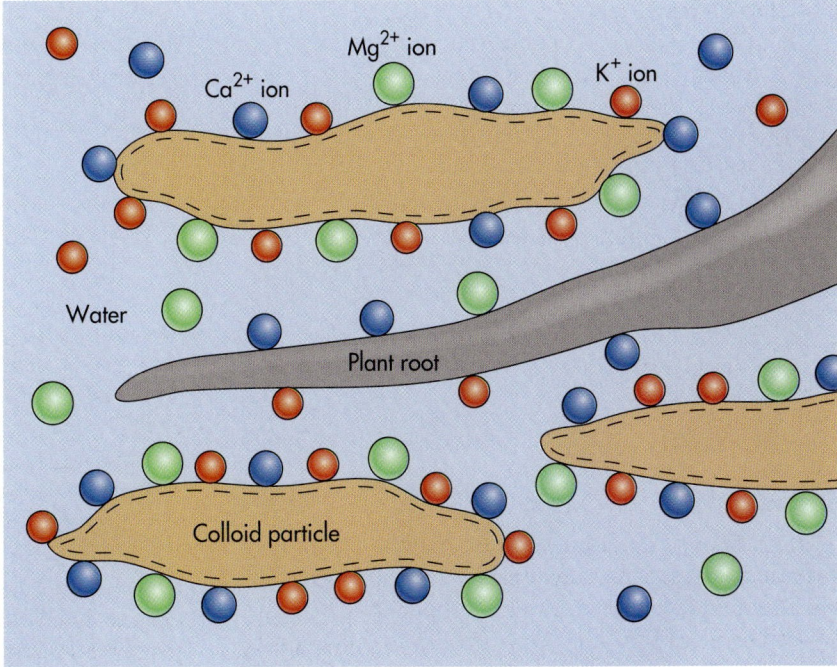

Figure 11.18 Microscopic view of colloids and soil fertility. Colloids are microscopic clays and organic particles that are negatively charged. Cations, which are positively charged atoms that exist in soil solution, are thus attracted to the colloid particles, which hold them until a plant root absorbs them.

Colloids *Very small (10 nanometers to 1 micrometer), evenly divided solids that do not settle in solution.*

Cations *Positively charged ions, such as sodium, potassium, calcium, and magnesium.*

Cation exchange capacity *The total amount of exchangeable cations that a soil can absorb.*

are typically the most fertile because they contain an abundance of colloids on which cations can be held and exchanged.

In the context of soil fertility, colloids are important because they keep minerals from being completely leached from the soil. Often, cations are held by the colloid until a plant root comes in contact with the colloid and absorbs the cation as a nutrient. Figure 11.18 illustrates how individual, negatively charged colloid particles (in this case clays) attract the positively charged cations surrounding them until a plant root "consumes" the stored nutrients.

Soil Profiles (Reading the Soil)

Now that we've seen some fundamental soil components, processes, and chemistry, let's examine the diagnostic characteristics resulting from the additions, transformations, translocations, and depletions that occur to form soil. The really nice thing about these basic processes is that they cause the soil to become progressively organized, through time, in a way that is easy to see by looking at a **soil profile.** Just as the profile of your face refers to a side view that ranges from the top of your head to your chin, a soil profile displays all of the soil characteristics in a vertical "image." A soil profile can be obtained by cleaning a natural exposure such as a creek bank, or, more commonly, by excavating a soil pit with a shovel or backhoe. Although it may not seem like it, this kind of work is really a lot of fun. Figure 11.19 shows examples of exposing a soil profile.

As soils form, they organize internally into distinct **soil horizons.** If the term "horizon" is confusing to you, simply think of the term as being equivalent to "layer" or "band" for the time being. Look again at the soil profile in Figure 11.19a. Do you see the distinct layers that occur in the profile? These layers are referred to as soil horizons because they typically blend into one another; in other words, it is very rare to see abrupt boundaries between different parts of the soil.

Soil profile *A vertical exposure in which all soil components can be seen.*

Soil horizons *The distinct layers within a soil that result from pedogenesis.*

(a)

(b)

Figure 11.19 Soil profiles. (a) After a soil pit is excavated you can then see the soil profile and its specific characteristics, such as the dark zone near the surface of this particular soil and the reddish zone beneath it.

(b) Sometimes soils are exposed naturally, such as along this stream cutbank. In these situations, they can easily be studied.

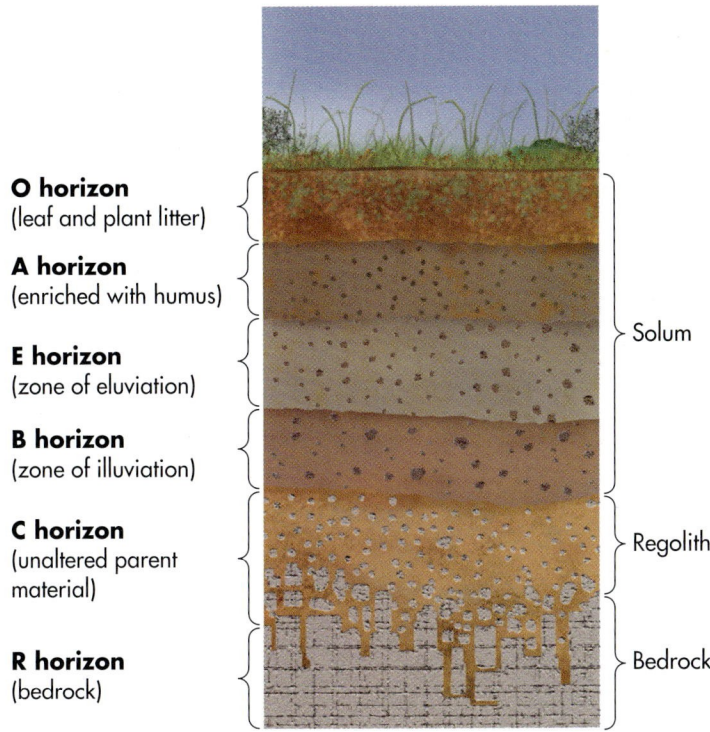

O horizon
(leaf and plant litter)

A horizon
(enriched with humus)

E horizon
(zone of eluviation)

B horizon
(zone of illuviation)

C horizon
(unaltered parent
material)

R horizon
(bedrock)

Solum

Regolith

Bedrock

Figure 11.20 Soil horizons in a fully developed soil on bedrock-dominated landscape. As you examine this figure, remember that the soil horizons evolve within the regolith.

The specific character of soil horizons varies dramatically between different parts of the world, but, in general, any well-developed soil contains, from top to bottom, O, A, E, B, C, and R horizons (Figure 11.20). These horizons form by different processes and have the following six characteristics.

1. **O horizon**—The O horizon is the uppermost horizon of the soil, existing at ground level, and consisting mainly of undecomposed organic matter. A good example of an O horizon is the leaf litter that accumulates on the forest floor during the fall season. Thus, an O horizon forms through the addition of organic matter.

2. **A horizon**—The A horizon is the uppermost mineral horizon of the soil. This horizon is distinctive because it contains abundant humus, which consists of

decomposed organic matter added to the soil. Given the relatively high humus content, A horizons are typically a darkish gray color. In general, A horizons have granular structure due to the high concentration of humus. In Figures 11.7b and 11.13, the dark horizons at the tops of those soil profiles are A horizons.

3. **E horizon**—The E horizon lies immediately below the A horizon (see Figure 11.20) and is the zone of eluviation; thus this horizon forms largely through the process of translocation. In other words, this horizon evolves due to the downward movement of minerals leaching from it. As a result of this eluvial process, the E horizon is usually relatively light in color compared to the horizons above and below it.

4. **B horizon** — The B horizon lies below the E horizon and is the zone of illuviation and thus forms largely through the process of translocation. In other words, this part of the soil evolves because minerals precipitate at this level. Recall that the zone of illuviation is deeper in more humid climates than in relatively dry ones, which is to say that the B horizon is deeper and thicker where precipitation is greater (see Figure 11.10). Given the accumulation of eluviated minerals in the B horizon, this horizon usually has stronger structure (such as "blocky") than the overlying horizons. The base of this horizon corresponds to the base of the soil. The full thickness of the soil, from the top of the A horizon to the bottom of the B horizon, is referred to as the **solum**.

5. **C horizon** — The C horizon consists of unaltered parent material that has yet to be affected by soil-forming processes. In other words, this horizon is regolith and is not part of the solum.

6. **R horizon** — Where soils form in landscapes dominated by bedrock, such as those you find in the Appalachian Mountains or parts of the Great Plains, an R horizon occurs. Very simply, the R horizon is the bedrock that is the source of the regolith/parent material, which is the stuff in which the soil forms.

Time and Soil Evolution

Although the profile in Figure 11.20 is typical, it is far from being the profile we see in every soil. In fact, not all horizons are present in every soil. Horizons also vary in color and configuration between different soils. These

O horizon *The uppermost soil horizon that consists of undecomposed plant litter.*

A horizon *The soil horizon that is enriched with humus.*

E horizon *The soil horizon that is progressively depleted of minerals through dissolution and translocation.*

B horizon *The soil horizon that forms below the E horizon because translocated minerals recrystallize.*

Solum *The A, E, and B horizons of a soil, which form through pedogenic processes.*

C horizon *Unaltered soil parent material.*

R horizon *Unweathered bedrock that underlies soil.*

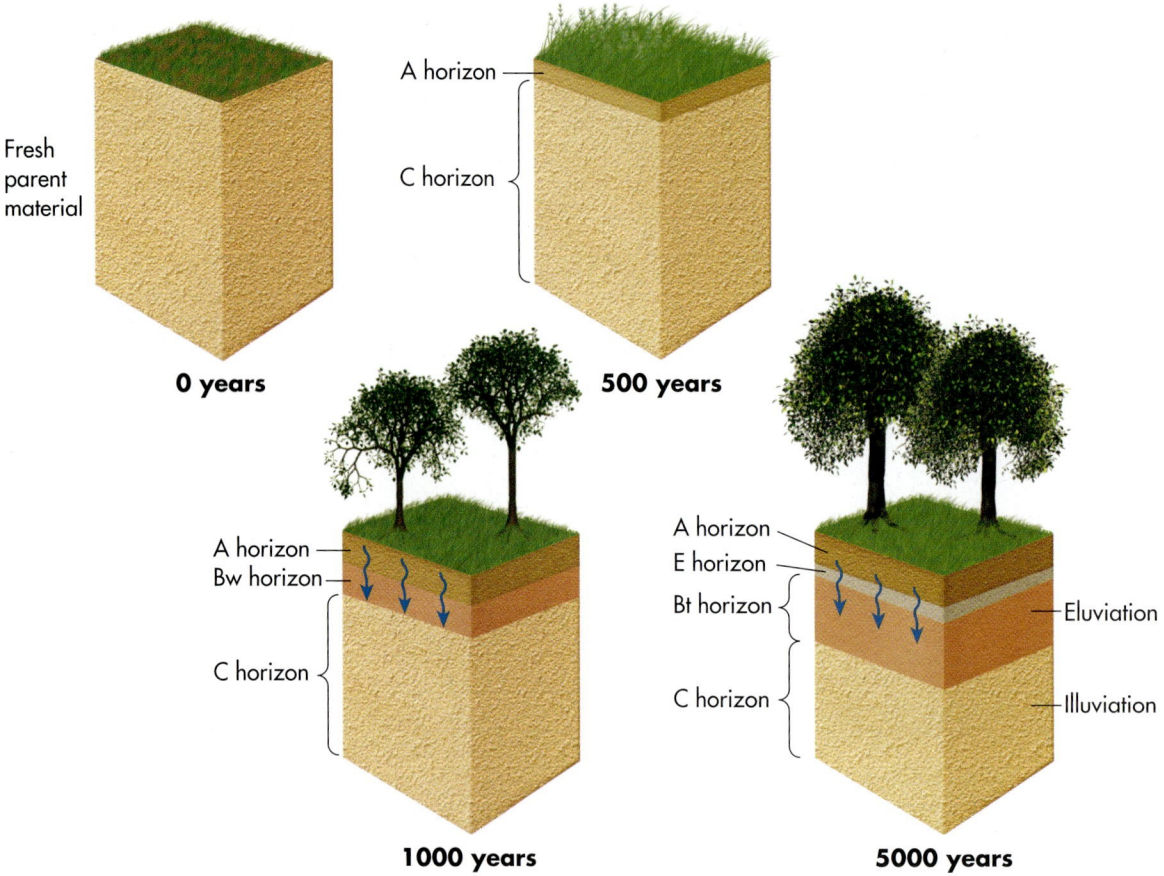

Figure 11.21 Hypothetical evolution of a generalized soil profile through time. In an idealized situation the development of A, B, and E horizons might take around 5000 years. Note the progression of horizon development at different periods of time.

differences are due not only to the five soil-forming factors discussed previously, but also to geographic variability in the way that soil processes function.

To see how soil horizons generally evolve, consider Figure 11.21. This figure presents the formation of a soil in a hypothetical profile observed at four separate times: today (0), 500, 1000, and 5000 years into the future. The idea is that at "time 0" a dump truck drops a big pile of sediment on your front yard. In the context of soils, this load of sediment can be thought of as fresh (transported) parent material in which a soil will form given the additions, translocations, transformations, and depletions that occur "in the future." Imagine what would happen if you excavated a soil pit and could then "visit" the site at future times.

If you could visit this site at three future times, you would see that the soil would evolve in a fairly predictable way. This is not to say that each soil would change in the same way with respect to horizon thicknesses or colors, or that it would evolve at the same speed given climate and textural differences. However, the overall pattern very well could be consistent. Following deposition of the fresh parent material, the first horizon that would become obvious in the first 500 years would likely be the A horizon. This horizon would develop due to the additions of organic matter that resulted from the plant and animal life at or near the surface. (Given your understanding of soil horizons, what would be the diagnostic characteristic of this horizon?)

At the same time that the A horizon is forming, translocations and transformations would also be occurring. The cumulative result of these combined processes might begin to be seen 1000 years in the future by the development of a Bw horizon immediately beneath the A horizon. The "w" means that the horizon is weakly developed, and will probably be identified because the structure has become slightly blocky due to the illuviation of clays and other minerals at this level of the soil. Another thing to notice is that the A horizon has become thicker because of the continual additions that have occurred over the millennia. It is probably darker as well.

If you returned to examine the soil pit 5000 years into the future you might very well discover that the soil had become fully developed. Remember, this development has occurred because plants, microbes, and animals have been interacting within this soil for a lengthy period of time. In addition, water has been percolating through the soil for the same length of time. As a result of these processes operating day by day, month by month, and year by year for 5000 years, the soil has a distinctive E horizon and a thick Bt horizon. The E horizon is the zone of the soil that has been extensively leached of its mineral and clay components. These minerals and clays moved downward and recrystallized in the evolving B horizon. In this case, the inclusion of the lower case "t" means that the horizon is texturally distinctive, through the illuviation of clay at that level, from the rest of the soil. Another way to make this distinction is to say that the B horizon is likely to be an **argillic horizon** with a strong blocky structure. This structure developed because abundant clay was added to this part of the soil through eluviation.

Soil Science and Classification

In the preceding section we discussed the basic model of soil formation and the various horizons that evolve over time. Given that model, we can now examine some of the soil variability on Earth, which is known because of the research conducted by scientists working in the field of **soil science.** This scientific discipline specifically deals with soils as natural resources, including their physical, chemical, and biological properties.

Soil science is a particularly relevant discipline to humans because we use soils in countless ways. Of particular importance is that soils are the medium in which we grow our food. Thus, understanding the properties of soils is absolutely essential to systematic agriculture and directly affects the quality and price of the food we buy. Soils are also the fabric in which we build our homes, buildings, highways, and a myriad of other structures. As a result, engineers must have a good understanding of soil characteristics in order to build structures that are properly supported. Soils are also important filters for surface and ground waters and are thus highly relevant to hydrology, which focuses on the way that water moves on Earth. People who study soils are called *soil scientists* (Figure 11.1). The goal of soil scientists is to improve our understanding of the Earth's soils in order to preserve and efficiently utilize them. In the U.S., many soil scientists are associated with the Natural Resource Conservation Service, which is a federal agency that works to manage and sustain natural resources in the country.

One of the ways in which soil scientists study and manage soils is in association with the concept of soil classification. In general, classification is a common practice in many scientific disciplines because it makes it easier to remember attributes about the phenomena under study. Classification also provides a basis from which one category can be compared with another. Given the predictable pedogenic processes and factors that operate on Earth, soils are particularly well suited for classification.

Although many different classification systems exist around the world, the one used in the U.S. is called **soil taxonomy.** This system is based on the existing properties of a soil, such as color, texture, structure, and mineral content, which can be measured. Many other soil classification systems are genetic; that is, the soil scientist attempts to reconstruct how the soil evolved, even though the initial environmental conditions may not be known. Because soil taxonomy is a generic, logically based system, the classification scheme is hierarchical in that there are several levels of general classifications, each with sublevels below it. With each successive lower level, fewer and fewer similarities appear between soil types.

The highest level within soil taxonomy is **soil order.** Twelve soil orders occur around the world, each one distinguished on the basis of diagnostic horizons (Table 11.1) that meet certain criteria. These diagnostic horizons are the **epipedon** (from the Greek word *epi*, meaning "over" or "upon"), which is the horizon formed at the soil surface, and the subsurface horizons formed by removal or accumulation of material.

Soil orders are subdivided into several categories, which are, from highest to lowest, *suborders, great groups, families,* and *series.* To give you an idea of how extensive this hierarchy is, 50 suborders and 180,000 soil series are recognized in the United States. It is well beyond the scope of this text to describe each of the soil series that have been identified; but we will outline the basic characteristics of the 12 soil orders. This will give you a good overview of how soils differ and how pedogenic processes and factors affect soil development.

The Twelve Soil Orders

This portion of the soils discussion focuses on the 12 soil orders that occur within the soil taxonomy classification system. These soils reflect the maximum development that occurs in these orders in association with various combinations of the soil-forming factors: (1) climate, (2) organisms,

---

Argillic horizon *A B horizon that is enriched in eluviated clay.*

Soil science *The study of soil as a natural resource through understanding of its physical, chemical, and biological properties.*

Soil taxonomy *The method of soil classification that is based on the physical, chemical, and biological properties of the soil.*

Soil order *A group of 12 distinctive soils differentiated at the most general level.*

Epipedon *The surface horizon of a soil profile.*

TABLE 11.1	Soil Orders, the Derivation of Their Names, and Characteristics	
Well-Developed Mineral Soils		
Order	**Name Derivation**	**Characteristics**
Alfisols	Nonsense syllable	Soils with high base content B horizon rich in eluviated clay. Found from low to subarctic latitudes.
Aridisols	Latin: *aridus* = dry	Soils in dry climates that are low in organic matter and frequently have subsurface horizons rich in calcium carbonate or soluble salts.
Mollisols	Latin: *mollis* = soft	Soils in semi-arid/subhumid grasslands in midlatitudes that have humus-rich A horizon and a B horizon that has a high base status.
Oxisols	French: *oxide* = oxide	Highly weathered soils in tropical environments that have low base status and subsurface horizon that has a high oxide concentration.
Ultisols	Latin: *ultimus* = last	Highly weathered soils in tropical and subtropical environments that have low base content and have a subsurface horizon rich in eluviated clay.
Spodosols	Greek: *spodos* = wood ashes	Soils in cool, moist environments that have a B horizon rich in eluviated sesquioxides.
Vertisols	Latin: *verto* = turn	Tropical and subtropical soils with high base status that contain an abundant amount of expandable clay that swells when wet and shrinks when dry.
Organic Soils		
Gelisols	Latin: *gelare* = freeze	Largely organic soils that form in extremely cold environments where permafrost is thick. Due to repeated freezing and thawing, soil horizons and surface expression are chaotic.
Histosols	Greek: *histos* = tissue	Very dark soils that consist mostly of organic matter. Typically found in cool/moist environments where organic decomposition is slow.
Weakly Developed Soils		
Andisols	Spanish: Andes	Weakly developed soils formed within glassy volcanic sediments ejected by active volcanoes.
Entisols	Nonsense syllable	Horizonless soils usually formed within recently deposited sediments.
Inceptisols	Latin: *inceptum* = beginning	Soils with poorly developed horizons, but which may evolve further.

(3) relief, (4) parent material, and (5) time. Note that two of these factors, climate and organisms, are distinctly environmental in nature. In other words, they reflect the specific climate and vegetation relationships in any given place or region. Given this close correlation, it should not surprise you that the global distribution of soil orders reflects these environmental factors and thus builds directly upon the previous two chapters. Figure 11.22 shows the geographic distribution of soil orders on Earth. Notice the similarity with respect to the patterns on this map and the global climate and vegetation maps seen earlier (see Figures 9.2 and 10.4, respectively). As you work your way through this discussion, try to keep these geographic patterns in mind.

As with the descriptions of climate and vegetation, we begin our discussion of soil orders by starting at the tropics and proceeding to higher latitudes. A few soil orders can occur anywhere under the proper conditions and will be dealt with at the end of the discussion. While you are studying the various soil orders, refer back to Figure 11.22, which shows the geographic distribution of soil orders and is similar to the global climate and vegetation maps in Figures 9.2 and 10.4, respectively. You might want to familiarize yourself with these illustrations again before you begin your tour of soil orders. Remember that many different variations of these soil orders arise, given the unique interactions of the five soil-forming factors; this discussion will give you only a basic overview of their character and distribution. The discussion of each order will be supported by a figure that includes a map of its global distribution, an example of a typical landscape associated with the order, and an example of a soil profile of that order.

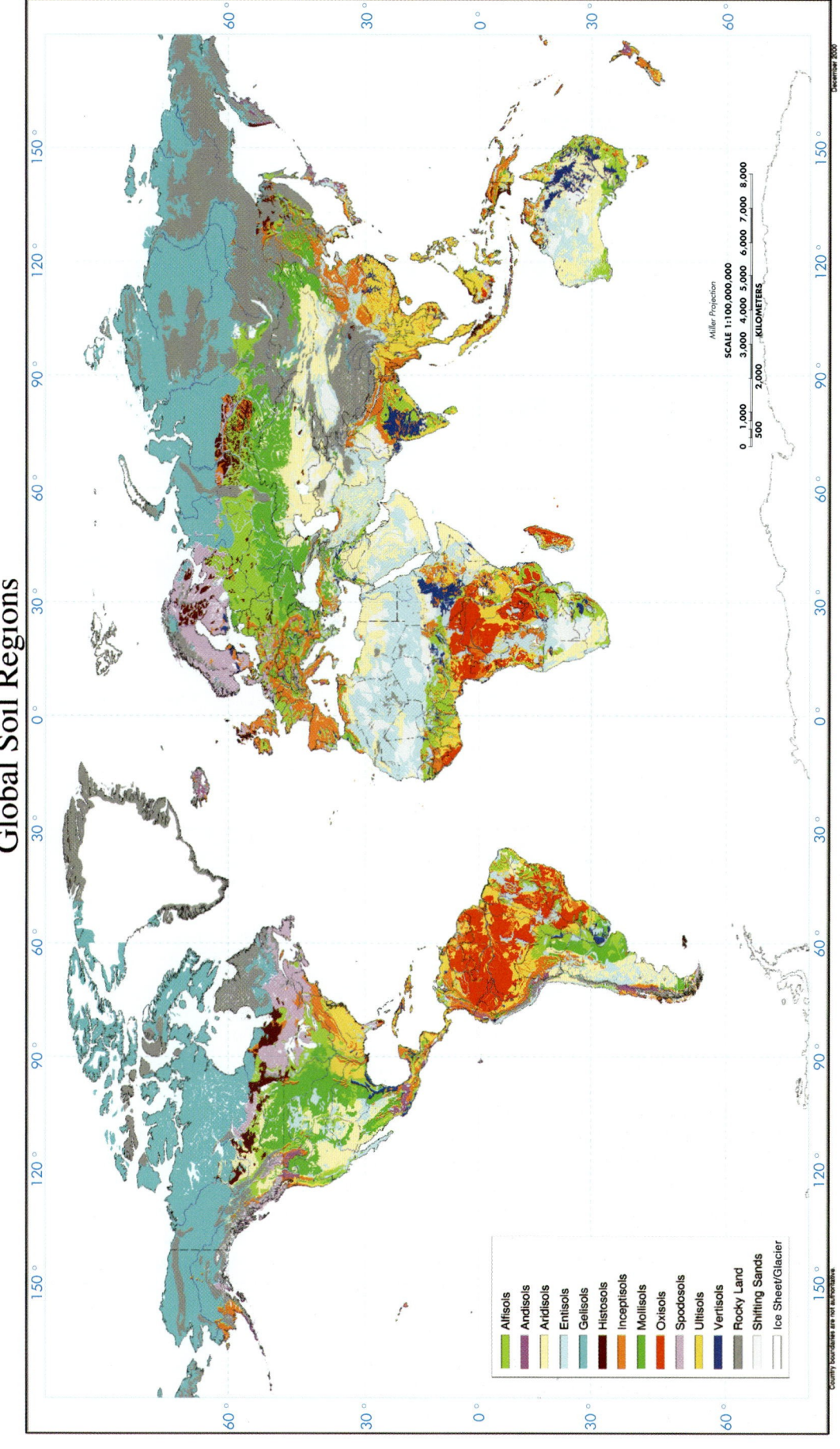

Global Soil Regions

Figure 11.22 The global distribution of major soil orders. This map is color-coded to reflect the geography of soil orders. Areas mapped as "Rocky Land," such as in central Asia, are places where large expanses of rock are exposed at the surface and soils have yet to form. Areas designated as "Shifting Sands," such as in the Sahara Desert, are places where the wind deposits sand in a way that precludes the development of soil. Areas mapped as Ice Sheet/Glacier, such as in Antarctica and Greenland, are places where ice covers the ground and soils cannot form. (Credit: U.S. Department of Agriculture)

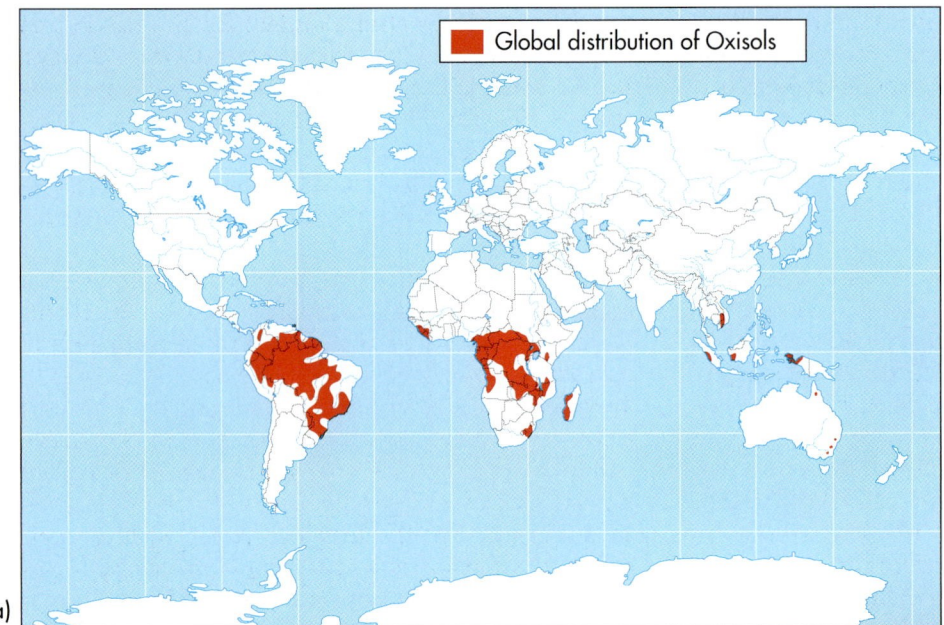

(a)

(b)

(c)

Figure 11.23 Oxisols. (a) Generalized map of ice-free Oxisols on Earth. These soils occur on about 8% of the Earth's ice-free area. (Credit: U.S. Department of Agriculture) (b) Oxisol landscape in Hawaii. Note the reddish color of the soil, which occurs due to extensive weathering in this tropical environment. These soils are used in the growth of sugarcane. (c) A typical Oxisol in Hawaii. The Ap horizon refers to the fact that the A horizon has been plowed. The "o" designation in the B horizons means that they contain oxidized iron and aluminum. These subhorizons are potentially distinguished from one another on the basis of texture, color, or structure.

Oxisols   Oxisols are soils that form in tropical environments and are found on about 8% of the ice-free land area on Earth (Figure 11.23a). These soils form through a specific soil-forming process associated with tropical environments called **laterization,** which means "to become brick-like." The essence of laterization is that the hot and wet environment causes a great deal of mineral weathering to occur within soils. Bases such as calcium, magnesium, and potassium ions are rapidly leached

Oxisols   *Mineral soils in tropical and subtropical environments that formed through laterization and thus have an oxic horizon within 2 meters of the surface.*

Laterization   *A regional soil-forming process in tropical and subtropical environments that results in extensive eluviation of minerals except for iron and aluminum.*

because they are easily mobilized and water consistently percolates through the soil. In fact, percolating water flow is so steady that even otherwise resistant minerals such as silica are leached. The only minerals that aren't leached are iron and aluminum sesquioxides (minerals containing three atoms of oxygen for every two atoms of the metal), which results in highly weathered A and B horizons that together are called an **oxic horizon** because the sesquioxides are oxidized (that is, atoms lose an electron). In addition, very little organic matter is added to the soil because decomposition of surface litter is rapid.

Oxisols are the most weathered of the soils formed by laterization. Given the oxidized character of these soils, they are usually red, yellow, or yellowish brown in color (Figure 11.23b and c). Well developed Oxisols typically have a very thin A horizon and a thick Bo horizon (Figure 11.23b), with the "o" reflecting the presence of oxidized sesquioxides. Although Oxisols are associated with rainforest, they actually have very low fertility because most nutrients have been weathered out of the system. In addition, Oxisols have low cation exchange capacity, which further reduces fertility. Given this fundamental soil infertility, plants derive most of their nutrients from standing and decomposing vegetation.

Farmers in Oxisol regions can temporarily improve soil fertility by cutting and burning the rainforest in the manner described in Chapter 10. Once the plant debris is burned, the resulting organic residue collects in the upper part of the soil, providing nutrients for crops. Unfortunately this method of soil fertilization is effective for only a few years in any given field, which requires that additional forest stands be cut to maintain crop production. This cause and effect is one of the principal driving forces leading to tropical deforestation.

Ultisols   Ultisols form in warm, moist, subtropical environments like those on the southeastern coasts of the U.S. and China. They also occur in the more humid parts within the tropical wet/dry climate regions. Overall they occur on about 9% of the ice-free land surfaces on Earth (Figure 11.24a). Recall from Chapter 10 that the vegetation in these tropical and subtropical areas consists of forest. As a result of this environmental association, these soils form through a variant of the laterization system and are similar to Oxisols in their reddish color (Figure 11.24b) and low base concentration.

The primary difference between Oxisols and Ultisols is that Ultisols are a bit more fertile because they are not as heavily leached. This relative lack of weathering happens because Ultisols occur in regions that experience a dry sea-

son. Ultisols also contain a distinct argillic (Bt) horizon (Figure 11.24c), resulting from the translocation of clays. In fact, some Ultisols have Bt horizons that contain as much as 70% clay. They support productive forests, but require the frequent addition of lime fertilizers if they are used for continuous agriculture. In this manner, Ultisols in the southeastern U.S. have been farmed for over 200 years.

Vertisols   A third soil order that frequently occurs in the tropical and subtropical regions is **Vertisols.** These soils are found on about 2% of the ice-free land surface on Earth, including some areas outside of the tropics (Figure 11.25a). The primary soil-forming factor associated with Vertisols is parent material because they contain a lot of expandable clays, which are clays that swell like a sponge when they are wet and shrink when they dry. Because these soils contain so much clay, fields that lie within Vertisol soils are typically dark with very strong structure, which you can see as the prominent irregular forms in Figure 11.25b.

One of the distinctive characteristics of Vertisols is that extensive cracks can form in the ground (Figure 11.25c) during long dry periods because the clays within the soil shrink so much. Because of the constant shrinking and swelling that occurs in these soils, they are constantly being mixed, with much of the mixing occurring when fragments of soil fall into the cracks. Within a soil profile you can frequently see the results of the shrinking and swelling through the presence of shiny clay skins on peds. These glossy surfaces, or *slickensides*, are created when the long-term expansion and contraction of the soil causes clays to be polished as they rub against one another. As a result, a diagnostic horizon of a Vertisol is a Bss horizon (Figure 12.25d), with the "ss" reflecting the presence of slickensides.

In fact, if you have ever noticed that cracks form in the ground during long droughts around where you live, then there is a good chance that the soils in your neighborhood are Vertisols, or closely related. In most places, Vertisols occur in isolated pockets that are too small to show up on the global soils map. Significant regions of Vertisols form in the tropical and subtropical regions of the world, however, including the Deccan Plateau of western India, along the Nile River in eastern Africa, and much of eastern Australia. Although these soils are often high in base content, they are difficult to farm because they are hard to cultivate.

If you ever enter the real-estate market to buy a home, try to avoid houses that are built on Vertisols because they often have weak foundations due to the constant shrinking and swelling that occurs within the surrounding soil.

Oxic horizon   *A diagnostic soil horizon in tropical and subtropical environments that is rich in iron oxides.*

Ultisols   *Mineral soils in subtropical environments that formed through laterization and thus are depleted of calcium and have an argillic horizon.*

Vertisols   *Soils that contain an abundance of expandable clay and thus swell and shrink during wet and dry cycles, respectively.*

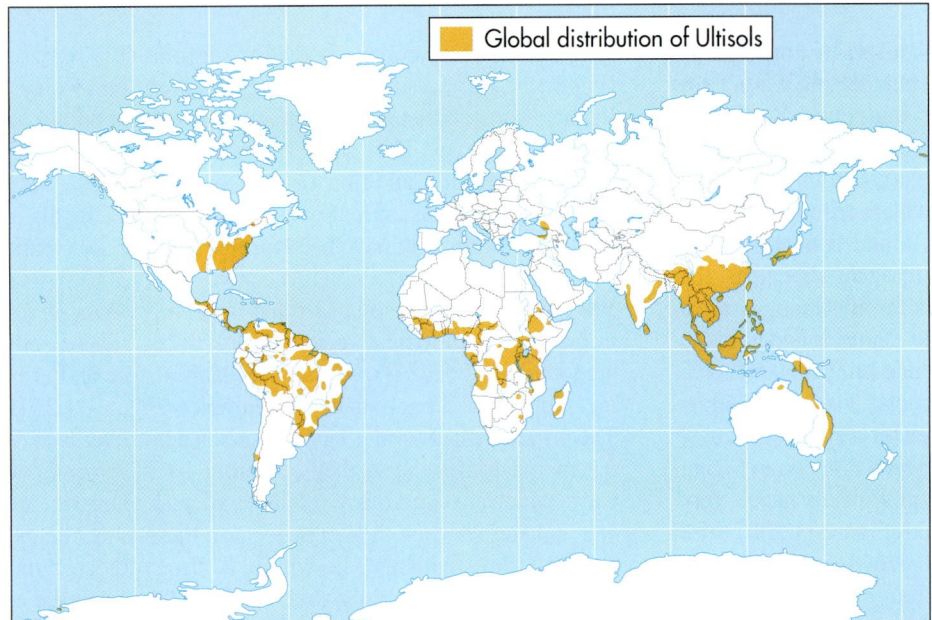

Global distribution of Ultisols

(a)

(b)

(c)

Figure 11.24 Ultisols. (a) Generalized map of Ultisols on Earth. These soils are found on 9% of the Earth's ice-free land area. (Credit: U.S. Department of Agriculture) **(b) A typical Ultisol landscape in central Brazil. Again, note the reddish color of this soil, which is due to extensive weathering in this subtropical environment. This region was once covered by forest and is now farmed. (c) A typical Ultisol from Arkansas. Note the E and Bt horizons, which indicate that abundant eluviation of clays has occurred.**

This process can be so strong that it can actually cause a basement wall to collapse, which costs thousands of dollars to repair. If you do purchase a home within Vertisol soils, make sure that some form of drainage system is present to move water away from the foundation.

Alfisols Alfisols form in cooler parts of the moist continental climate zones that lie poleward of the subtropical regions. These soils occur on about 10% of the Earth's ice-free land area over an extremely wide range of latitude, extending from as high as 60° N in North America to the

Alfisols *Soils generally found in seasonal midlatitude regions that formed through podzolization and have an alkaline argillic horizon.*

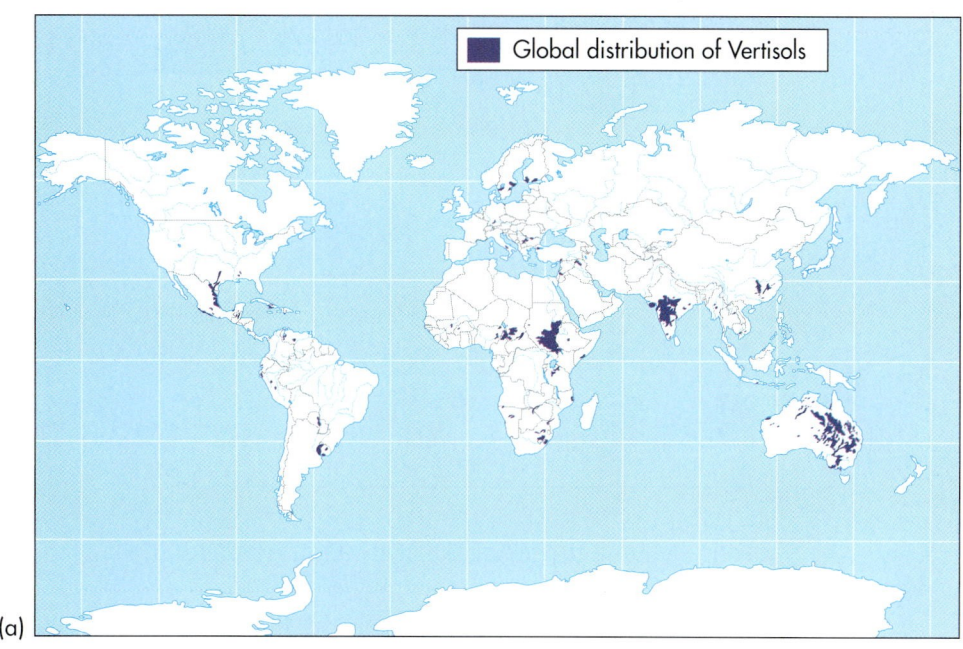
(a)

Global distribution of Vertisols

(c)

(b)

Ap

Ab

Bss

C

(d)

Figure 11.25 Vertisols. (a) Generalized map of Vertisols on Earth. These soils occur on about 2% of the Earth's ice-free land area. (Credit: U.S. Department of Agriculture) **(b) A Vertisol landscape in South Dakota. The small humps and ridges are known as "gilgai" and form due to frequent expansion and contraction of clays. (c) Ground cracks that resulted from contraction of expandable clays in a Vertisol. (d) A Vertisol in Puerto Rico that formed in clay-rich stream sediments. Note the distinctive Bss horizon.**

equatorial zone in South America and Africa (Figure 11.26a). Large areas of these soils occur in places like the American Midwest (Figure 11.26b), northern Canada, Europe, Asia, and the east coast of China. Smaller pockets are found in Central America, Australia, and Africa. Given the relatively high moisture of these regions, the natural landscape is largely covered with deciduous forest, but extensive areas of boreal forest occur as well.

Podzolization *A regional soil-forming process in cool, humid environments that results in the eluviation of iron, aluminum, and organic acids to form well-developed E and Bs horizons.*

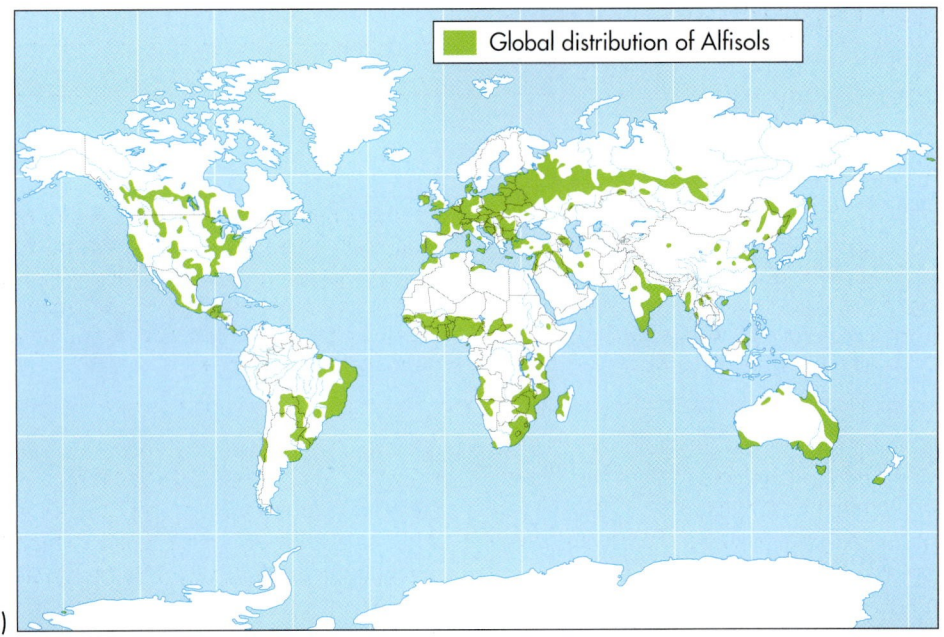

Global distribution of Alfisols

(a)

(b)

Ap

E

Bt

Bc

C

(c)

Alfisols form by a specific soil-forming process called **podzolization**. Podzolization is similar to laterization because both processes are associated with abundant leaching in their respective humid environments. The primary difference between the two processes, which is reflected in the character of the soils, is that organic matter decomposes slowly where podzolization occurs because the environment is cooler. Thus, a distinct O horizon forms in the regions where podzolization is the dominant pedogenic process. This O horizon consists mainly of deciduous leaves and coniferous needles. This litter layer is slightly acidic, which causes infiltrating soil water to become even more acidic. As a result, organic acids are incorporated into

Albic horizon *A diagnostic horizon of podzolization from which clay and free iron oxides have been removed, resulting in a light-colored E horizon.*

Figure 11.26 Alfisols. (a) Generalized map of Alfisols on Earth. These soils occur on about 10% of the Earth's ice-free land area. (Credit: U.S. Department of Agriculture) **(b) A typical Alfisol landscape in central Missouri. Prior to cultivation, the vegetation of this region was a mix of forest and tallgrass prairie. (c) A typical Alfisol from southern Michigan. The BC horizon contains elements of both the B and C horizons.**

AFRICAN CLIMATE, VEGETATION, AND SOILS

Now that you have examined the three tropical soils, you should review the relationship of climate, vegetation, and soil orders by going to the *GeoDiscoveries* website and selecting the module *African Climate, Vegetation, and Soils*. This module is a brief animation that illustrates how these variables are interrelated to one another to form a geographically distinct landscape. As you watch it, be sure to look at the spatial patterns that exist on the African continent. Once you complete the animation, be sure to answer the questions at the end of the module to test your understanding of this concept.

---

the upper part of the soil to form a distinct A horizon. From there, they are leached to the B horizon.

The leaching of organic acids causes translocation of the iron and aluminum sesquioxides, which are not moved where laterization occurs. In this environment, podzolization results in a very well-defined E horizon, which is ashy-gray in color and is thus called an **albic horizon** because it is so light. Below this horizon, a reddish B horizon forms where eluviated iron and aluminum collect. This horizon is called a **spodic horizon** because of the high sesquioxide content.

Alfisols are the soils that form in the warmer regions where podzolization occurs, specifically in the areas where the litter layer typically consists of deciduous leaves. Alfisols are similar to Ultisols in that they contain an argillic Bt horizon (Figure 11.26c) that is rich in clay. These soils are not as heavily weathered as Ultisols, however, which results in Alfisols generally being more fertile because they have higher concentrations of base ions. Nevertheless, Alfisols contain a distinct E horizon that is lighter, through extensive eluviation of minerals and clays, than the overlying A and underlying B horizons. Because the rate of decomposition is much less in this cooler environment, the A horizon of Alfisols is generally better developed and darker than in Ultisols.

Spodosols   Spodosols are acidic soils that are also associated with the podzolization process. These soils occur in about 3% of the ice-free land surfaces on Earth (Figure 11.27a). In contrast to Alfisols, Spodosols form in cool, humid regions poleward of places dominated by Alfisols. As a result, the vegetation in these regions is typically coniferous and hardwood forest (Figure 11.27b) and the litter layer is thus a mix of needles and leaves. As with Alfisols, this litter layer is acidic, which lowers the pH of percolating water sufficiently to mobilize organic acids and sesquioxides. In places where the parent material is sandy, soil formation occurs more rapidly because water percolates quickly through the solum.

Spodosols have a very distinct sequence of soil horizons (Figure 11.27c). The uppermost part of the soil usually contains a thin O horizon (needleleaf litter) over a thin A horizon, which is very dark due to the slow decomposition of organic matter. In well-developed Spodosols, the E horizon is very prominent because iron, aluminum, and base ions have all been eluviated, resulting in a distinct white or light-gray albic horizon. Directly beneath the E horizon is a particular type of spodic horizon, which is designated as a Bs horizon. This horizon is dark brown or reddish in color because it contains high amounts of illuviated organic matter, aluminum, and iron. If sufficient time or eluviation has transpired, the Bs horizon will actually become cemented with eluviated organic material called *ortstein*. In these instances, the B horizon will have a Bhs designation to reflect the presence of both humic acids and sesquioxides.

Histosols   Although Spodosols are the dominant soil in the cool and humid environments on Earth, another soil order is also specifically associated with these regions. This soil order, **Histosols**, is also commonly called *peat* or *muck*. These soils are found on about 1% of the Earth's ice-free land surfaces (Figure 11.28a).

As far as soils are concerned, Histosols are unique because they consist mostly of organic material, rather than mineral matter. Thus, the major horizons of Histosols are O horizons. These soils form in isolated shallow lakes, ponds, wetlands, and bogs (Figure 11.28b) where the decomposition of organic matter is very slow as

Spodic horizon *A mineral soil horizon characterized by the illuvial accumulation of aluminum, iron, and organic carbon.*

Spodosols *Soils in cool, humid regions that form through podzolization and contain a spodic horizon enriched in eluviated iron, aluminum, and organic carbon.*

Histosols *Organic soils that form in cool, wet environments where organic carbon decomposes very slowly.*

Figure 11.27 Spodosols. (a) Generalized map of Spodosols on Earth. These soils occur on about 3% of the Earth's ice-free land area. (Credit: U.S. Department of Agriculture) (b) A typical Spodosol landscape in the Great Lakes region. (c) A well-developed Spodosol in the Great Lakes region. Note the grayish E horizon and dark Bhs horizon. (d) Dwarf conifers in the Pygmy Forest along the coast of northern California. These trees, which grow to towering heights in other places, are stunted in this environment because the Spodosols here are so acidic and thus have low fertility.

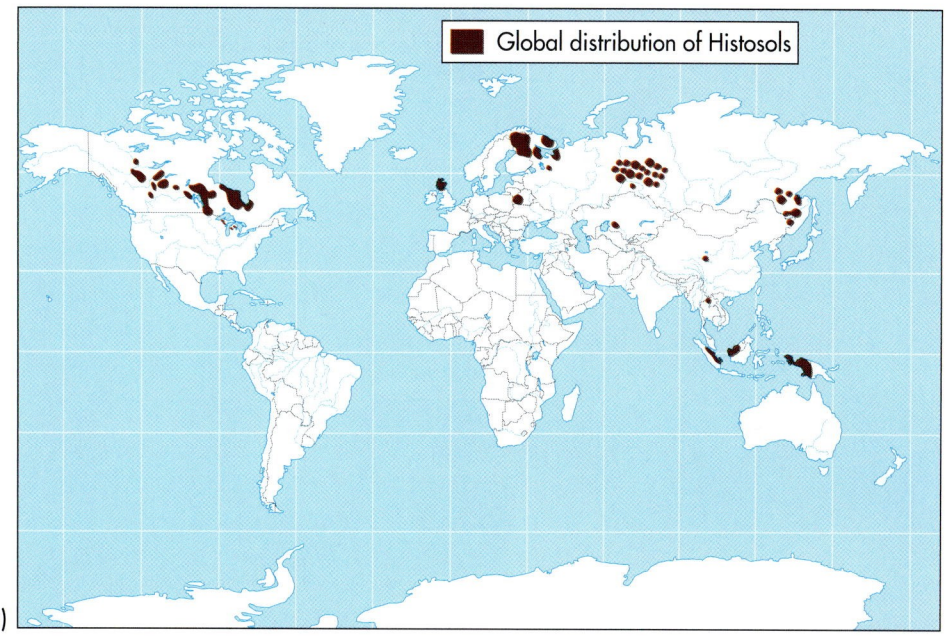
(a)

(b)

(c)

Figure 11.28 Histosols. (a) Generalized map of Histosols on Earth. These soils occur on about 1% of the Earth's ice-free land area. (Credit: U.S. Department of Agriculture) **(b) A typical Histosol landscape in Michigan forms in bogs like this where organic matter decomposes slowly. (c) A Histosol in Michigan. Note the O horizons, which reflect the high amount of organic matter. The Lma horizon is a calcium-rich clay called** *marl.*

a result of the cool environmental conditions. Given that these soils are largely organic, they are very dark in color (Figure 11.28c). Because these soils are very fertile they are actively cultivated for crops, such as mint or cranberries, in places where they are easily accessible. These soils are also used as mulch and a low-grade fuel source.

Gelisols In 1998 the soil taxonomy system was revised to include a new soil order: **Gelisols.** Gelisols develop in very cold climates at high latitudes or high mountain eleva-

tions and are estimated to occupy about 9% of the Earth's ice-free land area, including most of the state of Alaska (Figure 11.29a). Thus, these soils are frozen most of the year, resulting in permafrost within 2 meters of the surface. Because these soils are very cold, decomposition of

Gelisols *Soils in subarctic and arctic environments that contain permafrost within 2 meters of the surface.*

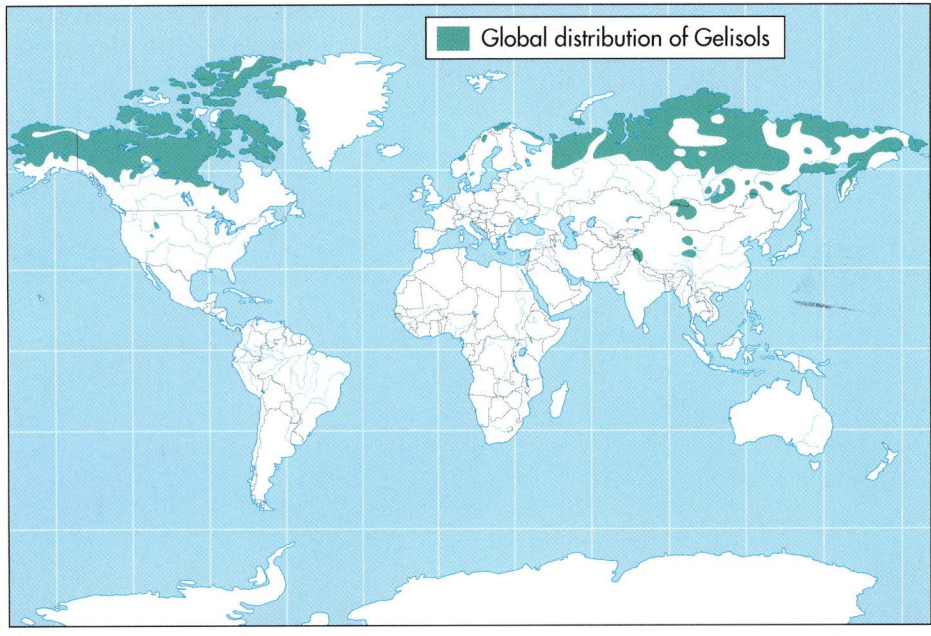

Figure 11.29 Gelisols. (a) Generalized map of Gelisols on Earth. These soils are found in about 9% of the Earth's ice-free land area. (Credit: U.S. Department of Agriculture) **(b). A typical Gelisol landscape in Alaska.**

The distinct polygons on the ground, called *patterned ground*, form as a result of repeated freezing and thawing. **(c) A Gelisol in Alaska. The Cf horizon is permafrost.**

organic matter proceeds even more slowly than where Histosols occur. As a result, Gelisols contain large amounts of organic carbon, some of which is in the form of muck or peat at the surface, and are thus dark in color (Figure 11.29c). Given that Gelisols regularly freeze and thaw, they are actively mixed and are frequently associated with deformed topography (Figure 11.29b).

Mollisols   Moving away from the humid zones, we next look at soils that form in more arid regions, beginning with **Mollisols.** These soils occur over about 7% of the Earth's soil network, including most of the American Midwest and Great Plains, a broad belt across Eurasia (including the Russian steppe), and a significant portion

of South America referred to as the Pampas (Figure 11.30a). Mollisols are closely associated with the more humid parts of the dry midlatitude climate and associated grassland vegetation (Figure 11.30b). They are the most extensive soil order in the U.S. and are found on 21.5% of the land area (Figure 11.30a). In the Great Plains, the presence of Mollisols is directly linked to the fact that the region lies within the Rocky Mountain rainshadow.

Mollisols   *Soils that form through calcification and have a mollic epidedon that overlies mineral matter that is more than 50% saturated with base ions.*

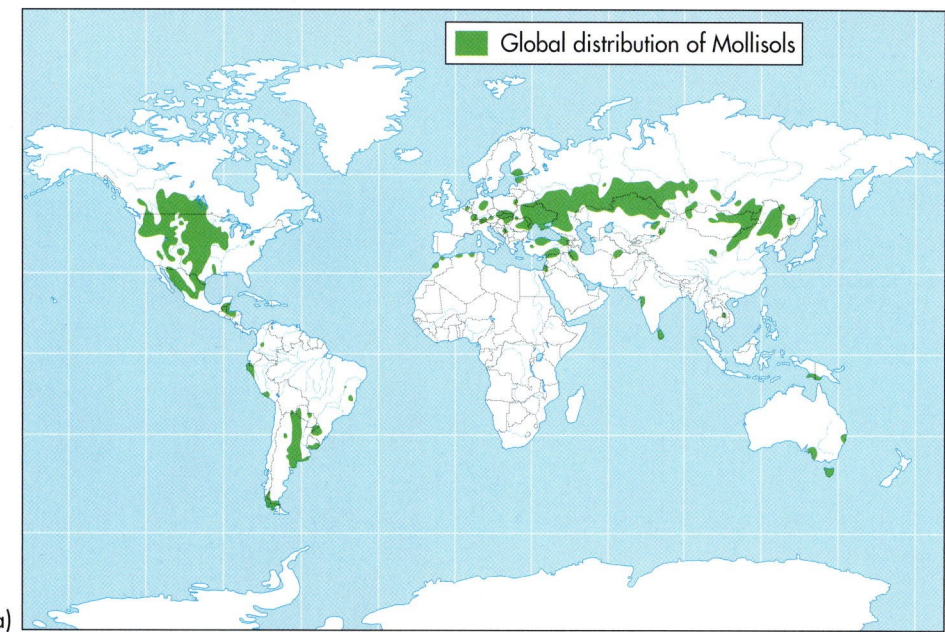

(a)

(b)

A

Bt

Btk

Bk

1

2

3

(c)

Figure 11.30 Mollisols. (a) Generalized map of Mollisols on Earth. These soils are found on about 7% of the Earth's ice-free land area. (Credit: U.S. Department of Agriculture) **(b) A typical Mollisol landscape in the central Great Plains. Note the tallgrass prairie. (c) A Mollisol in North Dakota. The Btk horizon reflects the presence of both eluviated clay and calcium. The Bk horizon contains eluviated calcium, which you can see as white nodules.**

Mollisols form in association with a specific pedogenic process called **calcification.** As with podzolization, vegetation plays an important role in the calcification process. Where podzolization occurs, deciduous and coniferous trees produce a needle-leaf litter layer that depresses the pH sufficiently to mobilize iron and aluminum sesquioxides. In grassland environments, where calcification takes place, the grasses play an important role in soil formation because they are base cyclers. The first part of the cycle occurs when grasses absorb bases as nutrients during the growing season.

Calcification *A regional soil-forming process in which calcium carbonate is cycled within the soil.*

Which one of the regional pedogenic processes is contributing to the development of this soil? What soil horizons are present? How can you tell? How did each of them form? Here's a hint: This soil is not fully developed, but the evolving horizons can be seen.

When the grass dies, however, the bases contained within the grass are cycled back into the soil as it decomposes.

Here is where the role of climate becomes important. If this base cycling somehow occurred in a humid environment, the bases would be leached completely out of the soil because they are so easily mobilized. Given the general moisture deficit that occurs in places like the Great Plains or the Russian Steppe, however, the bases are not completely leached. Instead, they are merely translocated to the zone of illuviation, or B horizon, just like iron and aluminum where podzolization occurs. Given that this part of the soil is rich in calcium, it is called a **calcic horizon**. This process can ultimately result in the complete cementation (solidification) of the B horizon by calcium carbonate, resulting in material called *caliche*.

Mollisols are also distinctive because they have thick, dark A horizons with high humus content (Figure 11.30c). This horizon forms in part because grasses decompose slowly in the relatively dry environment, therefore slowly and consistently adding humus to the soil.

Grass also has a dense underground root system that contributes to the high content of organic matter and calcium minerals in these soils. Soil organisms, like earthworms, ants, and moles, mix the soils, carrying organic matter deeper below the surface. Given the association with the calcification process, the B horizons of Mollisols are rich in calcium; in fact, the B horizon of a Mollisol is saturated at least to a level of 50% by bases such as calcium ions. It also often contains an abundance of translocated clays. As a result, the B horizon of a well-developed Mollisol is designated as a Btk ("k" meaning abundant calcium carbonate) horizon.

The combination of high amounts of organic matter and minerals make Mollisols the most fertile and productive soils on Earth because they contain abundant nutrients that plants need. In the U.S., Mollisols support a wide variety of agricultural crops, including corn, wheat, and soybeans. The next loaf of bread you buy, in fact, will most likely contain wheat grown in Mollisols.

Calcic horizon *A diagnostic soil horizon of calcification that is enriched in illuviated calcium carbonate.*

Aridisols *Mineral soils that form in arid environments and thus are poorly developed.*

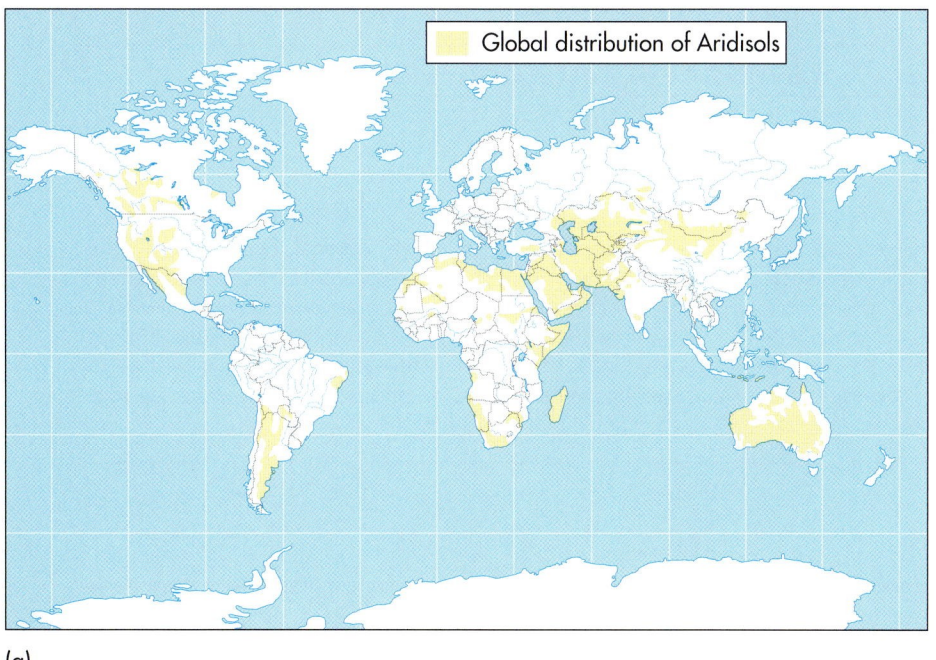

(a)

(c)

(b)

(d)

Figure 11.31 Aridisols. (a) Generalized map of Aridisols on Earth. These soils occur on about 12% of the Earth's ice-free land area. (Credit: U.S. Department of Agriculture) (b) A typical Aridisol landscape in the southwestern United States. (c) An Aridisol with a salic horizon in Nevada. Note the salts (white) at ground level. Three different parent materials were identified at this site, as indicated by the 2C2 and 3C3 horizons. This designation means that three distinct units of sediment have been identified here. (d) An Aridisol with a calcic horizon. Note the Bkm horizon, which reflects the dense concentration of eluviated carbonate.

REGIONAL PEDOGENIC PROCESSES

At this point we have completed our discussion of soils that generally form in association with regional environmental variables and pedogenic processes. These processes are particularly well suited for animation to more completely illustrate how they work. Go to the *GeoDiscoveries* website and select the module **Regional Pedogenic Processes**. This module illustrates how the soil-forming processes lateriza-tion, podzolization, calcification, and salinization influence the development of soils at a regional level. As you work your way through each of these processes, keep in mind that they reflect regional environmental variables and provide a good introduction to the global distribution of soils and soil classification. When you complete the animation, be sure to answer the questions at the end of the module to test your understanding of this concept.

Aridisols   As the name implies, Aridisols are soils that form in arid environments. These soils are found over about 12% of the Earth's ice-free land area, including the Sahara Desert and significant parts of Australia, South America, the Middle East, Asia, and the southwestern U.S. (Figure 11.31a). These dry environments have very sparse vegetation (Figure 11.31b), which means that Aridisols have poorly developed A horizons, if they have them at all (Figure 11.31c).

Aridisols can form either through the process of calcification or another process called **salinization**, which entails the movement of sodium (Na) in the soil. In the other pedogenic processes, such as laterization and podzolization, sodium is easily and completely leached from soils because it is one of the more soluble minerals and water is sufficiently abundant to move it out of the solum. In semi-arid to arid environments, however, sodium is not completely leached from soils and instead is moved only part way down the soil profile when infrequent rains do fall. When the soil dries out during the extended dry periods that occur in this environment, the partially leached minerals rise back up through the profile in a process called "wicking."

This wicking process is most active in enclosed basins where pools of water collect when it does rain. Sodium in these low spots can be leached many centimeters when the water percolates through the profile following a hard rain. When the site dries out, however, the salt rises back up and forms a salty crust at the surface called an Az horizon. If this crust reaches a thickness of 15 cm (6 in) and contains at least 20 g/kg of salt, it is called a **salic horizon**. These salts are highly toxic to most plants. In regions where calcification is the dominant process associated with Aridisols, the soils have thick Bk, or calcic, horizons that can become completely cemented with time. If they do, the horizon is designated as a Bkm horizon (Figure 11.31d).

We have now discussed each of the three major soil orders that occur within the midcontinental United States: Alfisols, Mollisols, and Aridisols. As a result, this juncture marks a good point to examine how soils vary along a study transect that extends from Ohio to central Colorado (Figure 11.32). Which side of this transect has the most moisture? Which side is drier? Do you know why? You should know that the eastern side of this transect is more humid due to the influx of mT air from the Gulf of Mexico. As you proceed westward, however, the climate generally becomes drier due to the influence of the Rocky Mountain rainshadow. The response of vegetation is that trees populate the eastern side of the transect. As you move westward, the vegetation shifts to tall grass and then to short grass.

How do soils respond along this transect? The primary change is that more leaching takes place on the eastern side of this transect than on the west. This geographic pattern should make sense to you because there is more moisture in Ohio than in eastern Colorado. As you can see in Figure 11.32, this distribution of soil water causes a distinct E horizon in Ohio, which fades away as you move farther west. The thickest A horizons are in the tallgrass region of the Mollisol belt because the biomass is so high, and thus decomposition and leaching are less than to the east. In this zone, the depth of illuviated carbonate (lime, $CaCO_3$) is the greatest and the B horizon is the thickest. As you move deeper into the rainshadow, in the western part of the Mollisol belt, the A horizon thins significantly and the depth of the B horizon becomes shallower. This pattern culminates within the zone of Aridisols with a very thin, or nonexistent A horizon and a very shallow, carbonate-rich

---

Salinization   *A regional soil-forming process in which soluble salts are cycled within the soil.*

Salic horizon   *The diagnostic horizon of salinization that forms due to the recrystallization of secondary salts.*

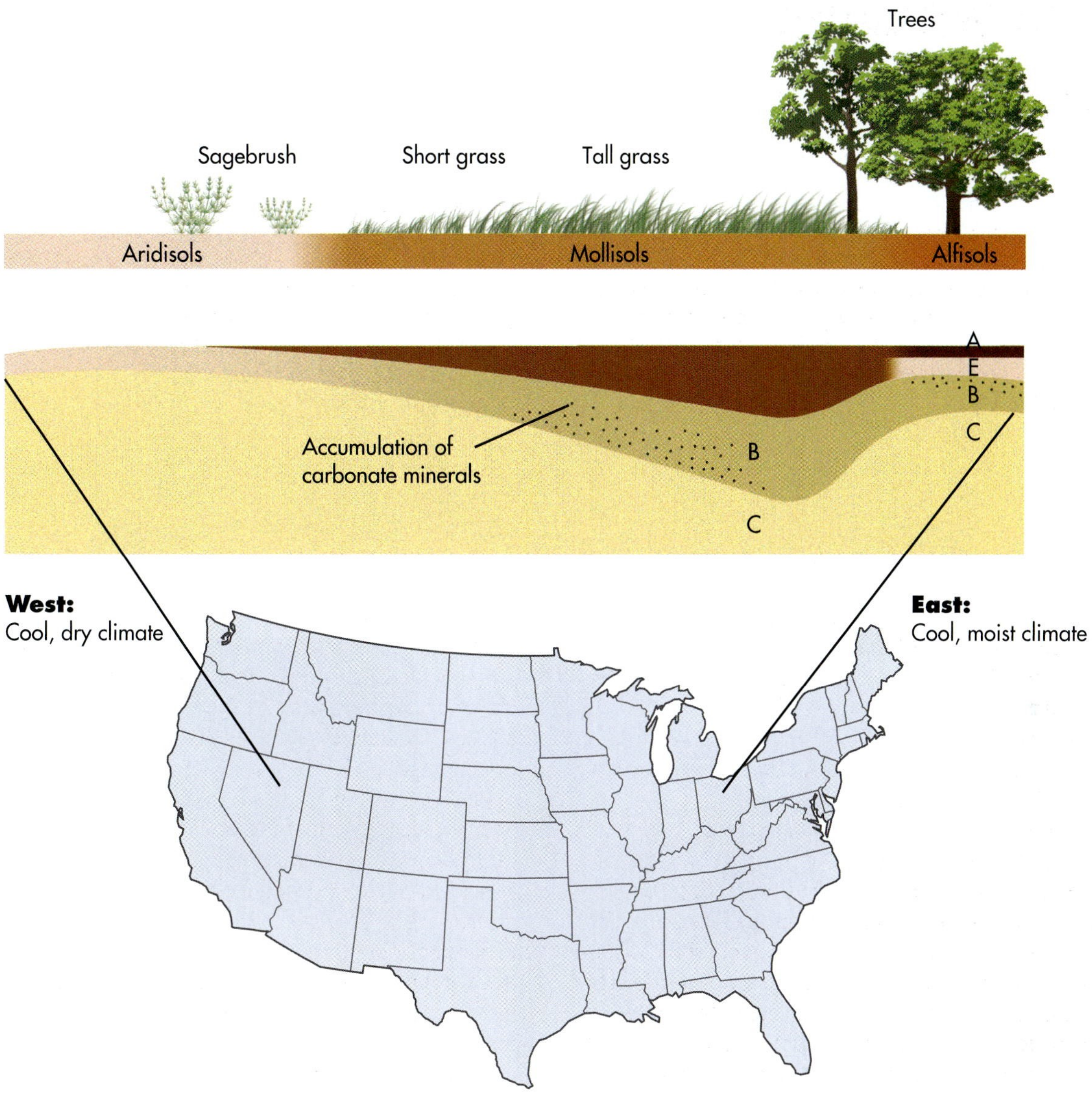

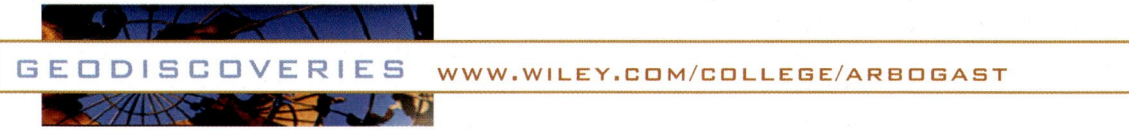

Figure 11.32 Midcontinent soil transect in the United States. **The change in soil characteristics is reflected by soil order, from central Nevada in the west to Ohio in the east.** (Credit: Modified from C. E. Millar, L. M. Turk, and H. D. Foth. *Fundamentals of Soil Science*. New York: John Wiley & Sons, 1990.)

GEODISCOVERIES WWW.WILEY.COM/COLLEGE/ARBOGAST

North American Climate, Vegetation, and Soils

At this point, you should have a basic understanding of mid-latitude soil orders and where they occur. In this context, it's a good time to review the close relationship between climate, vegetation, and soils in these regions. To do so, go to the *GeoDiscoveries* website and select the module ***North American Climate, Vegetation, and Soils***. This module contains a brief animation that illustrates how these variables are interrelated with one another to form a geographically distinct landscape. As you watch the animation, be sure to pay attention to the geographic patterns and their similarities. Once you complete the animation, be sure to answer the questions at the end of the module to test your understanding of this concept.

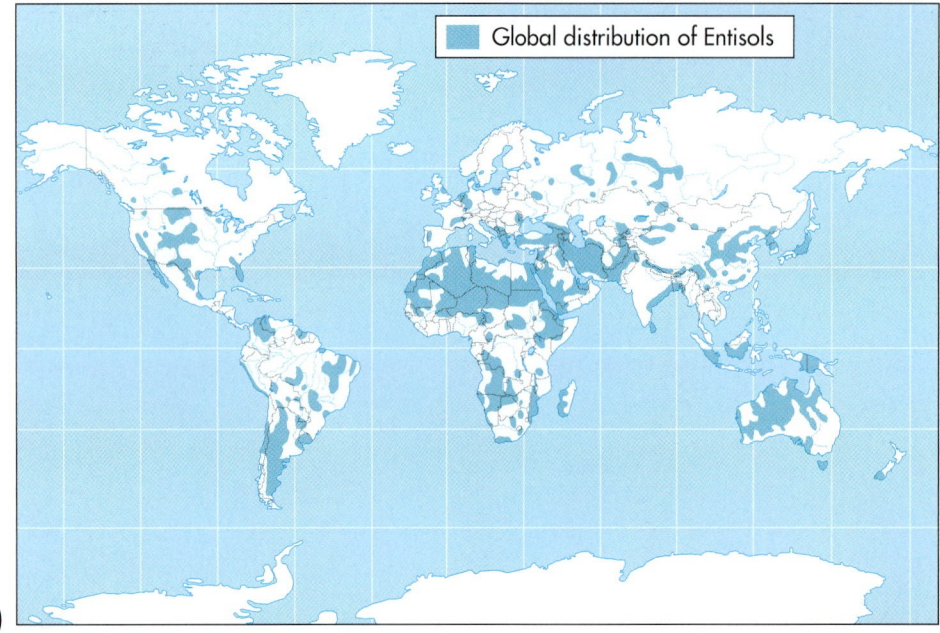

Global distribution of Entisols

(a)

(b)

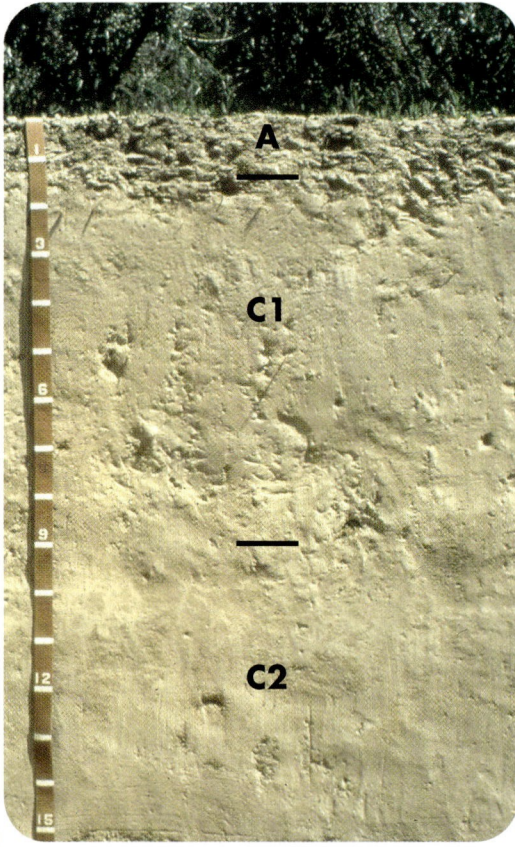

A

C1

C2

(c)

B horizon. This discussion is an excellent example of how many geographic variables come together to produce a distinctive pattern—in this case, regarding soils.

Entisols   Finally, we can examine the soils that are not related to any specific environmental variable or pedogenic process. In other words, these soils can be found in many different places on Earth and are a product of specific soil-forming factors. The most widespread of these random soils are **Entisols**, which means *young soils*. These soils occur on about 16% of the Earth's ice-free land area (Figure 11.33a).

Entisols   *Soils that are very weakly developed and thus have no distinct horizonation.*

Figure 11.33 Entisols. (a) Generalized map of Entisols on Earth. These soils occur on about 16% of the Earth's ice-free land area. (Credit: U.S. Department of Agriculture) **(b) An Entisol landscape in Michigan. The vegetation is growing on a recent deposit of wind-blown sand along Lake Michigan. (c) An Entisol in southern Idaho. Note the thin A horizon that directly overlies a C horizon.**

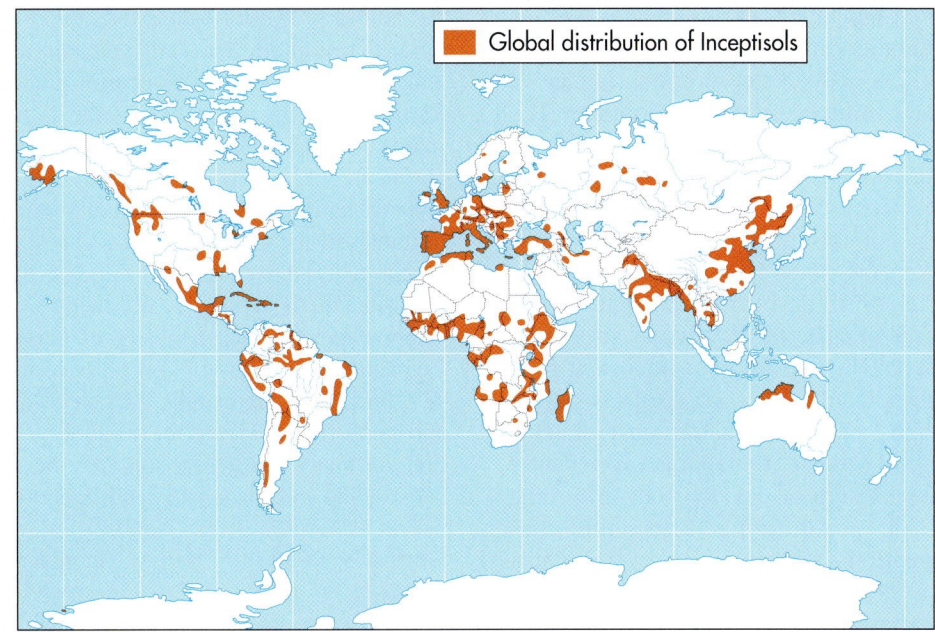

(a)

(b)

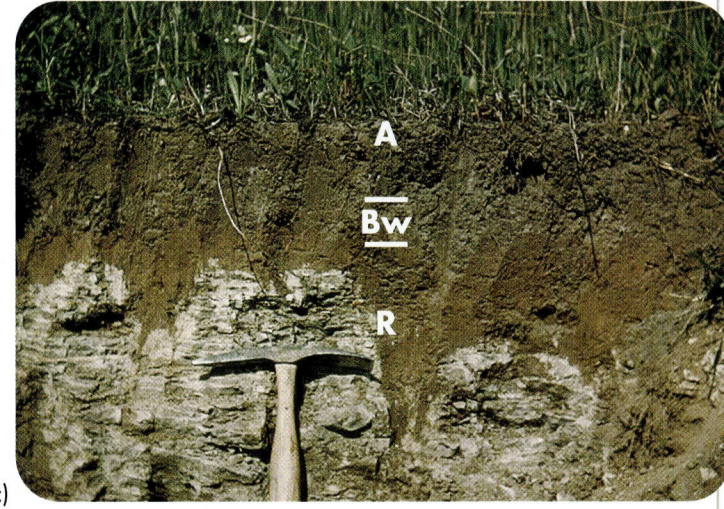

(c)

A

Bw

R

Figure 11.34 Inceptisols. (a) Generalized map of Inceptisols on Earth. (Credit: U.S. Department of Agriculture) These soils occur on about 10% of the Earth's ice-free land area. (b) An Inceptisol landscape. Soils on these steep slopes within the Appalachian Mountains have weakly developed horizons because they are easily eroded. (c) An Inceptisol in the Great Lakes region. This soil formed in glacial sediments that overlie bedrock (note the R horizon).

Although Entisols are soils in the sense that they can support plants, they are the least developed soil order because they have no diagnostic horizons such as an E or B horizon (Figue 11.33c). They may contain a thin A horizon. This overall lack of development can occur for several reasons: (1) the climate is extremely dry, making the processes of soil additions and translocations very weak; (2) the parent material is quartz sand, which is a resistant deposit that does not easily yield nutrients through weathering; or (3) the parent material has recently been transported, resulting in little time for a soil to develop (Figure 11.33b). Given that this range of soil-forming factors can occur just about anywhere, Entisols appear at all latitudes.

Inceptisols　Inceptisols are soils with weakly developed horizons. These soils are found on about 10% of the Earth's ice-free land area (Figure 11.34a), and

Inceptisols　*Soils that have one or more weakly developed horizons due to some alteration and removal of soluble minerals.*

can occur just about anywhere, even in places like the Appalachian Mountains (Figure 11.34b). It is useful to think of Inceptisols as being a step in formation above Entisols, which have limited horizon development. The primary soil-forming factor associated with Inceptisols is time. On the one hand, these soils are better developed than Entisols because they have had more time to develop. Thus, Inceptisols may have a Bw horizon (Figure 11.34c). On the other hand, insufficient time has elapsed for the soil to develop horizons that meet the specific criteria of a better developed soil order such as Mollisols or Ultisols. Another important soil-forming factor associated with Inceptisols is parent material because most of these soils form in sediments that have been recently transported, but not so recently that the soil shows no sign of development. As in the case of Entisols, the soil-forming factors that contribute to the formation of Inceptisols can occur just about anywhere.

Andisols   Andisols are soils that form in parent material that is more than half volcanic ash. Like Entisols and Inceptisols, Andisols occur at random places on the Earth, occupying about 1% of the Earth's ice-free land mass (Figure 11.35a). Most of Hawaii, for example, lies within areas classified as Andisols (Figure 11.35b). Given their volcanic origin, it is accurate to say that the most important soil-forming factor associated with Andisols is parent material because fresh deposits of volcanic ash are deposited on the landscape following a volcanic eruption. These deposits consist mostly of glass-like shards that frequently contain a high proportion of carbon. As a result, these soils are usually dark in color (Figure 11.35c) and quite fertile. Given the frequency of volcanic eruptions in many areas, Andisols may be poorly developed because the soils have had insufficient time to develop.

Andisols   *Soils formed in parent material that is at least 50% volcanic ash.*

KEY CONCEPTS TO REMEMBER ABOUT SOIL SCIENCE AND CLASSIFICATION

1. Soil science is a distinct subdiscipline of geography that focuses on the chemical, physical, and biological properties of soils.

2. One of the primary ways that soil scientists study and compare soils is through soil classification. Soils are classified (in the U.S.) through a hierarchical system called "soil taxonomy." The most general classification in soil taxonomy is the soil order, of which there are twelve. These orders are distinguished on the basis of measured physical properties and horizons.

3. Oxisols and Ultisols form as a result of laterization, which involves the intense weathering of soils in tropical and subtropical environments. These soils are typically reddish in color because minerals besides oxidized iron and aluminum are leached.

4. Alfisols and Spodosols form due to podzolization, which involves extensive translocation of organic acids and sesquioxides in cool, humid environments. Alfisols and Spodosols have distinctive E horizons that are light gray in color. Spodosols also have a distinctive Bs horizon that forms due to the illuviation of iron and aluminum.

5. Histosols are soils that are entirely organic because they form in cool, humid places where decomposition of organic matter is very slow.

6. Mollisols and Aridisols form in drier environments. Both can form because of calcification, which occurs where base ions are partially translocated in the soil. Aridisols can also form due to salinization, which involves the movement and surface recrystallization of sodium.

7. Andisols, Entisols, Gelisols, Inceptisols, and Vertisols are directly related to local environmental factors. Andisols form in volcanic environments, whereas Gelisols occur in subarctic regions where the soil is frozen most of the year. Entisols and Inceptisols can form anywhere and represent soils that are very weakly to slightly better developed, respectively. Vertisols are soils that contain abundant expandable clays that swell when wet and shrink when dry.

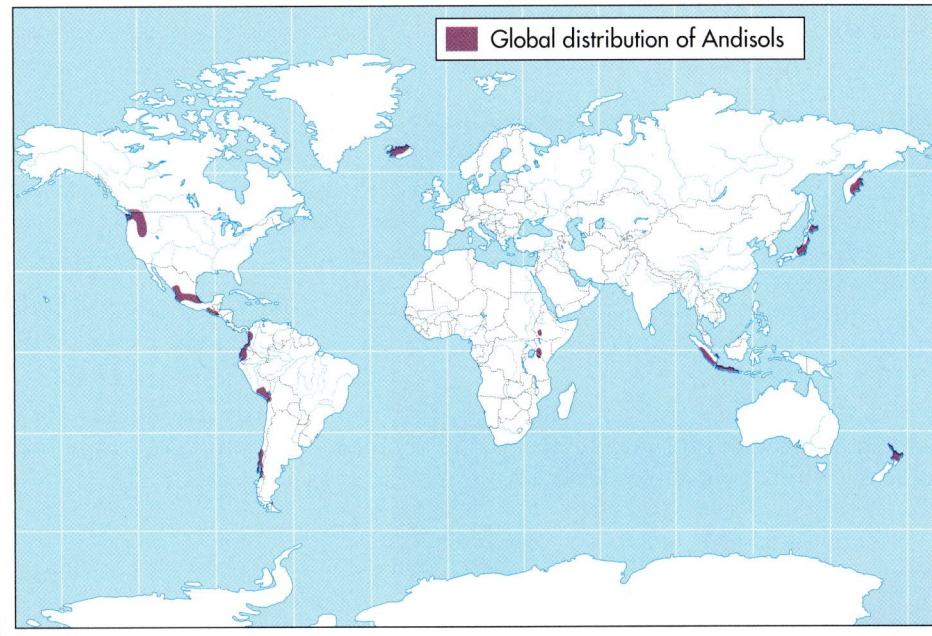

(a)

(b)

A & BW

C

(c)

Figure 11.35 Andisols. (a) Generalized map of Andisols on Earth. These soils are associated with volcanic regions and occur on about 1% of the Earth's ice-free land area. (Credit: U.S. Department of Agriculture) (b) Typical Andisol landscape in Hawaii. The Hawaiian Islands are covered in most places with volcanic sediments. (c) Andisol in Hawaii. Note the banding in this soil, which reflects periodic deposition of new volcanic parent material. The A&Bw horizon occurs because this part of the soil has elements of both an A and a Bw horizon.

Global distribution of Andisols

The Big Picture

Up to this point in this text, we have focused on geographic processes and patterns that are distinctly interrelated to one another. We began with a discussion of Earth–Sun geometry, which naturally led to a section on radiation and climate. These chapters provided the context for the global distribution and character of vegetation and soil. Now that we have concluded the chapter on soils, we turn our attention to processes associated with the solid Earth, or lithosphere. In other words, we're going to begin investigating the processes and variables that influence the shape of the landscape.

A good example of this landscape building is Teton Mountain range, which is located in northwestern Wyoming. This mountain range has evolved through the forces of internal processes within the solid Earth that have deformed rock on the surface. Although we have examined some relationships between the solid Earth and the atmosphere and oceans in the earlier chapters of the book, these relationships are incidental and even random in some instances. We begin exploring the lithosphere in more detail in the next chapter, which focuses on the nature of rocks on Earth, the configuration and character of the Earth's interior structure, and the topography (or geomorphology) of the Earth.

Summary of Key Concepts

1. Soil consists of the outermost layer of the Earth and forms through the complex interaction of additions, transformations, translocations, and losses. These processes occur in various combinations that depend on the five soil-forming factors: (a) climate, (b) organisms, (c) relief, (d) parent material, and (e) time.

2. Soils have a variety of distinctive characteristics, including color, texture, structure, pH, and cation exchange capacity, that can be measured and compared from one soil to another.

3. Soils are organized into horizons that form by distinctive processes. The O horizon is the uppermost horizon and consists of freshly added organic matter, such as leaves. The A horizon contains decomposed organic matter called humus. The E horizon forms through the eluviation of minerals, which recrystallize and collect through the process of illuviation in the underlying B horizon. The C horizon is unaltered parent material, which, in bedrock landscapes, is known as regolith.

4. Soils in the U.S. are classified on the basis of their genetic properties in a scheme called *soil taxonomy*. The highest level of this classification system is the *soil order*. Twelve *soil orders* are recognized. They are: Oxisols, Ultisols, Alfisols. Spodosols, Histosols, Vertisols, Gelisols, Mollisols, Aridisols, Entisols, Inceptisols, and Andisols.

Check Your Understanding

1. Why is regolith logically between bedrock and soil?

2. Describe the soil-forming factors and provide an example of how all of them, working together, can produce a soil.

3. Which texture will allow water to drain more rapidly: sand or clay? Why?

4. Which parent material will have a higher cation exchange capacity—one that is sandy, or one that is rich in clay? Explain your answer.

5. Which pair of horizons is created through the process of translocation? In what way does this evolution occur?

6. It is conceivable that a soil could be classified as an Entisol even though it is very "old." In what environment—tropical, subtropical, or arid— would this kind of weak development most likely occur? Why?

7. What is a soil profile and why is it useful for determining the characteristics of a soil?

8. How are Spodosols and Oxisols similar? How are they different?

9. Why is grass an important part of the calcification pedogenic model?

10. In which parent material would you most likely find an Inceptisol: a sand dune that is 500 years old, or one that is 1000 years old? Why?

ANSWERS TO VISUAL CONCEPT CHECKS

Visual Concept Check 11.1

The answer is *b*. The photograph shows an example of soil additions. When leaves fall to the ground, they slowly decompose and become incorporated into the soil.

Visual Concept Check 11.2

The answer is *d*; this soil is classified as a sandy clay.

Visual Concept Check 11.3

This soil is forming by the podzolization pedogenic process, which results in the eluviation of iron, aluminum, and organic acids. The soil contains developing A, E, and Bs horizons. You can tell the A horizon because it is dark, which occurs through the addition of humus. The E horizon is whitish and forms through the eluviation of iron and aluminum. The Bs horizon is visible beneath the E horizon as the light orange zone. This color reflects the illuviation of iron.

CHAPTER TWELVE

EARTH'S INTERNAL STRUCTURE, ROCK CYCLE, AND GEOLOGIC TIME

In the preceding chapter, we discussed soils, the very outer layer of the Earth. Now we turn our attention to the basic composition of the Earth and the deep layers that lie within it. Geographers are especially interested in the patterns and processes associated with the topmost layer of the Earth—the crust—because this is where modern landforms occur and people live. In this context, this chapter introduces the field of geomorphology, which is the study of landforms on Earth and how they

The Grand Canyon is one of the best known landscapes on Earth. In addition to being a beautiful place, the canyon is noteworthy because it contains rocks that span much of the Earth's history. This chapter focuses on the geologic history of the Earth, the evolution and composition of rocks, and the character of continents and ocean basins.

form. We will discuss these concepts within the framework of geologic time, a span of time so large that people often have difficulty understanding it and finding it relevant to their lives.

We begin our study by looking at the character of the Earth's inner structure. These inner layers are important because they drive the movements of the surface crust. After this discussion, we'll investigate the various kinds of rocks on Earth. In the last part of the chapter, we'll begin to examine Earth's landforms by looking at large-scale features—specifically the continents and ocean basins.

Earth's Inner Structure

Have you ever wondered what it's like inside the Earth? Science fiction writers have pondered this question for a long time; the most famous example is the 1864 book *Journey to the Center of the Earth* by Jules Verne. In this story, a professor leads a party of explorers to the center of the Earth through a volcano in Iceland. During their trip, the group travels through a maze of tunnels and encounters a wide variety of prehistoric animals and life-threatening hazards. Ultimately, however, they discover an underground world full of wild plants and animals to sustain them.

Although such an adventure is pure fiction, a lot of geologists would like to make the trip. However, it's not possible. In fact, the only way to actually see what the interior of the Earth looks like is to drill into it with a large machine and collect samples. Unfortunately, the deepest humans have ever been able to drill is about 12 km (7.5 mi) into the Earth. The distance to the center of the Earth is thousands of kilometers, however, which means that geologists must rely on indirect evidence to construct models about the Earth's interior.

Much of what we know about the Earth's structure is based upon how **seismic waves** travel through the Earth after earthquakes occur. Cooler areas within the Earth transmit these waves at higher speeds than do hotter zones because cooler regions are more rigid. Areas that are denser absorb seismic waves, whereas other variations in density cause seismic waves to be reflected or bent.

These differences have allowed scientists to determine the nature of Earth's interior and map its basic structure, somewhat like the explorers in the Jules Verne novel. Let's now discuss what we know about the Earth's interior structure.

The Major Layers

Looking at the inner Earth in cross section, you can see several major layers (Figure 12.1). Because it's difficult to imagine the depths and distances associated with these layers, let's compare them to the distances between places on a map of North America (Figure 12.2). For example, the distance to the center of the Earth's core—in other words, to the center of the Earth—is about 6370 km (3963 mi), which happens to be about the distance from Anchorage, Alaska, to Miami, Florida.

Earth's innermost layer is the **inner core.** This part of the Earth has a radius of about 1220 km (760 mi), which is about the distance from Anchorage to Prince Rupert, British Columbia (Figure 12.2). The inner core is composed mostly of solid iron, along with a little bit of nickel. The pressure in this part of the Earth is enormous, about 3 million times greater than the air pressure you feel at the surface, due to the weight of the overlying material of the rest of the Earth. This pressure explains why the inner core is solid even though the temperature there is well above the melting temperature of iron. In fact, geologists estimate that the temperature of the inner core ranges from about 3200° to 5200° C (5800° to 9400° F).

Seismic waves *Vibrations that travel through the Earth when stress is released in an earthquake.*

Inner core *The inner part of the Earth's core. This area is about 1220 km (760 mi) thick and consists of solid iron and nickel.*

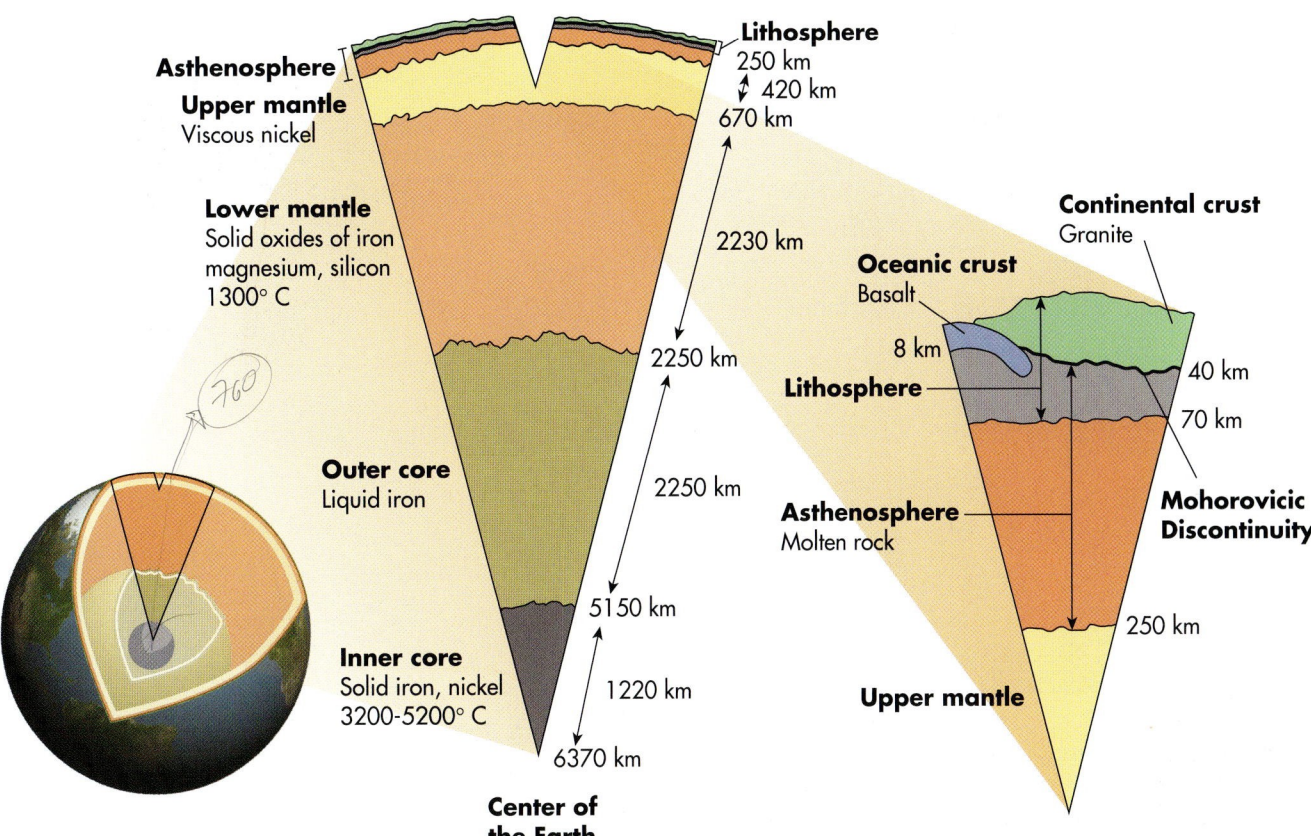

Figure 12.1 The interior of the Earth. The Earth's interior structure has several major layers, each with a distinct mineral composition and density. Note that the middle image is a blown-up diagram of the pie-shaped cross section in the left-hand diagram. The right-hand image, in turn, is a blown-up diagram of the pie-shaped cross section in the upper part of the middle image.

Surrounding the solid inner core is the **liquid outer core** (Figure 12.1). The top of this layer lies about 2900 km (1800 mi) below sea level. The liquid outer core is about 2250 km (1398 mi) thick, which is about the same distance as from Prince Rupert, British Columbia, to Laramie, Wyoming (Figure 12.2). In other words, it would take you 3 or 4 days to drive that distance. The outer core is composed of the same material as the inner core, except the iron is molten; that is, it's liquid. This variation occurs because the temperatures of the outer core approach those of the inner core, but there is less pressure. A key function of the outer core is that it generates at least 90% of the Earth's magnetic field and the resulting magnetosphere that protects the Earth from the solar wind. The magnetic field may exist because the Earth's inner core rotates 19 km (12 mi) per year faster than the rest of the planet. This rotational difference is

thought to cause circulation patterns in the outer core that generate electrical currents. These currents, in turn, may generate the magnetic field. Scientists are not certain of the exact processes that lead to the Earth's magnetic field, but circulation of the outer core is most likely involved.

As we continue to work our way out from the inner core, the next layer we encounter is the **mantle,** which surrounds the core (Figure 12.1). This part of the Earth is composed largely of solid iron, magnesium, and silicon oxides and is divided into two parts: the *lower mantle* and *upper mantle*. The lower mantle is about 2230 km (1385 mi) thick—in other words, about the same distance as from Laramie to Tallahassee, Florida (Figure 12.2). This distance would take you about 3 days to drive. Although temperatures in the mantle are still quite hot, over 1300° C (2370° F), they are much cooler than in the core. These cooler temperatures, coupled with intense pressure, cause

Liquid outer core *The outer part of the Earth's core. This area is about 2250 km (1400 mi) thick and consists of molten iron and nickel.*

Mantle *The layer of the Earth's interior that lies between the liquid outer core and the crust. This area is about 2900 km (1800 mi) thick and consists largely of silicate rock.*

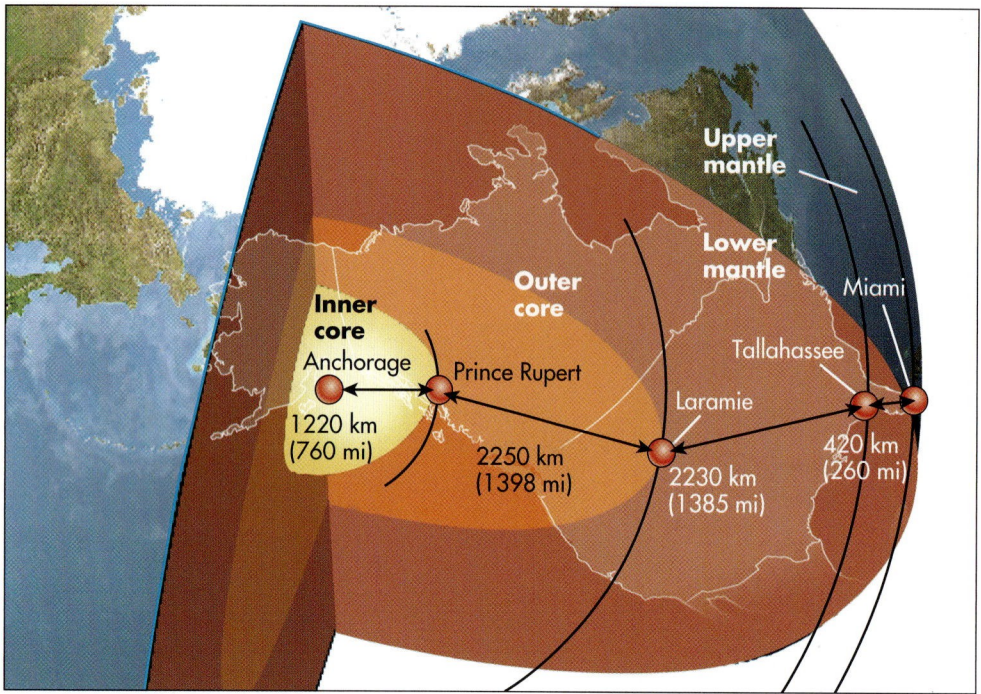

Figure 12.2 Map comparison with distance to the center of the Earth. The distance to the center of the Earth is about the width of the North American continent between Anchorage, Alaska, and Miami, Florida.

the lower mantle to be solid. This pressure lessens gradually, along with temperature, into the upper mantle.

The upper mantle is about 420 km (260 mi) thick, which is about the same distance as from Tallahassee to Miami, Florida. In contrast to the lower mantle, which is solid, the upper mantle consists mostly of viscous nickel; that is, it is a material like very thick syrup or a slowly flowing plastic. The upper part of the upper mantle is called the **asthenosphere,** which generally occurs between 40 to 250 km (25 to 105 mi) below the surface. This part of the inner Earth is the least rigid portion of the mantle because it contains scattered zones of high temperature due to radioactive decay. These zones comprise about 10% of the asthenosphere and consist of molten rock that moves very slowly by convection. This process is highly influential on the Earth's surface, causing earthquakes, volcanoes, and the deformation of rocks to form mountain chains.

The uppermost layer of the Earth is the **lithosphere** (Figure 12.1). This portion of the Earth extends from the surface into the uppermost part of the asthenosphere at a depth of 70 km (44 mi). This portion of the Earth contains the *crust*, which is the cool, stiff, and brittle exterior of the Earth. The boundary between the crust and the asthenosphere is very well defined and is called the **Mohorovicic Discontinuity** (pronounced Mo-ho-ro-vi-chich; "Moho" for short), named for the Yugoslavian scientist who first suggested its existence in 1909. This boundary is known because earthquake waves change speeds dramatically at the Moho level because of the dramatic density difference between the mantle and crust.

The crust is the thinnest of the major Earth layers, ranging in thickness from 8 to 40 km (about 5 to 25 mi). This thickness represents about 1% of the Earth's overall structure and is equivalent to driving from downtown Miami to a distant suburb in the metropolitan area. It floats on

---

Asthenosphere   *The layer of very soft rock that occurs in the upper part of the upper mantle. This region is about 40 to 250 km (25 to 105 mi) below the surface of the Earth. The soft character of this rock allows isostatic adjustments to occur.*

Lithosphere   *The outer, solid part of the Earth that is about 70 km (44 mi) thick and includes the uppermost part of the asthenosphere and the crust.*

Mohorovicic Discontinuity   *The boundary between the Earth's crust and the upper part of the asthenosphere; seismic waves change speed at this boundary.*

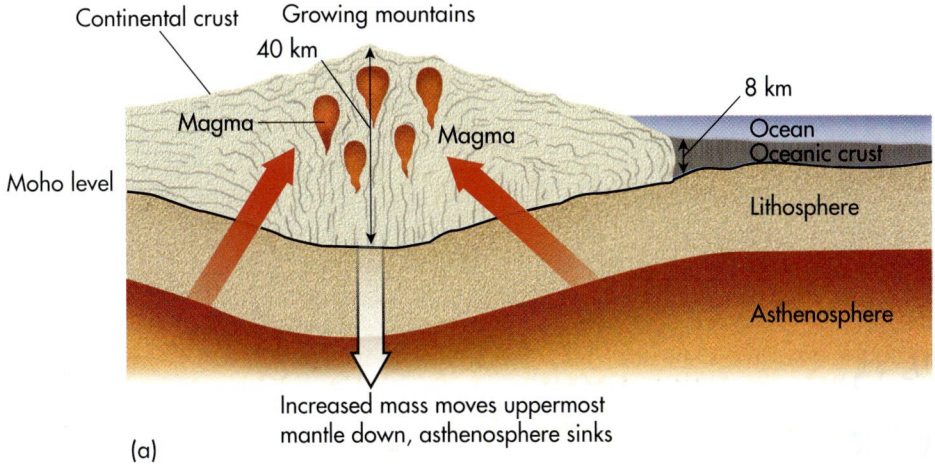

(a)

Continental crust
Growing mountains
40 km
8 km
Magma
Magma
Moho level
Ocean
Oceanic crust
Lithosphere
Asthenosphere
Increased mass moves uppermost mantle down, asthenosphere sinks

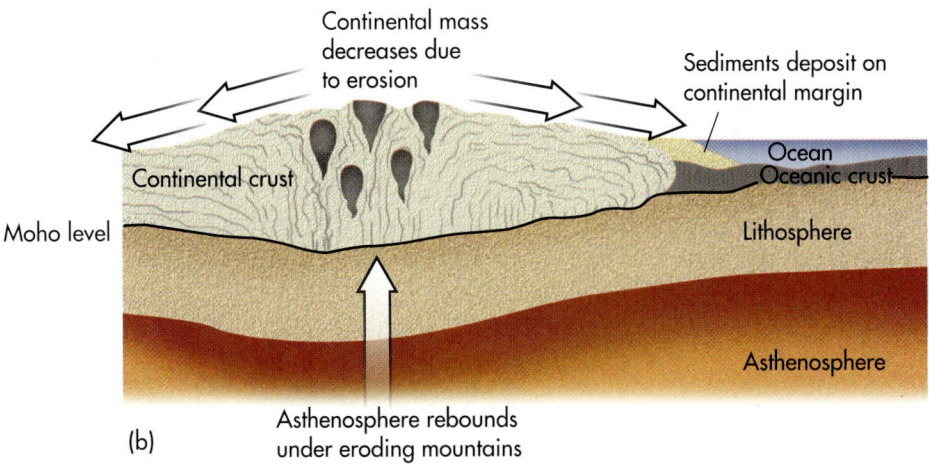

(b)

Continental mass decreases due to erosion
Sediments deposit on continental margin
Continental crust
Ocean
Oceanic crust
Moho level
Lithosphere
Asthenosphere
Asthenosphere rebounds under eroding mountains

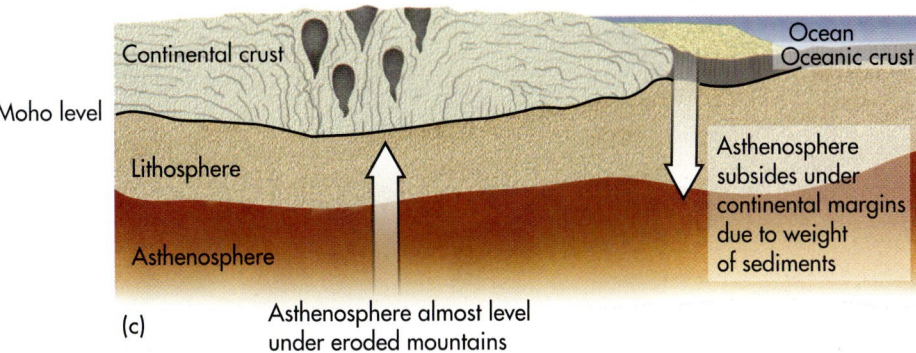

(c)

Continental crust
Ocean
Oceanic crust
Moho level
Lithosphere
Asthenosphere
Asthenosphere subsides under continental margins due to weight of sediments
Asthenosphere almost level under eroded mountains

Figure 12.3 Isostatic adjustment. (a) Continental crust is thickest where mountain ranges occur. Given the extreme weight of these locations, the underlying asthenosphere subsides relative to the area beneath the oceanic crust. (b) As mountains weather, the weight of the landmass decreases, allowing isostatic rebound of the asthenosphere. At the same time, the astheno- sphere beneath the continental margin begins to depress due to the increased weight of the freshly deposited sediments eroded from the continental interior. (c) As the continent weathers further, the isostatic rebound of the underlying asthenosphere continues. At the same time, the asthenosphere is further depressed due to the increased weight of margin sediments.

top of the mantle because its overall density is less. The crust contains two parts: oceanic crust and continental crust. **Oceanic crust** is about 8 km (about 5 mi) thick and consists mostly of a rock called *basalt*, which is very fine grained and high in silica, magnesium, and iron; thus, oceanic crust is often called *sima*, which is short for *si*lica and *ma*gnesium. **Continental crust,** in contrast, is composed largely of granite, which is a coarse-grained rock high in silica, aluminum, potassium, calcium, and sodium. As a result, continental crust is often called *sial*, short for *si*lica and *al*uminum. Continental crust is generally thicker and less dense than oceanic crust, averaging about 40 km (25 mi). It is thicker beneath mountain areas, where it can reach depths of 50 to 60 km (31 to 37 mi). In nonmountainous areas, in contrast, continental crust is only about 30 km (19 mi) thick. As you will see in the next chapter, the difference in composition between continental and oceanic crust is a critical part of the overall history of the Earth.

Now that we've examined all of the layers of the Earth, let's take a closer look at the relationship among the asthenosphere, lithosphere, and crust. Remember that these layers blend into one another, with the lithosphere being the transition between the asthenosphere and crust. Note in Figure 12.1 that the depth of the Moho level is generally greater beneath continents than under ocean basins. The reason for this pattern is fairly straightforward, as shown in the sequence of diagrams in Figure 12.3. Figure 12.3a represents a period of time during which intrusions of magma rise up from the mantle into the overlying continental crust. In this way, the overall elevation and mass of the continental crust increases, perhaps due to mountain building. This increased mass increases the pressure on the underlying asthenosphere, which is plastic, relative to the ocean basin. As a result, the depth of the Moho level increases beneath the continent.

Over time, the elevation of the continent reduces due to erosion (Figure 12.3b). This process transports sediments to the continental margin, where they are deposited. As a result, the mass of the continent slowly decreases and the underlying asthenosphere begins to bounce back (or rebound). At the same time, the asthenosphere beneath the continental margin begins to subside because of the weight of the new sediments. This process continues until the continent is nearly leveled (Figure 12.3c), which results in the near complete rebound of the underlying asthenosphere. At the continental margin, however, the asthenosphere subsides even more due to the increased weight of sediments transported from the landmass. This overall process of subsidence and rebound of the asthenosphere is called *isostatic adjustment*, or *isostacy* for short.

VISUAL CONCEPT CHECK 12.1

What two types of crust occur in landscapes that look like this? Which of the two is the most dense and why?

Oceanic crust *Basaltic part of the Earth's crust that makes up the ocean basins. Oceanic crust is also called sima because it consists largely of silica and magnesium and is about 8 km (about 5 mi) thick.*

Continental crust *Granitic part of the Earth's crust that makes up the continents. Continental crust is also called sial because it consists largely of silica and aluminum and averages about 40 km (25 mi) thick.*

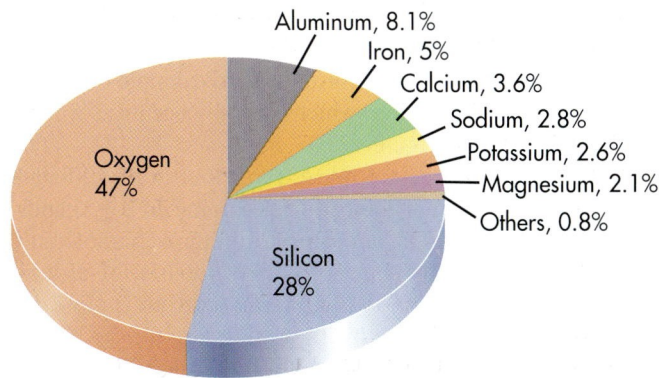

Figure 12.4 Elements in the Earth's crust. Percentages refer to the percentage by weight of each element in the Earth's crust.

Rocks and Minerals in the Earth's Crust

Have you ever picked up an interesting-looking rock when walking along the beach or hiking in the mountains. Have you ever noticed a fantastic exposure of rock along a highway? If you have, perhaps you wondered how the rocks formed, where they came from, or how old they were. In most cases the rocks were very old and contained elements that had been recycled many times throughout the Earth's history. Some of these minerals may have even been in the center of the Earth at one time. This part of the chapter examines the basic kinds of rocks on Earth and how they form.

Rocks are composed of a variety of Earth elements. Figure 12.4 shows the approximate percentage by weight of the various elements, including oxygen and silicon, with lesser amounts of aluminum, iron, calcium, sodium, potassium, and magnesium. These elements combine in various ways to form minerals. **Minerals** are naturally

Minerals *Naturally occurring substances with distinctive chemical configurations that usually manifest themselves in some kind of crystalline form.*

Figure 12.5 Quartz crystals. Quartz is an excellent example of a crystalline mineral. In small grains, it forms sediments in deserts, river deposits, and coastal areas. It is also a common part of most rocks formed in the continental crust.

occurring substances with distinctive chemical and atomic configurations that usually manifest themselves in some kind of crystalline form. A good example of a crystalline mineral is quartz, which typically occurs as a clear, six-sided prism (Figure 12.5).

When minerals are bonded together in a solid state they form **rock.** Although rocks come in all shapes, sizes, configurations, and ages, a common characteristic is that they usually (but not always) consist of at least two—and usually more—minerals that are bound together in some way. Most rocks are very old by human standards, with some being older than a billion years. Nevertheless, new rock is probably being created someplace at the very instant that you're reading this. Check and see if there is a volcanic eruption occurring on Earth right now that is spreading lava across the surface, or a large flood that is depositing mud in river valleys. If such an event is taking place, then new rock is being built there because material is accumulating that will one day become solid rock.

Because of the many different chemical elements that exist on Earth, an extremely wide variety of rock types exist. It goes way beyond the scope of this book to describe each one of the many different kinds of rocks. Instead, we'll focus on the three major rock classes—igneous, sedimentary, and metamorphic—and some fundamental kinds of rock that occur within these groups. After you complete this section of the chapter you should be able to identify these basic rocks, whether you find them lying randomly on the ground or exposed at the surface in a **rock outcrop.**

Igneous Rocks

One of the distinctive categories of rocks on Earth are those that are associated with liquid rock, known as **magma,** that originates in the mantle and moves upward into the Earth's crust or onto the surface (Figure 12.6a). When this material cools it forms **igneous rocks.** You are probably most familiar with magma that flows across the surface of the Earth as red-hot *lava* that erupts from volcanoes. This liquid rock ultimately cools to form **extrusive igneous rock,** which is so named because it results from magma that flows out of—or extrudes from—the Earth.

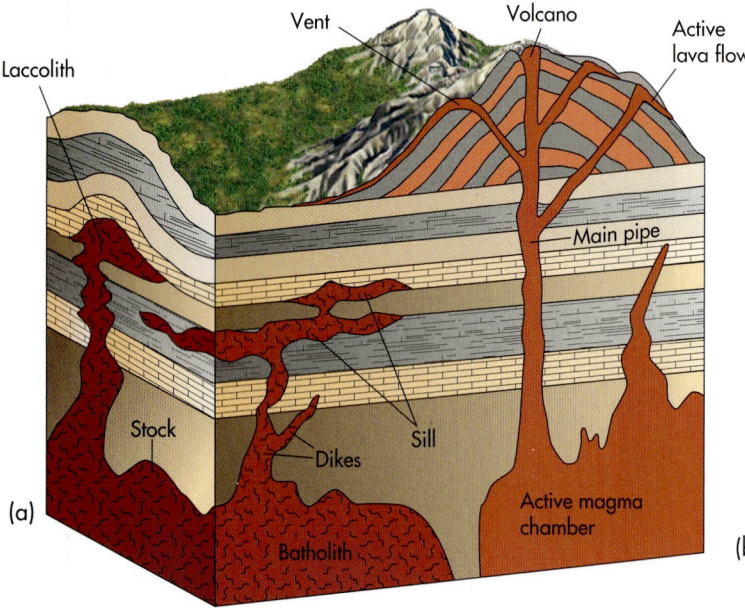

(a)

(b)

Figure 12.6 Various kinds of igneous environments. (a) This diagram illustrates the different ways that intrusive and extrusive igneous rocks form. (b) Lava flow at Mauna Loa Volcano in Hawaii. This material will ultimately cool and darken, consistent with the dark material in which it flows, forming extrusive igneous rock.

Rock *An amorphous mass of consolidated mineral matter.*

Rock outcrop *A place where rocks are exposed at the surface of the Earth.*

Magma *Molten rock beneath the surface of the Earth.*

Igneous rocks *Rocks that form when magma rises from the mantle and cools, either within the Earth's crust or on the surface.*

Extrusive igneous rocks *Rocks that form when magma cools on the Earth's surface.*

In contrast to the magma that flows onto the Earth's surface and cools, the majority of magma cools within the Earth to form **intrusive igneous rock,** so named because the magma intrudes into older rocks. This cooling process can occur in a variety of places beneath the Earth's surface (Figure 12.6b), with any body of intrusive igneous rock referred to as a **pluton**. The largest plutons are referred to as *batholiths*, which are features of no known depth that are up to several hundred kilometers long and 100 km (62 mi) wide. A smaller, somewhat rounded pluton that is attached to a batholith is called a *stock*.

In some instances, magma will move upward from a magma chamber into relatively small fissures within the overlying rock. One such feature is a *dike*, which is a pluton that consists of a wall-like feature that develops when

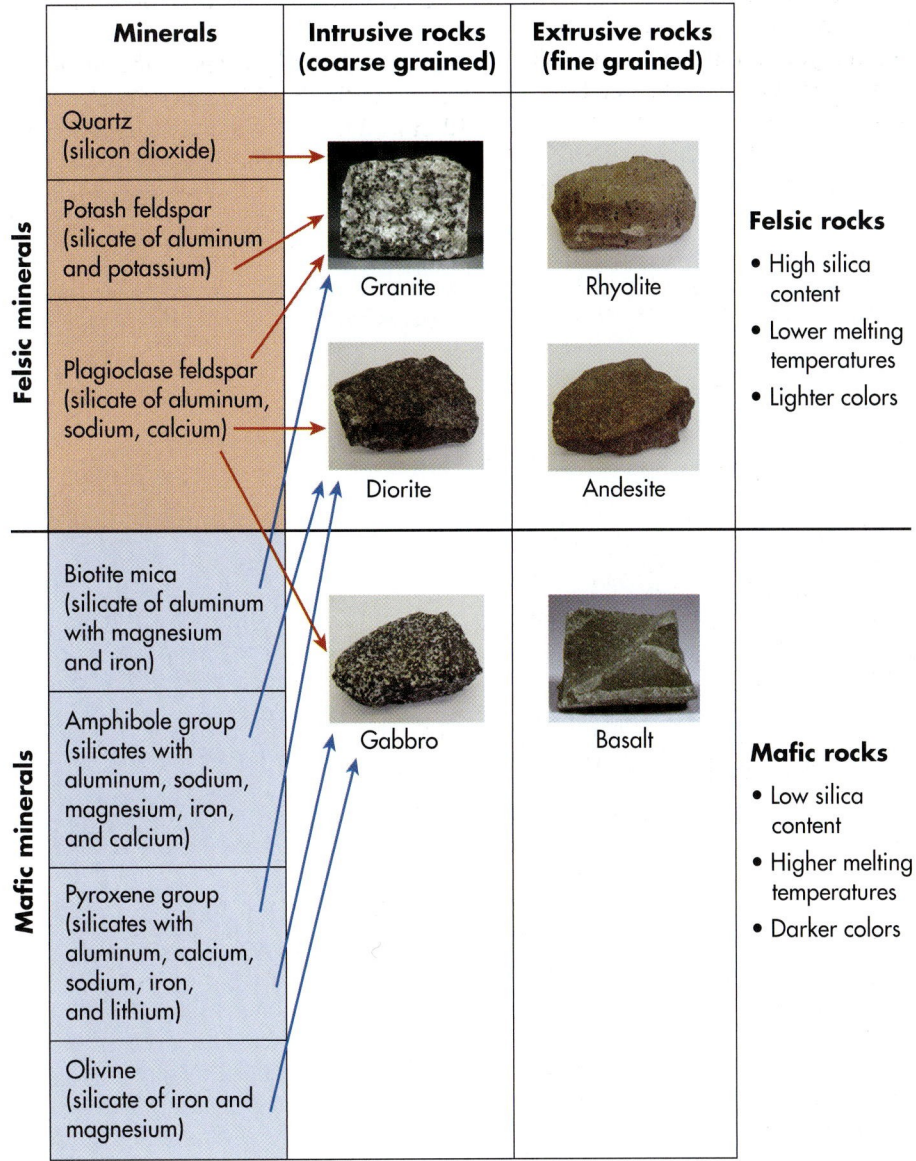

Minerals	Intrusive rocks (coarse grained)	Extrusive rocks (fine grained)	
Felsic minerals Quartz (silicon dioxide); Potash feldspar (silicate of aluminum and potassium); Plagioclase feldspar (silicate of aluminum, sodium, calcium)	Granite; Diorite	Rhyolite; Andesite	**Felsic rocks** • High silica content • Lower melting temperatures • Lighter colors
Mafic minerals Biotite mica (silicate of aluminum with magnesium and iron); Amphibole group (silicates with aluminum, sodium, magnesium, iron, and calcium); Pyroxene group (silicates with aluminum, calcium, sodium, iron, and lithium); Olivine (silicate of iron and magnesium)	Gabbro	Basalt	**Mafic rocks** • Low silica content • Higher melting temperatures • Darker colors

Figure 12.7 Silicate minerals and igneous rocks. This diagram shows the most important groups of silicate minerals, as well as some common igneous rocks. Notice that various combinations of minerals produce different rocks, both in extrusive and intrusive forms.

Intrusive igneous rocks *Rocks that form when magma cools within the Earth's crust.*

Pluton *An extremely large mass of intrusive igneous rock that forms within the Earth's crust.*

EXPOSED IGNEOUS INTRUSIONS

Two of the most dramatic natural features in North America consist of tall rock monoliths arising out of otherwise flat landscapes. They are not isolated mountains; rather, they are exposed igneous rock intrusions.

Shiprock, in the New Mexico desert, is sacred to the Navajo Indians, who call it *Winged Rock*. Standing 500 m (1700 ft) tall, the rock was once magma that cooled within the main pipe of a volcano that existed about 30 million years

Shiprock as it appears today. Note the prominent dikes (arrows) that radiate away from the central core.

ago. The surrounding body of the volcano and the associated rocks have disappeared due to erosion over the years, leaving the central solid rock standing by itself above the desert. Some of the dikes surrounding the volcanic pipe have also survived, forming radial rock fans about 3 m (10 ft) thick and 20 m (65 ft) high, extending as far as 3 km (2 mi) from Shiprock itself.

magma intrudes into a vertical rock fracture. Sometimes magma moves upward through such a fracture and then spreads horizontally within the crust along zones of rock weakness, forming a pluton called a *sill*. In other cases, upwardly moving magma spreads horizontally along a zone of rock weakness while at the same time pushing upward in such a way that the surface rocks are warped upward in a dome-like feature. When this warping magma cools, it forms a *laccolith*.

Regardless of whether the igneous rock is intrusive or extrusive, it consists largely of silicate minerals that contain crystalline chemical compounds dominated by sili-

con and oxygen atoms. Although there are many different kinds of silicate minerals, seven are most common (Figure 12.7). Three of these minerals are *felsic*, which means (among other things) that they have (a) high silica content; (b) relatively low melting temperatures; and (c) lighter colors. The remaining four of the seven silicate minerals are *mafic*, which essentially have opposite characteristics to felsic minerals.

Figure 12.7 indicates how silicate minerals can combine in different ways to form various kinds of igneous rocks. A good example on the mafic side of the diagram is the intrusive rock *gabbro*, which forms when plagioclase

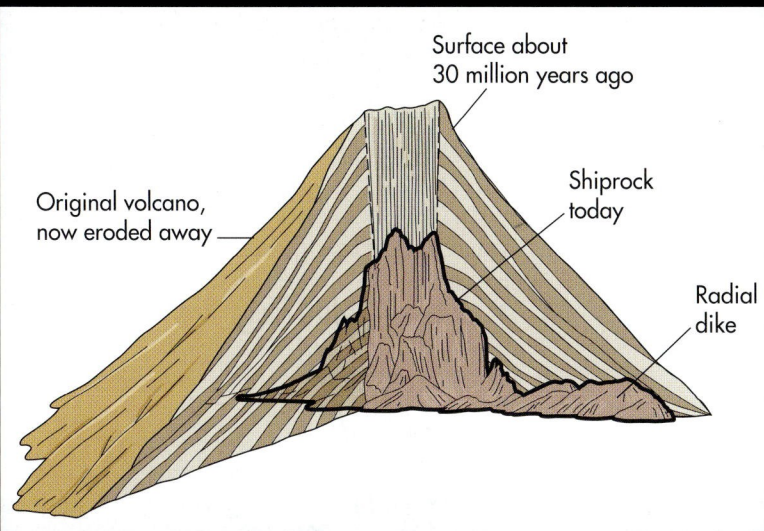

Diagram showing how the original volcano may have appeared before it eroded away. Try to imagine how much time must have occurred for the landscape to evolve in this way.

Another rock intrusion is Devil's Tower, located in the Great Plains of eastern Wyoming. Devil's Tower is 264 m (867 ft) tall and is thought to have formed when magma intruded into older rocks about 40 million years ago. Since that time the overlying rocks eroded, leaving the central core standing above the surrounding plains. The rocks of Devil's Tower have a unique columnar orientation as a result of contraction which occurred during the cooling of the magma.

The tall columns of Devil's Tower have made it a popular, if somewhat dangerous, route for serious rock climbers. It is also a sacred place for some Native American tribes and appeared in the popular movie *Close Encounters of the Third Kind* (1977) as a landing site for extraterrestrials. In 1906 it became the first National Landmark in the United States.

feldspar mixes with minerals from the pyroxene group and olivine. If the same minerals combine in an extrusive environment, the mafic rock is called *basalt*. On the felsic side of the diagram, a good example is the intrusive rock *granite*, which forms when potash feldspar combines with plagioclase feldspar, biotite mica, and quartz. In an extrusive setting, these minerals combine to form *rhyolite*. Note the other kinds of combinations that can occur.

You may be asking yourself, how can you tell the difference between extrusive and intrusive igneous rocks if they're made from the same materials? The answer to this question is that extrusive and intrusive rocks differ funda-

mentally in their appearance. Intrusive igneous rocks have coarse crystalline grains that you can see with your unaided eye. These coarse textures evolve because magma cools relatively slowly, perhaps over several thousand years, when it is trapped within the Earth's crust, such as in a sill, dike, or laccolith (see Figure 12.6a). A good example of a coarse-grained igneous rock is *granite* (Figure 12.7). Granite is a felsic rock that has distinctive dark grains consisting of the minerals *amphibole* and *biotite*, pinkish-tan grains of *feldspar*, and clear grains of *quartz*.

In contrast to intrusive igneous rocks, extrusive igneous rocks reach the surface and cool relatively

Figure 12.8 Obsidian. Obsidian is an extrusive igneous rock that develops when nongaseous lava cools very quickly, before crystals can develop. This rock was prized by prehistoric Native Americans because it has a rich, dark color and a smooth glassy surface.

These minerals can arise from any of the three major rock groups—igneous, sedimentary, or metamorphic (to be discussed later)—but their ultimate source is usually some form of igneous rock that was weathered in a way that liberated scores of individual particles such as clay (from feldspar) or sand (from quartz). These particles could then be transported as **sediment** by wind, water, or glaciers to a new place, where they subsequently accumulate as sedimentary deposits. Most sedimentary rocks are the remnants of sediments that accumulated in an oceanic environment. Many of these sediments were eroded from the continents in various ways and then carried to the ocean where they settled to the bottom.

After the sediment accumulates, all of the grains gradually become cemented to one another in a gradual process called **lithification** to form rock. This process occurs because water is squeezed out of the lower layers of sediments when they are compacted by the cumulative weight of the deposits above them. As a result of compaction and associated water loss, minerals such as calcium and silica recrystallize in the sediments and essentially glue them together to form rock.

Within the overall category of sedimentary rocks, there are three major classes of sediments: clastic, chemical, and organic. Table 12.1 lists the basic characteristics of these rocks.

quickly, sometimes in a matter of days. A good example of an extrusive igneous setting can be seen in Figure 12.6b, which shows a lava flow in Hawaii. Given that the lava cools relatively quickly in this setting, silicate minerals do not have time to congregate with one another, as they do in intrusive igneous rocks. The result is a fine-grained rock. A common kind of extrusive igneous rock is *basalt*, which is a dark mafic rock (Figure 12.7). Perhaps the most distinctive extrusive igneous rock is *obsidian* (Figure 12.8). Obsidian is also called *volcanic glass*, and forms when nongaseous lava cools so fast that crystals cannot form.

Sedimentary Rocks

Now let's turn to the second major rock class: **sedimentary rocks.** Although some forms of sedimentary rock have organic origins, most consist of vast quantities of formerly loose minerals that collected in some kind of depositional setting, such as a river floodplain, shallow sea floor, interior valley, lake, or marsh (Figure 12.9).

Clastic Sedimentary Rocks Perhaps the easiest type of sedimentary rock to understand is the category of clastic sedimentary rocks. These rocks contain mineral fragments derived from any of the three major rock groups that accumulate and lithify. The most common minerals in clastic sedimentary rocks are silicates, such as quartz and feldspar. Quartz is by far the most important clastic sediment, primarily because it is hard and very resistant to alteration. On the finer end of the textural scale, clays can be weathered from feldspars and carried by wind and water to some point where they are deposited.

There are several types of clastic sedimentary rocks. **Sandstone** is created when individual sand grains are deposited in thick layers by water along a beach or river, or by the wind within dunes. This collection of sand grains then lithifies, forming rock. When the sandstone also contains abundant rocks and pebbles of various sizes, it is called **conglomerate** (Figure 12.10).

Sedimentary rocks *Rocks that form through the deposition and lithification of small fragments or dissolved substances from other rocks, or, in some cases, marine animals.*

Sediment *Solid fragments of rocks that are transported to some location and deposited by wind, water, or ice.*

Lithification *The process whereby sediments are cemented through compaction to form rock.*

Sandstone *A sedimentary rock created when individual sand grains are deposited in thick layers by wind or water and lithify.*

Conglomorate *Sandstone that contains a wide variety of particle sizes.*

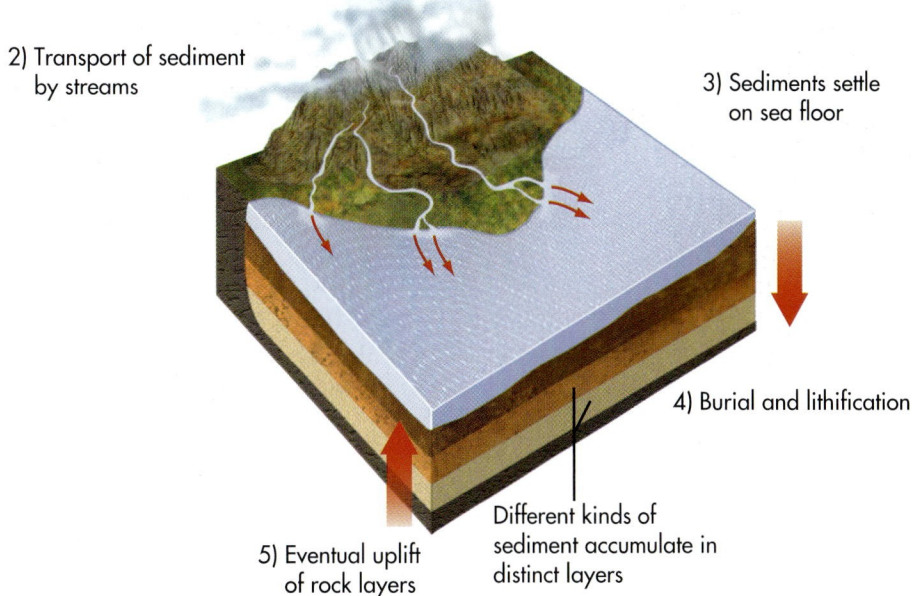

1) Sediments erode from continent

2) Transport of sediment by streams

3) Sediments settle on sea floor

4) Burial and lithification

Different kinds of sediment accumulate in distinct layers

5) Eventual uplift of rock layers

Figure 12.9 Marine sedimentary environments and rock formation. This generalized diagram illustrates how clastic marine sediments ultimately become rock, including (1) sediments erode from continents; (2) sediments are transported by streams into the ocean; (3) sediments settle to the ocean bottom; (4) the deposited sediments are buried and slowly lithify to become rock; and (5) the lithified sediments eventually uplift as new rock layers.

TABLE 12.1	Some Types of Common Sedimentary Rocks	
Class	**Rock Type**	**Composition**
Clastic: Formed from rock or mineral fragments	Conglomerate	Coarse-grained sandstone that contains rocks and pebbles of different sizes
	Sandstone	Sand grains cemented by minerals
	Siltstone	Silt particles cemented by minerals
	Claystone	Clay particles cemented by minerals
	Shale	Fine-grained particles that often contain fossils; most abundant sedimentary rocks; forms layers along which rock can split
Chemical: Formed from mineral precipitates from sea water or salty lakes	Limestone	Calcium carbonate, including that from animals and micro-organisms; forms rock layers
	Dolomite	Magnesium carbonates, mostly formed by chemical replacement in limestone
	Evaporites	Minerals (gypsum, rock salt, calcite) left in deposits after evaporation of water
	Chert, or flint	Extremely fine grains of silica (quartz) formed in layers or nodules within limestone
Organic: Formed from carbon-based organic matter	Coal	Carbon-based materials lithified by heat and pressure
	Oil (petroleum)	Liquid hydrocarbon trapped in sedimentary deposits; important fossil fuel
	Natural gas	Gaseous hydrocarbon trapped above oil in sedimentary deposits; important fossil fuel

Figure 12.10 The character of conglomerate. This rock is a sandstone that contains many different particle sizes. This range of particles may have accumulated in a stream environment.

An excellent place to see beautiful sandstone outcrops is the southwestern U.S., with one of the best known rock bodies being the Navajo Sandstone (Figure 12.11). The sediments comprising this rock accumulated in sand dunes approximately 175 million years ago when the region was a desert and sand was actively blown by the wind. Subsequently, the climate became more humid, resulting in rivers that buried the dunes with stream sediments. Still later, the region was covered by an ocean that deposited tremendous quantities of marine sediments. As a result of the compression and cementation of sands in the old dunes, they gradually became rock.

Recall from Chapter 11 that particle sizes smaller than sand are called silt and clay. When fine grains dominate a sedimentary deposit, it is referred to as *mud*. Silts and clays are easily suspended in and transported by moving water. Thus, mud typically accumulates in bodies of relatively still water, such as a coastal bay or lagoon (Figure 12.12a). After they are deposited in these settings, they lose a great deal of their original volume through compaction. The origin of these rocks and their associated names are very easy to remember: *siltstone* is dominated by silt-sized particles and *claystone* consists largely of clays. The most common fine-grained sedimentary rock is **shale,** which is a claystone that easily breaks into small flakes and has a platy appearance (Figure 12.12b).

Chemically Precipitated Sedimentary Rocks

When environmental conditions are favorable, thick deposits of mineral compounds can accumulate at the bottom of the ocean or inland lakes. In ocean environments, this process is associated with the precipitation and recrystallization of calcium carbonate from seawater. This

deposition can occur for a variety of reasons, including increase in water temperature, intense evaporation, upwelling of water, and bacterial decay, to name a few.

An excellent place to see the formation of chemically precipitated sedimentary rocks is the Bahama Islands, which lie off the eastern coast of Florida. Figure 12.13 is a satellite image that clearly shows the precipitation of calcium carbonate in this island chain. In this image, carbonate appears as the whitish plumes within the matrix of turquoise and bluish waters that lie between the islands. The turquoise and bluish colors appear because the water is shallow in the island complex and the satellite sensor can see the ocean bottom. Away from the islands, however, the ocean rapidly deepens, which is reflected by the dark ocean colors. Once the carbonates accumulate on the

Figure 12.11 The Navajo Sandstone at Zion National Park. Although it may be hard to believe, this landscape was once an active dune field. You can tell that the sands accumulated in dunes because of the distinct bedding lines exposed in the outcrop. Over a long period of time, the sand grains in the dunes lithified to form this rock body.

Shale *Sedimentary rock that consists of lithified clay-sized sediments.*

(a)

Figure 12.12 Formation and character of shale. (a) Mud collectively consists of individual clay particles, along with some silts, that settle out of suspension in places where water is very calm. If these sediments remain in place, they will gradually lithify to form shale. (b) Close up of shale. Note the flaky appearance of this rock type.

(b)

Figure 12.13 Carbonate depositional environment in the Bahama Islands. In this region, carbonate minerals are precipitating in the shallow waters of the island chain. You can see them as the whitish haze that surrounds the islands.

ocean floor, they subsequently lithify to form rock. The Bahama Islands, in fact, are made of calcium carbonate deposited in this manner.

In other situations, carbonate-bearing organisms such as clams, mussels, and oysters accumulate on the ocean bottom after they die. Subsequently, they become cemented when calcium carbonate precipitates and settles within the matrix of animal remains. The resulting rocks can contain abundant fossils (Figure 12.14), which enable geologists to reconstruct environmental conditions at the time the sediments were initially deposited. Regardless of the particular depositional process, whether it is entirely through carbonate precipitation or due to the accumulation of marine organisms, the resulting rocks are generally classified as **limestone** and can be incredibly thick, such as the famous White Cliffs of Dover, England. If the deposits also contain abundant magnesium, the resulting carbonate rock is called **dolomite.** Under more unusual conditions, silica can precipitate from solution at the bottom of the ocean to form *flint*, or *chert*.

In addition to forming in marine environments, chemically precipitated rocks also develop within continental lo-

Limestone *Sedimentary rock that consists of over 50% calcium carbonate ($CaCO_3$).*

Dolomite *Sedimentary rock that consists of over 50% calcium-magnesium carbonate ($CaMg[CO_3]_2$).*

Figure 12.14 Fossils in limestone. This rock formed millions of years ago when marine clams accumulated on the ocean bottom and calcium carbonate subsequently precipitated within the matrix of shells. The combined material lithified over time to form limestone.

cations when minerals evaporate from concentrated solutions (Figure 12.15a). These minerals collect at the surface and are called **evaporites.** As you can imagine, evaporites are most likely to evolve in semi-arid to arid geographic locations such as the southwestern U.S. (Figure 12.15b).

Organic Sedimentary Rocks As the name implies, organic sedimentary rocks consist of carbon-based materials that accumulate in thick deposits at the surface of the Earth. You saw a good example of this process in Chapter 11, when we discussed the accumulation of plant and organic matter as peat in cool and moist environments. Such thick accumulations of solid organic remains have occurred on a large scale over time, only to be buried by other sediments. When these organic deposits are buried progressively deeper, they slowly compact under the weight of the overlying sediments. In addition, they begin to be heated because temperature slowly rises with increased depth of burial. Over the course of millions of years, the amount of water and oxygen in the carbon progressively lowers and the deposit gradually solidifies, resulting in **coal.** Sometimes these deposits are "cooked" so much that they liquefy to form **petroleum** (crude oil). **Natural gas** comes from the remains of microscopic plants that live in the surface waters of the ocean. When they die, they settle to the bottom of the ocean where they decompose and form gas. This form of gas is typically about 85% methane and about 10% ethane, with smaller amounts of propane and butane.

(a)

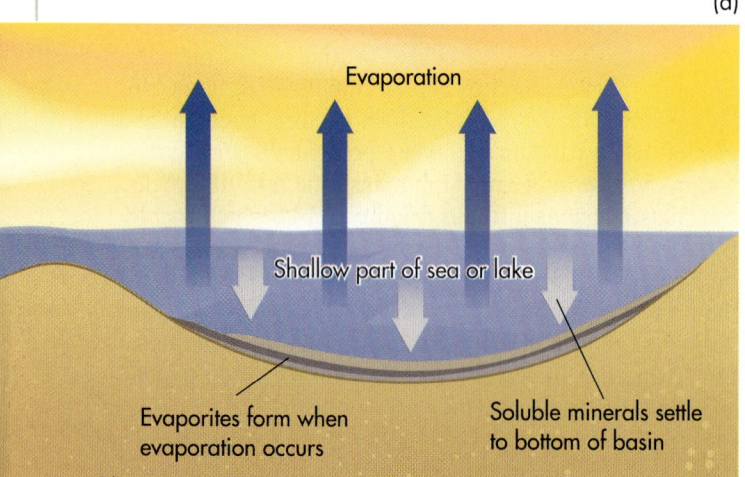

(b)

Figure 12.15 Formation of evaporites. (a) Evaporites form when minerals recrystallize on the exposed surface after waters evaporate from shallow oceans or inland lakes. (b) The whitish deposits in this image are salts and other highly soluble minerals that recrystallize as a crust when evaporation occurs in this extremely arid environment of Death Valley.

Evaporites *Surface salt residue that collects through the evaporation of water and the crystallization of sodium.*

Coal *A solid fossil fuel that consists of carbonized plants and animals.*

Petroleum *Naturally occurring, oily liquid that consists of ancient hydrocarbons.*

Natural gas *Naturally occurring mixture of hydrocarbons that often occurs in association with petroleum and is found in porous geologic formations.*

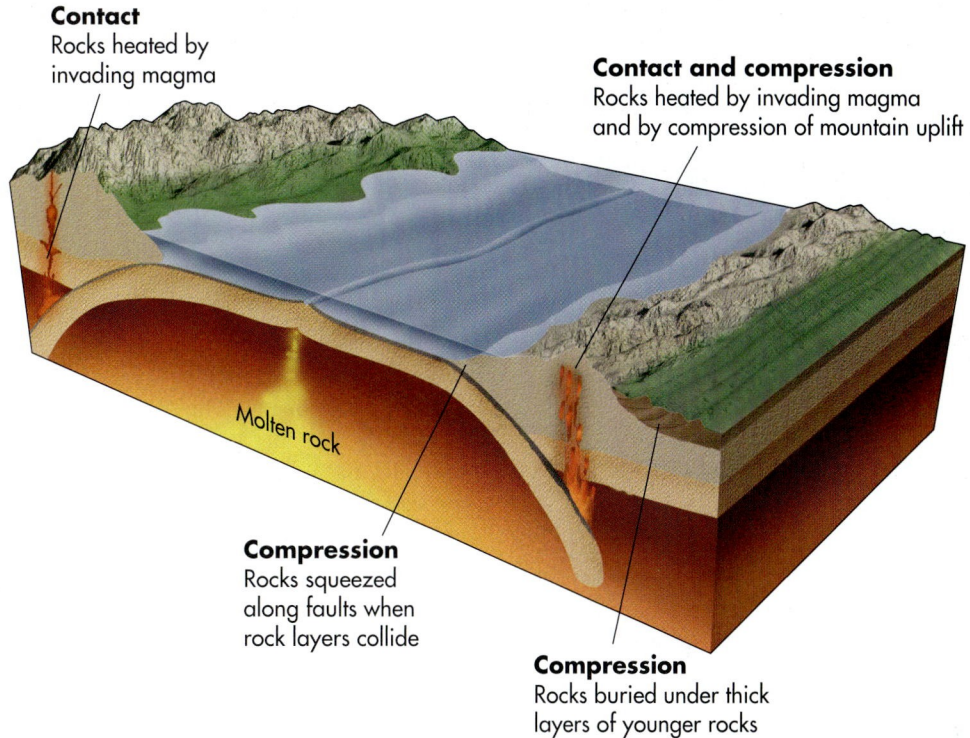

Contact
Rocks heated by
invading magma

Contact and compression
Rocks heated by invading magma
and by compression of mountain uplift

Molten rock

Compression
Rocks squeezed
along faults when
rock layers collide

Compression
Rocks buried under thick
layers of younger rocks

Figure 12.16 Formation of metamorphic rocks. Metamorphic rocks evolve due to intense pressure of overlying rocks or because rocks are heated dramatically when they come in contact with igneous bodies.

In places where organic sedimentary rocks occur, they are usually found in distinct beds that are separated by layers of limestone, sandstone, or shale. These deposits frequently accumulate in domes that form when rocks are bent due to compression for one reason or another; this process will be discussed more thoroughly in Chapter 13. Deposits of coal, petroleum, and natural gas are very important as fossil fuels to modern human civilization.

Metamorphic Rocks Igneous and sedimentary rocks can be altered after they develop. This alteration results in **metamorphic rock,** which occurs when a former igneous or sedimentary rock is subjected to intense heat or pressure within the Earth over millions of years. Thus, the name of the rock comes from a Greek word meaning "to change form." (A good example of metamorphosis is when a caterpillar changes into a butterfly.) Figure 12.16 shows the basic ways in which metamorphic rocks form. One way occurs when rocks are slowly compressed due to the thick accumulation of overlying sediments or because they are deeply buried by crustal processes. Another way that alteration occurs is through

contact metamorphism, which occurs when rocks are heated dramatically when they come in contact with invading magma in batholiths, dikes, or sills such as those illustrated in Figure 12.6. Whatever kind of alteration occurs, the result is that mineral components within the rock are rearranged to form different mineral varieties, or the minerals recrystallize into new mineral forms.

Although there are many different kinds of metamorphic rocks, we'll describe just a few here. *Slate* forms when shale is heated and compressed, forming a grayish, smooth rock that is quite hard. You may be familiar with slate because it is used as roofing shingles, patio flagstones, chalkboards, and the surface of the best pool tables. When slate is further heated and compressed, such that distinct planes and coarse texture develop, it becomes *schist* (Fig. 12.17a). Limestone that has undergone metamorphism becomes *marble,* which has a distinctive appearance because mineral impurities show up as swirling bands. Perhaps you are familiar with marble because it is frequently used in expensive tabletops and flooring. Finally, *gneiss* is a hard, banded metamorphic rock (Figure 12.17b) that forms when igneous or sedimentary rocks have been in close contact with intrusive magmas. The banding occurs because minerals become aligned during the metamorphic process. These kinds of rocks are described as being *foliated,* whereas those that lack banding are *nonfoliated* (Figure 12.17a).

Metamorphic rocks *Rocks that form when igneous, sedimentary, or other metamorphic rocks are subjected to intense heat and pressure.*

(a)

(b)

Figure 12.17 Some metamorphic rocks. (a) Schist forms when the metamorphic rock slate is further altered. (b) Gneiss evolves when igneous or sedimentary rocks are dramatically heated. Note the banding in this rock where minerals aligned under the intense heat.

The Rock Cycle

Recall from earlier chapters that distinct hydrologic and carbon cycles occur where carbon and water flow from one reservoir on Earth to another. There is also a distinct rock cycle (Figure 12.18). Although this cycle generally moves at a much slower pace than other cycles, it nevertheless comprises a distinct flow where rock gradually moves from one physical state or location to another.

The rock cycle works something like this. Let's begin at the top of the diagram in Figure 12.18 with the formation of an extrusive igneous rock such as basalt. The rock originates from magma that flows out of the ground, usually from a volcano. This magma cools and crystallizes, forming basalt that lies at or close to ground level. With time, the basalt gradually weathers due to various erosion processes, causing individual grains of the former rock mass to be transported and deposited someplace else. As the deposit of sediments thickens, those at the bottom of the stack become compressed, causing lithification and formation of sedimentary rocks. If these sedimentary rocks become deeply buried over time, or come in contact with the intense heat of an intrusive magma body, they will be altered to metamorphic rock because the original sediments will be recrystallized, recombined, or replaced. These metamorphic rocks, in turn, may be further heated such that they melt back to the magma state. If the sedimentary rocks remain near the surface, they too will be weathered, causing sediments to be loosened from the host rock mass. These sediments will then accumulate someplace else to form rock of a potentially different character. Remember that these are possible directions that the rock

Figure 12.18 Simplified view of the rock cycle. Over time, minerals move from one rock phase to another through a variety of processes, such as heating, melting, weathering, and recrystallization. This illustration does not contain all forms of movement within the cycle. For example, igneous rocks can melt and re-form as igneous rocks. Can you think of other forms not shown here?

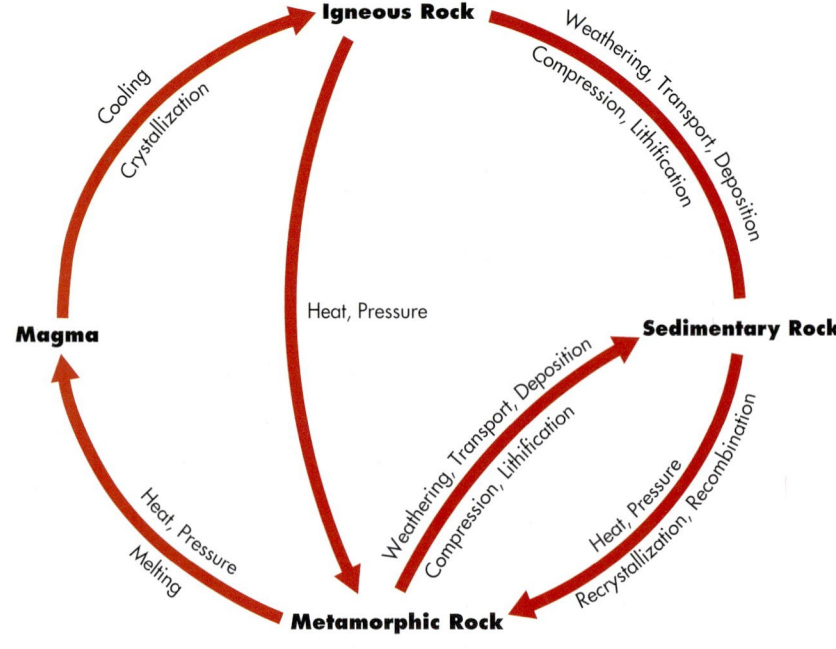

This image shows an outcrop of sedimentary rock. The lowermost rock unit (white arrow) is shale, whereas the one above it (black arrow) is limestone. Which one of the following choices best explains this combination of rocks?

a) They are igneous rocks.

b) They were modified by intense heat and pressure.

c) They collectively indicate fluctuating sea levels in the past.

d) They formed when sandy sediments accumulated along a beach.

cycle might take. It's possible that sedimentary rocks may be re-melted directly to magma or that metamorphic rocks may be weathered to form new sedimentary rock.

A good example of how the rock cycle works is the carbonate environment of the Bahama Islands (see Figure 12.13). Recall that in this area recrystallized calcium carbonate accumulates on the shallow ocean floor. In the context of the rock cycle, this carbonate likely originated from continental limestones that were eroded, and the resulting calcium carbonate was carried in solution by rivers to the ocean. Some of this carbonate is currently recrystallizing in the Bahama Islands to form new rock. Naturally, it has taken a long time for these combined processes to occur.

Geologic Time

In the context of the rock cycle, what does "a long time" mean? This is a very difficult question to answer in any kind of tangible way because it's not the kind of time you are used to thinking about. Instead, the rock cycle occurs within a time frame called **geologic time** (or *deep time*), which is a very different concept from what humans normally consider when thinking about time. Although you may have heard of this time scale before, it bears repeating here because it provides the context in which to view the rock cycle, as well as the overall evolution of the Earth.

GEODISCOVERIES: WWW.WILEY.COM/COLLEGE/ARBOGAST

THE ROCK CYCLE

Although the rock cycle moves very slowly, it's possible to animate the concept to capture its essence. Go to the *Geo-Discoveries* website and select the module **The Rock Cycle**. This module describes the basic ways in which rocks are created on Earth and how they move from one place to another. It begins with the formation of igneous rock and then de-

scribes how these rocks can be altered to form new kinds of rock, whether it be a new kind of igneous rock, sedimentary rock, or metamorphic rock. As you watch the animation, note the way rock moves from one reservoir to another through time. After you complete the animation, be sure to answer the questions at the end of the module to test your understanding of this concept.

Geologic time   *The period of time that encompasses all of Earth history, from its formation to the present.*

The basic premise of geologic time is that it is radically different from the human time scale. As human beings, we tend to think that anything 100 years old, or older, is ancient. For example, doesn't the American Civil War seem like a long time ago? After all, it occurred in the 1860s, long before your grandparents, or even great, great grandparents, lived. In the context of our lifetimes, 100 years is indeed a long time and the Civil War does seem like ancient history. As far as the history of the Earth is concerned, however, 100 years is a blink of an eye. Scientists accept that the Earth is a very old place (about 4.6 bil-

lion years old) that lies within a much larger and older universe (about 14 billion years old). Deep time is a concept that is, quite frankly, difficult to comprehend. It is important to understand the concept of deep time, however, because it provides a context within which we can study the Earth and helps us realize the rapid nature of environmental change associated with human impacts on the planet.

In an effort to make the concept of geologic time more understandable, geologists have devised a geologic time scale that is subdivided on the basis of major events in Earth history. Figure 12.19 shows this time scale and its

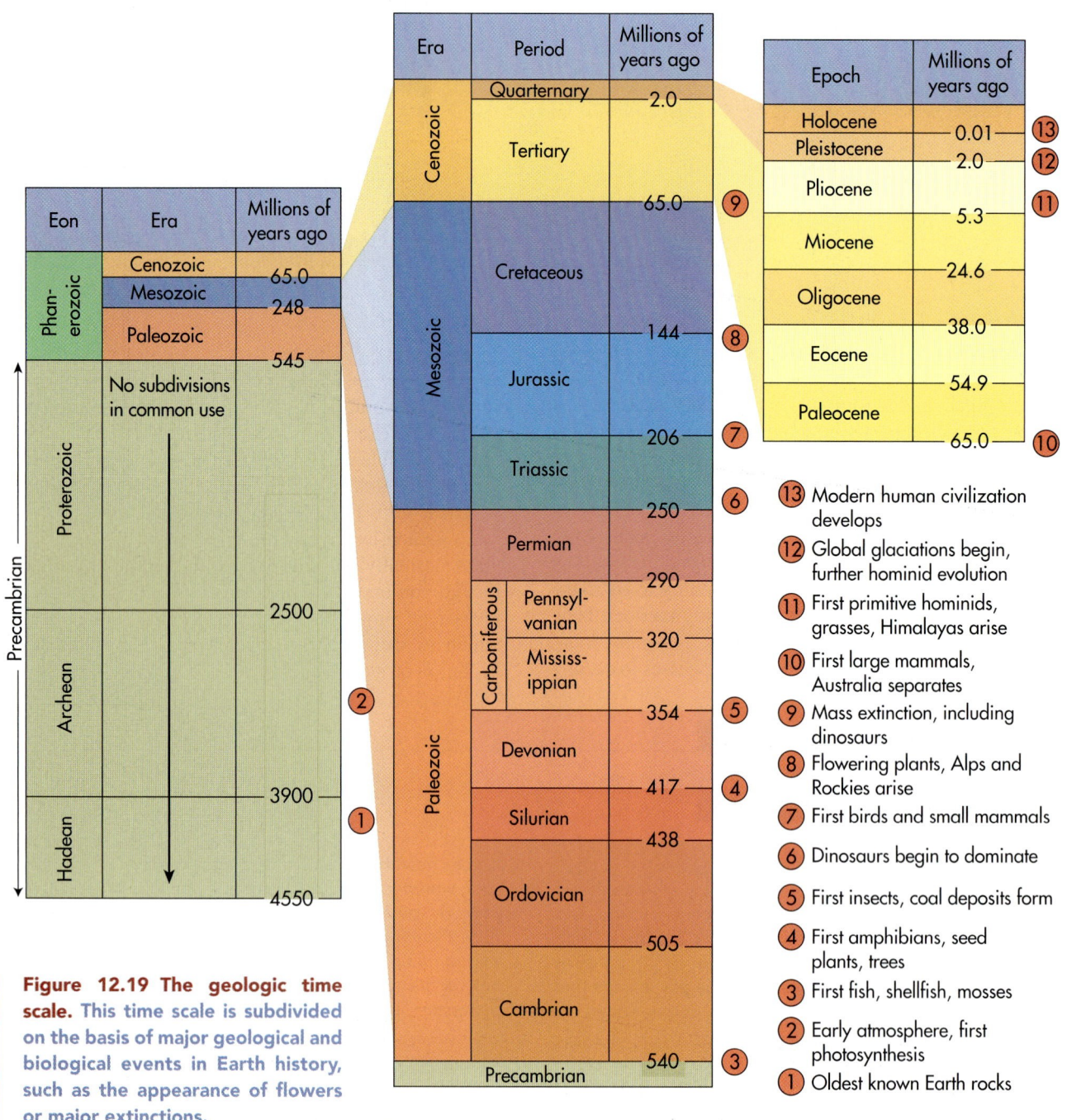

Figure 12.19 The geologic time scale. This time scale is subdivided on the basis of major geological and biological events in Earth history, such as the appearance of flowers or major extinctions.

various subdivisions. The time scale is categorized into (from largest time expanse to shortest) eons, eras, periods, and epochs. All time earlier than 540 million years (my) ago is lumped into the division of *Precambrian time,* which includes both the *Archean* and *Proterozoic* Eons. This interval of time represents 88% of Earth history, during which there was very little life. Notable events during early Precambrian time include the initial stages of atmospheric development, the emergence of the first cyanobacteria, and the onset of photosynthesis. Toward the end of Precambrian time the modern atmosphere evolved.

Beyond Precambrian time is the *Phanarezoic* Eon, which is subdivided into three major *eras*, the *Paleozoic*, *Mesozoic*, and *Cenozoic*. These eras are further subdivided into *periods*. The beginning of the Paleozoic Era (the *Cambrian* Period) is marked by the emergence of the first fish and shellfish about 540 million years ago, whereas the end of the Paleozoic Era is related to a major extinction during the *Permian* Period around 250 million years ago. Other significant events in the Paleozoic Era include the development of the first trees and amphibians at the beginning of the *Devonian* Period (~ 417 million years ago) and the first winged insects at the onset of the *Mississippian* Period (~ 354 million years ago).

The Mississippian Period is particularly noteworthy because organic deposits began to accumulate in sufficient quantity to ultimately form the extensive coal and other fossil fuel resources we use today. This period of massive organic deposition began because the average global temperature during the early Mississippian Period was quite warm, about 22° C (72° F). Thus, many parts of the Earth were covered by steamy swamps that contained copious amounts of organic carbon. As plants died, they slowly sank to the bottom of the swamps and began to decay. These deposits were later buried by thousands of meters of sediment; the resulting compaction and heating caused thick beds of coal to form and many petroleum and natural gas reserves. This period of massive organic deposition terminated at the end of the Pennsylvanian Period about 290 million years ago.

Following the Paleozoic Era is the Mesozoic Era. This era is a definable unit of deep time because it is the interval in which dinosaurs dominated life on the Earth, beginning around 250 million years ago and ending about 65 million years ago. *Isn't it amazing that dinosaurs existed for about 180 million years!* Nevertheless, the first birds and mammals evolved during this era of time, specifically at the onset of the *Jurassic* Period 206 million years ago. The *Cretaceous* Period is significant because dinosaurs reached their maximum dominance during this time, most notably in the form of the Tyrannosaurus rex, and the first flowering plants appeared. In addition, two major episodes of mountain building occurred in North America.

The end of the Cretaceous Period at 65 million years ago is marked by the extinction of the dinosaurs. Currently, most geologists think that this extinction may have been caused by the catastrophic impact of a large asteroid with Earth. This impact is thought to have dramatically influenced the Earth through a combination of numerous fires and high levels of atmospheric dust that may have blocked sunlight for months. As a result, the vast majority of animals that depended directly on plants, either through eating them or by eating the animals that ate plants, became extinct. In contrast, it is thought that animals such as small rodents and insects survived because they fed on the corpses and organic waste of the larger animals that were decimated.

The Cenozoic Era is significant because it represents all of post-dinosaur time, beginning 65 million years ago and continuing until the present. During this period most current landscape features such as mountain chains and river valleys on Earth developed, as well as life as we know it. The Cenozoic Era is subdivided into two periods, the *Tertiary* and *Quaternary*. Given the relative youth of these two periods, plus the fact that most current landforms formed within this interval, we know a great deal more about them than earlier periods. Thus, these periods

GEODISCOVERIES: WWW.WILEY.COM/COLLEGE/ARBOGAST

GEOLOGIC TIME

Understanding geologic time is difficult because it is hard to relate a diagram such as Figure 12.19 with the significant geologic events that are used to mark major intervals of time. Such events include major periods of mountain building, catastrophic extinctions, and the emergence of certain kinds of plants or animals in the fossil record. In an effort to facilitate this understanding, go to the *GeoDiscoveries* website and select the module **Geologic Time**. This module illustrates, with images and text, some of the major geologic events that are associated with past time intervals. The module's simulation is interactive because you can choose the time intervals that you want to "visit" and see the factors that make them unique. In this fashion, you will better understand why a major extinction occurred at the end of the Permian Period, why dinosaurs dominated the Mesozoic Era, and why the Quaternary Period is divided into the Pleistocene and Holocene Epochs. After you visit all of the time intervals within this simulation, be sure to answer the questions at the end of the module to test your understanding of the geologic time scale.

are further subdivided into *epochs*. The *Paleocene* Epoch, for example, is marked by the emergence of the first large mammals following the extinction of the dinosaurs 65 million years ago. Subsequently, the first primitive hominids developed at the beginning of the *Pliocene* Epoch 5.3 million years ago. In the context of human habitation of Earth, the *Pleistocene* and *Holocene* Epochs are significant because they are intervals of time during which most human evolution has occurred. We live in the Holocene Epoch, which spans the past 10,000 years.

Telling Geologic Time

You might be asking yourself, how do geologists really know the age of the Earth and when major events occurred? The answer lies in our understanding of the way in which rock elements radioactively decay through time. Rocks are composed of huge numbers of atoms, all of which contain protons and neutrons in their nuclei. Although many of these atoms remain stable indefinitely, some do not. These unstable atoms are called **radioactive isotopes.** In these atoms, particles within the nucleus break apart and the atom decays into a different element. It's as if you had a pet dog that suddenly woke up on Monday morning and had become a cat, then the following Monday changed again into a mouse.

In the context of calculating rock age, isotopic decay is important because radiation is emitted when this process (known as *radioactivity*) occurs. Each isotope decays at a different constant rate that is known by geologists. The reference time frame for the decay rate for any isotope is called

---

Radioactive isotopes *Unstable isotopes that emit radioactivity as they decay from one element to another.*

its *half-life*; this is the amount of time required for one-half of the isotopes in any given sample to decay. For example, thorium-232 requires 14.1 billion years to change through its decay series (including radium 228 and radon 220, among others) into the stable isotope lead-208. The half-life of the radioactive carbon isotope (carbon-14), in contrast, is 5730 years, during which time it converts to the stable isotope nitrogen-14. In this fashion, radioactive isotopes provide a running time clock for the history of the Earth and geologists use *radiometric dating* to calculate age.

If a geologist is interested in the age of a rock sample, he or she simply compares the amount of the original isotope with the amount of the decayed end product. Using radiocarbon dating, for example, a geologist can compare the amount of carbon-14 to the amount of nitrogen-14. If the ratio between the two is 1:1—that is, equal parts—then it means that one-half of the radioactive isotope has decayed. In other words, about 5730 years have passed since the carbon was deposited. You'll see in Chapter 18 how radiocarbon dating can be used to date the age of sand dunes.

Putting Geologic Time in Perspective

Although understanding radiometric dating provides confidence in the ages reported by geologists for the history of the Earth, it won't help you comprehend the enormity of the geologic time scale. If you want to put geologic time in some kind of perspective, consider that virtually all scientists believe that anatomically modern humans—that is, people looking basically like you—have lived on Earth for about 150,000 years. This sure sounds like a long time, doesn't it?

Here's another analogy that helps explain deep time. Imagine that all of geologic time is contained within a single calendar year, with time beginning on January 1 and extending until December 31, as shown in Table 12.2. In this

TABLE 12.2	The Geologic Time Scale Related to the Calendar Year	
Event	**Age***	**Time in Year**
Earth formed	4.6 by	Jan. 1
First single-celled organism	3.2 by	mid-April
Oxygen in atmosphere	2.0 by	mid-July
First cell with nucleus	1.0 by	Oct. 12
First vertebrates	625 my	Nov. 10
First land plants	340 my	Dec. 3
First reptiles	220 my	Dec. 13
First mammals	155 my	Dec. 18
Grand Canyon downcutting	10 my	Dec. 31, 5 A.M.
Early hominids	3 my	Dec. 31, 6 P.M.
Your birth	~ 18	Dec. 31 (0.05 seconds before midnight)

*by = billion years ago; my = million years ago

time context, each day is equivalent to 12.6 million years, each hour is 525,000 years, each minute is 8750 years, and each second is 146 years. The year begins with the formation of the Earth. From that point until the middle of April, no life forms inhabit the Earth. In other words, the Earth is a lifeless planet for the first 4.5 months of the "year". *Nothing* except single-cell organisms such as amoebas and simple bacteria exist from early May until late November, when the first vertebrate animals (fish) develop. Subsequently, the first land plants emerge on December 3. Reptiles develop on about December 13, followed by the first mammals on December 18. Dinosaurs dominate the Earth until late December, when they become extinct and mammals such as shrews and other small rodents emerge and more fully evolve. Toward the very end of the year, on December 31 at 5 A.M., the Grand Canyon begins to form. Early hominids don't evolve until 6 P.M. on December 31, which means that people looking generally like you have existed for an extremely short period of time within the year. Your life occurs within the last 0.05 seconds of the year. No wonder people have difficulty understanding the concept of geologic time!

Integrating Geologic Time and the Rock Cycle: The Spanish Peaks

Although analogies are fun to play with and help explain the concept of geologic time, it is possible to sense the essence of deep time directly when you travel. An excellent place to feel the depth of geologic time is the Spanish Peaks in southern Colorado, because this area has a complex geologic history that spans a relatively recent interval of time. In addition, you can see how the rock cycle fits within the context of the geologic time scale here. Let's take a closer look at this area and see how the concept of deep time relates to the rock cycle.

The Spanish Peaks lie on the eastern flank of the Sangre de Cristo mountain range (Figure 12.20a) and rise about 2134 m (7000 ft) above the Great Plains (Figure 12.20b), which lie immediately to the east. The evolution of this landscape began sometime before 75 million years ago when thick sedimentary deposits accumulated in a marine environment that fluctuated in depth. These deposits, consisting of alternating layers of sand, clay, and calcium carbonate may have originated from other rocks and slowly compressed and lithified over millions of years to form sandstone, shale, and limestone respectively, after the ocean receded.

Approximately 75 million years ago, during the late Cretaceous Period, the nearby Sangre de Cristo Mountains began to form (Figure 12.21). Over the course of the next 50 million years, until about 25 million years ago, the Sangre de Cristo Mountains slowly rose above the surrounding plains. The force of this uplift caused large fractures to appear in the preexisting layers of sedimentary rock. Between about 25 million and 14 million years ago, tremendous quantities of magma began to intrude into this portion of the Earth's crust,

Figure 12.20 The Spanish Peaks in southern Colorado. (a) This true color satellite image of Colorado and adjacent states shows the location of the Spanish Peaks. This image was acquired in September 2002 and shows the first snow of that year, which can be seen as white on the mountain crests. (b) Looking south at the Spanish Peaks. The crest of East Spanish Peak (left) is 3866 m (12,683 ft) high, whereas the top of West Spanish Peak (right) is 4153 m (13,626 ft) in elevation. Note the prominent radial dike that extends to West Spanish Peak. Several dikes like this one occur in the area.

South Dakota

Wyoming

Nebraska

Great Plains

Colorado

Utah

Sangre de Cristo Mts.

Spanish Peaks

New Mexico

Arizona

(a)

(b)

Radial Dike

AMAZING PLACES:

THE GRAND CANYON

The Grand Canyon is famous to many people because it's one of the most scenic places on Earth. From a geologic perspective, the Grand Canyon is significant because it contains rocks that span much of Earth history. As a result, the rocks tell us a great deal about the evolution of the Earth and the environmental changes that have occurred over time in what is now northwestern Arizona.

The oldest rocks are Precambrian in age and are found in the inner gorge of the canyon. The best known of these oldest rocks is the Vishnu Schist, which is a metamorphic rock about 2.0 to 1.7 billion years old. This rock layer was the foundation of an ancient mountain chain that was subsequently eroded between about 1.7

> The Grand Canyon is unique because it contains rocks that span about 1.5 billion years of Earth history and which accumulated in a variety of depositional environments.

and 1.25 billion years ago. Additional rocks of Precambrian age formed between about 1.25 billion and 825 million years ago.

After the last Precambrian sediments accumulated, another lengthy period of erosion occurred. This period lasted about 300 million years and is called the *Great Unconformity* because no rocks from this period exist. In fact, geologists believe that rocks were almost certainly there, but they were eroded by an advancing sea sometime late in that interval of time. The rocks above the Great Unconformity consist of alternating layers of sandstone, shale, and limestone that collectively accumu-

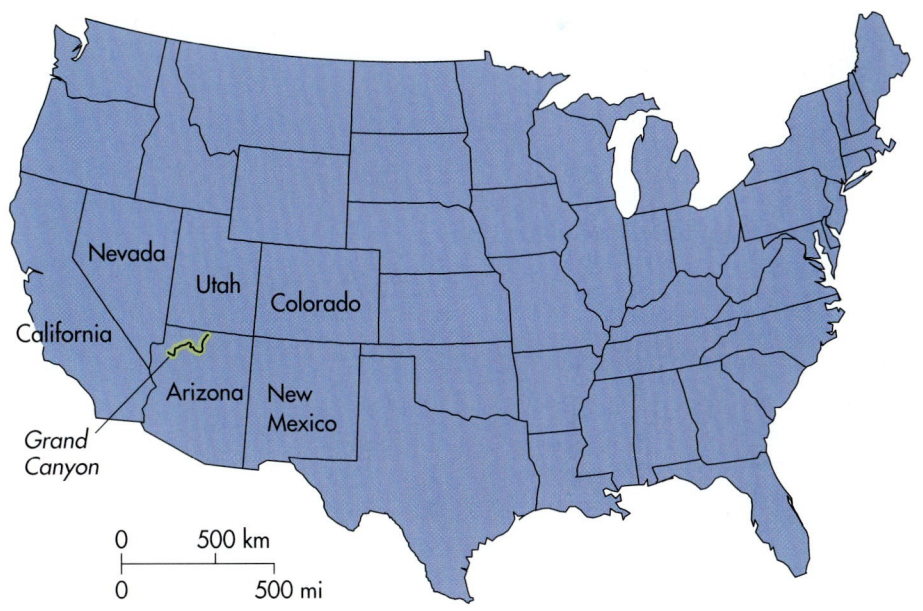

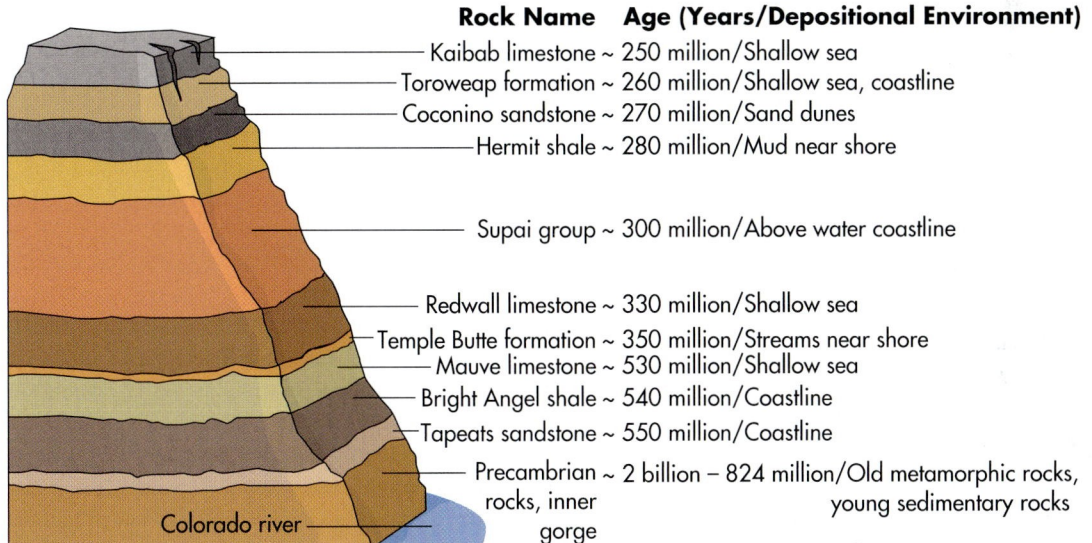

Rock Name	Age (Years/Depositional Environment)
Kaibab limestone ~	250 million/Shallow sea
Toroweap formation ~	260 million/Shallow sea, coastline
Coconino sandstone ~	270 million/Sand dunes
Hermit shale ~	280 million/Mud near shore
Supai group ~	300 million/Above water coastline
Redwall limestone ~	330 million/Shallow sea
Temple Butte formation ~	350 million/Streams near shore
Mauve limestone ~	530 million/Shallow sea
Bright Angel shale ~	540 million/Coastline
Tapeats sandstone ~	550 million/Coastline
Precambrian rocks, inner gorge ~	2 billion – 824 million/Old metamorphic rocks, young sedimentary rocks

Colorado river

Many different kinds of rocks are exposed in the walls of the Grand Canyon. The youngest rocks are found at the lip of the canyon.

lated between 550 and 250 million years ago. These rocks are testament to a fluctuating sea that resulted in the deposition of sandstone when the area was above sea level, shale when the area was a coastal bay or lagoon, and limestone when the region was beneath a shallow sea. These rocks have been progressively exposed in the past 5 million years because the Colorado River has been downcutting.

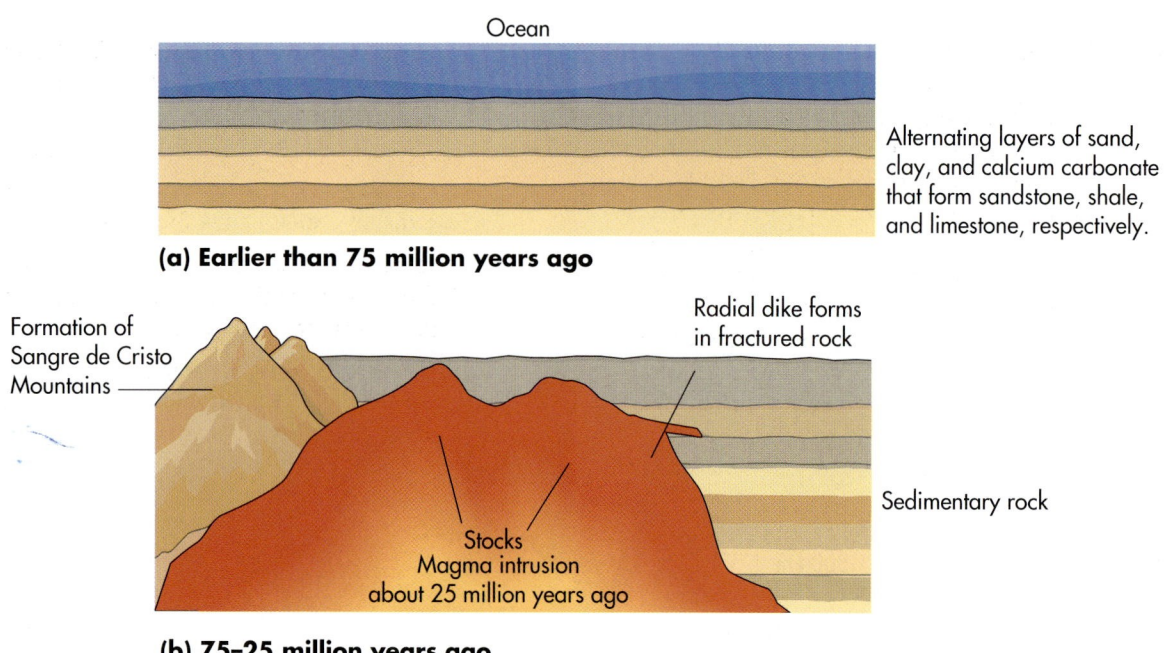

Ocean

Alternating layers of sand, clay, and calcium carbonate that form sandstone, shale, and limestone, respectively.

(a) Earlier than 75 million years ago

Formation of Sangre de Cristo Mountains

Radial dike forms in fractured rock

Sedimentary rock

Stocks
Magma intrusion about 25 million years ago

(b) 75–25 million years ago

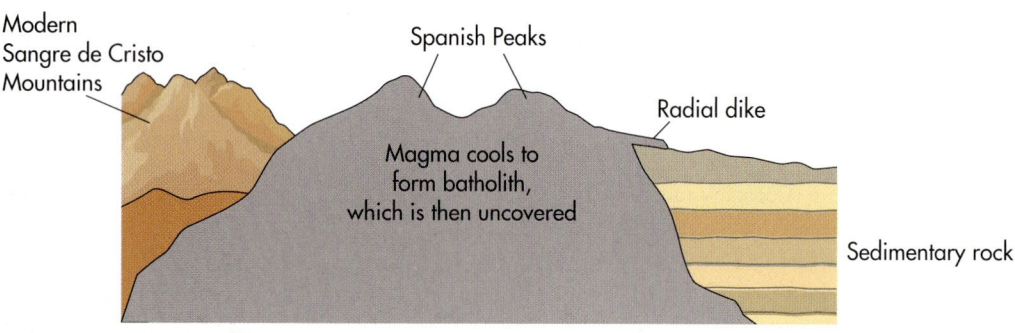

Modern Sangre de Cristo Mountains

Spanish Peaks

Radial dike

Magma cools to form batholith, which is then uncovered

Sedimentary rock

(c) 25 million years ago to present

Figure 12.21 Basic evolution of the Spanish Peaks. (a) Before 75 million years ago, the region was covered by a fluctuating ocean, which deposited alternating layers of sediment that became sandstones, shales, and limestones. (b) Formation of the Sangre de Cristo Mountains occurred between about 75 and 25 million years ago. The force of this uplift bent and fractured the sedimentary rocks to the east. Between about 25 and 14 million years ago magma intruded into these fractures. The magma subsequently cooled, forming a batholith, two stocks, and a number of radial dikes. During this same interval of time, the Sangre de Cristo range began to intensively erode and wear down. (c) In the past 5 million years the sedimentary rocks overlying the igneous features have been eroded, which has exposed the stocks and radial dikes seen in Figure 12.21b.

creating a giant chamber within the sedimentary rock body and filling smaller cracks in that same mass of rocks.

Sometime after 14 million years ago the intruded magma cooled to form a granite batholith with an associated pair of stocks at the top. Extending outward several kilometers from the stocks were **radial dikes** that also consisted of granite. In the past 5 million years, the sedimentary rocks that were once present were completely

eroded away from this part of Colorado, exposing the granite stocks and dikes on the surface of the Earth. The reason why the igneous bodies remained intact is that these kinds of rocks are much harder and more resistant to erosion than sedimentary rocks. The end result is that the Spanish Peaks now tower over the landscape.

Now that we've discussed how the Spanish Peaks formed, think about the broader questions related to geologic time and the rock cycle. In the context of the rock cycle, it's important to note that the sediments once contained within the layers of sandstone, limestone, and shale, which in turn covered the stocks and radial dikes, have been transported to other locations where new rock

Radial dikes *A long, wall-like feature of intrusive igneous rock that forms when magma is injected into thin cracks within older rock and cools.*

formations developed or are now developing. What is particularly impressive is that these layers of sedimentary rock must have collectively been greater than 2134 m (7000 ft) thick! Otherwise, the magma that formed the batholith, stocks, and radial dikes could not have been intruded within them. It is even more impressive that the process of erosion subsequently removed *all* of the sedimentary rock above the stocks. In the time since the stocks and dikes were exposed, they were being slowly worn down by erosion. The resulting liberated sediments have also been transported elsewhere to form new rock.

In the context of deep geologic time, the entire sequence of events that led to the development of what we now call the Spanish Peaks took over 75 million years, beginning with the deposition of more than 2134 m (7000 ft) of sand, calcium carbonate, and clay in a marine environment. These sediments slowly lithified to form rock. Next, the formation of the Sangre de Cristo Mountains lasted 50 million years and caused large fractures to develop in the sedimentary rocks. These fractures, in turn, filled with magma some time after 25 million years ago that intruded from a pair of developing stocks associated with an underlying batholith. This magma then cooled, forming granite. Then *all* of the sedimentary rocks were eroded away, exposing the granite stocks and dikes, and forming a pair of mountains. Imagine how much time *that* process took. And, perhaps the most impressive thing of all in this history is that the past 75 million years is encompassed within the last 2 weeks in the calendar analogy used to represent geologic time (see Table 12.2). In other words, the Spanish Peaks have existed for only a very short period of time within the geologic time scale. *Think about that!*

KEY CONCEPTS TO REMEMBER ABOUT ROCKS AND GEOLOGIC TIME

1. There are three kinds of basic rocks: igneous, sedimentary, and metamorphic.

2. Igneous rocks form when magma cools within the Earth (forming intrusive igneous rocks) or on the Earth's surface (forming extrusive igneous rocks).

3. Sedimentary rocks consist of large layers of formerly loose sediments that slowly compacted and lithified.

4. Metamorphic rocks form when igneous, sedimentary, or other metamorphic rocks are altered by intense heat and pressure.

5. The age of rocks can be confidently dated because radioactive isotopes decay at a known rate, called the half-life.

6. The geologic time scale requires that you think of time in a very different way than humans usually do.

Basic Geomorphology of Continents and Ocean Basins

As you now know, the Earth's crust fundamentally consists of two basic parts—the continental and oceanic crust. Figure 12.22 shows that oceanic crust is the dominant component of the Earth's crust in area, covering about 71% of the planet. This percentage is somewhat misleading, however, because relatively shallow oceans less than 150 m (500 ft) deep border most continents. Such an area is called a **continental shelf** and is frequently formed from sediments eroded from the nearby continent as described in Figure 12.3. These shelf areas have frequently been exposed landmasses in the past, approximately 1.6 million years ago, due to the effects of glaciation which lowered sea level. Their exposure increased the size of the continents to 35% of the Earth while decreasing the oceanic proportion to 65%. From the edge of the continental shelf, the ocean floor drops rapidly to depths of thousands of meters.

At this point, we can introduce a subdiscipline of physical geography called **geomorphology.** Geomorphology is the study of the formation, shape, spatial distribution, and evolution of landforms on the Earth—thus the name *geo* (Earth) *morphology* (shape). In contrast to a *landscape*, which is the overall appearance of a place in terms of its vegetation, topography, or human modifications, a **landform** is a distinct geographic feature, such as a mountain, river valley, coastline, or sand dune, to name but a very few. Figure 12.23 of the Teton Mountains in northwestern Wyoming shows two general kinds of landforms—specifically

---

Continental shelf *The part of the continental margin that extends from the lower water point seaward until the sea floor drops steeply.*

Geomorphology *The branch of physical geography that investigates the form and evolution of the Earth's surface.*

Landform *A natural feature, such as a hill or valley, on the surface of the Earth.*

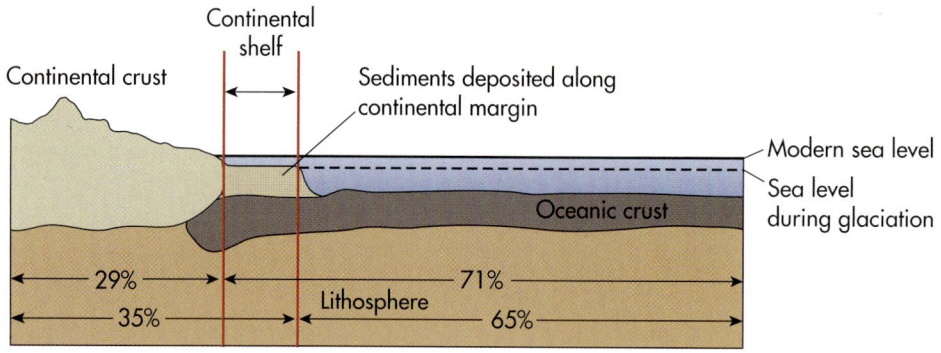

Continental crust · Continental shelf · Sediments deposited along continental margin · Modern sea level · Sea level during glaciation · Oceanic crust · Lithosphere

29% · 35% · 71% · 65%

Figure 12.22 Relative proportion of continents and ocean basins. Ocean basins are the largest geomorphic features on Earth, currently comprising 71% of the planetary surface. Continents, in contrast, make up 29% of the Earth's surface. Continents are some-what larger than they appear, however, because of the shallow and narrow transition zones to deep water. As a result, if sea level falls sufficiently, the relative proportion of the continents can increase, with a corresponding decrease of the ocean basins.

the mountains in the background and a river valley (the Snake River Valley) in the foreground.

Each of these landforms evolved due to distinct geomorphic processes, which will be discussed in later chapters. As you might imagine, the study of geomorphology is rooted strongly in geology because it requires a thorough understanding of how sediments are eroded, transported, and deposited. The discipline is also geographically based because geomorphologists are interested in the way landforms change across space and time. Geomorphology is relevant to everyday life because humans interact with landforms on the Earth's surface every day

in a wide variety of ways. For example, your house or apartment is built upon some kind of geomorphic landform, as is every other type of human structure on Earth. At this very instant, the surface of the Earth is being modified in countless places around the planet, which requires that engineers understand something about geomorphology. Sometimes, the Earth itself moves, through landslides, earthquakes, and other processes that directly impact people, sometimes with catastrophic results.

A basic concept within the discipline of geomorphology is *topographic relief*. Recall from Chapters 2 and 11 that *topography* is a cartographic term associated with the

Figure 12.23 The Teton Mountains and Snake River Valley in northwestern Wyoming. This dramatic landscape is an excellent example of geomorphic features. Geomorphology, a subfield of geography, focuses on the evolution of landforms such as these.

Figure 12.24 Location of major alpine chains on Earth. The Andes is the longest mountain chain, at 7200 km (4500 mi); the Rockies are not far behind, at 6000 km (3700 mi). However, the tallest mountains are in the Himalayas.

shape of the landscape and that *relief* refers to the vertical distance between the high and low places on that landscape. For example, the Teton Mountains is a region of high relief because the difference between mountain peaks and adjacent canyons is great. In contrast, a region such as the Snake River Plain is an area of low relief because it is relatively flat, at least compared to the nearby mountains.

Although the latter part of this book will focus on specific geomorphic processes and landforms, such as sand dune or beach formation, here we look at the basic geomorphic features of the continents and ocean basins. We'll begin by examining the geographic distribution of relief features on the continents.

In general, continents consist of two basic geomorphic regions: alpine chains and continental shields. As you study this portion of the chapter refer to Figures 12.24 and 12.25, which show the basic geomorphology of the continents.

Alpine Chains

Alpine chains are belts of active mountain building that occur either through the growth and eruption of volcanoes or because continental crust is deformed by tectonic processes. (These processes will be described in the next chapter.) Figure 12.24 shows the distribution of alpine chains in broadly curved sections that represent mountain arcs, including the Himalayan Mountains of south-central Asia, the Andes of South America, the Alps of southern Europe, and the Rockies/Cascades of North America. Most of the mountains in these ranges developed during the Cenozoic Era (see Figure 12.19) through tectonic processes, but significant amounts of volcanism have occurred as well, in places such as the Cascades and in Central America. Volcanism has been the dominant process in the numerous island arcs, such as Aleutian and Honshu arcs in the Pacific Ocean. These islands have literally grown out of the ocean through recurring volcanic activity over time.

Continental Shields

As you can see in Figure 12.24, mountain arcs occur over a relatively small area of the Earth's continents. The vast majority of the continents are fundamentally shields that are either exposed or covered by younger rocks (Figure 12.25). In contrast to the active mountain-building zones, continental shields are relatively inactive regions of low relief dominated by very old, stable rock, most of which is igneous or metamorphic. Many geologists like to think of shields as being the *basement* of continents because they are typically composed of the oldest rock (Precambrian) remaining on Earth and were the foundation on which subsequent sediments could be deposited to allow further expansion of the continents.

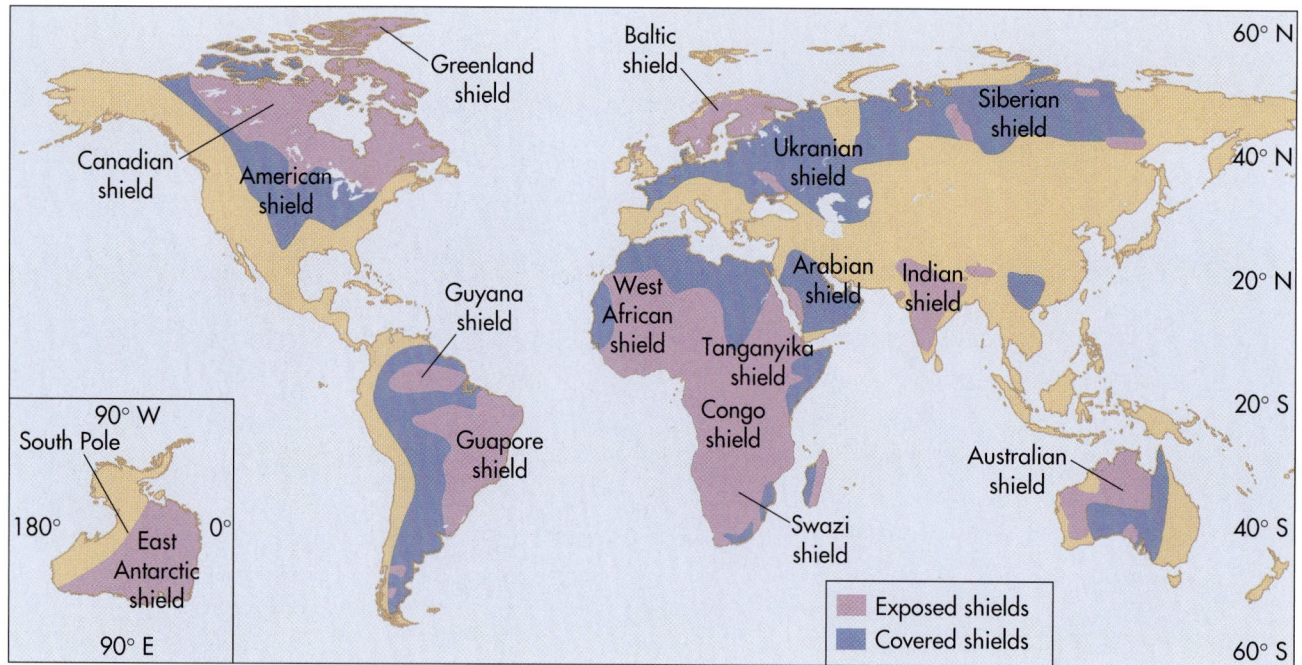

Figure 12.25 Geographic distributions of covered and exposed continental shields on Earth. Exposed shields tend to be better known because they are visible at the Earth's surface; thus, they are named here.

There are two classes of shields: exposed and covered. *Exposed shields* are broad regions, such as in northern Canada and much of Africa (see Figure 12.25), where the basement igneous and metamorphic rocks are exposed at the surface of the Earth. In the case of the Canadian Shield (Figure 12.26), it is exposed in large part because gigantic glaciers stripped away the younger, covering rock during the ice ages. *Covered shields*, in contrast, are areas like central North America and western Asia where the basement igneous and metamorphic rocks are buried by younger sedimentary layers that accumulated at times when shallow seas submerged the shields. These layers of marine sediments are frequently hundreds to thousands of meters (feet) thick and were lifted out of the oceans because the shield areas were broadly arched through tectonic activity.

Figure 12.26 The Canadian Shield. Precambrian igneous rocks are exposed at the surface in much of eastern Canada. Thick deposits of younger rock were stripped off due to continental glaciation that once covered these rocks.

KEY CONCEPTS TO REMEMBER ABOUT THE INTERIOR LAYERS OF THE EARTH

1. The Earth can be divided into continents and ocean basins. Oceans cover most of the Earth.

2. The crust is most dense in the ocean basins and less dense within the continents.

3. Geomorphology is the study of landscape evolution.

4. From a geomorphic perspective, the continents broadly consist of alpine chains and continental shields that are either covered or exposed.

Which one of the following statements is best associated with this image?

a) It is an image of an active extrusive igneous environment.

b) It is an environment where thick deposits of sedimentary rocks are apt to form.

c) It is a picture of oceanic crust.

d) It shows an area of high topographic relief.

The Big Picture

Now that we have discussed basic geology and the internal structure of the Earth, the next logical step is to examine how these concepts relate to landform evolution. In Chapter 13 we will focus on plate tectonics, continental drift, and the origin of mountain chains. These topics are directly related to Chapter 12 because they occur due to movements in the upper mantle. People experience these movements as earthquakes that rattle cities, and volcanoes, such as Mount Saint Helens, that eject material originating from deeper layers within the Earth. We will discuss how these processes operate and the landforms that result when they occur.

Summary of Key Concepts

1. Four major layers occur in the internal structure of the Earth, including (from lowest to highest): (a) inner core, (b) outer core, (c) lower mantle, (d) upper mantle (including the asthenosphere, lithosphere, and crust). These layers are identified on the basis of their chemical composition and solid or molten character. Of particular interest is the oceanic and continental crust. Oceanic crust consists largely of silica and magnesium (*sima*) and is denser than continental crust, which is composed of silica and aluminum (*sial*).

2. There are three basic kinds of rocks: igneous, sedimentary, and metamorphic. Igneous rocks form when magma cools within the Earth (forming intrusive igneous rocks) or on the Earth's surface (forming extrusive igneous rocks). Sedimentary rocks consist of large layers of formerly loose sediments that slowly compacted and lithified. Metamorphic rocks form when igneous, sedimentary, or other metamorphic rocks are altered by intense heat and pressure.

3. The rock cycle describes the ways in which sediments move from one place to another on Earth over time to become new kinds of rock, or how they are transformed to become new kinds of rock. A key concept in this cycle is that rocks gradually break apart as time goes by. The liberated rock fragments are then transported to another place, where they are deposited and gradually lithify to form new rock.

4. The Earth is about 4.6 billion years old. Geologic time refers to the length of time from the origin of the Earth to the present and is hierarchically subdivided into (from longest to shortest time) Eons, Eras, Periods, and Epochs.

5. Geomorphology is the study of landscape forms and their evolution. On a global scale, the Earth's geomorphology consists of continents and ocean basins. The geomorphology of continents can broadly be subdivided into alpine chains and shields.

Check Your Understanding

1. What are the four major layers of the Earth? What is the composition of each?

2. Describe the concept of isostatic depression and rebound. Why do each of these processes occur and why does this concept reflect the relationship between the crust and asthenosphere?

3. Describe the nature of the Earth's inner core and why it has the characteristics that it does.

4. What is the difference between intrusive and extrusive igneous rock? What are some examples of each kind of rock and how do you tell the difference between them?

5. What are the various kinds of sedimentary rock and why do they occur? How does the process of lithification fit into the context of sedimentary rock?

6. What are the two ways that rocks become metamorphic rocks?

7. Describe the flows that occur within the rock cycle. Where does the cycle begin? How do igneous rocks become sedimentary rocks? How do sedimentary rocks become metamorphic rocks? How do the processes of erosion, deposition, and lithification fit within the rock cycle?

8. How is the geologic time scale configured? Using the terms Epoch, Era, Period, and Eon, arrange them into the hierarchy of geologic time.

9. How does the concept of radioactivity fit within the context of calculating the age of ancient rocks? Why are the terms *radioactive isotope* and *half-life* relevant?

10. Describe the two types of crust found on Earth. Where is each type of crust found and what proportion of the Earth does each cover? What is the basic composition of each type of crust?

11. How does a covered shield differ from an exposed shield?

ANSWERS TO VISUAL CONCEPT CHECKS

Visual Concept Check 12.1

There are two types of crust present in this image—oceanic and continental crust. Oceanic crust, which is also called *sima*, is denser than continental crust because it consists largely of silica and magnesium. Continental crust, which is also called *sial*, is less dense because it consists largely of silica and aluminum.

Visual Concept Check 12.2

The answer to this question is *c*. The rocks collectively indicate fluctuating sea levels in the past. Limestone reflects the accumulation of carbonates in a shallow sea, whereas shale reflects the deposition of mud in a very shallow coastal bay or lagoon. So, at this site, the environment was once a shallow coastal bay or lagoon, and then changed to a shallow sea when sea level rose.

Visual Concept Check 12.3

The answer to this question is *d*. The image shows an area of high topographic relief because there is a dramatic difference in elevation between the high and low places on the landscape.

CHAPTER THIRTEEN

TECTONIC PROCESSES AND LANDFORMS

In the previous chapter we described the internal structure of the Earth, the way rocks form and evolve in the lithosphere, and the basic geomorphology of continents and ocean basins. We also discussed geologic time and how this time scale differs markedly from normal human time scales. Now we turn our attention to the ways in which interactions among the crust, mantle, and

Mountains are generally the most spectacular landforms on Earth, as this image of the Fitzroy Mountains in Argentina demonstrates. Mountains such as these form by tectonic processes, which consist of movements within the Earth's crust that deform rocks and cause earthquakes. This chapter focuses on those processes and the landforms that result.

asthenosphere shape the landscape to form mountains and volcanoes, as

well as cause destructive earthquakes. These processes fall under the topic

of plate tectonics and are associated with the kinds of major geologic events

that you frequently hear about.

Plate Tectonics

Recall from the last section of Chapter 12 that the Earth features a variety of alpine chains and continental shields that have a distinct geographic distribution. An important implication of this geography is that over the course of geologic time, mountains have grown and shield regions have gradually risen out of the sea. But how does a mountain grow? How is it possible that marine rocks such as limestones and shales can now be found at the top of mountains thousands of meters above the ocean?

In the first part of this chapter we'll examine these processes more closely by focusing on **plate tectonics,** which is the now-accepted theory that the Earth's crust consists of individual pieces (plates) that move individually and collectively. Understanding the theory of plate tectonics is important because it explains the geographic distribution of mountain chains and why earthquakes and volcanoes occur along plate boundaries. In the second part of the chapter we'll discuss the specific processes and landforms associated with volcanism and earthquakes.

Plate tectonics *The theory that the Earth's crust is divided into a small number of plates that move because they float on the asthenosphere.*

The Lithospheric Plates

A good place to begin our discussion of plate tectonics is the early 20th century, when a German geophysicist named Alfred Wegener noticed that many of the continents look as if they could have fitted together at one time. You can see this pattern yourself by examining Figure 13.1. Notice, for example, how the east coast of South America appears to fit into the southwestern edge of Africa. Wegener looked more deeply into the problem and discovered that portions of some continents share the same kind of rocks and fossils with other continents (Figure 13.1), which is precisely what one would expect if the continents were indeed connected at some point in the past. As a result of these studies, Wegener proposed in 1915 that a supercontinent existed about 300 million years ago that he named **Pangea** (meaning *whole Earth*). He further argued that the continents had drifted apart since that time in a process he called **continental drift.**

The reaction to Wegener's idea in the geologic community was openly hostile, as it directly challenged the

Pangea *The hypothetical supercontinent, composed of all the present continents, which existed between 300 and 200 million years ago.*

Continental drift *The theory that the continents move relative to one another in association with plate tectonics.*

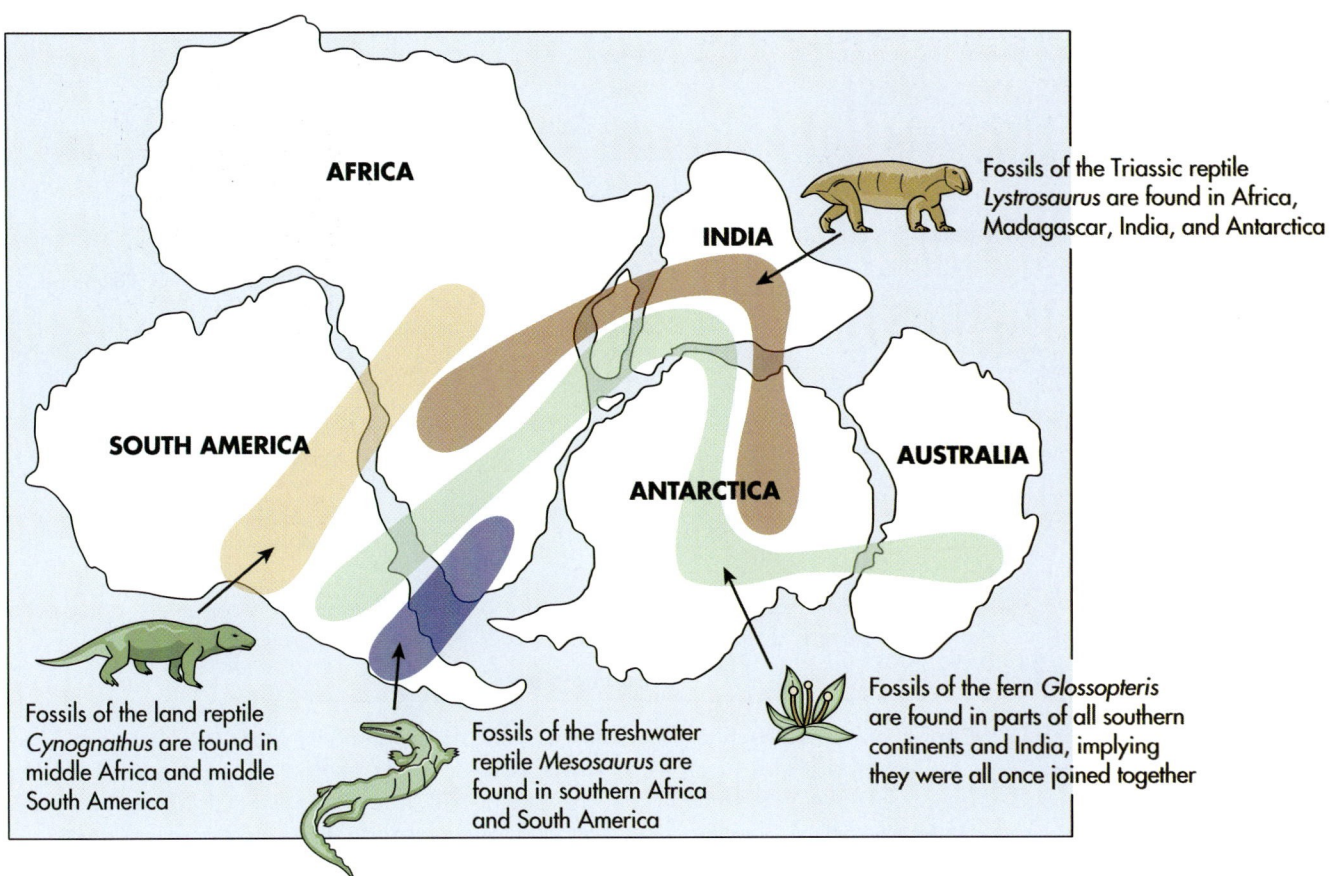

Figure 13.1 Fossils and continental drift. A strong line of evidence supporting the theory of continental drift is the occurrence of specific fossil plant and animal species on continents that are now far apart. (Credit: United States Geological Survey)

long-standing belief that the continents had been connected by a variety of land bridges at one time. In addition, his explanation for how the continents moved, specifically that they "plowed" through the denser oceanic crust, was physically impossible. As a result, Wegener's ideas were largely ignored until the 1950s, when a systematic study began of the ocean basins around the world. In addition, geologists began to investigate the cause of earthquakes more closely. These combined studies demonstrated that the Earth's crust consists of a series of interconnected plates that move (Figure 13.2). This new understanding was the foundation for the theory of plate tectonics that emerged in the 1960s. The theory of plate tectonics was revolutionary because it accurately explained a great deal of volcanic and earthquake activity on Earth, as you'll see later in this chapter. It also validated Wegener's basic premise of continental drift because it provided the mechanism for such migration. Unfortunately, Wegener perished on an expedition to Greenland in 1930 and thus did not live to see his theory vindicated.

Plate Movement

Given that continental drift has apparently occurred throughout time, the next question is *why* does it occur?

In general, the process occurs because the plates float on the asthenosphere and are moved by convection processes within it. Recall from Chapter 6 that convection occurs in the atmosphere when solar energy heats the ground surface and the warmed ground heats the overlying air. As the warm air rises, the cool air sinks. The same pattern of rising and sinking currents also happens inside the Earth. In contrast to atmospheric convection, which is driven by an outside source (the Sun), convection within the Earth occurs because radioactive decay of elements such as uranium, deep within the Earth, creates very high temperatures between $1000°$ and $2000°$ C ($1800°$ to $3600°$ F).

These high temperatures cause plumes of magma to rise slowly by convection within the mantle and into the asthenosphere (Figure 13.3). As these plumes reach the base of the crust, they spread horizontally and cool, moving segments of the Earth's crust in the process. To visualize this process, consider a mechanized people mover in the floor of an airport terminal or mall. This device works somewhat like a treadmill that continuously loops around. If you stand on it where the tread rises out of the floor, you're carried

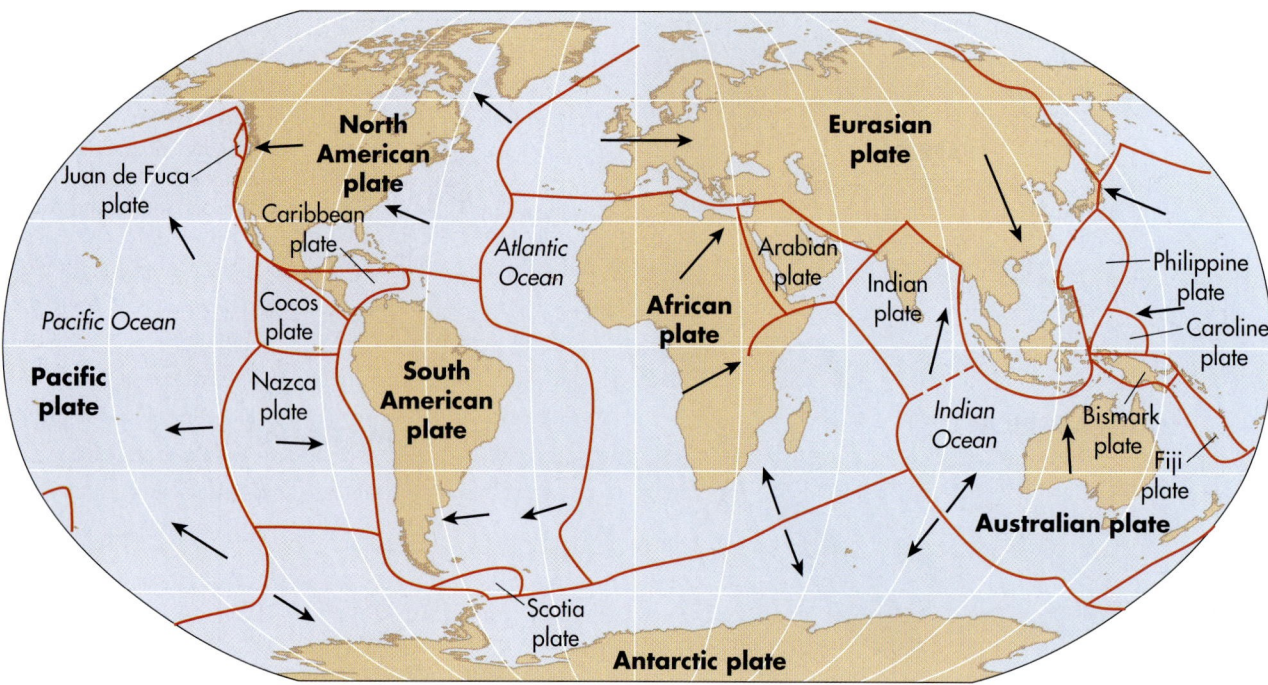

Figure 13.2 Geographic distribution of lithospheric plates. The seven major plates (boldface type) cover 94% of the Earth; about a dozen smaller plates cover the remaining 6% of the Earth. Notice the "jigsaw" appearance of the map and the way that the plates move (arrows) relative to one another. (The boundary between the Indian and Australian plates is uncertain; many sources refer instead to the "Indo-Australian" plate.)

forward. The process of continental drift works in somewhat the same way, with plates carried forward by the moving magma beneath. Ultimately, the magma slowly cools, solidifies, and sinks back into the mantle, where it melts again. In this sense, a magma convection current is somewhat similar to a tropical Hadley cell in the atmosphere.

How do we know that the plates are moving? For one thing, we can now measure the motion through remote sensing and global positioning systems. In California, for example, sensors placed on either side of major faults actually move relative to one another over the course of time, indicating that the underlying plates are shifting. Evidence of plate movement also appears in the age of the sea floor. Convection in the mantle brings magma up through fractures in the crust, where it extrudes onto the sea floor, cools, and forms new oceanic crust. Therefore, the ocean floor where spreading occurs is much younger than other locations. Figure 13.4 shows isochrons (lines of equal age) for the ocean floors around the world, and clearly indicates areas where oceanic plates have been spreading apart. You can see that in the Atlantic Ocean basin the oldest part of the oceanic crust is along the east coast of North America, where it is about 180 million

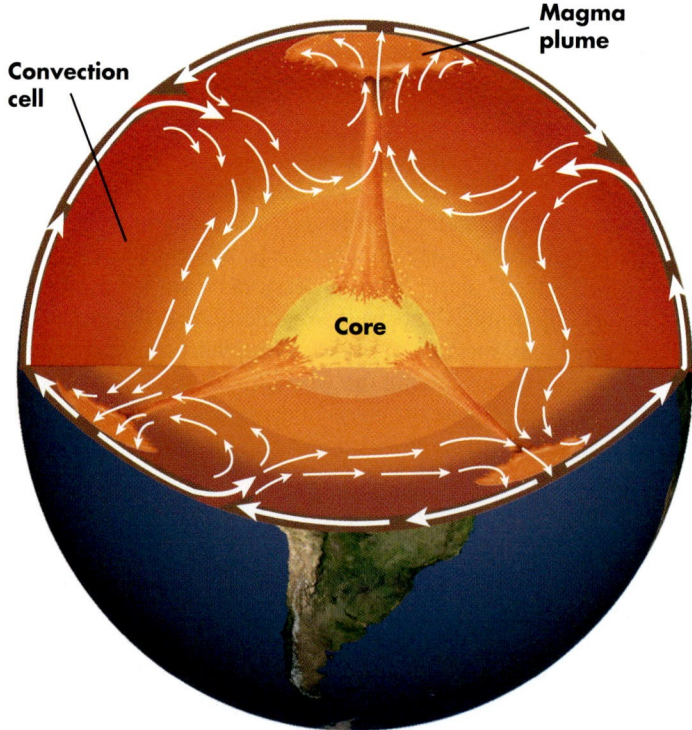

Figure 13.3 Convection plumes within the Earth. Magma plumes rise from deep within the Earth and spread apart at the crust. Rock created as the magma cools subsequently sinks back into the mantle, where it melts again.

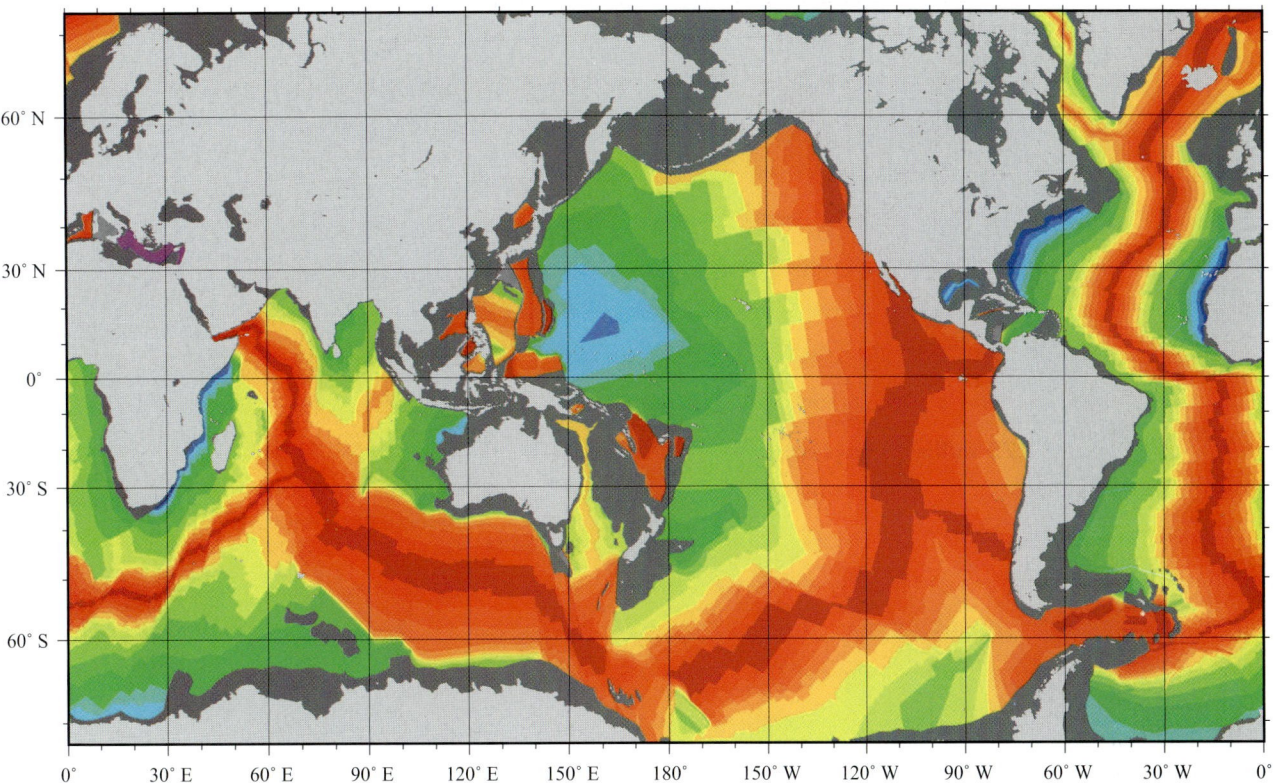

Figure 13.4 Sea floor ages. In this map colors are associated with sea floors, whereas the continents are gray. Notice how the youngest sea-floor rocks (in red) occur in the zones of sea-floor spreading, such as in the Atlantic Ocean, with progressively older rocks (yellow, greens and blues, respectively) away from these regions.

years old. In contrast, the sea floor in the center of the Atlantic Ocean is less than 10 million years old. Overall, the rate of crust movement appears to range from 1 to 15 cm (0.4 to 5.9 in) annually.

As a result of the new understanding that emerged in the 1960s about the behavior of oceanic crust and the age of the seafloor, it's been possible to reconstruct the patterns of continental drift through time. Geologists generally agree that the continents began to coalesce to form Pangea as early as three hundred million years ago, during the Pennsylvanian Period. If you recall from Chapter 12, this is the period of time when the Earth was much warmer and massive amounts of carbon began to accumulate in steamy swamps. Figure 13.5a shows how Pangea probably looked around 275 million years ago. Notice that South America and Africa were indeed joined at this time, as were the eastern coast of North America and Africa. Approximately two hundred million years ago, most of the continents began slowly to drift apart to reach their present positions. An important exception is the Indian landmass, which "raced" across what is now the Indian Ocean to collide with the southern side of the Asian continent. As we will see later, this collision created the Himalayan mountains.

Types of Plate Movements

Various things can happen on the Earth's surface when mantle convection moves segments of the Earth's crust around. Visualize it this way: if you and a friend are each in separate bumper cars, what are the possible ways you can move in relation to each other? You could move away from each other, you could move side by side past each other, or you could move toward each other and eventually crash. The same movements happen with crustal plates on Earth. In this section we describe the various types of plate movements and explain what we would expect to see on the Earth's surface from each type of movement.

Passive Plate Margins

The simplest kind of tectonic interaction occurs at **passive plate margins.** As the name implies, passive plate margins are relatively stable localities from a

Passive plate margin *A place where the continental crust and the oceanic crust are on the same tectonic plate and thus do not move relative to one another.*

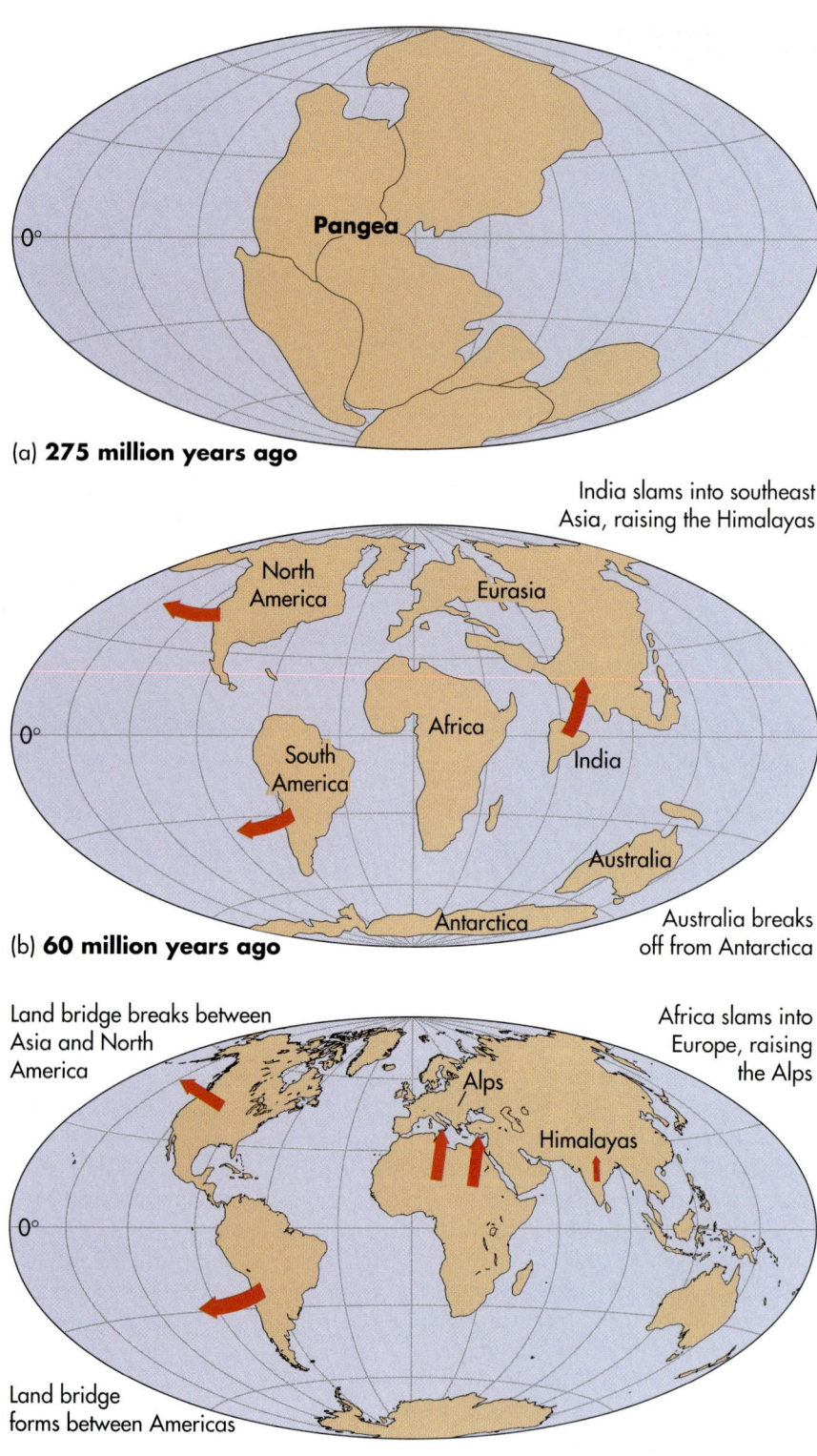

(a) 275 million years ago

India slams into southeast
Asia, raising the Himalayas

(b) 60 million years ago

Australia breaks
off from Antarctica

Land bridge breaks between
Asia and North
America

Africa slams into
Europe, raising
the Alps

Land bridge
forms between Americas

(c) Today

Figure 13.5 Continental drift in the past 275 million years. (a) Geography of the Pangea supercontinent during the Permian period about 275 million years ago. At this time, all the continents were connected to form one giant landmass. (b) Geography of continents at the beginning of the Cenozoic Era about 60 million years ago. Note the relative direction of continental movement. (c) Configuration of continents today. The continents continue to move around the surface of the Earth and will eventually change the world's appearance again.

CONTINENTAL DRIFT

To see how the plates have drifted over time, go to the *GeoDiscoveries* website and access the module **Continental Drift**. This animation shows the position of continental landmasses during different intervals of geologic time. Use the controls in the animation to play it, to step forward or backward in time, or to rewind to the beginning of the animation. You can also put the individual continental positions into the context of the geologic time scale by accessing specific time periods such as the Triassic or Cretaceous. After you complete the animation, be sure to answer the questions at the end of the module to test your understanding of continental drift.

tectonic perspective. These regions frequently occur in places where the continental crust and the bordering oceanic crust are actually on the same tectonic plate. One such place is the eastern seaboard of North America, which has a passive margin with the oceanic crust in the western part of the Atlantic Ocean basin. Even though the continental crust that forms the North American landmass is a separate rock body from the adjoining oceanic crust in the Atlantic Ocean, they both lie within the North American plate (Figure 13.6) and thus have a passive margin in the sense that it is not geologically active.

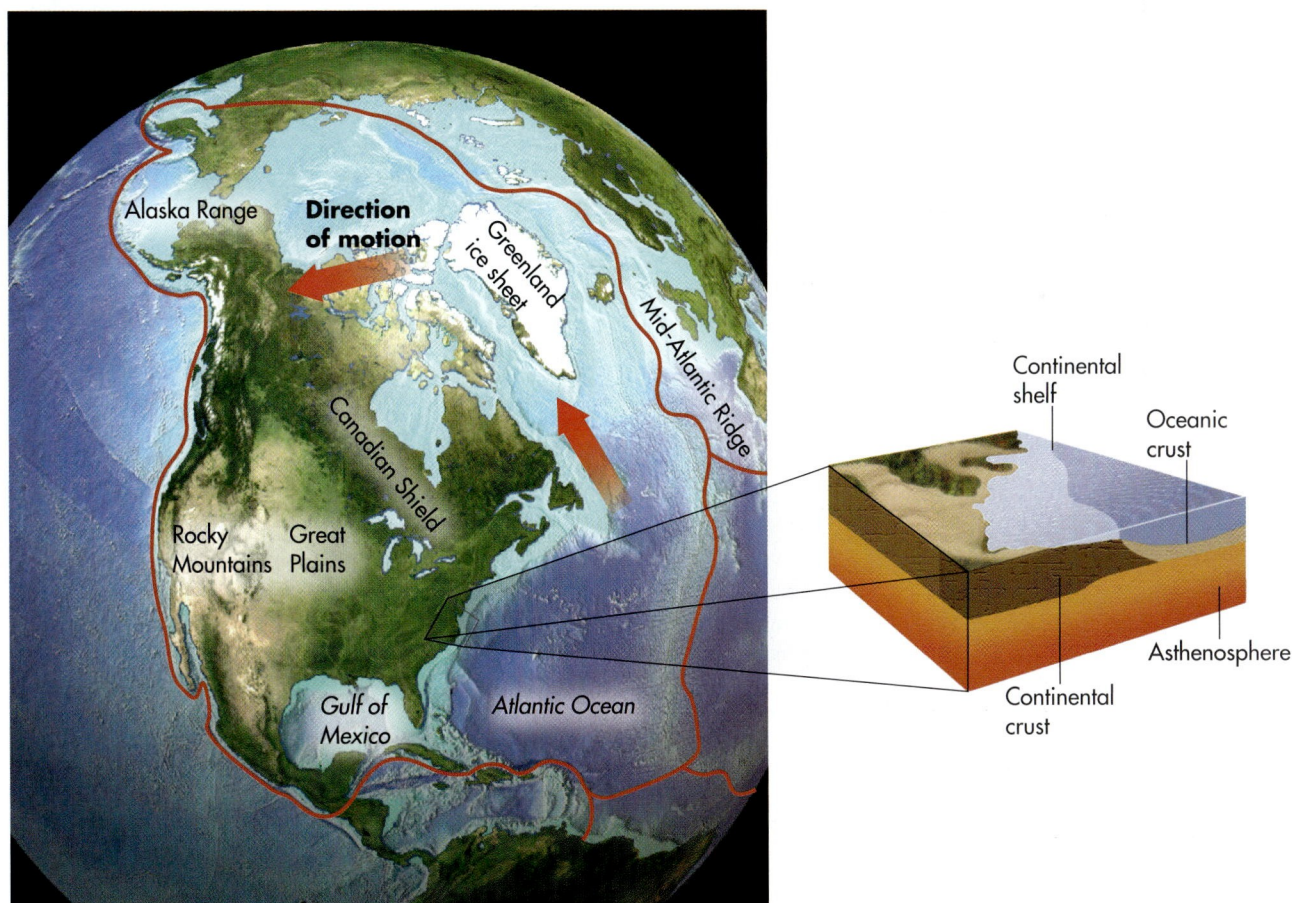

Figure 13.6 Passive margin along the eastern edge of North America. The continental crust of North America meets the oceanic crust of the Atlantic Ocean. Nevertheless, they are part of the same tectonic plate and are slowly moving as one body to the northwest.

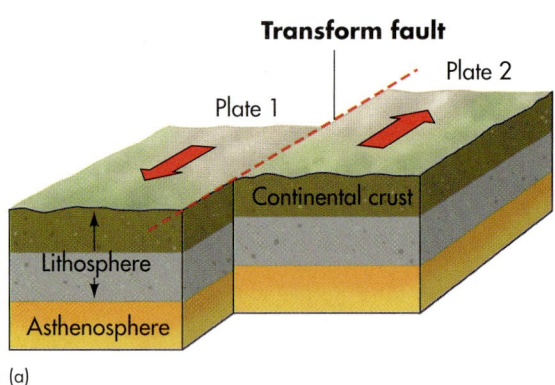

(a)

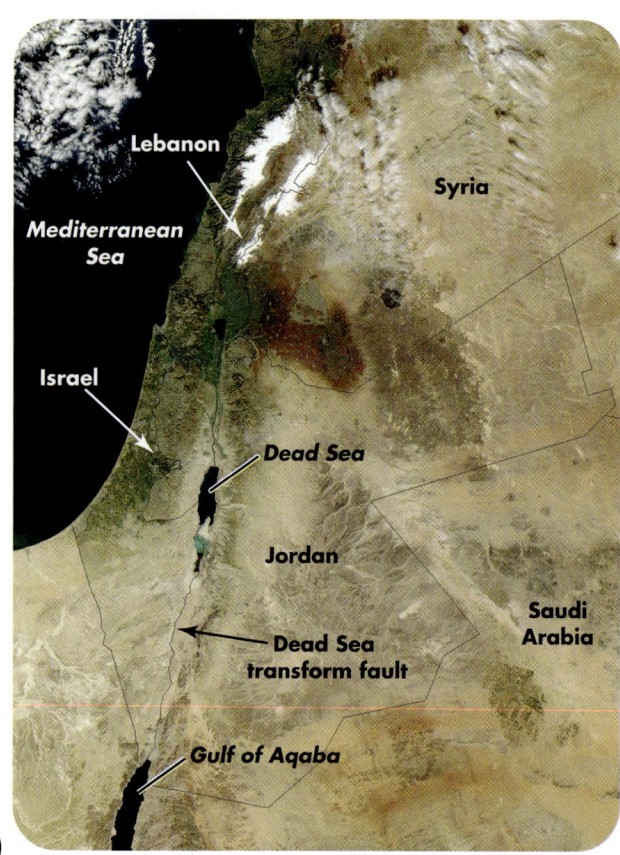

(b)

Figure 13.7 Transform plate margin. (a) Plates that meet at transform margins move horizontally past each other along transform faults. (b) Satellite image of the Dead Sea Fault. This transform fault occurs along the boundary of the Arabian and African plates and extends from the Gulf of Aqaba on the southern tip of Israel through the Dead Sea and north into Lebanon. The Arabian plate is generally moving to the north, whereas the African plate is moving in a southerly direction in this region.

Transform Plate Margins

Another type of plate boundary is the **transform plate margin.** In contrast to passive plate margins, transform boundaries are places where plates slide horizontally past each other (Figure 13.7a). At transform margins, the plane of motion is along a nearly vertical break (or fault) that extends through the entire lithosphere. The San Andreas Fault in California and the Dead Sea Fault along the border of Israel and Jordan (Figure 13.7b) are excellent examples of transform faults. We'll examine the San Andreas Fault later in the chapter when we discuss earthquakes.

Plate Divergence

Plate divergence is the process by which lithospheric plates move away from one another. This type of movement occurs in regions where rising magma plumes within the Earth move upward and outward between plate fractures, spreading the Earth's plates apart in a process called **rifting** (or divergence). A great place to see the process of rifting is on the sea floor. Recall from the earlier discussion about the evidence for plate tectonics that the youngest sea-floor ages

occur where magma extrudes through fractures in the oceanic crust (Figure 13.8). This extrusion of magma produces a ridge-like feature, appropriately called a **mid-oceanic ridge,** that lies parallel to the rift zone. The Mid-Atlantic Ridge is perhaps the best-known oceanic ridge and is located, as its name suggests, at the middle of the Atlantic Ocean along the Atlantic rift zone. If these plates are diverging, what do you suspect is happening to the width of the Atlantic Ocean? The answer is that it's slowly increasing every year. This process began when Pangea began to break apart about 200 million years ago and continues to this day.

Plates can also diverge within continents, causing a split in the landmass. One of the best-known locations of continental plate divergence is in eastern Africa. In this region, continental rifting is occurring in three places that merge at a single place known as a *triple junction*. Figure 13.9 shows that this triple junction is in close geographic association with the Arabian, African, and Indian plates. Two of the rifts are oceanic, with one opening the Red Sea and the other causing enlargement of the Gulf of Aden on the southern side of Saudi Arabia. On land, the third rift, the East African Rift, extends in a southerly direction from the juncture of the

Transform plate margin *A plate boundary where opposing plates move horizontally relative to each other.*

Rifting *The spreading apart of the Earth's crust by magma rising between fractures in the Earth's plates.*

Mid-oceanic ridge *A ridge-like feature that develops along a rift zone in the ocean due to magma upwelling.*

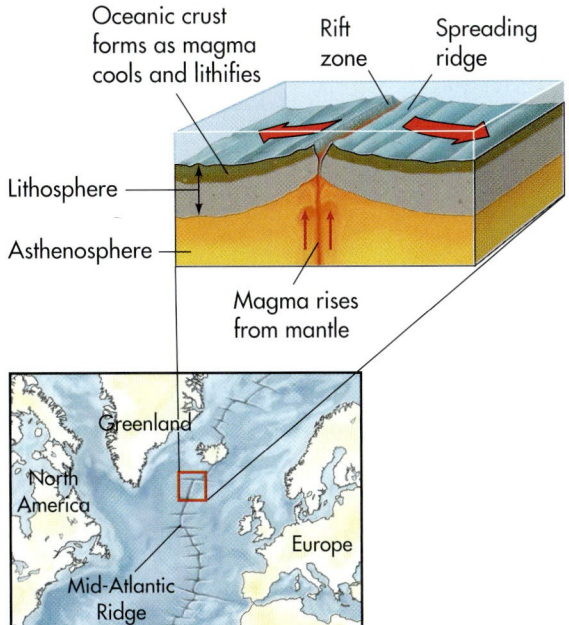

Figure 13.8 Oceanic rifting, sea floor spreading, and plate divergence. In ocean basins, the sea floor spreads where plumes of magma interact with the crust, creating a rift. One of the best examples of sea-floor spreading exists in the middle of the Atlantic Ocean.

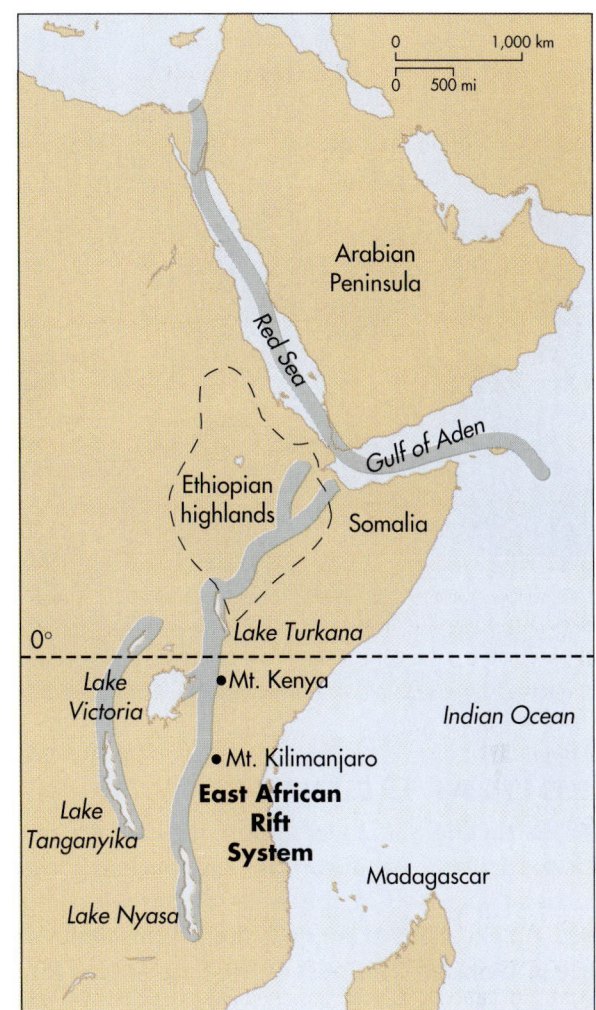

(a)

(b)

(c)

Figure 13.9 The Great Rift system in eastern Africa and the Middle East. (a) The triple junction of the Gulf of Aden, the Red Sea, and the East African Rift is the meeting place of three plate boundaries. (b) Satellite image of the triple junction with view to east/south-east. The Red Sea is to the upper left, the Gulf of Aden is to the upper right, and the East African Rift is in the foreground. (c) The Suguta Valley in Kenya is part of the East African Rift; note the steep bluffs bordering the valley.

AMAZING PLACES

OLDUVAI GORGE

In reconstructing human evolution, the East African Rift is particularly significant because ancient rocks have been exposed in the walls of the widening chasm. An excellent example is the famous Olduvai Gorge in Tanzania. This part of the East African Rift

This cliff exposure at Olduvai Gorge contains rocks up to two million years old and consists of alternating layers of volcanic, lake, and river sediments.

system is bordered by high cliffs exposing rocks that are about two million years old. These rocks contain numerous hominid fossils, such as the *Australopithecus boisei* skull pictured on the following page, that are important pieces in the human evo-

Red Sea and the Gulf of Aden. Along this rift, the eastern part of the African continent is slowly splitting in two, resulting in several large lakes, such as Lake Victoria.

Transform plate margins and plate divergence do not necessarily occur separately; they can also happen at the same time and place. Because plate divergences usually occur along jagged lines, plates often strike against each other as they diverge. Where diverging plates rub against each other, a transform margin occurs.

Plate Convergence

We have just seen what happens when plates diverge from one another at rift zones. Now let's turn to the processes in which plates are coming together in association with plate convergence. Plate convergence occurs through the processes of collision, when plates directly impact each other, or subduction, when one plate is forced beneath another.

Collision Plate collision results when two converging plates meet. This process usually occurs when two plates of continental crust meet because they have similar density and thus one does not ride over or below the other. In other words, they smash together like two cars in a head-on collision. The crust, like the cars, crumples, causing folding of the formerly horizontal bedrock through compression (Figure 13.10). Try this for yourself with a piece of notebook paper. Start by laying it flat on a desk or table. Then compress the paper by pushing the left and right edges of the sheet toward one another. What happens? You should see that one side of the paper is folded up, whereas the other side is folded down. The same thing happens to rocks when the compressive force is large enough.

You can see evidence for past compression and folding of rocks by looking at the orientation of rock

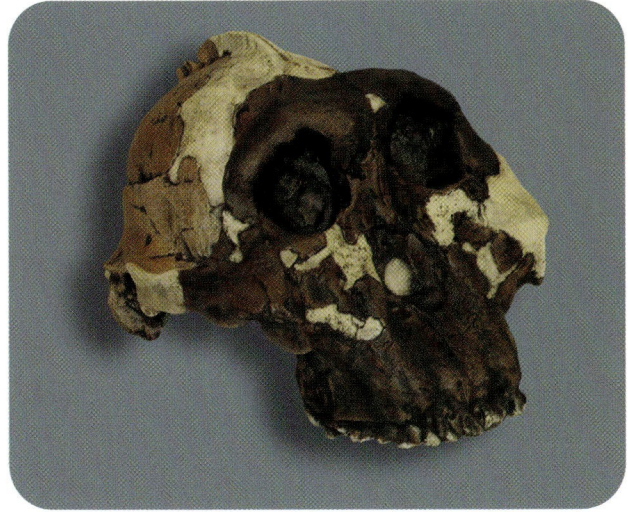

This *Australopithecus boisei* skull was discovered in the cliff walls of Olduvai Gorge and is approximately 1.8 million years old. Fossils like this one are significant clues about the evolution of human beings and might not have been discovered if the rift system had not been present.

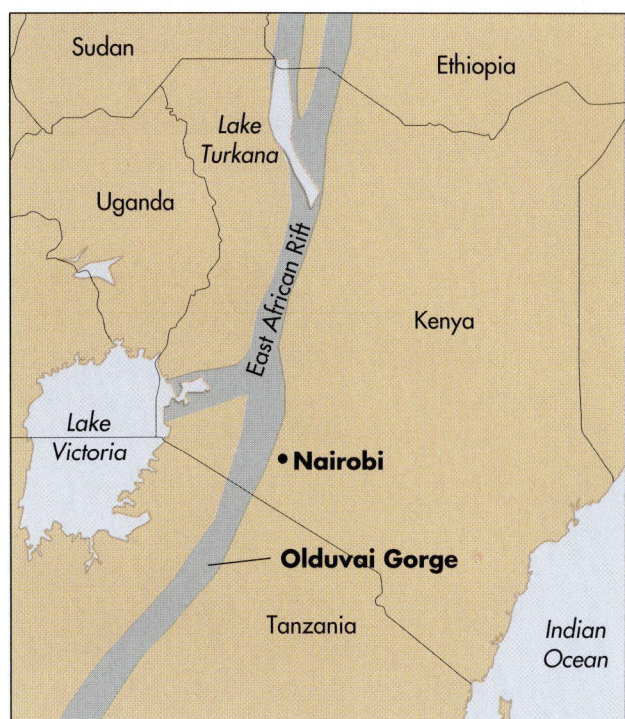

Olduvai Gorge in East Africa. Olduvai Gorge, located in the northern part of Tanzania, is part of the East African Rift system.

lutionary puzzle. The age of these fossils has been determined by dating the volcanic rocks (by potassium-argon or argon-argon dating) that lie immediately above or below the sediments in which the fossil remains lie. *Australopithecus boisei*, for ex-

ample, lived about 1.8 million years ago. If it weren't for the cliff exposures created by the rift, these fossils might not have been discovered.

layers, or **rock structure.** Figure 13.10 illustrates how rock layers are folded and the various features that result. In general, a close correlation exists between the amount of compression and the nature of the fold. A **monocline** is a one-sided slope where beds of horizontal rock are inclined in a single direction over a large area. This kind of fold occurs when the amount of compression is relatively low. An excellent example of a monocline occurs in southeastern Utah, near the town

of Mexican Hat (Figure 13.10c). More extensive compression causes anticlines and synclines to form (Figure 13.10a). An **anticline** is a portion of the fold where the rock layers arc upward to form a concave arch along the fold axis. One-half of such a fold is called a *limb*. In contrast, a **syncline** is the portion of the fold where rock layers dip downward to form a convex trough. If the collision is especially intense, the rocks can be folded so much that an **overturned** (or

Rock structure   *The internal arrangement of rock layers.*

Monocline   *A geologic landform in which rock beds are inclined in a single direction over a large distance.*

Anticline   *A convex fold in rock in which rock layers are bent upward into an arch.*

Syncline   *A concave fold in rock in which rock layers are bent downward to form a trough.*

Overturned fold   *A structural feature in which the fold limb is tilted beyond vertical, which results in both limbs inclined in the same direction, but not at the same angle.*

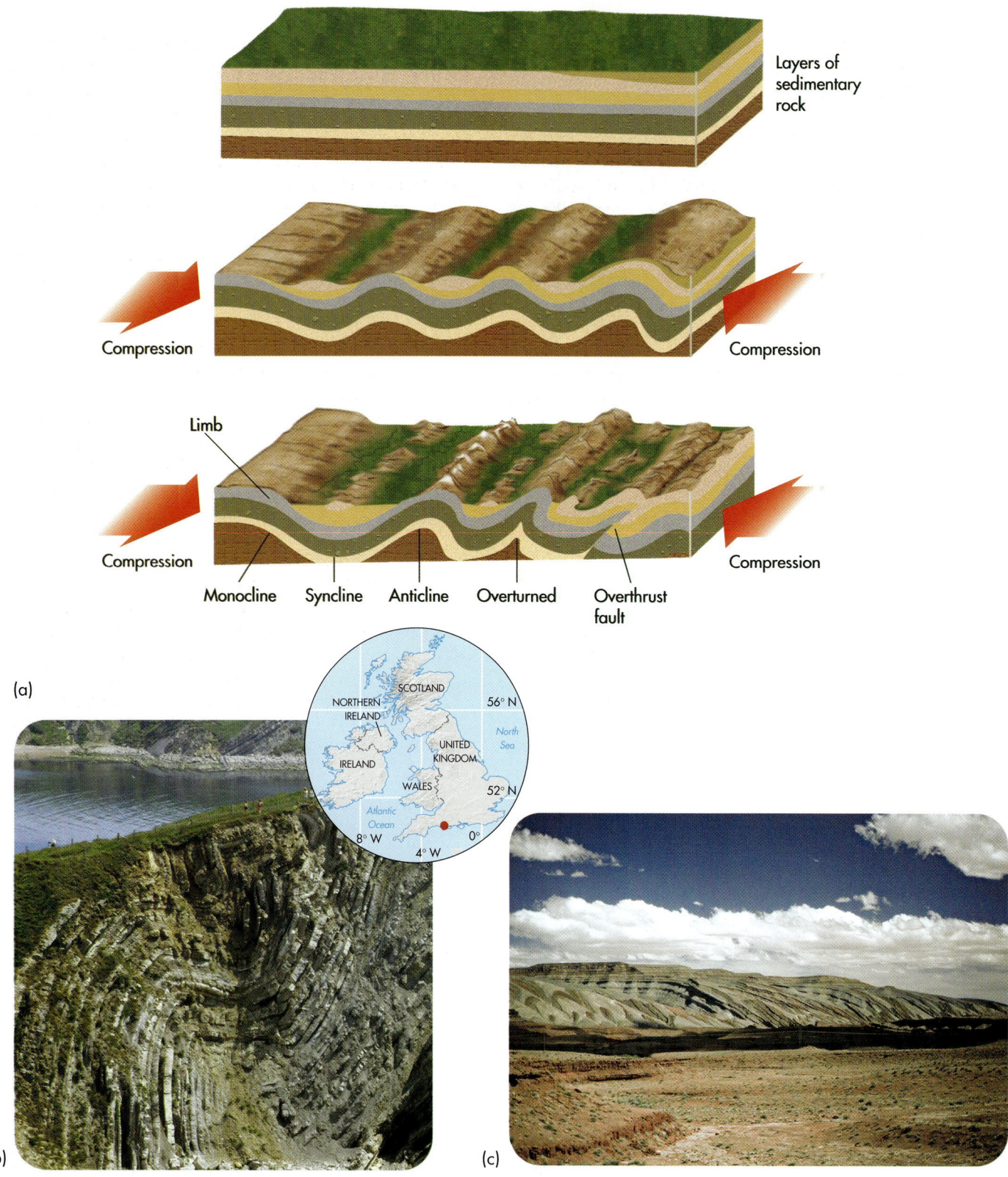

Layers of sedimentary rock

Compression — Compression

Limb

Compression — Compression

Monocline Syncline Anticline Overturned Overthrust fault

(a)

SCOTLAND
56° N
NORTHERN IRELAND
North Sea
IRELAND
UNITED KINGDOM
WALES
52° N
Atlantic Ocean
8° W
4° W
0°

(b)

(c)

Figure 13.10 The folding process. (a) The vast majority of rocks lithify within horizontal beds. If these rocks are subsequently compressed by plate collison they are deformed in some way. In many places, folded rocks are associated with mountains and valleys. (b) Folds in a rock outcrop in the United Kingdom. These rocks were once horizontal but have been bent in several different directions by compressional forces. (c) The Comb Ridge Monocline near Mexican Hat, Utah. Compressional forces produced a single fold here that connects two beds of relatively horizontal sedimentary rock. One of these beds of horizontal rock is on the top of the mountain, whereas the other underlies the landscape in the foreground.

recumbent) **fold** results. Still more compression results in an **overthrust fold,** which occurs when one part of the rock mass is shoved up and over the other.

Evolution of the Appalachian Mountains

Large-scale compression of rocks is frequently associated with mountain building. Such an event usually occurs within a distinct period of geologic time called an **orogeny.** There are many excellent examples on Earth of how continental collision and compression cause orogenies in which rocks are intensively folded. One such example is the Appalachian Mountains, which extend approximately 2500 km (~1600 mi) along the eastern United States. Several phases of compressional mountain building have occurred in this region, with the earliest being the *Taconic Orogeny* during the Orovician Period (~450 million years ago). Following this period of mountain building, another interval of uplift called the *Acadian Orogeny* occurred during the Devonian Period (~375 million years ago). Both the Taconic and Acadian Orogenies were centered in the northeastern part of the Appalachians and resulted in mountain ranges such as the Adirondacks and the Catskills in New York.

Following the Acadian Orogeny, mountain building in the Appalachians shifted to the area from Pennsylvania to Alabama. This area was intensively folded when the northwest corner of Africa collided with the east coast of North America at the time Pangea formed. This period of mountain building is known as the *Allegheny Orogeny* and occurred during the Permian Period between 290 and 248 million years ago. Deformation of rocks during this orogeny resulted in rugged mountains that may have been between 5000 m and 8000 m (~16,400 and 26,200 ft) high. In other words, if you could go back in time you would have seen mountains in what is now Virginia that looked much like the modern Rocky Mountains in Colorado, or maybe even the modern Himalayas in Tibet. Many of the mountains within the Appalachians are underlain by overthrust faults (see Figure 13.10a) where portions of the African plate were shoved up and over part of the North American plate. Over the past 200 million years the towering mountains that once existed have been eroded to well-rounded landforms, such as those seen in the Great Smoky Mountains in North Carolina, that are less than 2000 m (6550 ft) high.

The Appalachian Mountains are subdivided into several different subregions. Perhaps the best known of these regions is the Ridge and Valley Province, which extends along much of the length of the Appalachian range. As the name implies, the landscape here consists of a series of distinct ridges and valleys that parallel one another. Figure 13.11a shows a satellite view of this pattern of alternating ridges and valleys. A characteristic ground-level view is seen in Figure 13.11b.

Given the topography of the Ridge and Valley Province, it's tempting to assume that the ridges are anticlines because they arch up and the valleys are synclines because they bow downward. In fact, the structural/topographic relationships are much more complex than you might think. To see the complexity of these structures and how they relate topographically, examine Figure 13.12, which shows a general time sequence illustrating how the landscape evolved. Part (a) shows what the topography might have looked like after it was initially folded during the Allegheny Orogeny. Note the well-developed anticlinal ridges and synclinal valleys across the landscape. During and after the orogeny, a parallel network of streams developed on the landscape that began to vigorously erode the uplifted rocks. However, these streams flowed on rocks of variable resistance (or hardness), so that portions of the landscape were eroded more easily than others. In addition, erosion focused on the highest points, which were the crests of the original anticlines.

As a result of the preferential erosion that took place, the relationship between surface topography and underlying structure became less clear. Look at the ridge marked *A* in Figure 13.12b, for example. This ridge was initially underlain by an anticline (Figure 13.12a). Erosion removed the crest of the ridge, however, forming a valley parallel to the underlying structure. The ridges on each side of the valley are remnants of the limbs of the anticline. Such a ridge is called a **hogback ridge** because one side of the ridge is steeper than the other. Also note that the valley formed between the two hogbacks at point *A* In Figure 13.12b is underlain by the original anticline; that is, the rocks arch up there. In other words, the stream eroded into the core of the anticline, forming a valley within it along its length. Such a topographic feature is called an **anticlinal valley.** A **synclinal valley** lies immediately to the right of the hogback ridge-and-valley network at point *A* and is what you'd expect given the shape of the

---

Overthrust fold *A structural feature where one part of the rock mass is shoved up and over the other.*

Orogeny *A period of mountain building, such as the Alleghany Orogeny.*

Hogback ridge *A ridge underlain by gently tipped rock strata with a long, gradual slope on one side and a relative steep scarp or cliff on the other.*

Anticlinal valley *A topographic valley that occurs along the axis of a structural anticline.*

Synclinal valley *A topographic valley that occurs along the axis of a structural syncline.*

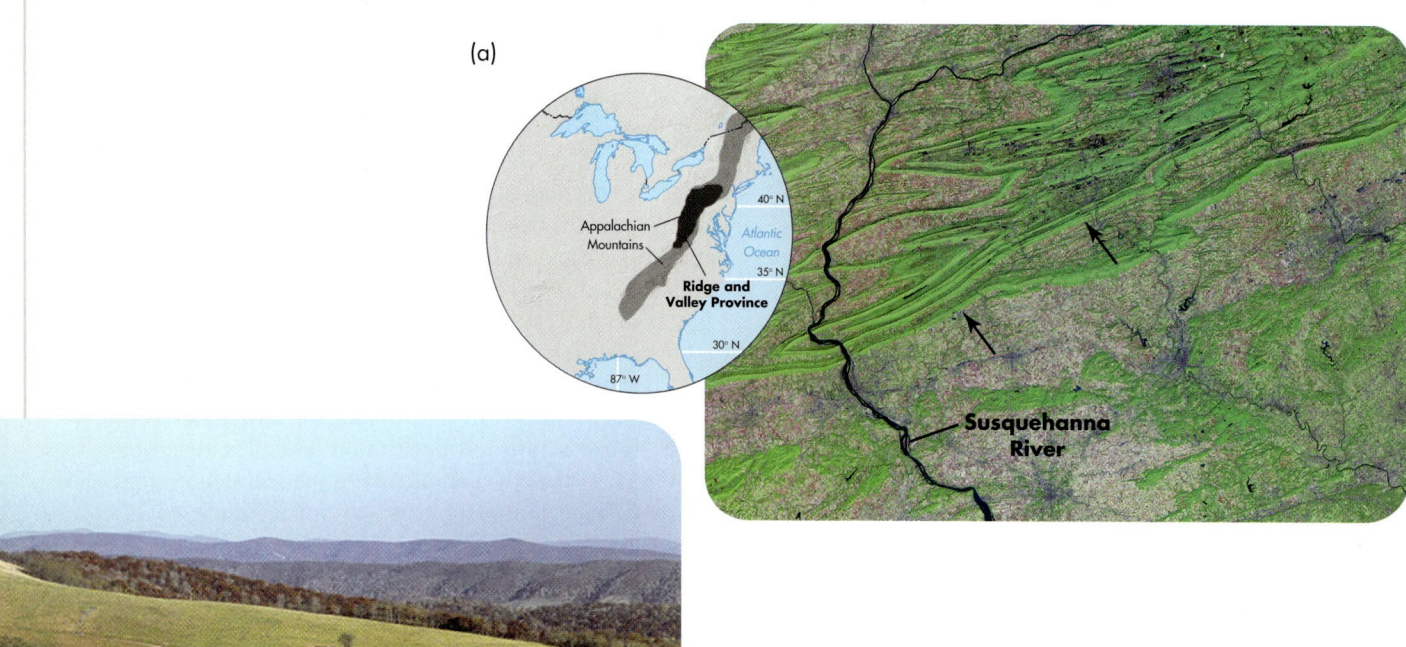

(a)

(b)

Figure 13.11 Appalachian Ridge and Valley. (a) Infrared satellite image of a portion of the Ridge and Valley Province in central Pennsylvania. The ridges (arrows) are covered by forest, which makes them stand out particularly well. (b) Typical panoramic view in the Ridge and Valley Province.

low ground there. To the right of that valley is another hogback ridge, which borders yet another anticlinal valley. See how the valley is cut into rocks that arch up?

Now look at point *B* in Figure 13.12b. A ridge exists at the surface here, but it's actually underlain by a syncline. See the U-shape of the rock stratum that caps the ridge? This is the same rock body that forms the hogback ridge to the left and now caps the ridge here, because much of the former anticline to the left was removed by erosion. In other words, the topography at point *B* has been inverted so that what was once the valley floor (point *B*, upper image) is now the crest of a ridge (point *B*, lower image). This type of inversion could occur only if the rock that formerly was in the valley bottom (upper image) is very hard and resistant to erosion relative to the rock that capped the anticline at the beginning of the sequence. Such complex topographic/structural relationships demonstrate that landscapes are sometimes more complex than they appear. They also illustrate how much *time* must be involved for this kind of landscape to evolve. First there had to be uplift and then *substantial* amounts of erosion. Overall, it took about 300 million years!

Although the Appalachian Mountains are a great example of how collision produced a landscape, it's certainly not the only example. Crustal collision continues to uplift

mountains in several prominent places around the world. One place you've probably heard of is the famous Alps in southern Europe. This mountain range forms the border between Austria and Italy and began to form during the Cretaceous Period (144–65 my ago) when the continental plate of Africa collided with the southern side of the European plate. A second orogeny occurred during the Tertiary Period (65–2 my ago). An interesting aspect of the Alps geology is that part of the mountain range consists of rock material from the African plate. In other words, part of the European landmass is actually rock from Africa.

Another mountain range that has formed largely through continental collision is the Himalaya Mountains (Figure 13.13), which border India and China. You're probably aware that Mt. Everest is the highest mountain in the world. What you may not have known is that this peak, as well as the entire Himalayan chain, has been uplifted because the northeastern part of the Indian plate has been slamming into the southern side of the Eurasian plate for about the past 30 million years. As a result of this collision the mountains have shot out of the ground, at least as far as geologic time is concerned, uplifting metamorphic and sedimentary rocks (including marine fossils) thousands of meters above sea level.

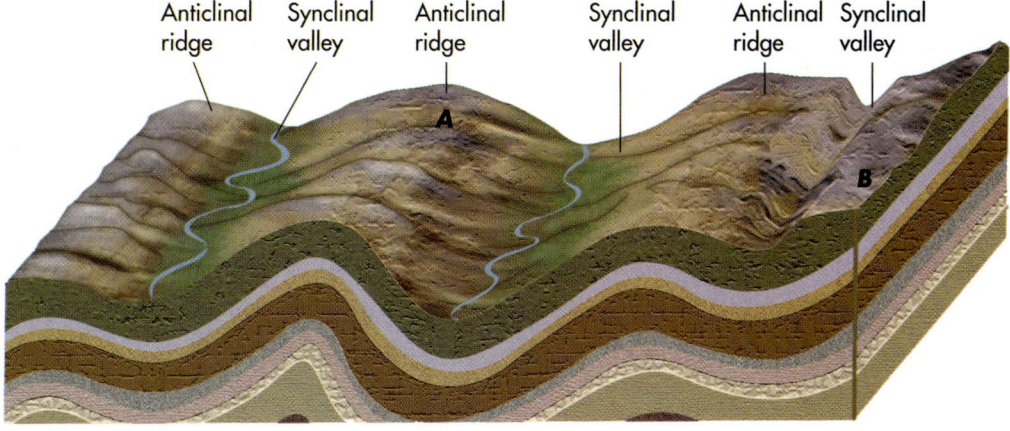

(a) Initial folding during Allegheny Orogeny

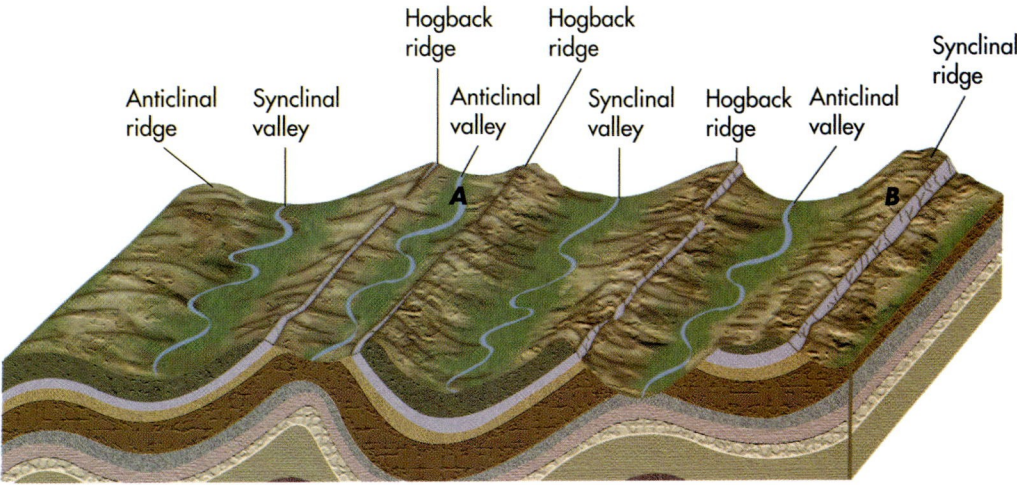

(b) After erosion of softer rock layers

Figure 13.12 General evolution of the Ridge and Valley Province. (a) The landscape shortly after folding. Anticlines form the high ground, whereas valleys occur in synclines. (b) As time progresses, erosion modifies the landscape. Anticlines at *A* and *B* are eroded by streams, forming valleys in upward-arching structures. Hogback ridges exist on both sides of the anticlinal valley at *A*, where limbs of the former anticline form high ground. To the right at point *B*, extensive erosion has inverted the topography so that the ridge is underlain by a syncline.

Figure 13.13 The Himalaya Mountains. Mount Everest is the tallest mountain in the world at 8850 m (29,035 ft). The 25 tallest mountains in the world are all in the Himalayas, including all 14 peaks that are over 8000 m (~26,250 ft). This mountain chain extends in an arc about 2400 km (about 1500 mi) long and forms the border between India and Tibet. The Himalayas formed due to collision of the Indian plate with the Asian landmass.

FOLDING

A pair of animations on the *GeoDiscoveries* website nicely illustrates the process of crustal folding. The first one is the module **Folding**, which animates the fundamental process of rock deformation by compression. As you watch this animation, be sure to think about geological structure and note how anticlines, synclines, and other folds develop.

After you watch the *Folding* module, turn your attention to the module **The Folded Appalachians**. This animation focuses specifically on the Appalachian Mountains and will give you a better feel for that landscape. As you watch this animation, think about the process of folding, and try to visualize how this landscape evolved and the amount of time it must have taken. Pay particular attention to changes that have occurred on this landscape since the last orogeny in the region. After you complete each of these animations, be sure to answer the questions at the end to test your understanding of these concepts.

Subduction   We've just described what happens when bodies of continental crust slam into each other in a collision zone. Now let's examine the processes that take place when continental crust and oceanic crust converge. In these regions, the dominant tectonic process is **subduction,** where one lithospheric plate is forced beneath another (Figure 13.14). Subduction is initiated when oceanic plates diverge due to sea floor spreading. As noted previously, this spreading forces oceanic crust to move horizontally away from the spreading zone and toward continental margins. When the oceanic crust converges with the continental crust, the denser oceanic crust material is slowly forced beneath the less dense continental crust and down into the upper mantle. When the oceanic crust encounters the upper mantle, it melts and is recycled into magma as part of the rock cycle.

It's important to note that the process of subduction is not continuous, but is instead episodic. In other words, the oceanic plate does not flow smoothly beneath the continental crust. A good analogy for the subduction process is imagining that you're trying to jam a piece of wood into an opening that is slightly narrower than the wood body. It can be done, but only by applying excessive force, perhaps with a hammer. When the wood is struck, it moves a bit through the opening; otherwise, it remains in place. Subduction is somewhat the same in that the oceanic crust repeatedly stops and starts in its path below the continental crust. It moves only when the buildup of stress behind it—which is ultimately derived from the zone of seafloor spreading—exceeds the force of friction holding it in place beneath the continental crust. This stress causes significant deformation of the rocks above the subduction zone, resulting in the construction of mountain ranges.

Subduction   *The process by which one lithospheric plate is forced beneath another. This usually happens when oceanic crust descends beneath continental crust.*

VISUAL CONCEPT CHECK 13.1

Which one of the following statements is accurate with respect to the geologic structure in this roadcut?

a)   **The rocks are horizontal.**

b)   **This is a nice example of a rift system.**

c)   **The structure shows the limb of an anticline.**

d)   **The structure shows the base of a syncline.**

5. Volcano erupts

4. Magma rises through cracks in continental crust

1. Oceanic crust moves toward continental crust

2. Subduction of dense oceanic crust below lighter continental crust

3. Oceanic crust melts into magma

Figure 13.14 The subduction process. Subduction occurs where dense oceanic crust is forced beneath less dense continental crust. This process causes the oceanic crust to move down into the upper mantle and melt. When the magma produced rises to the surface through cracks in the overlying crust, it frequently causes volcanoes.

GEODISCOVERIES   WWW.WILEY.COM/COLLEGE/ARBOGAST

PLATE TECTONICS

To review the process of plate tectonics and the movements that occur along plate boundaries, and to see how their movements actually look, you should access a pair of animations on the *GeoDiscoveries* website. The first module is **Active and Passive Margins**, which compares the movements that occur on unstable and stable tectonic boundaries. The second module is **Tectonic Plate Boundary Relationships**, which focuses on subduction, transform, and collision boundaries. As you watch these animations, be sure to associate the boundary process with the type of landscapes that were discussed in the preceding text. After you complete each animation, be sure to answer the questions at the end of the media to test your understanding of plate tectonics and plate margins.

Earthquakes

The previous discussion described the various ways in which plate margins move relative to one another and how mountains can form in these areas. Now we're going to turn our attention to more specific processes associated with plate boundaries—vulcanism and earthquakes. Let's look first at earthquakes because they occur both as independent events and frequently in association with volcanoes.

Earthquake Processes

You've no doubt heard about a major earthquake that happened someplace, perhaps one that killed hundreds or even thousands of people. In May 2006, for example, an earthquake occurred in Java, Indonesia, that killed over 5000 people. An **earthquake** occurs when the sudden release of accumulated tectonic stress results in an instantaneous movement of the Earth's crust. While this type of movement is usually associated with plate boundaries, it can also occur due to the rapid movement of magma within a volcano. Such movements produce shock waves that radiate through the lithosphere, causing the ground to shake in sometimes dramatic ways. Earthquakes are most frequently associated with plate boundaries because these are locations where stress builds as plates grind against each other. This stress creates a fracture between adjoining plates called a **fault,** along which the adjacent rock bodies are displaced relative to each other. Faults and earthquakes can also occur in the middle of plates, but this association is less common. Although most large earthquakes occur along major plate boundaries, smaller earthquakes can occur whenever a small fault occurs within a rock body.

To see the relationship of earthquakes and plate boundaries, look again at Figure 13.7a, which shows the plate movements along a transform margin. The important thing to remember is that these plates do not slide by one another gradually. Instead, these masses of rock are locked by friction for long periods of geologic time until the stress builds up to a critical point and the fault ruptures along the fault plane. The place within the lithosphere where the fault breaks is called the *focus* and usually occurs a few kilometers deep within the ground (Figure 13.15). In California, for example, the average focal depth is between 6.4 and 9.5 km (4 to 6 mi). The point on the surface directly above the focus is known as the epicenter. When the rocks finally break free of each other, they do so in an abrupt movement and release powerful seismic waves into the surrounding crust.

Locating the Epicenter When an earthquake occurs, the first goal of geologists is to determine the location of the epicenter. This is done through the process of triangulation, in which the distance to the epicenter is compared among seismographs stationed at three separate locations. This methodology is based on the fact that there are two kinds of seismic waves, *P waves* (or *primary waves*) and *S waves* (or *secondary waves*), that travel at different speeds. P waves are compressional waves that cause the Earth's crust to expand and contract rapidly in a horizontal manner as the waves radiate out from the epicenter. P waves typically move at about 1.5 to 8 km (~1 to 5 mi) per second through the crust. S waves, in contrast, move about 60% to 70% slower than P waves and shake the ground in a vertical fashion that is similar to sending a wave pulse along a rope or string when you whip it up and down. In an earthquake, S waves cause the ground to vibrate in a rolling fashion that can be quite noticeable and frightening. The magnitude of both these waves is measured at observation stations with a device called a seismograph (Figure 13.16).

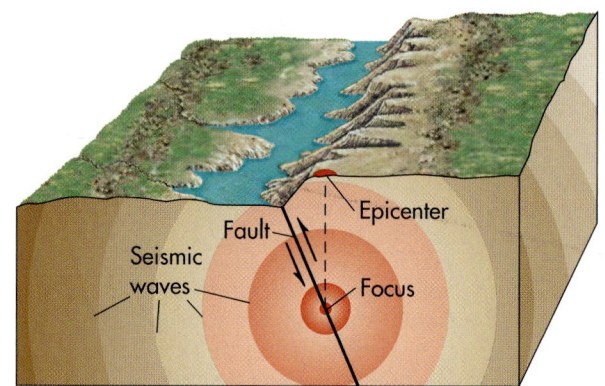

Figure 13.15 Idealized earthquake processes. Earthquakes occur when tension is abruptly released along a fault, causing opposing rock bodies or plates to move in opposite directions relative to one another. This movement generates seismic waves that radiate through the crust. Uplift along a fault can result in a distinct cliff or bluff along a fault, known as a fault scarp. The motion of plates that causes an earthquake can come from stress buildup associated with a transform fault, subduction, and collision.

Earthquake *Shaking of the Earth's surface due to the instantaneous release of accumulated stress along a fault plane or from underground movements within a volcano.*

Fault *A crack in the Earth's crust that results in the displacement of one lithospheric plate or rock body relative to another.*

Epicenter *The point on the Earth's surface that lies directly over the focus of an earthquake.*

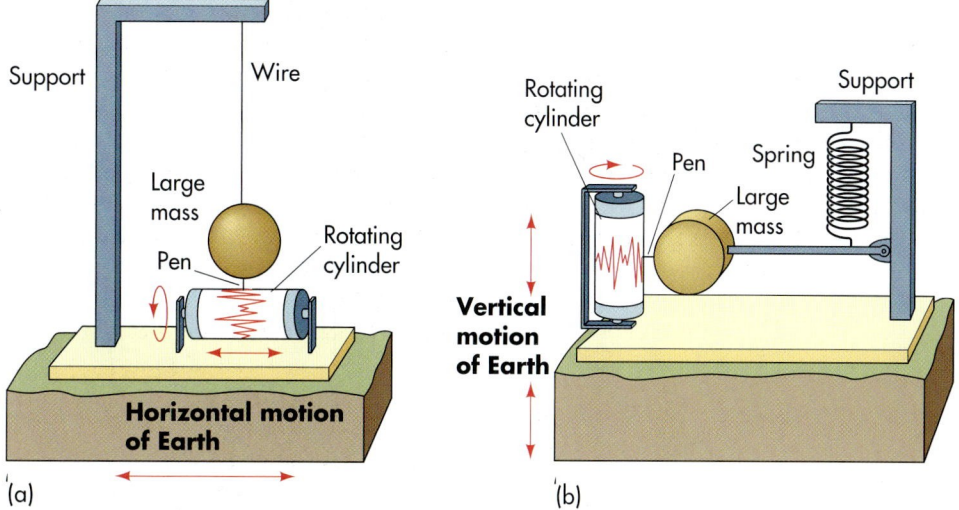

(a)

Support   Wire

Large mass

Pen

Rotating cylinder

Horizontal motion of Earth

(b)

Rotating cylinder

Pen   Spring

Large mass

Support

Vertical motion of Earth

Figure 13.16 Measuring seismic waves. Seismic waves associated with an earthquake are measured with a seismograph. When the Earth shakes, a platform moves back and forth beneath a stationary pen that records the movement on a continuous roll of pa-

per. Primary waves (P waves) consist of compressional waves that cause vibrations parallel to their travel direction. In contrast, secondary waves (S waves) cause vertical vibrations to develop. P waves move more quickly than S waves and are thus the first to be felt.

Given the known speed difference between P waves and S waves, it is possible to calculate the distance to an earthquake epicenter at any observation station by noting the time lag between the passing of each wave. When the passing of seismic waves is viewed in a regional context, it's logical to assume that a wave will require more time to reach an observation station that is farther away from the epicenter than one that is closer. With these time patterns in mind, geologists can determine the earthquake

epicenter by triangulating the time lag between three separate observation stations. The epicenter is located at the point where the respective distance measurements overlap (Figure 13.17).

Measuring Earthquake Magnitude

The strength of earthquakes is measured on the **Richter scale,** which is related to the amplitude of seismic waves moving through the Earth's crust as determined

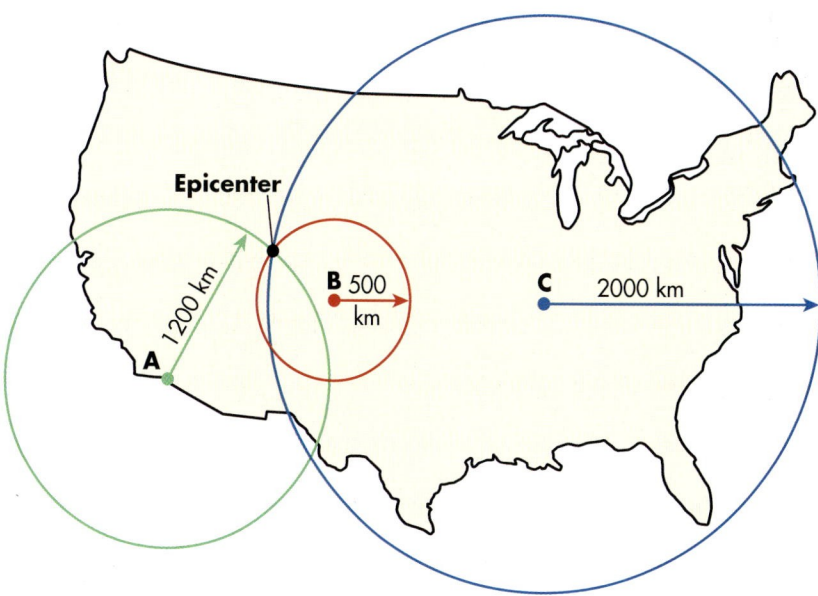

Epicenter

A 1200 km

B 500 km

C 2000 km

Figure 13.17 Seismic triangulation. The epicenter of an earthquake is the point where the distance measurements from three seismic observation stations overlap.

Richter scale   *The logarithmic scale used to measure the strength of an earthquake.*

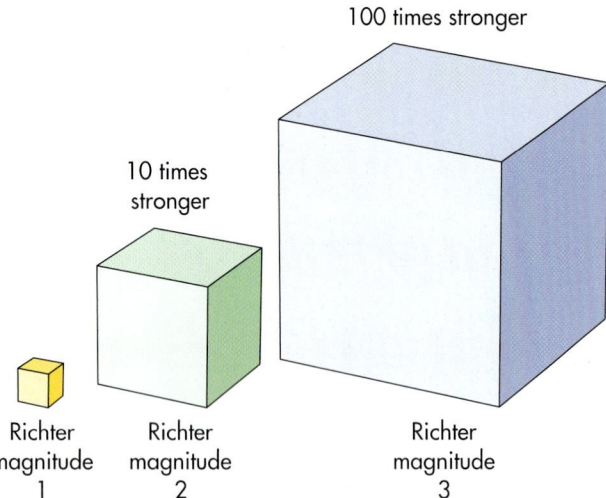

Figure 13.18 The Richter scale. These cubes show the logarithmic nature of the Richter scale, with the size of each cube representing relative power. For example, a magnitude-2 earthquake is 10 times stronger than a magnitude-1 earthquake. Similarly, a magnitude-3 earthquake is 100 times stronger than a magnitude-1 earthquake.

100 times stronger

10 times stronger

Richter magnitude 1

Richter magnitude 2

Richter magnitude 3

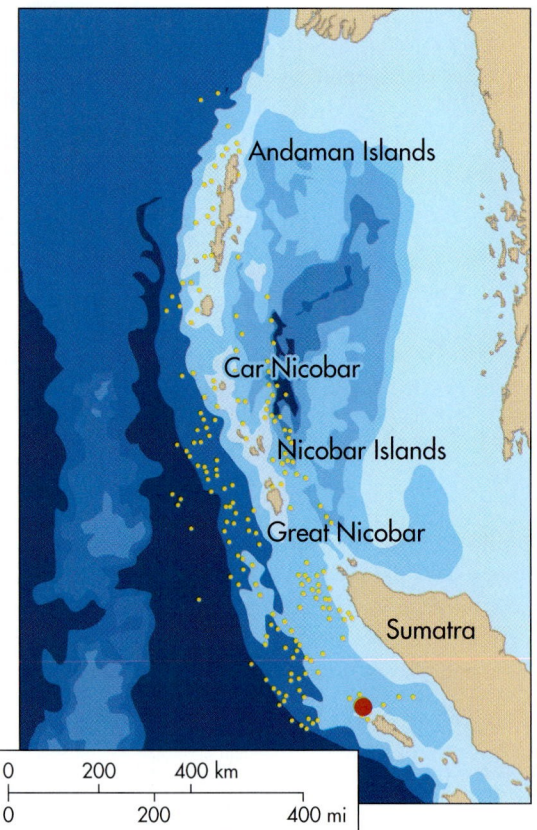

Figure 13.19 Map showing the epicenter of the Sumatra-Andaman earthquake in December, 2004. The initial earthquake (at red dot) measured 9.1 on the Richter scale and was the second most powerful quake on record. As a result of this earthquake, a massive tsunami was created that killed over 200,000 people along the coasts of the Indian Ocean. The yellow dots show the location of aftershocks.

by a seismograph. The Richter scale is logarithmic and is represented by whole numbers and decimal fractions. In general, an increase of one magnitude unit (such as 4 to 5) corresponds to ten times greater ground motion; an increase of two magnitude units corresponds to 100 times greater ground motion; and so on (Figure 13.18). The magnitudes of earthquakes range between 0 (weakest) and 9 (strongest) on the Richter scale. The actual ground motion for, say, a magnitude-5 earthquake is about 0.04 millimeters at a distance of 100 kilometers from the epicenter; it is 1.1 millimeters at a distance of 10 kilometers from the epicenter.

The hazards associated with severe earthquakes were clearly seen when the second strongest earthquake ever recorded occurred near Sumatra, an Indonesian island in the eastern part of the Indian Ocean (Figure 13.19) in December, 2004. This earthquake, known officially as the Sumatra-Andaman earthquake, measured a staggering 9.1 on the Richter Scale and lasted for 10 minutes. This is a very long period of time when you consider that most earthquakes last no more than a few seconds. The length of the rupture along the Sumatra fault associated with this quake was 1200–1300 km (720–780 mi), making it the longest such rupture ever observed. The quake was so strong that the *entire* Earth shook, with ground motions of at least 1.3 cm (.5 in) occurring *everywhere* on the globe as seismic waves spread from the epicenter. According to the United States Geological Survey, for example, the ground motion associated with the quake caused the water level in a well in Virginia suddenly to increase 9 m (3 ft)! In the four days after the initial earthquake, more than

150 aftershocks greater than magnitude 4.0 occurred. This earthquake caused the massive tsunami that killed over 200,000 people along coasts of the Indian Ocean. The formation of tsunamis, including this one, will be described in Chapter 19.

Types of Faults In addition to the production of seismic waves, earthquakes cause deformation of the rocks both within the crust and on the Earth's surface. This deformation is usually associated with the various kinds of faults that occur when opposing rock bodies move relative to each other in association with plate tectonics (Figure 13.20).

A **normal fault** is a vertical fault in which one slab of the rock is displaced up and the other slab down (Figures 13.20a and 13.21). This kind of fault is created by tension forces acting in opposite directions. Where such

Normal fault *A steeply inclined fault in which the hanging rock block moves relatively downward.*

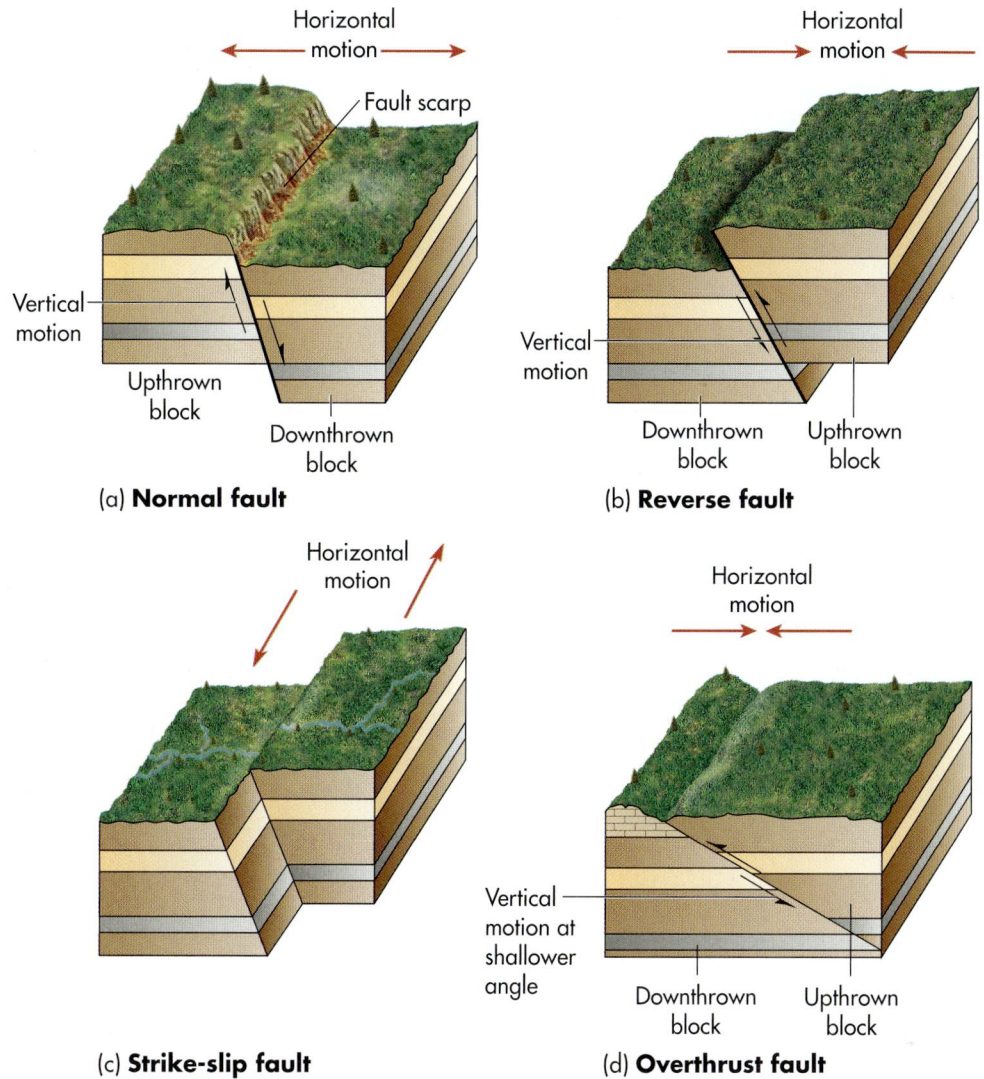

Figure 13.20 Types of faults. (a) Normal faults and (b) reverse faults result when one block moves up while the other moves down. (c) Strike-slip faults result when blocks move horizontally relative to one another. (d) Overthrust faults occur when the upthrown block also slides over the downthrown block.

faults occur, the opposing blocks are pulled away from one another by gravity, which causes one of the fault blocks to slip up relative to the fault plane as an upthrown block while the other slips down as a downthrown block. The exposed side of the upthrown block forms a cliff-like feature known as a **fault escarpment** (or *scarp* for short).

Another kind of fault is a **reverse fault** (Figure 13.20b). Although a reverse fault looks very similar to the normal fault, the cause and nature of the movement differ. Whereas a normal fault entails movement of blocks away from each other, a reverse fault results when rock blocks move toward each other, causing one block to ride up steeply over the other.

As noted previously, overthrust faults result when one rock body is thrust up and over another, usually in association with folding. These faults differ from normal and reverse faults because their fault planes are usually at a shallower angle in comparison.

A spectacular example of the effects of normal faulting at the surface is the famous Basin and Range Province in the southwestern United States. This region extends from the Wasatch Mountains in eastern Utah to the Sierra Nevada Mountains along the border of Cali-

Fault escarpment   *A step-like feature on the Earth's surface created by fault slippage.*

Reverse fault   *A steeply inclined fault in which the hanging rock block moves relatively upward.*

Figure 13.21 A normal fault. This roadcut exposure shows a small normal fault in surface rocks near Death Valley, California. The fault plane is clearly visible as the diagonal contact between the two blocks of rock.

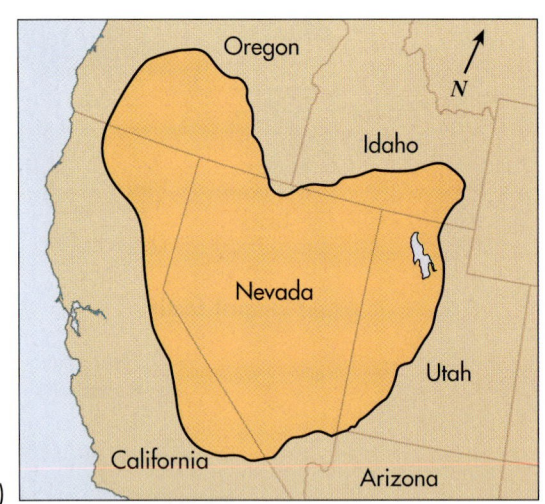

(a)

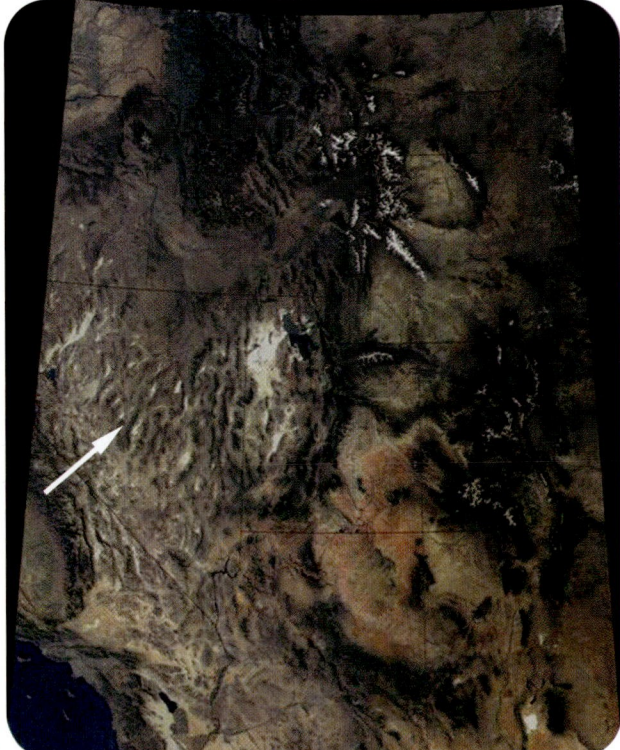

(b)

(c)

Figure 13.22 The Basin and Range Province. (a) The Basin and Range Province extends over much of the American southwest. (b) Satellite image of the Basin and Range Province (arrow) showing the semi-linear network of mountain ranges (darker shades) and intervening basins. (c) Panoramic view of a portion of the Basin and Range Province. Note the distinct mountain ranges and the intervening basins.

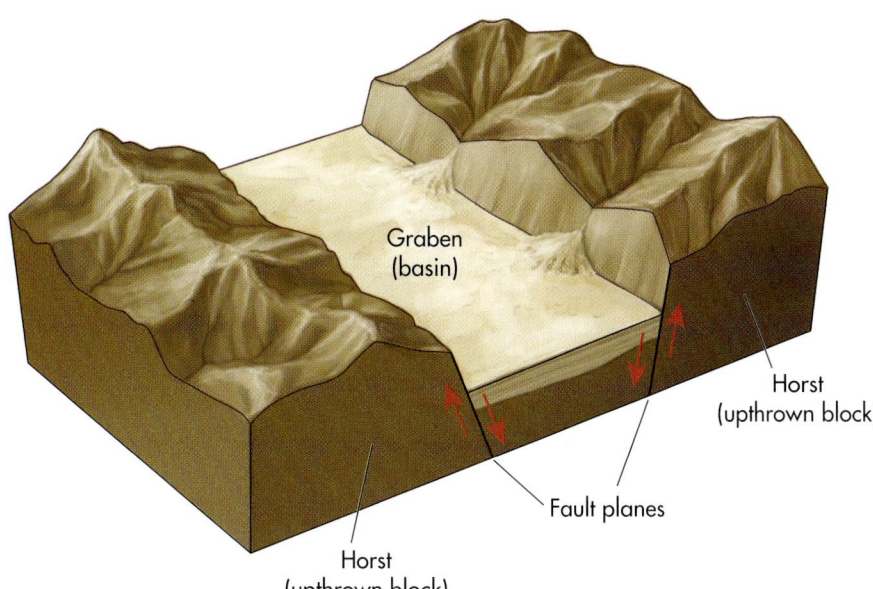

Graben
(basin)

Horst
(upthrown block)

Fault planes

Horst
(upthrown block)

Figure 13.23 Formation of the Basin and Range. The Basin and Range consists of a series of normal faults produced by stretching of the continental crust over the past 20 million years. After uplift, erosion attacks the mountain ranges (horsts). The resulting sediments are then deposited in the intervening basins (grabens), slowly filling them with sediment.

fornia and Nevada (Figure 13.22a). It is characterized by numerous mountain ranges and intervening valleys that are generally aligned from north to south (Figure 13.22b). The relief between the crests of mountain ranges and the adjacent basins can be as much as about 3050 m (10,000 ft).

The Basin and Range Province began to form about 20 million years ago when the North American crust began to stretch in this part of the continent due to pressure applied by an underlying magma plume. As a result of this stretching, the crust thinned, cracked, and pulled apart, creating numerous normal faults that have planes inclined at about a 60° angle (Figure 13.23). This faulting created the series of alternating basins and ranges that characterize the area. From a structural perspective, an upthrown block is called a **horst,** whereas the downthrown block is called a **graben.** The most impressive of these upthrown blocks is the Sierra Nevada Mountains, which form the western boundary of the Basin and Range Province. As you approach the range from the east, the mountain range forms an abrupt escarpment that rises over 2700 m (~9000 ft) above Owens Valley, the graben to the east (Figure 13.24). Given that the Sierra horst is tilted to the west, the slope toward the west is much more gradual.

Return to Figure 13.23 for a moment and notice the way that the bedrock of the graben is covered with layers of sediment. These sediments are derived from the adjacent ranges, which have been the focus of intense erosion because slopes there are steep and stream energy, when streams do flow, is very high. The eroded sediments from the ranges are subsequently deposited in the grabens, producing a relatively level landscape that is underlain by sediments that originated in

the nearby mountain ranges. We'll examine some of these deposits more closely in Chapter 16.

Of the four types of faults illustrated in Figure 13.20, three are directly associated with some kind of uplift—specifically the normal, reverse, and overthrust faults. In

Figure 13.24 The Sierra Nevada Mountains. The Sierra Nevada Mountains form the western boundary of the Basin and Range Province. Uplift and westward tilting of the horst has formed an approximately 2700-m (~9000-ft) escarpment into the Owens Valley graben (foreground).

Horst *An upthrown block of rock that lies between two steeply inclined fault blocks.*

Graben *A downthrown block of rock that lies between two steeply inclined fault blocks.*

and North American plates, with the Pacific plate moving in a northwesterly direction relative to the North American plate at an average rate of about 5 cm (2 in) per year. Figure 13.25 shows the fault geography, including the major subsidiary faults in the San Francisco region. Many of these faults, such as the Hayward and Garlock faults, are strike-slip faults. In many areas, the San Andreas Fault can be seen as a distinct linear trough on the Earth's surface, as pictured in Figure 13.26. In this photograph, you can see the Pacific plate is in the foreground, whereas the North American Plate is the hilly ground in the background. You can also see how incremental the movement of the fault has been by looking at the stream that flows from the North American plate onto the Pacific Plate. If you look closely, you can see that the stream bends sharply at the fault and then flows directly on it for a short distance before turning onto the Pacific plate. In other words, the Pacific plate has moved sufficiently slow to allow the stream to bend with it. Geologists believe that the total accumulated displacement along this fault is at least 350 miles since it came into being about 15–20 million years ago. You've probably heard that one day Los Angeles will move to the location of modern San Francisco. It probably will, but it won't reach that point for thousands of years.

Given the immense size of the San Andreas Fault, the earthquakes associated with it are sometimes quite violent and destructive. The largest earthquake on the San Andreas in recorded history is the 1906 San Fran-

Figure 13.25 California fault systems. The San Andreas Fault is a transform fault that forms the boundary between the North American (to the east) and Pacific (to the west) plates. Movement along this fault is horizontal.

Figure 13.26 The San Andreas Fault. In southern California, the Pacific plate (foreground) is moving toward the northwest (left) relative to the North American plate. Note the change in stream direction that occurs at the fault boundary. This distinct pattern evolved because the stream flows onto the Pacific plate from the North American plate. Because the Pacific plate has been gradually moving in a northwest direction relative to the North American plate, the stream has slowly changed direction as well.

contrast to these types of faults, the **strike-slip fault** entails purely horizontal movement of the two plates past each other. Strike-slip faults and transform faults are closely related because they share the same kinds of horizontal movement. In fact, the primary difference between transform and strike-slip faults is that transform faults are associated with large, tectonic plate boundaries, whereas strike-slip faults occur where small rock blocks move horizontally relative to one another.

As indicated previously, one of the best-known transform faults is the famous San Andreas Fault in California. This massive fault occurs along the boundary of the Pacific

Strike-slip fault *A structural fault along which two lithospheric plates or rock blocks move horizontally in opposite directions and parallel to the fault line.*

This view of the San Andreas Fault is in central California. Which one of the following choices explains why there are hills along each side of the fault?

a) There is subduction occurring at the fault.

b) The San Andreas fault is a normal fault.

c) There is rifting occurring along the fault.

d) There is compression of the rocks on either side of the fault.

cisco earthquake, which had a magnitude of 7.9 on the Richter scale and ruptured along 470 km (290 mi) of the fault. This extensive earthquake caused as much as 8.5 m (28 ft) of near instantaneous slip in some places and resulted in extensive damage in San Francisco and the surrounding area. Between the effects of the earthquake and the catastrophic fire that occurred afterward, 700 to 800 people were killed and about 28,000 buildings were destroyed. With respect to future earthquake activity, geologists now believe there is a 70% chance that an earthquake of magnitude greater than 6.0 will occur in the San Francisco region by 2030.

GEODISCOVERIES   WWW.WILEY.COM/COLLEGE/ARBOGAST

TYPES OF FAULTS

To review the various kinds of faults, go to the *GeoDiscoveries* website and access the module ***Types of Faults***. This animation contains a video presentation that illustrates the fault types discussed in this chapter and will demonstrate the various ways that faults move and the stresses that cause this movement. As you watch the media, try to relate the various fault movements with the character of the landscape. After you have completed the media, be sure to answer the questions at the end of the module to test your understanding of fault types.

KEY CONCEPTS TO REMEMBER ABOUT FAULTS AND EARTHQUAKES

1. There are four major kinds of faults: normal, reverse, overthrust, and strike-slip faults.

2. Earthquakes occur when moving plates are stuck along fault planes, which causes stress to increase until a rupture occurs along the fault.

3. The place within the Earth where a fault breaks is called the focus. The location on the surface directly above the focus is called the epicenter.

4. Earthquakes are measured with a seismograph, which measures the intensity of p and s waves that radiate outward from the earthquake focus.

5. The strength of earthquakes is measured on the logarithmic Richter scale, which indicates the magnitude of seismic waves passing through the Earth's crust.

Volcanoes

We now turn our attention to volcanoes, which are another major feature associated with plate tectonics. As we briefly discussed in Chapter 12, a **volcano** is a mountain or large hill containing a conduit that extends down into the upper mantle, through which magma, ash, and gases are periodically ejected onto the surface of the Earth or into the atmosphere (Figure 13.27). In most cases, volcanoes lie dormant for some time and erupt only when the pressure of material rising from the mantle becomes excessive. The length of dormancy varies dramatically between volcanoes. Some volcanoes lie dormant for hundreds or thousands of years before they erupt, whereas others are in a near constant state of eruption.

Although there is no hard and fast classification scheme, most geologists distinguish three basic kinds of volcanoes: cinder cones, composite volcanoes, and shield volcanoes. This classification depends in large part on whether the eruption is explosive or fluid. We'll discuss the volcanoes associated with explosive eruptions first.

Explosive Volcanoes

A **cinder-cone volcano** is the easiest kind of volcano to understand, as it evolves through the accumulation of solidified magma fragments, rock debris, and ash that are ejected from a central vent. These volcanoes usually form quickly, have steep sides (~30°), and are small relative to composite volcanoes and shield volcanoes. A great example of a cinder-cone volcano is Mt. Capulin (Figure 13.28), which is located in northeastern New Mexico and erupted only once about 62,000 years ago.

Figure 13.27 Volcano eruptions. (a) Ash and steam cloud from Augustine volcano in Alaska. (b) Magma eruption at the Pu'u 'O'o volcano in Hawaii.

(a)

(b)

Figure 13.28 Cinder-cone volcanoes. Cinder-cone volcanoes are relatively small volcanoes that build up by the accumulation of solidified magma fragments, rock, and ash over a short period of time. Mt. Capulin is a typical cinder-cone volcano. This volcano is about 305 m (1000 ft) high.

Volcano *A mountain or large hill containing a conduit that extends down into the upper mantle, through which magma, ash, and gases are periodically ejected onto the surface of the Earth or into the atmosphere.*

Cinder-cone volcano *A small, steep-sided volcano that consists of solidified magma fragments and rock debris that may form in only one eruption.*

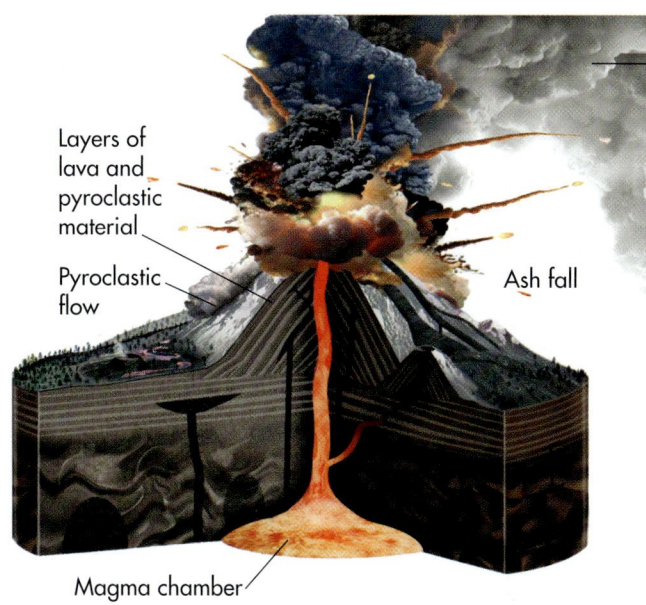

Spreading cloud of smoke and ash

Layers of lava and pyroclastic material

Pyroclastic flow

Ash fall

Magma chamber

(a)

(b)

Figure 13.29 Composite volcanoes. (a) Composite volcanoes consist of layers of lava and pyroclastic material that pile up around one or more conduits to subterranean magma chambers. (b) Mt. Fuji in Japan, reverently called *Fuji-san* by the Japanese people, is a composite volcano.

Composite volcanoes build up through the deposition of alternating layers of lava, which is magma flowing on the surface, and fragmented rock debris called **pyroclastic material** (or *tephra*), such as volcanic ash, cinders, and boulders (Figure 13.29a). Also called *stratovolcanoes* because they contain strata or layers of volcanic debris, these volcanoes typically have moderately steep cones with a semi-horizontal top containing the crater. These volcanoes are much larger than cinder cones, perhaps over 3000 m (10,000 ft) high, and may erupt several times, usually violently, over the course of their active history. A beautiful example of a composite volcano is the famous Mt. Fuji in Japan (Figure 13.29b).

Volcanic Arcs at Plate Boundaries Mt. Fuji is part of the **Pacific Ring of Fire,** which is a chain of volcanoes that follows most of the outline of the Pacific plate

(Figure 13.30). This chain arcs northeast from New Zealand, along the eastern coast of Asia, and north across the Aleutian Islands of Alaska. From there, it continues south along the western coast of North and South America. This region contains about 75% of the world's volcanoes, including Mt. Ruapehu in New Zealand, Mt. Pinatubo in the Philippines, Mt. Fuji in Japan (Figure 13.29b), Kliuchevskoi Volcano on Russia's Kamchatka Peninsula, Augustine volcano in Alaska, Mt. Garabaldi in Canada, and Cotapaxi in Chile, just to name a few. Most of these volcanoes are composite and have formed because subduction is occurring at many places along the Pacific plate. Given the tectonic stress associated with subduction, numerous earthquakes occur along this plate boundary as well.

Composite volcanoes, such as those along the Pacific Ring of Fire, typically lie dormant for long periods of time between eruptions. In New Zealand, for example, the last major eruption of Mt. Taranaki occurred in 1755. When composite volcanoes do erupt, they tend to explode violently, sending clouds of ash high into the atmosphere. This kind of eruption occurs because the magma within composite volcanoes is rich in silicas, which are minerals containing silicon (Si), and therefore highly viscous (meaning sticky and slow flowing). As a result, gases are trapped in the magma and build up pressure within the volcano until it explodes violently. Such an eruption may literally blow the top off of a mountain. Subsequently, the

Composite volcano *A large, steep-sided volcano that grows through progressive volcanic eruptions, which are usually explosive, and consists of layers of volcanic debris.*

Pyroclastic material *Fragmented rock materials resulting from a volcanic explosion or ejection from a volcanic vent.*

Pacific Ring of Fire *The chain of volcanoes that occurs along the edge of the Pacific lithospheric plate.*

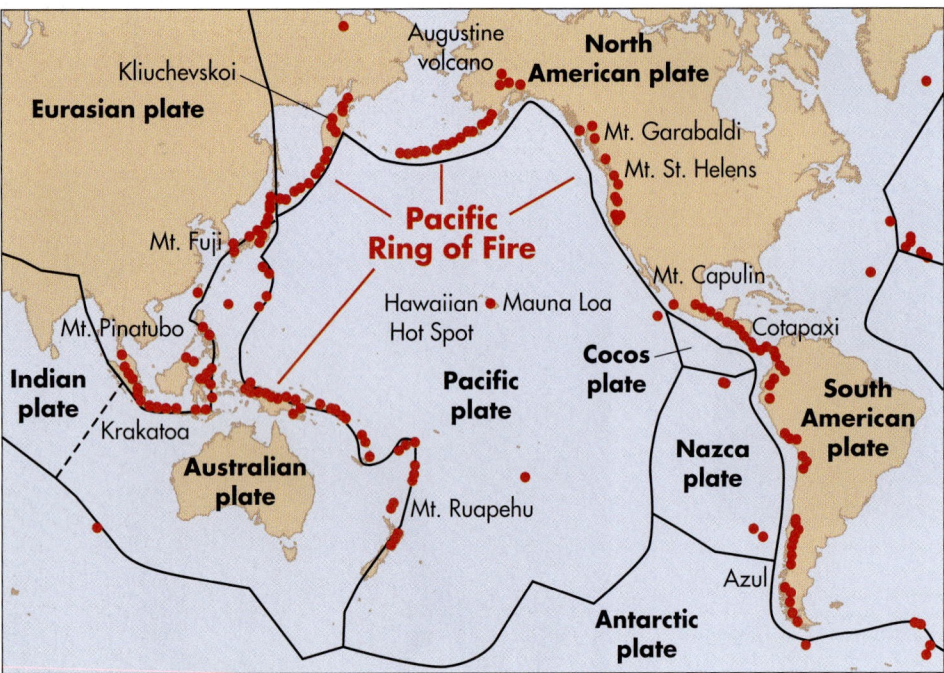

Figure 13.30 The Pacific Ring of Fire. Subduction along the Pacific plate and subsidiary plates (such as the Nazca and Cocos plates) has produced many prominent composite volcanoes. The locations of a few of them are shown here.

left-over crater may partially fill with a **lava dome,** which consists of a steep-sided mound built by highly viscous magma that slowly oozes from the central vent. Although these features sometimes occur alone and are thus classified as a particular type of volcano by many geologists, the vast majority of the recently active lava domes occur in association with composite volcanoes.

An excellent example of an explosive eruption occurred at Mt. Pinatubo in the Philippine Islands in 1991. This volcano began to erupt on June 7 of that year when non-gaseous magma (like soda-pop gone flat) reached the surface of the mountain and oozed out to form a lava dome. A spectacular eruption occurred on June 12 when millions of cubic meters (yards) of volcanic material exploded from the volcano. On June 15, a still greater eruption occurred, one that was catastrophic in nature. This eruption ejected more than 5 cubic kilometers (1 cubic mile) of volcanic material, resulting in an ash cloud that rose 35 km (22 mi) into the air. Some of this ash fell as far away as the Indian Ocean, and satellites tracked the ash cloud several times around the globe. Fortunately, due to active monitoring, geologists were able to predict the nature and general timing of the eruption in a way that allowed the 60,000 people who lived near the volcano to

evacuate safely before it exploded. Much of this warning was based on the fact that numerous earthquakes occurred in the vicinity of the mountain prior to the eruption, suggesting that significant activity was happening within.

Explosive eruptions are also common in the Pacific Northwest of North America along a string of volcanoes in the Pacific Ring of fire known as the Cascade Volcanic Arc. This volcanic arc is present because the Juan de Fuca plate, which is located off the west coast of North America (see Figure 13.2), is subducting beneath the North American plate to the east. The best known of the Cascade volcanoes is Mt. St. Helens, which lies in the southern part of the state of Washington. This volcano is so well known because the most recent major volcanic eruption in the mainland United States occurred here, in May, 1980.

Prior to the eruption, Mt. St. Helens had a classic conical shape (Figure 13.31a) and, as a result, it was frequently called the Mt. Fuji of North America because it was such a picturesque volcano. Mt. St. Helens had been dormant for 123 years when it slowly began to reactivate in the spring of 1980. Several small eruptions occurred throughout April and early May of that year that opened a new crater and sent periodic clouds of volcanic debris onto the surrounding landscape. On May 18, the volcano erupted catastrophically (Figure 13.31b) in a manner similar to that observed more recently at Mt. Pinatubo. Much of the volcanic cone of Mt. St. Helens was blown off in this eruption and the surrounding landscape was devastated (Figure 13.31c).

Lava dome *A steep-sided volcanic landform consisting of highly viscous lava that does not flow far from its point of origin before it solidifies.*

(a)

(b)

(c)

Figure 13.31 Eruption of Mt. St. Helens. (a) The near-perfect conical shape of Mt. St. Helens before the May, 1980 eruption. (b) The catastrophic eruption on May 18, 1980. (c) Mt. St. Helens after the eruption. The blast was so explosive that much of the peak was blown off, reducing the height of the mountain from 2950 m (9677 ft) to about 2550 m (8364 ft) and leaving a wide crater where the cone once existed. Note the lava dome (arrow) in the center of the crater.

Subsequently, a nice example of a lava dome formed in the center of the crater. Evidence for this eruption, such as the new pattern of streams and the partial filling of nearby Spirit Lake with volcanic debris, can still be seen on satellite imagery.

Mt. St. Helens provided a recent reminder that it remains the most active volcano in the Cascade Volcanic Arc, apparently because it continues to be more consistently fed with magma from the subducting Juan de Fuca plate than the other volcanoes in the region. Following a short period of increased seismic activity, Mt. St. Helens erupted again in October, 2004, creating a moderate ash plume. Volcanic activity continued sporadically into May, 2005, including low levels of seismic activity, low emissions of steam and volcanic gases, and minor productions of ash. None of this activity was catastrophic in nature and was merely associated with enlargement of the lava dome within the crater left by the 1980 eruption.

Given the active history of Mt. St. Helens in the past 4000 years, this volcano continues to be closely monitored by the United States Geologic Survey, as are other large volcanoes such as Mt. Rainier and Mt. Adams. It is very possible that one of these volcanoes, or another one in the Cascade Volcanic Arc, will erupt in a violent way during the course of your life. If such an eruption were to occur on Mt. Rainier, for example, it would truly be a catastrophic event. In addition to the destruction caused by the blast, the eruption would cause ancient glaciers on the mountain summit suddenly to melt, resulting in a massive flood of volcanic debris and water (called a *lahar*) that would sweep

CRATER LAKE

Crater Lake in Oregon is the second-deepest lake in North America, measuring 389 m (1934 ft) deep. It is roughly 14.5 km (9 mi) wide, almost perfectly circular in shape, and its water is a remarkably pure blue. However, if you look closely at the island in the middle of the lake and the steep rock walls surrounding it, you can begin to realize an even more remarkable fact—Crater Lake is the site of a dormant volcano.

Geologists have pieced together the story of how the crater formed when the top of the volcano collapsed into the partially empty magma chamber after eruption. The resulting crater is called a *caldera* and is particularly

Crater Lake in Oregon is the site of a dormant volcano, that exploded over 6800 years ago. Note Wizard Island, which formed during the most recent volcanic activity about 4000 years ago.

deep, with sheer rock walls. The mountain itself, called Mt. Mazama, formed about 420,000 years ago and erupted only 6850 years ago, in one of the biggest explosions of the past 10,000 years. Ash from the mountain has been traced as far as Alberta, Canada, and pyroclastic flows extend up to 40 miles from the central vent. Approximately 4000 years ago, a relatively small eruption created Wizard Island, a small volcanic cone that protrudes out of the water to form an island. At some point

toward the populated areas near Tacoma, Washington. This flood would likely cause numerous deaths and massive property loss.

Fluid Volcanoes and Hotspots

In addition to cinder-cone and composite volcanoes, a third kind of volcano is called a **shield volcano** (Figure 13.32). When compared to cinder-cone and composite volcanoes, shield volcanoes are very broad and are underlain by magma that contains relatively little silica. Because of the low silica, these magmas have a relatively low viscosity and flow easily as broad sheets and rivers of molten rock that typically cool to form basalt. Through successive eruptions, the volcanoes gradually build up-

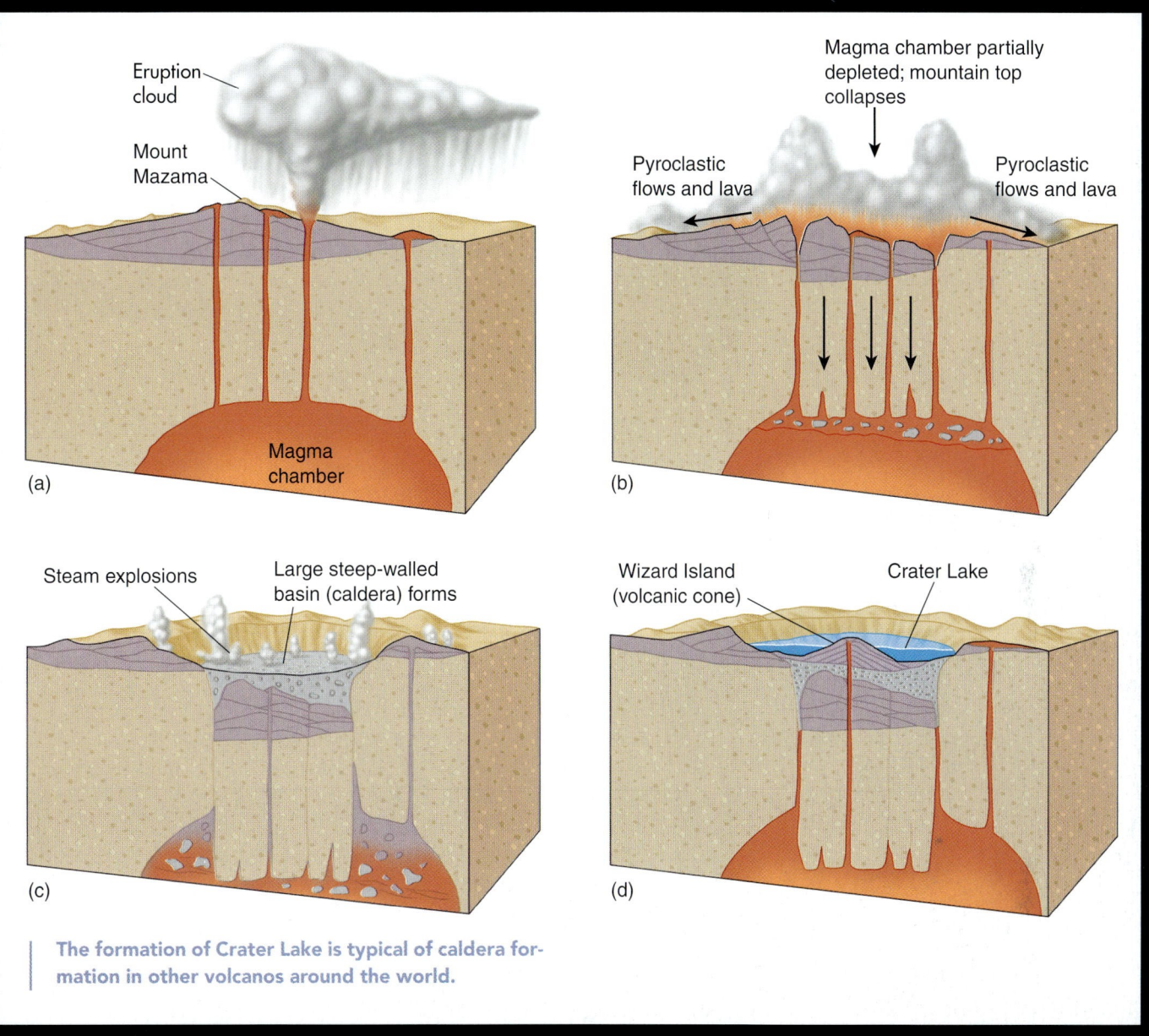

Eruption cloud

Mount Mazama

Magma chamber

(a)

Magma chamber partially depleted; mountain top collapses

Pyroclastic flows and lava

Pyroclastic flows and lava

(b)

Steam explosions

Large steep-walled basin (caldera) forms

(c)

Wizard Island (volcanic cone)

Crater Lake

(d)

The formation of Crater Lake is typical of caldera formation in other volcanos around the world.

following the major eruption the caldera filled with water, forming Crater Lake. If you ever visit Crater Lake you'll find the water to be incredibly blue. This occurs because there is no stream supplying water to the lake, so there is very little sediment washing into it.

ward over a wide area when compared to composite volcanoes. Notice that these volcanoes also have much lower relief than composite volcanoes.

The island state of Hawaii is an example of the geologic history and construction of shield volcanoes. Although Hawaii is not the only place where shield volcanoes occur—they are also found in the Galapagos Islands and many other places—it is a great place to see how this type of volcano shapes the landscape. Note that the Hawaiian Islands are in the middle of the Pacific

Shield volcano *A very broad volcano with shallow slopes that forms in association with nonviscous lava flows.*

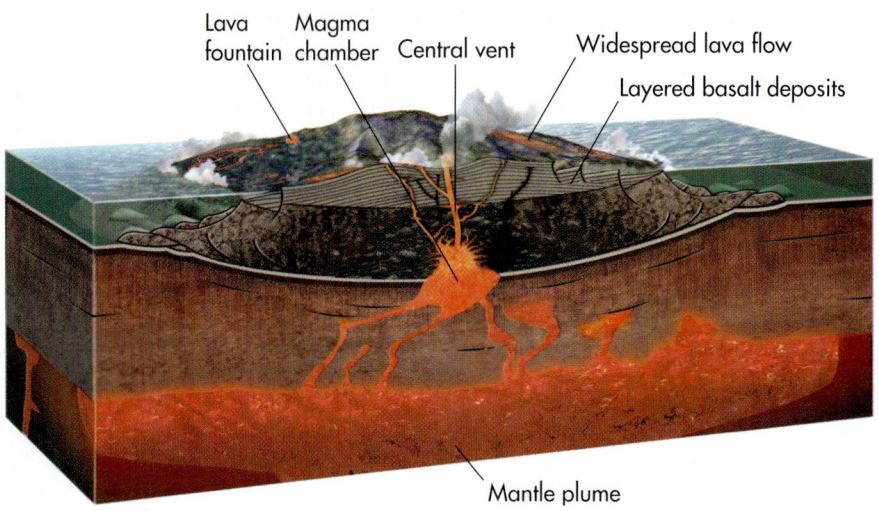

Lava fountain Magma chamber Central vent Widespread lava flow Layered basalt deposits

Mantle plume

Figure 13.32 Shield volcanoes. Shield volcanoes are broad, low features that surround a central magma chamber. Lava associated with shield volcanoes typically has a low viscosity, which allows it to flow freely in broad streams of molten rock.

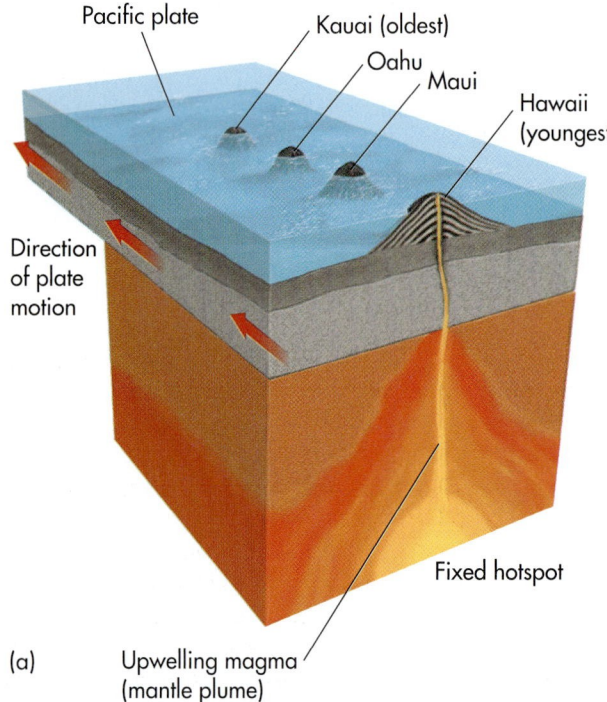

Pacific plate
Kauai (oldest)
Oahu Maui
Hawaii (youngest)

Direction of plate motion

Fixed hotspot

(a)
Upwelling magma (mantle plume)

plate. Given the previous discussion about composite volcanoes and their relationship to plate boundaries, you might wonder how Hawaiian volcanoes can exist where no subduction is taking place. The difference is that the Hawaiian Islands exist over a geologic feature called a **hotspot.** Hotspots are stationary zones in the asthenosphere where upwelling magma from a mantle plume is released at the surface through volcanic eruptions. One

Hotspot *A stationary zone of magma upwelling that is associated with volcanism within the interior of a crustal plate.*

Figure 13.33 The Hawaiian Hotspot. (a) This hotspot is fixed because it is at the top of a mantle plume. The Hawaiian island chain has evolved through progressive movement of the Pacific plate over the hotspot. (b) Mauna Loa is the largest volcano on Earth. (c) The island of Hawaii consists of five separate shield volcanoes.

(b)

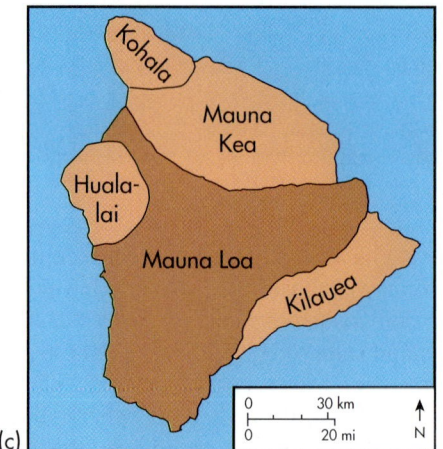

Kohala
Mauna Kea
Huala-lai
Mauna Loa
Kilauea

0   30 km
0   20 mi
N

(c)

of the most interesting hotspots on Earth is the Hawaiian hotspot, underlying the island of Hawaii, the easternmost island in the chain (Figure 13.33a). This hotspot is releasing highly fluid magma that has built a variety of shield volcanoes, such as Mauna Loa on the island of Hawaii (Figure 13.33b). This volcano is the largest volcano on Earth and is one of five that collectively comprise the island (Figure 13.33c).

This Hawaiian hotspot has apparently been active for a long part of geologic time, and provides further evidence supporting the theory of continental drift. Figure 13.34 shows a trail of islands and submerged highlands called *seamounts* that extend to the northwest from the island of Hawaii. This line of features exists because the oceanic crust in the Pacific basin has rotated through time, beginning about 70 million years ago, over the Hawaiian hotspot. Do you see the sharp turn in the island/seamount chain? The features north of that point, beginning with the Emperor Seamount Chain, must have formed when the plate was moving generally to the north as it flowed over the hotspot. Shortly after the Emperor Seamount Chain formed, approximately 43 million years ago, the plate began moving more toward the northwest. This movement has continued to the present day and explains why the state of Hawaii includes a string of islands that project northwest of the island of Hawaii. Each of these islands was formed when that particular part of the Pacific plate was over the hotspot, in the same manner as the island of Hawaii. Now, project yourself into the geologic future for a minute. Where would you expect the next island of the Hawaiian Islands to develop—to the northwest of Hawaii or to the southeast? Why?

Hotspots occur not only in ocean basins, but deep within continents as well. Such a continental hotspot is the Yellowstone hotspot, which lies beneath Yellowstone National Park. Similar to the Hawaiian hotspot, the Yellowstone hotspot is a fixed zone of upwelling magma, in this case beneath the North American plate. Geological studies indicate that the apparent location of the Yellowstone hotspot has moved over time (Figure 13.35a), reflecting the west/southwestern migration of the North American plate. About 16 million years ago the hotspot was located in the southeastern corner of Oregon. Between 12 million and about two million years ago it migrated across what is now southern Idaho.

The Yellowstone hotspot has generally been in its present location for approximately the past two million years, with three major eruptions occurring during this time at 2.2 million years ago, 1.3 million years ago, and about 630,000 years ago. These eruptions were cataclysmic, with each forming a giant caldera (Figure 13.35b). The Yellowstone caldera—that is, the crater from the most recent eruption—is over 50 km (~32 mi) wide. As you can imagine, eruptions of this magnitude must have ejected enormous amounts of volcanic debris. The first eruption in the area, which is the largest of the three, ejected 2500 km³ (~1550 mi³) of material. By comparison, only 1–2 km³ (~.6–1.2 mi³) of material

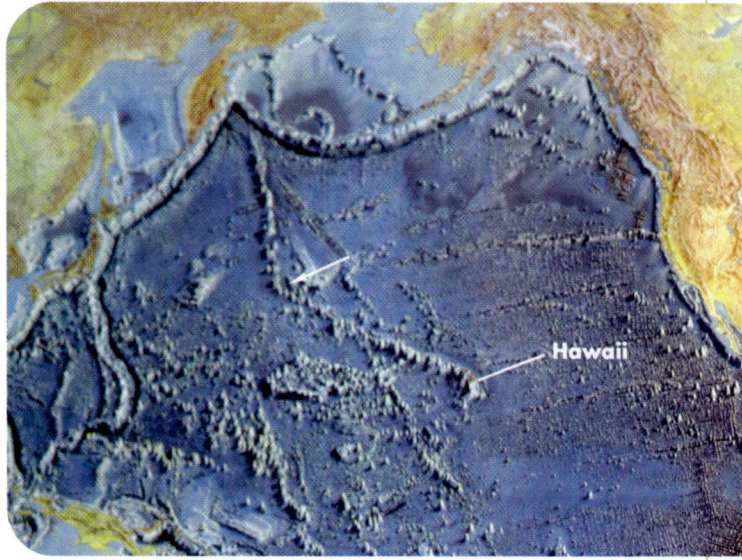

Figure 13.34 Impact of the Hawaiian hotspot over time. In addition to creating the modern Hawaiian Islands, the apparent migration of the hotspot formed the Emperor Seamount Chain (arrow). Note that the chain contains a sharp turn that reflects a shift in the drift of the Pacific plate.

was ejected by Mt. St. Helens in 1980. Imagine what the Yellowstone eruptions must have been like!

If you've been to Yellowstone National Park, you know that it contains abundant evidence that the underlying hotspot remains active. Geothermal features such as geysers, mudpots, and fumeroles are very common within the park. A **geyser** is a superheated fountain of water that suddenly sprays into the air on a periodic basis (Figure 13.36). This process occurs because boiling water beneath the Earth is constricted as it rises through a subterranean passageway. When the pressure builds sufficiently, the water bursts into the sky. When the geyser has used all the water in its reservoir, it dies away until the water is replenished from the surface or underground springs. A **mudpot** consists of a bubbling mixture of gaseous mud and water (Figure 13.36c). These systems form where hot water is limited and hydrogen sulfide gas is present, creating sulfuric acid. This acid dissolves the surrounding rock into fine particles of

Geyser *A superheated fountain of water that suddenly sprays into the air on a periodic basis.*

Mudpot *A bubbling mixture of gaseous mud and water at the Earth's surface that is associated with geothermal activity.*

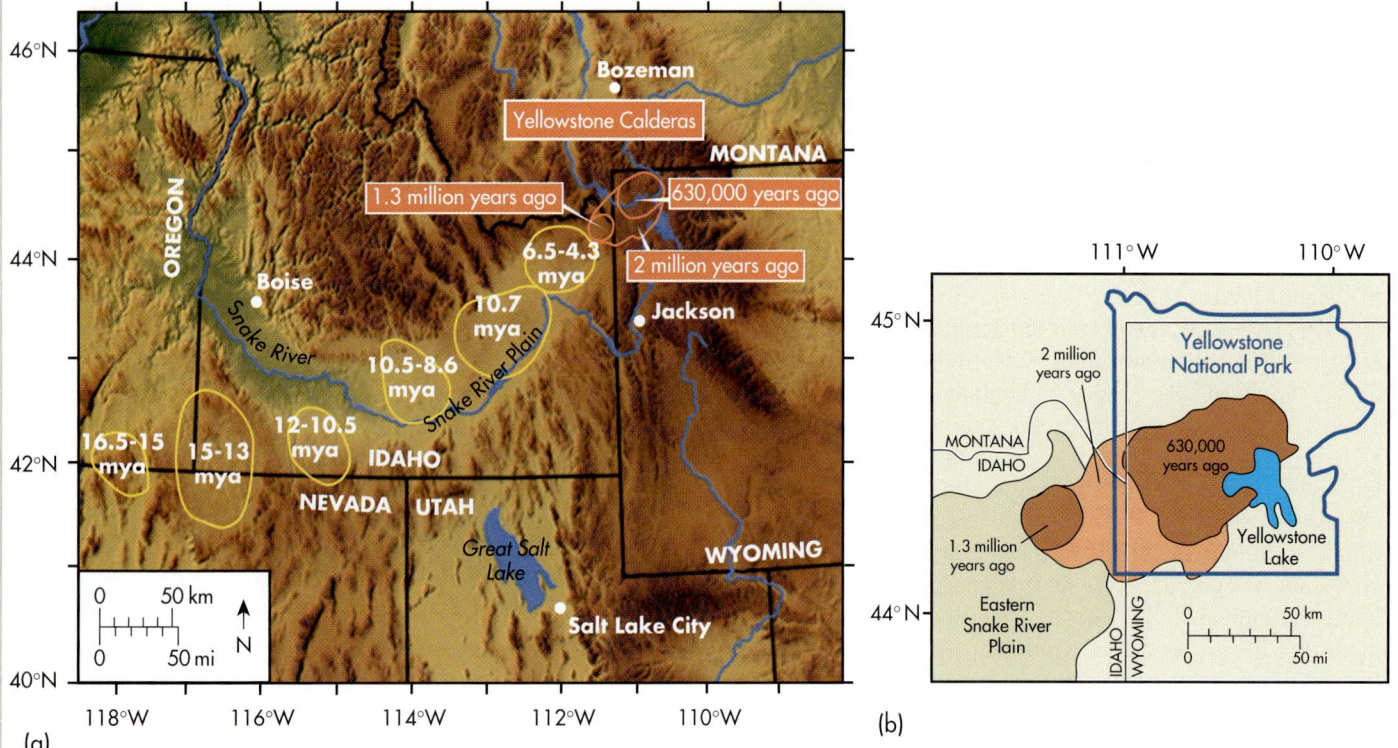

(a)

(b)

Figure 13.35 The Yellowstone hotspot. (a) Migration of the Yellowstone hotspot between 16 million and 600,000 years ago. Although it appears that the hotspot has moved, it is the North American plate, in fact, that has migrated toward the southwest. (b) Calderas in the vicinity of Yellowstone National Park in northwest Wyoming. Each of these calderas formed during individual eruptions.

silica and clay that mix with what little water there is to form the mudpot. A **fumerole** is a steam vent that results because underlying groundwater is boiled away before reaching the surface.

Fumerole *A steam vent that results because underlying groundwater is boiled away before reaching the surface.*

(a)

(b)

Figure 13.36 Geothermal features at Yellowstone National Park. (a) Schematic illustration of geyser processes. Water below the Earth's surface is super-heated and rises in small cracks in the overlying rock. If a constriction exists in this rock, the heated water backs up until the growing pressure causes it to blast toward the surface where it erupts. (b) Old Faithful Geyser is probably the most famous system of its kind in the world. It derives its name because it erupts on average every 91 minutes.

VOLCANOES

Now that we've discussed volcanoes thoroughly, you can see their associated processes in animated form. Go to the *GeoDiscoveries* website and access the module **Volcanoes**. This module contains a video that compares shield and composite volcanoes and shows various examples of each kind. As you watch the video, note how the concept of plate tectonics fits into the evolution of volcanoes, where they occur, and how they erupt. After you complete the animation, be sure to answer the questions at the end of the module to test your understanding about volcanoes.

VISUAL CONCEPT CHECK 13.3

Given your understanding of volcanoes, describe the volcano pictured here. Is it a cinder-cone volcano, composite or shield volcano? How can you tell? What kind of eruption will most likely occur when it does erupt?

KEY CONCEPTS TO REMEMBER ABOUT VOLCANOES

1. Volcanoes occur most frequently on plate boundaries. In particular, most of the world's volcanoes are found along the Pacific Ring of Fire.

2. Volcanoes are most frequently associated with the process of subduction because oceanic crust melts and then rises through cracks in overlying continental crust. The Cascade volcanic arc is an example of this relationship.

3. Some volcanoes form over hotspots, which are places where upwelling magma reaches the surface. Examples of hotspot activity are the Hawaiian Islands and Yellowstone National Park.

4. Volcanic eruptions can be broadly classified as explosive or fluid. Explosive eruptions occur when the magma is viscous, forming composite volcanoes (also called stratovolcanoes) such as those in the Cascades. When the magma is not viscous, it flows freely, resulting in broad rivers of lava that collectively form shield volcanoes such as those in the Hawaiian Islands. Cinder cone volcanoes evolve through the accumulation of solidified magma fragments, rock debris, and ash that are ejected from a central vent.

The Big Picture

Now that we have described how rocks form and mountains grow, we can next examine how these features are modified over geologic time. For instance, in our discussion in this chapter about the Appalachian Mountains, we stated that these mountains have been severely eroded from much higher peaks, resulting in today's well-rounded mountains. One of the ways in which this erosion occurred was through the process of rocks falling under the force of gravity, such as those pictured below. This image illustrates one of the many different ways in which rocks and landforms are modified after they form. The next chapter begins this investigation by focusing on weathering, which is the process through which rocks and landscapes are altered incrementally or suddenly by water and gravity.

Summary of Key Concepts

1. The Earth's crust is not a single sheet, but rather a jigsaw puzzle of interconnected plates. A variety of evidence (location of fossil remains, apparent magnetic polar wander, and actual measurement of motion) strongly supports the theory of continental drift, indicating that plates have shifted through geologic time.

2. Plate margins can be passive, transform, converging, or diverging. At diverging plate margins, magma plumes from the asthenosphere cause rifting, either on continents or on the sea floor. At converging plate margins, plates collide or subduct, causing geomorphic features such as alpine chains and volcanoes.

3. Folding is the process by which rocks are compressed and deformed. Several kinds of geologic structure are associated with folded topography, including (1) monoclines, (2) synclines, (3) anticlines, (4) overturned folds, and (5) overthrust faults.

4. There are four major kinds of faults: (1) normal, (2) reverse, (3) overthrust, and (4) strike-slip faults. Earthquakes occur when moving plates are stuck along a fault, which increases stress until a rupture occurs along the fault plane. The strength of earthquakes is measured on the logarithmic Richter scale, which indicates the magnitude of seismic waves passing through the Earth's crust.

5. A volcano is a mountain that erupts periodically when the buildup of pressure beneath it passes a critical threshold. Most volcanoes occur along subduction zones. There are three primary kinds of volcanoes: (1) cinder-cone volcanoes, (2) composite volcanoes, and (3) shield volcanoes. The type of volcano depends on the character of the magma associated with the system.

Check Your Understanding

1. Describe the two types of crust found on Earth. Where is each type of crust found and what proportion of the Earth does each cover? What is the basic composition of each type of crust?

2. How does fossil evidence support the theory of continental drift?

3. Explain how magnetic declination is used to reconstruct continental drift.

4. What is the difference between a converging and diverging plate boundary? Why do diverging plate boundaries occur?

5. How does the age of the Atlantic seafloor support the theory that the mid-Atlantic plate boundary is a rift?

6. How does continental collision cause mountains to grow? What are the two processes through which alpine chains evolve?

7. Describe the geologic structure of the Ridge and Valley Province in the Appalachian Mountains. How can it be that some mountains are underlain by structural synclines, whereas others are underlain by anticlines?

8. Why do large earthquakes occur along plate boundaries?

9. In the context of the Richter scale, how much stronger is a magnitude-5 earthquake than a magnitude-2 earthquake?

10. Why are volcanoes associated with subduction zones?

11. What are the differences between stratovolcanoes and shield volcanoes? Where do such volcanoes occur?

12. Why is Yellowstone National Park located in Wyoming? In the context of your answer, explain how the migration of the Yellowstone hotspot fits into the picture.

ANSWERS TO VISUAL CONCEPT CHECKS

Visual Concept Check 13.1

The answer is *c*; the structure shows the limb of an anticline. You can recognize this as an anticline because the rock structure rises rather steeply.

Visual Concept Check 13.2

The answer is *d*; there is compression of the rocks on either side of the fault. Movement along the fault produces incredible strain that is causing deformation of the rocks on either side of the fault.

Visual Concept Check 13.3

This volcano is a composite volcano, specifically Mt. Ranier in Washington. You can tell the nature of this volcano because it is much too large to be a cinder cone volcano. It's not a shield volcano because it is tall with moderately steep sides. Mt. Ranier will probably erupt violently because the magma is highly viscous; thus, a great deal of pressure will build before the eruption occurs.

CHAPTER FOURTEEN

WEATHERING AND MASS MOVEMENT

For the remainder of this book, we'll discuss the various geomorphic

processes that shape the surface of the Earth on a smaller scale than the

mountains, volcanoes, and earthquakes described in Chapter 13. These

smaller-scale processes produce the rolling hills, valleys, and floodplains that

are common in most parts of the world. But first we need to examine the

mechanisms that break apart rock and loosen sediment so that they can be

subsequently transported through the process of erosion to some other place

and deposited. Keep in mind that the topics we've studied so far—especially

Monument Valley is one of the many beautiful places in the southwest U.S. These monuments are remnants of a solid rock mass that once covered the area. Year after year, piece by piece, rock fragments broke off from this rock mass and were carried away by erosion to create a stunning landscape. This chapter focuses on the way that rocks break apart and wear down through the process of weathering.

climate, vegetation, and soil formation—affect these mechanisms and contribute to the variations you see in different landscapes.

It's useful to think of the various geomorphic processes described in the rest of the text as part of a continuum of landscape evolution that begins with tectonic processes, which tend to elevate large blocks of land above sea level and provide topographic relief. Once landforms have been uplifted, other forces begin to reduce the overall relief and ultimately smooth the landscape over a long period of time, a process called *landscape denudation*. Two major components of this process are weathering and mass movement.

Weathering

Recall from Chapter 11 that **weathering** is the process by which rocks break down or decay into smaller fragments. Two primary kinds of weathering processes take place: *mechanical* (also called *physical*) and *chemical*. Each of these weathering processes can operate alone, but they often work together slowly to break rocks apart. Figure 14.1 shows the relationship between these weathering processes and climate. Notice that physical weathering is more pronounced in colder/drier environments, whereas chemical weathering is related more to warmer/wetter climates. The reasons for these environmental relationships will become clear as you work through the chapter.

An important factor that influences the amount or pattern of weathering is the relative resistance of different kinds of rock at an outcrop. Some rocks are simply harder than others and are thus more resistant to weathering and erosion. Limestone, for example, is generally more resistant to physical weathering than shale. Within the limestone rocks, the dolomite units are more resistant than chalky rock units, which sometimes dissolve easily due to the chemical interactions discussed in Chapter 15. It is also possible that certain parts of an individual rock unit are more resistant to weathering and erosion than others within the same unit due to variations in the firmness of the cementing materials. This kind of differential cementation is particularly common in sandstone. In places where this kind of variability exists within a rock outcrop

or within an individual rock unit, long-term weathering can have a noticeable impact and can even result in very unusual forms (Figure 14.2). Such a weathering pattern is called *differential weathering* because rock units weather at different rates. If a more resistant rock layer overlies a less resistant layer, it protects the underlying rock somewhat from erosion and is thus known as **caprock.** As you study the various weathering processes, keep in mind that they can be influenced by the relative difference in resistance between and within individual rock units.

Mechanical Weathering

Mechanical weathering, also called *physical weathering*, involves the destruction of rocks through physical stresses. When this kind of weathering occurs, rocks do not change their chemical composition. Instead, they simply break into progressively smaller pieces of the same kind of rock. As they do so, the overall surface area of the rock progressively increases, which facilitates even more weathering because more area within the rock mass is available for additional weathering to occur. This portion of the chapter describes the various kinds of mechanical weathering.

Frost Wedging   The most common type of mechanical weathering is **frost wedging.** It begins when water finds its way into fractures in the rock. A good example of such a rock fissure is a **joint** (Figure 14.3). Joints can form when rock repeatedly expands

---

Weathering   *Physical or chemical modification of rock or sediment that occurs over time.*

Caprock   *A unit of relatively resistant rock that caps the top of a land form and thus protects underlying rocks from erosion.*

Mechanical weathering   *The breakup of a body of rock into smaller rocks of the same type.*

Frost wedging   *Expansion and contraction of water in rock cracks due to freezing and thawing.*

Joint   *A crack or fissure along horizontal or vertical planes in a rock mass that divides the rock into large blocks.*

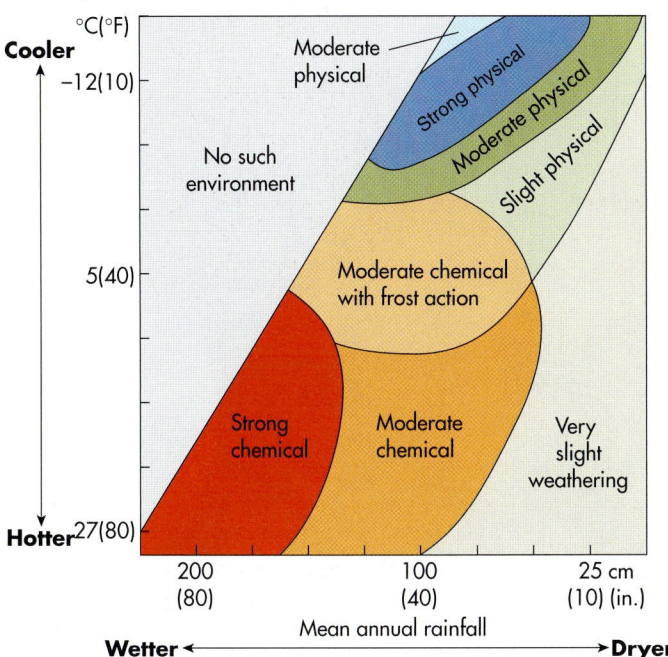

Figure 14.1 Weathering environments. Physical weathering dominates in colder, drier environments, whereas chemical weathering tends to prevail in warmer, wetter regions (although some overlap occurs).

Figure 14.2 Differential weathering in Arches National Park. Differential weathering occurs when rock resistance varies between or within individual rock units. This form of weathering often leaves unusual rock forms called *hoodoos*, such as this one in Utah. In this particular example, the caprock is a sandstone that is slightly more resistant to weathering than the underlying rock layer.

and contracts with heating and cooling, respectively, causing fractures to develop along horizontal and vertical planes that essentially break the rock into large blocks. Joints can also form due to regional tectonic forces that produce stress within the rock mass. Although joint features may look like faults to you, they are not because there is no relative movement on either side of the fracture.

Once joints develop, they allow water to flow deeper into the rock mass at different points. If the rock body lies in a region that has distinct winter and summer seasons, the water will freeze and thaw, respectively, over the course of the year. When water freezes, it expands its volume as much as 9%; you can see this by noticing how water in an ice cube tray freezes into cubes that are slightly larger than the liquid water level. When water thaws back to liquid form, it contracts. (By the way, water is one of the very few substances that act this way; most materials are larger in volume as a liquid than as a solid.) Over the course of time, the stress produced by this freeze/thaw cycle

Figure 14.3 Joints in rocks. Joints are fissures that develop along horizontal or vertical planes within rock. These vertical joints appear in a rock outcrop in the Grand Canyon.

| The Atlas Mountains are well-rounded mountains in northwest Africa.

THE LONG-TERM EFFECTS OF WEATHERING AND EROSION ON MOUNTAINS

Geographers can often tell the relative age of landforms by looking at the general impact that weathering and erosion have had on them over time. This kind of comparison is particularly valid in mountain ranges. Compare a typical scene in the Atlas Mountains in Morocco with the scene from the Andes Mountains, which extend the length of the

causes the joints progressively to enlarge so that boulders slowly break away from each other (Figure 14.4). In places where frost action is most active, such as high mountain peaks or arctic environments, large fields of fractured boulders form (Figure 14.5) and are called *felsenmeer* (German for "rock sea").

In addition to its effects on rock, the freezing and thawing of water can also cause modifications in soil. These modifications are most pronounced where soils are fine-textured with horizontal planes between silt and clay particles. Under these conditions, water collects in the planes and then expands upward upon freezing, causing the soil to rise upward in a process called **frost heaving.** If you happen to live in the northern part of the United

States or in Canada you can see evidence of frost heaving all around you. One place to see it is in agricultural fields in the region. In these areas, individual stones and cobbles (large, rounded rock fragments) can be slowly brought to the surface over time when water collects beneath them and freezes. After each winter, farmers pull any freshly surfaced rocks out of their fields to facilitate plowing and stack them in a pile at the edge of the lot. The piles can grow larger with time as more rocks are brought to the surface each year due to frost heaving. You can also see

Frost heaving *Upthrust of sediment or soil due to the freezing of wet soil beneath.*

west coast of South America. Note that the Atlas Mountains are well rounded, whereas the Andes are very angular with sharp jagged peaks. Which one of the mountain ranges looks older to you? Why?

Based on the relative appearance of the two mountain ranges alone, a geographer would conclude that the Atlas Mountains are probably older than the Andes Mountains and that they have gradually been worn down through a variety of weathering and erosion processes. In fact, the Atlas Mountains were created between 290 and 248 million years ago during the same Allegheny Orogeny that formed the Appalachian Mountains in North America. The Andes Mountains, in contrast, are only about seven million years old. This dramatic age difference means that weathering and erosion have been active for at least 240 million years longer in the Atlas Mountains than in the Andes Mountains.

the effects of frost heaving on northern paved roads when cracks develop in them because they are uplifted somewhat during the winter due to the effects of frost. Building contractors are acutely aware of the frost-heaving process. In fact, to avoid problems with structures such as decks settling in uneven ways, builders are required to install support footings below the depth of the frost line. This line becomes progressively deeper with higher latitude or altitude.

Two other weathering processes have the same effect as frost wedging in that they break rocks down into progressively smaller fragments. One is related to the impact of plants when their roots grow into cracks and joints in rock. As the plants and associated roots grow, they can literally cause rock to split farther apart (Figure 14.6). The second process takes place in association with temperature changes that occur over the course of seasons, between day and night, or due to changes in the direction the Sun faces a rock over the course of the day. These temperature fluctuations can cause rock to expand and contract when heated and cooled, respectively. Over the course of countless such cycles, the rock will gradually break into progressively smaller pieces. This kind of mechanical weathering is very prominent in arid regions (Figure 14.7) or at high elevations where intense solar radiation occurs during the day and radiational cooling prevails at night.

Figure 14.4 Frost wedging. Frost wedging occurs when water gets into cracks and joints within rocks and subsequently expands and contracts upon freezing and thawing, respectively. This process causes cracks to widen over time. Frost wedging can cause even large boulders to fracture in two, such as these in Great Smoky Mountain National Park.

Figure 14.5 Rock sea, or felsenmeer. This field of boulders in western Canada was created through the long-term process of frost action at high altitudes.

Salt Crystal Growth Another kind of physical weathering process is **salt-crystal growth,** which involves the buildup of salts on rock surfaces. Salt is a dispersing agent and its concentration subsequently causes the minerals that cement the sediment grains within the rock to weaken their bonds; this, in turn, loosens sediment so that it can then be eroded. This process is similar to frost action in that it requires the presence of water moving within rock.

An excellent place to see the process and effects of salt-crystal growth is in the steep cliffs of the southwestern United States, where permeable sandstones often overlie relatively impermeable shales (Figure 14.8). Water stored within the sandstones tends to move downward fairly easily because these rocks have large pore-spaces; we'll discuss the process more fully in Chapter 15. This moving water easily dissolves salts such as halite (sodium chloride), calcite (calcium carbonate), or gypsum (calcium sulfate). When this downward-moving water meets the shale, it can no longer move downward and thus flows horizontally toward the cliff wall, where it seeps out and rapidly evaporates.

As a result of this evaporation, salts build up on the rock surface at the contact between the sandstone and shale. This high concentration of salts dissolves the ce-

Salt-crystal growth *A weathering process that involves the buildup of salts on rock surfaces through evaporation. These salts weaken the cement that bonds rock particles, allowing them later to be washed or blown away.*

Figure 14.6 Plant roots and mechanical weathering. Plants can slowly break rocks apart, especially when roots work their way into cracks and crevices such as these.

menting mineral bonds, which, in turn, causes the sand grains within the rock slowly to separate from one another. These loosened grains are then washed or blown away by the water or wind, respectively, forming a niche at the contact between the two rock types. Although salt-crystal growth is a small-scale process, it has contributed greatly to the evolution of landscapes in the canyons of the desert southwest by creating large niches and cave-like recesses in cliff walls.

Exfoliation Another way in which rock physically weathers is when it expands through the process of *unloading*. Unloading occurs when deep rocks are slowly uncovered because overlying rock gradually erodes off of them. As the overlying rock mass is gradually removed, the amount of compression on the deeper rocks progressively lessens. If unloading progresses to the point where the formerly deep rocks become surface rocks, they may expand so much as the pressure lessens that curved joints develop parallel to the surface. These joints are called *sheeting structure* and result in the flaking of loosened rock from the main rock mass in a process called **exfoliation** (Figure 14.9). Although unloading may be the dominant cause of exfoliation, the process is probably enhanced by daily and seasonal temperature cycles that cause the rock to expand and contract.

Exfoliation is most active in igneous and metamorphic rocks, but also occurs in some sedimentary rocks. A great place to see exfoliation on a large scale is Yosemite

Figure 14.7 Temperature changes and mechanical weathering. These multicolored rocks in Death Valley National Park in California are slowly weathering because they have expanded and contracted countless times due to significant temperature changes.

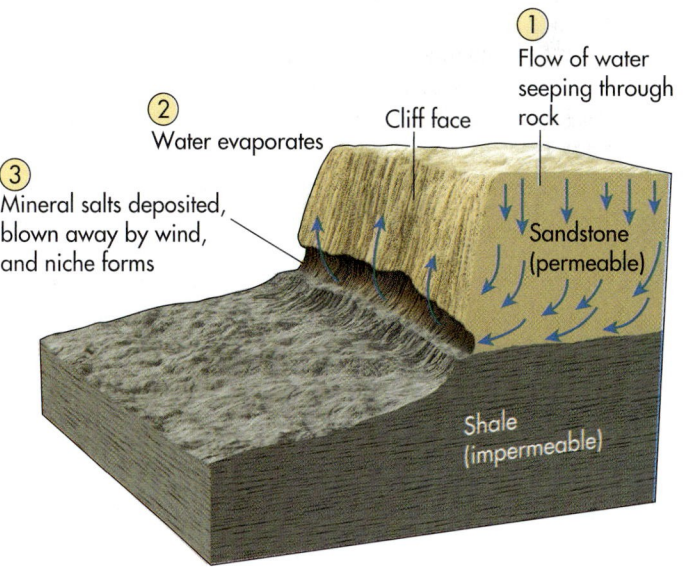

Figure 14.8 Salt-crystal growth and niche formation. Salt-crystal growth is most prominent where permeable sandstones overlie relatively impermeable shales. As a result, water seeps out at the contact between the two rock types, and salt-crystal growth in the sandstone creates a niche.

① Flow of water seeping through rock

② Water evaporates

Cliff face

③ Mineral salts deposited, blown away by wind, and niche forms

Sandstone (permeable)

Shale (impermeable)

Exfoliation *Form of physical weathering in which sheets of rock flake away due to seasonal temperature changes or by the expansion of the rock due to unloading.*

ANASAZI CLIFF DWELLINGS

The Anasazi were a distinct cultural group of prehistoric Native Americans, ancestral to the modern Pueblo Indians, who lived in the southwest region from about A.D. 1 to A.D. 1300. Although the Anasazi had very distinctive pottery and art, they are best known for their cliff dwellings that dot the region. For example, Mesa Verde National Park in southwestern Colorado is known for its numerous Anasazi cliff dwellings. These buildings are found in niches within the thick sandstones of the Mesa Verde Group, which overlies

Cliff Palace ruin at Mesa Verde National Park. Ruins like this are well preserved in this semi-arid environment and are scattered throughout the park.

the relatively impermeable Mancos Shale. Note how this sequence of rock compares with the diagram in Figure 14.8. If you are ever traveling in this part of the country, be sure to visit this spectacular park and see ruins such as Cliff Palace and Cliff Tree House, which are wonderfully preserved in the semi-arid climate. Similar ruins can be seen at Bandelier National Monument and Canyon de Chelly, among many other places.

National Park in California. Here, several large exfoliation domes exist where granite monoliths are exposed at high altitudes. The most famous of these landforms is *Half Dome*, which towers over the adjacent Yosemite Valley (Figure 14.9b). A close look at this peak reveals the curved sheeting structure associated with exfoliation.

Chemical Weathering

Another way in which rocks are altered is by **chemical weathering.** In contrast to physical weathering, which causes rock simply to break apart into smaller fragments,

chemical weathering decays the rock by changing its chemical composition. This decomposition occurs through the interaction of air, water, biological processes, and rock minerals in various ways. In particular, water is very important as a weathering agent because of its abil-

Chemical weathering *The decomposition and alteration of rocks due to chemical actions of natural physical and biological processes.*

Ceremonial kiva in cliff niche at Bandalier National Monument. Archaeologists believe that Anasazi priests climbed into structures like this one to perform religious ceremonies.

years, probably more. Nevertheless, the niches are virtually *unchanged* in that time span. We know that because the ruins that lie within the niches are still in phenomenal condition. What happened to the Anasazi? Why did they leave their dwellings and disappear from the canyons? Nobody knows for sure. The best scientific guess is that their culture collapsed due to a combination of climate change and their own poor land-use practices.

As you visit such places, think about geologic time, and how much time must have elapsed for these landscapes to form. First, sediments accumulated in a sea or ocean that varied in depth; subsequently, they slowly lithified to become rock. Then, over millions of years, streams slowly cut the canyons in the area. At the same time, the niches slowly evolved due to salt crystal growth. Then, prehistoric people moved into the area. You can get a sense of the time when you consider that no one has lived in these dwellings for at least 700

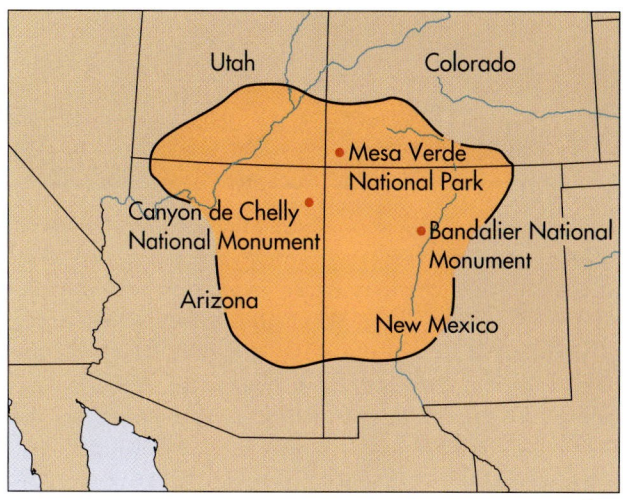

The area occupied by the Anasazi Indians is in the southwestern United States. Ruins from this cultural group are preserved in a number of places, including Mesa Verde National Park, Bandalier National Monument, and Canyon de Chelly National Monument.

ity to dissolve minerals. As a result, chemical weathering is typically more active in warm, humid environments (see Figure 14.1) where abundant water is present and various chemical reactions occur freely. Several chemical weathering processes are common, each having a unique combination of environmental variables that cause rock to decompose.

Hydrolysis Hydrolysis is a weathering process that decomposes silicate minerals within rocks. This form of weathering occurs when silicate molecules are split after the addition of hydrogen and hydroxyl ions derived from water. In granitic rocks, for example, hydrolysis causes feldspars to weather into clays, such as kaolinite and silica, that may be subsequently washed or blown away and deposited somewhere in a soil or stream system. If a sufficient

Hydrolysis *A chemical weathering process that results in the decomposition of silicate molecules within rock through the reaction of hydrogen and hydroxyl ions in water.*

(a)

(b)

Figure 14.9 Exfoliation. Exfoliation occurs when overlying rocks are removed through unloading and when temperature changes cause expansion and contraction of surface rocks. When these fluctuations occur, sheeting structure develops and rock flakes off the surface. (a) Exfoliation on Enchanted Rock, an exposed batholith in central Texas. Note how flaky the surface appears. (b) The peak of Half Dome mountain (arrow) in Yosemite National Park is also composed of granite. Its unique shape gives rise to its name. The peak shows the sheeting structure and jointing associated with the process of exfoliation. These sheets gradually flake off, revealing underlying rock that will then itself expand.

amount of clay is removed in this fashion, more resistant quartz grains are exposed and, with further weathering, liberated from the host rock, to be transported by a stream or blown into sand dunes.

You can see the effects of hydrolysis in several different ways. The most prominent action of hydrolysis on the landscape is that the sharp edges and corners of rocks become rounded, a process called *spheroidal weathering* (Figure 14.10). This process has the biggest impact where numerous joints occur in rocks, providing more corners and edges on which decomposition can occur.

Oxidation Another form of chemical weathering is called **oxidation.** In contrast to hydrolysis, which requires water, oxidation occurs when oxygen is added to chemical compounds and causes electrons within the compounds to be lost. You're probably most familiar with this process from seeing tools and household equipment rust when left outside for a few days. Very simply, what is occurring is that oxygen reacts with iron to produce iron

oxides (rust), which are reddish-colored and flake off with time. The iron underneath is then exposed to the air and forms rust in its turn. The same process happens in rocks that contain abundant iron, such as those in the southwest U. S. These rocks have been heavily oxidized, resulting in their orange appearance (Figure 14.11).

Carbonation A third kind of chemical weathering occurs when water containing carbon dioxide causes minerals to dissolve and wash away in solution. This form of weathering is called **carbonation** (because carbon is reacting with minerals) and occurs because carbon dioxide is easily dissolved by atmospheric water vapor, forming carbonic acid. As a result, precipitation contains this carbonic acid, which is sufficiently acidic to dissolve many minerals, especially limestone (calcium carbonate). When rainfall collects on the rock surface, it dissolves the limestone and forms pits (Figure 14.12). If you're interested in the weathering rates where you live, take a trip to your local cemetery and compare the decomposition in

Oxidation *A form of chemical weathering in which oxygen chemically combines with metallic iron to form iron oxides, resulting in the loss of electrons.*

Carbonation *A type of chemical weathering caused by rainwater that has absorbed atmospheric carbon dioxide and formed a weak carbonic acid that slowly dissolves rock.*

(a)

Edges
Water can act on
two surfaces

Hydrolysis

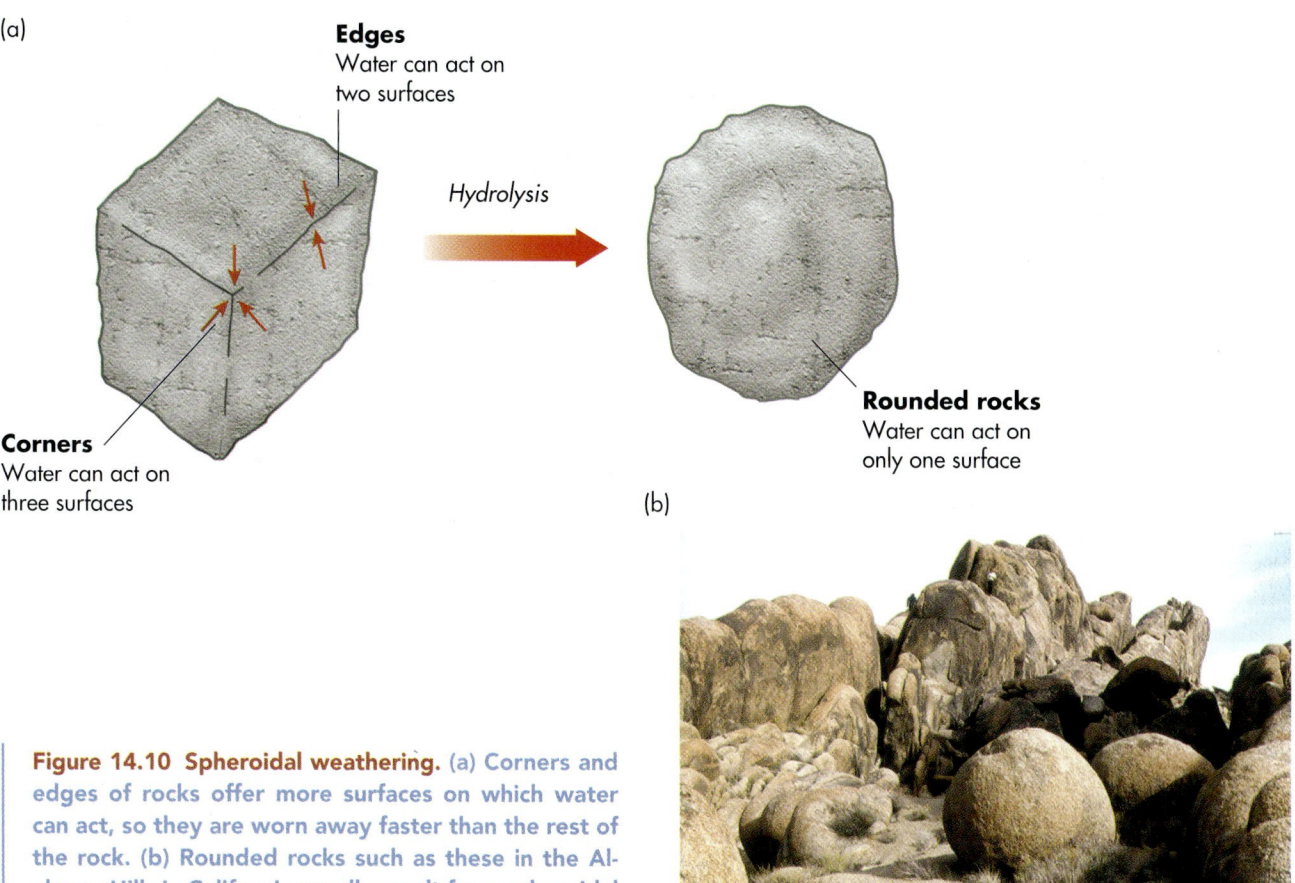

Corners
Water can act on
three surfaces

Rounded rocks
Water can act on
only one surface

(b)

Figure 14.10 Spheroidal weathering. (a) Corners and edges of rocks offer more surfaces on which water can act, so they are worn away faster than the rest of the rock. (b) Rounded rocks such as these in the Alabama Hills in California usually result from spheroidal weathering. Note the people for scale.

limestone versus marble or slate tombstones. Since marble and slate are much more resistant than limestone to weathering, the difference can be dramatic.

Acid Rain As just described, carbonation is the natural process through which rocks are chemically weathered when they come in contact with acidic water. In a similar way, by-products of human industrial activity cause chemical weathering of rocks through the process of acid rain. **Acid rain** is the broad term applied to the way in which industrial acids fall out of the atmosphere in association with precipitation.

In North America, acid rain is a particular problem in the northeastern part of the United States and eastern Canada. This area is geographically distinct because it lies downwind of large industrial cities, such as Chicago, Detroit, Cleveland, and Pittsburgh, where factories emit chemical by-products into the atmosphere. In particular,

---

Acid rain *The precipitation by rain, fog, or snow of strong mineral acids, primarily sulfur dioxide and nitrogen oxides, that originated in factories.*

sulfur dioxide and nitrogen oxides are the primary acids linked to acid rain. According to the United States Environmental Protection Agency, about 2/3 of all sulfur dioxide and 1/4 of all nitrogen oxides in the atmosphere come from electric power generation plants that rely on burning fossil fuels such as coal.

After these acids are emitted from factories in the industrial Midwest, they are carried aloft and eastward by the prevailing westerly winds. Subsequently, they are deposited on the surface of the Earth through precipitation or within fog. As a result, the pH of rainfall in the eastern part of the United States is about 4, whereas it is generally above 5 in the western half of the country. Similar patterns have been identified in other parts of the world, particularly Eastern Europe.

The impact of acid rain on the landscape and as a weathering agent has been significant. With respect to chemical weathering, acid rain dramatically accelerates the weathering process. Although some of this weathering would have occurred by the natural process of carbonation, it has certainly been enhanced by acid rain. In addition to the acceleration of chemical weathering, acid rain is also associated with several negative environmental impacts,

Figure 14.11 Oxidation. Oxidized sandstones in the southwest United States. Scenes like this are very common in this part of the country because iron-bearing rocks crop out in many places due to the arid climate and lack of vegetation.

Figure 14.12 Effects of carbonation in limestone. Acidic rainwater causes calcium carbonate on rock surfaces to dissolve, forming irregular depressions and pits such as these on this dolomite on the shore of Lake Huron in Ontario, Canada.

especially upon forests and lakes. Forests that lie within the belt of acid precipitation are often stressed (Figure 14.13). This stress occurs in part because supporting soils are acidified, which accelerates the leaching of nutrient cations before the trees absorb them. There is also evidence that acid deposition on the needle leaves of conifers reduces their tolerance to cold. Similarly, the pH of lakes and ponds can be lowered due to acid rain, causing stress to the plants and animals that live within those water bodies. Many lakes in New England, for example, do not support fish because they contain toxic levels of inorganic acids.

Although acid rain is a serious environmental problem, recent studies indicate that environmental regulations imposed in 1970 are having a positive effect, at least as far as the United States and Canada are concerned. Prior to the major amendments to the Clean Air Act in 1970, the amount of industrial acids being emitted into the air had been essentially unchecked since the onset of the industrial revolution in the 19th century. These emissions peaked at 28.8 million metric tons in 1973 in the United States alone. With the amending of the Clean Air Act in 1970, which specified a variety of industrial regulations designed to curb acid emissions, these rates in the U.S. decreased 32% to 19.6 million metric tons in 1998. Although nitrogen oxide emissions remain unchanged, we appear to have at least turned the corner with respect to acid rain in North America. Major problems remain in the former Soviet bloc of Eastern Europe, however, where virtually no environmental regulations were in effect during the Communist era. As a result, the relatively new democratic governments are struggling with how to cope with the serious environmental degradation in these areas.

Figure 14.13 Forest stress due to acid rain. Acidic rainfall is concentrated in the northeastern part of the United States. Forests in these areas are affected because soils are acidified, which accelerates leaching of nutrient cations before the trees can absorb them. In addition, trees become less tolerant to cold weather.

This image is of a rock in the southwestern United States. Given what you see, and the location of the rock, which one of the following weathering processes produced the rock pattern visible here?

a) Hydrolysis

b) Carbonation

c) Freeze-thaw activity

d) Thermal expansion

e) Oxidation

GEODISCOVERIES WWW.WILEY.COM/COLLEGE/ARBOGAST

WEATHERING

This point in the chapter is a good time to review the various weathering processes that we've discussed so far. Given that these processes involve movement of water and the wearing down of the landscape through time, it is particularly appropriate to review them in an animated format. Go to the *GeoDiscoveries* website and open the module **Weathering**. This module reviews the various mechanical and chemical weathering processes that we've discussed so far. It shows, for example, how frost wedging gradually causes rock to

split apart and how the surface area of a rock mass increases with progressive mechanical weathering. The animation also reviews the various ways that rock decomposition occurs as a result of chemical weathering processes such as hydrolysis and carbonation. As you watch this animation, think about how weathering contributes to the wearing down of rock masses and the time involved for this erosion to occur on a noticeable scale. After you complete the animation, be sure to answer the questions at the end of the module to test your understanding of weathering.

KEY CONCEPTS TO REMEMBER ABOUT WEATHERING

1. Two general kinds of weathering processes take place: mechanical (physical) and chemical. Mechanical weathering causes rock bodies to break into smaller rock fragments, whereas chemical weathering causes the rock to change chemical composition.

2. Although mechanical weathering and chemical weathering can occur anywhere, most mechanical weathering occurs in cold environments, whereas chemical weathering is more pervasive in warm/humid climates.

3. The three major kinds of mechanical weathering are frost wedging, salt-crystal growth, and exfoliation. Frost wedging involves water that expands and contracts upon freezing and thawing in joints and

cracks in rock. Salt-crystal growth occurs when water seeps out of rock in arid environments. Exfoliation happens when rock expands because of unloading and temperature changes.

4. The three major kinds of chemical weathering are hydrolysis, oxidation, and carbonation. Hydrolysis occurs when water interacts with rocks, causing chemical reactions to occur. Oxidation occurs when oxygen combines with metals to form oxides. Carbonation takes place when carbonic acid in precipitation accumulates on rocks and causes them slowly to dissolve.

5. Humans have increased the amount of chemical weathering downwind of industrial cities through the process of acid rain. Most of these acids come in the form of sulfur dioxide and nitrogen oxides derived from coal-burning power plants.

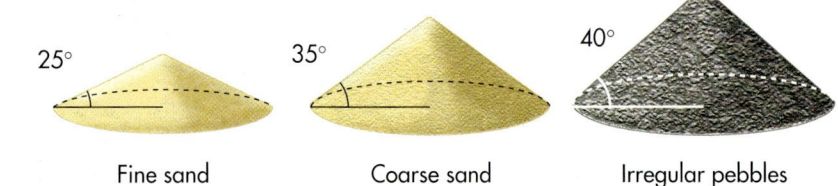

25° Fine sand

35° Coarse sand

40° Irregular pebbles

Figure 14.14 The angle of repose. The angle of repose is the natural maximum angle at which sediments of a given size can rest before they begin to slide under the force of gravity. In general, coarse sediments have a steeper angle of repose than finer sediments.

Mass Wasting

Once sediment is liberated from rock as a result of the weathering process, it can be moved and deposited by gravity, wind, waves, or flowing water and glaciers. In this way, the Earth's landscape is modified and shaped to form features such as hills, valleys, beaches, and sand dunes. The remainder of this chapter will focus on the effect of gravity and the role it plays in landscape evolution by moving large volumes of sediment through the process of **mass wasting.**

Mass wasting occurs specifically on hill slopes, which are land surfaces inclined at various angles from the horizontal. This process occurs when the downward force of gravity overcomes the resisting force of the slope material's shearing strength—in other words, its cohesiveness and internal friction. The point at which sediment becomes unstable is called the *threshold point* and depends a great deal upon the material's angle of repose. This angle is the natural maximum slope that any deposit of a particular kind of sediment—for example, sand—can achieve without moving under the force of gravity. The angle of repose ranges between about 25° for fine sands to approximately 40° for angular pebbles and cobbles. You can see the angle of repose for yourself when you see piles of sediment deposited at construction sites or sand and gravel quarries. After the sediment is dumped, its slope naturally moves to the angle of repose (Figure 14.14).

Generally speaking, most landscapes contain a variety of hill slopes that grade downward to a stream of some size (Figure 14.15). The *local relief* of the area—that is, the vertical distance between the top of the hills and the stream down slope—may have been created either by uplift of the landscape or through downcutting by the stream. In some cases, both processes occur.

Regardless of how the relief was created, the resulting hill slopes contain many distinctive components. In all probability, the core of the slope is some form of bedrock that dates either to an igneous or sedimentary process that occurred tens of millions or hundreds of millions of years ago. In relatively humid areas, the solid core is mantled by regolith, which, if you recall, is formed when rock is weathered. Also recall that regolith is further modified through various pedogenic processes at the surface to form soil. Depending upon the steepness of the hill slope and the regional environmental conditions, the soil will either be well developed and thick, or poorly developed and thin. Most mass-wasting processes occur within the mantling regolith and soil, although where relief is particularly steep, they may also impact outcropping bedrock. Sediment that moves down the hillslope due to some mass-wasting process is called **colluvium.** Colluvium tends to be deposited at the base of hillslopes because the gradient

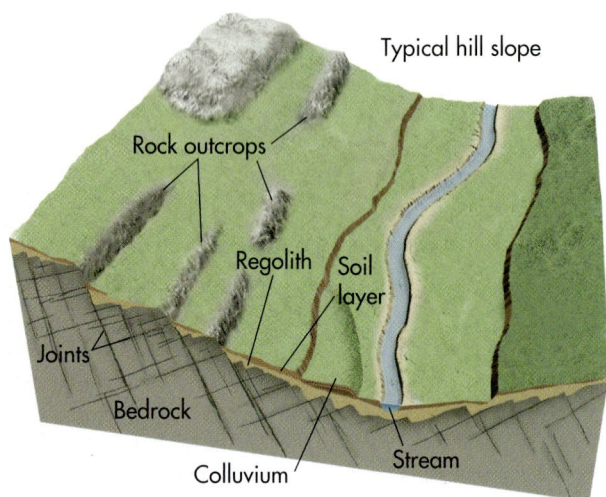

Typical hill slope

Rock outcrops

Regolith Soil layer

Joints

Bedrock

Colluvium Stream

Figure 14.15 Generalized view of a typical hillslope. Note the relationships of bedrock, regolith, soils, and valley stream.

Mass wasting *Movement of rock, sediment, and soil down slope due to the force of gravity.*

Colluvium *Unconsolidated sediment that accumulates at the base of a slope.*

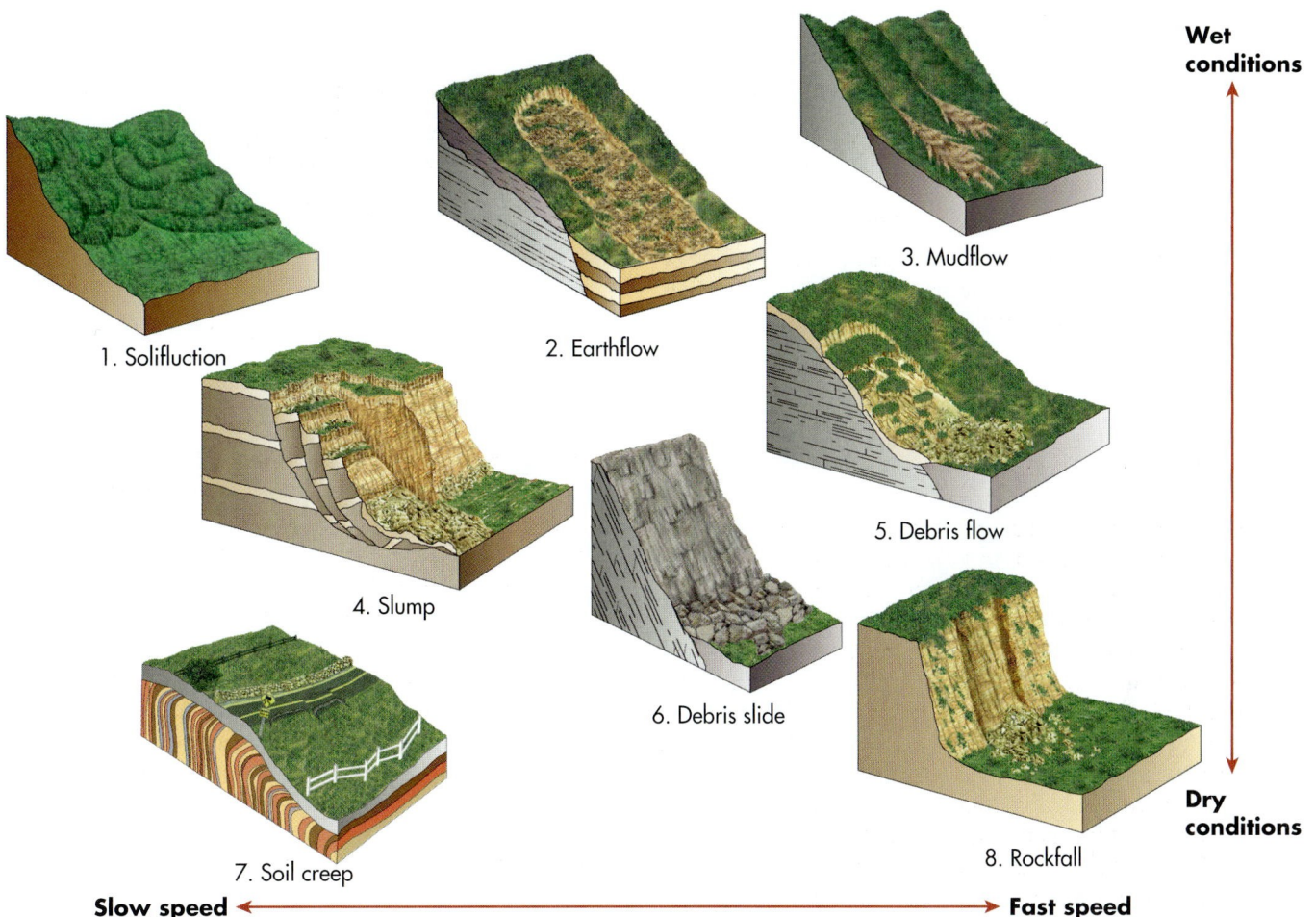

Wet conditions

1. Solifluction

2. Earthflow

3. Mudflow

4. Slump

5. Debris flow

6. Debris slide

7. Soil creep

8. Rockfall

Dry conditions

Slow speed ← → **Fast speed**

Figure 14.16 Major processes and forms associated with mass wasting. Movements to the right of this figure are relatively fast, those to the left are relatively slow. Movements near the top of this figure are triggered by wet conditions, such as heavy rain; those near the bottom of the figure can occur under dry conditions.

becomes less at those locations and the energy to transport sediment is lost.

There are several different kinds of mass wasting, as shown in Figure 14.16. Although these variations tend to grade into one another, there is a general association between specific environmental settings and individual mass-wasting processes. These processes are subdivided on the basis of:

1. The kinds of Earth materials being moved, such as rock or regolith, and their association with the presence of water

2. Their physical properties, including whether they are hard, plastic, or fluid

3. The kind of motion, for example, whether the material is falling or flowing

4. How quickly the process occurs

The following discussion focuses on the processes outlined in Figure 14.17.

Rockfall

The easiest kind of mass-wasting process to envision is **rockfall.** As the name implies, rockfall occurs when rocks fall quickly down a hillslope under the force of gravity. Rockfall is usually most prominent in places where extensive rock outcrops occur on steep hillslopes or canyon walls (Figure 14.17a). Frost wedging and large temperature changes contribute significantly to rockfall because these processes gradually split outcropping rocks apart from one another. Eventually, the gravity threshold is passed and stones and boulders break free to fall onto the lower part of the hillslope. Rockfall also occurs in association with differential weathering in places where relatively resistant rocks such as limestone and sandstone overlie softer rocks like shale. In these situations, the shale surfaces gradually

Rockfall *A mass-wasting process in which rocks break free from cliff faces and rapidly tumble into the valley below.*

(a)

(b)

Figure 14.17 Rockfall. (a) A typical rockfall in a steep-walled canyon. Rockfall occurs when rocks in canyon walls are gradually loosened from the host rock mass by frost wedging and/or large temperature changes and thus quickly fall under the force of gravity. (b) Talus cones, such as these in Banff National Park in Canada, form along canyon walls and mountain escarpments where ravines and gullies funnel boulders into the adjacent valley.

retreat into the cliff through a combination of weathering and erosion, leaving the resistant rock as a ledge that juts out over the shale. Sooner or later the force of gravity exceeds the strength of the rock and it collapses. This process is very common in places like the Grand Canyon.

Extensive deposits of fallen rock, called **talus**, can accumulate through rockfall. These deposits can collect in relatively small amounts along the base of individual, small hillslopes or in larger quantities along high cliffs. Typically, talus slopes have a relatively high angle of repose, ranging between 35° and 40°, because boulders become lodged together and thus support one another. In most cases, talus accumulates in cone-shaped bodies called *talus cones* (Figure 14.18b). This pattern occurs because most canyon walls are interspersed with ravines and gullies that funnel boulders down where they accumulate in distinct masses. In places where numerous talus cones lie adjacent to each other, the resulting landform is called a *talus apron*.

Soil Creep

The slowest mass-wasting process is called **soil creep.** Soil creep occurs on virtually every soil-covered slope, as the perpetual force of gravity slowly pulls surface particles downhill. In this fashion, the soil behaves as a plastic substance that is gradually molded over time. This process is enhanced by cycles of wetting, drying, freezing and thawing in the soil, as well as from disturbances caused by burrowing animals and even people. The result of soil creep is that the surface deposits of a hill slope, and even artificial features such as fenceposts and telephone poles, tend to shift slightly downhill (Figure 14.18). Although this process does not produce any distinctive landforms, it causes the angles of hillslopes and the elevation of hilltops to become slowly reduced.

Although soil creep is a general process that occurs virtually everywhere hillslopes exist, a distinctive kind

Talus *A pile of rock fragments and boulders that accumulates below a cliff due to rockfall.*

Soil creep *The gradual downhill movement of soil, trees, and rocks due to the force of gravity.*

Figure 14.18 Evidence of soil creep. Soil creep occurs when hillslopes gradually move downhill under the force of gravity. Notice here that a portion of a fence line is leaning downhill, which is a good sign that soil creep is occurring.

Figure 14.19 Solifluction in Kyrgyzstan. The distinctive lobes on this hillslope have evolved because unfrozen soil becomes saturated in the summer and sags downhill over underlying permafrost.

of creep called **solifluction** (meaning "soil flowage") occurs above the tree line in tundra landscapes. This process is very straightforward and involves the impact of seasonal temperature changes on the landscape. Most of the year, the soil is frozen from the surface to some depth. During the short summer months, the upper part of the soil thaws. Water within this part of the soil (the active layer) can't percolate very deeply, however, because the sediment beneath it is continuously frozen as permafrost. As a result, the overlying (unfrozen) soil becomes saturated and slowly sags downhill in a very uneven way, producing distinct lobes that overlap one another (Figure 14.19).

Landslides

Another form of mass-wasting process is called a **landslide.** As the name implies, a landslide is a mass of rock, regolith, and soil that flows downhill. This movement is initiated when an event such as an earthquake or period of heavy rain causes the downward force of gravity to overcome the sediment's cohesiveness and internal friction. In general, slope-shear strength depends upon the frictional resistance and overall cohesion of the underlying sediment. When slope shear strength is reduced for some reason, a portion of a hillslope will often move downhill at a rapid rate, either as a debris slide or a slump.

A **debris slide** occurs when slope failure occurs along a plane that is roughly parallel to the surface. A famous example of a massive debris slide is the *Madison slide*, which occurred in Montana near Yellowstone National Park on August 17, 1959. Late in the evening of that day, a magnitude-7.1 earthquake rocked the region for about 30 seconds. This earthquake initiated a massive landslide (Figure 14.20) on a mountain that borders the Madison River, which flows through the area. Approximately 28 million cubic meters (37 million cubic yards) of rock was instantly detached from the mountain in an area that measured about 600 m (about 2000 ft) in length and 300 m (984 ft) in thickness. During the ensuing landslide, the rock mass slid down the mountain slope, crossed the Madison River, and, in the process, killed 28 people who had been camping in the vicinity. The landslide was so large that it dammed the Madison River, creating a new lake, now known as *Earthquake Lake*, which is nearly 100 m (330 ft) deep.

Solifluction *A form of soil creep that occurs in arctic environments where freeze-thaw processes result in lobes of soil moving gradually down slope.*

Landslide *An instantaneous movement of soil and bedrock down a steep slope in response to gravity, triggered by an event such as an earthquake.*

Debris slide *A mass-wasting process in which slope failure occurs along a plane that is roughly parallel to the surface.*

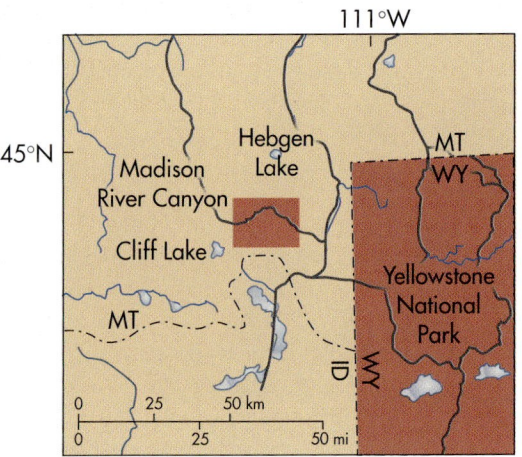

Figure 14.20 The Madison slide. When the threshold point is passed, a portion of a slope can fail instantly, causing tons of rock and sediment to move rapidly down the hill. The massive Madison slide was prompted by a large earthquake in the region. It blocked the Madison River, forming Earthquake Lake (arrow).

In contrast to debris slides, which flow parallel to the surface, a **slump** occurs when rock and sediment rotates and moves down the slope along a concave plane relative to the surface. Slumps typically occur when impermeable clay-rich deposits such as shale underlie more porous rock or sediment. Under such circumstances, water will flow through the upper deposits and then parallel to the clay surface upon encountering

it. This water lubricates the contact between the rock layers and thus reduces frictional force, causing the rock and sediment mass to rotate as it slides down the hill (Figure 14.21a).

At the top of the slump, a wall-like escarpment is produced where the slump block breaks away, leaving a steep face. The sediment in the middle of the slump usually moves more quickly than near the scarp, resulting in a series of step-like features called *slump terraces*. At the base of the slump, the sediment flows more slowly and bulges out from the middle section of the flow as a distinctive *toe*. Although slumps

Slump *A mass-wasting process in which rock and sediment rotates and moves down the slope along a concave plane relative to the surface.*

(a)

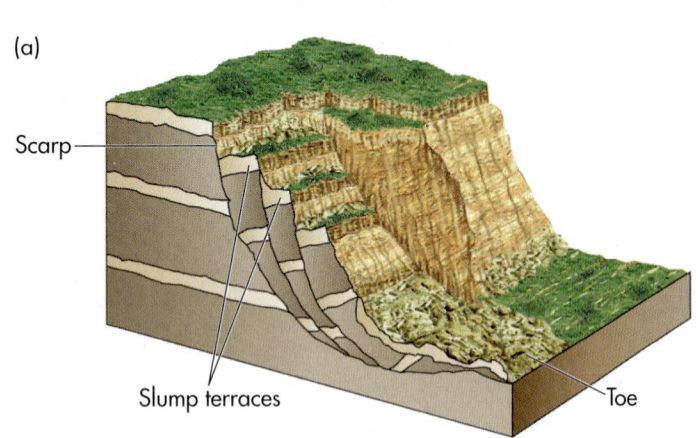

(b)

Figure 14.21 Slumps. (a) A typical slump includes a scarp, slump terraces, and a toe. (b) Slumps like this frequently occur in north-central Kansas because a thin shale (and associated regolith) at the base of hills becomes lubricated during heavy spring rains. Repeated slumps like this one have gradually caused the hillslopes to retreat throughout the area, forming these small hills.

happen virtually everywhere, they are by far more common in humid regions because the sediments are more frequently saturated there. They usually occur in isolated places, such as in Figure 14.21b, with very little impact on humans. Occasionally, however, a large slump occurs in a residential zone and causes severe damage to personal property.

Flows

As the name implies, flows are mass-wasting events that involve sediments that are very wet. Such wet conditions occur when rainfall is high for a period of time, or when humans have stripped hillslopes of protective vegetation. These processes are catastrophic forms of mass wasting and are a serious environmental hazard in many parts of the world. Flows are common in southern California, for example, and are especially dangerous in suburban areas where homes are built on steep hillslopes. In fact, every two or three years some form of major flow occurs in this area that causes extensive property loss and even the occasional human death. Three basic kinds of flow occur: 1) earthflows; 2) mudflows, and 3) debris flows.

An **earthflow** is a slow-to-rapid type of mass movement that involves soil and other loose sediments, some of which may be coarse. This kind of flow may move downhill across a fairly broad surface. Although a **mudflow** is very similar to an earthflow, it differs because it consists of fine-

textured sediments and is extremely fluid; thus, it moves more quickly. Mudflows are most frequently associated with steep canyons in mountainous regions and are initiated in humid zones when heavy rains fall on already saturated slopes. In arid zones, they occur when heavy rains fall on hill slopes that are unprotected by vegetation. Regardless of the environment, mudflows begin when large volumes of water begin to run directly off the surface down the canyon. As the water flows, it picks up increasing amounts of mud from the underlying soil, resulting in a thick, viscous flow of water-laden sediment that has tremendous power. In large mudflows, the power is so great that large boulders are frequently transported great distances (Figure 14.22) and human structures such as homes and bridges can be destroyed. Once the mudflow reaches the valley, it often spreads out to form an apron-like landform (see Figure 14.16).

Another kind of flow associated with heavy rainfall is a **debris flow.** Debris flows differ from mudflows because, in addition to mud, they also contain a mix of boulders, trees, and even buildings as they flow rapidly downhill. These rapidly moving and powerful flows can be triggered by extremely heavy rains in mountainous regions or by a volcanic eruption that instantly melts summit glaciers.

Debris flows are very serious environmental hazards that can impact people catastrophically. An unfortunate example was the massive debris flow that swept down a steep hill slope into the coastal community of La Conchita, California, in January, 2005 (Figure 14.23). This debris flow was triggered by a three-day storm that produced nearly 25 cm (10 in) of rain in the region.

The slopes above La Conchita consist of ancient debris flow and landslide deposits that had failed to a smaller extent in 1995 when a large slump destroyed nine homes and blocked a county road. When these deposits became saturated again in early 2005 they failed on a much larger scale, releasing approximately 14,000 m^3 (45,900 ft^2) of material that flowed more than 100 m (330 ft) into the community. This debris flow destroyed 13 homes and killed 10 people. Given that slope failures frequently occur in California during winter, this disaster demonstrates why it is important to understanding how mass-wasting processes operate and how they can be triggered by passing weather systems.

Avalanches

Another mass-wasting process that you've probably heard of is an avalanche. An **avalanche** is a large mass of snow or rock that suddenly slides down a mountainside (Figure 14.24a). Avalanches are most commonly associated with

Figure 14.22 Mudflow deposit. This mudflow deposit is located in the Never-Summer Range in Colorado. Note the lack of coarse fragments such as rocks and boulders.

Earthflow *A slow-to-rapid type of mass movement that involves soil and other loose sediments, some of which may be coarse.*

Mudflow *A well saturated and highly fluid mass of fine-textured sediment.*

Debris flow *A rapidly flowing and extremely powerful mass of water, rocks, sediment, boulders, and trees.*

Avalanche *A large mass of snow or rock that suddenly slides down a mountainside.*

N 34° 21.736' W 119° 27.668' 297 m   03/16/2005, 9:36:35 AM

Figure 14.23 The La Conchita debris flow. This massive debris flow along the coast of southern California was triggered by heavy rains over the course of three days. Thick deposits of mud, trees, and other debris swept into the community of La Conchita and unfortunately killed several people.

snow and typically occur when thick deposits of snow accumulate at steep angles on mountain slopes. These steep-angled snow deposits can become unstable when a subsequent large storm produces another layer of snow of lighter density than the snow it buries. Under these conditions, an avalanche will occur because the two layers of snow are not firmly bound together to form a uniform mass of snow. When the uppermost layer is somehow destabilized, a distinct section will break free as a *slab avalanche*. Such an avalanche may release up to 228 million cubic meters (300 million cubic yards) of snow, which is equivalent to about 20 football fields covered with about 3 m (10 ft) of snow.

Avalanches have three major components (Figure 14.24b). They begin in a place called the *starting zone*, which is usually located on the steepest slopes of a mountain where deep snows lie at precarious angles. After the avalanche begins, it typically follows a chute or ravine system on the mountain slope called a *track*. The track is where most destruction takes place, either through trees being knocked down or through the movement of boulders and other debris. As you travel through the mountains in summer, you can often see the track of a recent avalanche by looking for avalanche scars, which are generally treeless zones extending down the slope of a mountain. Ultimately, an avalanche loses its energy when it reaches the base of the mountain and the slope lessens. This area is called the *runout* and is the place where snow and debris are deposited.

VISUAL CONCEPT CHECK 14.2

What type of mass wasting process is pictured here?

a) Debris flow

b) Mudslide

c) Rockfall

d) Slumping

e) Avalanche

(a)

(b)

starting zone

track

runout zone

Figure 14.24 Avalanches. (a) Avalanches are usually associated with the sudden failure of enormous quantities of snow that cascade down steep mountain slopes. (b) Avalanches begin at the starting zone, travel down the track, and end at the runout.

GEODISCOVERIES   WWW.WILEY.COM/COLLEGE/ARBOGAST

WEATHERING AND MASS MOVEMENTS

To better understand the process of mass movements, go now to the *GeoDiscoveries* website and access the module *Weathering and Mass Movements*. This module allows you to visualize the various mass-wasting processes in action. As you work through this module you will see how these processes operate, how they differ from each other, and how they shape the landscape. In addition, you will better comprehend the impact that these hazards have on people and the structures in which they work, live, and use. Once you complete the animation, be sure to answer the questions at the end of the module to test your understanding of this concept.

KEY CONCEPTS TO REMEMBER ABOUT MASS WASTING

1. Mass wasting refers to the downslope movement of sediment, soil, and rock under the force of gravity.

2. Mass-wasting processes are differentiated on the basis of how quickly they move and the relative saturation of sediments.

3. Rockfall, creep, solifluction, and landslides are mass-wasting processes that occur under the force of gravity. Rockfall and landslides occur when rocks and sediments break free for some reason, such as lubrication of underlying clays or earthquakes.

4. Flows occur when sediments become saturated and then move under the force of gravity. Flow processes include earthflows, mudflows, and debris flows.

5. Avalanches occur in mountainous areas when steep snow packs break free.

The Big Picture

Now that we've discussed how rocks can be broken apart and sediment transported by mass wasting, we'll turn to some other geomorphic processes that produce landforms. The next chapter focuses on the processes and landscapes associated with groundwater, which, as the name implies, is water stored within the ground. Much of Chapter 15 focuses on karst landscapes, which are created when rocks dissolve. In this sense, the next chapter logically flows from Chapter 14 because chemical weathering and the formation of karst landscapes are processes that are related to one another. A cave, such as the one illustrated, is an example of a karst feature. If you've been to a cave, you know that karst landscapes can be some of the more mysterious features on Earth. The next chapter will describe how these features form, as well as the ways in which water moves and is stored in the ground.

Summary of Key Concepts

1. Weathering refers to the decay and breakup of rocks on the surface of the Earth. Two general kinds of weathering processes take place: mechanical (physical) and chemical. Mechanical weathering causes rock bodies to break into smaller rock fragments, whereas chemical weathering causes the rock to change chemical composition. These general processes are closely associated with environmental conditions and may vary locally due to the relative resistance of different rock units.

2. The three major kinds of mechanical weathering are: frost wedging, salt-crystal growth, and exfoliation. The three major kinds of chemical weathering are: hydrolysis, oxidation, and carbonation.

3. Mass wasting refers to the downslope movement of sediment, soil, and rock under the force of gravity. Mass-wasting processes are differentiated on the basis of how quickly they move and the relative saturation of sediments.

4. Rockfall and soil creep are mass-wasting processes that are associated with unsaturated sediments.

5. Flows occur when sediments become saturated and then move under the force of gravity. Flow processes include solifluction, landslides, earthflows, mudflows, and debris flows.

Check Your Understanding

1. Why is the term *weathering* appropriate for the process it describes?

2. What does the term *differential weathering* mean and why does it occur?

3. Why is chemical weathering more likely to occur in warm/wet environments, whereas frost action mostly occurs in colder regions?

4. Why does frost wedging cause physical or mechanical weathering? How is it similar to the effect that plant roots and large temperature changes have on rock? Why do these processes increase the likelihood of further weathering of a rock mass?

5. A prominent cultural feature in the American southwest is cliff dwellings. How did the niches and recesses that contain these structures evolve?

6. How does the process of unloading lead to exfoliation?

7. How are the processes of hydrolysis and carbonation similar? How are they different?

8. What is acid rain and why is it most closely associated with the northeastern part of the United States and Canada in North America?

9. Why is the process of soil creep considered to be a mass-wasting process?

10. Describe the process of solifluction and its environmental context.

11. Why is slumping more likely to occur in humid regions than arid regions? What particular rock type—sandstone or shale—is most commonly associated with slumping? Why?

12. What is the difference between a mudflow and a debris flow? How would the process of human-induced deforestation make a mudflow or debris flow more likely?

ANSWERS TO VISUAL CONCEPT CHECKS

Visual Concept Check 14.1

This weathering pattern was most likely produced by thermal expansion. It's certainly the result of mechanical weathering because the rock is merely breaking up into smaller fragments. Freeze-thaw is a mechanical process, but is unlikely to have much of an effect in southern Texas.

Visual Concept Check 14.2

The mass-wasting process pictured here is rockfall. The rocks at the base of the cliff broke free from the exposure above and fell to their current location.

CHAPTER FIFTEEN   GROUNDWATER AND KARST LANDSCAPES

Most people interact with water a great deal every day. Think about all the

many things you do that involve water. You drink it, cook with it, play in it,

wash with it, and countless other things. Water is absolutely critical to life on

Earth in many different ways, yet we often take it for granted. How often do you

think about where water comes from and the various places where water is

stored? Or even about what it would be like to run out of water? In many

Groundwater is a very important part of the Earth and for human beings because much of our water comes from this source. It can also shape the ground below you, forming fascinating subterranean chambers such as this cave in Malaysia. This chapter focuses on the characteristics of groundwater and the impact that it can have on the landscape, both above and below the Earth's surface.

regions of the Earth, water is a precious commodity and, as a result, is the focus of many political battles as people strive to acquire the water they need.

Recall from Chapter 6 that water covers about 70% of the Earth. In previous chapters we examined how water moves in the air and within the ocean. Now we focus on the very small portion of the water on Earth that is stored in the ground as groundwater.

Movement and Storage of Groundwater

In most circumstances, water that falls to the surface is absorbed by the soil, where it is stored in the *soil-water belt* (Figure 15.1). Remember from Chapter 11 that water is absorbed in soil by infiltration, which occurs through pore spaces between sediment grains (Figures 11.3 and 11.4), along pathways associated with the soil structure, or down plant roots. When there is very little water in the soil, it is held tightly to sediment grains as

hygroscopic water that is largely unavailable for plant absorption (Figure 15.2). The wilting point is the threshold at which, for any given soil, water is no longer available for plant uptake. With increasing precipitation, the pore spaces begin to fill with *capillary water*. Some of this water is lost to the atmosphere through evaporation, but plants absorb a great deal of it before it is returned to

Hygroscopic water *Soil water held so tightly by sediment grains that it is unavailable for plant use.*

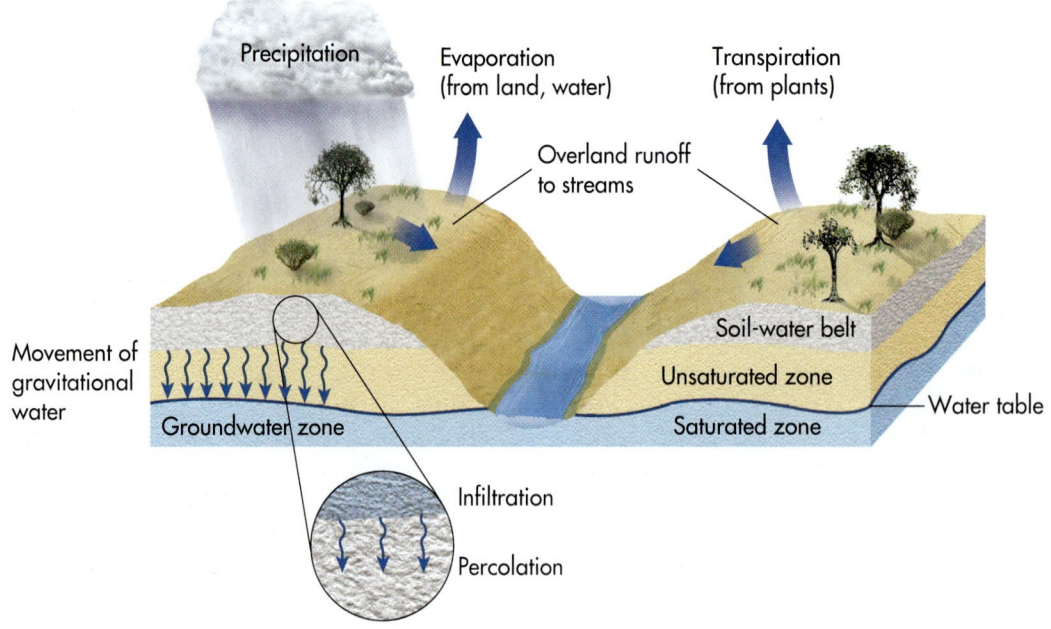

Figure 15.1 The groundwater model. Precipitation is initially stored in the soil-water belt until it becomes saturated. After that occurs, water can either run off the surface in streams or percolate through the unsaturated zone to the saturated zone.

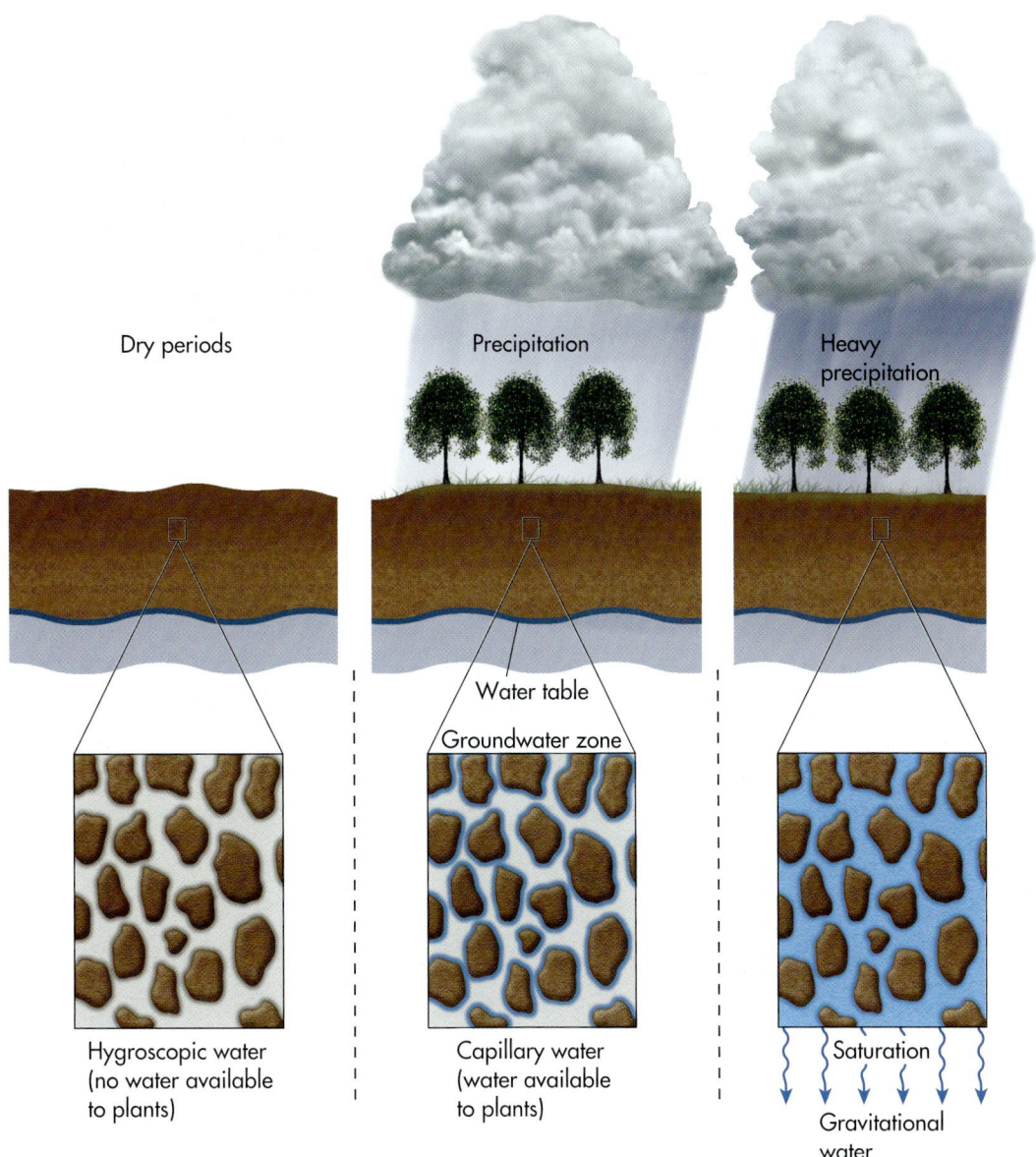

Dry periods

Precipitation

Heavy precipitation

Water table

Groundwater zone

Hygroscopic water
(no water available
to plants)

Capillary water
(water available
to plants)

Saturation

Gravitational
water

Figure 15.2 Stages of soil water. During dry periods, soil grains tightly hold onto hygroscopic water. With increasing rainfall, the pores begin to fill with capillary water. Once field capacity is reached, pores are full and excess water can run off the surface and flow downward as gravitational water.

the air through the process of transpiration (see Figure 15.1). The combined processes of evaporation and transpiration are often referred to as evapotranspiration.

If precipitation continues to the point where soil pore spaces become completely filled with water, the soil is considered to be at its **field capacity.** Any excess water is free to move more deeply in the sediment or rock, under the force of gravity, as gravitational water. As this water moves downward, it first passes through an area called the **unsaturated zone** (see Figure 15.1 again), which is the portion of the rock or sediment mass where pore spaces may be partially filled with water, but also contain oxygen and other gases.

At some depth, the water ceases to move downward and collects, so that sediment pore spaces are filled. This cessation of gravitational movement of water usually results when the infiltrating water encounters a rock layer such as shale or dense limestone that is impermeable; in other words, the pore spaces are so small that water will not flow through them. Such an impermeable layer is

Field capacity   *The amount of water remaining in the soil after the soil is completely drained of gravitational water.*

Unsaturated zone   *The area between the soil water belt and the water table where pore spaces are not saturated with water.*

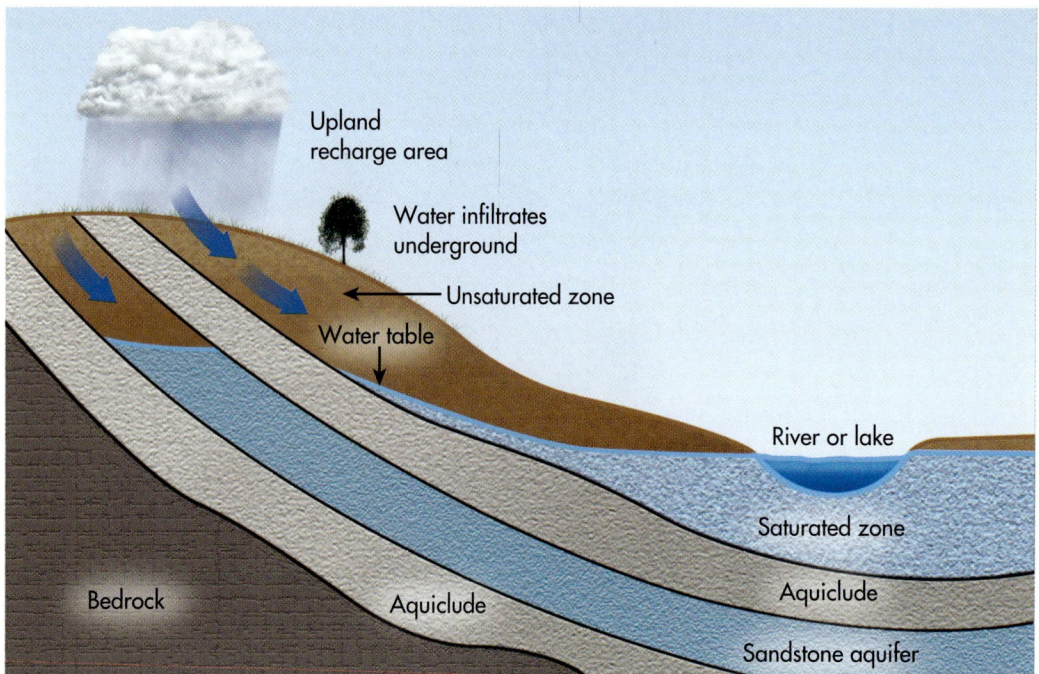

Upland recharge area

Water infiltrates underground

Unsaturated zone

Water table

River or lake

Saturated zone

Aquiclude

Bedrock

Aquiclude

Sandstone aquifer

Figure 15.3 Water tables and aquifers. A water table is the top of the saturated zone of rock and soil. Depending on the kind of rock deeper in the ground, water may collect in saturated zones called aquifers, enclosed by layers of impermeable rock called aquicludes. An aquifer is a saturated zone that holds enough water to serve as a source for irrigation or drinking water.

referred to as an **aquiclude** and causes water to collect in the sediments or rock above it because, although it may contain water, it does not allow transmission of water through it. Over long periods of time, the pore spaces within the sediments that overlie the aquiclude become filled with water and a **saturated zone** develops (Figure 15.3). Zones of saturation can be small and localized by some unique rock structure or sediment mass, or they can be regional in their extent. If a saturated zone is sufficiently large to be a significant source of water for people, it is called an **aquifer.** Regardless of its spatial extent, the top of the saturated zone is referred to as the **water table** and marks the boundary between the groundwater below and the unsaturated zone above. In most places, the water table is not visible because it's beneath the ground. Sometimes, however, topographic depressions or stream channels dip below the level of the water table, as you can see in Figure 15.3. Where such topographic relationships occur, a stream or lake will be present.

The High Plains Aquifer

The most extensive aquifers occur in regions where abundant sandy sediments or sandstone rocks are found across a broad geographical area. Aquifers are one of the most important natural resources on Earth. Perhaps the best-known and most heavily utilized aquifer in North America is the *High Plains Aquifer.* Also called the *Ogallala Aquifer,* this groundwater reservoir is found in the Great Plains region of the United States and underlies portions of eight states, ranging from South Dakota south to Texas (Figure 15.4). The High Plains Aquifer is largely formed in the thick, porous deposits of unlithified sand and gravel of the Ogallala Formation, which consists of sediments that weathered and washed out from the Rocky Mountains between 19 and five million years ago. These sediments blanketed the western part of the Great Plains and slowly filled with water from rainfall and snowmelt during the ensuing years. Much of this groundwater accumu-

Aquiclude *An impermeable body of rock that may contain water but does not allow transmission of water through it.*

Saturated zone *The zone of rock below and including the water table where pore spaces are completely filled with water.*

Aquifer *A geological formation that contains a suitable amount of water to be accessed for human use.*

Water table *The top of the saturated zone.*

Figure 15.4 Distribution of the High Plains Aquifer. This vast underground reservoir of water underlies eight states in the central United States.

lated during the ice ages, between about 1.6 million to 10,000 years ago, when the regional climate was much wetter than it is today.

As you can see in Figure 15.4 and Table 15.1, the size of the High Plains Aquifer is staggering. It covers an area of about 280,000 km² (174,000 mi²) and contains 1,315,000 hectare feet (3,250,000 acre feet) of water. In other words, the aquifer contains sufficient water to cover over 1,000,000 hectares (3,000,000 acres) to a depth of about 30 cm (12 in). This volume is approximately equivalent to one of the larger Great Lakes, such as Lake Huron. As you can see from Table 15.1, the thickness of the saturated zone varies dramatically across the region. In Nebraska, for example, the average saturated thickness is currently 104 m (342 ft). In Colorado, on the other hand, the average saturated thickness

of the aquifer is only 24 m (79 ft). This variation exists because the sediments are sandy all the way to the surface in Nebraska, resulting in rapid infiltration of water, whereas in Colorado they are covered in most places by relatively impermeable deposits.

The Great Plains is one of the most important agricultural regions on Earth, producing much of the world's wheat, corn, and soybeans. One of the primary reasons for this high productivity is that the soils are generally Mollisols, which, as we saw in Chapter 11, have thick, organic-rich A horizons and contain B horizons laden with nutrients such as calcium and magnesium. Despite the semi-arid environment, these fertile soils attracted large numbers of farmers to the region in the late 1800s and early 1900s during the period of western settlement. These farmers quickly learned, however,

TABLE 15.1	Data Associated with the High Plains Aquifer*									
Characteristic	Unit	Total	CO	KS	NE	NM	OK	SD	TX	WY
Area underlain by aquifer	mi^2	174,050	14,900	30,500	63,650	9,450	7,350	4,750	35,450	8,000
% of total aquifer area	%	100	8.6	17.5	36.6	5.4	4.2	2.7	20.4	4.6
% of each state underlain by aquifer	%	—	14	38	83	8	11	7	13	8
Average area-weighted saturated thickness in 1980	ft	190	79	101	342	51	130	207	110	182
Volume of drainable water in storage in millions 1980	acre-ft	3250	120	320	2130	50	110	60	390	70

* Modified from Gutentag, E.D., et al., 1984, *Geohydrology of the High Plains Aquifer in Parts of Colorado, Kansas, Nebraska, New Mexico, Oklahoma, South Dakota, Texas, and Wyoming*, U.S. Geological Survey Professional Paper 1400-B, p. 63.

that the region could be an unforgiving place for agriculture, with frequent droughts that decimated crops. They knew that a large groundwater supply existed, one that could potentially be used to irrigate crops, but they could not access it in any significant way with the technology of the time.

The early settlers saw abundant evidence for a large aquifer in the Great Plains. First, many artesian wells were scattered throughout the region. An **artesian well** occurs in places where an aquifer is sandwiched between two aquicludes (Figure 15.5a). When this relationship occurs, water naturally rises to the surface at springs (Figure 15.5b) due to the pressure applied by incoming water elsewhere in the aquifer. Second, settlers also noticed that many streams had a fairly consistent supply of water in them, even during periods of relative drought. This consistent flow resulted because stream channels intersected with the water table, providing a steady flow of water. These patterns are common not just to the High Plains Aquifer, but to any place where the water table is high and intersects depressions and streams.

For the first half of the 20th century, access to the High Plains Aquifer was limited and largely confined to artesian and hand-excavated wells. However, everything changed in the early 1950s with the advent of a new technology called *center-pivot irrigation*. These systems consist of massive pumping units (Figure 15.6a) that tap into the aquifer and distribute the water to farm fields through self-rotating sprinkler systems. Such a system may slowly cover a circular area approximately 64 hectares (160 acres) in size (Figure 15.6b).

Development of center-pivot systems was a major advancement for the agriculture-based economy of the region, enabling farmers to alter the semi-arid environment by providing a consistent supply of water to high-yield crops such as corn that would otherwise not consistently grow there. In fact, these systems have been so wildly successful that tens of thousands of them are currently in operation around the region. The next time you fly across the country, be sure to notice this important part of the nation's cultural landscape. It's significant to you because much of your food is produced in this fashion.

Center-pivot technology is used to access not only the High Plains Aquifer, but also other aquifers around the world. An example of this technology elsewhere can be seen in the Sahara Desert in Egypt. Remember from Chapter 9 that the Sahara Desert is an incredibly dry place due to the influence of the subtropical high-pressure system. Thus, it's not a place that naturally supports large-scale agriculture. Instead, agriculture is generally confined to isolated oases, such as Siwa Oasis (see the following Amazing Places feature). At various times in the geological past, however, most recently between about 20,000 and 10,000 years ago, the Sahara Desert was a more humid place, resulting in the storage of vast quantities of groundwater. With population pressure increasing, farmers are now moving from the oases into the desert and tapping previously inaccessible groundwater with center-pivot irrigation systems. Without these systems, it would be impossible to grow crops in

Artesian well *A well in which water from a confined aquifer rises to the surface through natural pressure.*

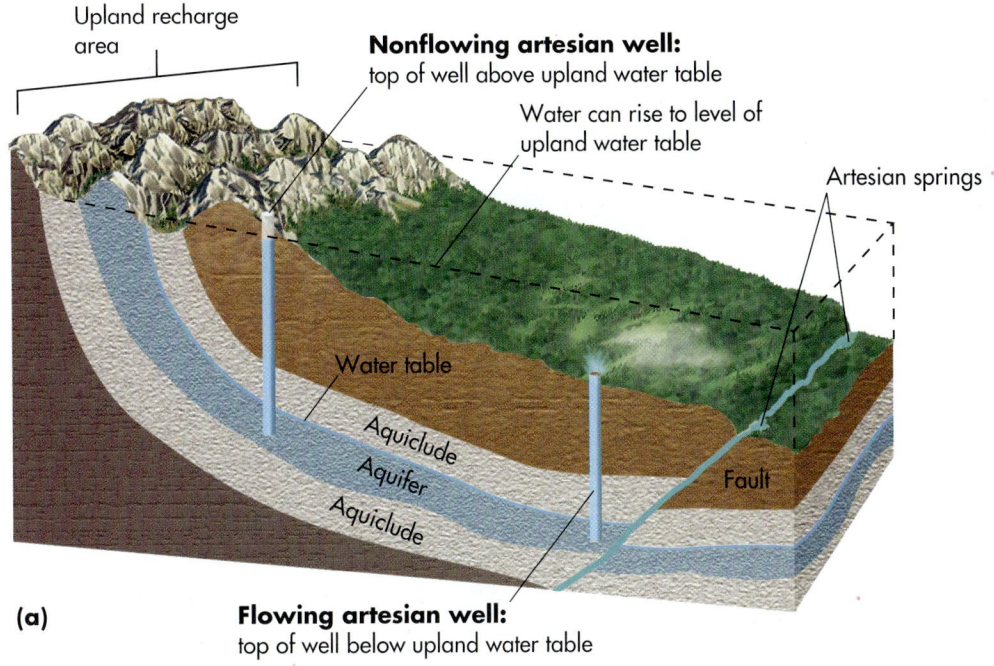

Upland recharge area

Nonflowing artesian well:
top of well above upland water table

Water can rise to level of upland water table

Artesian springs

Water table

Aquiclude

Aquifer

Aquiclude

Fault

(a)

Flowing artesian well:
top of well below upland water table

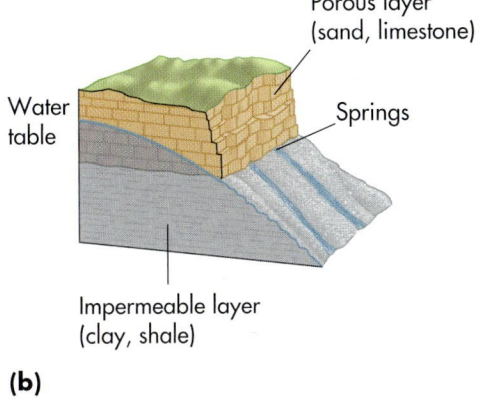

Porous layer
(sand, limestone)

Water table

Springs

Impermeable layer
(clay, shale)

(b)

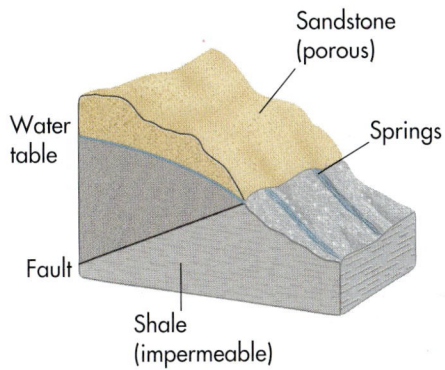

Sandstone
(porous)

Water table

Springs

Fault

Shale
(impermeable)

Figure 15.5 Artesian wells and springs. (a) When aquifers are confined between aquicludes, water may rise to the surface in an artesian well due to pressure applied by incoming water elsewhere in the aquifer. Where depressions or channels intersect the water table—for instance, at a fault—lakes and streams de-velop. (b) A spring is a place where water naturally seeps out of the ground. This may occur because the groundwater lies between aquicludes, as in an aquifer, or water may flow horizontally along the contact be-tween overlying coarse-textured sediments and an aquiclude.

this expansive desert where only a few cm (in) of pre-cipitation falls every year.

Although the rapid expansion of center-pivot irriga-tion has enabled farmers all around the world to expand their operations into marginal regions, or to grow crops such as corn that otherwise could not be cultivated in a semi-arid or arid environment, the long-term environmen-tal consequences of extensive groundwater mining are

signficant. The most obvious impact of extensive ground-water extraction is that the water table has dropped in many places. This decline can be seen at the local level with any individual well as a **cone of depression** in

Cone of depression *The cone-shaped depression of the water table that occurs around a well.*

Figure 15.6 Aerial view of center-pivot irrigation. Center-pivot pumping systems such as these in Nebraska were developed in the mid-20th century and are almost ubiquitous now in parts of the Great Plains. Look for them when you fly over the country.

which water is drawn down in a circular pattern from the sediments surrounding the well (Figure 15.7a). If pumping from the well ceases, the cone will fill back in, both by recharge from the surface and by water seeping into the empty pore spaces from the surrounding saturated zone.

In many regions, however, groundwater extraction is increasing because more wells are being installed and water is used year-round for farming and ranching. As a result, the drain on underground water has become serious at a regional level in many parts of the world. Figure 15.7b shows the impact that two wells—a deep well and a shallow, seasonal well—may have on the groundwater supply in a particular area. In the Great Plains of North America drawdown has occurred in thousands of places, causing a significant decline in the elevation of the water table in many locations. In areas within south-

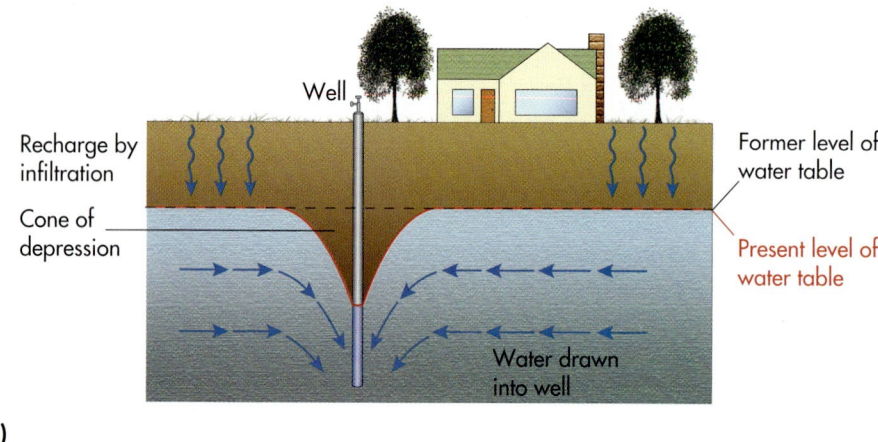

(a)

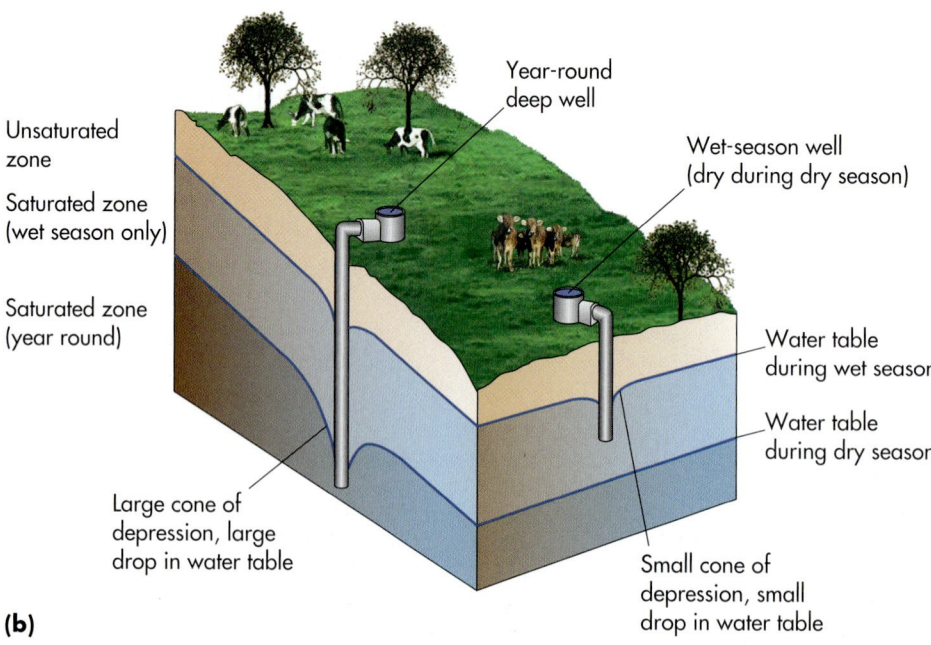

(b)

Figure 15.7 Groundwater depletion. (a) When water is drawn from an aquifer, the water table is depressed in a conical shape. (b) Deep wells draw water from zones that are saturated year round, draining far more water than shallower, seasonal wells.

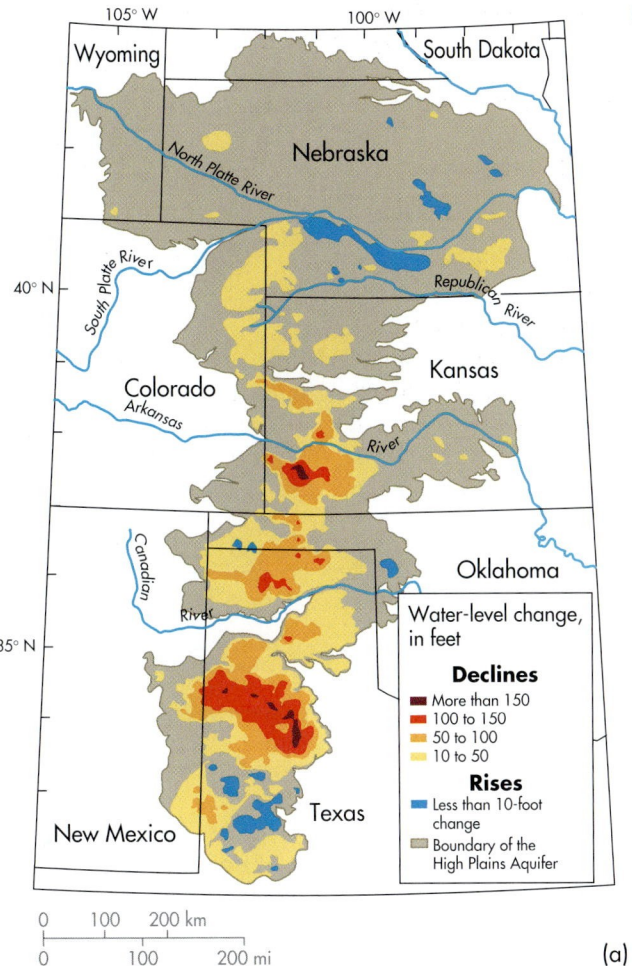

Figure 15.8 Drawdown of the High Plains Aquifer. (a) Due to excess groundwater mining, the water table has dropped significantly in an area that ranges from southwestern Kansas to the Texas panhandle. (Source: United States Geological Survey) (b) The effect of aquifer depletion has been significant on many streams. For example, prior to aggressive groundwater extraction in the region, Rattlesnake Creek in south-central Kansas was a perennial, spring-fed stream. Now, it is dry most of the time, as shown here, because the water table is far below the level of the streambed. The students in the foreground are studying the sand dunes that are exposed along the old stream bed.

western Kansas and the panhandle of Texas, for example, the average drop in the level of the water table was more than 12 m (40 ft) between 1980 and 1996 (Figure 15.8a). This rapid decline has occurred in these regions because the withdrawal rate is as much as 22 times greater than the recharge rate, which averages only 0.5 cm (0.2 in) per year. Signs of this drop in the water table can be seen throughout the region. For example, many of the streams are beginning to dry up (Figure 15.8b) because their channels no longer intersect the water table.

As a result of the heavy reliance on groundwater mining, the High Plains Aquifer may disappear in most places by the middle of the 21st century. This estimate is based on the amount of groundwater extracted from the aquifer compared to its recharge rate due to infiltration of precipitation. Estimates are similar for the depletion of other aquifers around the world. In the Sahara Desert, for example, estimates suggest that aquifers there may last 50 years. Put another way, natural resources that took hundreds of thousands of years to evolve will disappear essentially within one century due to human impact.

In response to this important environmental issue, many farmers in the Great Plains are shifting to dryland agriculture, which is a crop-rotation system designed to conserve soil moisture by leaving fields fallow (or crop-free) every other year. Other farmers are attempting to maximize short-term profits by mining the aquifer until it disappears in their region, at which time they will shift to dry-land agriculture or leave farming entirely. In the Sahara Desert, it's conceivable that oases will revert to desert because regional water tables will drop significantly.

Subsidence

The High Plains Aquifer demonstrates what happens to water table elevation when groundwater is tapped excessively. In addition to these changes, which occur deep within the ground, adjustments sometimes take place on the surface when large amounts of groundwater are removed, either for irrigation or to supply drinking water in heavily populated regions. The most obvious surface change occurs when the land sinks due to the removal of

OASES IN THE SAHARA DESERT

When you think of the Sahara desert you typically think of an incredibly dry and hot place where it is impossible to grow crops and it would be difficult to live. Although this is a generally accurate picture of the region, it is not entirely so

An oasis exists where groundwater comes to the surface in the desert.

because groundwater reaches the surface at springs and artesian wells at some places. Such a place is called an *oasis* (thought to be derived from the ancient Egyptian word "wah," meaning "fertile place in the desert") where relatively dense vegetation flourishes. About 75 percent of

water from underlying sediments. This type of response is called **subsidence.**

Subsidence is a particularly significant problem in and around large cities. For example, Venice, Italy, is approximately 1400 years old and was constructed on a landscape consisting of over 100 low-lying islands in a lagoon in the northern part of the Adriatic Sea. Underlying the islands are about 1000 m (about 3300 ft) of weakly cemented deposits of sand, gravel, clay, and silt. As the city expanded and the associated buildings became larger and more numerous, the underlying deposits were compacted by the weight of the structures above. This caused the ground to sink so that homes are now at the shoreline. This problem has been aggravated by groundwater withdrawal, both to provide drinking water and for industrial purposes. Similar problems have also recently occurred in China and California. In the

Subsidence *The settling or sinking of a surface as a result of the loss of support from underlying water, soils, or strata.*

San Joaquin Valley of California, for example, groundwater removal for irrigation has caused the ground to subside as much as 4.8 m (16 ft) in places over a 35-year period.

Groundwater Contamination

Another major environmental problem associated with the relationship between humans and groundwater is contamination by pollutants. Our industrial society produces prodigious amounts of solid and liquid waste. Until recently, no regulations governed the disposal of this waste. As a result, much of it was simply carted off to large holes in the ground, some excavated and some natural, where it was perhaps burned and buried. Infiltrating rainwater would interact with chemical compounds in the waste and carry pollutants into the saturated zone, where they would then flow along the various groundwater paths to wells and streams (Figure 15.9). Given that drinking water is often obtained from groundwater, contamination can be a significant health risk.

The location of Siwa Oasis in western Egypt.

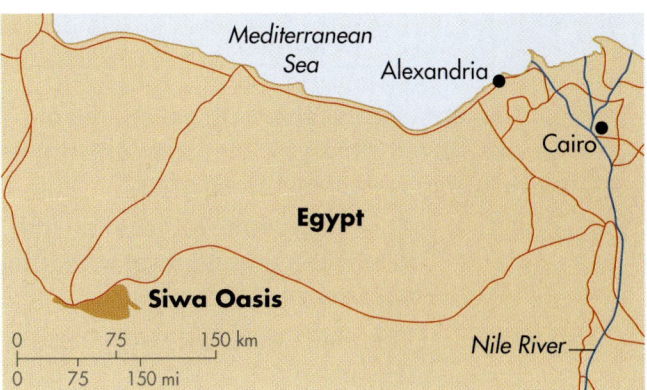

City dwelling before a palm forest at Siwa Oasis.

the people living in the Sahara are clustered around oases, where they cultivate dates, corn, and barley, among other crops.

A fascinating example of a Saharan oasis is *Siwa* (pronounced: sewa) *Oasis* in western Egypt. This oasis is about 560 km (350 mi) west-southwest of Cairo and is one of the five major oases in Egypt. It is about 82 km (51 mi) long, ranges in width from 2 to 20 km (1.2 to 12.4 mi), and has about 200 springs. Indigenous North Africans first inhabited the oasis around 12,000 years ago. It has been occupied since that time, under several different names, and subsequently became a regional religious center in 550 B.C. when the Greek temple of Amun was constructed. In fact, Alexander the Great visited the temple in 331 B.C., seeking confirmation from the temple oracles that he was the son of the god Zeus.

Currently about 23,000 people live at Siwa Oasis, with most being Berber-speaking Sudanic peoples who live in mud-brick structures. The economic base of the oasis is agriculture, with about 300,000 date palms and 70,000 olive trees. Although the oasis is still relatively isolated, it is now connected to the outside world by a road system and is thus increasingly becoming a major tourist destination.

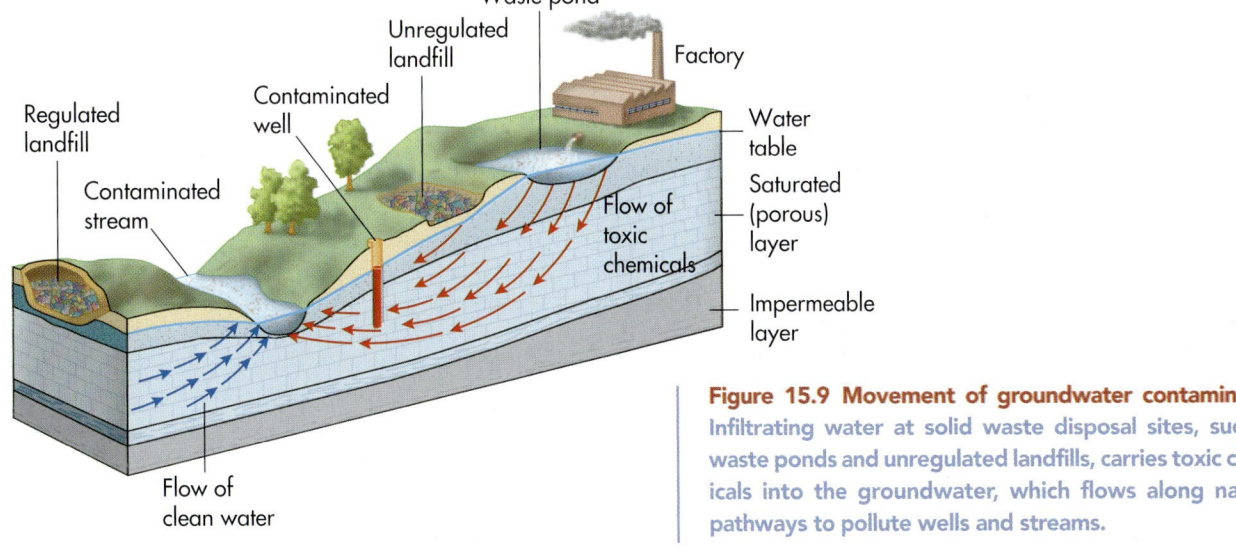

Figure 15.9 Movement of groundwater contaminants. Infiltrating water at solid waste disposal sites, such as waste ponds and unregulated landfills, carries toxic chemicals into the groundwater, which flows along natural pathways to pollute wells and streams.

This photograph shows the shoreline of a natural lake. Given what you see here, which one of the following statements is accurate?

a) The shoreline represents the place where the ground level meets the local water table.

b) The water table lies below the level of the lake bottom.

c) The local water table is very low.

d) A tremendous amount of groundwater mining has occurred in the area.

Since the establishment of the Environmental Protection Agency (EPA) in the middle 1970s, a more systematic effort has been made to control the disposal of solid waste in the United States. One important way in which solid waste disposal is intensely regulated by the EPA occurs in highly engineered landfills. In contrast to the traditional "hole-in-the-ground" approach used previously, modern engineered landfills are designed to eliminate the flow of toxic chemicals and other contaminants into the groundwater supply. For example, landfill engineers can line the base and sides of the landfill with impermeable material so that groundwater contamination does not occur. Any toxic chemicals and contaminants produced in the landfill are pumped to a settling pond where the waste is oxidized and neutralized. In spite of these measures, landfills and other hazardous waste facilities remain very controversial.

KEY CONCEPTS TO REMEMBER ABOUT GROUNDWATER

1. Groundwater refers to water that is stored within the ground.

2. Some water is stored in soil pore spaces to form the soil-water belt. When the soil is dry, water is nevertheless held tightly as hygroscopic water. With increasing precipitation, water fills more of the pores as capillary water. If the soil and associated pores are saturated, it reaches field capacity and water is free to move downward under the force of gravity.

3. When water percolates downward, it first encounters the unsaturated zone. At some point in its downward flow, water collects and fills the sediment pore spaces and a saturated zone develops. The sediments in this zone usually consist of sand and gravel that are underlain by an aquiclude of impermeable clay or shale.

4. The top of the saturated zone is called the water table. In places where the groundwater supply is sufficiently large to be a source of water for people, this zone is called an aquifer.

5. The High Plains Aquifer is the largest aquifer in North America, underlying much of the Great Plains. This aquifer is being extensively mined to irrigate agricultural crops. Consequently, a significant decrease in water table elevation is occurring in many places.

HYDROLOGIC CYCLE AND GROUNDWATER

Now that we have discussed the basic components associated with the hydrologic cycle and groundwater, go to the *GeoDiscoveries* website and open the module **Hydrologic Cycle and Groundwater**. This animation allows you to visualize the processes that have been described so far in this chapter. As you view the animation, notice the relationship among evaporation, precipitation, and the way that groundwater is stored and moves. Review how the water table fluctuates and the way that water is delivered to streams when their channels intersect the groundwater reservoir. The last part of the animation discusses how humans are interacting with the groundwater supply in the form of contamination associated with landfills and hazardous waste sites.

Karst Landforms and Landscapes

Have you ever been to Mammoth Cave in Kentucky, Carlsbad Caverns in New Mexico, or any other cave system? If so, you were exploring how groundwater can shape the subterranean landscape over a long period of time. These effects are visible not just beneath the ground, but can also be seen at the surface in the form of depressions and, sometimes, even hills.

A region that contains a lot of caves and other soluble rock features is said to be a landscape with **karst topography.** Karst topography is most closely associated with extensive, thick deposits of limestone that typically contain more than 80% calcium carbonate. Rocks with this high carbonate content dissolve easily through the weathering process of carbonation. This erosion produces a complex system of joints that allows water to flow from the surface into the underlying rock. Vegetation also plays a role because it provides organic acids that enhance the solution process.

Karst landscapes are widespread in places of high rainfall, such as the eastern part of the United States and southeastern China (Figure 15.10). They also occur in re-

Karst topography   *Terrain that is generally underlain by soluble rocks, such as limestone and dolomite, where the landscape evolves largely through the dissolution of rock.*

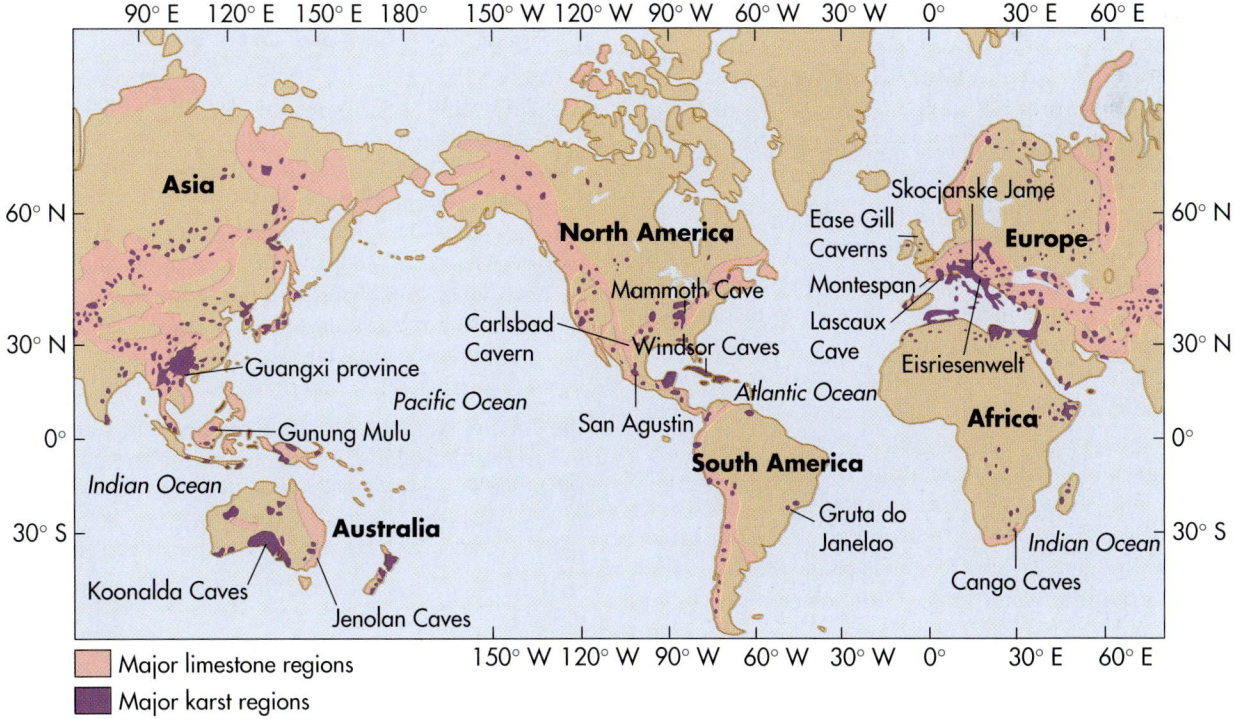

Figure 15.10 Location of major limestone and karst regions on Earth. Large areas of limestone and karst occur on nearly every continent, except for Africa.

gions that are now arid but which were humid at some point in the geologic past, such as New Mexico and north-western Africa.

Caves and Caverns

Perhaps the best-known karst landforms are **caves,** which consist of underground voids in rock that are sufficiently large for people to somehow enter. Although most caves are small, some cave systems are enormous and contain numerous passageways and large chambers. Such a large cave system is called a *cavern.* To give you an idea of how large chambers can be, consider that The Big Room at Carlsbad Caverns in New Mexico is about 110,000 m² (1,184,030 ft²) in area.

People have been aware of caves and caverns for a long time. In Europe and Asia, for example, caves provided shelter for prehistoric people during the Pleistocene Epoch when gargantuan glaciers covered much of the landscape. Evidence for these occupations can be seen in Europe in the number of beautiful paintings (known as cave art) that adorn the cave walls. These paintings, as well as burials deep within the caves, have given archaeologists important clues about the way people lived and the animals that they hunted.

Caves and caverns form due to the complex interaction of climate, groundwater, and local streams over time. Figure 15.11 illustrates the basic evolution of these landscapes. Cave formation begins during a time (Stage 1) when the water table is high because the stream has not yet downcut a deep valley by erosion. We'll discuss this concept of downcutting in more detail later, but for now, suffice it to say that one of the things that streams do as they evolve is carve out deep valleys. In the context of karst topography, the important relationship is that the water table mirrors the behavior of the stream because groundwater flows to the lowest point on the landscape— in other words, the river channel.

During Stage 1, the process of carbonation is concentrated in the deposits of limestone that occur just below the water table. The resulting dissolved sediments are then carried by underground streams to the surface channel. These kinds of streams evolve because water concentrates along joints in the rocks where it can move freely. These cracks slowly enlarge over time through dissolution and the water can thus flow more quickly as an underground stream. Over time, the process of carbonation and associated groundwater flow creates a myriad of passageways and caverns at the water table boundary. In Stage 2, the stream has cut a valley into the rock, lowering the water table. Groundwater cuts large channels and caverns into the rock as it flows through to

Cave *A cavity in rock, produced by the dissolution of calcium carbonate, that is large enough for someone to enter.*

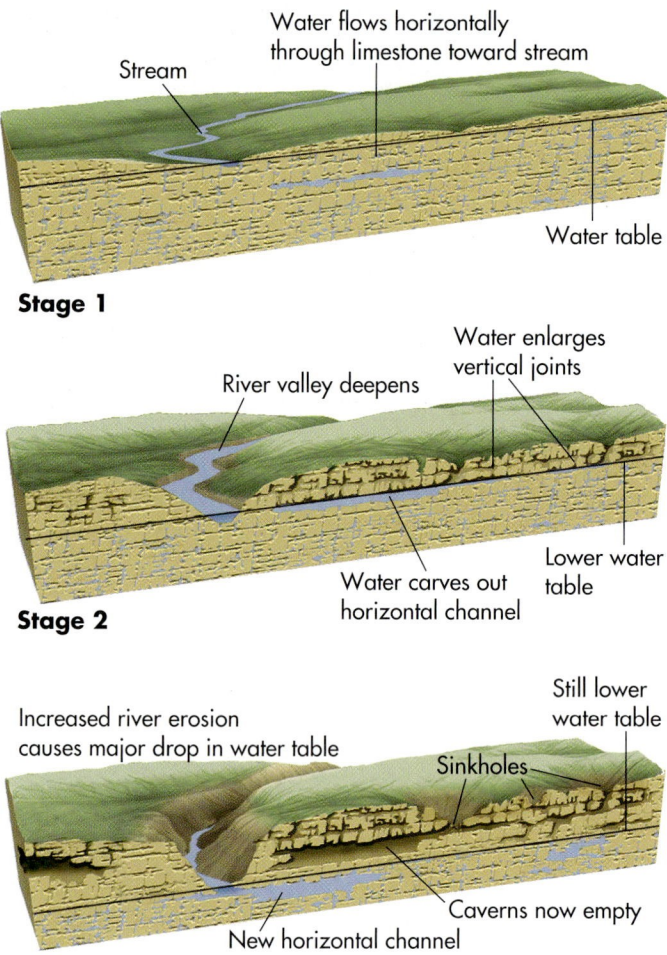

Stage 1

Stage 2

Stage 3

Figure 15.11 Cave and cavern evolution. In Stage 1, limestone below the water table dissolves, forming submerged caverns. As the stream downcuts in Stage 2, the water table drops in response, causing water to carve out channels in the limestone as the water flows to the river. In Stage 3, the water table drops farther, enough so that the formerly submerged caverns are now located in the unsaturated zone. Surface water percolates into these open features, resulting in the formation of stalagmites and stalactites.

the stream. At this time, these features are submerged in groundwater. Finally, in Stage 3, as the stream continues to downcut and deepens its valley, the water table also lowers still farther. As a result of this lower water table, the formerly submerged passageways of Stage 1 are now in the unsaturated zone and thus become open caves and caverns.

Once Stage 3 is complete, the dominant process in some caverns shifts to the formation of stalactites (hanging rods), stalagmites (upward-pointing rods), columns, and drip curtains (Figure 15.12). These unusual forma-

Figure 15.12 Drip formations at Carlsbad Caverns in New Mexico. Here, dripping water enriched in calcium carbonate has formed numerous stalactites, stalagmites, and columns in the Papoose Room. This form of deposition occurs when percolating groundwater dissolves limestone and drips, leaving calcium carbonate behind.

tions evolve in caves where large quantities of surface water percolate downward into the roof of the open cavern. As this water moves through the overlying rock, the solution becomes enriched in dissolved carbonates before it slowly drips into the cave. Stalactites in the roof of a cave form in much the same way that an icicle lengthens downward during a winter thaw. Stalagmites, in contrast, develop when water drips to the cave floor. This water evaporates, leaving the calcium behind in a deposit that slowly grows upward over time.

Karst Topography

Karst landforms can also evolve at the surface. The most common surface landform is a **sinkhole,** which is a depression that occurs in regions of cavernous limestone (Figure 15.11). Sinkholes form when caves enlarge so much that their ceilings collapse due to the force of gravity, causing the surface rocks and sediment to sink, often suddenly. Sinkholes frequently enlarge for a period of time after their initial formation because surface water tends to funnel into them. Sometimes this surface flow becomes a **disappearing stream,** leaving the sinkhole and entering the underground channels. Often, this downward-flowing stream causes a pipe to evolve that connects

(a)

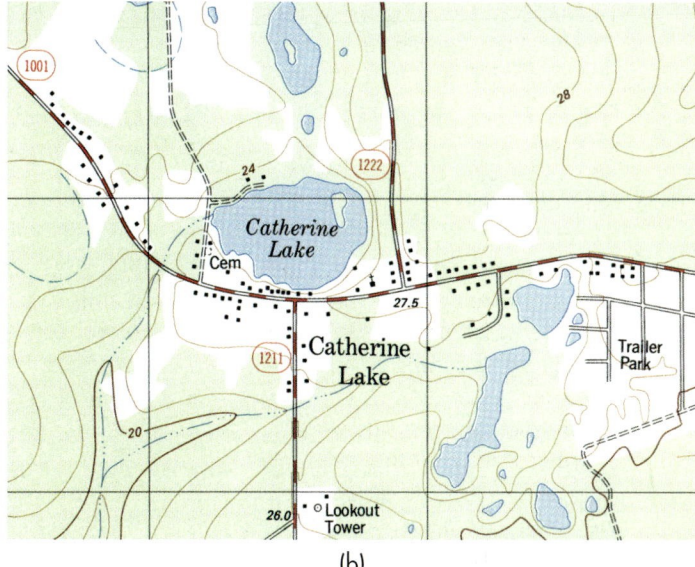

(b)

Figure 15.13 Sinkhole topography. (a) Sinkholes that fill with water form ponds. (b) Sinkholes show up clearly on topographic maps as ponds, where they are filled with water, and as depressions, where contour lines form circular features.

the sinkhole with the caverns and passageways beneath. At other times sinkholes remain detached from the underlying cavern and simply fill with water, appearing as ponds on the landscape and in topographic maps (Figure 15.13). In regions where karst activity occurs in densely populated regions, sinkholes are a significant geological hazard because they can unexpectedly swallow homes or portions of highways.

Sinkhole *A topographic depression that forms when underlying rock dissolves, causing the surface to collapse.*

Disappearing stream *A surface river or stream that flows into a sinkhole and subsequently moves into an underground river system.*

AMAZING PLACES

TOWER KARST IN CHINA

Another form of karst topography occurs in humid subtropical areas where thick beds of limestone and high water tables exist. This kind of karst is called *tower* (or *haystack*) *karst* and results in spectacular landscapes. In order for tower karst to evolve, karst processes must have operated for a long period of time, during which a variety of subterranean caves and passageways developed. Over time, these features gradually enlarge so that sinkholes and various other collapse structures form. If enough of these karst features exist, the effect is to produce a number of towering hills that are the unaffected remnants of the block of solid limestone that once existed. Streams also play a role in the development of tower karst because they cut steep valleys into the regional rock mass.

An excellent place to see fantastic tower karst is in southeastern China, near the city of Guilin in the northern part of Guangxi province. This area lies within the humid subtropical climate region and is affected by the southeastern monsoon. The Guilin area is covered by carbonate rocks that may reach a thickness of 4600 m (~15,000 ft). Two major surface karst landforms occur in the region: (1) the peak-cluster depressions (Fengcong) and (2) the peak-forest plain (Fenglin). These areas collectively cover an area 2429 km^2 (938 mi^2) in size. The Fengcong landscape consists of a group of peaks separated from each other by collapse depres-

Tower karst near Guilin, China, is a surreal landscape. Remember that a solid block of limestone was once here and most of it has dissolved, leaving these isolated peaks.

sions, whereas the Fenglin areas are a group of isolated peaks separated by flat ground created by streams. Taken together, these hills literally tower over the landscape, reaching heights of 200 m (660 ft) above the surrounding lowlands and giving the region a beautifully surreal quality. In the regions close to the United States, the only place to see tower karst is the northern part of Puerto Rico.

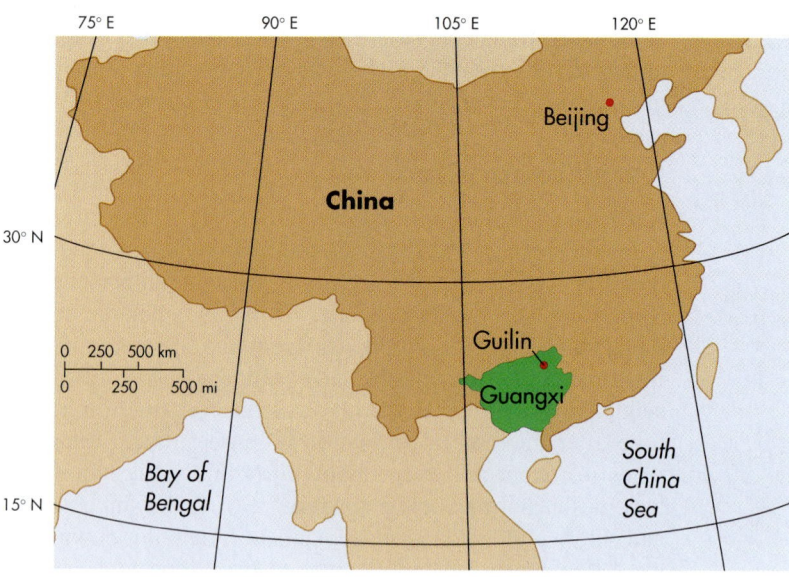

The Guangxi province is located in southeastern China. Spectacular tower karst exists near the city of Guilin.

Given what you see in this photograph, which one of the following statements is accurate?

a) The local water table intersects with the ground level at this place.

b) This photograph shows the formation of tower karst.

c) The top of a cave collapsed, forming a sinkhole.

d) The local water table was much higher at some point in the past.

KEY CONCEPTS TO REMEMBER ABOUT KARST LANDSCAPES

1. Karst landforms and landscapes form in association with groundwater and include features such as caves and sinkholes.

2. Caves begin to form during a period of high water table when flowing groundwater dissolves carbonate-rich limestones through the process of carbonation. Caves become open when the water table falls, usually in association with stream downcutting or, perhaps, climate change.

3. Once caves are opened, drip processes form features such as stalagmites and stalactites.

4. Sinkholes are surface depressions that result when underlying caves collapse, causing subsidence.

The Big Picture

Water is stored and moves around the Earth in a variety of ways. This chapter focused on the different ways in which water moves and is stored within the ground. In the next chapter we focus on how water is stored at the surface of the Earth in streams and lakes. Although this form of water makes up a tiny part of the hydrologic cycle, rivers are responsible for most of the sculpting that occurs within the Earth's landscapes. In addition, a river is one of the few places where you can see non-atmospheric water actually flow on Earth as part of the hydrologic cycle. Over the course of your lifetime, water issues will become increasingly more important as human demands on water increase. Not only will this demand impact groundwater resources, as you have seen, but it will impact rivers and streams as well. To fully comprehend some of the critical events that occur in association with stream systems, or thoroughly enjoy a canoe ride down a nice river, you must recognize how streams function and what landforms they produce. The next chapter will provide you with that background.

Summary of Key Concepts

1. Groundwater refers to water that is stored within the ground. Some water is stored in soil pore spaces to form the soil-water belt. If the soil and associated pores are saturated, water is free to move downward under the force of gravity. When water percolates downward, it first encounters the unsaturated zone.

2. At some depth the sediment pore spaces are saturated with water, forming the saturated zone. The sediments in this zone usually consist of sand and gravel that are underlain by an aquiclude of impermeable clay or shale. The top of the saturated zone is called the water table. In places where the groundwater supply is sufficiently large for people to use for irrigation or drinking water, the saturated zone is called an aquifer.

3. Groundwater can become contaminated when chemicals from old landfills or dumps slowly move into the system.

4. Karst landforms and landscapes form in association with groundwater and include features such as caves and sinkholes.

5. Caves begin to form during a period of high water table when flowing groundwater dissolves carbonate-rich limestones through the process of carbonation. Caves become open when the water table falls, usually in association with stream downcutting or, perhaps, climate change.

6. Tower (or haystack) karst topography forms when numerous caverns and subterranean passageways collapse, leaving individual high hills.

Check Your Understanding

1. What is the soil water belt and how do the terms hygroscopic water, capillary water, and field capacity relate to this term?

2. What is the unsaturated zone and where is it located in the groundwater model?

3. What role does an aquiclude have in the formation of a saturated zone?

4. Why do sandy and gravelly deposits make good aquifers?

5. Describe the relationship among topography, the water table, and the development of natural lakes.

6. What is a cone of depression and how is it related to declining water tables in places like the Great Plains?

7. Why are many streams going dry in places where aggressive extraction of groundwater occurs?

8. Why are modern landfills more environmentally safe than unregulated disposal sites of years past?

9. What is karst topography and how do caves evolve?

10. How does the formation of a sinkhole relate to the process of land subsidence?

ANSWERS TO VISUAL CONCEPT CHECKS

Visual Concept Check 15.1

The answer to this question is *a*: the shoreline represents the place where the ground level meets the local water table. Such a natural lake exists when depressions dip below the level of the water table.

Visual Concept Check 15.2

The answer to this question is *d*: the local water table was much higher at some point in the past. This cave formed when carbonate rocks dissolved when the water table was higher. After the water table fell, the cave was left behind.

CHAPTER SIXTEEN

FLUVIAL SYSTEMS AND LANDFORMS

In Chapter 15 we examined the processes and landforms associated with groundwater. Now we will study water that flows across the surface of the Earth in stream channels. Such bodies of water are called *fluvial systems* and include rivers, which are large streams of flowing water, and creeks or brooks, which are smaller. Regardless of their size or what they are specifically called, the common variable shared among all streams is that they are bodies of water that flow downhill in channels due to the force of gravity. In the process of flowing in this manner, they shape the landscape in very distinct ways.

Flowing water is the most important geomorphic agent on Earth, shaping the landscape in a variety of fascinating ways. Humans also interact with streams in numerous ways, using them for transportation arteries, supplies of drinking water, and energy production, to name a few. This chapter focuses on the way that water flows across the Earth's surface in streams, the landforms that are produced by these processes, and some of the ways that people interact with them.

In most parts of the world, streams are a visible indication of the more distant geological past. In humid regions, streams have likely been flowing consistently along the same basic course for thousands of years, shaping the landscape gradually through erosion and deposition. Where streams occur in semi-arid to arid regions, in contrast, they may flow only during the wet season, or only every few years. When they do flow, however, they may contain so much water and have so much power that they change the landscape dramatically in a very short time.

Overland Flow and Drainage Basins

Streams are an important part of human history because they are conduits that connect the continental interior with the ocean. All flowing water ultimately returns to the sea in some fashion as part of the hydrologic cycle. Streams have long served as transportation corridors that link widely spaced groups of people economically. Rivers such as the Rhine in Europe, the Nile in Africa, the Mekong in Southeast Asia, the Amazon in South America, and the Hudson and Mississippi in the United States, to name but a very few, are important physical features in the evolution of human societies. The famous Lewis and Clark expedition, for example, used the Missouri River as its pathway deep into the American west in 1803. Figure 16.1 shows the location of major rivers around the world.

Figure 16.1 Important rivers on Earth. Rivers consist of flowing water that connect the continental interior with the ocean. As a result, these rivers, and streams like them, have been used as transportation arteries over the course of human history.

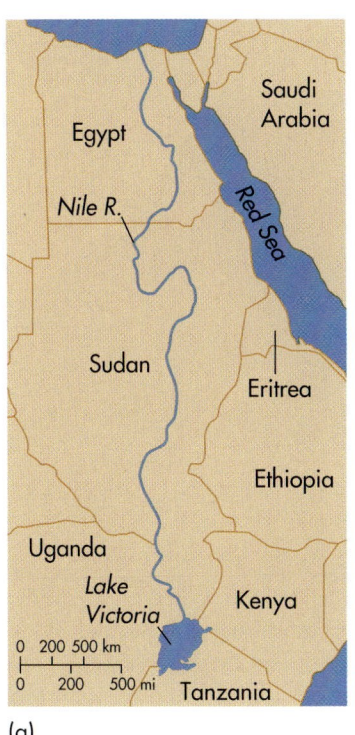

(a)

(b)

Figure 16.2 Source of the Nile River. (a) The Nile River begins at the north end of Lake Victoria and flows northward for about 10,730 km (6670 mi) before it ultimately reaches the Mediterranean Sea in Egypt. (b) This image shows the outlet where water flows out of Lake Victoria into the Nile River.

Origin of Streams

Where does the water in streams come from? Some streams flow all year long and are called *perennial streams*, whereas others contain water for only a short time and are thus *ephemeral streams*. You might initially think that the water comes directly from precipitation that falls into the stream channel, but that's really only a negligible contribution. Some stream water comes from melting glaciers, which provide water for streams that flow away from the ice front. A more important source of stream water is water that seeps out of the ground where stream channels intersect the water table. We discussed this concept in Chapter 15 (see Figure 15.3). In many areas, this source may be the most important contributor of stream water.

Still another important source for streams is relatively static bodies of water stored in lakes and ponds, which are fed both by precipitation and groundwater. If the water level in such a lake reaches a sufficient depth, it may spill over at a low spot on the landscape, called an *outlet*, and become the source of a river. A good analogy to this natural scenario is if you fill a bathtub beyond its holding capacity. At some point, the water begins to spill over the edge of the tub at the rim's lowest spot and flow across the floor. So long as you continue running water in the tub, it will stay full and water will flow out of it.

Many rivers start at an outlet from a major lake. An excellent example is Lake Victoria in east central Africa, which is the source of the Nile River. The lake straddles the borders of Uganda, Kenya, and Tanzania

(Figure 16.2a) and lies in a depression created by the East Africa Rift system (Figure 13.9). Approximately 67,850 km^2 (26,197 mi^2) in size, Lake Victoria is the second largest freshwater lake in area on Earth (behind Lake Superior in North America). Excess water spills over an outlet at the north end of Lake Victoria (Figure 16.2b) and marks the beginning of the Nile River. The Nile River is about 10,730 km (6670 mi) long, making it the longest river in the world, and flows northward through Sudan and Egypt to the Mediterranean Sea. The best-known river in North America, the Mississippi, also begins at a lake outlet, Lake Itasca in Minnesota. At this place, you can literally walk across the Mississippi River!

Another significant source of stream water is surface runoff. Runoff occurs most commonly in association with wet periods when soils are saturated and pore spaces can no longer absorb additional precipitation. As we discussed in Chapter 15, these periods are times of groundwater recharge because water is free to move from the soil downward through the force of gravity. At the same time that water is moving downward through the soil, water runs off the surface and toward the stream as overland flow (Figure 16.3). You can see the process of runoff clearly in any parking lot during and shortly after a strong storm. The asphalt keeps water from soaking into the soil beneath; as a result, the water flows across the surface in sheets toward a drain of some kind. This type of runoff, called *sheet runoff*, also occurs in areas where slopes are very steep and rainfall simply does not have a chance to soak into the ground.

Figure 16.3 Storm runoff. During periods of heavy rain, soils become saturated and water flows across the surface. This kind of flow is an important source of stream water.

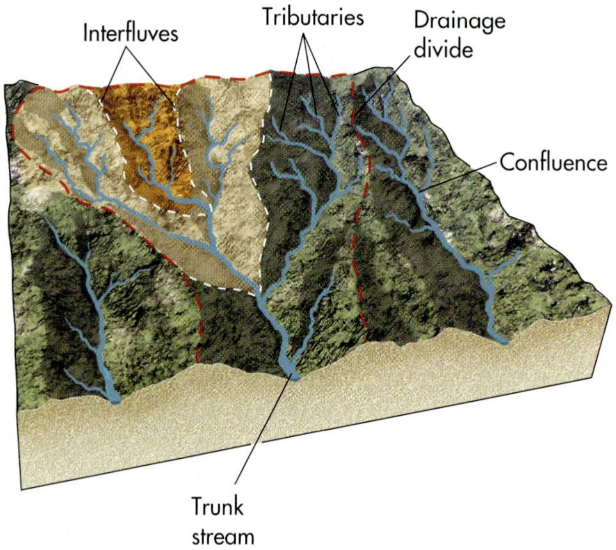

Drainage Basins

Before we examine the ways in which streams flow and the landforms they produce, we first need to describe how stream systems are organized and how they are topographically distinct from each other. We begin this discussion with the concept of the drainage basin. Simply put, a **drainage basin** is the geographical area that contributes groundwater and runoff to any particular stream. Another term frequently used to define the same area is *watershed*.

To visualize better the concept of a drainage basin, let's use an extreme example by comparing a stream in Massachusetts and another in California. Naturally, you wouldn't expect rainfall on the east coast to run directly off into a stream on the west coast. Thus, streams in California lie in a separate drainage basin or watershed from those in Massachusetts. With this extreme example in mind, let's turn now to how streams are organized within the same geographic region.

We begin by considering how streams are topographically distinct from one another and establishing some basic terminology. Streams in any two watersheds are separated by a topographic feature called a **drainage divide**, which is an area of elevated terrain that forms a kind of rim around any given basin (Figure 16.4). Below this rim, the topography slopes downward into the core of the basin. As a result, all runoff and groundwater flows into streams that progressively funnel into the **trunk stream,**

Figure 16.4 Drainage basins. Also called watersheds, drainage basins are separated from one another by divides, which are high points on the surrounding landscape. Each large drainage basin contains numerous tributaries, which themselves are separated from each other by interfluves.

which is the largest stream in the drainage basin. Each stream that flows into the trunk stream, or a stream that flows into another stream, is called a **tributary.** Tributaries are themselves separated from one another by relatively small topographic high points called **interfluves.** The point where a tributary joins the trunk stream, or any stream for that matter, is referred to as a **confluence** (Figure 16.4).

Drainage basins range from several hectares to thousands of square kilometers in size, with numerous watersheds contained within larger ones in a nested fashion, as you can see in Figure 16.4. Notice, for example, that the small drainage basins portrayed in tan is nested within the drainage basin highlighted in yellow. In other words, water that drains into the tan basin ultimately flows into the larger stream contained within the yellow basin at a confluence. This stream ultimately flows into the trunk stream at another confluence, like the one pictured in Figure 16.5. This image shows the confluence of the two major tributaries of the Kansas

---

Drainage basin *The geographical area that contributes groundwater and runoff to any particular stream.*

Drainage divide *An area of raised land that forms a separating rim between two adjacent drainage basins.*

Trunk stream *The primary stream of a drainage basin.*

Tributary *A stream or river that flows into a larger stream or river.*

Interfluves *Topographic high points in a drainage basin that separate one tributary from another.*

Confluence *The place where two streams join together.*

Figure 16.5 A typical river confluence. The Smoky Hill River flows into this image from the upper left, whereas the Republican River enters from the upper right. They join at a confluence in the center of the photograph to form the Kansas River, which is the trunk stream of the Kansas River basin, and flows toward the bottom of the image.

River, which is the trunk stream of the Kansas River basin in the central United States. To see the geographic relationship of these streams, refer back to the map in Figure 15.4. In this map note that the Republican River drains the northern portion of Kansas and southern Nebraska, whereas the Smoky Hill River drains western Kansas. These two streams meet in eastern Kansas to form the Kansas River.

Let's put this discussion of drainage basins into the context of major watersheds in the United States. The map in Figure 16.6 not only shows the geographic position and size of the largest drainage basins, but also indicates which ocean the trunk stream of any particular basin flows into. The largest watershed in North America is the Mississippi River basin (bold outline), which drains most of the United States and directs water to the Gulf of Mexico. The Mississippi watershed contains several other large basins, including the Ohio, Arkansas, and Missouri, which are named after their trunk streams. Another way to describe these trunk streams is that they are major tributaries of the Mississippi River. Each of these trunk streams, in turn, contains a number of tributaries. A major tributary of the Missouri River, for example, is the Kansas River shown in Figure 16.5. In other words, it's appropriate to say that the Smoky Hill and Republican rivers are tributaries not only of the Kansas, but also of the Missouri and Mississippi rivers.

The western divide of the Mississippi basin is the Rocky Mountains. This divide is the famous *continental divide* because it separates the Mississippi watershed and its drainage to the Gulf of Mexico from streams that deliver runoff into the Pacific Ocean, such as the Colorado and Columbia Rivers. The eastern divide of the Mississippi basin is the Appalachian Mountains, which is also a continental divide because it

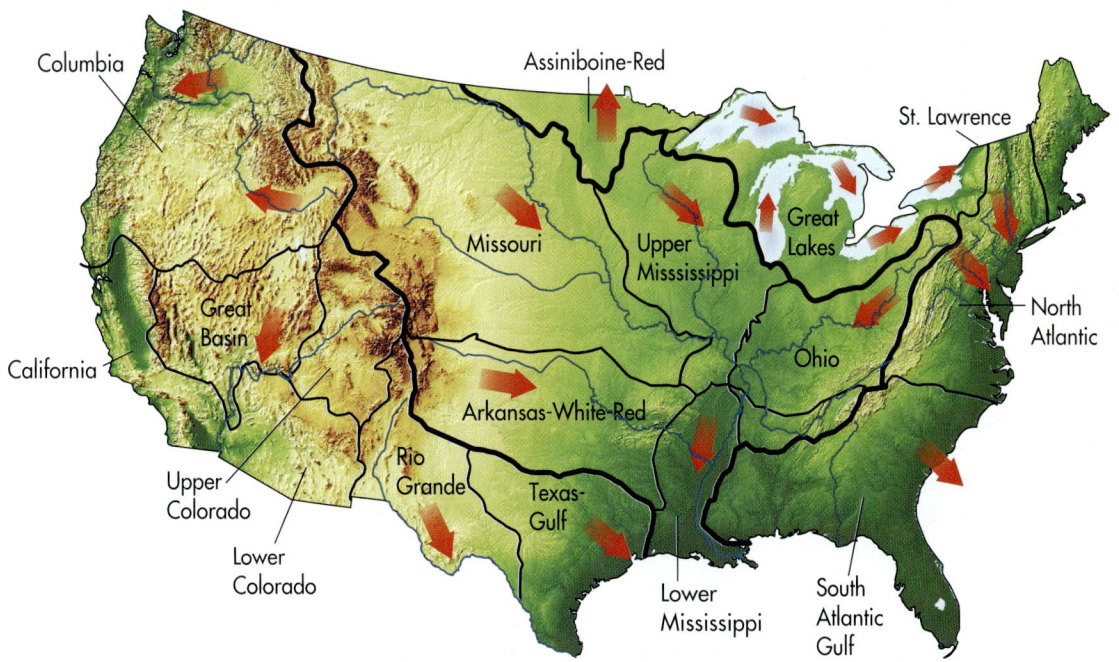

Figure 16.6 Major United States watersheds. Several large drainage basins occur in the United States. Each of these basins contains an array of nested drainage basins like those shown in Figure 16.4. Note the direction of flow for each trunk stream, The Mississippi River is the largest drainage basin in the country and is outlined in bold.

separates Atlantic-flowing drainages such as the Delaware and Hudson basins from the Gulf-flowing waters of the continental interior.

Drainage Patterns, Density, and Stream Ordering

Drainage networks can be classified in a couple of different ways. One method is to look at the drainage pattern, which is the way that the various streams within a drainage basin are spatially organized. The four primary drainage patterns are shown in Figure 16.7.

1. **Dendritic**—A branching, tree-like drainage pattern evolves in areas of uniform rock resistance and structure, with little distortion by folding or faulting. Notice in Figure 16.7a that the streams develop a random branching network similar to a tree.

2. **Rectangular**—A rectangular drainage pattern occurs when joints and faults steer streams at right angles to one another. This pattern occurs because water flows preferentially to these zones of weakened bedrock where the water more freely erodes. Notice in Figure 16.7b the angular nature of the drainage pattern.

3. **Trellis**—A trellis drainage pattern is a system of streams that develops in areas such as the Ridge and Valley Province in the Appalachian Mountains where rocks are folded. In this area, major streams tend to flow parallel to one another in adjacent valleys within the folded mountain belts. Minor tributaries flow into larger streams at right angles.

4. **Radial**—In a radial drainage pattern, streams radiate outward from a central point, forming a spoke-like pattern of rivers. This kind of pattern tends to evolve where streams flow away from rounded upland areas such as a volcano.

Another way to characterize a drainage basin is by calculating the drainage density of a watershed. **Drainage density** refers to the relative density of natural drainage channels in a given area. This value is calculated by dividing the total length of all streams in the basin by the area of the basin:

$$\text{drainage density} = \frac{\text{total length of all streams}}{\text{area of drainage basin}}$$

Drainage density *The measure of stream channel length per unit area of drainage basin.*

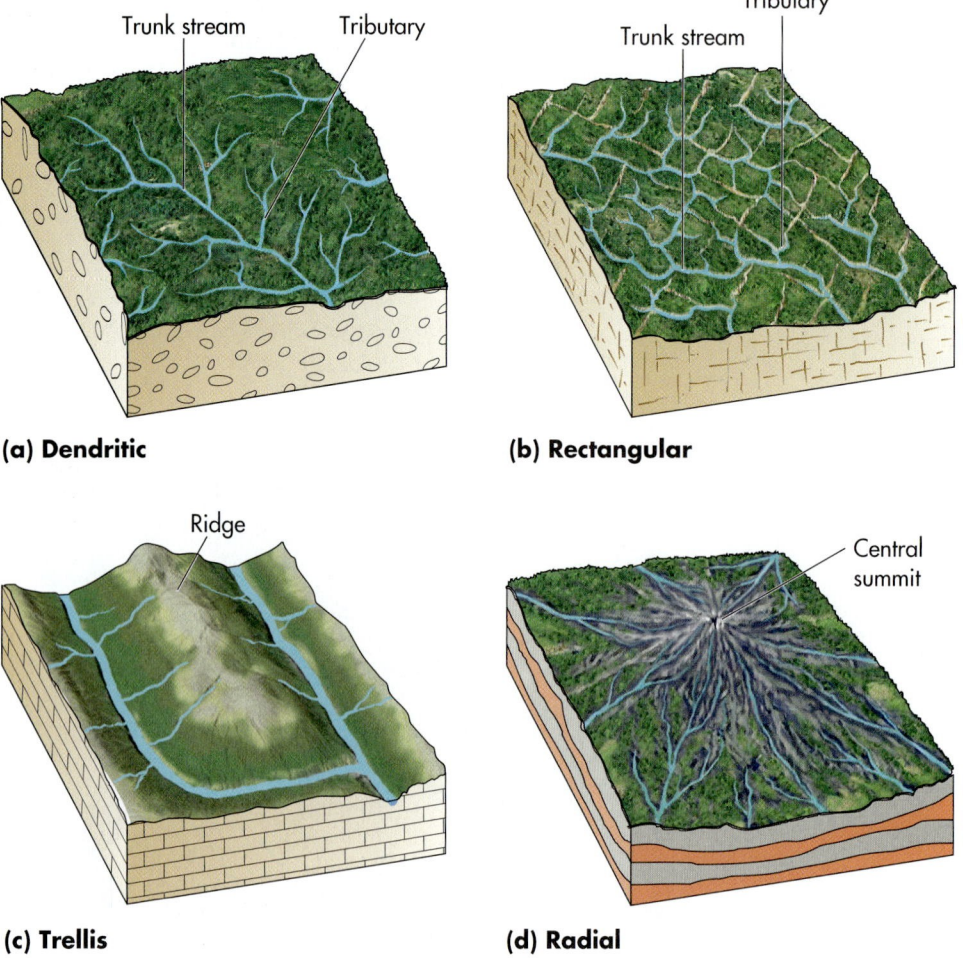

(a) Dendritic

(b) Rectangular

(c) Trellis

(d) Radial

Figure 16.7 Drainage patterns. Stream networks are configured in four primary drainage patterns: (a) dendritic, (b) rectangular, (c) trellis, and (d) radial.

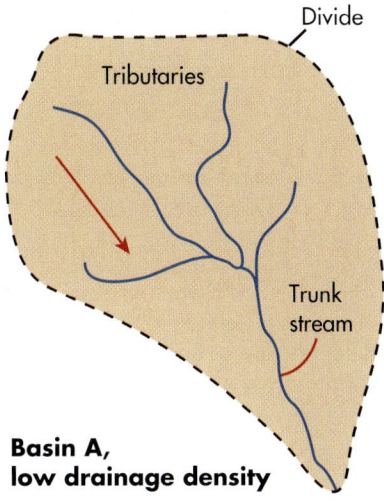

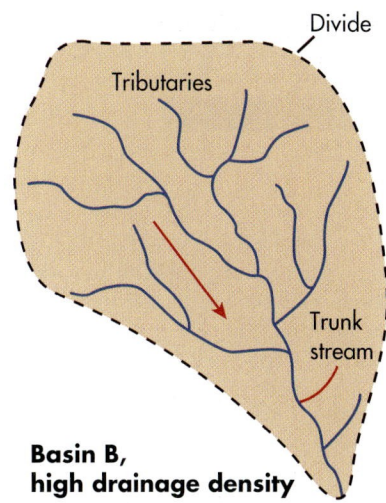

Basin A,
low drainage density

Basin B,
high drainage density

Figure 16.8 Drainage density differences. Basin A has a lower drainage density than basin B. This may result from differences in climate or type of underlying rock or sediment. The arrows indicate the general direction of flow in each watershed.

Figure 16.8 shows an example of differences in drainage density. This diagram shows two watersheds (A and B) of the same size that clearly contain a variable number of streams. Which area, basin A or B, has the higher drainage density? The answer is basin B. Why might such a difference exist? One reason might be that basin B occurs in a more humid climate, so there are more streams to handle the extra runoff of this region. A second possibility is that basin A occurs on a landscape lying over porous sands that allow water to seep rapidly into the ground, whereas watershed B occurs in a region of relatively impermeable clay that causes water to flow overland. The measurement of drainage density provides a hy-

drologist or geomorphologist with a useful numerical measure of runoff potential and how much streams have shaped the landscape.

A third way to characterize a drainage network is through the process of stream ordering. Streams within any drainage network can be arranged into a hierarchy based upon their size. Stream ordering is useful because it provides a good relative measure of a stream's place in a basin hierarchy, which is typically a function of how many tributaries occur in any given watershed. Ordering the streams of any basin is a simple task, as shown in Figure 16.9. An important rule to remember when calculating stream order is that it changes whenever two streams of the *same order*

Figure 16.9 Stream ordering. Stream ordering classifies the hierarchy of channels in a drainage basin. Stream order changes only where two streams of the same order merge. There is a positive relationship between stream size and stream order, with streams of progressively higher order portrayed by progressively thicker lines on the diagram.

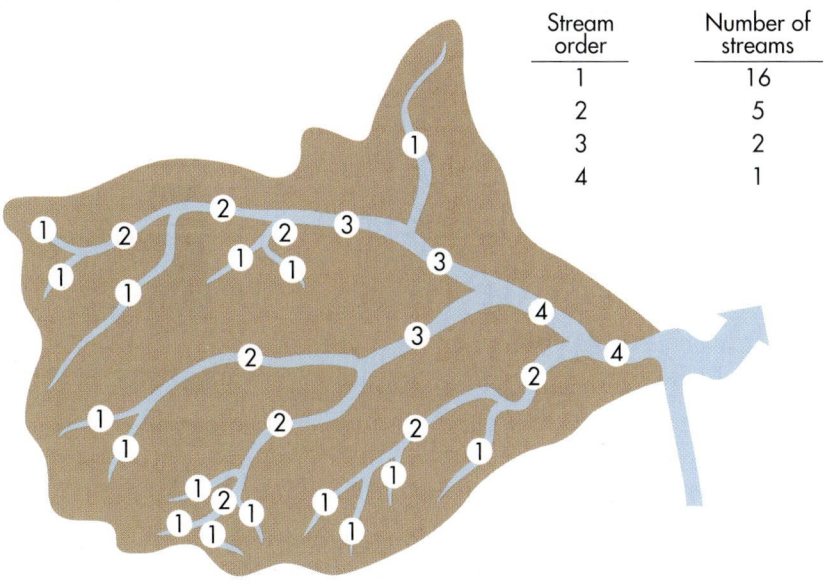

Stream order	Number of streams
1	16
2	5
3	2
4	1

THE RHINE RIVER

The next section of the chapter focuses on hydraulic geometry and channel flow. You can preview this material by going to the *GeoDiscoveries* website and accessing the module **The Rhine River**. This video follows the course of the famous Rhine River in Europe from its origin in the Swiss Alps to where it meets the Atlantic Ocean in Holland. As you watch the video, notice the various landscapes through which the Rhine flows. Listen for the technical terms, such as graded stream, rapids, and valley, and try to understand their meaning. Although these terms won't have much meaning for you now, they will be thoroughly described later in the chapter. As you continue through the chapter, keep this video in mind as an excellent example of a stream. Once you complete the module, be sure to answer the questions at the end of the module to test your understanding of this concept.

join at their confluence. The smallest tributaries of a basin are first-order streams. At the confluence of the first-order tributaries, a second-order stream forms. Subsequently, two second-order streams join to form a third-order stream.

You may be tempted to think that the ordering of a stream changes *whenever* it meets another stream, regardless of the second stream's order. For example, Figure 16.9 shows several places where first-order streams flow into streams of a higher order, such as a second- or third-order stream. You can also see a confluence where a second-order stream meets a third-order stream. It is important to remember that stream order changes *only* at the confluence of streams of the same order. For example, just because a first-order stream meets a second-order stream, that doesn't mean the second-order stream becomes a third-order stream. It doesn't; rather, it remains a second-order stream. The second-order stream becomes a third-order stream *only* where it meets with another second-order stream.

A last point to make about stream ordering is that there is a positive relationship between increased stream order and stream size. In other words, you can expect a second-order stream in a drainage basin to be larger than a first-order stream. Similarly, a third-order stream is bigger than a second-order stream. This increase occurs because a stream of a higher order contains the combined flow of the lower ones. Thus, higher-order streams logically contain more water than lower-order streams do. In Figure 16.9, this geographical relationship is portrayed by using progressively thicker lines to represent streams of progressively higher order.

KEY CONCEPTS TO REMEMBER ABOUT RUNOFF AND DRAINAGE BASINS

1. Streams consist of channels in which water flows downhill. Runoff refers to water that runs off land surfaces into streams.

2. Streams are fed by runoff or groundwater, or may have a large lake as their source.

3. A drainage basin is the area that contributes runoff or groundwater to any given stream. Drainage basins are separated from one another by divides, which are topographic rims consisting of relatively high ridges and hills.

4. Drainage density refers to the relative density of streams in comparable areas and is influenced by climate, underlying bedrock, or both.

5. Streams are classified by their order, which ranges from low (such as 1 or 2) to high (such as 5 or 6). Stream order changes only at the confluence of two streams of the same order.

Hydraulic Geometry and Channel Flow

Streams are very important natural systems on Earth for several reasons. They supply us with water for drinking and irrigation, a transportation system, energy for industry, and a waste-disposal system. In times of flooding, streams can have a devastating effect on all of these systems and on residential areas. Streams are also responsible for shaping much of the landscape. Because streams have such wide influences on our lives and on the landscapes they produce, it is important to understand how they behave.

The study of flowing water on the Earth's surface is called *stream hydrology* and focuses on the geometrical attributes, or *hydraulic geometry*, of river channels. Streams have a measurable width, depth, velocity, slope down which they flow, and discharge. Hydraulic geometry characterizes the cross-sectional components of any given stream's channel form (Figure 16.10) by measuring these simple variables:

- w = channel width (how wide the channel is in which the stream is flowing).
- d = depth (how deep the channel is).
- v = velocity (how fast the water is moving in the channel).
- s = slope (how steep the slope [also called *gradient*] is on which the stream is flowing).
- Q = discharge (how much water is flowing in the channel).

These variables are intimately related to one another. For example, the velocity of a stream closely correlates to the channel slope, with streams on steep gradients moving more quickly than those on shallower slopes. With increases in width, depth, and velocity, stream discharge becomes greater.

Stream Discharge

The most important hydraulic variable is stream discharge (Q), which refers to how much water is flowing in the channel. This value is calculated by the simple equation

$$Q = w \times d \times v$$

Discharge is measured with a device called a stream gauge that can either be portable or fixed in one place in a gauging station. Stream discharge is typically expressed as cubic meters per second (m^3/s) or cubic feet per second (ft^3/s). With increases in width, depth, and velocity, stream discharge becomes greater.

In most streams, discharge varies over the course of the year, depending on cycles of rainfall and relative drought. The amount of discharge that serves as the point of

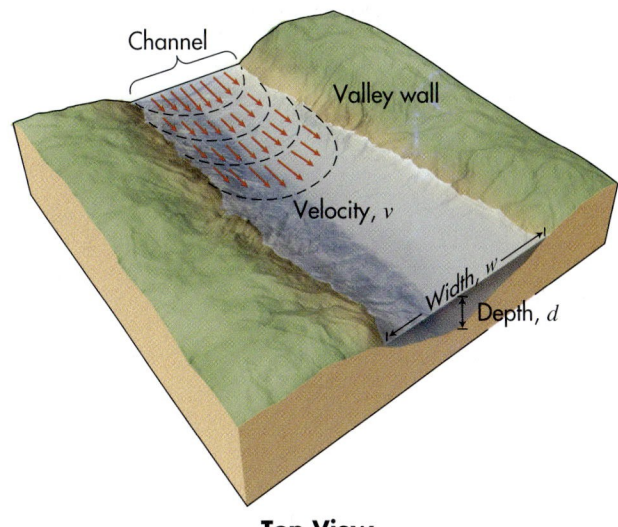

Top View

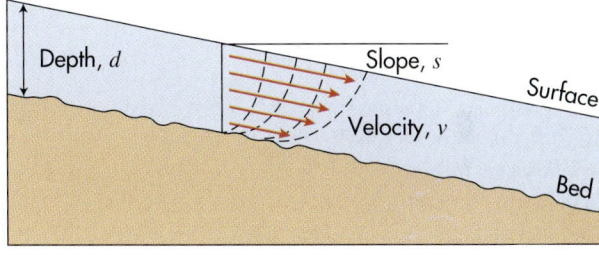

Side View

Figure 16.10 Hydraulic variables. Streams have a measurable width, depth, velocity, slope (or gradient) down which they flow, and discharge. Velocity is least on the sides and bottom of the channel due to the effects of friction.

VISUAL CONCEPT CHECK 16.1

The famous Danube River, shown here, is the second-longest river in Europe, flowing from Germany to the Black Sea. Which one of the following statements is correct and can accurately be determined?

a) The Danube is a high-order stream.

b) The discharge of the Danube is low.

c) The watershed of the Danube is small.

d) The slope of the Danube is very steep.

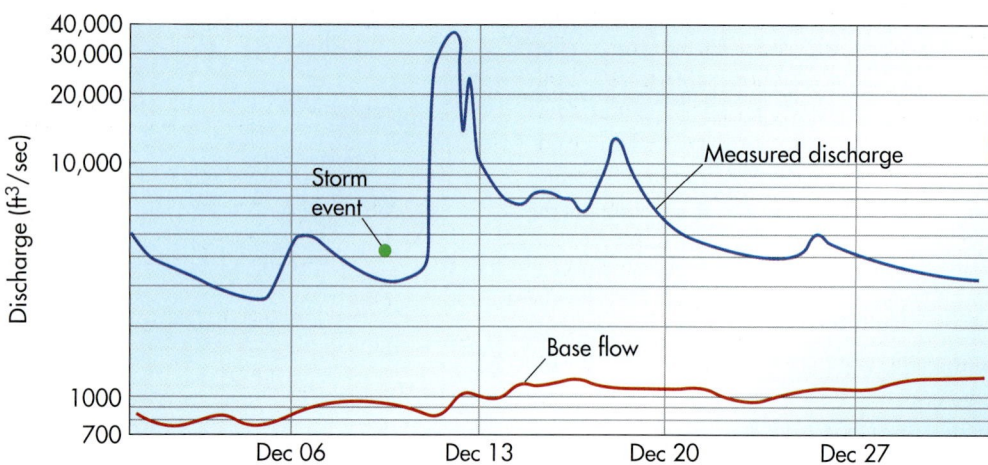

Figure 16.11 Rappahannock River hydrograph. This hydrograph illustrates how discharge varied during December, 2003 (blue line), in the Rappahannock River near Fredricksburg, Virginia. Base flow (red line) indicates the rate of flow maintained mostly by groundwater influx. Note the peak that occurs in association with increased runoff caused by a storm event. (Source: United States Geological Survey)

comparison for any stream at any given place and time is **base flow,** which is the flow rate that is sustained solely by groundwater influx. During periods of heavy precipitation, the amount of runoff to the stream increases, causing stream discharge to increase above the base flow amount. If the amount of water increases such that the channel is full of water, then the stream is said to be at **bankfull discharge.**

Figure 16.11 is a stream hydrograph that shows how stream discharge varies as groundwater influx and runoff change. A **stream hydrograph** is a graph that shows the fluctuation in stream discharge over the course of time. This particular hydrograph is from the Rappahannock River in Virginia and was collected from a measuring station near the city of Fredricksburg. An important thing to note from this hydrograph is that stream discharge does not immediately increase during a storm event. Instead, a time lag occurs between the period of heaviest rainfall and maximum stream discharge because it takes a period of time for the water to flow overland and reach the channel.

Flooding When stream discharge increases so that water begins to spill out of the channel and over the adjoining ground, the river is said to be at **flood stage.** In humid regions, most high-order streams are bordered by low, relatively level terrain generally referred to as the *floodplain.* Although we'll discuss floodplain formation

in greater detail later in this chapter, it's enough to say now that the stream creates this surface, which stands only a few meters above the channel water level at base flow. During the wet season, the combination of significant surface runoff and groundwater influx can cause the floodplain to become inundated by excess discharge. In most humid locations floods are annual to semi-annual events and are part of the natural evolution of river systems. In addition, they are important in the maintenance of the flora and fauna that live along a river (which, if you recall from Chapter 10, is called the *riparian zone*).

Occasionally, periods of extremely heavy rains occur in a watershed, causing the streams to rise to spectacular levels that result in great damage to human property and even sometimes significant loss of life. Fortunately, these kinds of floods are rare, with a close statistical relationship between the *return period*—in other words, the frequency of the event—and its magnitude. To put it more simply, more extreme floods happen less frequently than do lower-magnitude floods. Hydrologists who monitor flood frequency may describe a particular large flood as being, for example, a 30-year or 50-year event; in other words, it statistically occurs once every 30 or 50 years. They also refer to the probability of a particular discharge being exceeded by an even larger one.

To see how flood magnitude, return period, and statistical probability can be used to predict stream behavior,

Base flow *The amount of stream discharge at any given place and time that is solely the product of groundwater seepage.*

Bankfull discharge *The amount of discharge at which the stream channel is full.*

Stream hydrograph *The graphical representation of stream discharge over a period of time.*

Flood stage *The level at which stream discharge begins to spill out of the channel into the surrounding area.*

examine Figure 16.12, which shows data from the Skykomish River at Goldbar, Washington. The Skykomish has a base flow of about 200 m³/s (about 7,060 ft³/s), which statistically occurs every year and has a 99% probability of being equaled or exceeded on an annual basis. According to the graph, the 10-year peak discharge in this stream is about 1700 m³/s (~60,000 ft³/s) and this flow has a 20 % chance of being equaled or exceeded each year. A more unlikely event is the 50-year flood, which has a discharge of approximately 2500 m³/s (about 88,300 ft³/s). This discharge has only a 2% probability of being equaled or exceeded on a yearly basis. This is not to say that a 50-year flood or even a larger one could not happen in successive years—it could. It's just not very likely from a statistical perspective.

1993 Mississippi Basin Floods   Perhaps the best example of a major flood in the recent past within the United States is the Mississippi River flood of 1993, which affected a large part of the upper Midwest. This flood was at least a 100-year event and has been classified by some hydrologists as a 500-year flood. It dominated the national media during June and July of that year because of its catastrophic impact on people living along the Mississippi River and its tributaries. The causes of this flood and the basin response provide an excellent example of how environmental variables are interrelated. Understanding such relationships may help you deal with a major flood in the future where you live.

The stage for this massive flood was actually set during the preceding winter when unusually heavy rain and snow saturated the soils in the region. Following the wet

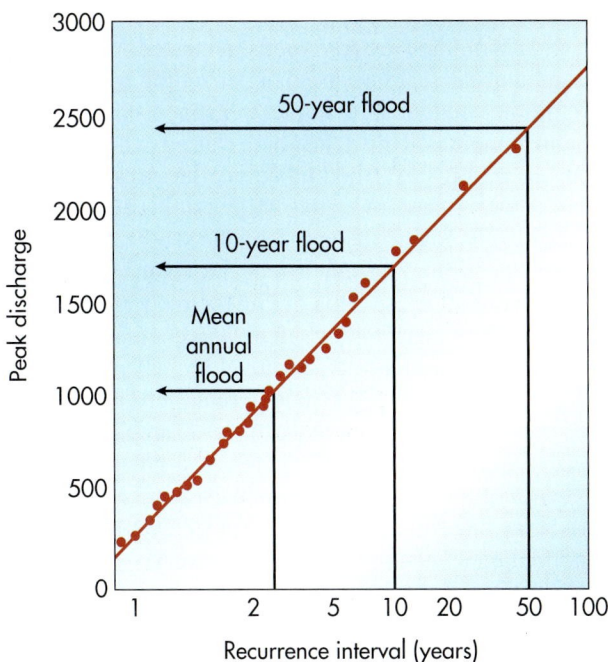

Figure 16.12 Flood frequency data for the Skykomish River at Goldbar, Washington. Each data point is the measured maximum discharge over a 53-year period.

winter, an unusual weather pattern developed during June and July in which a strong subtropical high-pressure system formed and stalled over the southeastern United States (Figure 16.13). Recall from Chapter 8 that atmospheric pressure systems in the midlatitudes rarely stay in one

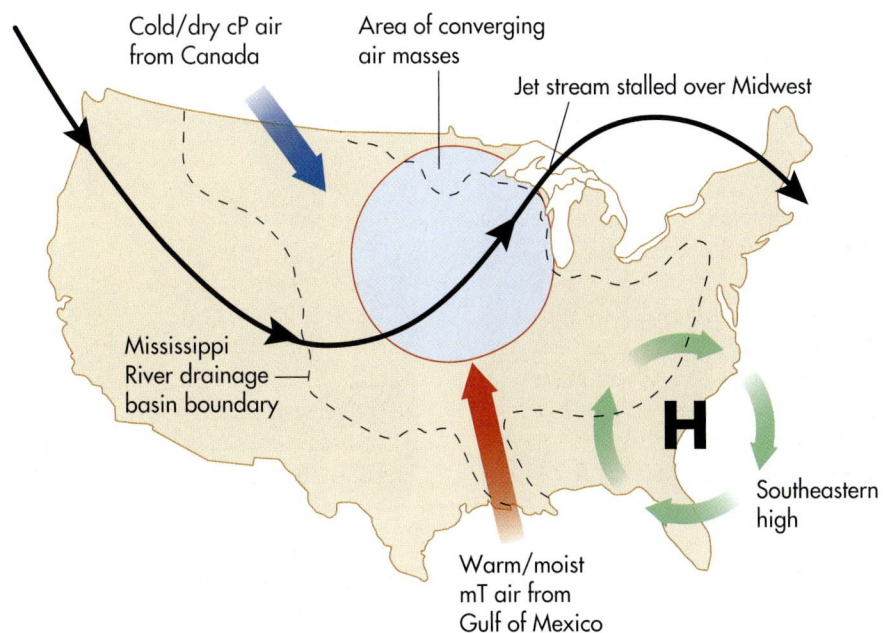

Figure 16.13 Meteorology of 1993 flood. A high-pressure system stalled over the southeastern United States and persistently pumped humid mT air into the upper Midwest. This water vapor condensed and fell as rainfall along the stalled frontal boundary in the region. This intense rain caused widespread flooding.

place but instead typically migrate from west to east in association with the westerly winds. Nineteen ninety-three was an unusual year because a high over the southeastern U.S. did not migrate and, in fact, remained stationary for about two months. This stalled high caused two significant things to occur. First, the midlatitude jet stream became "stuck" in a sense over the Midwest because it was blocked from moving to the east by the southeastern high. Second, the clockwise rotation of the southeastern high pumped continuous streams of Gulf water vapor into the Midwest. Given that cool/dry cP air was continuously colliding with warm/moist mT air along the stalled jet stream in the upper Midwest, prodigious rains fell during June and July. During July 4th and 5th, for example, between 10 to 15 cm (4 to 6 in) of rain fell in a broad swath of southeastern Iowa. This kind of rain event was a common occurrence that summer as the region received over *200%* of normal precipitation for that time of year!

Given that the soils of the region were already saturated from the wet winter, most of the rainfall could not be absorbed into the soil. As a result, flooding was widespread, affecting most of the major streams and tributaries in Iowa, southeastern Wisconsin, western Illinois, northern Missouri, eastern Nebraska, and northeastern Kansas. Since all of these streams progressively merge and ultimately flow into the Mississippi River, flooding along this stream was intense (Figure 16.14). Overall, this flood caused $15 billion of damage and was responsible for at least 50 deaths.

(a)

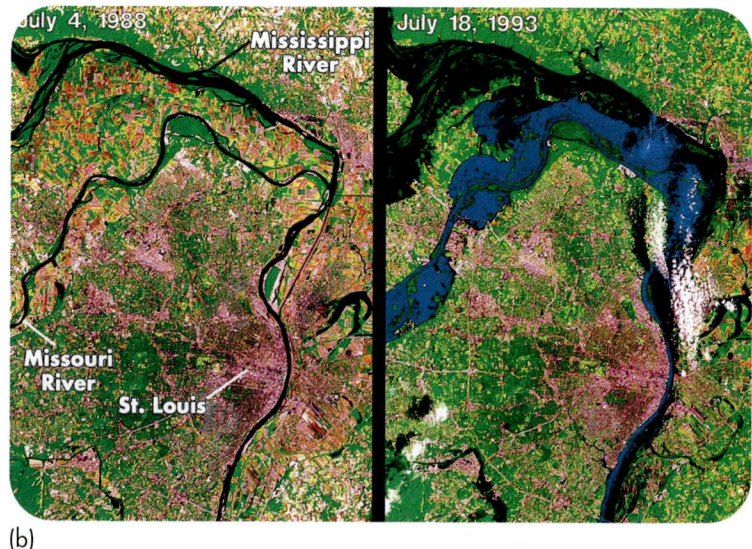

(b)

Figure 16.14 Flooding along the Mississippi River in 1993. (a) Aerial photograph of flooding along the Mississippi River in 1993. (b) Landsat images of the confluence of the Missouri and Mississippi Rivers north of St. Louis during a normal year (left) and during 1993 (right). In these images, vegetation and urban areas appear in green and pink tones, respectively. The extensive flooding in 1993 appears as blue and black areas, indicating submerged floodplains.

Fluvial Processes and Landforms

Now that we have seen how water flows across the surface of the Earth, let's examine the landforms that streams produce. Landforms created by fluvial processes are called *fluvial landforms*. Although processes such as glaciation, wind, and ocean waves also act to shape the Earth's surface (and will be covered in subsequent chapters), running water is the single most important geomorphic agent on Earth. This process is driven by gravity, which is the most important geomorphic force on Earth.

Erosion and Deposition

All landforms on Earth are products of either erosion or deposition. Sometimes both processes act in concert to shape a landscape. Simply put, erosion occurs when sediment is removed from one place by a geomorphic process such as slumping, stream flow, or glaciation. Erosion is usually more intense in areas of high relief because the force of gravity enhances the ability of geomorphic processes to do work by increasing their power. Other factors that influence the process of erosion are vegetation and climate. Erosion tends to be less in areas of dense vegetation because the plants protect hill slopes from the effects of running water. Stream erosion is also limited in the driest deserts because very little running water occurs there. In fact, the most intense erosion tends to occur in semi-arid to sub-humid zones where plant cover is relatively sparse but heavy rains that can move sediment often occur.

In contrast to erosion, the process of deposition occurs when sediments that are being transported (after being eroded from someplace else) stop moving and are dropped. Deposition occurs when the transporting agent (such as wind or running water) simply loses the power to carry the sediment, which can happen for a variety of reasons. A good analogy is if you pick up something heavy, perhaps a large rock, and try to carry it some great distance. At some point, you don't have the strength to carry the rock any farther and must drop it.

Although erosion and deposition often work in tandem to shape a landscape, one or the other process tends to dominate in any particular place. Thus, it's possible to classify landforms generally as being either erosional or depositional in their nature. Erosional landforms are created when sediment, soil, or rock is stripped away by some geomorphic process. Depositional landforms, in contrast, form when sediment accumulates after being dropped.

Figure 16.15 shows a simplified example of these basic categories. Here, the mountain slopes have largely been shaped by erosion due to the high energy created by the steep relief. This process creates some distinctive landforms. The most prominent feature is a *peak*, which is the highest point on any given mountain. Peaks are typically separated by a lower landform called a *saddle*. As streams cut into the mountain

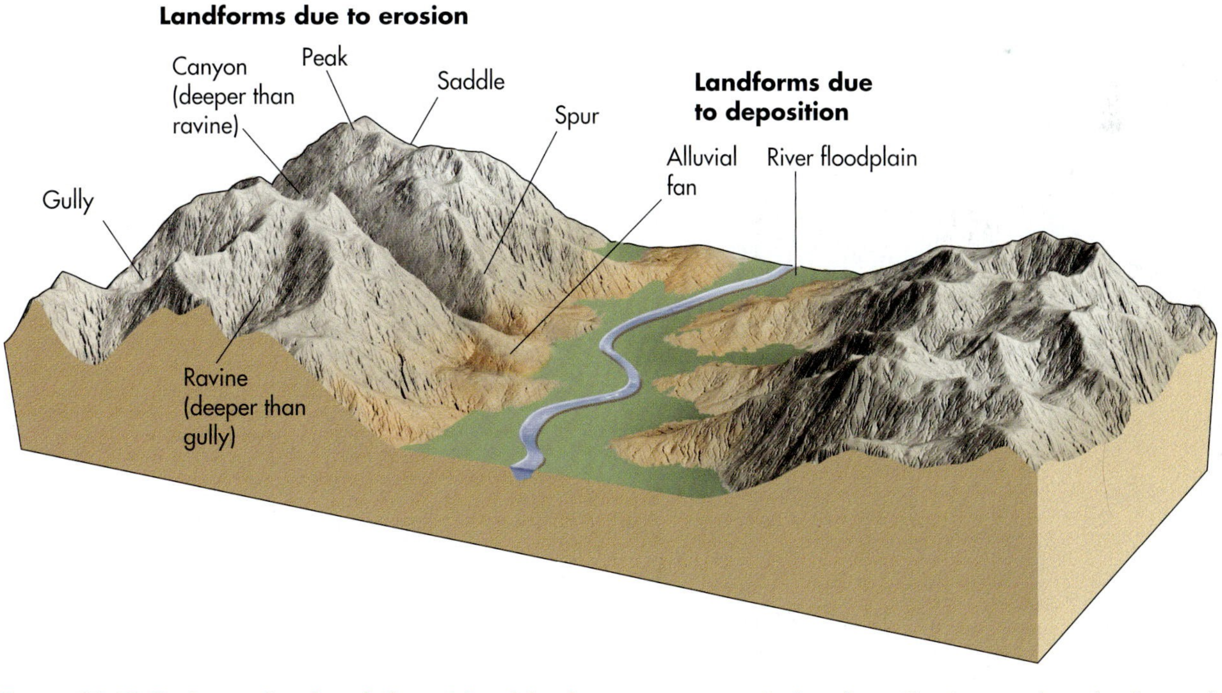

Landforms due to erosion

Canyon (deeper than ravine), Peak, Saddle, Spur, Gully, Ravine (deeper than gully)

Landforms due to deposition

Alluvial fan, River floodplain

Figure 16.15 Basic erosional and depositional landforms. Although erosion or deposition can occur anywhere, one of these processes tends to dominate in any particular place. Erosion tends to be focused on areas of high relief, whereas deposition typically dominates in more level terrain.

slopes, they first create a shallow *gully*, which can enlarge to become a *ravine* and then, if sufficient time and erosion later occur, a deep and broad *canyon*. These features are separated from one another by a relatively high ridge called a *spur*, which is, in effect, a drainage divide. Over time, the eroded hill-slope sediments are transported into the valley below, where they may be deposited on more level terrain within an *alluvial fan* or *river floodplain*. Here the relief lessens and geomorphic processes lose their power. We will discuss these terms in more detail later in the chapter.

Fluvial Erosion on Hill Slopes   Let's begin our discussion of stream erosion by focusing on the part of the landscape that is most intensely eroded by running water: hill slopes. Hill slopes are the most active zones of fluvial erosion because, as indicated before, the force of gravity is greatest in areas of high relief, which, in turn, causes running water to flow more quickly and thus with more energy. Sediment transport begins on hill slopes as soil erosion, initiated when the fluid drag associated with overland flow picks up sediment. In humid regions, dense vegetation protects slopes, and erosion occurs sufficiently slowly that the soil maintains all of its horizons. However, the process of slope erosion accelerates in more arid regions, deforested areas, and agricultural fields because raindrops fall directly on bare soil. When this occurs, sediment is loosened, lifted, and dropped into new positions through a process called *splash erosion* (Figure 16.16).

As water continues to flow across the surface, the first erosional landforms to form are rills. **Rills** consist of

Rills   *Small drainage channels that are cut into hill slopes by running water.*

small furrows in the ground, a few centimeters wide and deep, that develop when overland flow becomes concentrated along preferential pathways (Figure 16.17a). As rills begin to merge, stream power and erosion intensify so that larger and deeper ditch-like features develop called gullies (Figure 16.17b).

Sediments eroded from hill slopes by overland flow are either deposited quickly or are carried to a stream where they can be transported great distances. Deposition of eroded slope sediments often occurs at the base of the hill where the slope becomes gentler, the effect of gravity lessens, and there is less power to carry sediment. Recall from Chapter 14 that sediment deposited in this manner is called *colluvium* and can form an apron-like landform that may be difficult to distinguish in humid regions of dense vegetation.

(a)

(b)

Figure 16.17 Formation of rills and gullies. (a) Rills are small channels that evolve when heavy rains fall on hillslopes. (b) This gully formed due to the enlargement and merging of rills as a result of rapid runoff.

Figure 16.16 Splash erosion. Erosion occurs because large raindrops create a small crater when they impact on bare soil.

Stream Gradation

Now that we've discussed the processes and landforms associated with overland flow on hill slopes, we can examine how landscapes evolve along larger, high-order streams. Although these streams do not vigorously erode like rills and gullies, they nevertheless shape the landscape in subtle ways that have profound effects over a long period of time. In fact, in most higher-order streams the combined processes of runoff, erosion, and deposition have been occurring for thousands of years.

Think of a stream in your neighborhood or the city in which you live. Although its basic appearance (for example, its width, depth, or valley width) probably hasn't changed much in the past few thousand years, especially if it lies in a typical watershed in the humid eastern United States, it was likely a very different system tens of thousands (or even millions) of years ago, when it responded to major climate changes or tectonic forces. Perhaps this ancestral stream had a much steeper gradient and narrower valley than it currently does. In fact, what you likely see now is a stream that is thoroughly adjusted to the environmental setting in which it occurs—specifically, the amount of precipitation that falls, the size of the drainage basin, the density of vegetation on the landscape, and the rock and sediment types found there.

The significant point is that streams evolve in a predictable way that allows them to *just* carry the average amount of sediment that they receive from hill slopes and tributary channels in the drainage basin. This sediment can be carried in a variety of ways (Figure 16.18):

1. **Dissolved load**—Mineral ions that are carried in solution and are invisible during transport in a manner similar to dissolved sugar in water.

2. **Suspended load**—Sediment that floats along in the stream. This load usually consists of clays and silts that are held up within the water by turbulence, which means that the water flows in a very irregular and disorderly way.

3. **Bed load**—Larger particles such as sand and gravel that roll, slide or bounce along the channel bed in a process called *saltation*. This form of transport dominates in mountain streams where slopes are steep and stream velocity is high.

It is usually possible to determine the dominant kind of sediment a stream is carrying by noting the basic channel characteristics. If a stream is transporting mostly suspended sediment, it will typically have a single, deep channel that gradually curves from side to side across the landscape

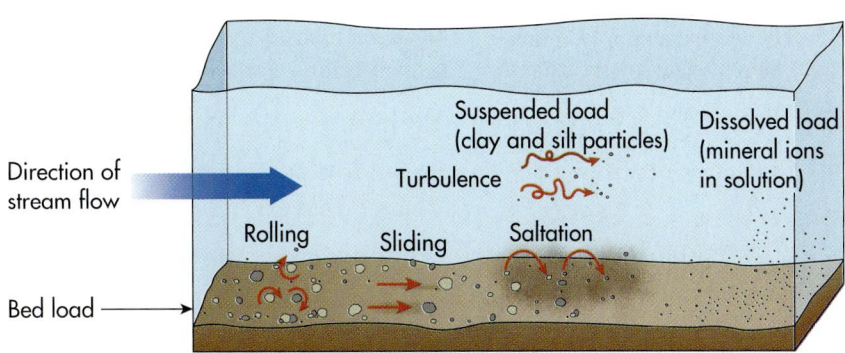

Figure 16.18 Sediment load. Streams carry their sediment load as dissolved, suspended, and bed load.

(a)

(b)

Figure 16.19 Stream load and channel characteristics. (a) A meandering stream generally consists of a single channel that winds back and forth across the landscape. (b) Braided streams such as this one in Alaska consist of a maze of wide, shallow channels because the dominant sediment load is sand and gravel.

(Figure 16.19a). Such a stream is called a **meandering stream** and often contains muddy-looking water due to the high concentration of suspended sediment that was eroded from deposits containing abundant silt and clay. In the Mississippi River, for example, approximately 90 percent of the sediment carried within the river is suspended load. No wonder this river is frequently called *Big Muddy*! If a stream's dominant load is bed load, then it will have a **braided stream** pattern (Figure 16.19b), which consists of a maze of interconnected, wide, shallow channels that look like the braids in someone's hair. These streams have this form because the channel banks consist of easily eroded deposits of sand and gravel. These deposits fall into the river during high discharge, allowing the stream to spread over a broad area.

Let's turn now to the concept of stream gradation. Think of a stream as being capable of internal adjustment so that it can maintain an equilibrium state with the surrounding environmental variables. Imagine, for example, if the vegetation on hill slopes were suddenly removed, as in a forest fire—how would the stream respond? Rill and gully erosion would increase and more sediment would be delivered to streams because vegetation would no longer be there to protect the hill slopes from heavy rain. In an effort to adjust to this higher sediment supply, much of the sediment would initially be deposited on the chan-

nel bottom (because the stream couldn't carry it) in such a way as to increase the overall stream gradient. This kind of deposition, which raises the elevation of the channel bed, is called **aggradation.** The increased gradient caused by aggradation would result in higher flow velocity and power, which would, in turn, enable the stream to carry the increased sediment load. If hill-slope vegetation was re-established and erosion slowed through time, the stream would require less energy to carry the reduced sediment load. As a result, the stream would gradually erode its channel bed in a process called **degradation** (or *downcutting*) until a new gradient was established that enabled the stream to be able to carry just the lessened supply of sediment. A stream in this equilibrium condition is referred to as a **graded stream** because it can effectively carry its average sediment load.

Let's examine more closely how a stream might evolve to reach a graded condition. Figure 16.20a shows several stages in the development of a hypothetical landscape. The figure traces the evolution of a stream's **longitudinal profile** (or simply "profile"); the profile shows the change in the stream's gradient along its length. The beginning point for this developmental sequence could be a period of rapid uplift that elevated the upper part of the basin, or, conversely, this sequence could be triggered by a major climate shift or a major environmental change

Meandering stream *A river or small stream that curves back and forth across its valley.*

Braided stream *A network of converging and diverging stream channels within an individual stream system that are separated from each other by deposits of sand and gravel.*

Aggradation *The progressive accumulation of sediment along or within a stream.*

Degradation *The topographic lowering of a stream channel by stream erosion.*

Graded stream *A stream that is capable of transporting the average sediment load provided to it over time.*

Longitudinal profile *A graph that illustrates the change in stream gradient in cross section along a stream from its source to its mouth.*

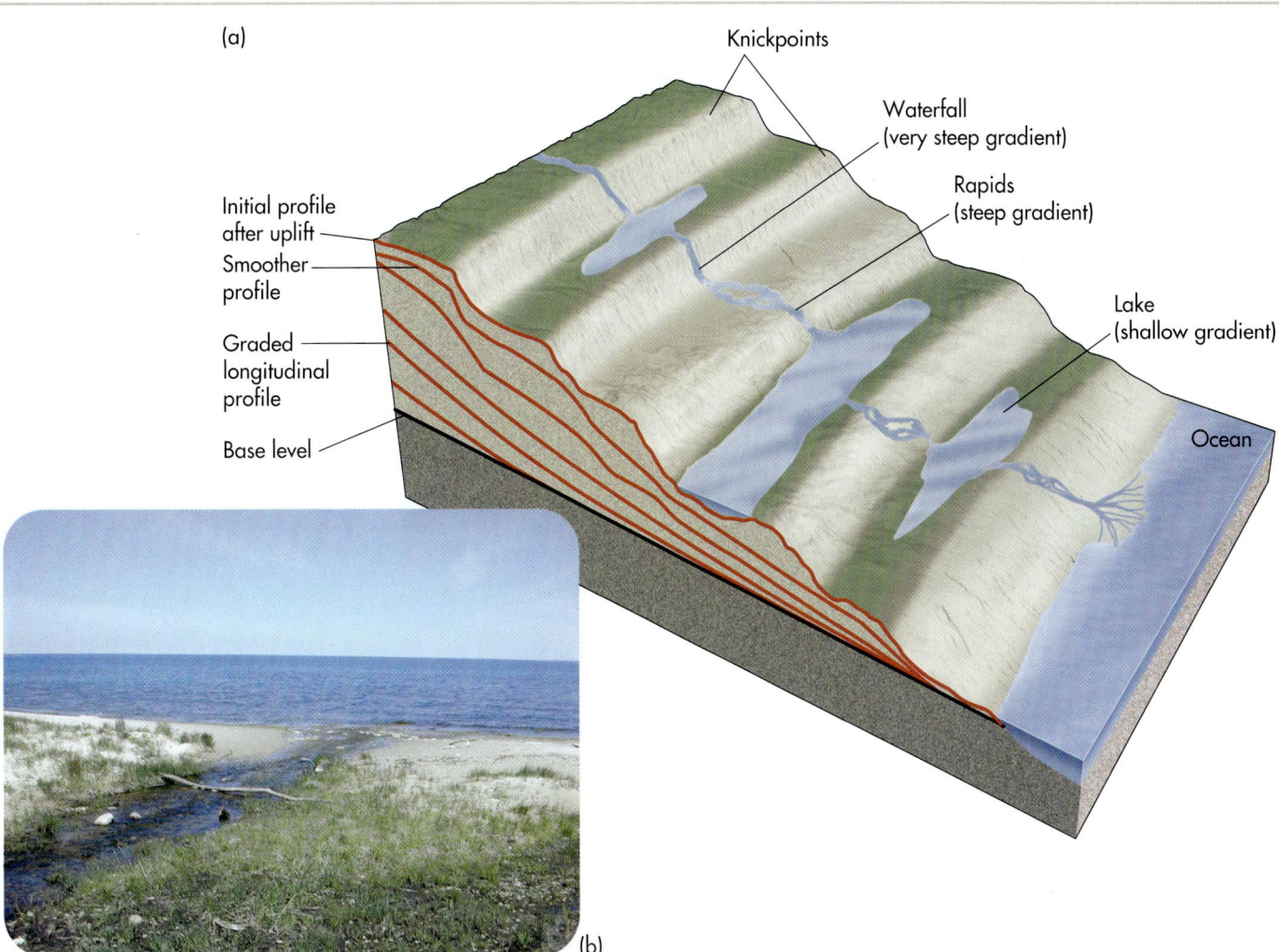

(a)

Knickpoints

Waterfall
(very steep gradient)

Rapids
(steep gradient)

Lake
(shallow gradient)

Initial profile
after uplift

Smoother
profile

Graded
longitudinal
profile

Base level

Ocean

(b)

Figure 16.20 Evolution of a graded stream profile. (a) Following landscape uplift, the stream profile is very uneven. With time, however, it gradually smooths into an ideal longitudinal profile. From this point forward, the gradient of the channel progressively lowers. (b) A major control of stream behavior is base level, which is the point at which a stream flows into the ocean, lake, or another stream. This example shows the location where Antrim Creek flows into Lake Michigan. In other words, Lake Michigan is base level for Antrim Creek.

such as the uncovering of the landscape following deglaciation. For simplicity, let's imagine that the sequence begins with some kind of uplift. Notice in Figure 16.20 that several profiles can be seen in the side of the illustration. Each one of these profile lines represents the gradient of the channel bed at a distinct period of time in the evolution of the system. For reference, imagine that the entire sequence of events takes about one million years to complete.

The top profile line represents the stream gradient following the initial uplift of the landscape above sea level. Soon after the landscape was uplifted, a stream would begin to develop to handle the runoff associated with heavy rains and groundwater flow. Given the nature of streams, runoff on this landscape would naturally flow downhill toward the sea and would follow the path of least resistance. The gradient would naturally be very steep in some places and very shallow in others. The

steepest gradients would most likely be associated with resistant rock layers because stream erosion would be relatively slow in these locations compared to the areas of less resistant rock. As a result, the stream gradient is greater where resistant rocks occur and less in areas of softer rock. A place where the stream gradient is significantly steeper than other places within the stream is called a *knickpoint*. Knickpoints are visible within streams as rapids and, where gradients are especially steep, as waterfalls. Where the gradient is low, overland flow would collect in lakes. Assuming that climate and vegetation cover remained essentially the same, the stream would spend perhaps its first 300,000 to 400,000 years adjusting to the variations in gradient that occur across the landscape. Note, however, that between the top two profile lines, the overall stream gradient becomes smoother.

One of the reasons why the stream gradient smooths is that the knickpoints slowly retreat upstream as they are worn

AMAZING PLACES

WATERFALLS

Niagara Falls is one of the best-known waterfalls in the world, pouring 195,000 cubic feet of water every second over a drop of 49 m (160 ft), extending in two sections that measure 790 m (2592 ft) long (Horseshoe Falls) and 305 m (1001 ft) long (American Falls). However, it is far from being the only spectacular fall of water along a river. Iguacu Falls, near the common borders of Argentina, Paraguay, and Brazil, extends some 4 km (2.5 mi) long and is about 82 m (270 ft) high. The falls are

Iguacu Falls in Brazil is one of the largest waterfalls in South America.

divided into roughly 275 separate channels, pouring about 5012 m^3 (177,000 ft^3) of water per second over the rocks. Like Niagara Falls, Iguacu Falls is receding by erosion of the underlying rock, as is evident from the aerial view of the falls and the Iguacu River.

The world's tallest free-standing waterfall is Angel Falls in Venezuela (see Chapter 2, Amazing Places). Angel Falls is located in the Guayana highlands, where it plunges off of the edge of an

down by stream erosion. A good example of how this process occurs is at a great waterfall like Niagara Falls at the Ontario–New York border (Figure 16.21). Niagara Falls is on the Niagara River, which flows from Lake Erie into Lake Ontario over a resistant layer of limestone that overlies a layer of easily eroded shale. This limestone was scraped clean and covered by a huge glacier until about 12,000 years ago, at which time the ice receded into Canada and an ungraded landscape was left behind. Since the ice left, the falls have retreated more than 11 km (6.8 mi) at a rate of about 1.3 m (4.3 ft) per year. The primary reason for this retreat is that the force of the water striking the plunge pool erodes the shale at the base of the falls. As the shale erodes headward over time, the overlying limestone collapses under its own weight and the falls retreat. This cycle has been repeated countless times in the past 12,000 years and continues to this day.

We noted earlier that as knickpoints are slowly removed, the stream profile gradually smooths. By perhaps 500,000 years into our stream's history, the stream becomes graded; in other words, it can carry the average

sediment load in the basin. At this time, the stream gradient has the ideal longitudinal profile (see Figure 16.20a), with a steeper slope in the upper reaches of the basin and a shallower slope downstream. From this point forward, this graded profile is maintained and gradually lowered as the landscape is further eroded.

Notice in Figure 16.20a that the point where the stream enters the sea never changes in the vertical sense; in other words, it doesn't rise or lower even though the gradient upstream fluctuates with time. The reason for this consistent position is that *all* streams that flow into the sea do so precisely at sea level. In this case sea level is the stream's **base level,** which is the lowest level at which a stream can erode its channel bed. This concept is

Base level   *The lowest level at which a stream can no longer lower its bed, because it flows into the ocean, a lake, or another stream.*

Victoria Falls in Zambia is considered by many to be one of the wonders of the natural world.

extensive plateau. It then falls as a single cataract over 730 m (2400 ft) to the river below.

The largest single curtain of falling water in the world is Victoria Falls along the Zambezi River. It is over 1707 m (5600 ft) long and 100 m (328 ft) tall (on average) and well deserves its African name of Mosi-oa-Tuya, or "the smoke that thunders." Discharge of the river varies widely with time of year and rainfall, but has been measured at roughly 295,000 cubic feet per second.

(a)

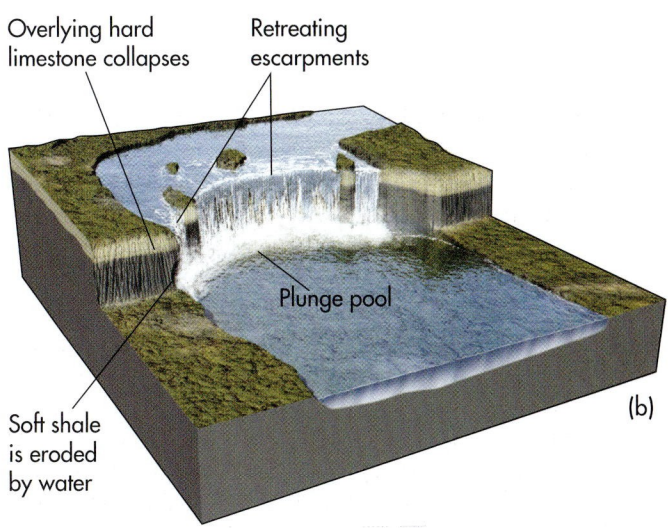

(b)

Figure 16.21 Niagara Falls. (a) Niagara Falls is a 51-m (167-ft) waterfall on the Niagara River. It actually consists of two waterfalls, the Horseshoe Falls on the Canadian side (to the right) and the American Falls (to the left). The eastern side of Lake Erie lies in the background. (b) The Niagara River flows over a resistant layer of limestone that overlies relatively soft shale. This shale is eroded headward in the plunge pool, which results in collapse of the overlying limestone and retreat of the falls. Niagara Falls has retreated more than 11 km (6.8 mi) in the past 12,000 years.

THE GRADED STREAM

Now that we have discussed the basic components associated with stream gradation, let's watch the process in action by viewing an animation. Go to the *GeoDiscoveries* website and access the module **The Graded Stream**. This animation nicely illustrates the basic process of stream evolution and gradient adjustment. As you watch this animation, you will be looking at an initial longitudinal profile similar to the uppermost one in Figure 16.20a. Notice how this profile smooths as time progresses through the combined process of knickpoint reduction and filling of lakes and ponds with sediment. Ultimately, the stream will obtain a slope that enables it just to carry the sediment that is provided through hill slope erosion. Once you complete the animation, be sure to answer the questions at the end of the module to test your understanding of this concept.

---

easy to visualize where a stream meets the ocean or large lake because the stream cannot erode its bed below the level of the water body in which it flows (Figure 16.20b). If it did, it would be flowing uphill to meet the ocean or lake, which is physically impossible. The concept of base level also applies on a more local level—for example, where one stream flows into another at a confluence. As at the ocean, the elevation of a tributary's stream bed must be the same as the channel bed of the stream in which it flows, otherwise it too would be flowing uphill.

Let's return to Figure 16.20a to see how base level relates to stream behavior. In this particular diagram, base level did not change over the one million years of hypothetical stream evolution and thus the point at which it enters the sea remained stationary. In reality, however, sea level—in other words, base level—would probably change frequently over such a period of time and would thus be a major control in the behavior of the stream. If sea level fell, then a wave of erosion would proceed upstream as the stream eroded down to the new base level. On the other hand, if sea level rose, then the stream would begin to aggrade its bed so that it could rise to the new base level. The stream would also begin to erode if uplift occurred in the source area of the stream, which would cause the channel slope to increase and result in more energy. If such a scenario occurred, the stream system would be said to be *rejuvenated*; that is, it would begin the developmental cycle all over again.

KEY CONCEPTS TO REMEMBER ABOUT STREAM GRADATION

1. Streams carry sediment as dissolved load, suspended load, or bed load.

2. A graded stream is a stream that can just carry the average sediment load in the basin.

3. An example of an ungraded landscape is one that was recently uplifted. This uplift creates a very steep gradient. As a result, a stream is able to carry more sediment than is provided to it from eroded hillslopes. This stream will downcut its bed through a process called degradation.

4. Stream aggradation occurs when the sediment load is greater than the stream's capacity to carry it.

5. Base level is the point where a stream flows into the ocean and is a major control of a stream's behavior. When base level rises, streams deposit sediment to match that elevation. In contrast, when base level falls, streams erode their channel beds to cut down to sea level.

Evolution of Stream Valleys and Floodplains

We now turn to the landforms that evolve in association with the long-term process of stream gradation. As you work your way through this section of the chapter, refer to Figure 16.22. This diagram shows the four-stage evolution of a stream and its associated valley.

For simplicity, let's begin by imagining that a landscape was uplifted at some point in the distant past. With this uplift in mind, compare Figure 16.22 with the top stream profile in Figure 16.20a. Lakes and waterfalls are present in both illustrations. Recall that lakes occur in areas where the slope is shallow, whereas waterfalls are associated with knickpoints where the rocks are resistant and the gradient

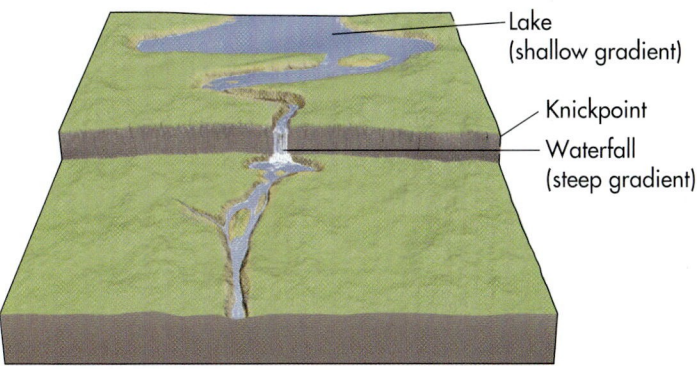

(a) Initial uplift

Lake (shallow gradient)

Knickpoint

Waterfall (steep gradient)

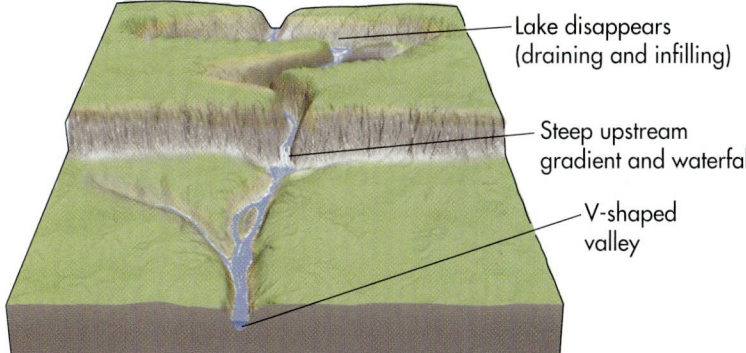

(b) Vigorous downcutting

Lake disappears (draining and infilling)

Steep upstream gradient and waterfall

V-shaped valley

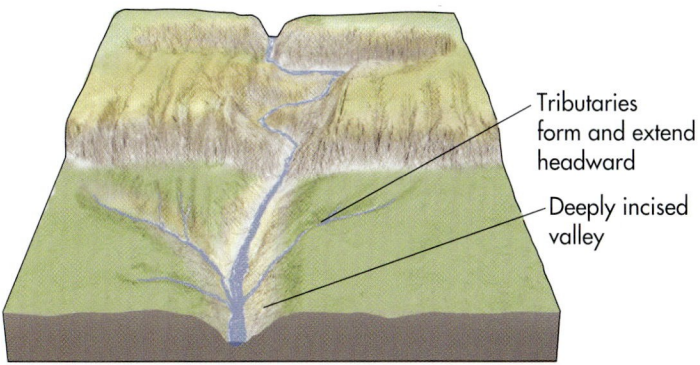

(c) Reaching equilibrium

Tributaries form and extend headward

Deeply incised valley

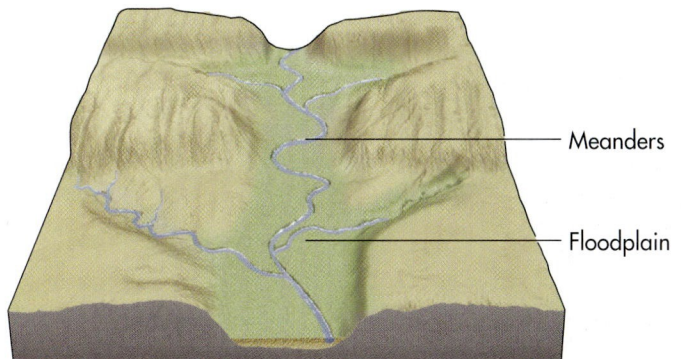

(d) Horizontal expansion

Meanders

Floodplain

Figure 16.22 Evolution of a graded stream and associated valley. During the first stages of valley evolution, the stream downcuts vigorously because it has a steep gradient and excess energy. When the stream reaches a graded state, it begins to migrate horizontally, resulting in a wide valley.

is steep. In this early stage of gradation, the stream has a capacity to carry more sediment than is being delivered to it. As a result, the stream begins to downcut vigorously in an attempt to reach a state of equilibrium. In so doing, the stream carves out a deep V-shaped valley (Figure 16.22b) that is consistent with the Grand Canyon of the Yellowstone River (Figure 16.23).

As time progresses (Figures 16.22b and c), the lake disappears through a combination of infilling by eroded upstream sediments and draining because the river cut back into it after eroding through the resistant knickpoint downstream. During this interval of time, the gradient becomes smoother, as in the second curve in Figure 16.20a. Turning back to Figure 16.22b and c, the stream begins to extend its valley farther upstream (or *headward*) because the gradient is steepest in this part of the system. Because of this steep gradient, the stream naturally flows at a greater speed and thus has more energy to erode sedi-

Figure 16.23 The Grand Canyon of the Yellowstone River. Here, the Yellowstone has yet to reach a graded condition, as indicated by the deep V-shaped canyon and waterfall. Compare this photograph with Figure 16.22b.

ments. In Figure 16.22c, you can see that tributaries begin to develop and also extend their valleys upstream by headward erosion. At the same time as this headward erosion is occurring, the canyon continues to be more deeply incised.

Look at Figure 16.22d. The stream valley looks different from its appearance in the previous three images. Now, the valley is no longer V-shaped, but is quite wide. This change is evidence that the stream has reached a graded condition and has the graded longitudinal profile seen in Figure 16.20a. Recall that during the early stages of stream valley evolution the stream erodes vertically because it has proportionately more energy than sediment load. When a stream becomes graded, however, it no longer needs to downcut throughout the entire system because it has reached a gradient on which it can trans-

port the average load that is provided from the hillslopes in the upper reaches of the system. Although downcutting may continue in the upper reaches of a watershed where the gradient is steep, the stream begins to expend its energy in a horizontal fashion in the lower reaches where the gradient is less.

Stream Meandering As noted previously, streams migrate horizontally through a process called meandering, in which numerous pronounced arcs and curving bends develop in the valley (Figure 16.22d). Stream meanders are fascinating places because both erosion and deposition occur simultaneously. As a result, stream meanders migrate back and forth across the valley over time. Figure 16.24 shows the processes associated with a typical stream

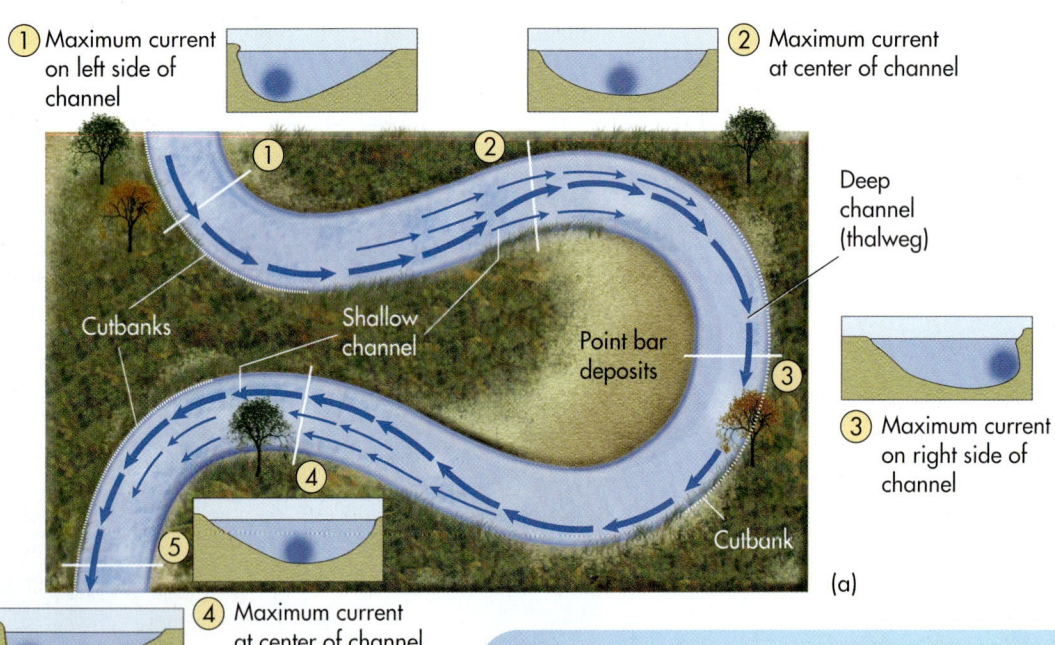

Figure 16.24 Features of stream meanders. (a) Deposition of sediment, forming a point bar, occurs on the inside of a meander because stream velocity is low. In contrast, the stream erodes the cutbank on the outside of the meander because velocity is high. (b) Note the distinct point bar and cutbank in this stream meander along the Kansas River.

STREAM MEANDERING

To view an animation on the meandering process, go to the *GeoDiscoveries* website and access the module **Stream Meandering**. This animation illustrates how meander bends form over time in an initially straight river and also shows the formation of an oxbow lake. As you watch this animation, relate it back to the evolution of a stream valley. Once you complete the animation, be sure to answer the questions at the end of the module to test your understanding of this concept.

meander. As the river flows around the meander bend, the centrifugal force associated with its current carries the main line of flow of the channel, or *thalweg,* to the outside of the curve. Thus, the channel is deepest and the flow velocity is highest in this part of the stream. Given the high stream power that exists on the outside of the meander, this part of the arc is a zone of aggressive erosion. As a result, the surface leading away from the river is undercut, which causes the overlying sediment to collapse into the stream, forming a steep slope called a *cutbank*. At the same time that erosion shapes the cutbank, deposition occurs on the inside of the meander bend because the channel is shallow in this part and energy is low. As a result, alluvium accumulates as a long, curving deposit of sediment called a *point bar* that is only about 1 m (about 3.3 ft) above the level of the stream.

Evidence that streams are actively meandering comes in the form of oxbow lakes. An **oxbow lake** is a water-filled former meander bend that was detached from the main channel in a stream valley. This lake develops when the stream erodes the cutbank as water migrates

Oxbow lake *A portion of an abandoned stream channel that is cut off from the rest of the stream by the meandering process and is filled with stagnant water.*

downstream. At the same time that cutbank erosion is occurring, deposition of sediment takes place at the corresponding point bar. Given the combined impact of these processes, the neck between meander loops gradually becomes narrower. Ultimately, one or more meander loops can be entirely cut through when the stream floods. This neck cutting is part of the stream's continuous process of gradient adjustment and results in a temporary lake (Figure 16.25). With time, an oxbow will usually fill with sediment derived from major floods along the active channel. In the first stages of sediment infilling, the oxbow will become a swamp that contains plants such as reeds and cattails. As sediment continues slowly to accumulate with successive

(a)

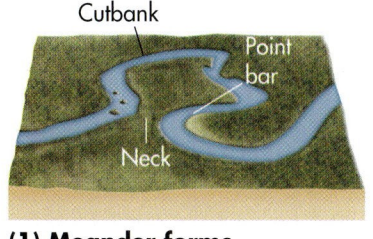

(1) Meander forms

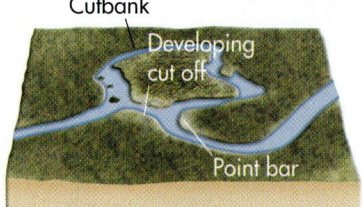

(2) Cutoff develops

(3) Oxbow lake forms

(b)

Figure 16.25 Meander evolution and formation of an oxbow lake. (a) Meander cutoff in the East River in Colorado. (b) Schematic views of the evolution of an oxbow lake on the East River. Diagram 1 illustrates the possible appearance of the meander before a breach of the neck occurred. Diagram 2 shows the landscape in its present condition. Diagram 3 illustrates how the oxbow lake may look when the meander ends seal with sediment.

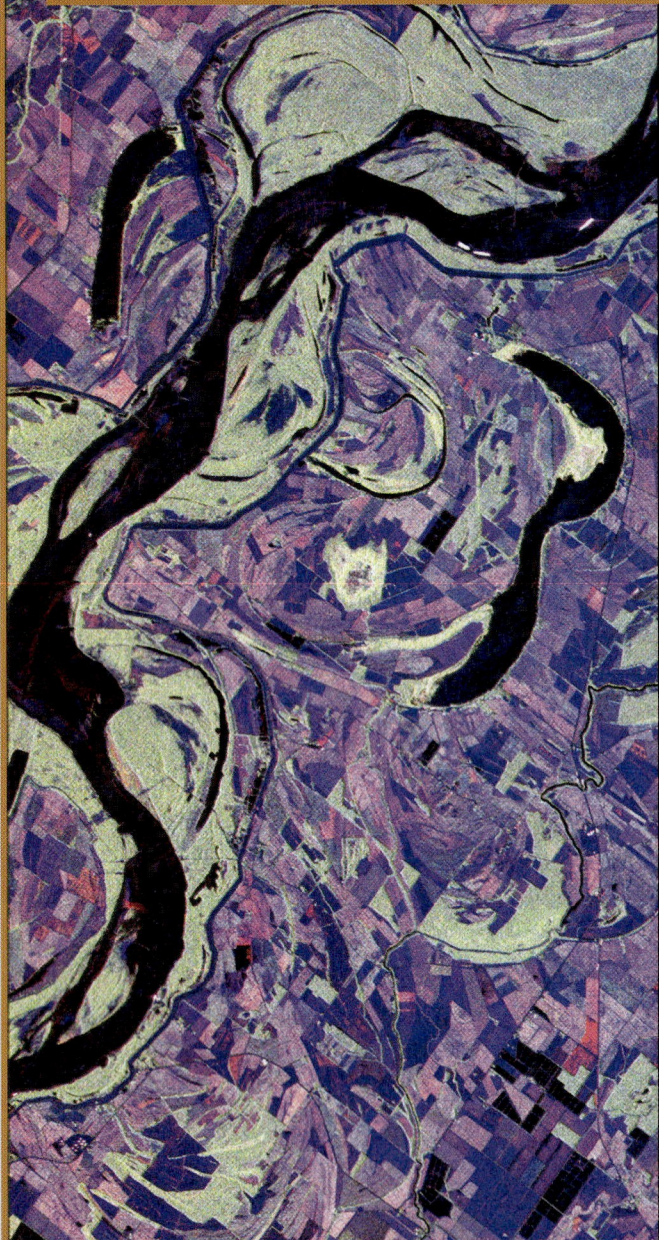

MISSISSIPPI RIVER MEANDERS

You can learn much about a river by examining satellite images. This image provides an excellent example. It's a synthetic aperture radar image of the Mississippi River that was acquired on the space shuttle in October, 1994. The image is centered at about 32.75° N, 90.5° W and covers an area of about 23 km by 40 km (14.2 mi by 24.8 mi). North is toward the upper right of the image. On this image, water is black. The accompanying diagram shows the location of the oxbow lakes and some of the meander scars. See if you can find some more.

As you can see, the Mississippi River is actively meandering in this area, with numerous oxbow lakes and meander scars visible. By looking at such an image, a geographer can determine where the fastest and slowest velocities are within the river. Note that two such places are identified on the diagram. In addition, the geographer can determine the areas that are prone to flooding. For example, the elevation between the current channel and the modern floodplain boundary is probably low, which means that the area is likely to flood during high discharge. A geographer also knows that the numerous signs of meandering mean that the Mississippi River must be close to base level in this area and that the stream is in, or close to, a graded condition.

The meandering Mississippi along the borders of Mississippi, Arkansas, and Louisiana. Note the numerous oxbow lakes and meander scars highlighted in the diagram.

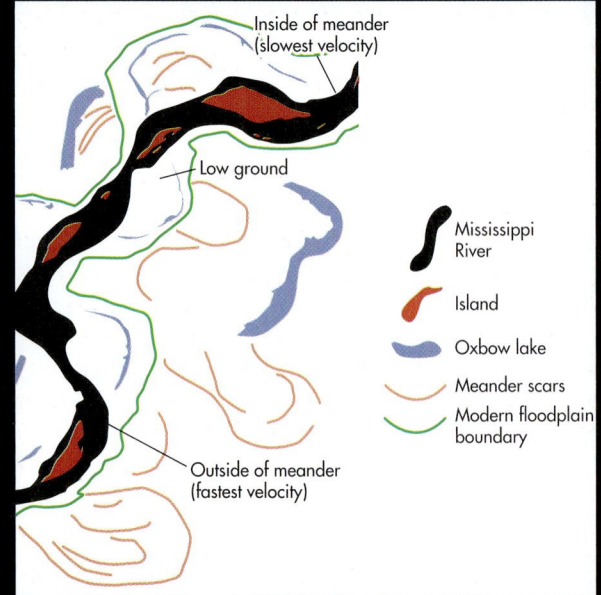

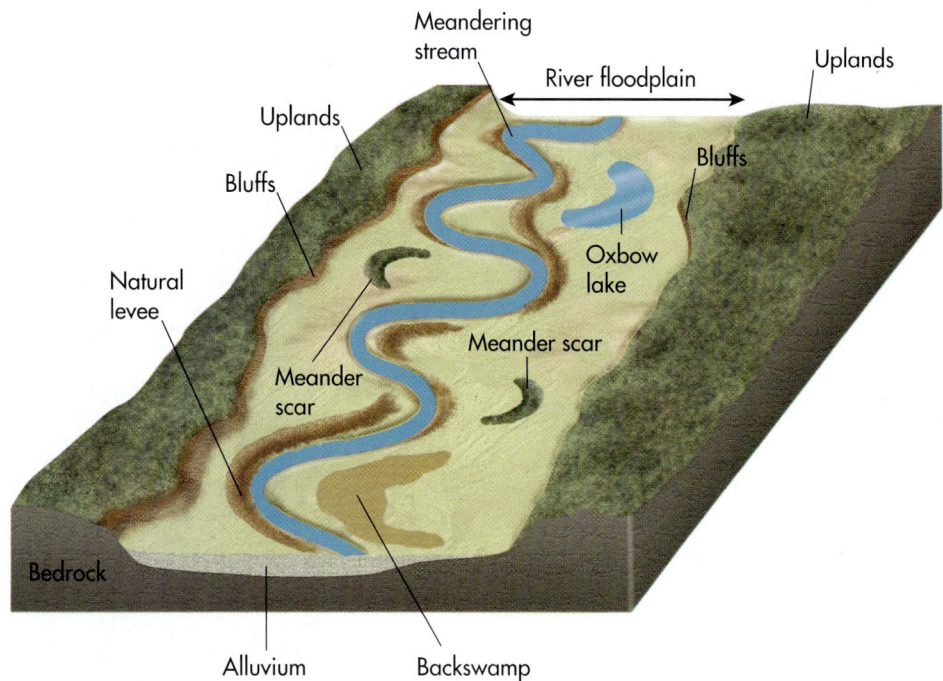

Meandering stream

River floodplain

Uplands

Uplands

Bluffs

Bluffs

Oxbow lake

Natural levee

Meander scar

Meander scar

Bedrock

Alluvium

Backswamp

Figure 16.26 Floodplain features of a graded stream. A graded stream has well developed meanders, natural levees, oxbow lakes, and backswamps.

floods, the swamp will ultimately fill completely with alluvium. Nevertheless, evidence that an oxbow lake was once present will remain in the form of a *meander scar*.

Valley Widening and Floodplain Formation

In the context of stream meandering, it's important to realize that the processes of cutbank erosion and point bar deposition occur simultaneously. This results in stream meanders that literally migrate across the valley over time. This lateral migration, in turn, is important because meanders slowly cut into the side slopes, or *bluffs*, that flank the channel, causing stream valleys gradually to widen, as you can see in Figure 16.22d. Note in this figure that the stream valley is now broad, essentially flat, and bordered by steep, rocky bluffs..

A stream valley usually contains thick deposits of alluvium that were deposited by the combined processes of point bar development and overbank deposition during large floods (Figure 16.26). During floods, the coarsest sediments (usually sand grains) are deposited along the channel margin because they can't be carried far away from the stream due to their size. This near-channel deposition results in a ridge-like landform along the river called a **natural levee** that may be 1 to 2 m (3.3 to 6.6 ft) high. Finer silt and clay-sized particles are carried farther away from the stream during floods and, when deposited,

Natural levee *A small ridge that develops along the channel of a stream through the deposition of relatively coarse sediment when flooding occurs.*

form low **backswamps** because they are poorly drained, for reasons outlined in Chapter 11.

Overall, however, the combined processes of meander development, point-bar formation, and overbank deposition produce the essentially flat topography of the stream valley. This part of the landscape is generally called the *river floodplain* because it can be submerged during large floods, even though such an event may be rare. The river floodplain differs slightly from the *active floodplain*, which is the low, frequently flooded surface that results mostly from point-bar formation on the inside of meander bends.

Backswamps *Marsh floodplain landforms that develop behind natural levees in which fine-grained sediments settle after a flood.*

> ### KEY CONCEPTS TO REMEMBER ABOUT GRADED STREAMS AND THE EVOLUTION OF STREAM VALLEYS
>
> 1. When a landscape is recently uplifted, the resulting ungraded streams degrade by cutting V-shaped valleys that contain waterfalls.
>
> 2. Waterfalls occur in places where rock is resistant to erosion and progressively erode headward until the gradient is smoothed.

This image shows a stream in an alpine setting. Given what you see here, which one of the following statements accurately reflects the stream in this landscape?

a) The stream has a wide valley.

b) The stream is actively meandering.

c) The stream has high discharge.

d) The stream is not graded.

e) The stream is a high-order stream.

3. As streams approach a graded state, they begin to erode horizontally through the process of meandering. This process produces a wide valley because bordering bluffs are eroded.

4. In the process of meandering, erosion occurs on the outside of the meander at the cutbank, whereas deposition is focused at the point bar on the inside of the meander.

5. You can tell that a stream is actively meandering (in present time) if it has well developed oxbow lakes and meander scars.

Entrenched Meanders   As previously discussed, a meandering stream develops when a stream reaches a graded condition. These conditions don't last forever, however, and usually a stream must periodically adjust to new environmental conditions such as tectonic uplift, a drop in base level, or climate change. The landscape response to these changes is usually preserved in stream valleys as distinct landforms.

For example, when a graded, meandering stream responds to a new period of landscape uplift associated with a major tectonic event or base level drop, the gradient of the stream is dramatically increased. Such a stream once again has a capacity to carry more sediment than is delivered to it from the hill slopes. As a result, the stream is said to be *rejuvenated* and will begin a new period of downcutting and channel-bed erosion.

When a graded stream is rejuvenated, entrenched meanders develop, such as the well known Goosenecks of the San Juan River in Utah (Figure 16.27). Prior to uplift

Figure 16.27 Goosenecks of the San Juan River in Utah. These entrenched meanders formed as a result of massive uplift of the Colorado Plateau in the past 10 million years, which increased the stream gradient of the San Juan River. The gorge is about 370 m (1400 ft) deep.

of the Colorado Plateau region, the San Juan River was a meandering stream and the channel bed was somewhere near what is now the lip of the gorge. In other words, no gorge was present at that time. In the past 10 million years the entire Colorado Plateau has been uplifted about 1.5 km (0.9 mi). This period of uplift caused the San Juan River to begin eroding its channel bed while preserving its meandering pattern (because it was no longer eroding horizontally). This same period of uplift also initiated the downcutting by the Colorado River that created the Grand Canyon in Arizona.

Alluvial Terraces

Entrenched meanders illustrate the kind of landscape evolution that can occur when massive steepening of a stream's slope happens due to large-scale regional uplift. Most of the time, however, the environmental adjustments within a drainage basin are subtler, consisting of relatively minor fluctuations in base level or in the regional climate. Although these adjustments do not result in dramatic canyons and gorges, such as the Goosenecks of the San Juan River, they nevertheless create distinctive landforms.

The most prominent landform created by smaller-scale fluvial adjustments is an alluvial terrace. An **alluvial terrace** is a broad, flat surface that occurs within the stream valley and is elevated with respect to the river so that it floods much less frequently than the topographically lower point bars. Although an alluvial terrace is generally considered to be part of the floodplain, it is also useful to think of an alluvial terrace as being an abandoned floodplain in the sense that it is no longer frequently flooded.

Floodplain abandonment can occur in one of two ways. One way is that, through repeated flooding, thick deposits of alluvial sediment (or **alluvium**) accumulate on the active floodplain in a process called *alluviation*. As these sediments gradually thicken with repeated flooding, the active floodplain slowly becomes elevated with respect to the stream so that increasingly higher discharges are required to flood the surface.

Another way in which floodplain abandonment and terrace formation occurs is when a stream downcuts following a period of meandering. This downcutting may be associated with either a subtle period of regional uplift, base level drop, or regional cimate change. When such an event occurs, the stream responds to the new ungraded conditions by downcutting through its own alluvium. This downcutting elevates the former active floodplain, creating terraces on either side of the river that are at the same elevation. Such terraces are called *paired terraces*.

Alluvial terrace *A level, step-like landform that forms when a stream erodes its bed so that an essentially horizontal surface is raised relative to the channel.*

Alluvium *Sediment deposited by a stream.*

If the stream is able to continue meandering as it downcuts, it may remove a portion of the alluvium on the outside of meander bends through cutbank erosion, leaving surfaces of unequal elevation called *unpaired terraces* on either side of the river. Many streams have numerous terraces that progressively step down to the river (Figure 16.28). Whenever you see such a sequence of surfaces, it indicates that the stream has had a very complex history of flooding and downcutting.

Integrating Concepts into a Graded Stream Model At this point in the chapter, we have discussed the concepts of stream order, hydraulic variables, stream behavior, and landform evolution. Now, let's try to integrate all of these concepts into a conceptual framework that links them within the graded stream model. Remember that once a stream reaches a graded condition it has the graded longitudinal profile seen in Figure 16.20a. Once this condition is reached, a very close and predictable relationship arises among stream order, hydraulic variables, and the type of landforms you would see along the stream's profile (Figure 16.29). Stream gradient is highest in the upper reaches of the system, which translates to higher velocity in these areas. As a result, the stream expends its energy vertically, and well-developed gullies and canyons appear in this part of the watershed. Given that the streams here are low-order systems, however, their width and depth are relatively small because they contain little water. This means that discharge is also low in streams in this part of the drainage basin.

As we move farther down the profile into the middle reaches of the system, you can see that slope decreases, which causes stream velocity to drop correspondingly. As a result, the stream has less energy to expend downcutting and it begins to meander. This shift results in the stream valley becoming wider in this part of the system. Width, depth, and discharge also increase because streams in this reach have a higher order due to the progressive funneling of low-order streams into the system. With higher discharge the potential of flooding increases, which will result in deposition of alluvium on the active floodplain because the valley is wider than it is in the higher reaches. Deposition of alluvium means that stream terraces may also occur in this part of the profile, especially if some waves of downcutting progressed through the area in response to a drop in base level.

As we continue into the lowest reaches of the profile, the pattern seen in the middle part of the system becomes even more pronounced. Here, the trunk stream is fully developed, which naturally has a relatively high order because it contains virtually all of the water that lower-order tributaries can provide. Stream gradient is very low here, resulting in low stream velocity and the formation of well-developed meanders and a correspondingly wide valley. Discharge is high, however, reflecting a progressively greater width and depth. Given that all of the runoff

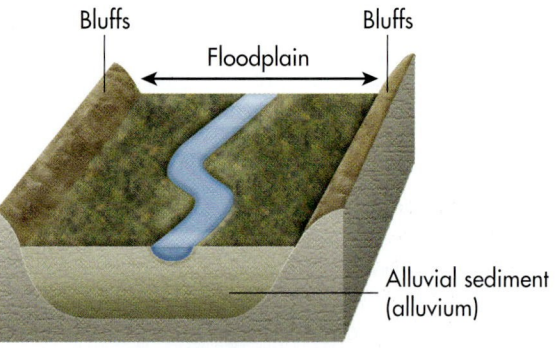

Bluffs — Floodplain — Bluffs

Alluvial sediment (alluvium)

(a) Original floodplain

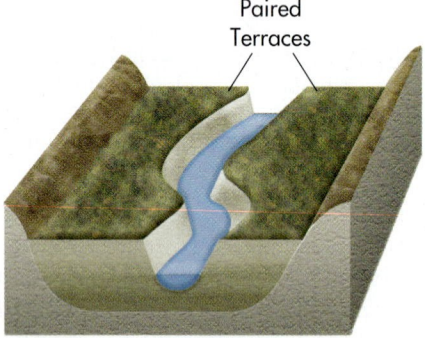

Paired Terraces

(b) River downcutting due to regional uplift, base level drop, or climate change

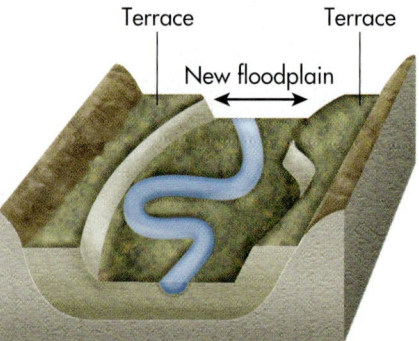

Terrace — New floodplain — Terrace

(c) New floodplain

(d)

Figure 16.28 Formation of alluvial terraces. Alluvial terraces form through a combination of meandering, flooding, and downcutting in a river valley. (a) As the stream meanders, it creates an active floodplain that is underlain by alluvial sediment. (b) During a period of downcutting, the formerly active floodplain is elevated with respect to the river; thus it rarely floods and a terrace forms. (c) A new active floodplain forms at a lower elevation when the stream begins to meander again. (d) Alluvial terraces along the Cave River on the South Island of New Zealand. Notice that two terraces (arrows) occur along this valley, which means that at least two periods of downcutting have occurred.

in the drainage basin now flows through this central conduit, the potential exists for major flooding in this part of the system. Consequently, the alluvium is probably very thick and well-developed terraces are likely present. These landforms may be very prominent if significant base level fluctuations have occurred through time.

Other Depositional Landforms Associated with Streams

So far in this discussion of alluvial processes, we have focused on landforms created in association with the evolution of graded stream systems and their valleys. These landforms, such as point bars, cutbanks, oxbow lakes, and terraces, evolve through the combined processes of erosion and deposition. Now, let's turn our attention to a pair of alluvial landforms that are essentially depositional in character and are not related to the stream valley per se. These landforms are alluvial fans and deltas.

Alluvial Fans   An **alluvial fan** is a landform created by alluvial aggradation in areas of high relief where bedload-dominated streams flow out of mountainous or hilly areas onto an adjacent plain. Very simply, as the stream travels down the steep gradient it can carry an abundance of coarse sediment supplied by the mountain or hill. However, when the stream reaches the plain at the base of the hill or mountain, it loses energy because the gradient there is much lower. As a result of this reduced gradient, stream power is lost and aggradation occurs. Sediment

Alluvial fan   *A fan-shaped landform of low relief that forms where a stream flows out of an area of high relief into a broad, open plane where the gradient is less and deposition occurs.*

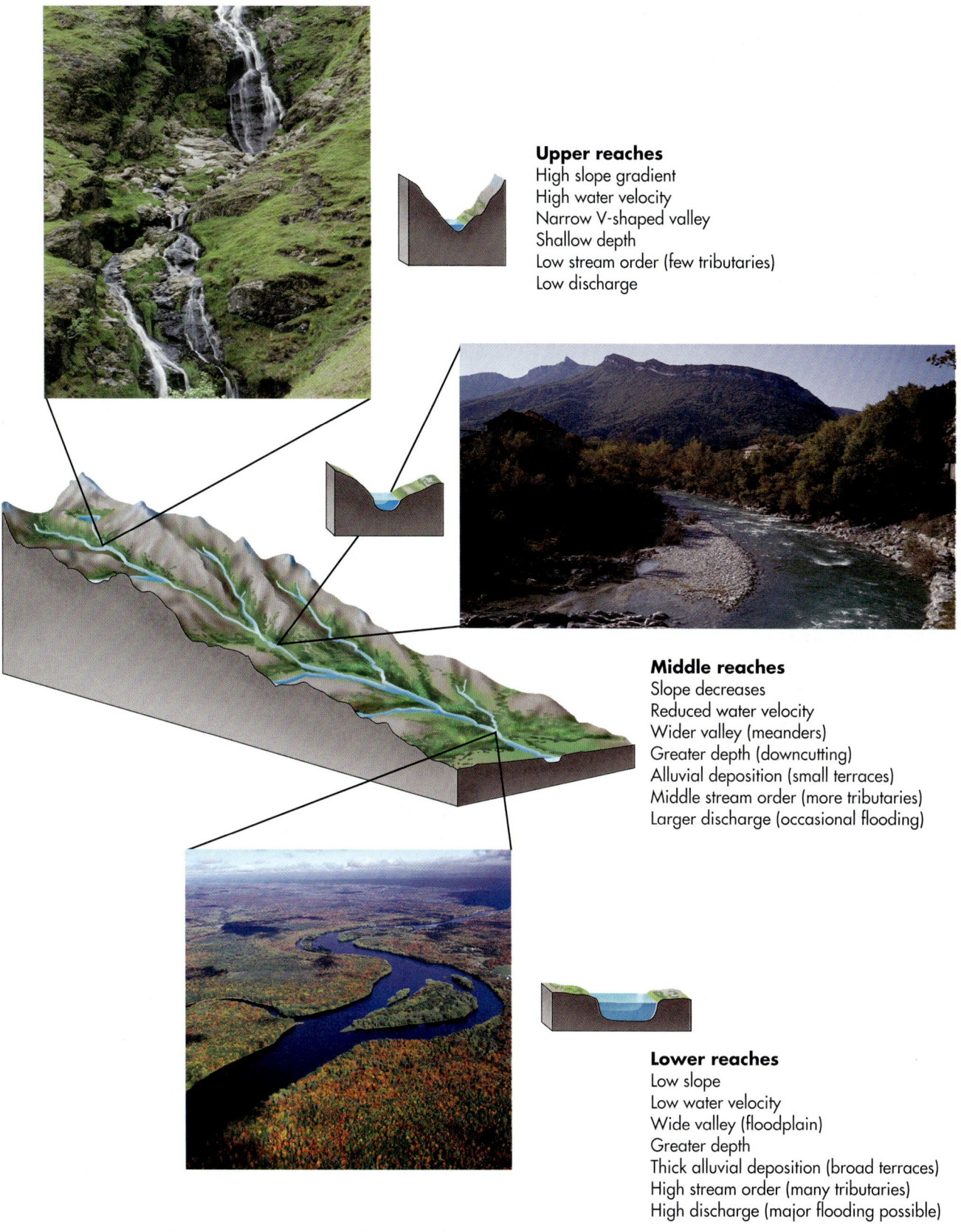

Upper reaches
High slope gradient
High water velocity
Narrow V-shaped valley
Shallow depth
Low stream order (few tributaries)
Low discharge

Middle reaches
Slope decreases
Reduced water velocity
Wider valley (meanders)
Greater depth (downcutting)
Alluvial deposition (small terraces)
Middle stream order (more tributaries)
Larger discharge (occasional flooding)

Lower reaches
Low slope
Low water velocity
Wide valley (floodplain)
Greater depth
Thick alluvial deposition (broad terraces)
High stream order (many tributaries)
High discharge (major flooding possible)

Figure 16.29 Relationship of stream order, valley form, and hydraulic variables along a graded stream. Streams in the upper reaches of the profile typically have the highest slopes and velocity, but the lowest widths, depths, and discharge. With distance along the profile, stream order and discharge increase, and gradient and velocity correspondingly decrease.

Figure 16.30 Alluvial fans in the southwestern United States. This typical alluvial fan in Death Valley, California, has a fan-like shape where sediment is deposited by streams periodically flowing out of the mountains.

deposition begins at the fan apex, which is the point where the stream leaves the mountain. Below the apex, the braided stream sweeps back and forth through time, and in so doing creates a semicircular landform that looks much like a hand-held fan from above.

Although alluvial fans can occur virtually anywhere that streams flow abruptly from areas of high relief to surfaces that have lower slopes, the most pronounced fans occur in deserts such as those in the southwestern United States. Here, alluvial fans occur along many of the mountain fronts within the region (Figure 16.30) and are sometimes many kilometers wide and have steep gradients. These fans are so large that they often overlap with one another along the mountain front. Many of these fans are complex in that they contain interbedded deposits of mud, sand, and gravel. The sand and gravel layers indicate deposition by a bedload stream, whereas the fine sediments accumulate during major mudflows that periodically occur. Given that the mud layers function very well as aquicludes, the coarse deposits above them frequently make excellent aquifers.

Deltas   As described previously, streams eventually flow to the sea, their ultimate base level. This place, also called the *river mouth*, is a zone of intense sediment deposition for reasons somewhat similar to the processes associated with alluvial fan development. With alluvial fans, aggradation occurs because discharge velocity is dramatically reduced when the stream leaves the steep mountain front. Similarly, when a stream encounters the relatively still waters of the ocean, its forward velocity slows significantly. As an analogy, think about what happens when you run from the beach into the ocean or a

large lake. You're able to keep moving forward for a while after you enter the water, but at some point, you slow down because the water stops your momentum. The same thing basically happens to a stream's velocity when it enters the ocean at the mouth. When you run into the ocean you ultimately fall down. For the same reason, a stream drops its sediment when it enters a relatively still body of water.

The landform created by sediment deposition at a river mouth is called a **delta** because, from above, it usually has a triangular shape similar to the Greek letter delta (Δ). An example of this shape is the Nile River delta (Figure 16.31). A delta consists of a nearly flat plain of sediment with finer particles located farther away from the river mouth. In other words, the coarsest sediments, such as sand and gravel, are deposited closest to the mouth because they fall out and are deposited as soon as the stream begins to lose energy. As stream energy further weakens away from the river mouth, the silt-sized fraction of the sediment load is then deposited. The clay fraction is carried the farthest in suspension and ultimately accumulates far from the river mouth.

Figure 16.31 The Nile River delta. This true-color image shows the triangular shape of the Nile River delta where the Nile meets the Mediterranean Sea. The color green represents dense foliage and crops that are being supplied with water and nutrients from the fertile soils of the Nile River valley and delta. Note the surrounding desert.

Delta   *A low, level plain that develops where a stream flows into a relatively still body of water so that its velocity decreases and alluvial deposition occurs.*

(a)

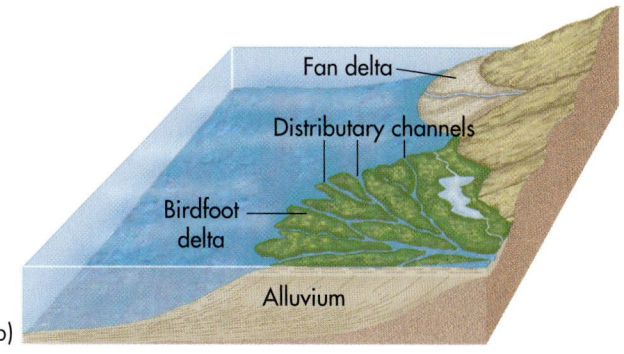

(b)

(c)

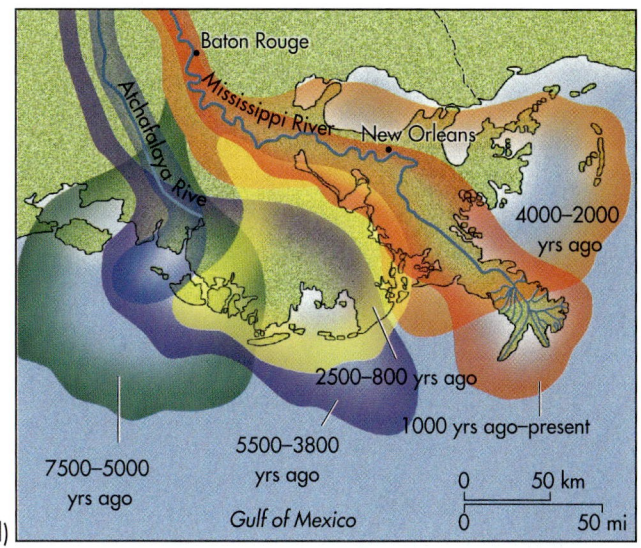

(d)

Figure 16.32 The Mississippi River delta. (a) This satellite image beautifully shows the Mississippi River's birdfoot delta. The delta consists of a distributary network that funnels sediment (whitish hues) into the Gulf of Mexico. (b) The Mississippi River delta is a complex system of distributary channels that is underlain by copious amounts of sediment derived from the continental interior. (c) The Mississippi Delta is a landscape of very low relief, consisting of numerous swamps and marshes that lie within a network of distributary stream channels. (d) Migration of the delta locations over the past 7500 years.

This natural process of segregating sediment particles by size over distance is called *sorting*.

Another excellent example of a river delta is the Mississippi River delta in Louisiana, where the Mississippi River meets the Gulf of Mexico. This delta is a classic *birdfoot delta* because it appears to contain a number of individual toes that can easily be seen in the satellite image (Figure 16.32). Each one of these "toes" is in fact a distinct river channel that is part of a well-defined distributary network (Figure 16.32b and c). Unlike the tributary system in the upper part of the drainage basin, which funnels water into the trunk stream, the distributory network systematically drains water away from the main channel into the ocean.

Such a network develops because the delta gradient is very low and new channels form wherever a breach occurs in the natural levee along the stream. Wherever these breaks occur, fresh water pours into the bay of the distributary, which subsequently fills with alluvial sediment. When the bay ultimately fills, it seals off and a levee breach occurs at another place in the delta. This cycle occurs simultaneously at several places, which is why the delta has the distinct distributary network. The entire delta is underlain by enormous quantities of sediment

(Figure 16.32b) that originated in places ranging from the Appalachian Mountains to the Rocky Mountains.

The Mississippi delta has a complex history that spans the past 120 million years. During this period of time it has constructed a variety of depositional lobes that fundamentally formed the southern part of Louisiana. Each of these lobes formed during a specific interval of time when the Mississippi entered the ocean at a particular place along the coast. Wherever that occurred, tremendous quantities of sediment poured into the ocean, forming a delta. When the sediment load ultimately became sufficiently thick to actually form new land that stood above sea level, the active delta moved to a new location where the stream gradient was steeper and sediment could thus be transported. The current delta is about 1000 years old and is the seventh of the series built in the past 7500 years (Figure 16.32d).

This most recent delta lobe is apparently full of sediment, and the main body of the Mississippi River could move to the channel now occupied by the Atchafalaya River to the west (Figure 16.32d). This route would shorten the distance between central Louisiana and the Gulf of Mexico to less than one-half the present distance. However, such a move would be economically disastrous for New Orleans, because this busy port city is accessed through the current mouth of the Mississippi River. In an effort to combat this problem, an artificial control system currently separates the Mississippi River from the Atchafalaya River at the Old River Control Project upstream from Baton Rouge. It's only a matter of time, however, before a major flood occurs and the Mississippi circumvents the barrier to flow down its new course. Pay attention, it may occur in your lifetime!

VISUAL CONCEPT CHECK 16.3

Which one of the following choices accurately describes the evolution of the alluvial landform pictured here?

a) It formed due to deposition and intense downcutting, which created alluvial terraces.

b) It formed because the streams are graded streams that flow on a perennial basis.

c) It formed because deposition of alluvial sediment occurs where the stream gradient lessens.

d) It consists of a point bar that developed during the process of stream meandering.

KEY CONCEPTS TO REMEMBER ABOUT ENTRENCHED MEANDERS, TERRACES, ALLUVIAL FANS, AND DELTAS

1. Entrenched meanders form when the slope of a graded, meandering stream is increased through extensive regional uplift, base level drop, or climate change.

2. Alluvial terraces are abandoned floodplains that have been elevated through the process of either alluviation or stream downcutting so that they are no longer frequently flooded.

3. An alluvial fan has the appearance (from above) of a hand-held fan and forms when a bedload-dominated stream flows down a mountain front and aggrades in a radial pattern in the adjacent valley.

4. A delta is a low-lying plain that forms at the mouth of rivers where they meet the ocean. The sediment contained within a delta is well sorted, with coarse deposits accumulating at the river mouth and fine sediments transported to the delta front.

5. Deltas and alluvial fans are similar in that they form when stream velocity is reduced such that sediment can no longer be carried. This decreased power occurs at an alluvial fan because the channel slope reduces between the mountain and valley. At a delta, stream power is lost because the stream flows into the relatively still ocean water.

Human Interactions with Streams

The attempt to control the direction of the Mississippi River, as just described, is a nice example of how humans interact with rivers to meet societal needs. In this particular case, if (and when) the Mississippi River moves to the Atchafalaya channel, the city of New Orleans may no longer have direct access to the Gulf of Mexico and its important status as a port city may be jeopardized. That's why a lot of money has been invested in the Old River Control Project to maintain the current course of the Mississippi River. Although this problem is a major issue in southern Louisiana, it is only one of countless examples of human interaction with river systems in the world. Here we will look more closely at some of the ways in which people are attempting to control and adjust fluvial systems.

Urbanization

People have a big impact on fluvial systems through the process of urbanization. In the early part of this chapter, we explained the process of runoff by analogy with water draining off an impervious parking lot. The reality is that with increasing urbanization, parking lots and other urban structures are covering greater amounts of ground around the world. For example, think about the sizes of the parking lots associated with the malls and shopping centers where you live. In many cases, entire drainage basins now lie within urban areas where the ground cover is no longer vegetation and soil, but mostly asphalt and concrete.

The impact of urbanization on stream behavior is significant, especially following a heavy rainfall. Figure 16.33 shows a hydrograph that illustrates the theoretical response of a stream to a heavy storm prior to urbanization (green line) and after urbanization (blue line). Before the city was constructed, the lag time between the storm peak (when it was raining hardest) and peak stream discharge was slow. This long lag time existed because the soil-water belt had first to be saturated before the stream could be supplied with runoff or water from a recharged aquifer.

After urbanization, however, the response of stream discharge is markedly different in two specific ways. Can you see them in the stream hydrograph? First, the stream response is much faster; in other words, the lag time between the storm peak and peak discharge is less. This decreased lag time occurs because rain water is no longer absorbed first into the soil, but runs off the urban parking lots and other concrete surfaces directly into the stream. Second, the peak discharge in an urbanized setting is *greater* after a given storm event than it was following a storm of similar magnitude during pre-urban times. Very simply, this increased discharge occurs in the urbanized stream because more water runs off into the stream from the various asphalt and concrete areas rather than being absorbed by the soil. Given these relationships, urban streams are often described as being *flashy* because they react more quickly than their rural counterparts.

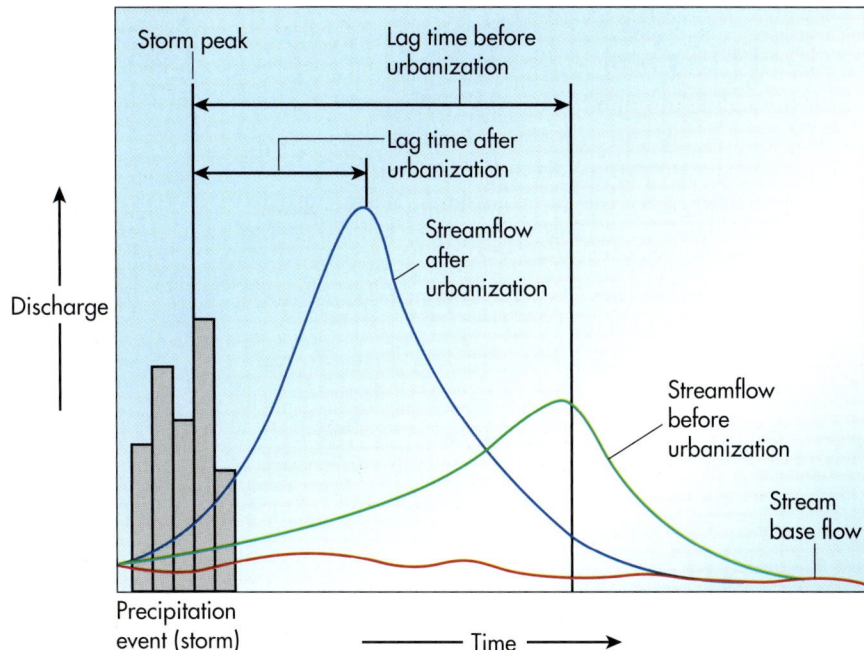

Figure 16.33 Stream hydrograph before and after urbanization. The response of the urbanized stream is much faster and larger than during pre-urban conditions.

Figure 16.34 Flash flooding. A flash flood is a brief but intense flood that occurs more frequently in urban areas where stream response is rapid. About half of flood-related fatalities occur when people try to drive across flooded roadways.

These conditions make *flash floods* much more likely in urbanized watersheds than in rural drainage basins if the same type of storm event occurs in both places. In contrast to river floods, which are slow-evolving events such as the 1993 Mississippi basin floods that lasted for weeks, flash floods are brief but intense (Figure 16.34), and can result in both significant property damage and loss of life. One way you can protect yourself during such an event is *never* to drive across a flooded roadway, even if the water appears to be shallow. After all, many automobiles become buoyant and can be carried away in only about 0.6 m (2 ft) of water. Of all of the human deaths attributed to flooding, approximately half occur when people are in their vehicles.

Artificial Levees

Given the increasing number of people living on floodplains around the world, major river and flash floods can be economically disastrous. As noted previously, for example, the 1993 floods in the Mississippi Basin caused $15 billion in damage. In the context of this kind of cost, humans have engineered a variety of control structures to mitigate the effects of flooding. One kind of flood control structure is an artificial levee.

Recall that a natural levee is a small ridge that forms along the stream channel where coarse sediments accumulate during a flood. This ridge increases the height of the stream bank so that the next flood must have a greater

Artificial levee *An engineered stucture along a river that effectively raises the height of the river bank and thus confines flood discharge.*

discharge, and thus be deeper, in order to reach flood stage. An **artificial levee** is based on the same principle as a natural levee, with the difference that an artificial levee is designed by engineers and is usually built to a height such that a given city or valuable farmland is protected from a major event such as a 50-year or 100-year flood (Figure 16.35). In the United States, most of the large artifical levees along major rivers such as the Mississippi, Ohio, and Missouri are built and maintained by the United States Army Corps of Engineers. For example, in the Memphis district of the Corps of Engineers, levees are built to a height of about 7.6 m (25 ft) above the natural elevation of the Mississippi River bank.

Although levees certainly provide a societal benefit with respect to flood protection, they also have negative side effects. One problem with levees is that they keep the river from replenishing the sensitive wetland ecosystem that borders most rivers. This ecosystem is intimately related to the frequent but low-level floods that naturally occur along the stream and bring a fresh supply of sediment and nutrients into the wetlands. This relationship is lost, however, along a stream that is heavily protected by levees. Another problem with levees is that they may actually increase the intensity of flooding when major floods do occur. Under natural conditions during a major flood, a stream will slowly spill out of its banks and spread across its floodplain as discharge increases. Although this slow spread can certainly be damaging, the gradual nature of the flood decreases the stream power and has a cushioning effect on locations downstream. In a levied section of river, however, the

Figure 16.35 Engineered levees. This artificial levee along the Mississippi River in Guttenberg, Iowa, kept the Mississippi River from flooding the town of Guttenberg during a minor flood in 2001. The U.S. Army Corps of Engineers built and maintains this structure, as well as many others like it, along the Mississippi.

water cannot slowly spread horizontally because it is confined between the levees. As a result, the depth of the water increases beyond what it naturally would (if it were allowed to slowly spill out of its banks) and the associated stream power grows. Much of this energy is transferred to the levee system, which may then fail during especially large floods like the 1993 Mississippi River flood. If such a failure occurs, water catastrophically pours through the breach in the levee, resulting in tremendous damage to farmland and human structures where the break occurs.

Dams and Reservoirs

Another way that humans interact with rivers is through the construction of dams, which are among the largest human structures in the world. A **dam** is an engineered obstruction that is built across a river to control its flow. When you were young, you may have built dams by piling dirt across small streams to control them. In large dams, this control is obtained through hydraulic gates in the dam that can be adjusted to allow a specific discharge of water to proceed downstream. Dams typically have behind them a large lake or reservoir that evolves through the process of impounding stream water that would otherwise flow downstream. A well-known example of a dam is Hoover Dam, which straddles the Colorado River on the border of Arizona and Nevada (Figure 16.36). Formed behind this dam is Lake Mead, which is one of the largest artificially cre-

Figure 16.36 Hoover Dam on the Colorado River. This dam was constructed in the early 1930s to provide flood control and hydroelectricity for the region. Lake Mead is the artificial lake upstream of Hoover Dam.

ated bodies of water in the world, with an area of 603 sq km (233 sq mi).

Historically, dams have been built for three primary reasons: to provide 1) hydroelectric energy, and 2) flood control, and 3) to enhance river navigation. Dams improve navigation because the amount of water in the river downstream of a dam can be kept at a constant level by controlling the amount of water released from the reservoir, except during the most extreme droughts or floods. For example, Figure 16.37 is a hydrograph

---

Dam *A barrier that blocks or restricts the downstream movement of a stream.*

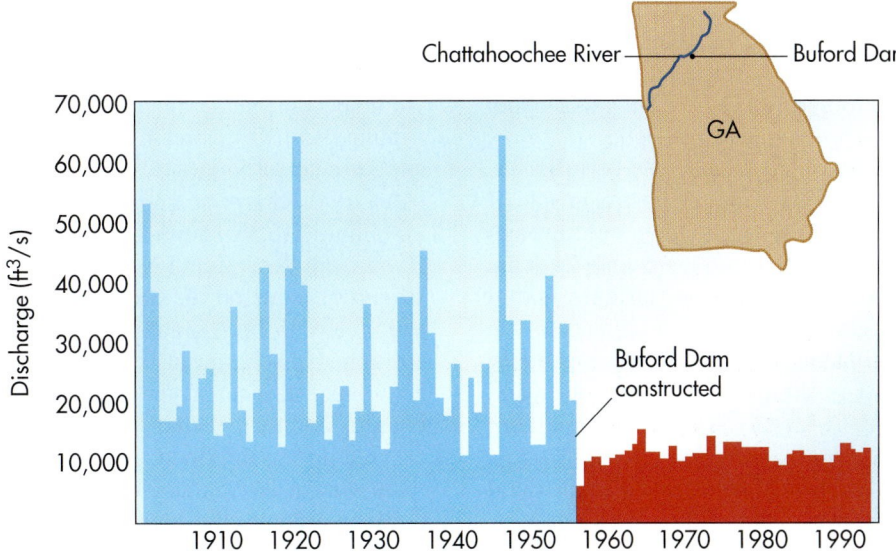

Figure 16.37 Annual flows for the Chattahoochee River downstream of Buford Dam in Georgia. Notice the change in the variability of annual discharge before and after the construction of Buford Dam in 1956. (Source: United States Geological Survey)

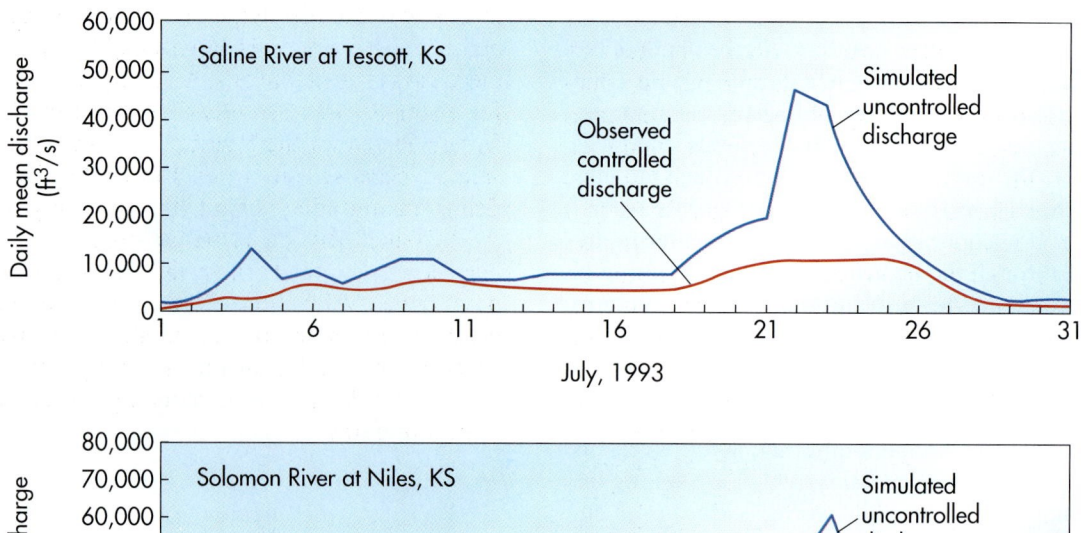

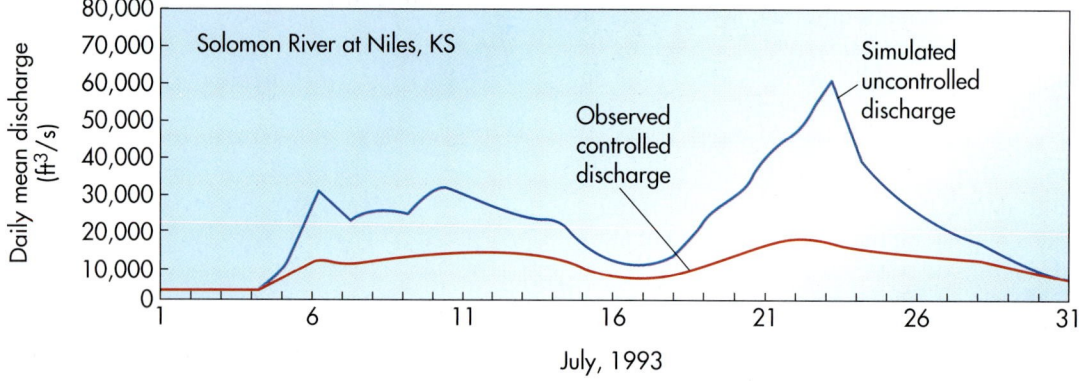

Figure 16.38 River hydrographs from Tescott and Niles, Kansas during the 1993 flood in the Mississippi watershed. Note the difference between simulated peak discharge (blue line) if flood-control dams had not been present and actual peak discharge (red line) resulting from controlled outflow from the dams upstream of this pair of towns. (Source: United States Geological Survey)

from the Chattahoochee River at Narcross, Georgia, which lies downstream of Buford Dam. Note how the discharge of the stream was held much more constant following dam construction than would have occurred naturally before. As a result of this controlling effect, rivers such as the Ohio, Mississippi, and Missouri are important transportation corridors because water depth is maintained at more or less the same depth.

With respect to hydroelectricity, this form of energy is produced when water from the upstream side of the dam (from the reservoir) flows down and through tunnels in the dam to its outlet into the channel. As the water flows through the dam, it spins turbines within generators that produce electricity. The hydroelectric generators at Hoover Dam, for example, are capable of supplying nearly 1.5 million kilowatts of power and provide electricity to Arizona, Nevada, and southern California.

The third major function of dams is to provide flood protection for areas downstream of the structures. Very simply, dams reduce the chances of a flood because they can be used to store excess runoff in the reservoir by controlling the outflow of water at the dam during periods of heavy rainfall. In theory, the gates of a dam can be shut completely during a flood, which would cause all of the stream discharge upstream of the dam to be trapped in the reservoir. Upstream of the dam, the lake would become larger and progressively fill the trunk and tributary valleys upstream. Below the dam, however, the stream channel would be dry until it began to receive tributary discharge someplace downstream.

Figure 16.38 shows hydrographs during the 1993 flood from Tescott and Niles, Kansas, which lie downstream of dams on the Saline and Solomon Rivers, respectively. In both graphs the blue line represents the simulated discharge at each town if the dams hadn't been present, whereas the red line shows the actual discharge. Although actual peak discharge certainly increased during the flood interval, it was about 80% less than what it is simulated to have been. As a result of these two dams, the towns of Niles and Tescott were protected from the flood.

Negative Environmental Effects of Dams Although dams have a clear benefit to society in that they control flooding, produce electricity, and enhance shipping, they also have a variety of societal and environmental costs associated with them. The most basic societal cost is that dams are extremely expensive, a cost which, in the United States, has been paid for largely through tax contributions. One example of extraordinary dam cost is the Itaipu Dam on the border of Paraguay and Brazil. This structure is currently the largest hydroelectric dam in the world, producing up to 12,600 megawatts of electricity, and provides 78% of the electricity used in Paraguay and 26% of that consumed in Brazil. Construction of this dam began in 1975 and was finished in 1983. It cost between $20 billion to $25 billion U.S. dollars to complete and was such a large financial drain that it contributed to the downturn of the Brazilian economy in the 1990s. Similar costs are projected for construction of the Three Gorges Dam in China, which is currently being constructed on the Yangtze River and is expected to be fully operational in 2009.

Another negative effect associated with dams is specifically related to the reservoirs they impound. In the course of their creation, reservoirs flood large areas upstream of the dam site. If this flooding occurs in heavily populated areas, many people need to be relocated. Upstream of the Three Gorges Dam, approximately 632 km^2 (244 mi^2) of land will ultimately be inundated. As a result, up to 1 million people have had to move to avoid being flooded. In a related vein, ancient cultural artifacts are sometimes lost in this way. Upstream of the Aswan Dam in Egypt, which was constructed across the Nile River, impoundment of Lake Nasser resulted in the loss of numerous classical Egyptian artifacts.

In addition to the societal costs of reservoirs, they also have negative environmental impacts. These impacts are most closely related to the loss of ecosystems and spectacular scenery. Prior to construction of the Itaipu Dam, for example, the Parana River flowed through a spectacular gorge that contained a variety of beautiful waterfalls. Many of these waterfalls, along with much of the habitat unique to the plants and animals of the region, are now submerged beneath the reservoir. Similar scenery and habitat loss occurred in the canyonland region of the southwestern United States when dams such as Glen Canyon Dam and Hoover Dam (Figure 16.36) were constructed.

In the context of ecosystem impacts, dams also have negative environmental costs because of the way they regulate the discharge of streams, when, in fact, stream flow is naturally highly variable over the course of the year. The riparian ecosystem is adapted to this variable discharge and is now being pressured because stream flow is largely held constant to meet the various societal needs. A variety of animal species, for example, are currently threatened by dams in the United States. In the Pacific Northwest, dams impact salmon because they often form a barrier that stops fish from migrating upstream to spawning sites. Along the Missouri River, the pallid sturgeon is threatened because it requires high spring flows to trigger its reproductive cycle; these flows are moderated significantly by the numerous dams along the river. At this point in time, environmentalists and government agencies are working toward reducing the environmental impact of dams.

KEY CONCEPTS TO REMEMBER ABOUT HUMAN IMPACTS ON STREAMS

1. Most streams in the world have been impacted by humans in some way.

2. Asphalt and concrete structures in cities increase storm runoff. As a result, the lag time of a stream decreases, peak discharge increases, and flash floods are more likely.

3. Artificial levees are ridges built along stream channels to increase the height of the stream bank so that protected areas are rarely flooded. These systems work, but increase the overall power of the stream at high discharge by confining it to a narrow channel rather than allowing gradual flooding.

4. Dams are used to control downstream discharge on rivers by impounding water (in reservoirs) and releasing it slowly. Dams are used for flood control and hydroelectric production, and to facilitate channel navigation.

5. Although dams provide a definite societal need, they have significant environmental impact because the natural rhythm of seasonally variable stream flow is lost. In addition, they are a barrier for migrating river species.

The Big Picture

In this chapter we focused on the way that flowing water moves and shapes the landscape. Next we'll turn to the way that frozen water flows across the Earth's surface in the form of glaciers. Remember that most fresh water on Earth is stored in glaciers and polar ice sheets. Similar to stream systems, glaciers actually flow; in fact, you can think of a glacier as being a river of ice. Also similar to streams, glaciers have had a large role in shaping the Earth's landscape. The next chapter will focus on the way that glaciers evolve, how they behave, and the landforms they produce. Our discussion will incorporate concepts related to climate, the hydrologic cycle, and Earth–Sun geometry, and should give you a much greater appreciation for the role that ice has played on Earth.

Summary of Key Concepts

1. Streams consist of channels in which water flows downhill under the force of gravity. They are fed by runoff or groundwater or may have a large lake as their source and are separated topographically from other streams by bordering high ground called divides.

2. Stream flow in a channel is defined by its hydraulic variables, including width, depth, velocity, slope, and discharge. These variables are interrelated with one another. For example, stream discharge is calculated by the equation $Q = w \times d \times v$.

3. In the course of their evolution, streams gradually sculpt the landscape so that the slope of the stream is sufficiently steep to carry the average discharge in the river. During this shaping process, the slope of the stream decreases. Once the stream reaches this stage of evolution, it is said to be "graded." The behavior of the stream is controlled in large part by base level and climate.

4. As streams approach a graded state, they begin to erode horizontally through meandering. This process produces a wide valley because bordering bluffs are eroded.

5. Streams produce a variety of erosional and depositional landforms, including gullies, cutbanks, point bars, deltas, and terraces.

6. Given the importance of streams to human societies, they are heavily managed.

Check Your Understanding

1. How does the concept of runoff differ from groundwater percolation?

2. Define the concept of *drainage basin* and describe one of the major watersheds in North America. What are the boundaries of this drainage basin and what are some of the major tributaries?

3. Imagine that there are two drainage basins in a humid environment. One of the watersheds has high drainage density whereas the other one does not. Assuming that each drainage basin lies within a region of similar relief, what environmental variable most likely accounts for the difference in drainage density? Why?

4. What is the minimum number of tributaries that a fourth-order stream has? How many stream confluences are there in this watershed?

5. Name four primary hydraulic variables. Provide an example of how one variable influences another.

6. Although the terms *base level* and *base flow* sound similar, they have quite different meanings. Define each term and describe how they differ from one another.

7. Which landscape is more likely to contain graded streams, one that has recently been uplifted or one that has been stable for a long period of time with a consistent climate? Why?

8. Name and describe three landforms associated with a stream that is striving to reach a graded condition.

9. Name and describe three landforms associated with a graded stream.

10. How can deposition and erosion both occur along a stream meander?

11. Imagine that you encountered a stream with well developed meanders that are deeply entrenched. Describe two reasons why this landscape would evolve.

12. Why is a terrace an abandoned floodplain? Describe the two ways that terraces form.

13. How are alluvial fans and deltas similar to one another?

14. Why is peak discharge greater in streams contained within urban areas than those in the surrounding countryside?

15. Describe how a dam provides flood control for a city that lies downstream of the structure.

ANSWERS TO VISUAL CONCEPT CHECKS

Visual Concept Check 16.1

The correct answer is *a*; the Danube is a high-order stream. You can tell that the Danube is a high-order stream because it is large, with high discharge. This appearance results from the numerous lower-order tributaries that feed into the river.

Visual Concept Check 16.2

The correct answer is *d*; the stream is not graded. Remember that graded streams have shallow slopes that allow a stream to migrate horizontally. This stream has a very steep slope, which means that most of its energy will be used to erode vertically.

Visual Concept Check 16.3

The correct answer is *c*. This landform is an alluvial fan, which forms when sediment is deposited at the mountain front where a stream with a steep slope enters the valley. At this location, sediment is deposited because the stream velocity slows, resulting in the development of this fan-shaped landform.

CHAPTER SEVENTEEN

GLACIAL GEOMORPHOLOGY: PROCESSES AND LANDFORMS

In the preceding chapter, we discussed the processes and resulting landforms associated

with liquid water that flows across the Earth's surface in streams. Now we examine the

processes associated with flowing *ice* and the landforms that result. About 77% of the

fresh water stored on land occurs as ice, and, although it may not seem possible, this

form of water moves, becoming a river of slowly flowing ice called a glacier. Like rivers,

glaciers shape the landscape in predictable ways through erosion and deposition.

Flowing ice has had a huge impact on the Earth throughout history. Although glaciers such as this one in Alaska's Denali National Park move slowly, they have tremendous power and are capable of eroding and depositing prodigous quantities of sediment. Much of the landscape seen here has been shaped in this way. This chapter focuses on glacial processes and the landforms that result from flowing ice.

At the present time, ice covers about 11% of the Earth. However, as little as 18,000 years ago, glaciers covered about 30% of the planet, including much of North America and Europe. As the climate warmed, the ice retreated to progressively higher latitudes and new landscapes created by glacial processes were uncovered. Although these landscapes have been modified since that time, much of the topography in northern North America and Europe is the direct result of glacial processes. If you live in Canada or any of the northern states such as Wisconsin, Michigan, New York, Washington, or Illinois, there is a good chance your home is built on deposits left by glaciers.

Development of a Glacier

A **glacier** is a slowly moving mass of dense ice (Figure 17.1) formed by the gradual thickening, compaction, and recrystallization of snow and water over time. Glaciers form in regions where heavy snowfall does not melt completely during the summer. In other words, a key factor in glacial formation is that summer temperatures must be sufficiently cool that some snow is left over to serve as the foundation for the next year's snowpack. Only then can snow depth slowly increase so that it can be transformed to a river of flowing ice that grinds and shapes the landscape.

The Metamorphosis of Snow to Glacial Ice

How does snow transform into a flowing glacier? Recall the discussion in Chapter 11 about certain kinds of metamorphic rocks and how they form because of high pressure due to compaction from overlying sediments. The process of snow transformation to glacial ice is very similar. The first thing to remember is that this snow-to-ice metamorphosis takes a lot of time, often hundreds or even thousands of years. Over this course of time, snow accumulates in distinct layers that are very similar to sedimentary rock deposits (Figure 17.2). Given this similarity, some glaciologists think of glacial ice as being a form of sedimentary rock.

When snow first falls, it is approximately 80% air. With time, this snow is progressively buried by new annual layers of snow. As a result, it gradually compacts and recrystallizes due to the weight of snow above it. In addition, some of the snow might melt and refreeze. Water may percolate downward into the compacting snow from summertime rain that falls on the surface of the snow pack and from the limited snowmelt that occurs within it. As a result of the combined processes of compaction and percolation of liquid water, the lowest deposits of snow are transformed into a compact, granular substance called **firn** (Figure 17.3). Firn is somewhat similar to the crunchy snow that you can almost walk on top of after the snow melts a bit during the daytime and then refreezes at night. With additional compaction, pockets of air in the firn further decrease in size or are squeezed out entirely, and the ice crystals become aligned with one another. The resulting dense material is known as glacial ice. At this time, the mass is 90% solid and begins to flow slowly under its own weight, much like maple syrup across a pancake. Once this movement begins, the ice mass becomes a glacier. In

Glacier *A slow-moving mass of dense ice.*

Firn *The compact, granular substance that is the transition stage between snow and glacial ice.*

Figure 17.1 A typical glacier. Valley glaciers like this one in Alaska's Chugach National Park look just like a frozen river, slowly flowing down from the mountains to the lowlands.

most mountain glaciers, the full transformation to active glacier begins when the snow and ice mass reaches a depth of about 40 m (about 130 ft).

The Glacial Mass Budget

Like streams, glaciers are natural systems that depend on inputs and outputs to function. In the case of streams the input is liquid water in the form of precipitation, runoff, and groundwater. The outputs are stream discharge and water vapor (through evaporation). With glacial systems, the input is snow and the outputs are ice, meltwater, and again, water vapor. In the context of a glacial system, it is useful to think of the relationship between input and outputs as a budget, analogous to your bank account. This analogy is often useful to show how natural systems achieve balance; we used it in Chapter 4 to describe the flow of terrestrial radiation. The input you have in this account is money that's either earned, borrowed, or been given to you. These funds are subsequently spent on outputs such as housing, transportation, food, clothing, and entertainment. If you make more money than you spend, your bank account grows. Conversely, if you spend more money than you earn, your bank account shrinks.

A glacier works on the same basic principle as your bank account. The input to the glacial system is snow that accumulates at the upper end of the glacier (Figure 17.4). This area is fittingly called the **zone of accumulation**

Figure 17.2 Annual snow layers in a glacier. A key factor that determines if a glacier will form is whether snow from the preceding winter melts in summer. If the snow doesn't melt, then it can accumulate in layers like these exposed in the Quelccaya Ice Cap in Peru. The average thickness of these layers is about .75 m (2.5 ft). Note the similarity between these layers and those found in non-deformed sedimentary rocks.

and is where the addition of annual snow exceeds what is lost through melting and evaporation. Glacial outputs occur primarily down the valley toward the front of the glacier, where the temperatures are higher. This area is called the **zone of ablation** (or *wastage*) and is the place where more annual melting and evaporation of ice takes place than snow accumulation. Somewhere between the ablation and accumulation zones is the **equilibrium line**, which marks the place where system inputs and outputs are in balance.

The overall budget of the glacial system is continually in flux, much like your bank account, with variable amounts of input and output over the course of time.

Zone of accumulation *The geographical region where snow accumulates and feeds the growth of a glacier.*

Zone of ablation *The part of a glacier where melting exceeds snow accumulation.*

Equilibrium line *The place on a glacier where snow accumulation and melting are in balance.*

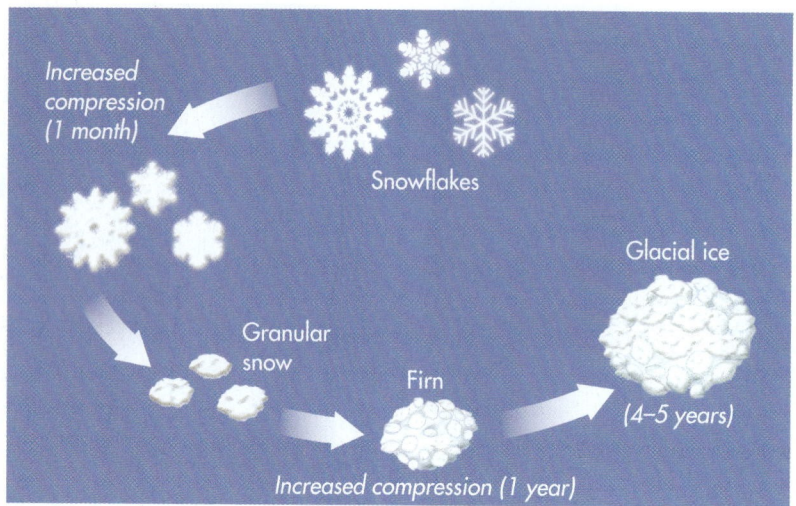

Figure 17.3 The transformation of snow to glacial ice. When snow first falls it is 80% air. As it compacts, its density increases to the point that it becomes granular. When these grains of snow become more compact they collectively form firn. With additional compression, the firn becomes glacial ice. In general, the most dense and compressed layers of ice are at the bottom of a glacier, while the top layers are relatively recent snowfall.

During periods when there is more accumulation of snow in the ice cap or ice field than there are outputs at the zone of ablation, the equilibrium line moves down the valley and the glacier grows and advances. Conversely, when there is more wastage in the zone of ablation than there is accumulation of snow in the ice cap or ice field, the equilibrium line moves up the valley and the glacier shrinks and retreats.

Glacial Movement

As we've just noted, glaciers actually move. Contrary to what you may think, a glacier is not just a block of ice that happens to slide down a hill. Rather, glacial movement is caused by the ability of ice to flow under its own weight.

Most glacial movement occurs as a result of *internal deformation* within the ice. This internal movement is related to the fact that the interior of a glacier behaves like a plastic that can be shaped and manipulated to some degree. Given the friction that exists between the ice and the rock on which it flows, the ice moves most quickly in the core of the glacier (Figure 17.5). Thus, ice-flow dynamics are quite similar to a stream system. However, in contrast to streams, which flow quickly, average glacial flow is on the order of 10 to 30 cm (about 4 to 12 in) per day.

It is important to note that this internal deformation continues even if the glacier is melting, shrinking, and retreating. In other words, the interior of the glacier continues to move forward *even though* the front of the glacier may actually be retreating up the mountain valley as a result of fluctuations within the overall mass budget (Figure 17.6).

Not all of the ice is plastic, however. In fact, the upper part of the ice is quite brittle and breaks easily. When this part of the glacier fractures, prominent **crevasses** can develop that extend from the surface of the ice to some depth within it. Crevasses usually form when the ice flows over a small ridge in the bedrock below (Figure 17.7). This ridge causes compression to build in the ice on the up-valley side of the ridge. On the downstream side of the ridge, the ice stretches, causing tension to build and crevasses to form. Crevasses are most closely associated with the equilibrium line because that is the place where they become visible as you move down the glacier (see Figure 17.4). These giant cracks in the ice are visible in the zone of ablation because they are not covered by fresh snow. In the zone of accumulation, however, they are not as visible because fresh snow tends to hide them.

In addition to the flow associated with internal deformation, glacier movement also occurs through a process called *basal slip*. This process results when high amounts of meltwater accumulate at the base of the glacier, thereby providing a lubricant between the underlying rock and the ice mass. This lubricant can cause a glacier to move relatively rapidly through a process called *surging*, with movement accelerated to greater than 30 cm (12 in) per day.

Crevasse *A deep crack in a glacier.*

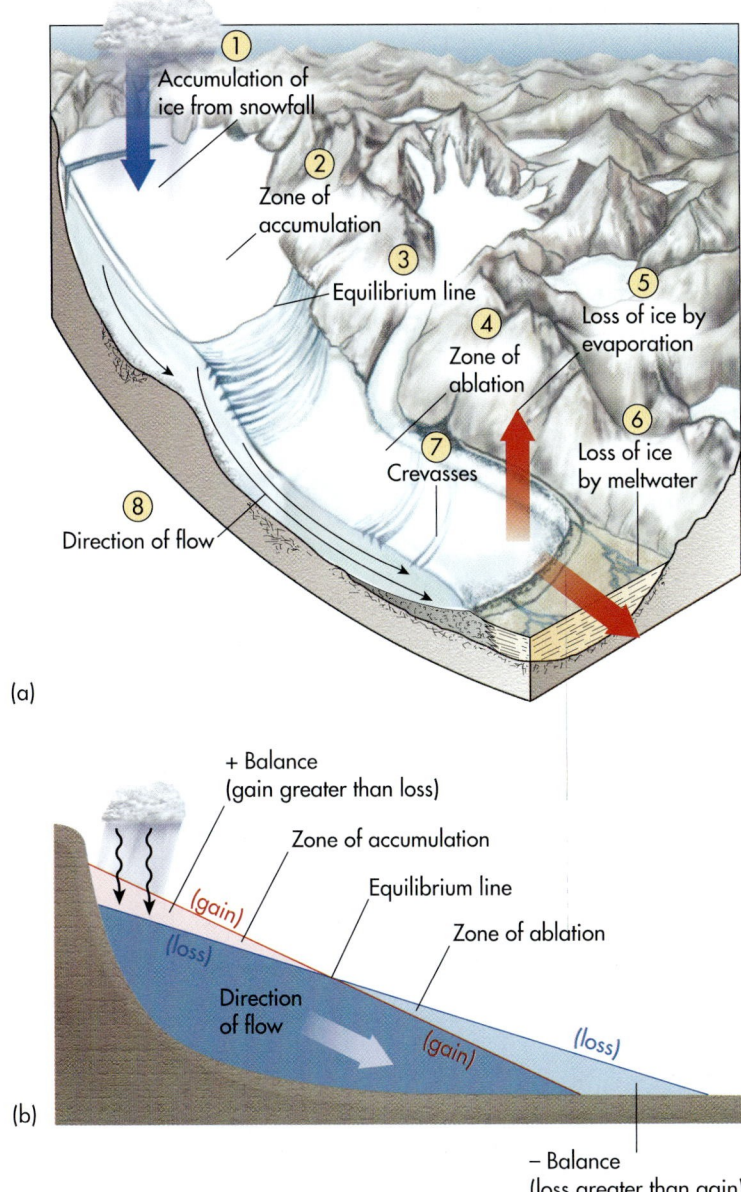

(a)

(b)

Figure 17.4 The glacial mass budget. (a) The glacial mass budget is the balance between snowfall in the zone of accumulation and melting and evaporation at the zone of ablation. (b) Side view of the glacial mass budget. Note the geographical relationship of the zones of accumulation and ablation and the way in which ice flows.

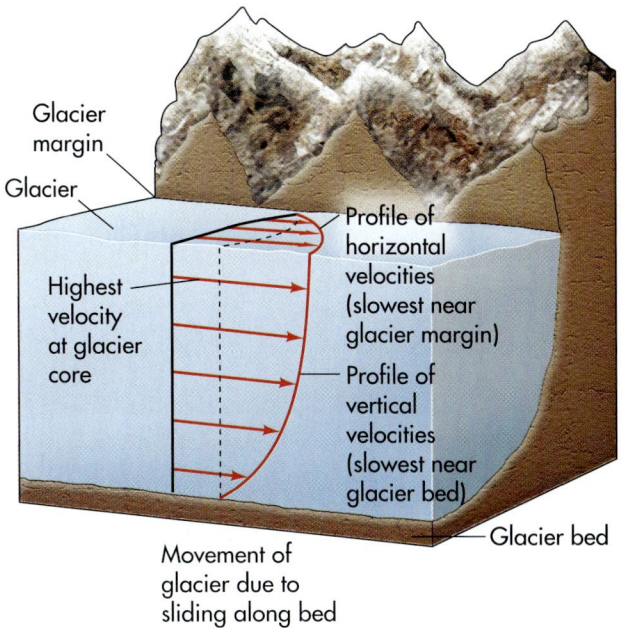

Figure 17.5 Internal flow anatomy of a glacier. Glacial ice is a plastic substance that flows. Because of friction, the ice flows more slowly on its edge and base where it is in contact with bedrock. Velocity gradually increases toward the core of the glacier because it can move more freely.

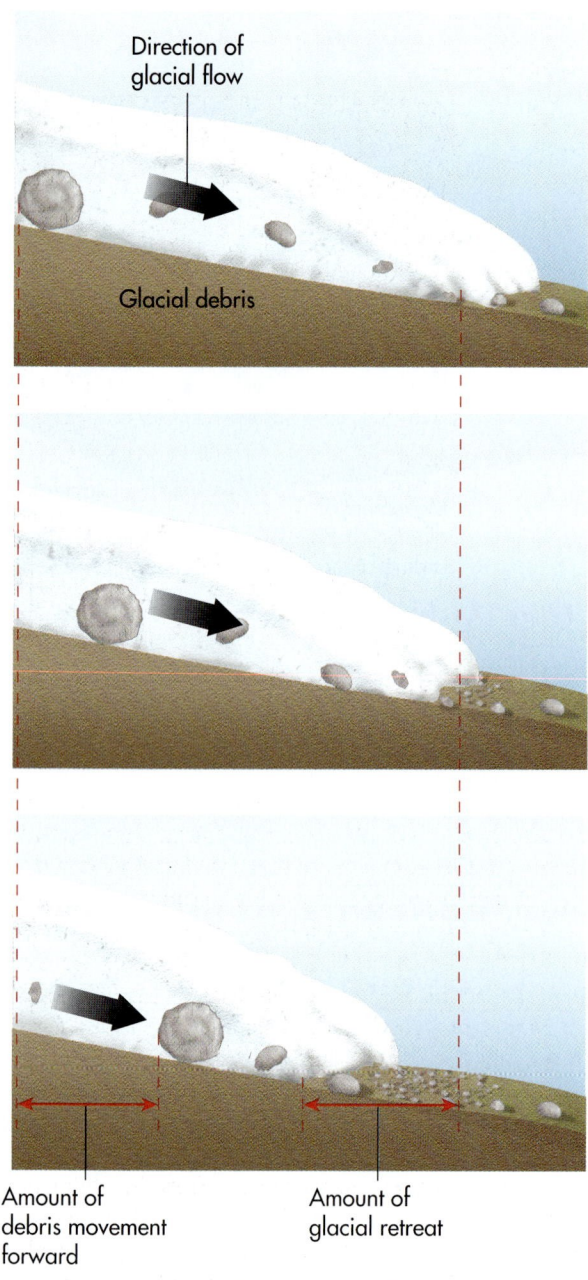

Direction of glacial flow

Glacial debris

Amount of debris movement forward

Amount of glacial retreat

Figure 17.6 Internal movement of a glacier. The interior of a glacier continues to flow forward, carrying large rocks and other debris, even though the position of the glacier's front may retreat due to fluctuations in the glacial mass budget.

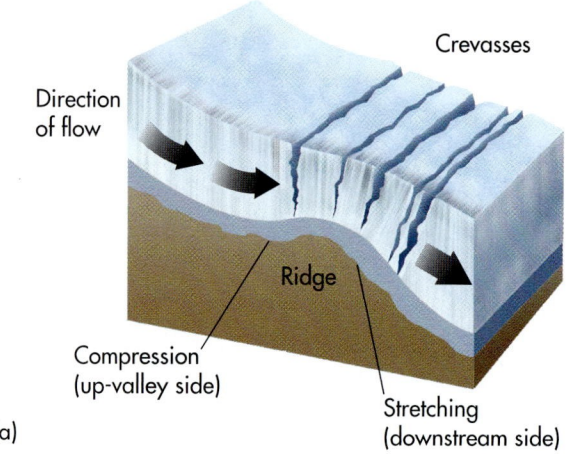

Direction of flow

Crevasses

Ridge

Compression (up-valley side)

Stretching (downstream side)

(a)

(b)

Figure 17.7 Formation of a crevasse. (a) A crevasse is a deep fissure that extends from the surface to some depth within the ice. Crevasses form when ice is compressed on the upstream side of an underlying bedrock ridge. On the downstream side of the ridge the ice stretches, tension is high, and crevasses develop. (b) A typical crevasse can be 27 m (~90 ft) deep or more, causing great danger to skiers or hikers attempting to cross a glacier.

Once you understand the glacial mass budget and what a glacier looks like in that context, it's possible to look at an alpine glacier and interpret where features such as the equilibrium line and zones of ablation and accumulation are located. Look at this image and see if you can identify the zones of accumulation and ablation. Where is the equilibrium line? How can you tell?

Types of Glaciers

Now that we've seen the basic processes associated with glacier movement, let's discuss the different kinds of glaciers by examining the environments in which they form.

Glaciers in Mountainous Regions

Perhaps the most obvious place that glaciers form is in high mountainous environments. These regions are a logical place to find large masses of ice because mountain ranges are places where heavy orographic snows can accumulate. Another important factor in these high altitudes is that temperatures are usually cold to very cool and snow is thus slow to melt.

Three basic kinds of ice masses occur in mountainous areas. The largest are ice caps and ice fields (Figure 17.8). An **ice cap** is a continuous sheet of ice, roughly circular in shape, that covers the entire landscape. In contrast, an **ice field** is more linear in form and is relatively discontinuous, burying all but the tallest mountains in ice that may be hundreds of meters thick. Where these mountains rise above the ice, they are called *nunataks*, which means "lonely stones."

Ice caps and ice fields are frequently the source area for the third kind of ice mass found in mountains, **alpine glaciers**, which are glaciers that flow down valleys extending away from the high country. Such glaciers exist in many mountain ranges around the world, including the Southern Alps of New Zealand, the Swiss Alps in Europe, the Canadian Rockies, the Himalayan Mountains, and the Andes Mountains in South America. Several alpine glaciers occur in the United States, with the majority in Washington and Alaska (Figure 17.9). As you can see from that figure, most of these glaciers are quite small.

In areas where such glaciers flow in steep-sided valleys, they are called *valley glaciers* (Figure 17.10) because they occupy landscapes originally carved by streams. If a valley glacier extends entirely out of a valley into the lowland beyond the mountain front, it is called a *piedmont glacier*. Some valley glaciers even terminate at the ocean. These glaciers are called *tidewater glaciers* and are particularly common along the coast of Alaska. These glaciers are highly prized as a tourist destination because the landscape is a unique interaction of ice and the ocean. One of the particularly interesting aspects of tidewater glaciers is that large blocks of ice break off the front of the glacier in a process called *calving*. The calved

Ice cap *A large ice mass in mountainous regions that is approximately circular in form.*

Ice field *A large ice mass in mountainous regions that is generally linear in form.*

Alpine glacier *A glacier in mountainous regions that flows down pre-existing valleys.*

blocks then fall into the ocean where they become *icebergs*. Such an iceberg, of course, was responsible for the tragic sinking of the Titanic in 1912.

In addition to ice caps and ice fields, another source area for an alpine glacier is a **cirque**, which is a small, bowl-like depression on a mountain flank formed by glacial erosion. A good example of an environment where numerous cirque glaciers have formed is the Cascade Mountains in the Pacific Northwest. Recall from Chapter 7 that this region receives abundant orographic precipitation from moisture-laden air flowing off the Pacific Ocean to the west. This precipitation

---

Cirque *A bowl-like depression that serves as a source area for some alpine glaciers.*

is especially high on the numerous volcanoes that arc along the range. Given that the peaks of the volcanoes are usually greater than 3000 m (9800 ft) in elevation, summer temperatures near the mountain crests are typically cool.

Figure 17.11a shows the effect that these cool temperatures have on snow cover. Note the extent of the snow cover on the flanks of Mount Hood in Oregon during late summer. Snow cover of this magnitude has resulted in the formation of several cirque glaciers on this mountain, as well as on other volcanoes in the range. On Mount Hood, a variety of glaciers flank the mountain, extending from near the crest at about 3400 m (about 11,200 ft) down to an elevation of about 1800 m (6000 ft). These glaciers ring the mountain on all sides, as shown in Figure 17.11b.

Figure 17.8 Ice caps and ice fields. (a) The Vatnajökull ice cap in southeastern Iceland. (b) The Patagonia ice field in the Andes Mountains of southern Chile. This long ice mass has numerous glaciers flowing out of it. (c) Nunataks rise above the ice in the Juneau ice field in Alaska.

(a)

(b)

(c)

AMAZING PLACES

THE VATNAJÖKULL ICE CAP

As far as large ice masses are concerned, Iceland's Vatnajökull ice cap is particularly amazing because it covers the Grímsvötn volcanic system, which includes the 10-km-wide Grímsvötn caldera. Located on the eastern side of the island, this volcanic system is related to extrusion of magma associated with seafloor spreading along the Atlantic Mid-Oceanic Ridge, which extends through the center of Iceland. This system has been very active historically, most recently with the eruption of the caldera in 1998. This eruption was visually fascinating because it ejected volcanic debris *through* the overlying ice cap. The intense heat from the eruption caused a significant amount of the ice to melt instantaneously, producing a giant and destructive river called a jokulhlaup (pronounced "yokalup") that burst out from beneath the ice cap.

Eruption of the Grimsvötn volcano in November, 2004. This volcano has the very unique distinction of erupting through an overlying ice cap.

Iceland is an island in the northern Atlantic Ocean that has formed astride the Mid-Atlantic rift zone. Given this tectonic activity, several active volcanoes occur on the island, including two that underlie the Vatnajökull ice cap.

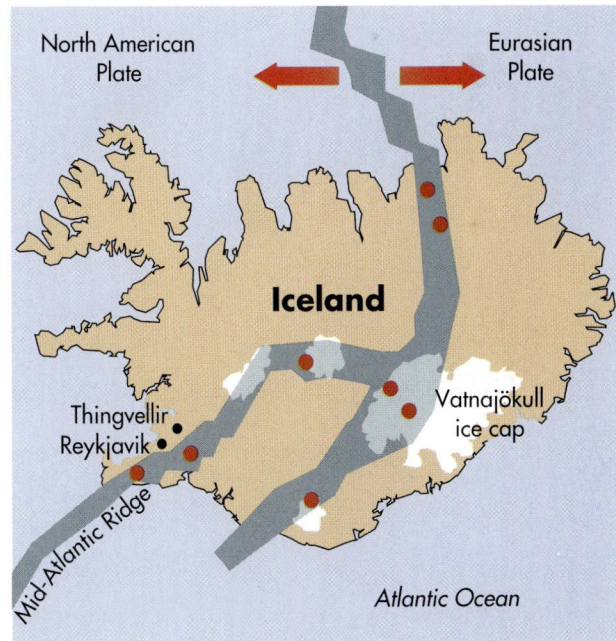

Washington
950 glaciers
160 sq. mi.

Montana
200 glaciers
16 sq. mi.

Oregon
60 glaciers
8 sq. mi.

Idaho
20 glaciers
<1 sq. mi.

Wyoming
100 glaciers
19 sq. mi.

Nevada
5 glaciers
<1 sq. mi.

Utah
1 glacier
<1 sq. mi.

Colorado
25 glaciers
<1 sq. mi.

California
290 glaciers
19 sq. mi.

Pacific
Ocean

0   100   200 km

0   100   200 mi

Alaska
Thousands
of glaciers
29,000 sq. mi.

Arctic Ocean

Pacific
Ocean

Figure 17.9 Alpine glaciers in the Pacific Northwest and Alaska. Most glaciers in the continental U.S. are relatively small, as you can see by the area they cover. However, Alaska has some large and impressive glaciers.

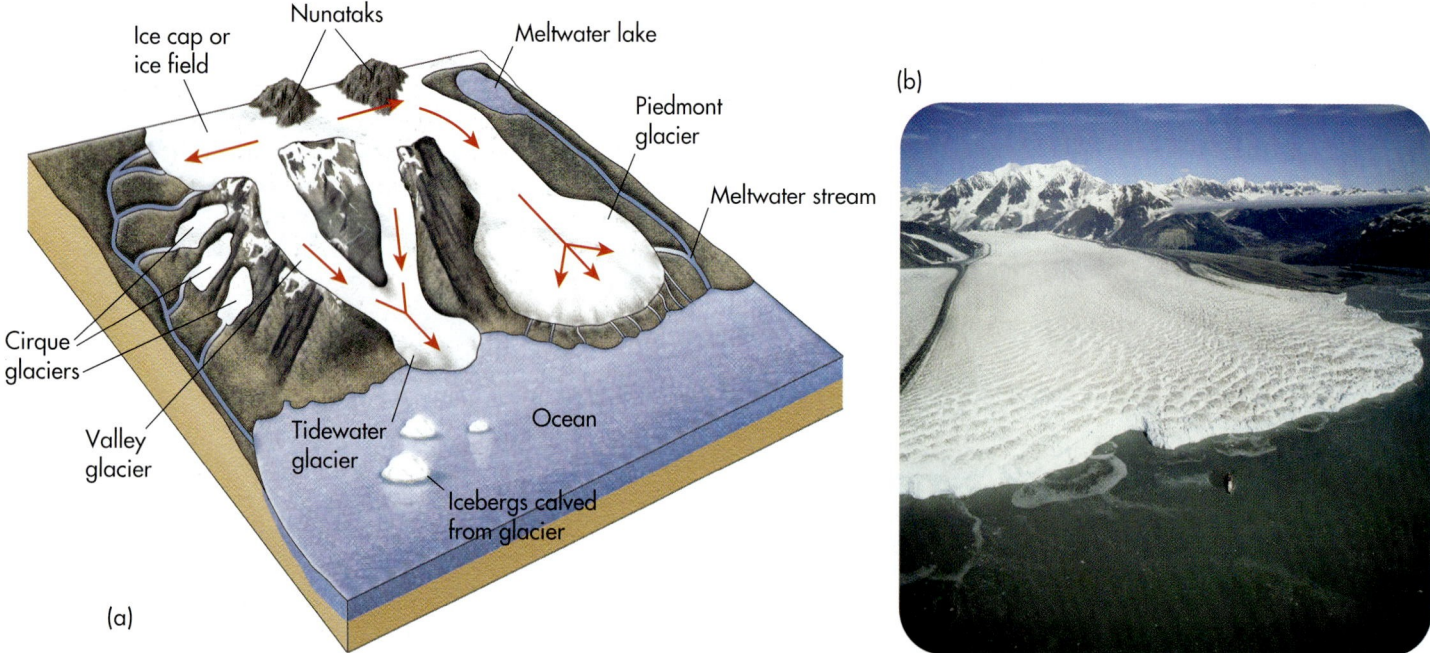

Ice cap or ice field

Nunataks

Meltwater lake

Piedmont glacier

Meltwater stream

Cirque glaciers

Valley glacier

Tidewater glacier

Ocean

Icebergs calved from glacier

(a)

(b)

Figure 17.10 Valley glaciers. (a) Valley glaciers frequently originate at ice fields and ice caps and flow down terrain initially carved by streams. They become piedmont glaciers if they expand beyond the valley front. (b) Tidewater glaciers occur where glacier ice meets the sea. In these situations, such as seen here at the terminus of the Hubbard glacier in Alaska, ice calves off the front of the glacier and falls into the water to form icebergs. Note the cruise ship in front of the glacier for scale.

(a)

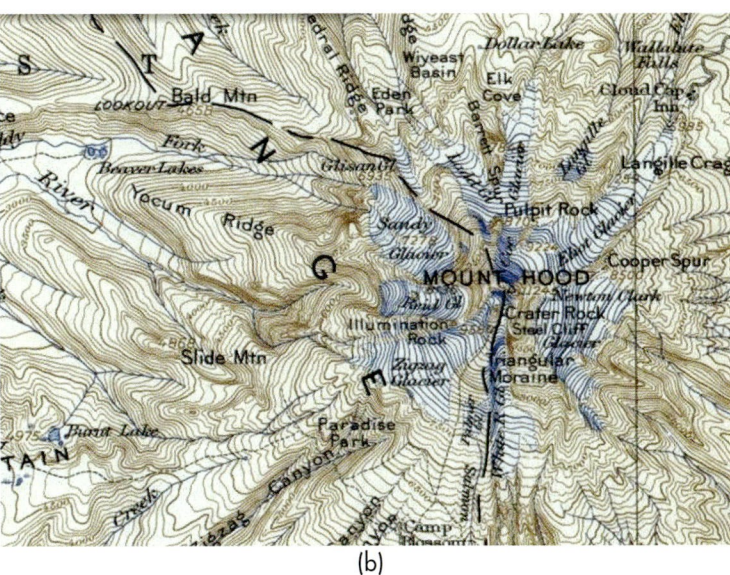
(b)

Figure 17.11 Cascade alpine glaciers. (a) Mount Hood is a composite volcano that reaches an elevation of over 3400 m (11,200 ft) and thus is an excellent place for the development of alpine glaciers. (b) This topographic map illustrates the various glaciers that ring the flanks of Mount Hood.

Continental Glaciers

In addition to mountain glaciers, another form of glacier is called a continental glacier. As the name implies, a **continental glacier** is a huge ice mass that covers a large part of a continent or large island. Also called *ice sheets*, continental glaciers once covered much of North America and Europe. However, they are now largely confined to the island of Greenland and the continent of Antarctica (Figure 17.12) because summer temperatures are cold at these high latitudes.

Continental glaciers contain an enormous volume of ice that is more than 3000 m (10,000 ft) deep in places. The Antarctic ice sheet covers 90% of the continent and alone contains 91% of all the glacial ice on Earth (amounting to 13.9 million km^3, or 3.3 million mi^3). Similarly, the Greenland ice sheet covers 80% of that island and is about 1.8 million km^3 (0.43 million mi^3) in size. In both Greenland and Antarctica, the tremendous weight of these extensive ice sheets is so great that the underlying lithosphere has been pushed down into the upper part of the asthenosphere to some degree by a process called *isostatic depression* (see Figure 12.3).

Continental glacier *An enormous body of ice that covers a significant part of a large landmass.*

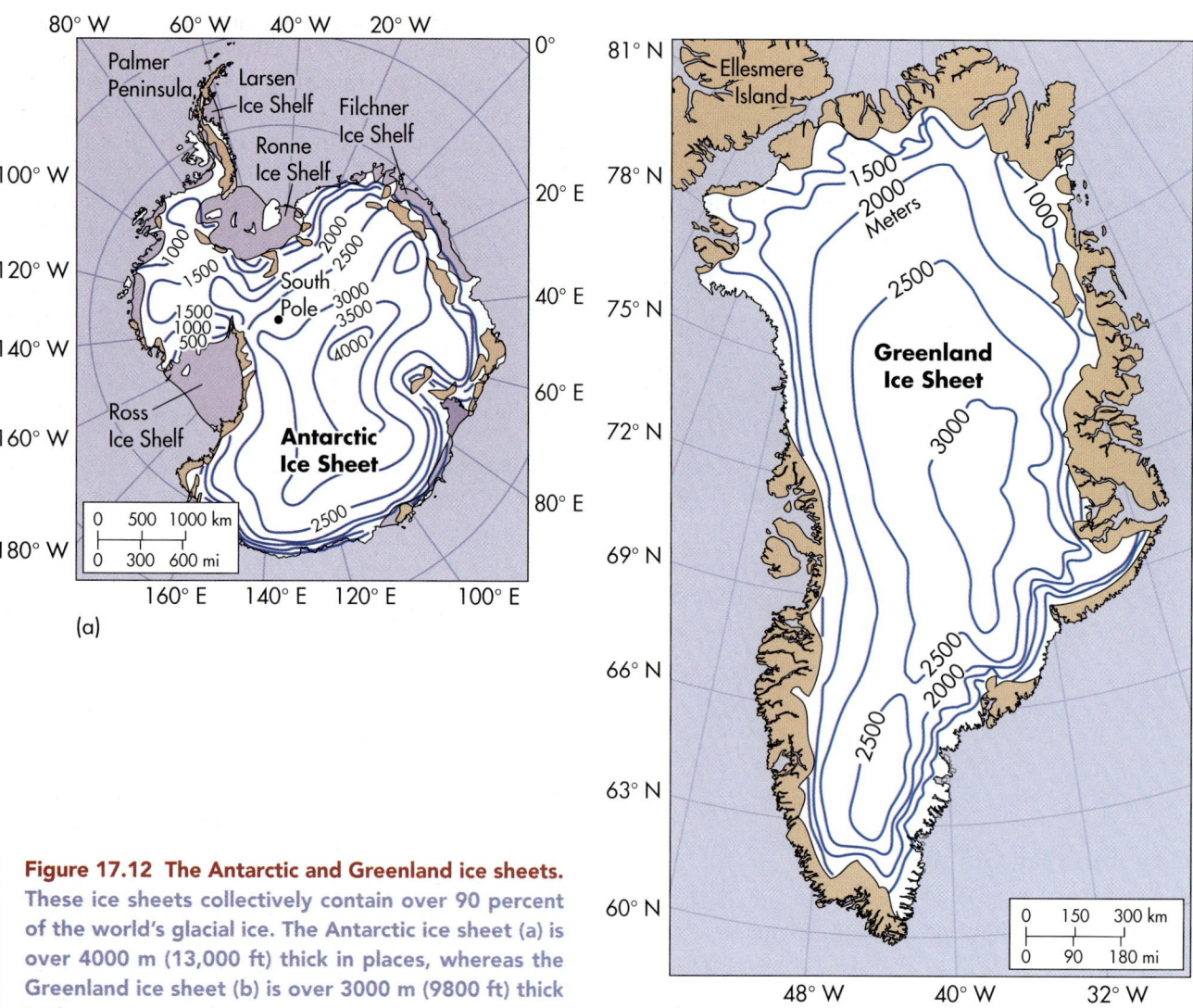

Figure 17.12 The Antarctic and Greenland ice sheets. These ice sheets collectively contain over 90 percent of the world's glacial ice. The Antarctic ice sheet (a) is over 4000 m (13,000 ft) thick in places, whereas the Greenland ice sheet (b) is over 3000 m (9800 ft) thick in the center.

Glacial Landforms

Like stream systems, glaciers create landforms through the processes of erosion and deposition. Put simply, as the glacier advances it erodes the Earth's surface by picking up rocks and other debris that it will later deposit at some other place. In this section, we divide glacial landforms into two groups: 1) landforms created mainly by erosion and 2) landforms created mainly by deposition. We also discuss the actual processes of erosion and deposition in some detail.

Landforms Made by Glacial Erosion

Erosion by glaciers occurs mainly through the processes of abrasion and plucking. The glacier carries along with it particles of sand or rock fragments that are frozen to the bottom

of the moving ice. In **glacial abrasion**, these fragments scratch and gouge the underlying bedrock as the glacier moves. Abrasion results in **glacial striations**, or scratches that indicate the direction of glacial movement (Figure 17.13). When the scratches are particularly deep and well expressed, they are called **glacial grooves**.

Glacial abrasion   *An erosional process caused by the grinding action of a glacier on rock.*

Glacial striations   *Scratches in rock produced by glacial abrasion.*

Glacial grooves   *Deep furrows in rock produced by glacial abrasion.*

Figure 17.13 Glacial striations. Glacial striations such as these in Minnesota are linear scratches in bedrock produced when rocks at the bottom of a glacier grind across the bedrock.

In **glacial plucking**, a glacier rips large rocks or boulders from the ground as it moves. These boulders are subsequently carried within the ice or are frozen to the bottom of the glacier, where they act as tools for abrasion. When the glacier retreats, it drops the boulders in random locations as it melts. These random boulders on the landscape are known as **glacial erratics** because they appear to be out of place in their current location. In other words, these boulders now occur in places where that particular rock does not crop out. Often these erratics are far away from their original source area.

A great example of glacial plucking and long-distance transport can be seen in the Sioux Quartzite in the midwestern United States. This rock unit is a Precambrian quartzite that crops out in southwestern Minnesota at several places, including Pipestone National Monument. This area was overridden and smoothed during the ice ages when a massive continental glacier stretched at times from northeastern Canada to southern Illinois. (We will describe this history in more detail later in the chapter.) One such advance of the ice plucked rocks from the Pipestone region and carried them to southeastern Nebraska and northeastern Kansas, where they were dropped as glacial erratics (Figure 17.14b).

The combined processes of glacial abrasion and plucking generally scrape and round the hills. The most distinctive landform produced by these combined processes is the roche moutonnée. A **roche moutonnée** is a rounded bedrock hill that has a gradual slope on one side, called the stoss side, and a steep slope on the other, the lee side (Figure 17.15a). The stoss side faces the direction from which the glacier came and was smoothed by abrasion as the ice moved over it. The lee side, in contrast, is steeper and more irregular with respect to its surface because the ice plucked rock from this side of the hill.

Although glacial erosion occurs in both alpine and continental environments, it is more common in mountainous regions and results in very dramatic landscapes that are easy to interpret. In general, alpine glaciers modify a landscape by rounding parts of it at the same time that other aspects are sharpened. To see how a pre-glacial mountainous landscape is transformed by glaciation, refer to Figure 17.16 during the following discussion.

The basic landform created by glacial erosion in alpine regions is a cirque, which, as described earlier, is a broad amphitheater that forms on the flanks of a mountain. Cirques typically have very steep side and head walls and a floor that is flat to shallow sloping (Figure 17.17a). As noted previously, cirques are source areas for alpine glaciation and are enlarged through the combined processes of glacier plucking and mass wasting of the adjacent walls. Sometimes the ice in a cirque melts completely away, leaving water in the quarried depression. Such a small lake is called a **tarn** (Figure 17.17b).

In many cases, cirques exist on several sides of an individual mountain. Over time, these cirques may progressively enlarge and steepen their respective side and head-walls so that the ridges between them narrow and become quite steep and thin, with serrated crests (Figure 17.18). Such a thin, steep ridge is called an **arête**, the French word for "fishbone." If a mountain has three or more arêtes resulting

Glacial plucking *An erosional process by which rocks are ripped out of the ground by a glacier.*

Glacial erratics *Large boulders that have been plucked and transported a great distance before they are deposited.*

Roche moutonnée *A landform produced by glacial abrasion and plucking that has a shallow slope on one side and a steep slope on the other side.*

Tarn *A small lake that forms within a glacial cirque.*

Arête *A sharp ridge that forms between two glacial cirques.*

(a)

Figure 17.14 Plucking and erratics of Sioux Quartzite. (a) The Sioux Quartzite is Precambrian bedrock that crops out in southwestern Minnesota. Rocks from this outcrop were plucked by glacial ice about 600,000 years ago and carried as far south as north-

(b)

eastern Kansas. (b) This Sioux Quartzite erratic lies in northeastern Kansas. It is highly resistant to erosion, which is why it remains a prominent landscape feature even though it's been resting there for about 600,000 years.

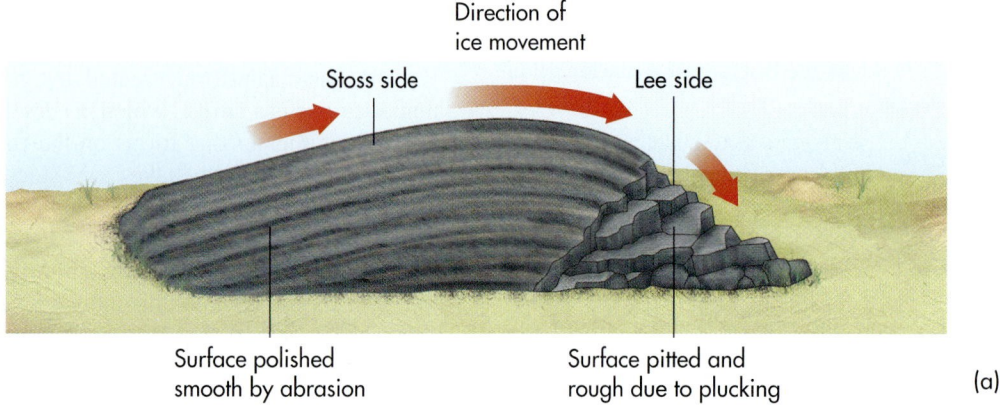

Direction of ice movement

Stoss side　　　Lee side

Surface polished smooth by abrasion

Surface pitted and rough due to plucking

(a)

(b)

Figure 17.15 Roche moutonnée. (a) A roche moutonnée is created through the combined processes of abrasion on the stoss side and plucking on the lee side. (b) A roche moutonnée formed in the Manhattan Schist in New York's Central Park. The stoss side is on the right-hand side of the image.

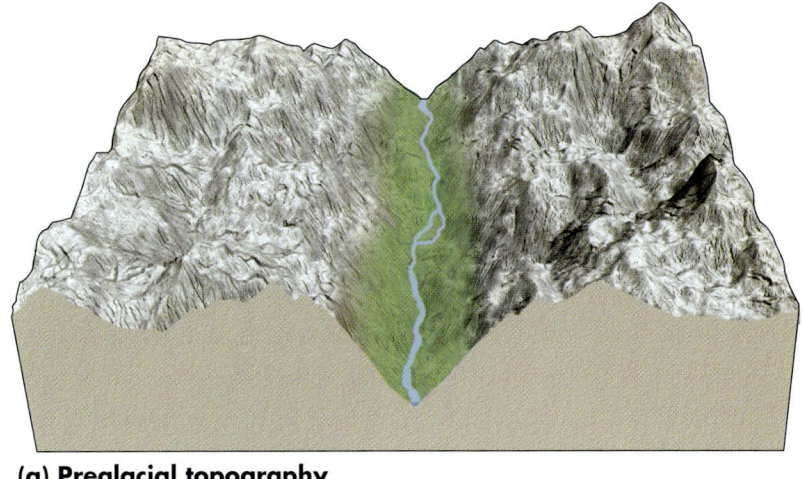

(a) Preglacial topography

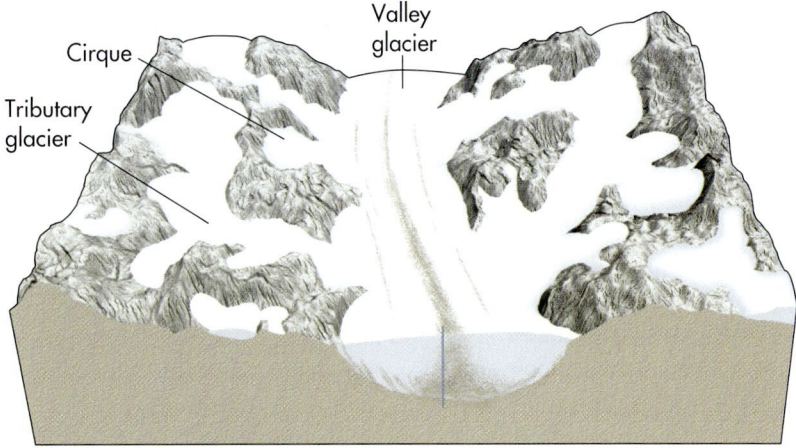

Cirque

Tributary
glacier

Valley
glacier

(b) Maximum glaciation

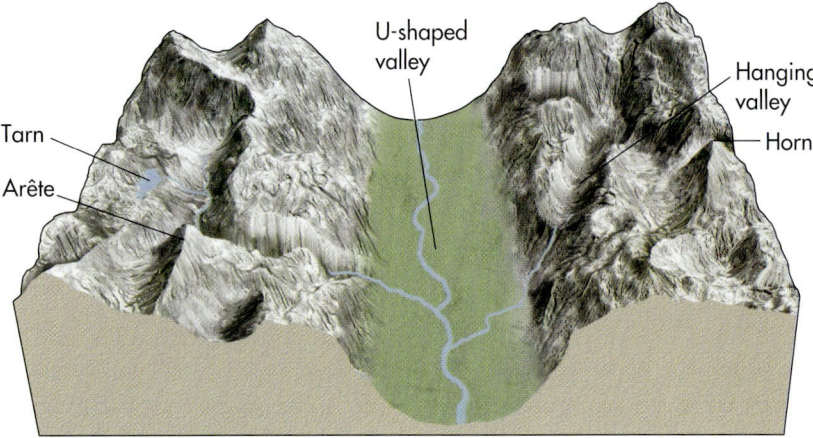

U-shaped
valley

Hanging
valley

Tarn

Horn

Arête

(c) Postglacial topography

Figure 17.16 Modification of mountainous land-scapes by glaciation. (a) The preglacial topography consists of rounded mountain slopes and a V-shaped stream valley. (b) During maximum glaciation, well-defined cirque glaciers shape mountain slopes and a thick valley glacier fills the former stream valley. The valley glacier is fed by tributary glaciers flowing into it. (c) The postglacial topography is much sharper on the mountain slopes and the former V-shaped stream valley is rounded to a U-shaped valley.

(a)

(b)

Figure 17.17 Glacial cirques and tarns. (a) Glacial cirques, such as this one in the Canadian Rockies, are bowl-shape depressions created by the scouring action of glacial ice. (b) A tarn is a meltwater lake that fills the depression within a cirque. This one is located in the North Cascades of Washington.

from intersecting cirques, then the mountain is called a **horn**. The most famous example of a horn is the Matterhorn in the Swiss Alps.

Glacial erosion occurs not only on high alpine peaks, but also within the valleys that slope away from the accumulation zones of snow. Sometimes glaciers never leave their cirque source areas, presumably because snow accumulation is not sufficient to cause glacial advance. If snow accumulation is high, however, a glacier will advance out of the cirque basin and move down the adjoining valley. When this type of advance occurs, the glacier modifies the preexisting V-shaped valley into a U-shaped valley called a **glacial trough** (Figure 17.19). These landforms are *very* distinctive and easy to identify the next time you're in the mountains. The deepest and best developed glacial troughs are formed by the trunk glacier in a particular mountain system (see Figure 17.16 again). Smaller, tributary glaciers also carve U-shaped troughs, but not to the same depth as the trunk valleys into which they flow. As a result, a distinctive **hanging valley** is produced at the intersection of the tributary and trunk valleys when the glaciers fully recede. A hanging valley is often a place where a waterfall develops, such as the famous Yosemite Falls, as the tributary stream flows toward the trunk stream.

Horn *A mountain with three or more arêtes on its flanks.*

Glacial trough *A deep, U-shaped valley carved by an alpine glacier.*

Hanging valley *An elevated U-shaped valley (with respect to a glacial trough) formed by a tributary alpine glacier.*

Figure 17.18 The Matterhorn in Switzerland. This mountain is 4478 m (14,692 ft) tall and has been dramatically shaped by alpine glaciation.

Figure 17.19 A glacial trough. This well-developed U-shaped trough in Montana was carved by a trunk valley glacier.

With an understanding of glacial processes, it's possible to visualize how a landscape has evolved through time. Given what we have discussed in this chapter so far, can you describe how glaciation has modified this landscape? What is the evidence for that modification?

KEY CONCEPTS TO REMEMBER ABOUT GLACIAL EROSION AND RESULTING LANDFORMS

1. Glaciers erode through the processes of abrasion (grinding) and plucking (ripping).

2. Glacial striations and grooves are produced through abrasion.

3. A roche moutonnée is a streamlined bedrock landform produced by the combined processes of abrasion (on the upstream side) and plucking (on the lee side).

4. Extensive glacial erosion occurs in alpine regions. In general, a preglacial landscape is converted to one that is more angular after glaciation. Resulting landforms include cirques, tarns, arêtes, horns, troughs, hanging valleys, and waterfalls.

Deposition of Glacial Drift and Resulting Landforms

Now that we have examined the landforms created by glacial erosion, let's consider those constructed through glacial deposition. As you study this section of the text, refer frequently to Figure 17.20, which shows most of the landforms discussed as well as the basic geographical relationships of these features.

Glacial Till   Sediment deposited by a glacier is called **glacial drift** and occurs in two basic forms: glacial till and glacial outwash. Let's first discuss **glacial till**, which is sediment deposited directly by a glacier. This form of glacial drift is relatively unsorted, which means that any given deposit likely contains many different particle sizes, ranging from clay to sand and even cobbles and small boulders.

Basal till (or lodgement till) is a form of glacial till deposited at the base of the glacier. An analogy for how sediments are deposited in this way is how you smear peanut butter on bread with a knife. Peanut butter sticks to the knife blade and is then smeared onto the bread with force. A glacier deposits basal till in somewhat the same way and can do so either as the glacier advances or retreats. Because of the intense pressure at the base of the glacier, due to the weight of the ice, basal till is usually quite dense and relatively fine grained because it has been ground and crushed.

Another form of glacial till is *ablation till*, which is sediment that is carried within or on top of the ice and deposited when the glacier melts. Ablation till usually contains coarser fragments such as boulders because they are not crushed at the base of the glacier. As a result, ablation till tends to be even less well sorted than basal till.

Glacial drift   *Sediment deposited indirectly or directly by a glacier.*

Glacial till   *Sediment deposited directly by a glacier.*

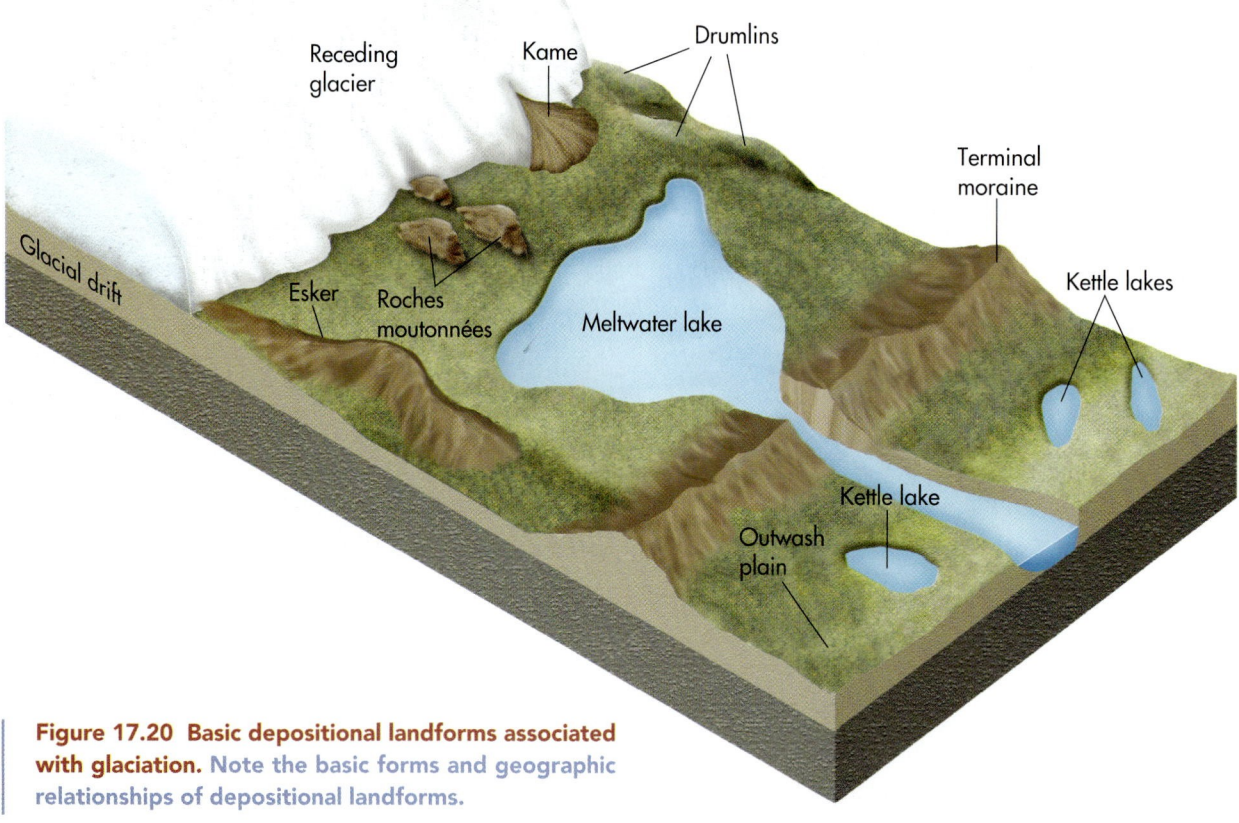

Figure 17.20 Basic depositional landforms associated with glaciation. Note the basic forms and geographic relationships of depositional landforms.

Deposition of glacial till can result in very distinctive landforms. The most distinctive type of till landform is a **moraine**, which is a winding ridge-like feature that forms at the front or side of a glacier, or between two glaciers. A glacial moraine is a very poorly sorted deposit, ranging from clays to boulders, that appears when sediments are brought forward by the conveyor action within a glacier until they are dropped out of it (Figure 17.21). In this fashion, deposition of moraine till is somewhat analogous to when a dump truck deposits dirt at a construction site. However, instead of the single pile of sediment that a dump truck produces, a moraine is a distinct ridge because it forms at the front or side of a slowly moving glacier. It is important to note that a well-developed moraine forms only when the ice front or side is stationary for a relatively long period of time. Only then can the persistent internal motion of the ice continually bring forth enough deposits for a visible ridge to grow.

Moraine   *A winding ridge-like feature that forms at the front or side of a glacier or between two glaciers.*

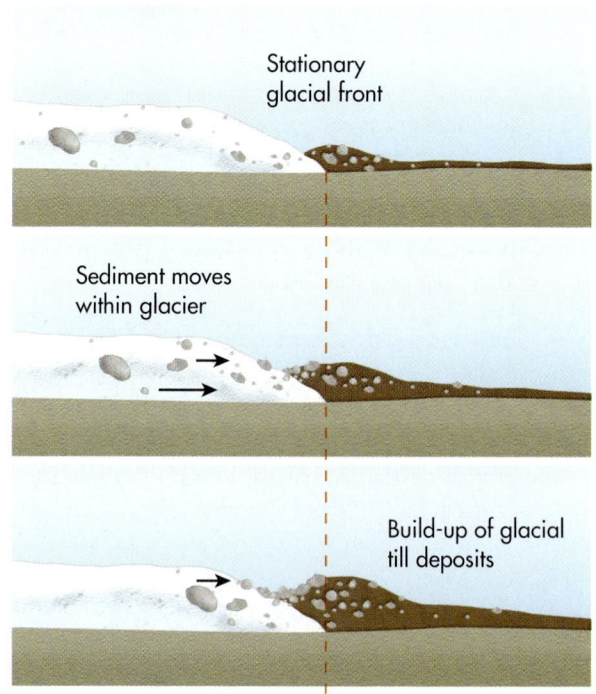

Figure 17.21 Formation of a moraine. A moraine forms when the position of the ice front is stationary for a long period of time. This allows till to be deposited in one place due to the conveyor-like action within the interior of a glacier.

Lateral moraines

Medial moraines

Terminal moraine

Recessional moraines

Terminus of glacier

(a)

(b)

Figure 17.22 Types of moraines. (a) Lateral, medial, and end moraines in an alpine glacier system. (b) The hill in the background is a portion of the Port Huron moraine in northwestern Lower Michigan. This moraine can be traced around much of northern Lower Michigan.

The specific classification of any given moraine depends upon its spatial relationship to the glacier that deposited its till. Figure 17.22 shows examples of the different kinds of moraines. A *lateral moraine* forms on the edge of a glacier and is usually best expressed along the sides of alpine glacial troughs. When a lateral moraine is sandwiched between two glaciers, it's called a *medial moraine*. A *terminal moraine* (also called an *end moraine*) forms at the front of the ice and marks the long-term stationary position at the farthest advance of a glacier. Another form of moraine that forms at the ice front is a *recessional moraine*. Although a recessional moraine may look exactly the same as a terminal moraine, it differs from a terminal moraine because it forms when the position of the ice front becomes stationary for a time during an overall retreat of the glacier. A final kind of moraine is called *ground moraine*. This kind of moraine forms when the retreat of the ice is slow but steady, allowing an irregular pattern of deposition that creates a hummocky landscape consisting of small hills and depressions (Figure 17.23).

After till is deposited, it can subsequently be modified by direct and indirect glacial processes, leaving distinctive landforms. One such landform is a **drumlin**, which is till that has been streamlined in the direction of ice flow (Figure 17.24). In this sense, a drumlin is somewhat similar to

Drumlin *A streamlined landform created when a glacier deforms previously deposited till.*

a roche moutonnée, but, in this case, the blunt end faces into the direction of ice flow. The formation of drumlins is not fully understood. Some may form because a glacier flows over previously deposited till that has uneven resis-

Figure 17.23 Ground moraine on the Coteau du Missouri in northwestern North Dakota. This rolling landscape was created because the glacier was retreating slowly at the same time it was depositing sediment in front and beneath it.

Figure 17.24 A drumlin field in Wisconsin. These streamlined hills are distinct landforms that develop beneath the ice.

tance, perhaps due to scattered boulder deposits or because some parts of the till have a frozen core. As the glacier flows over the resistant till, it may streamline less resistant sediment downstream of the core. Also, meltwater at the base of the glacier may have a role in the formation of some drumlins by streamlining sediments when periodic dramatic discharge bursts occur. Drumlins can range in size from 100 to 5000 m (300 ft to over 3 mi) and can be as much as 200 m (650 ft) high. They often occur in "swarms" and are particularly common in areas of Wisconsin, Michigan, and New York where continental glaciation was intense during the last ice age.

Glacial Outwash   We have just discussed the deposition of glacial till, which is sediment that accumulates while in direct contact with an ice mass. Now we turn to sediment that is carried and deposited by water that flows out and underneath a glacier as it melts. This kind of sediment is called **glacial outwash** and results from *glaciofluvial processes*, which are the combined effects of glaciers and streams.

Glacial outwash differs fundamentally from glacial till in that it is well sorted, which means that most of the sediments are roughly the same size, namely sand and gravel. This higher degree of sorting results from the fact that flowing water naturally segregates sediments on the basis of their size (see Chapter 16). For example, flooding stream systems deposit sand near the channel (resulting in natural

levees) and clays toward the valley edge (forming wetlands and swamps). When glaciers melt, they produce prodigious amounts of meltwater that flows underneath and out in front of the ice. These flowing bodies behave according to basic hydraulic principles and therefore sort sediments in the same way as "normal" rivers do. These sediments are also stratified, which means that distinctive layers of sand, silt, or gravel can be traced horizontally for some distance.

Deposition of glacial outwash creates some distinctive landforms. The most common and extensive glaciofluvial landform is an outwash plain. As the name implies, an **outwash plain** is a relatively flat landscape (Figure 17.25) that was created through the deposition of sediments carried by meltwater flowing out in front of a glacier. The associated meltwater streams are usually choked with coarse sediment and are therefore braided. Recall from the previous chapter that braided streams are associated with aggradation, which, in the case of outwash plains, can result in very thick deposits of water-laid sediment in front of the ice.

Another type of landform associated with glacial meltwater is a **kame,** which is a large mound of sediment deposited along the front of a slowly melting or stationary glacier (see Figure 17.20). A kame can form either as an alluvial fan at the ice margin or as a deltaic deposit if the ice borders a lake. Kames usually contain stratified deposits of sand and gravel that progressively bury older deposits of a similar kind. Kame landscapes typically consist of rolling and irregular hills that are quite large, so large, in fact, that many ski resorts in the upper Midwest are built on these landforms.

Figure 17.25 Outwash plain at Jackson Hole, Wyoming. Note the broad, horizontal surface that dominates the center of this image. This landform evolved because meltwater streams deposited the sediment in front of a valley glacier.

Glacial outwash   *Sediment deposited by meltwater streams emanating from a glacier.*

Outwash plain   *A broad landscape of limited relief created by the deposition of glacial outwash.*

Kame   *A large mound of sediment deposited along the front of a slowly melting or stationary glacier.*

A landform that is somewhat similar to a glacial kame in its development is an esker. An **esker** is a winding ridge (Figure 17.26) of coarse sediment deposited by a stream flowing *under* the ice. In this instance, the cross section of a sub-glacial stream channel is inverted with respect to a normal river channel; that is, the channel looks like an up-side-down U in cross section. The flowing water is confined by the ice at the top and sides of the tunnel and by the ground beneath it, and so is the deposition of sediment. When the ice eventually melts, a meandering ridge of well-sorted and stratified glacial sediment is left behind. Most eskers are discontinuous because the conditions required for sub-glacial fluvial deposition are difficult to maintain over long distances. Nevertheless, eskers can be over 30 m (100 ft) high and tens of kilometers long.

Still yet another kind of landform associated with glacial outwash is called a **kettle lake.** A kettle lake forms when a large block of ice falls off the front of a glacier and is subsequently buried by glacial outwash as the ice retreats. At some point, long after the glacier is gone, the buried block of ice melts. In the process of melting, the block creates its own lake, first by forming the depression that results from the subsiding sediments, and second through providing the water to fill the depression. This lake can subsequently be maintained at its original size if the

Figure 17.27 A kettle lake. Kettle lakes form when ice blocks calve off of the front of a glacier, are subsequently buried by glacial sediment, and then melt, forming a depression and lake. This air photo shows two beautiful kettle lakes in Wisconsin.

depression intersects the water table. Kettle lakes are very common in heavily glaciated terrain and are randomly scattered on the landscape, as shown in Figure 17.27. If you happen to live in or near Minnesota, you know that the state is known as the *Land of 10,000 Lakes.* The vast majority of these lakes are kettle lakes. The next time you camp at a small lake in this area or any place in the northern tier of the United States, there's a good chance that the lake evolved because a former ice block was buried and then melted. Try to imagine the process the next time you're there.

Figure 17.26 An esker. Eskers form when meltwater streams flow in winding tunnels beneath a glacier, depositing sediment that forms a ridge when the ice melts. This esker forms a prominent ridge in Blue Lake, Minnesota.

Esker *A winding ridge formed by a stream that flows beneath a glacier.*

Kettle lake *A lake that forms when a block of ice falls off of the glacial front, is buried by glacial drift, and then melts, forming a depression that fills with water.*

1. Two primary kinds of glacial deposits occur: till and outwash. Glacial till is deposited in direct contact with the ice, whereas glacial outwash accumulates through meltwater streams flowing in front of the ice.

2. Glacial till is relatively unsorted and accumulates either as basal (lodgement) till smeared under the bottom of the glacier or as ablation till laid down as the ice melts.

3. A distinctive depositional landform created by glaciers is a moraine (end, lateral, or medial), which is a ridge of till that forms when the ice front or margin is in one place for a relatively long period of time. Drumlins are streamlined till landforms created by the weight and pressure of the overlying flowing ice.

Glaciers construct landforms in many different ways, both in direct contact with the ice and in close proximity to it. Given your understanding of landscape evolution and glaciers, what kind of landform is being constructed in this image? How can you tell? What will the relief of this landform be like when it's fully developed?

4. Glacial outwash is relatively well sorted because it is deposited by flowing meltwater in front of the ice. The associated streams are typically braided and create a broad, flat surface known as an outwash plain. Sometimes glacial outwash buries a block of ice that broke off the front of a receding glacier. When this ice block subsequently melts, it forms a water-filled depression called a kettle lake.

5. Kames and eskers are distinctive meltwater landforms. A kame is an irregularly shaped hill that basically consists of an alluvial fan or deltaic deposit that forms in contact with the ice. Eskers are winding hills that develop when sediment accumulates in a stream that flows in a tunnel at the base of the ice.

History of Glaciation

The many examples we have seen of landforms caused by glaciers shows that in the past, ice sheets were an important and widespread feature around the world. The presence of drumlins in places like New York and Michigan, for example, indicates that ice sheets existed where they no longer do. You might wonder if it's possible for such conditions to return at some point in the future. To answer this question, it's logical to examine the history of glaciation on Earth because it provides a context within which to view present glacial landforms and the likelihood that future glaciation may occur.

Before we begin, it's important to establish some general terminology. Periods of glacial advance, when glaciation is a dominant world-wide process, are called *glacials*. During glacials, temperatures are cooler on average, ice sheets expand and become larger, and sea level is lower because a great deal of water is stored in the ice sheets.

Periods of glacial retreat, such as we are in now and have been for about the past 10,000 years, are called *interglacials*. During interglacials, temperatures are warmer on average, ice sheets are smaller, and sea level is higher because meltwater from the glaciers has returned to the sea.

Although the evidence is extremely vague, the first major glacial period on Earth appears to have occurred during the Precambrian period about 2.3 billion years ago. This glaciation was in response to a major fluctuation in the composition of the early atmosphere. Prior to this period, the atmosphere was composed mostly of methane, a major greenhouse gas. Sometime about 2.3 billion years ago the amount of oxygen in the atmosphere increased dramatically, which in turn cooled the Earth so that much of the planet was covered by ice. In fact, so much of the Earth was glaciated that this period is often referred to as the *snowball Earth*. During this glacial interval, many of the world's oceans were frozen to a depth of about 0.8 km (0.5 mi). Similar "snowball" glacials occurred later in the Precambrian period, at about 750 million and 600 million years ago, and are most closely associated with decreases in carbon dioxide, another greenhouse gas.

Although much of the Earth was heavily glaciated during these so-called snowball glaciations, these periods are neither the most recent nor the best-known glacial intervals. Instead, the most significant and best-understood glacial periods are those associated with the great ice ages of the Pleistocene epoch in the recent geological past. During this epoch of time, alpine and continental glaciation occurred on such a large scale that 30% of the Earth was covered by glacial ice at its maximum extent (Figure 17.28). For perspective, remember that contemporary glaciers now extend over only 11% of the Earth.

To put the ice age in the proper chronological context, we must turn back to the concept of geologic time for a moment. Recall that the Cenozoic era is the period of time that encompasses the past 65 million years.

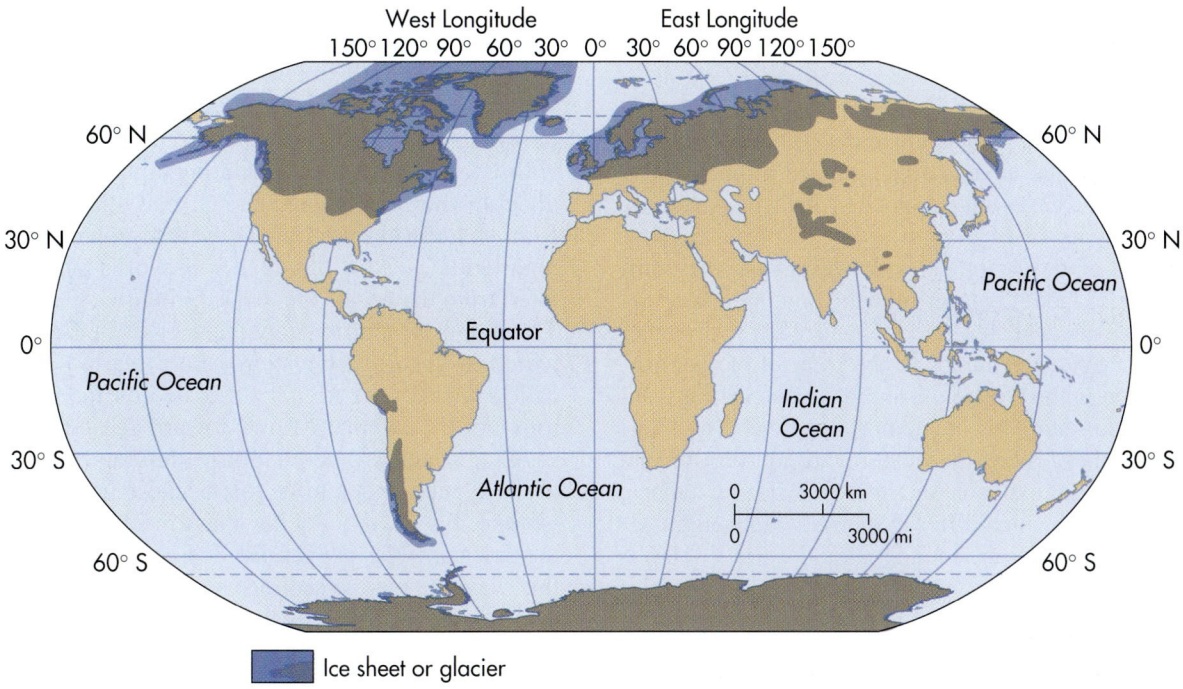

Figure 17.28 Maximum extent of glaciation during the Pleistocene Epoch. During this period of time, glaciers covered 30% of the Earth, compared to 11% today.

Within this time frame are two major periods: the Tertiary, from 65 million to 1.8 million years ago, and the Quaternary, from 1.8 million years ago to the present time. The Quaternary period is further subdivided into two epochs: the Pleistocene and Holocene. As we just noted, the Pleistocene epoch is considered to be the time of the ice ages and lasted from 1.8 million years ago to about 10,000 years ago. We happen to be living in the Holocene epoch, which includes the past 10,000 years.

When geologists and geomorphologists talk about the ice age, they are basically referring to the Pleistocene epoch. This is not to say that glaciers were consistently advancing during the entirety of this epoch; they were not. In fact, several different glacials and interglacials occurred during the Pleistocene, each lasting tens of thousands of years. Although geologists typically think of the ice age as ending at the end of the Pleistocene, about 10,000 years ago, this designation is arbitrary in a glacial sense because our current epoch, the Holocene, may be nothing more than an interglacial period that is generally consistent in length and character with previous interglacials.

How Many Pleistocene Glaciations Were There?

To give you a feel for the magnitude of Pleistocene climate rhythms and ice sheet response, consider the number of glacial periods that occurred. Until recently, the chronology of Pleistocene glaciation was reconstructed solely through geological studies in which stratified tills in Europe and North America were investigated. According to these early studies, four major advances of glacier ice occurred on Earth during the Pleistocene. In North America, these "classic" advances were generally named after the locations in which the ice advanced to its farthest point and where prominent evidence was found in the form of distinctive glacial drift. These glacial periods were given the following names (and approximate ages):

1. Nebraskan—1 million years ago

2. Kansan—625,000 years ago

3. Illinoisan—300,000 years ago

4. Wisconsin—35,000 to 10,000 years ago

In each of these advances, a thick continental glacier extended from the Arctic to a point deep within the American midwest. East of the Rocky Mountains, this glacier is typically referred to as the **Laurentide Ice Sheet** because it covered the Laurentian Mountains in eastern Canada. Thick ice caps also covered much of the Rockies and extended west to the Pacific coast and south into Washington. Collectively, these ice caps are referred

Laurentide Ice Sheet *The continental glacier that covered eastern Canada and parts of the northeastern United States during the Pleistocene Epoch.*

to as the **Cordilleran Ice Sheet,** which, at its maximum extent, merged with the Laurentide Ice Sheet in central Canada. According to classical theory, the Laurentide Ice Sheet reached its southernmost point during the so-called Kansan glaciation about 600,000 years ago. This is the glacial advance that stopped at present-day northeastern Kansas and deposited the glacial erratic shown in Figure 17.14.

Recent evidence indicates, however, that the traditional glacial model is too simplistic and that major periods of ice advance and retreat occurred at many other times. The evidence comes in the form of oxygen isotopes stored in the Greenland and Antarctic ice caps and from marine sediments in equatorial regions. The foundation of these new theories regarding Pleistocene glaciation is that oxygen contains two primary isotopes, O-16 and O-18, which differ in their atomic weight. Although both isotopes are found in liquid water, O-16 water is almost 500 times more common and generally more easily evaporated than its heavier counterpart, O-18 water. This variation depends somewhat on the temperature of the water at the time it evaporates. When water is warmer, for example, proportionately more O-18 water is evaporated than when the water temperatures are colder. Thus, it is possible to use the ratio of O-16 to O-18 in annual layers of ice in ice caps and yearly accumulations of marine sediments to indirectly reconstruct climate cycles in the Pleistocene.

To see how this method of climate reconstruction works, examine Figure 17.29, which illustrates the way in which O-18 and O-16 move relative to each other within the hydrologic cycle during a glacial period. The diagram shows a hypothetical air mass that flows away from its tropical source area toward a growing glacier at high latitudes. Note that relatively more O-16 is evaporated from the ocean than 0-18 during this glacial cycle. This leaves proportionately more O-18 in the seawater, which is absorbed by microscopic marine organisms called *benthic foraminifera* that live on the ocean bottom. As the air mass flows toward the center of the growing glacier, precipitation occurs and O-18 is the first oxygen isotope to be lost because it is the heavier isotope. By the time the air mass reaches the center of the ice mass, the oxygen isotopes are mostly O-16. In this way, the marine sediments (which contain the benthic foraminifera) become relatively enriched in O-18 (because O-16 is preferentially evaporated) at the same time that thickening glacial ice contains proportionately more of the O-16 isotope (because most O-18 was left behind or precipitated earlier).

So how do scientists reconstruct this record of climate change? It's very simple; they take cores of the ma-

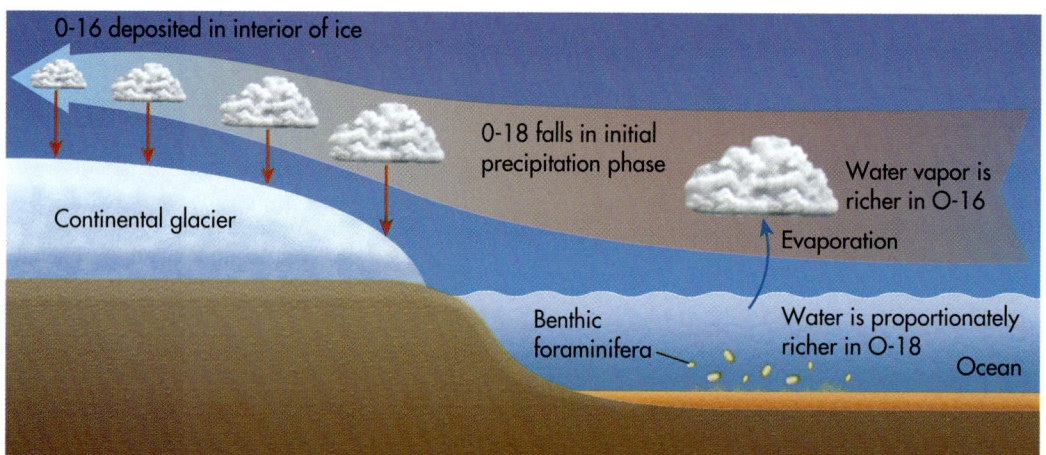

Figure 17.29 Using oxygen isotope ratios to reconstruct glacial and interglacial cycles. Seawater contains both O-16 and O-18 isotopes, which are preferentially evaporated (O-16) and precipitated (O-18) relative to one another during glacial cycles. Thus, microscopic marine organisms (benthic foraminifera) absorb relatively more of the heavier O-18 isotope during glacial cycles. At the same time, moisture-laden air flows toward growing ice sheets, where O-18 is precipitated first because it is heavier than O-16. By the time the air mass reaches the center of the ice sheet, it is enriched in O-16 relative to O-18.

Cordilleran Ice Sheet   *The ice cap that covered much of the mountains in the northwestern part of North America during the Pleistocene Epoch.*

(a)

(b)

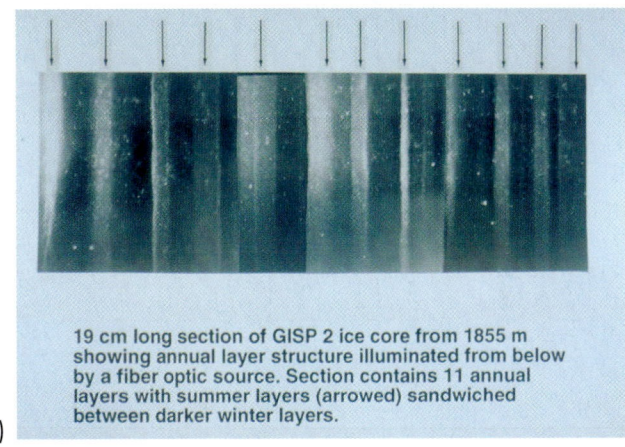

19 cm long section of GISP 2 ice core from 1855 m showing annual layer structure illuminated from below by a fiber optic source. Section contains 11 annual layers with summer layers (arrowed) sandwiched between darker winter layers.

(c)

Figure 17.30 Ice cores. (a) A coring station on the Greenland Ice Sheet. The long tube extending above the tent is used to extract a core (cylinder of ice) from the glacier. (b) This typical ice core was obtained with the tube shown in part (a), which penetrated numerous annual layers of ice. (c) An example of annual ice layers, showing distinct bands within the ice. Each of these bands represents an annual accumulation of snow that transformed to glacial ice. Each annual layer contains a sample of the atmosphere at the time the snow fell, including the relative amount of O-16 and O-18.

rine sediments and of large ice masses. Both the marine sediments and glacial ice masses gradually thicken with time and the annual records of this growth are preserved. To see how this record of annual growth is preserved in a glacier, look again at the annual layers of ice shown in Figure 17.2. Each of these annual layers contains a sample of the prehistoric atmosphere at the time the snow accumulated, that is, the ratio of O-16 to O-18 each year that snow fell.

Although the coring process is essentially the same in both ocean and ice settings, it's much easier to visualize on a glacier (Figure 17.30). Members of a coring expedition pick a place where the ice is expected to contain the best annual sequence of isotopic data. Once they arrive at that location they set up a station (Figure 17.30a) where a large tube can be drilled vertically into the glacier, through a large number of annual ice layers, to a great depth. With this core tube, they can extract a cylinder of ice (Figure 17.30b) that contains a sample of all the annual snowfalls (and thus the annual ratio of

O-16/O-18) they can penetrate. Once the core is extracted, scientists merely have to count the annual layers (Figure 17.30c) to record the length of time represented in the cylinder of ice and then determine changes that occur in the ratio of O-16 and O-18 over time. The process is basically the same in an ocean setting, with the primary difference being that a large boat is obviously required in this environment and the core penetrates sediment rather than ice.

In certain cores, the annual layers of oxygen-containing sediment or ice extend in time throughout the entirety of the Pleistocene, recording a sequence of distinct **oxygen isotope stages** that reflect specific interglacial or glacial time intervals. Overall, the oxygen isotope record

Oxygen isotope stages   *Periods of time that have distinct O-18/O-16 ratios, which are used to reconstruct prehistoric climate change.*

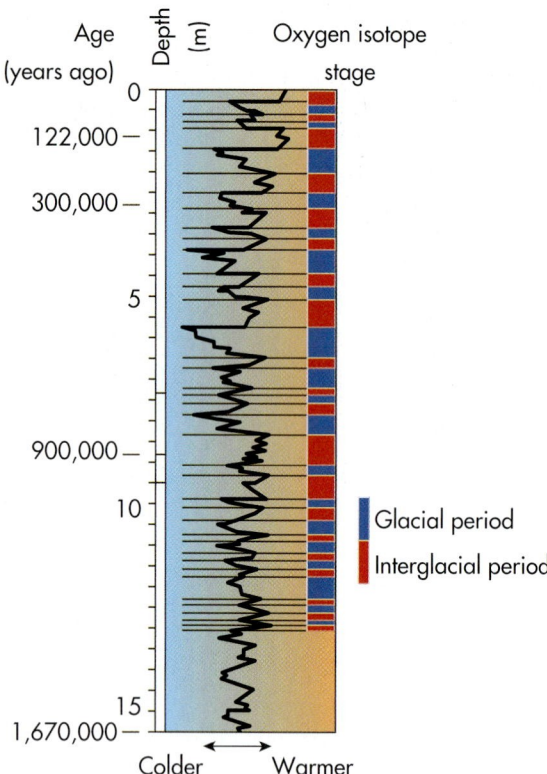

Age (years ago) — Depth (m) — Oxygen isotope stage

| | |
| Glacial period |
| Interglacial period |

Colder ⟷ Warmer

Figure 17.31 Oxygen isotope stages during the Pleistocene Epoch. Each oxygen isotope stage reflects a glacial or interglacial period, as indicated through oxygen isotope ratios in marine sediments. This record indicates that glacial periods have been quite common during the Pleistocene Epoch. (Source: Adapted from Shackelton, N.J. and Opdyke, N. D.,1976. *Oxygen-Isotope and Paleomagnetic Stratigraphy of Pacific Core v28–239, Late Pliocene to Latest Pleistocene.* Geological Society of America Memoir 145).

demonstrates that there were perhaps 18 glacial periods during the Pleistocene (Figure 17.31). In this revised context, the only names maintained from the traditional glacial model are the Illinoisan and Wisconsin advances because the deposits associated with these most recent glaciations are easily recognizable. All other glaciations are lumped into one category called *Pre-Illinoisan*.

The Wisconsin Glaciation and the Evolution of the Great Lakes

Given that the Wisconsin glacial advance is the most recent, its effects are the most studied of all Pleistocene glaciations. A good place to examine the Wisconsin glaciation and its impact on the landscape is the Great Lakes region. Following the post-Illinoisan interglacial period, the Laurentide ice sheet began to expand again about 110,000 years ago. For the next approximately 90,000 years ice generally advanced south, reaching its most southerly position in central Ohio, Indiana, and Illinois by about 20,000 years ago. At this time, *all* of what is now known as the Great Lakes and the state of Michigan were buried beneath a sheet of ice (Figure 17.32a) that was probably hundreds of meters (feet) thick. This ice subsequently melted from the Great Lakes region between about 16,000 and 10,000 years ago.

Although the most recent deglaciation of the Great Lakes region left an abundance of glacial landforms, including moraines (such as Figure 17.22b), kames, eskers, outwash plains, and drumlins, the most prominent landscape feature is the Great Lakes themselves (Figure 17.32b). The evolution of these lakes is fascinating and is a testament to the power of glacial ice and how rocks of different resistance can influence the shape of a landscape. Prior to Pleistocene glaciation, the areas now occupied by the Great Lakes were large river valleys that were cut into soft deposits of shale. Very resistant granites, limestones, and sandstones underlie the areas that are now regional landmasses.

As the Laurentide Ice Sheet oscillated back and forth across the region during the Pleistocene, the glacier preferentially eroded the softer rocks. In this way, the former river valleys gradually evolved into enormous glacial troughs. Because these troughs progressively enlarged and deepened during subsequent glaciations, they became the preferential pathways of the ice front whenever it advanced back into the region. More resistant rocks were much less eroded and became the foundation for landforms such as the Upper and Lower Peninsulas of Michigan, the Bruce Peninsula in Ontario, the Keweenaw Peninsula in Upper Michigan, and a variety of islands.

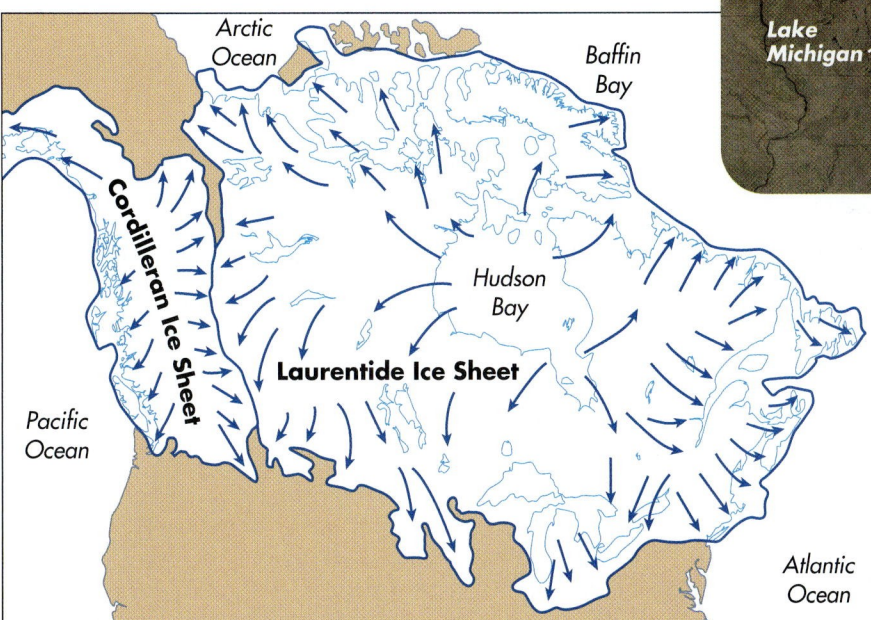

(a)

(b)

Figure 17.32 Extent and impact of the Wisconsin glaciation. (a) During the most recent ice age, the Laurentide Ice Sheet and Cordilleran Ice Sheet collectively covered most of northern North America. (b) Satellite image of the Great Lakes acquired in April, 2005. These lakes were formed when the advancing ice further enlarged and deepened pre-existing glacial troughs that had been scoured out of relatively soft rocks. Note the extent of snow cover, frozen lakes, and the suspended sediment in Lake Erie.

KEY CONCEPTS TO REMEMBER ABOUT THE HISTORY OF GLACIATION ON EARTH

1. Several periods of intense glaciation occurred during the Precambrian period in which most of the Earth was covered by ice. Such a period is referred to as a snowball Earth.

2. In the context of geologic time, the most recent period of major continental glaciations is the Pleistocene epoch, which is a subset of the Quaternary period that lasted from approximately 1.6 million to 10,000 years ago.

3. Until recently, it was believed that four major glaciations occurred during the Pleistocene epoch. Analysis of oxygen isotopes (O-16 and O-18) contained within ice cores and marine sediments indicates that potentially 18 major glacial periods occurred during the Pleistocene. The best understood of these glaciations are the Illinoisan and Wisconsin periods, which occurred most recently. All other glacial periods are lumped into one category called the Pre-Illinoisan.

4. Most glacial landforms in the upper Midwest, northeastern United States, and Canada are products of the Wisconsin glaciation, which occurred between about 110,000 and 10,000 years ago. The most distinctive large-scale landforms are the Great Lakes, which are glacial troughs cut into soft bedrock.

The horizontal lines visible in this hill side in Missoula, Montana, are old shorelines created by wave action of glacial Lake Missoula during the most recent ice age. To imagine the depth of the water, note the school building in the foreground for scale.

THE CHANNELED SCABLANDS IN EASTERN WASHINGTON

The Channeled Scablands in eastern Washington is known for the numerous and randomly spaced dry canyons that occur throughout the region. In the early 20th century, the geologist J. Harlan Bretz hypothesized that these landforms were created by catastrophic floods caused by melting glaciers. He was ridiculed for most of his life about this idea because most geologists thought that Earth-changing processes due to glaciers had to be slow and gradual. However, by the time Bretz died in 1981, his hypothesis had been accepted. In fact, he received geology's highest honor, the Penrose Medal, in 1979.

Geologists now believe that a lobe of the Cordilleran Ice Sheet—the Purcell Trench Lobe—advanced far enough into Montana that it blocked a valley through which glacial meltwater flowed west. This blockage created an ice dam that allowed a deep lake to form behind it, spreading back into the mountain valleys to the east. This lake is known to geomorphologists as Glacial Lake Missoula because it formed shorelines that are still visible around Missoula, Montana. These shorelines indicate that the lake was as big as today's Lake Ontario and Lake Erie combined. As the lake deepened, it eventually began to flow over the top of the ice dam, finally causing the dam to break catastrophically.

Torrential floodwaters, up to 300 m (1000 ft) deep, subsequently burst onto the plains of what is now eastern Washington, flowing at a rate of ten times the total of all the rivers

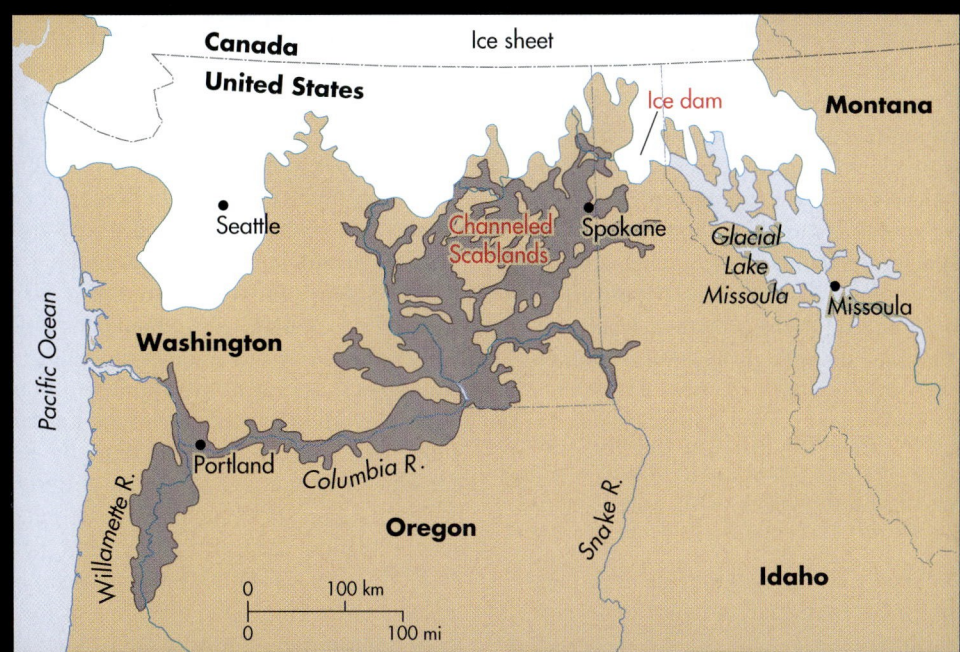

The Channeled Scablands cover a large area that includes most of southeastern Washington.

on Earth and emptying Lake Missoula in perhaps as little as 48 hours. This type of catastrophic discharge may have happened up to 100 times and created the enormous valleys carved into the loess and basalt that comprise this landscape. The network of now dry channels is so well defined that it can be seen on satellite imagery.

The Channeled Scablands show up beautifully as light brown channel forms (arrows) in this satellite image that focuses on eastern Washington, northeast Oregon, and western Idaho.

What Causes Glaciation?

Given that we know numerous glaciations took place in the past 1.6 million years, the next logical question is: what causes this remarkable process? The obvious place to look is climate change, specifically fluctuations in average temperatures in the glacial source areas at high latitudes. Remember, the key factor in glacial periods is not necessarily how cold it gets in the winter, but rather how cool it stays in the summer at the glacial source area—the place where the glacier begins to grow. Even seemingly slight decreases in average temperatures occurring for long periods of time can lead to the growth of glaciers.

Climate fluctuations are related to several possible factors:

1. Variations in solar radiation, possibly caused by periodic increases in dust and gases in the space through which the Earth passes or by variations in sunspot cycles.

2. Periods of reduced carbon dioxide in the atmosphere, allowing more longwave radiation to escape, cooling the Earth.

3. Increased volcanic activity, leading to increased levels of volcanic ash and dust in the atmosphere that decrease the amount of solar radiation reaching the Earth's surface.

4. Changes in the Earth's orbit around the Sun, the tilt of the Earth's axis, and the rotation of the Earth, each occurring over long periods of time and affecting the amount of solar radiation received at the Earth's surface.

Glacial Periods and Earth-Sun Geometry

Although adjustments in the first three climate-forming factors may be linked in some way to specific glacial periods, scientists now believe that the dominant variable is Earth/Sun geometric change. Astrophysicist Milutin Milankovich first proposed this theory in the 1920s and argued that fluctuations in the Earth's tilt, rotation, and orbit best explained cycles of continental glaciation. This theory was largely ignored until the middle 1970s when, as discussed earlier, isotopic analyses of marine cores revealed that there were many more glacial periods than previously believed. Most important, scientists discovered a close correlation between the orbital fluctuations proposed by Milankovitch and the timing of glacial and interglacial periods. Thus, the **Milankovitch theory** is now widely accepted as best explaining the Pleistocene glacial history. Let's take a closer look at this important theory by discussing its three components.

1) *Orbital Eccentricity*. The *eccentricity* of the Earth's orbit refers to the variation in the shape of the orbit over time (Figure 17.33). Recall from Chapter 3 that the Earth's orbit is not perfectly circular in its current configuration, but is instead slightly elliptical. The elliptical

nature of the orbit results in the Earth being about 5 million km (about 3.1 million mi) closer to the Sun at perihelion on or about January 3 than it is during aphelion on or about July 4. This difference amounts to about a 6% increase in incoming solar radiation from July to January. Over the course of cycles that last approximately 90,000 to 100,000 years, however, the orbit becomes even more elliptical. When the orbit is more elliptical, the amount of insolation received at perihelion is about 20% to 30% greater than at aphelion. This kind of variability would result in a substantially different climate from what we experience today, with cooler summers in the high latitudes of the northern hemisphere than currently occur.

2) *Tilt Obliquity*. The term *obliquity* refers to the variations that occur with respect to the axial tilt of the Earth. As you know, the axis of the Earth is tilted 23.5 degrees from its orbital plane around the Sun. This tilt fundamentally explains the seasonal differences that we experience. During a cycle that averages about 40,000 years, the tilt of the Earth's axis varies between 22.1° and 24.5° (Figure 17.34). As tilt increases, the seasonal contrast becomes greater, so that winters are colder and summers are warmer. A tilt of 24°, for example, results in about 8% more solar radiation received at high latitudes.

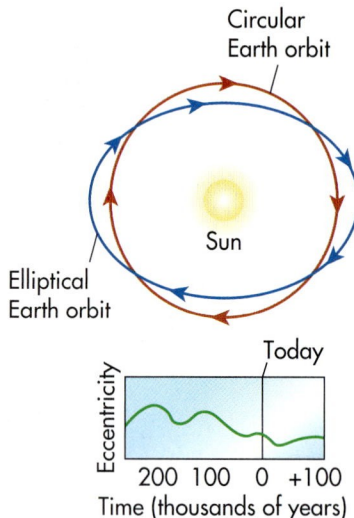

Figure 17.33 Orbital eccentricity. When the Earth's orbit is highly elliptical, we receive more radiation at perihelion than at aphelion. This orbital variation occurs on cycles that last about 100,000 years and causes the Earth to be about 11.35 million miles closer to the Sun during orbital perihelion than at aphelion.

Milankovitch theory *The theory that best explains Pleistocene glacial/interglacial cycles through long-term variations in the Earth's orbital eccentricity, tilt, and axial precession.*

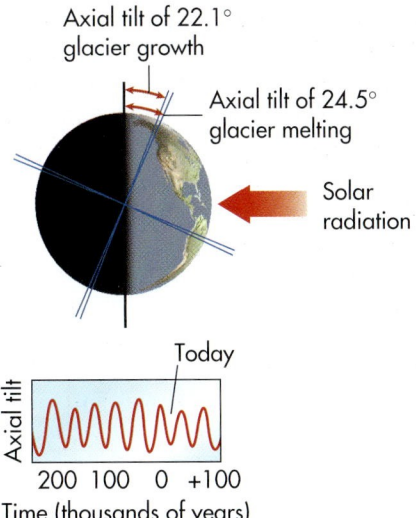

Axial tilt of 22.1°
glacier growth

Axial tilt of 24.5°
glacier melting

Solar radiation

Today

Axial tilt

200 100 0 +100
Time (thousands of years)

Figure 17.34 Tilt obliquity. The amount of Earth's axial tilt varies over 40,000-year cycles, ranging from 22.1° to 24.5°. When tilt is less, glacial source areas experience cooler summers, allowing glaciers to grow. Seasonality is greatest when axial tilt is high.

In the context of continental glaciation, more tilt means warmer summers, which makes it less likely that glaciers can become established. On the other hand, conditions are more favorable for glacier growth when tilt is less because summer temperatures are cooler at high latitudes.

3) *Orbital Precession.* The *precession* of the Earth's orbit refers to the Earth's wobbles on its axis as it orbits the Sun. The Earth's axis is not fixed in space. Over the course of an approximately 23,000-year cycle, the axis slowly migrates, making it appear as if the planet is wobbling as when a spinning top slows down (Figure 17.35). This change alters the orientation of the Earth with respect to perihelion and aphelion. If a particular hemisphere is pointed toward the Sun at perihelion, that hemisphere will be pointing away at aphelion, and the difference in seasons will be more extreme. This seasonal effect is reversed for the opposite hemisphere. Currently, northern summer occurs near aphelion.

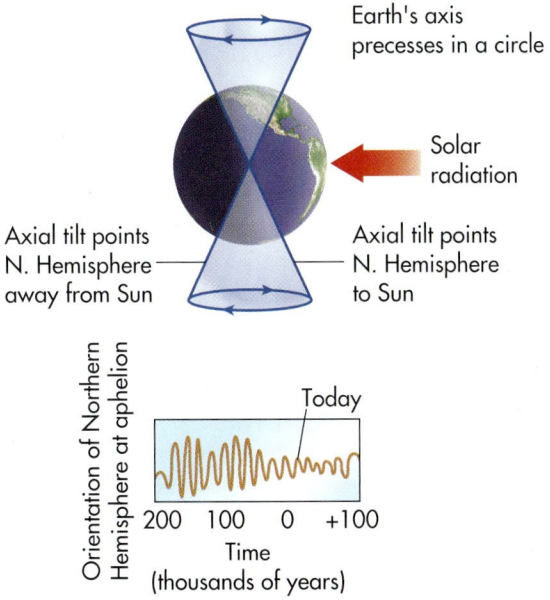

Earth's axis precesses in a circle

Solar radiation

Axial tilt points N. Hemisphere away from Sun

Axial tilt points N. Hemisphere to Sun

Orientation of Northern Hemisphere at aphelion

Today

200 100 0 +100
Time (thousands of years)

Figure 17.35 Orbital precession. Over the course of a 23,000-year cycle, the Earth's axis slowly wobbles. The impact of this wobble is that a gradual change occurs with respect to the hemisphere that is tilted toward or away from the Sun at aphelion and perihelion.

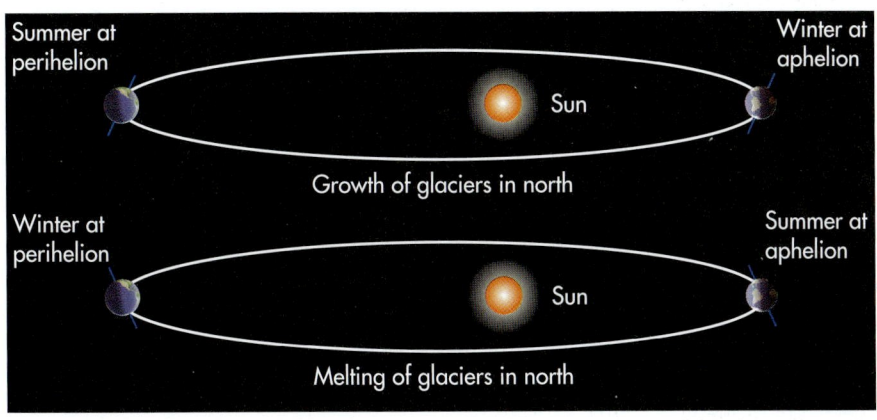

Summer at perihelion

Winter at aphelion

Sun

Growth of glaciers in north

Winter at perihelion

Summer at aphelion

Sun

Melting of glaciers in north

THE MILANKOVITCH THEORY

To see how the Milankovitch theory looks in animation, go to the *GeoDiscoveries* website and open the module **The Milankovitch Theory**. This module shows the way that the Earth-Sun geometric relationship changes through time and how it contributes to glaciation. As you watch the animation, notice how variations in orbital eccentricity, axial tilt, and orbital preces-

sion result in cooler summers at high latitudes where continental glaciers begin to grow. An excellent animation of orbital precession can also be seen at http://earthobservatory.nasa.gov/Library/Giants/Milankovitch/milankovitch_2a_high.html. After you complete the *GeoDiscoveries* animation, be sure to answer the questions at the end of the module to test your understanding of this concept.

As discussed previously, isotopic data from marine cores indicate a close correlation between orbital variations and periods of glaciation. Expansion of glaciers seems to occur on 23,000-, 46,000-, and 100,000-year cycles, with one, two, or all of the orbital variables playing a major role within any given glaciation. The 23,000- and 46,000-year cycles are minor fluctuations and are apparently most related to eccentricity of orbit and precession of the equinoxes, in other words, the time of year when a particular hemisphere is tilted toward or away from the Sun. Major glaciation seems to occur every 100,000 years, and although this 100,000-year cycle mirrors the length of the orbital eccentricity cycle, there appears to be no direct relationship between the two. Instead, the 100,000-year rhythm seems to be most closely related to fluctuations in axial tilt and to the difference between the long time it takes a continental glacier to grow and the relatively short time it takes for the glacier to melt. When ice began to expand about 110,000 years ago, for example, axial tilt was 22°. Subsequently, it took 80,000 or 90,000 years for the Laurentide Ice Sheet to reach its maximum size during the Wisconsin glaciation. About 15,000 years ago, tilt had increased to 24° and the high latitudes received much more summertime radiation. In the ensuing 10,000 years, the great mass of ice melted almost entirely.

important factor appears to be Earth-Sun geometric fluctuations.

3. Earth-Sun geometric fluctuations include orbital eccentricity, axial tilt variability, and changes in orbital precession. These variables appear to change over the course of 100,000-, 40,000, and 23,000-year cycles, respectively.

4. Although some expansion of continental glacier ice appears to occur in 23,000- and 46,000-year cycles, major glaciation occurs every 100,000 years. The primary variable causing major glaciation appears to be variability in axial tilt.

Impact of Global Climate Change on Glaciers

Recall from Chapter 9 that global warming is a major environmental issue facing the world. Although this period of warming may be entirely part of a natural cycle, most climatologists believe that it is associated with increased levels of atmospheric carbon dioxide (and other greenhouse gases) caused by human industrial activity. Regardless of the cause, the Earth appears to be in a distinct warming phase that has been linked to the increased incidence of drought, species migration and loss, and sea level rise. Global warming is also having a significant impact on glaciers around the world.

As previously discussed in this chapter, glacial advance and retreat is controlled by fluctuations in the glacial mass budget (see Figure 17.4). When snowfall in the source area exceeds melting in the zone of ablation, glaciers thicken and expand, either down valleys in alpine settings or across landmasses in continental glaciers. Glaciers naturally recede during periods when the rate of melting in the zone of ablation exceeds the accumulation of snow in the source area. During the Pleistocene Epoch, the budget of global ice

> ### KEY CONCEPTS TO REMEMBER ABOUT THE CAUSES OF GLACIATION
>
> 1. The most important variable with respect to glaciation is cool summers, when the previous winter's snow does not melt.
>
> 2. Although variables such as sunspot cycles, carbon dioxide levels, and volcanic dust may play a role as causes of glaciation, the most

(a)

(b)

(c)

Figure 17.36 The melting snows of Kilimanjaro. (a) Although Mt. Kilimanjaro lies near the Equator, it has long had a substantial ice cap due to its high elevation. (b) Landsat image of the ice cover on Mt. Kilimanjaro in February, 1993. (c) Landsat image of Mt. Kilimanjaro in February, 2000. Notice the extent of melting that occurred over only seven years.

masses was controlled largely by fluctuations in Earth-Sun geometry, as described in the Milankovitch Theory. More recently, the last significant period of glacial advance occurred during the Little Ice Age, a period of significant global cooling that occurred between about 1500 and 1850 A.D. Since that time, glaciers have generally been retreating around the world, in many instances at rates faster than previously recorded. This rapid rate of glacial retreat has been attributed by many scientists to human-induced global warming.

Although some of the world's glaciers have actually advanced in the past few decades, such as the Perito Moreno Glacier in Argentina, the vast majority are melting at rapid rates. Glacier retreat has been very noticeable in most alpine environments. In the Swiss Alps, for example, the Rhône glacier has retreated about 2.5 km (1.5 mi) in approximately the past 140 years. In Peru, the Ururashraju Glacier retreated 500 m (1640 ft) between 1986 and 1999 alone! Another excellent example of rapid alpine glacial retreat is on Mt. Kilimanjaro, which is the largest mountain in Africa at an elevation of 5895 m (19,336 ft). Although it lies near the Equator, this volcano on the East African Rift has had a substantial ice cap for more than 11,000 years. This ice cap, which is seen in Figure 17.36a, inspired the famous Hemingway novel, *The Snows of Kilimanjaro*. In this past century the extent of the ice mass at Mt. Kilimanjaro has shrunk about 80%. Significant glacial melting was observed in Landsat imagery between 1993 and 2000 alone (Figures 17.36b and c). If the rate of melting continues, it is estimated that the entire glacier will disappear by 2015.

Rapid melting of glaciers has also been observed on the Greenland and Antarctic ice sheets. On the Greenland Ice Sheet, the area on the edge of the glacier affected by melting increased by 16% between 1979 and 2002. You can see in Figure 17.37 how the area affected by melting changed between 1992 and 2002. On the Antarctic Ice Sheet, the rate of glacial melting is even more impressive. An example of this melting is the collapse of the Larsen B Ice Shelf that occurred over a 35-day period beginning January 31, 2002. An ice shelf is a plate of glacial ice that extends from the landmass into the water; many of these features extend off the Antarctic landmass. During the Antarctic summer of 2002, about 3500 km^2 (1,352 mi^2) of the Larsen B Ice Shelf disintegrated. For scale, consider that the entire state of Rhode Island is 2717 km^2 (1049 mi^2) in size. Figure 17.38 shows the location of the ice shelf and the progression of its breakup in two satellite images. The breakup is believed to be due to the rapid warming (about 0.5 °C [0.9° F] per decade) that has occurred since the 1940s. This warming has had a profound impact, causing the extent of seven ice shelves on the Antarctic peninsula to decline by a total of about 13,500 km^2 (5212 mi^2) since 1974.

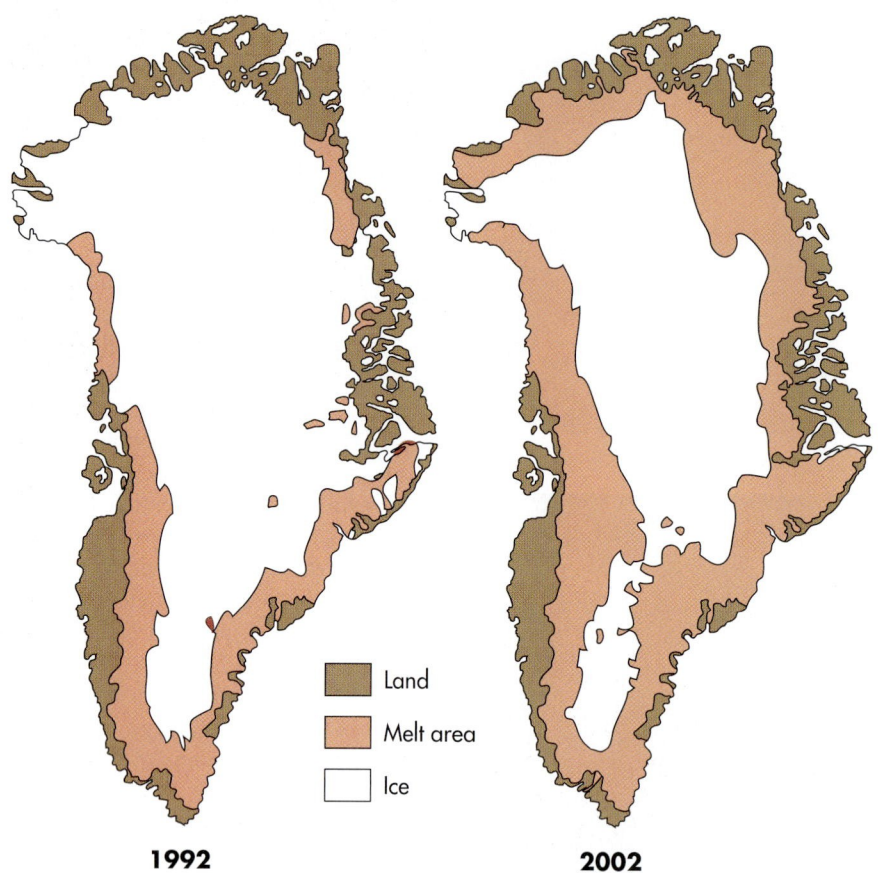

1992   **2002**

Land
Melt area
Ice

Figure 17.37 Change in melt area between 1992 and 2002 on the edge of the Greenland Ice Sheet. The Greenland Ice Sheet appears to be melting at a rapid rate. Notice, for example, how the area affected by melting in 2002 is much greater than it was in 1992. (Credit: NASA)

Periglacial Processes and Landscapes

Although we are clearly living in an interglacial period, one that may even become significantly warmer if global warming models prove to be accurate, much of the Earth remains very cold over much or all of the year. Naturally, these very cold regions are located at high altitudes and latitudes and are associated with polar-type climates. Associated with these climates is a range of specific geomorphic processes referred to as **periglacial processes;** these processes occur in a near-glacial environment and involve substantial amounts of frost action, specifically the effects of continual freezing and thawing on the landscape. Periglacial processes are closely related to physical weathering, mass movement, and soil modification and affect approximately 20% of the Earth's surface.

Permafrost

As the name implies, **permafrost** is ground that is permanently frozen. These conditions typically develop when soil or rock temperatures remain below 0° C (32° F) for at least two years. Permafrost may develop because the area is covered by a glacier or the area may be in close proximity to the ice margin and thus periglacial.

From a geographical perspective, regions of permafrost can be subdivided into two primary categories: 1) continuous permafrost and 2) discontinuous permafrost. Figure 17.39 shows the geographical distribution of these regions in the Northern Hemisphere. Large areas of permafrost also occur in Antarctica but are not described here. Regions of continuous permafrost in the Northern Hemisphere are associated with the most frigid temperatures, which basically occur poleward of the −7°C (19°F) mean annual isotherm. In areas

Periglacial processes *The suite of processes involving frost action, permafrost, and ground ice that occurs in arctic environments or along the margins of ice sheets.*

Permafrost *Ground that is permanently frozen.*

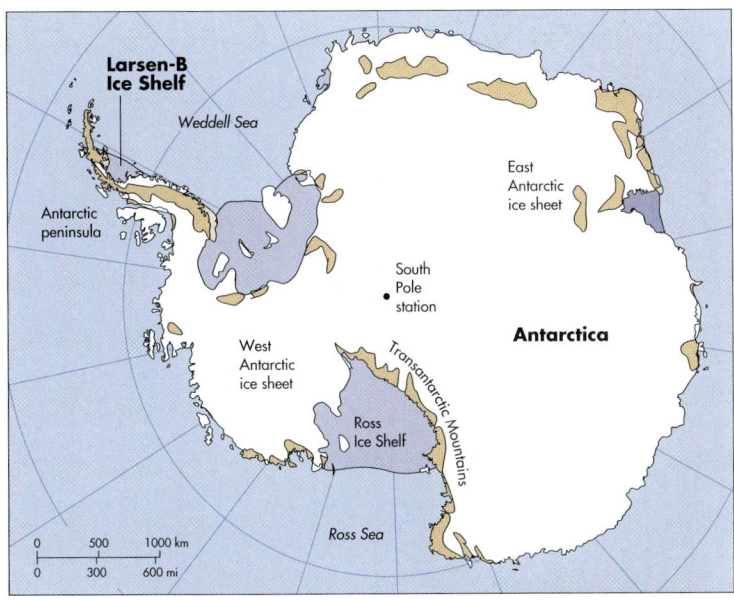

(a)

(b)

(c)

Figure 17.38 Disintegration of the Larsen B Ice Shelf in Antarctica in 2002. (a) Location of the Larsen B Ice Shelf in Antarctica. (b) The ice shelf when breakup began on January 31, 2002. The dark spots on the shelf are patches of meltwater. (c) Status of breakup on March 5, 2002. The largest ice blocks are hundreds of km² in size.

of continuous permafrost, all surfaces are frozen except those insulated beneath deep lakes or rivers. Although the average depth of permafrost in these regions is about 400 m (1300 ft), it can reach the impressive depth of 1000 m (3300 ft).

South of the region of continuous and deeply frozen ground, the mean annual temperatures become warmer

and permafrost is discontinuous and shallower. This region of discontinuous permafrost basically extends in a broad area bounded by the −7° C (19° F) and the −1° C (30.2° F) isotherms. Permafrost in this region is discontinuous where slopes face south toward the Sun and in places where the ground is insulated by snow. Overall, permafrost of some kind occurs in about 50% of Canada and 80% of Alaska. To see how the depth and spatial extent of permafrost varies across the continuous and discontinuous zones, examine Figure 17.40, which shows a typical cross section across the high latitudes.

Permafrost Processes Figure 17.40 also shows how permafrost impacts the landscape. For example, notice the areas denoted as *active layer*. The active layer is the part of the soil that melts and refreezes on a daily or seasonal basis. This zone lies above the subsurface permafrost layer and predictably varies with latitude, ranging in thickness from up to 2 m (6.6 ft) in areas of discontinuous permafrost to only 10 cm (4 in) in northern areas of continuous permafrost.

As you can imagine, the active layer is sensitive to climate change. Although the response is slow, colder temperatures gradually cause the thickness of the active layer to decrease, whereas warmer temperatures result in permafrost melting and an increased thickness of the active layer. Abundant evidence indicates that Arctic permafrost is rapidly thawing in the Earth's current warming

Ground ice   *Distinct zones of frozen water that occur in permafrost regions.*

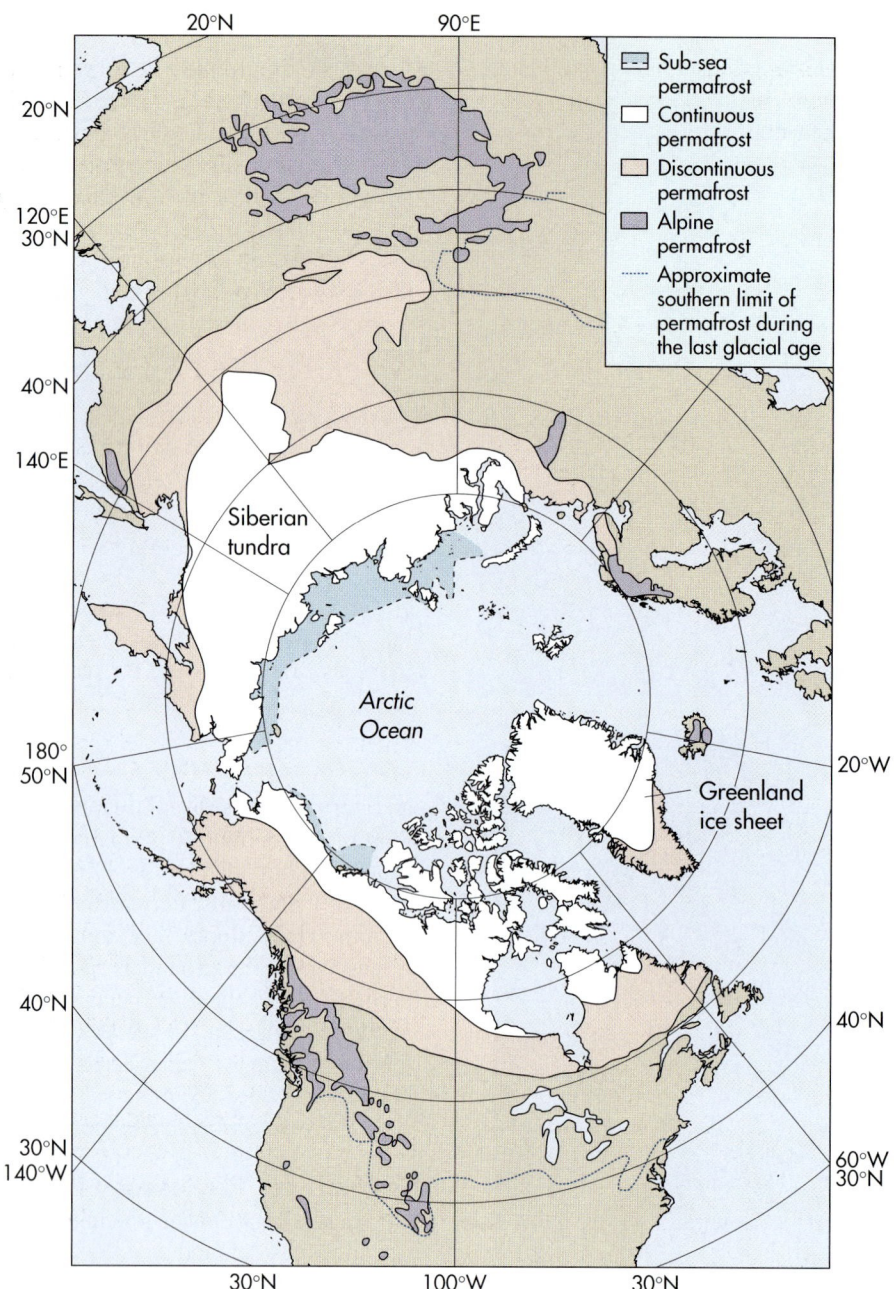

Figure 17.39 Geographical distribution of permafrost in the Northern Hemisphere. Note that permafrost can exist in the oceans as well as on land.

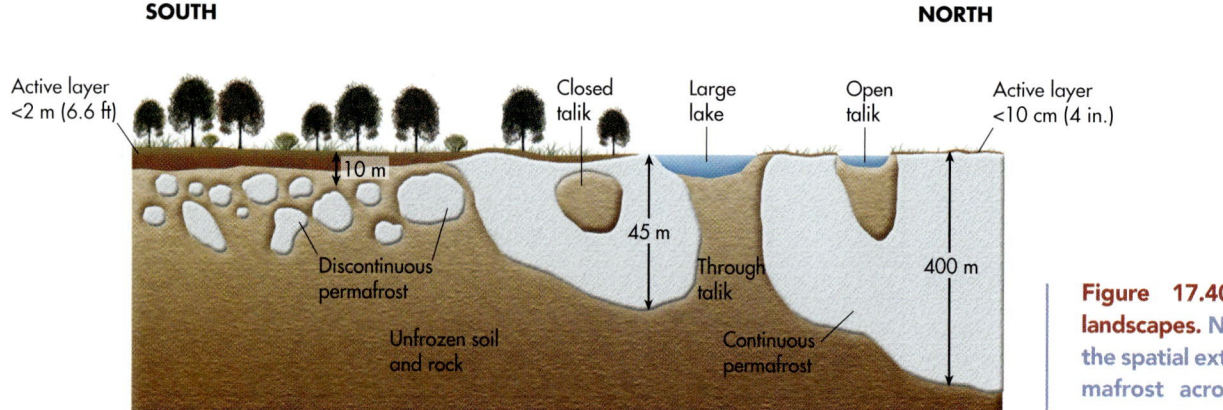

Figure 17.40 Typical periglacial landscapes. Note the differences in the spatial extent and depth of permafrost across this hypothetical cross section (not to scale).

phase. This thawing is releasing tons of CO_2 stored in the ground into the atmosphere, which should contribute to even more warming in the future.

(a)

(b)

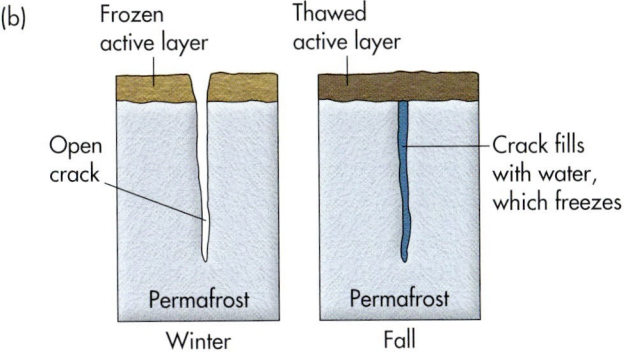

After 500 years

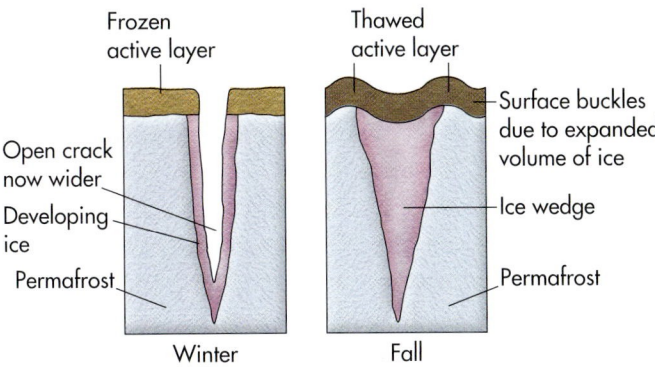

Figure 17.41 Formation of an ice wedge. (a) An ice wedge in northern Canada. (b) Sequence of steps in the formation of an ice wedge. These features form when water penetrates an open crack in the ground and freezes. Over time, frequent freezing and thawing slowly causes the crack to expand, forming a wedge. (Modified from Lachenbruch, 1962. "Mechanics of thermal contraction and ice-wedge polygons in permafrost," *Geological Society of America Special Paper 70*)

In addition to the active layer, another important feature in a permafrost landscape is a talik (see Figure 17.40 again). A *talik* is a body of unfrozen ground that occurs within a zone of discontinous permafrost or beneath a lake or river in the continuous region. Taliks are especially important in discontinuous permafrost zones because they are a link between the active layer and groundwater, forming a kind of conduit through which water can move freely.

Ground Ice and Associated Landforms

Within permafrost regions, distinct zones of frozen water occur within the ground. These areas are generally referred to as **ground ice** and contain highly variable amounts of water, ranging from near zero to completely saturated. Areas of ground ice expand and contract along freezing fronts, which are the boundaries between frozen and unfrozen ground. As these areas expand and contract, they cause physical weathering of the landscape through the frost action processes described in Chapter 14. Remember that when water freezes, it expands by about 9%. This expansion causes rocks to break along joint lines. In addition, water expansion can cause soil to move vertically or horizontally through the processes of *frost-heaving* and *frost-thrusting*, respectively.

Ground ice comes in many forms. The most basic kind of ground ice is *pore ice*, where water freezes in soil pore spaces. Ground ice also occurs as *horizontal lenses* and *veins* of ice that extend in random directions. A third kind of ground ice is an *ice wedge*, which results when water enters a crack in the ground (Figure 17.41). *Segregated ice* is ground ice that is buried but grows when additional water is added. Finally, *intrusive ice* is ground ice that forms when water is injected under pressure, perhaps because it lies between a pair of aquicludes. Where intrusive ice occurs the surface of the ground can bulge up to 60 m (200 ft) upward, forming a landform called a *pingo* (Figure 17.42).

Figure 17.42 A pingo. A pingo forms when water is injected into the ground under some form of hydraulic pressure. This pressure forces the ground to bulge upward.

Figure 17.43 Patterned ground. The distinctive polygons on this landscape form when frost action preferentially brings large stones and rocks to the surface.

In addition to pingos, frost action can result in a variety of other landforms. A *palsa* (Spanish for "elliptical") is a rounded or elliptical mound of peat that is similar to a pingo in that it is thrust upward. It differs from a pingo, however, because it is only 1–10 m (3–30 ft) high and contains ice lenses rather than a solid core. Another distinctive periglacial landform is *patterned ground* (Figure 17.43). As the name implies, patterned ground is a landscape that has evolved to contain distinctive shapes. These shapes are usually polygons that form when frost action preferentially brings coarser materials—stones and boulders—to the surface.

The Big Picture

The last two chapters have focused on the processes and landforms associated with flowing water and ice. In the next chapter we turn to another kind of flow, moving air. You will discover that wind can have a dramatic impact on the landscape through the erosion and transport of wind-blown sediment. These processes shape the landscape in predictable ways that produce landforms such as sand dunes and vast plains covered by wind-blown silt. Eolian processes are most closely associated with dry places, so the chapter will begin with the geomorphology of arid environments and will integrate the elements of climate, weathering, and rock structure discussed earlier.

Summary of Key Concepts

1. Glaciers form when snow accumulates annually and does not melt during the summer months. In this way, the snow mass gradually thickens with each passing year and the snow gradually changes into glacial ice through compression.

2. A growing ice mass becomes a glacier when it begins to flow under its own weight. Whether a glacier advances or retreats depends upon the glacial mass budget, which is the balance between snow accumulation in the source area and melting at the zone of ablation. Melting and accumulation are in balance along an equilibrium line.

3. Ice fields and ice caps are large ice masses that spread horizontally across mountainous landscapes. Valley glaciers flow down pre-existing stream valleys in alpine settings. Continental glaciers, such as the Antarctic Ice Sheet, are thousands of meters thick and spread across continental landmasses.

4. Glaciers erode the ground beneath them through the combined processes of abrasion and plucking. In alpine settings, these processes change the landscape from generally rounded to much more angular, with sharp features such as arêtes, horns, and cirques.

5. Glaciers deposit sediment either directly as till or indirectly from meltwater streams as outwash. These processes produce landforms such as moraines, kames, and outwash plains.

6. During the Pleistocene Epoch (between about 1.6 million and 10,000 years ago), glacial ice covered up to 30% of the Earth, including significant parts of North America, Europe, and Asia. The most recent period of significant glacial advance

was the Wisconsin glaciation, which ended about 10,000 years ago. Glaciers now cover about 11% of the Earth. These tremendous fluctuations in ice volume are intimately related to gradual changes that occur with respect to Earth-Sun geometry.

7. Evidence for the Pleistocene ice ages is found in many places in the northern part of the United States and Canada. The Great Lakes, for example, were created because large ice lobes enlarged pre-existing stream valleys.

8. Permafrost refers to ground that is perpetually frozen, forming features such as ice wedges and pingos. Areas of current permafrost conditions occur at high latitudes and altitudes.

Check Your Understanding

1. Why are summer-time temperatures a critical variable with respect to the formation of a glacier?

2. Describe the steps involved in the metamorphosis of snow to glacier ice.

3. How does a glacier flow? Where does the ice move the fastest and slowest? Why?

4. Describe the difference between glacial abrasion and glacial plucking. What landform is produced by a combination of these two processes?

5. The next time you're in the mountains, how will you be able to tell whether or not the landscape has been glaciated? What are some of the characteristic landforms that should be present if the landscape was glaciated?

6. How does a hanging valley form?

7. What is the difference between valley and cirque glaciers?

8. In what two regions of the Earth are large continental glaciers currently found?

9. What is the difference between glacial till and glacial outwash? What are the diagnostic characteristics between each deposit?

10. How does the formation of a recessional moraine differ from that of a terminal moraine?

11. Describe the logic behind the use of oxygen isotopes to reconstruct the glacial chronology of the Pleistocene. How does this chronology differ from the traditional model?

12. Describe the three components of the Milankovitch Theory. How does a change in each of these variables enhance the possibility of glaciation?

13. Which process—chemical or physical weathering—is most likely dominant in periglacial environments? Why?

14. What is the difference between continuous and discontinuous permafrost? What is their geographical distribution and why does it make sense given your understanding of permafrost processes?

15. What is ground ice? Describe two landforms created by this phenomenon.

ANSWERS TO VISUAL CONCEPT CHECKS

Visual Concept Check 17.1

The glacial mass budget refers to the balance between snow accumulation in the source area and melting in the zone of ablation. The point where these two process are in balance is known as the equilibrium line. Below the equilibrium line, crevasses are visible, whereas above the equilibrium line they are covered with snow.

Visual Concept Check 17.2

Glaciers modify alpine landscapes in a very intense way. Prior to glaciation, alpine landscapes are generally rounded, with streams flowing in V-shaped valleys. Through the combined processes of abrasion and plucking, alpine glaciers cause mountainous regions to become more angular, forming features such as cirques, arêtes, and horns. Valley glaciers enlarge and deepen former stream valleys, changing them to U-shaped valleys.

Visual Concept Check 17.3

This photograph of a stream system in the Southern Alps of New Zealand is a nice example of how outwash plains form. Several lines of evidence point to this conclusion. The snow-capped peaks in the background suggest that alpine glaciers are present, which must be delivering coarse sediment to the stream in glacial meltwater. You can tell that the sediment is coarse because the stream is braided. Given that fluvial deposition is occurring, you can surmise that the landscape will have very limited relief and will thus be consistent with an outwash plain.

CHAPTER EIGHTEEN

ARID LANDSCAPES AND EOLIAN PROCESSES

In this chapter we focus on arid geomorphology and the way that flowing air moves sediment and shapes the landscape. Wind processes are typically referred to as *eolian processes*, which produce eolian landforms. (The term *eolian* is derived from Aeolus, the Greek god of the winds.) Although the ability of wind to shape the landscape is small relative to flowing water in streams or flowing ice in glaciers, a fascinating array of erosional and depositional landforms can

Some of the most beautiful landforms on Earth are produced by the wind, such as this sand dune in the Sahara Desert. In contrast to mountains, stream terraces, and glacial moraines, eolian landforms are fascinating because they can easily develop and be noticeably modified during the course of a human lifetime. This chapter focuses on the way wind shapes the land and the kind of landforms that occur in arid environments.

nevertheless result over time. It is important to understand the role that

wind plays with respect to landform evolution because more than

one-third of the land on our planet is characterized as arid or semi-arid.

In addition, many of these landscapes are agriculturally important and

can be adversely affected by the wind. Given that the impact of eolian

processes is most pronounced in these dry places, we'll begin the

chapter with a general discussion of arid environments.

Arid Landscapes

Let's begin the discussion of eolian processes by looking more closely at the geography and character of arid environments. This discussion incorporates many of the topics that were covered earlier in the text—atmospheric circulation, plant geography, soils, geology, and fluvial processes. The goal of this part of the chapter is to provide you with a good understanding of desert environments and why they occur where they do.

Figure 18.1 shows the geography of arid and semi-arid regions on Earth. Such climates occur in both

warm and cold settings, and collectively they form the largest climate region on the planet, one that covers perhaps as much as 35% of the Earth's land surface. Within the Köppen climate system (Figure 9.4), these climate regions fall within the arid desert (BW) and semi-arid steppe (BS) climate categories. The spatial distribution of these dry regions is based on one of three factors:

1. *Dominance of Subtropical High Pressure Systems—* In areas located between about 15° and 35° N and S latitude, strong high-pressure systems dominate the weather for all or significant por-

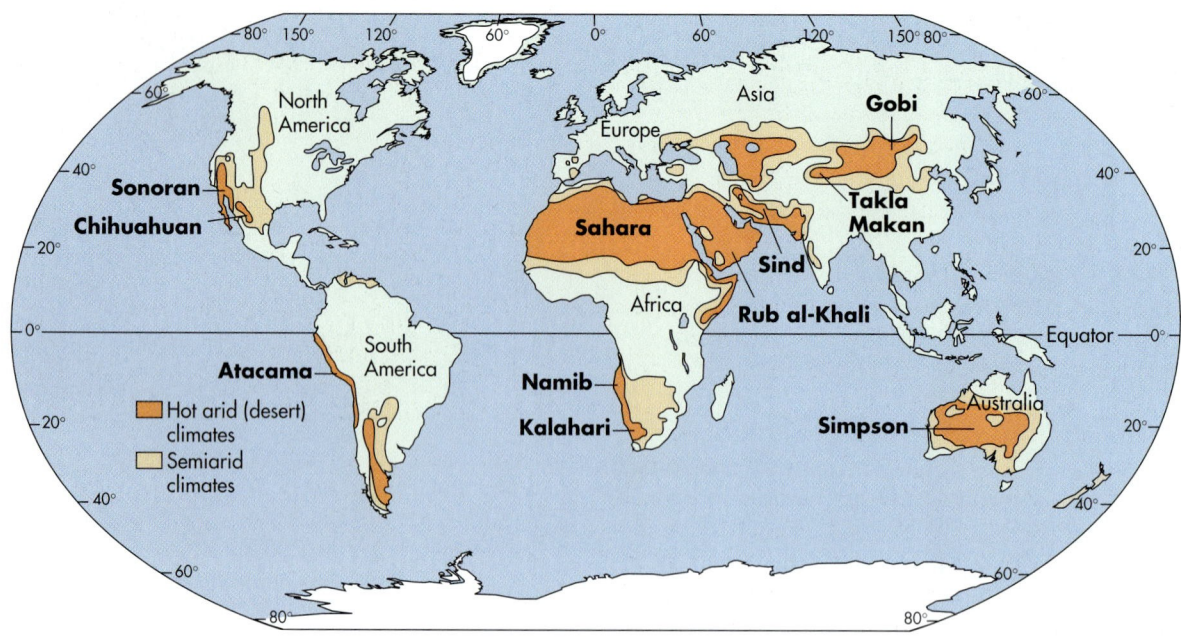

Figure 18.1 **Deserts of the world.** Arid and semi-arid deserts collectively cover about one-third of the Earth's land surface.

tions of the year. These descending air masses are components of the Hadley cells (shown in Figure 6.16) that circulate within the atmosphere in the tropical latitudes. The sub-tropical high-pressure systems are very closely associated with the largest deserts on Earth, such as the Sahara and Kalahari in Africa, the Simpson Desert in Australia, and the Chihuahuan Desert in Mexico (Figure 18.1).

2. *Rainshadow Effect*—Another environmental factor that is associated with deserts is the rainshadow effect. If you recall from Chapter 7, a rainshadow occurs on the lee side of mountain ranges where descending air warms adiabatically. This process was outlined in Figure 7.24 and is responsible for the large zone of semi-arid climate in the High Plains of the U.S. It also is related to the presence of the Atacama Desert along the west coast of South America. This desert was discussed as an Amazing Place in Chapter 10.

3. *Distance from Large Water Bodies*—The last environmental variable that is related to dry regions is the proximity to a water body such as an ocean. Locations deep within continents tend to receive relatively small amounts of moisture-laden air and can therefore be quite dry. An excellent example of this kind of situation is the Taklimakan and Gobi deserts in Asia, as well as the other large areas of arid and semi-arid lands that lie deep within that continent.

Desert Geomorphology

A key element that all deserts and semi-arid regions share is a relative lack of vegetation when compared to more humid areas. As a result, descriptive terms such as *sparse* and *barren* are frequently used to describe dry regions. This relationship of climate and vegetation was thoroughly explored in Chapters 9 and 10 and can be seen in numerous photographs of desert envirnoments, such as Figures 9.12 and 10.15.

Although the barren nature of these environments may at first be unattractive or even foreboding to some, desert lanscapes can be beautiful, with spectacular rock outcrops and panoramic vistas. In the context of geomorphology, deserts are great places because you can actually *see* important elements like rock structure, differential weathering, and the entire outline of landforms clearly. Although landforms in humid regions can be thoroughly investigated, it is somewhat more complicated because thick stands of vegetation often mar the view. In contrast, return to the photograph of the alluvial fan in 16.30 and notice how clearly you can see that particular landform.

The Southwestern United States An excellent place to see desert geomorphology at a place you may visit is the southwestern U.S. This region lies in the rainshadow of the Sierra Nevada Mountains and encompasses a number of previously discussed arid places such as the Sonoran Desert (Figure 10.l4), the Grand Canyon (see *Amazing Places* in Chapter 12), and the Basin and Range Province (Figure 13.22), to name a few. Each of these places is noteworthy from a geomorphic perspective because a predictable array of landforms occurs throughout the region and are readily visible to anyone passing through the area.

One of the most important factors that has influenced landform evolution in the southwestern U.S. is flowing water. Although it might seem that the influence of water would be minimal in dry lands where most streams are ephemeral, the fact is that streams are very important in these landscapes with respect to landform evolution. The reason is that vegetation cover is minimal. As a result, when it does rain in an arid environment, erosion by flowing water is extensive because hillslopes are not protected by plants. Flash flooding in these areas can easily occur and the landscape may be shaped dramatically. A particularly distinctive landform in the western U.S. formed by flooded streams is an **arroyo,** which is a steep-sided gully cut into alluvial sediments (Figure 18.2).

Figure 18.2 Arroyos in the Southwest. Arroyos are steep-sided gullies cut into alluvium. A flash flood recently passed through this arroyo in Arizona and deposited sediment on the gully floor.

Arroyo *A deep, steep-sided gully that is cut into alluvium.*

Another variable that strongly influences landforms in the West is rock structure. We saw a great example of this relationship in the Basin and Range (Figures 13.22 and 13.23), which has formed due to the effects of normal faulting that has produced a number of upthrown and downthrown blocks called horsts and grabens, respectively. The most dramatic example of an upthrown block is the Sierra Nevada Mountains, which are pictured in Figure 13.24.

The importance of rock structure can also be seen where the rocks are horizontal and have not been deformed (Figure 18.3). In these areas, which are very common in the west, the rocks usually consist of alternating layers of limestone, sandstone, and shale. The limestones and sandstones are generally the most resistant rocks to erosion, whereas the shales are relatively soft and easily eroded. Where these patterns occur, the hard rocks usually form a resistant caprock that protects the underlying shale to some extent from erosion. Notice in Figure 18.3 that shale tends to form shallow slopes, whereas the more resistant rocks are associated with steep slopes. Shale tends to erode backward gradually due to slope wash when it does rain. This gradual slope retreat ultimately undermines the caprock above and causes it to collapse due to rockfall (see Figure 14.17a) and the cycle begins again.

The net effect of this kind of landscape evolution is that large parts of the western U.S. appear to be cut up, with upright islands of various sizes in some apparent relationship to one another. This landscape really represents nothing more than a slow progression of erosion in which the landscape has been gradually dissected. Figure 18.3 illustrates the landforms associated with this progressive

evolution. It's important to notice how each of the rock layers can be traced from one landform to another. This occurs because the rock was once a uniform mass that had broad regional extent. Through a combination of stream erosion, differential weathering, and mass wasting, the landscape was slowly dissected, leaving a succession of flat-topped remnants that tower throughout the area.

The landforms produced by these long-term processes are easy to see if you travel in the Southwest. A **plateau** is a broad platform tens of kilometers across that is elevated up to perhaps 300 m (about 980 ft) above the surrounding terrain. Dissection of a plateau begins when a stream downcuts vigorously into the underlying rock strata to form a **canyon**, which is a deep, narrow valley with very steep sides (Figure 18.4). The most spectacular canyon, of course, is the Grand Canyon, but numerous smaller canyons dot the landscape. This system of canyons is largely associated with the uplift of the Colorado Plateau about 30 million years ago. This uplift increased the slope of streams dramatically. Since that time, streams and their tributaries have been vigorously downcutting to reach a new graded state in a manner consistent with the discussion in Chapter 16 (Figure 16.20a).

With the system of canyons in place, further stream erosion and mass wasting processes begin to attack the slopes of the dissected plateau. These combined processes gradually leave isolated remnants of the plateau standing alone in progressively smaller and smaller landforms. You can easily visualize this progression in Figure 18.3. The largest of these plateau fragments is a **mesa**, which can be up to several kilometers across. A somewhat smaller remnant is a **butte**, which

Figure 18.3 Prominent desert landforms associated with horizontal rock structure. Over time, a plateau is dissected into progressively smaller landforms that give the region a distinct appearance.

Plateau *A very broad, horizontal surface that is upheld by resistant caprock.*

Canyon *A very steep-sided valley that is cut into bedrock.*

Mesa *A broad horizontal surface, smaller than a plateau, that is upheld by resistant caprock.*

Butte *A steep-sided hill or peak that is often a remnant of a plateau or mesa.*

Figure 18.4 A typical canyon in the western U.S. Canyons such as this one are very common in the region and evolved due to vigorous stream incision. Exposed in the canyon walls are layers of sedimentary rock.

Figure 18.5 Landforms in Monument Valley, Utah. This photograph nicely illustrates the concept that the mesa, buttes, and pinnacles were once part of a uniform rock mass that was heavily eroded over a long time.

Figure 18.6 Playa in Death Valley, California. Lakes form in basins and on plains in the American West because they collect storm runoff. They subsequently dry out during drought periods, leaving this distinctive whitish feature behind.

gradually wears down to form a tower-like landform called a **pinnacle.** A great place to see these kinds of landforms in close proximity to one another is Monument Valley in Utah (Figure 18.5).

The net effect of the progressive dissection and shrinking of plateaus, mesas, buttes, and pinnacles is that the landscape is gradually worn down. This evolution ultimately produces broad plains that are scattered throughout the region. A common feature in these intervening plains, as well as the grabens of the Basin and Range, is a dried lake bed called a **playa** (Figures 18.3 and 18.6). These features are associated with closed topographic depressions that may temporarily fill with water during wet periods. They subsequently dry out during extended droughts, leaving salty evaporites on the surface. In many cases, these old lake beds contain a lengthy record of climate change and sedimentation in the region.

Pinnacle   *A steep-sided, narrow tower that is the final remnant of a plateau, mesa, or butte.*

Playa   *A dried lake bed that forms when runoff collects in closed topographic depressions in arid regions.*

KEY CONCEPTS TO REMEMBER ABOUT ARID LANDSCAPES

1. Approximately one-third of the Earth's land surface is a mix of arid and semi-arid deserts.

2. Arid lands are associated with Subtropical High pressure systems, rainshadows, and locations that are deep within continents.

3. Deserts and other dry lands are excellent places to study geomorphology because the vegetation cover is relatively thin and thus features such as rock structure and landforms are easy to see.

4. In many places of the American West, a distinct progression of landforms is associated with horizontal rock structure. This progression includes plateaus, canyons, mesas, buttes, pinnacles, and plains.

5. A playa is a dried lake bed that occurs in plains and grabens. These lakes temporarily fill with water during wet periods and subsequently dry out during periods of drought.

Eolian Erosion and Transport

In addition to the progression of erosional landforms that occurs in many deserts, another prominent geomorphic agent in these areas is **eolian processes**. These processes are most active in desert regions because 1) strong winds are common, 2) there is often a large supply of sand and silt that can be blown, and 3) vegetation cover is minimal and the wind is thus free to erode sediment. This section of the chapter focuses on the way that moving air behaves and the way it shapes the landscape.

The Fluid Behavior of Wind and Sediment Transport

Like water, wind behaves according to specific physical laws associated with fluids. From this perspective, the primary difference between wind and water is that water is much denser than wind. For example, think of how much more buoyant you are in water than in air. Because of this variation in buoyancy, rock fragments appear to be much heavier in air than in water. Thus, to move a particle of sediment, winds must have speeds almost 30 times greater than the speed of water currents to move the same particle.

Both water and wind behave as a fluid with a predictable velocity gradient in the vertical dimension. When air flows over a bare, unvegetated surface, there is a very thin boundary layer where the wind velocity is zero. Above this layer, to a height of about 3 cm, velocity initially increases rapidly, but then slows with additional height.

The ability of wind to erode sediment is similar to that of water in that sediment particles begin to move once a critical threshold velocity is passed. This threshold varies depending on the size of any given particle, with small grains more easily moved than larger ones (Figure 18.7). To see how wind moves sediment, consider Figure 18.8. Let's focus for a minute on a single particle of sediment and how it would begin to move. Initial movement of the particle begins when the threshold velocity is crossed, which causes the sediment to oscillate at the surface beneath the wind. If the particle is less than 0.06 mm—in other words, classified as silt or clay—it will suddenly fly nearly vertically into the air and be carried in suspension for great distances. These fine grains are kept in the air by turbulence (Figure 18.9) and may travel hundreds of kilometers before they settle back to Earth.

In contrast to silts and clays, which are carried in suspension, larger particles such as sand either bounce across the ground through the process of *saltation* or roll along the surface in a process called *creep* (see Figure 18.8). Saltation begins when a sand grain flies into the air. Because this grain is large relative to silts and clays, it is carried downwind only a short distance and then falls back to the surface under the force of gravity. When the grain strikes the ground, it transfers momentum to other sand

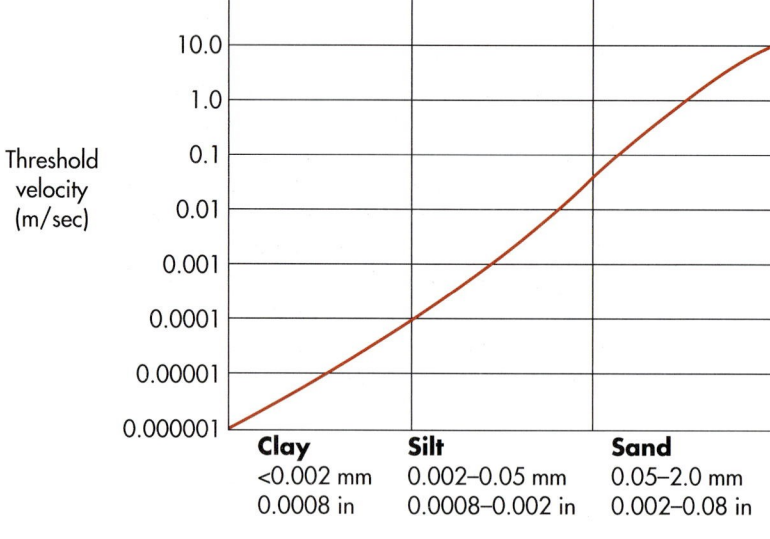

Threshold velocity (m/sec)

	Clay	Silt	Sand
	<0.002 mm	0.002–0.05 mm	0.05–2.0 mm
	0.0008 in	0.0008–0.002 in	0.002–0.08 in

Figure 18.7 Threshold velocity required for wind to carry various-sized particles. The largest sand particles require a wind velocity of 10 m/sec (~33 ft/sec) to be moved by the wind.

Eolian processes *Geomorphic processes associated with the way that wind erodes, transports, and deposits sediment.*

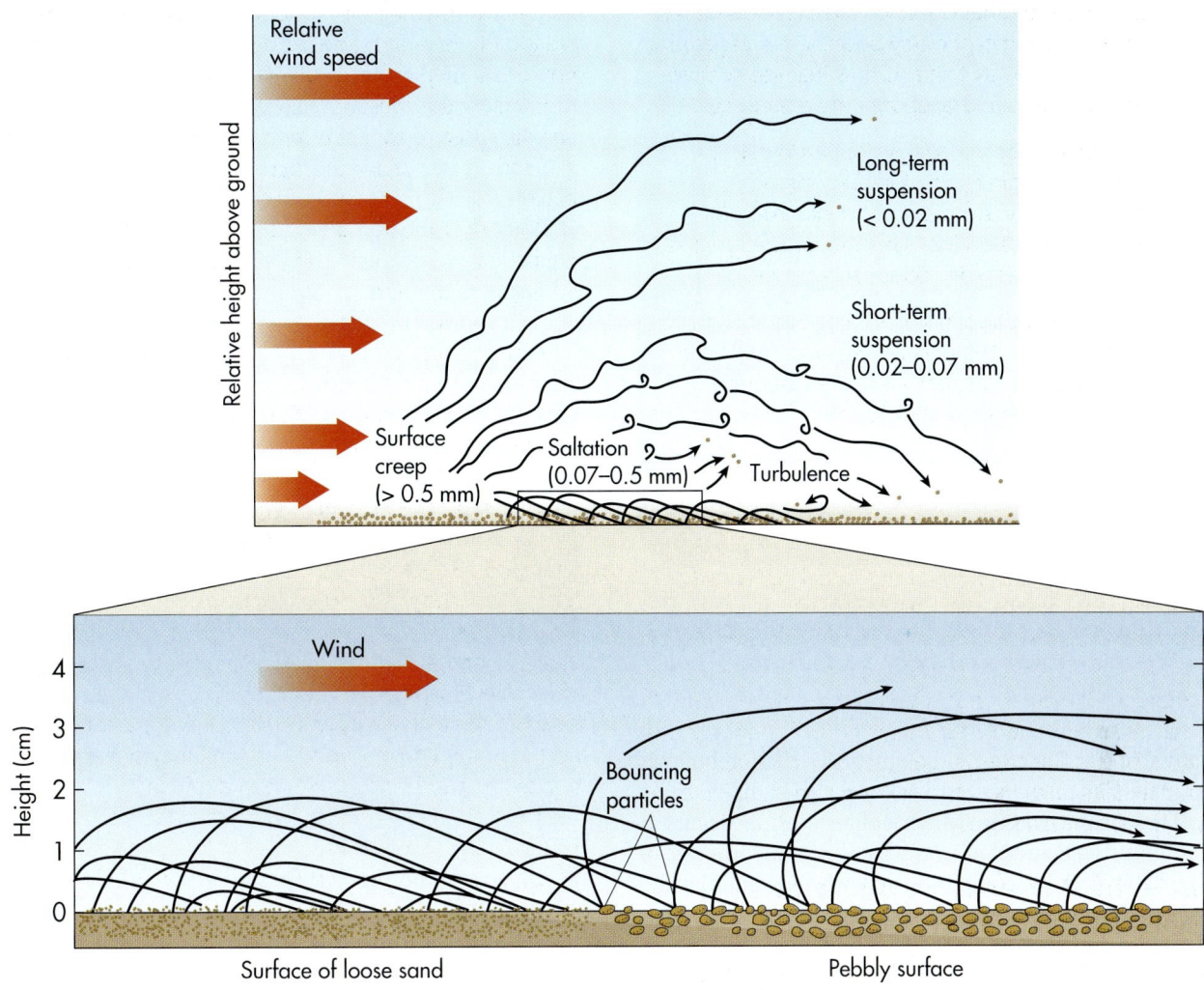

Figure 18.8 Ways in which eolian sediment is transported. Wind-blown sediment can move in suspension, by saltation, or by surface creep.

grains that, in turn, fly vertically into the air to repeat the process farther downwind. Creeping grains are larger, such as small pebbles, and consistently maintain contact with the surface. Like saltating grains, however, they cause movement in other grains through momentum transfer.

Eolian Erosional Landforms

Let's examine the process of eolian erosion on a larger scale and the landforms that result. Wind erosion occurs by either one of two processes: deflation or abrasion. When **deflation** occurs, turbulent air blows loose soil particles away. If this process persists over time across a

broad surface, all fine particles are ultimately removed, leaving a concentration of coarser pebbles and cobbles that collectively form a surface called **desert pavement** (Figure 18.10). These surfaces are usually a mix of dark colors, collectively referred to as *desert varnish,* that is a coating of fine clays and bacteria that forms on rock surfaces over many decades. Although desert pavement protects underlying layers of fine particles from further deflation by capping them, the surface can easily be disturbed by human activities. When deflation is more localized, a landform called a deflation hollow may form. As the name implies, a **deflation hollow** is a basin formed by the

---

Deflation   *Removal of sediment from a surface by wind action.*

Desert pavement   *A resistant, pavement-like surface created when fine particles blow away and coarse sediment such as pebbles and gravel are left behind.*

Deflation hollow   *A depression created by wind erosion.*

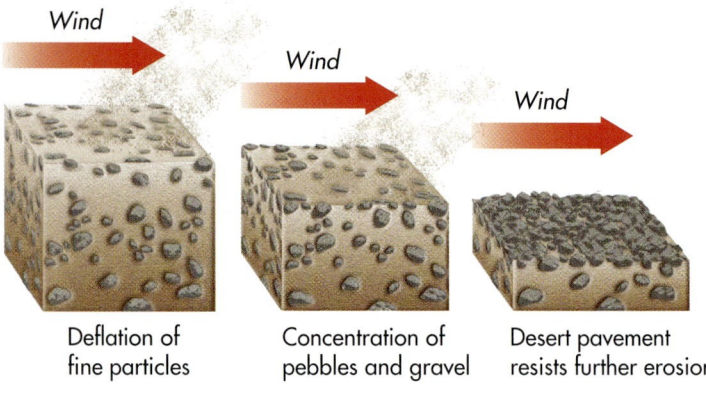

Deflation of fine particles

Concentration of pebbles and gravel

Desert pavement resists further erosion

(a)

(b)

Figure 18.9 Eolian dust in suspension. Silt and clay particles can be carried far in the air by turbulent flow, as demonstrated here in north central Iowa.

Figure 18.10 Desert pavement. (a) Desert pavement forms when fine particles progressively blow away, concentrating the coarse particles into a continuous gravel surface. (b) A typical desert pavement. Note the very coarse texture of this surface.

removal of fine material. Although most deflation hollows are small, some may exceed 1.6 km (1 mi) in diameter.

The second way in which eolian erosion occurs is through **abrasion.** You can visualize the process of eolian abrasion by comparing it with the sandblasting conducted by work crews. As you've probably seen, sandblasting is a method used to clean buildings and bridges by blowing sand in a compressed stream of air to polish and abrade a surface. The process of eolian abrasion works much slower than mechanical sandblasting and is confined to the space immediately above the ground, but nevertheless has the same effect over a long period of time. Many factors influence how rapidly abrasion occurs, including the strength and persistence of the wind, the hardness and angularity of the blowing sand grains, and the resistance of the rock being abraded.

Individual rocks that have been influenced by abrasion are called **ventifacts** ("artifacts of the wind") and tend to be pitted, grooved, or polished (Figure 18.11a and b). In addition, they are typically aerodynamically shaped in the direction of the prevailing winds. Sometimes an entire rock outcrop or series of rock outcrops is streamlined by abrading winds. When this larger-scale abrasion occurs, elongated, wind-sculpted ridges called **yardangs** form (Figure 18.11c). Yardangs are similar to glacial

roches moutonées in that they form as a result of two distinct processes, with abrasion dominating on the windward side of the feature and deflation on the leeward side. Yardangs range from a few meters to several kilometers in length and are very prominent in desert regions such as those in Egypt and Iran. These landforms have also been identified on Mars, indicating that this planet also has an active eolian landscape.

Abrasion *Erosion that occurs when particles grind against each other.*

Ventifact *An individual rock that is pitted, grooved, or streamlined through wind abrasion.*

Yardangs *Ridges that are sculpted and streamlined by wind abrasion and deflation.*

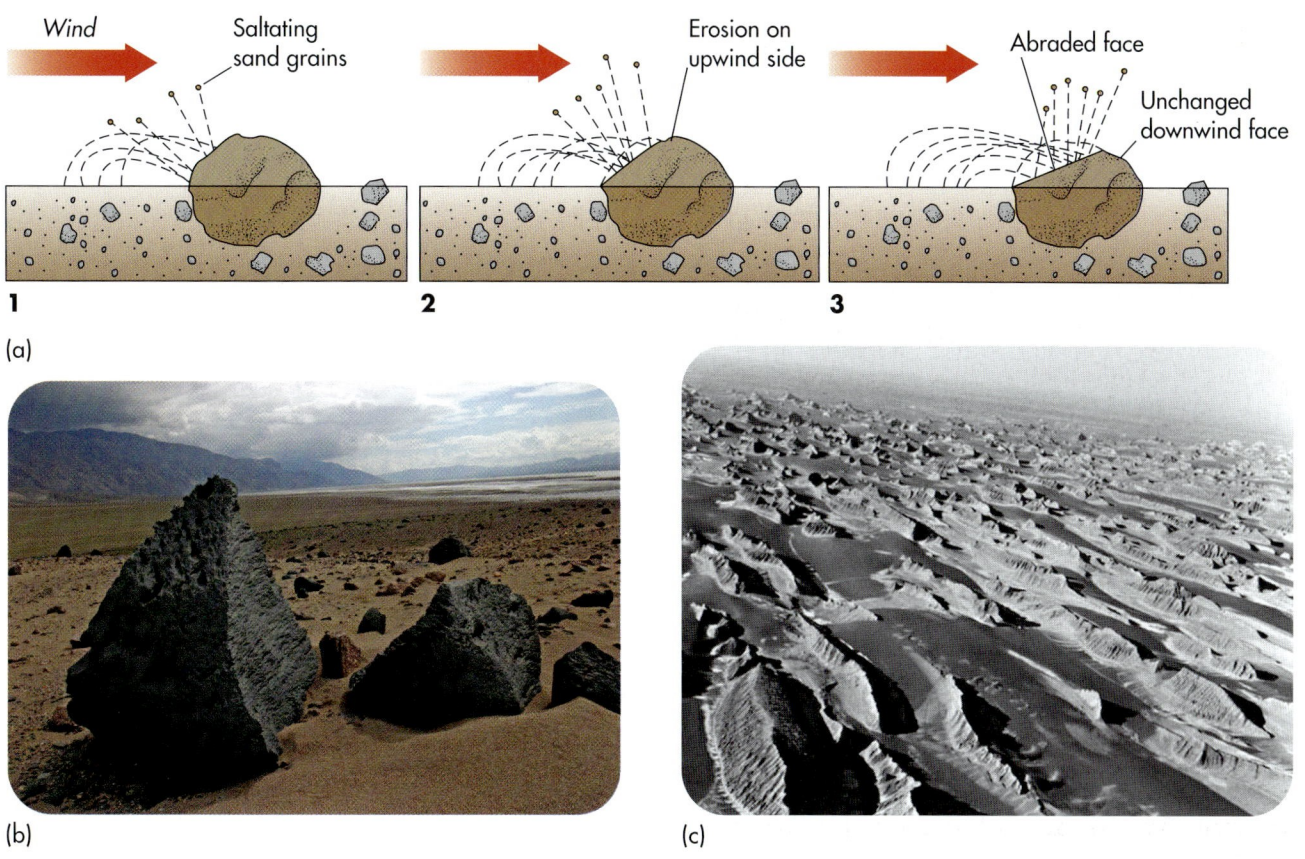

(a)

(b)

(c)

Figure 18.11 Erosional features created by the wind. (a) Ventifacts are individual rocks that have been streamlined by the wind. (b) Ventifacts such as these in Death Valley form when wind sandblasts rocks. (c) These yardangs in southeastern Iran are up to 80 m (~ 263 ft) high and are some of the largest on Earth.

VISUAL CONCEPT CHECK 18.1

Imagine that you're on a vacation in the Sonoran Desert in the southwestern United States. As you travel around this region, you frequently see landscapes such as this one, covered with small stones and pebbles. How have these landscapes formed?

KEY CONCEPTS TO REMEMBER ABOUT EOLIAN PROCESSES

1. Like water, wind is a fluid that behaves in a predictable physical way with respect to flow and velocity.

2. As a fluid, wind is much less dense than water, which means that it must flow at a larger velocity (than that of water) to generate enough force to carry a particle of a given size.

3. Sediment is moved by the wind in three ways, depending on particle size: suspension (silt and clay), saltation (sand), and creep (pebbles and, in strong winds, gravel).

4. Wind erosion works through the processes of deflation and abrasion. Deflation occurs when sediment is physically moved from one place to another. Abrasion is a natural form of sandblasting.

Eolian Deposition and Landforms

In the preceding section we discussed how wind behaves as a fluid and how it erodes and transports sediments. These processes produce distinct erosional landforms, such as desert pavement and yardangs. However, the sediment removed from one place must accumulate someplace else,

forming depositional landforms. This part of the chapter focuses on the landforms produced by eolian deposition.

How Sand Dunes Form

A multitude of landforms can develop through deposition of eolian sediments. Of these landforms, you're probably most familiar with those associated with windblown sand. The largest depositional features related to eolian sand are sand sheets and sand seas. Sand sheets are horizontal to semi-horizontal bodies of sand that exhibit little or no surface topography. In contrast, sand seas are vast regions where enormous quantities of sand result in a wide variety of dune types. The best known sand sea is in the Sahara Desert in Africa (Figure 18.12a). Sand seas occur elsewhere in the world, however, such as the Namib sand sea in the Namib Desert on the southwest coast of Africa (Figure 18.1). The largest sand sea in North America is the Nebraska Sand Hills, which is over 32,000 km² (about 12,350 mi²) in size. This sand sea is largely stabilized at present due to an extensive cover of grass (Figure 18.12b). Research indicates, however, that it has been very active at various times within the Holocene Epoch (the past 10,000 years) when intensive droughts reduced the vegetation cover.

Of all desert and wind phenomena, sand dunes have probably received the most scientific attention. Although a variety of dune forms are recognized, their forms sometimes grade into one another with no clear distinctive feature. Dunes typically exhibit a characteristic profile consisting of

(a)

(b)

Figure 18.12 Sand seas. (a) The Sahara Desert is best known for the massive sand dunes that have formed in this very dry environment, which is dominated by the Subtropical High pressure system. (b) The Nebraska Sand Hills is the largest sand sea in the Western Hemisphere and contains a wide variety of dune forms. Although isolated areas of blowing sand occur in the area, the dunes are mostly stabilized by an extensive cover of grass.

three components: the **backslope** or windward surface, where erosion dominates; the **crest** or highest point of the dune; and the **slip face** or lee slope, where deposition occurs (Figure 18.13). The backslope has the shallowest slope, ranging between 10° and 15°. In contrast, the slip face commonly stands between 30° and 34°. This slope is near the angle of repose for sand, which means that it's the maximum slope at which the sand will not slump under the force of gravity.

Deposition on the slip face occurs when saltating sands on the backslope move up and over the crest onto the lee slope, where they are protected from the wind. Another way in which slip face deposition occurs is when the slope becomes too steep, exceeding the angle of repose, and sands slump to a new slope position under the influence of gravity. The effect of these dual processes of backslope saltation and slip-face deposition is that dunes move across the landscape.

Backslope   *The gradual slope of a dune that faces the prevailing winds.*

Crest   *The highest point of a dune.*

Slip face   *The steep slope that lies on the leeward side of a sand dune.*

The extent of sand dune migration depends upon the balance between erosion on the backslope and deposition on the lee slope. As a result, maintenance of the dune form depends upon the forward movement of the entire feature because the amount of backslope erosion is balanced by the amount of deposition on the lee slope. In reality, variations in fluid dynamics and other factors result in complex cross-sectional characteristics in most dunes.

Overall, the most important factors that influence the shapes of dunes are wind speed, the amount of stabilizing vegetation, and the sand supply—that is, how much sand is present for the wind to move. The amount and kind of vegetation is a critical variable because plant roots hold deposits of eolian sand together. If a sand dune is well vegetated, it tends to remain stable—that is, it stays in one place—unless it is significantly disturbed through human impact, drought, or a major storm. If such an event occurs, the dune becomes active and begins to migrate.

Types of Sand Dunes

Despite the intricacies in dune morphology, several attempts have been made to categorize dunes based upon their appearance in plan view, that is, from above. Such classification schemes largely rely on some potentially

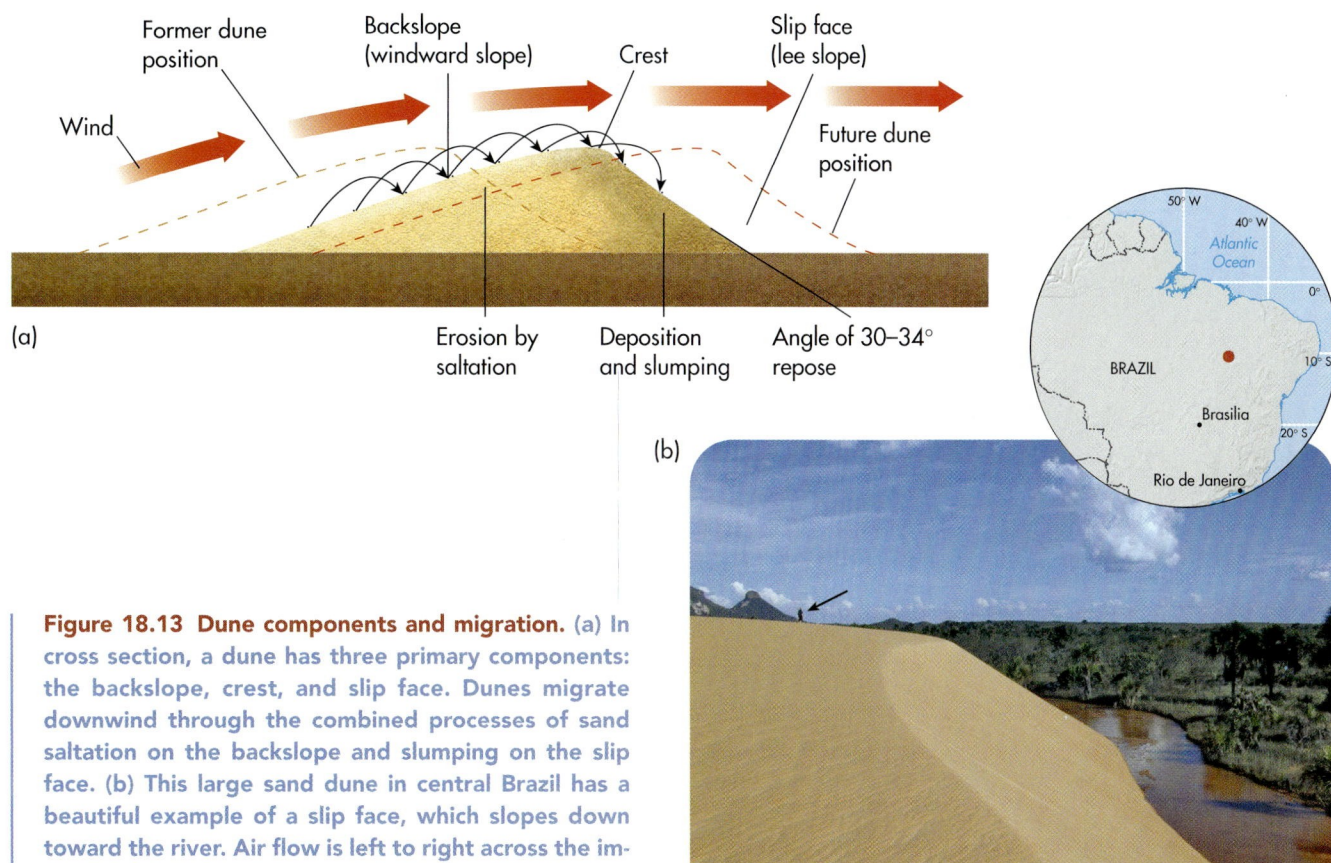

Figure 18.13 Dune components and migration. (a) In cross section, a dune has three primary components: the backslope, crest, and slip face. Dunes migrate downwind through the combined processes of sand saltation on the backslope and slumping on the slip face. (b) This large sand dune in central Brazil has a beautiful example of a slip face, which slopes down toward the river. Air flow is left to right across the image. Note the pinnacle in the background and the person (arrow) for scale.

distinguishing factors, including shape, slip-face orientation, wind type, and surface conditions. At least eight major classes of dune morphology have been identified, with each type occurring individually or in compound and complex forms. Figure 18.14 illustrates these classes. Compound dunes occur when dunes of the same type are superimposed on one another. Complex dune fields are regions where more than one type of dune appear.

Several types of dunes form in landscapes with no vegetation. Three dune classes—*barchan, barchanoid*

ridges, and *transverse ridges*—develop under these conditions, but their forms vary according to sand supply and wind strength and direction. Barchan dunes are the classic eolian landforms in sandy landscapes. These dunes look like a crescent in plain view and have a gently inclined windward slope with steep lee side, around which tapering cusps of the dune project downwind (Figure 18.15). If persistently strong winds occur or if multi-directional winds persist, the crest of a barchan may be truncated and the lee slope flattened, resulting in *dome* dunes.

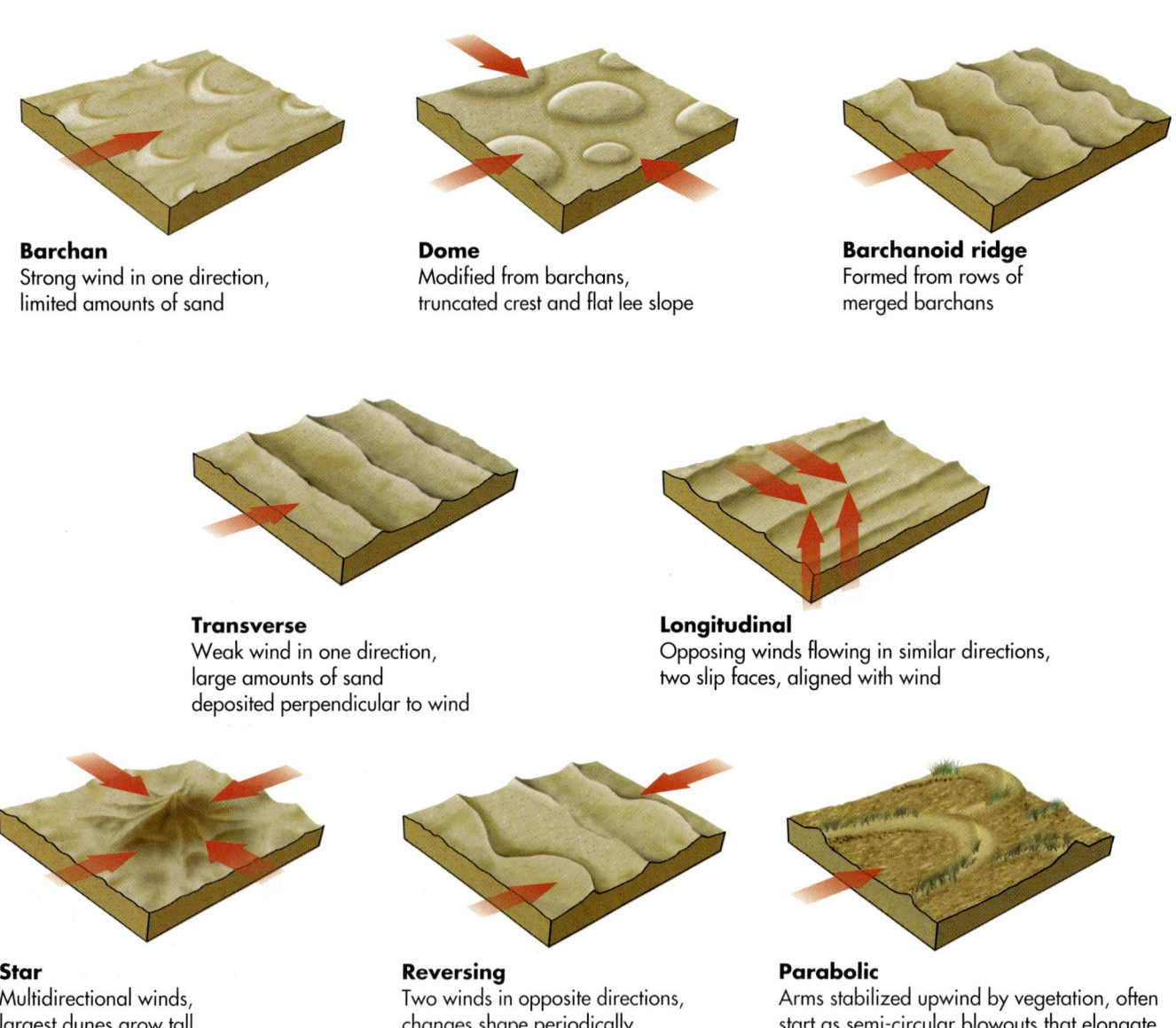

Barchan
Strong wind in one direction, limited amounts of sand

Dome
Modified from barchans, truncated crest and flat lee slope

Barchanoid ridge
Formed from rows of merged barchans

Transverse
Weak wind in one direction, large amounts of sand deposited perpendicular to wind

Longitudinal
Opposing winds flowing in similar directions, two slip faces, aligned with wind

Star
Multidirectional winds, largest dunes grow tall instead of moving

Reversing
Two winds in opposite directions, changes shape periodically

Parabolic
Arms stabilized upwind by vegetation, often start as semi-circular blowouts that elongate

Figure 18.14 Sand dune classification. Eight major varieties of sand dunes occur, with form depending on wind direction, sand supply, and the amount of vegetation cover. Arrows represent the direction of prevailing winds.

Figure 18.15 **Barchan dunes.** Barchan dunes in Namibia. The prevailing winds are from right to left, resulting in dune migration toward the left-hand side of the image.

Figure 18.17 **A typical blowout.** Blowouts like these in the Nebraska Sand Hills form when vegetation thins and strong winds deflate the core of a sandy deposit.

Usually, barchans are isolated dunes that form in areas of strong wind and relatively small amounts of sand. Where winds are weak, however, and massive quantities of sand are present, barchanoid ridges may form. Barchanoid ridges are sinuous, asymmetrical ridges that are transitional to a transverse ridge. Transverse ridges are primarily a conse-

(a)

(b)

Figure 18.16 **Types of dunes.** (a) Longitudinal dunes such as these in western Australia form as a result of two dominant wind directions. (b) Star dunes such as these in the Sahara desert form when winds are multi-directional.

quence of unidirectional winds that result in a linear deposit of sand perpendicular to the prevailing wind direction.

Another kind of dune that forms in barren landscapes is a *longitudinal dune* (Figure 18.16a). Longitudinal dunes have the same linear form as transverse ridges, but differ because they develop along a line between two prevailing wind directions. In areas of multidirectional winds with very little vegetation, *star dunes* may form (Figure 18.16b). Star dunes can grow to be very large. In the Sahara, for example, star dunes may be several hundred meters in height and many kilometers in diameter. Rather than migrate, star dunes simply grow in height. A transitional dune between star dunes and transverse ridges is the *reversing dune*, which forms where two winds from nearly opposite directions are balanced with respect to strength and duration. As a result, a second slip face periodically develops.

Dunes that form in the presence of vegetation and/or moisture include *parabolic* and *blowout* dunes (Figure 18.17). Both of these dune types are essentially deflation

GREAT SAND DUNES NATIONAL MONUMENT

One of the most impressive dune fields in North America is Great Sand Dunes National Monument in south central Colorado. This dune field lies within the San Luis Valley, between the San Juan and Sangre de Cristo Mountains. Overall the dune field is about 63 km² (24 mi²) in size and contains dunes that are up to 213 m (700 ft) high, making them the tallest dunes in North America.

The tallest sand dunes in North America occur in the Great Sand Dunes National Monument. This photograph illustrates the beautiful juxtaposition of the dunes in the foreground and the Sangre de Cristo Mountains in the background.

The dune field has formed over the past several thousand years because westerly winds blow over the San Juan range to the west and then rapidly sink into the rainshadow of the San Luis Valley, often reaching speeds greater than 64 km (40 mi) per hour. These dry winds then pick up

features with slip faces that slope in many directions. A typical blowout is a circular bowl, whereas parabolic dunes have U- or V-shaped arms that point upwind. These arms are usually stabilized by vegetation, which causes the dune to elongate as it migrates downwind. Parabolic dunes frequently originate as blowouts that subsequently take on the parabolic form upon further migration.

Loess

Another form of eolian deposit is called **loess** (a German word that, in the United States, is usually pronounced "luss"), which consists of fine-grained, wind-blown silt. Loess is typically highly calcareous, which means that it contains a lot (as much as ~40%) of calcium and is yellowish-tan (buff-colored) in color. Often, loess originates within the same region as the sand that forms dune fields. However, because silt can be carried far in the air by suspension, it becomes sorted out from the sand and is subsequently deposited great distances from its original source, leaving the dune fields behind (Figure 18.18). Loess also originates in stream valleys after silt is deposited on floodplains

Loess *Wind-blown silt.*

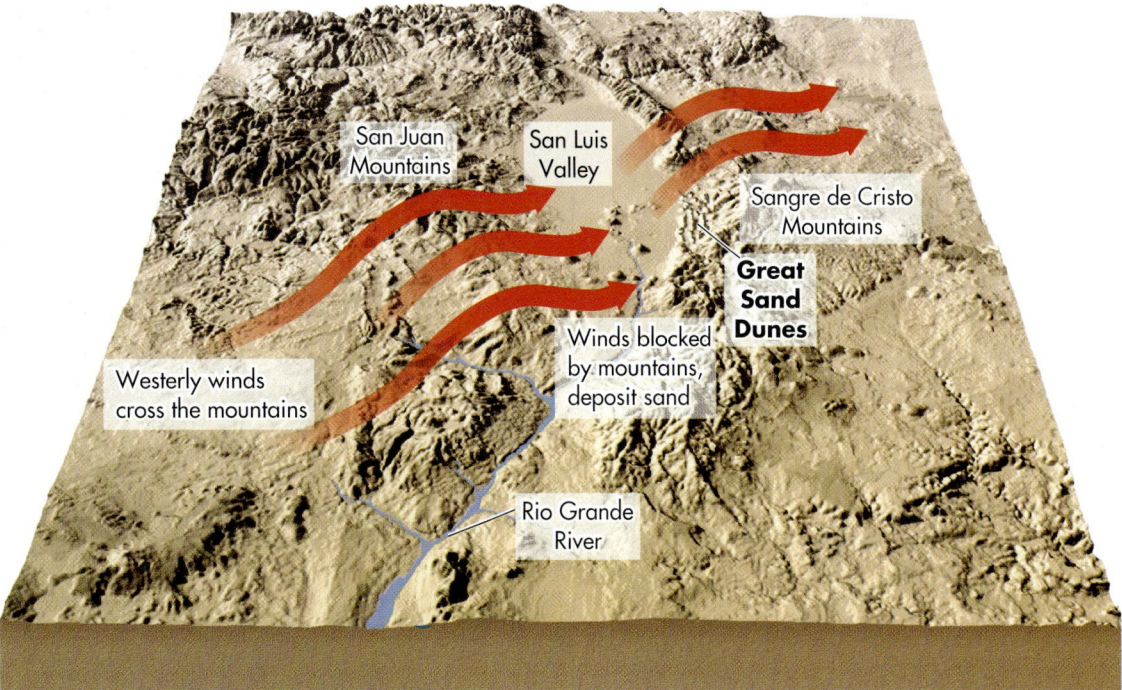

The sand dunes in Great Sand Dunes National Monument formed due to a unique combination of atmospheric and geomorphic variables. After westerly winds cross the San Juan Mountains, they descend into the San Luis Valley. As the wind blows across the valley, it picks up sands deposited by the Rio Grande River and transports them to the mountain front, which blocks the sands from moving further.

floodplain sands, which are mostly weathered igneous sediments from the San Juan Mountains that were carried to the valley by the Rio Grande River. The winds transport these sediments eastward, where they are blocked by the Sangre de Cristo Mountains. Here, the sand deposits accumulate to form dunes.

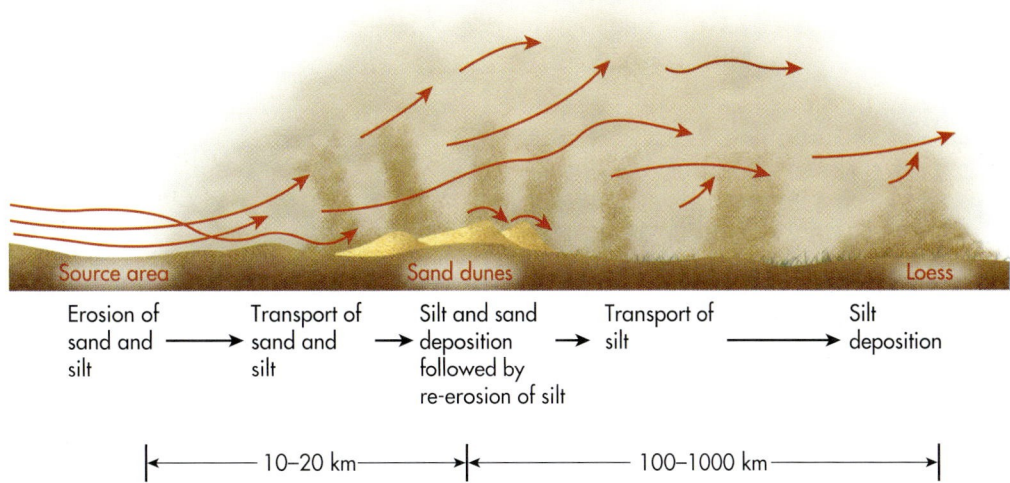

Figure 18.18 The process of eolian sorting. After sediments are deflated from a source area, silts and clays are transported great distances in suspension. Coarse sands, in contrast, travel a relatively short distance.

If you understand how different kinds of dunes form, you can easily interpret the dune formations in arid landscapes. For example, this photo shows barchan dunes near the Skeleton Coast in Namibia. Assuming that north is at the top of the photograph, what does the orientation of the dunes tell you about the direction of prevailing winds? What must the sand supply be like at this place—high or low? Can you identify the backslopes and slip faces of the dunes?

during floods. A great deal of loess also owes its origins to the thick deposits of glacial till and outwash deposited during the Ice Age. Following the deposition of these sediments by the glacier, the ice retreated, exposing surfaces that were unvegetated and easily deflated by strong winds. These winds picked up the silts contained within these deposits and carried them downwind.

Loess is a distinctive deposit for two primary reasons. One is that deposits of loess occur as thick blankets of sediment that are regional in their extent, with approximately 10% of the Earth's land surfaces covered in this way. For example, extensive deposits of loess cover much of the central United States, Argentina, Russia, and China (Figure 18.19). These areas are some of the most productive agricultural regions on Earth because loess can absorb a lot of water and the

GEODISCOVERIES   WWW.WILEY.COM/COLLEGE/ARBOGAST

EOLIAN PROCESSES AND LANDFORMS

Now that we have discussed the basic components of eolian processes and associated landforms, let's visualize these concepts in an animated format. To do so, go to the *Geo-Discoveries* website and open the module *Eolian Processes and Landforms*. This module describes the fluid flow of air

and how it shapes the landscape. In this animation, you'll see how abrasion shapes rock outcrops; how the variables of sand supply, wind direction, and vegetation influence the formation of dunes; and how loess develops. After you complete the animation, be sure to answer the questions at the end of the module to test your understanding of this concept.

SAND DUNES ON MARS

You may already know that a lot of research is being conducted to learn about the environment and geologic history of Mars. The reason for this intensive research program is that Mars and Earth are about the same size, are relatively close to one another, and have a somewhat similar geologic history. Evidence suggests that flowing water once existed on Mars, which means that the planet may have had an atmosphere somewhat similar to Earth's at one time. It also means that some form of life may have once existed there, or may still exist in some obscure place.

One of the ways that Mars is being studied is with satellite remote sensing. This program is revealing much about the character and history of Mars, including what you can see in this image. Do you recognize these features? They are barchan dunes. To a geographer, the presence of barchan dunes on Mars means that wind blows sand on the planet. It also means that the amount of sand available to blow is somewhat limited. Of particular interest on these dunes are the randomly spaced dark patches. If these dunes were on Earth, you might assume that these spots are patches of vegetation. But, of course, vegetation does not exist on Mars. So what are they? The best guess so far is that the dunes are covered with a light frost in the Martian winter and that the spots represent places where the frost sublimates (turns to vapor) as summer approaches. Fascinating, isn't it?

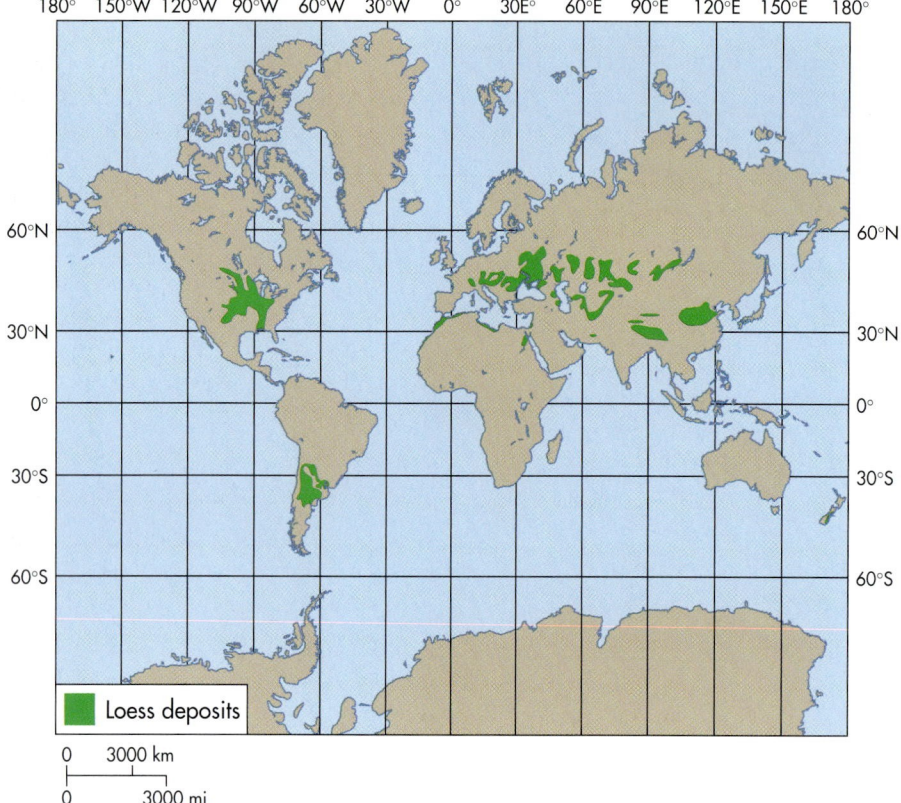

Figure 18.19 Significant loess deposits in the world. Loess covers about 10% of the Earth's landmass and about 30% of the United States. Much of the loess in the U.S. originated from glacial sediments that were subsequently deflated.

deposits are usually calcareous. A second distinctive aspect of loess is that it is capable of maintaining nearly vertical faces upon exposure (Figure 18.20). This durability occurs because the silt grains are strongly attracted to one another and are bound together by the calcium carbonate contained within it.

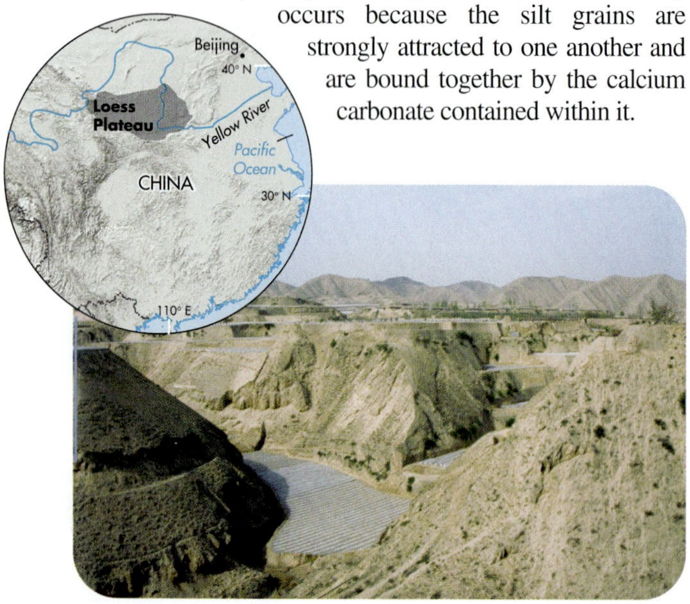

Figure 18.20 Loess deposits near the Yellow River in China. This entire landscape is covered by thick deposits of loess. Given that loess is highly calcareous, it holds vertical faces very well, which is why agricultural fields can be carved into the sediments.

KEY CONCEPTS TO REMEMBER ABOUT EOLIAN LANDFORMS

1. The best-developed depositional eolian landforms are sand dunes. Sand dunes occur in a variety of forms, determined by wind direction and strength, sand supply, and the amount of stabilizing vegetation.

2. The wind is an efficient sorting mechanism, which means that sediments are separated by size. Clay-sized particles travel the farthest from the source area. Silts move an intermediate distance, and pebbles and gravels remain in place because they are too large to move by wind.

3. Another eolian deposit is loess, which is windblown silt that is carried great distances in suspension before it settles to the ground. Loess covers much of the Earth and is very fertile.

Human Interactions with Eolian Processes

In Chapter 16 we examined how humans interact with stream systems. Humans can also impact the eolian landscape in many ways. In several examples around the world, people have contributed to the degradation of a landscape through the process of **desertification;** that is, transforming a formerly vegetated landscape into one that is relatively barren and highly susceptible to wind erosion. This process is enhanced in marginal, semi-arid landscapes where the vegetation density is al-

Desertification *The process through which a formerly vegetated landscape gradually becomes desert-like.*

ready low. Figure 18.21 shows extensive regions of the Earth that are susceptible to desertification. Until recent times, people have generally avoided these kinds of landscapes because rainfall to sustain crops and animal herds is unpredictable. However, with increasing population pressure, more and more people have been moving into these formerly uninhabited places in search of land that can be cleared for farmland. In the process of this development, the impact on the landscape has been enormous. As you will see in the following discussion, human-induced desertification can have catastrophic consequences.

Desertification in the African Sahel

One place where human-induced or human-enhanced desertification has occurred is the Sahel region of Africa. The Sahel is a narrow band of latitude that lies between

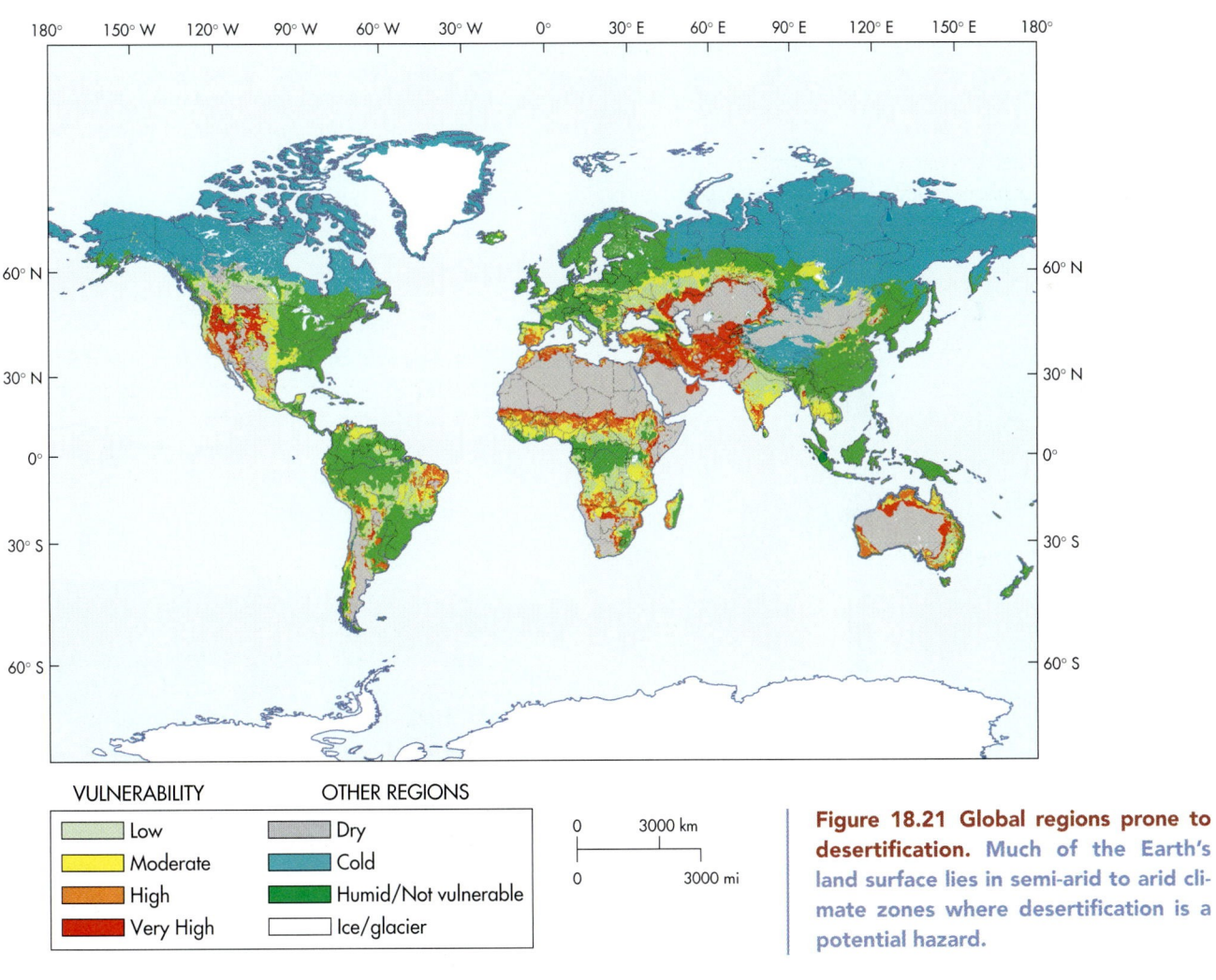

Figure 18.21 Global regions prone to desertification. Much of the Earth's land surface lies in semi-arid to arid climate zones where desertification is a potential hazard.

VULNERABILITY

- Low
- Moderate
- High
- Very High

OTHER REGIONS

- Dry
- Cold
- Humid/Not vulnerable
- Ice/glacier

0   3000 km
0   3000 mi

about 15–18° N and extends from the country of Senegal on the west coast to the country of Sudan on the east coast (Figure 18.22a). Recall from Chapter 9 that this part of Africa has a distinct wet/dry precipitation cycle that depends upon the seasonal passage of the (ITCZ). Consequently, the wet season occurs from late June to mid-September when the ITCZ is in the region. The remainder of the year, in contrast, is quite dry because of the dominance of the Subtropical High pressure system. In the context of vegetation, the region lies between the Sahara Desert to the north and the tropical savannah (grass and open forest) to the south. In other words, this region is an **ecotone** because it lies between two distinct ecosystems and serves as the transition from one to the other. The photograph in Figure 18.22b gives you a feel for the character of the place.

Prior to the 19th century, the Sahel was largely home to nomadic pastoralists and small-scale sedentary farmers. The nomads lived mostly in the driest (northern) part of the region, which bounds the Sahara Desert to the north, whereas the farmers lived in the humid southern part of the region. The nomads maintained small herds of cattle and goats and followed a practice of moving their herds to find grass that their cattle could graze. Given the

seasonal movement of the ITCZ and associated rainfall, their lifestyle evolved into a pattern of annual north-south migrations across the region; in other words, they moved their cattle to follow the rain. The farmers, in contrast, tended to stay in one place and had evolved a subsistence strategy that allowed them to be successful in this marginal environment. These strategies included staggering their plantings to have a continuous source of food, growing grains that mature quickly, and farming near a source of water such as river valleys.

Overall, the people in the Sahel had evolved a lifestyle that was compatible with the carrying capacity of the landscape; that is, the number of people living there did not exceed what the landscape could support without degradation. This sensitive relationship began to change in the late 1800s when Great Britain and France began to organize the region into distinct colonial entities. With this organization, boundaries arose that became barriers to the annual cycle of nomadic migration. In addition, the economic focus changed from living within the regional carrying capacity to exploitation of raw materials for consumption in Europe. Although the colonial era ended in the mid-20th century with the advent of African nationalism, the state boundaries, and their effect on human behavior, remained in place. People became increasingly sedentary, which resulted in the cultivation of fragile soils, overgrazing, and the removal of trees for firewood. In addition, overall population increased because people

Ecotone *The transition between two distinct ecosystems that contains species of each area.*

(a)

(b)

Figure 18.22 The Sahel region in Africa. (a) This region is an ecotone that lies between the grass and forested regions to the south and the Sahara Desert to the north. (b) This typical shot of the Sahel shows the character of the landscape. Note the widely spaced trees and intervening barren areas.

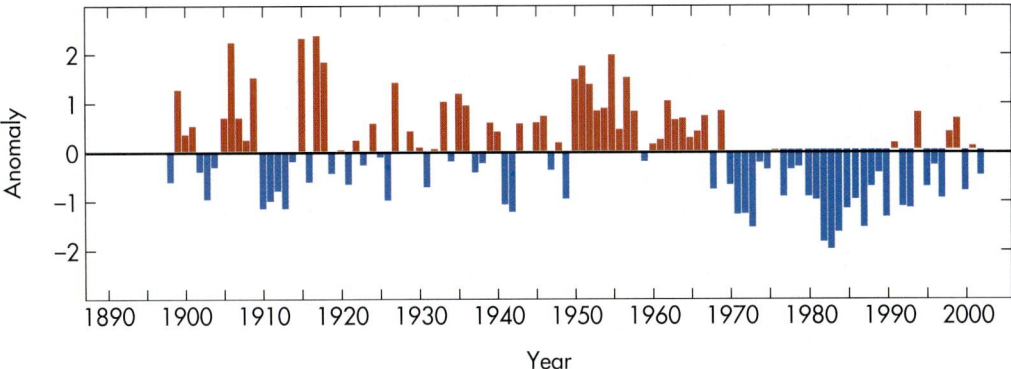

Figure 18.23 Annual rainfall from June through October in the Sahel, 1898–2002. In this graph, 0 represents the mean amount of rainfall during this period. Positive numbers above the mean are annual amounts of rainfall that are above the mean. Conversely, negative values below the mean represent below-average amounts of precipitation. Until the late 1960s, the region was relatively wet. Drought conditions have dominated since that time (Source: World Meteorogical Surface Station).

from more humid regions to the south migrated into the Sahel in search of places to cultivate.

As a result of the combined factors of boundary establishment and increasing overall population, the landscape of the Sahel began to degrade in the early 20th century because the carrying capacity was greatly exceeded. Landscape degradation and desertification of the Sahel reached a catastrophic level during the late 1960s and early 1970s. As you can see in Figure 18.23, this period coincides with a major period of drought that has basically lasted to the present time. Coupled with the preexisting human-induced degradation that had already occurred on the landscape, this drought resulted in widespread desertification in the region.

In association with this drought, hundreds of thousands of people and countless numbers of livestock died of starvation; in fact, it has been estimated that approximately 100,000 people died in 1973 alone. The amount of wind erosion has also increased over the region, frequently resulting in the transportation of large amounts of atmospheric dust (Figure 18.24). Some researchers believe that this increased eolian dust is a direct result of the human impact on the landscape. Others believe, however, that the dust is a result of increased occurrence and intensity of tropical waves (low-pressure systems) in the region.

Desertification in the Great Plains of the United States: The Dust Bowl

People in the United States sometimes think that catastrophic environmental impacts such as those in the Sahel occur only in poor countries where most people have little education and economic opportunity. This view of the world is inaccurate. In fact, many examples have occurred within the so-called "developed world" of human impacts on the landscape resulting in dire environmental consequences. An excellent example in

Atlantic Ocean

Africa

Figure 18.24 Atmospheric dust off the northwest coast of Africa. The combination of drought and human environmental impacts in the Sahel may have increased the incidence of dust blowing off of the African continent.

the United States of human-induced desertification on a large scale is the region known as the *Dust Bowl*, which became part of the national consciousness in the 1930s.

The dust bowl region is located in the western Great Plains and includes parts of Colorado, Kansas, Nebraska, New Mexico, Oklahoma, and Texas (Figure 18.25). As discussed previously, the climate of this region is semi-arid due to the rainshadow of the Rocky Mountains. In response to this climate, the natural vegetation is predominantly short-grass prairie.

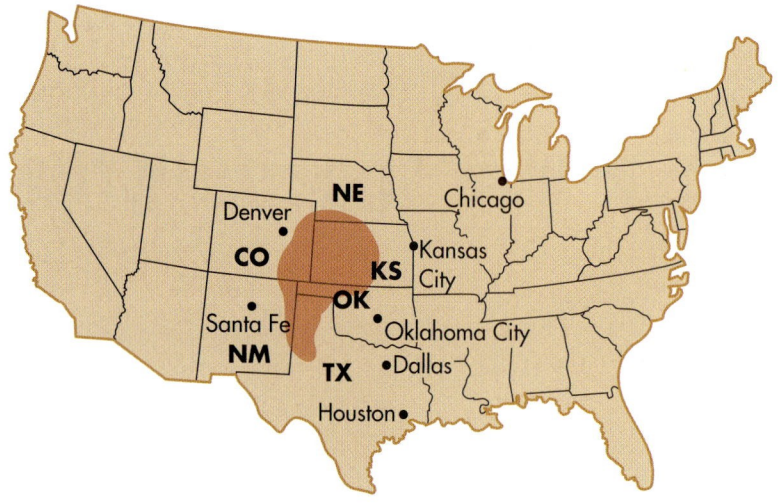

Figure 18.25 The dust bowl region. Long-term, severe drought in the 1930s impacted a large area of the Great Plains, including parts of Nebraska, Kansas, Colorado, Oklahoma, New Mexico, and Texas.

Until the later part of the 1800s, this region was largely uninhabited, except for small numbers of nomadic groups of Native Americans such as the Comanche and Pawnee. Given the hunting and gathering subsistence of these groups, they had very little impact on the landscape because they moved frequently from one place to another. However, beginning in the 1870s, large numbers of European Americans from the eastern United States began to settle in the western Great Plains in association with the Homestead Act of 1862. In this context, each family of homesteaders was given a free quarter section, or about 65 hectares (160 acres), of land with the understanding that the land would be developed in some way. Given the high fertility of the Mollisols found in the region, the vast majority of this land was plowed and subsequently farmed. In this way, most of the native grassland in the region was destroyed over the next 50 years.

Although the region is known for recurring drought because of its rainshadow location, the first few decades of European settlement were sufficiently wet to give farmers reason for long-term optimism regarding their economic future in the area. A drought in the early 1890s caused some concern, but was not overly extreme. In the middle 1930s, however, a period of especially intense drought occurred that decimated the landscape and thus the regional farm economy. Undoubtedly, this drought would have severely impacted the landscape even if it had occurred during the pre-settlement period when the ground was covered with grass. However, since most of the grass had been plowed under in the preceding few decades, the soils were exposed and easily deflated by the very high winds that accompanied the drought.

One of the lasting images from the dust bowl era is the massive dust storms that developed when tons of topsoil were blown away. Figure 18.26 shows an immense wall of windblown dust from one of these storms. They were so intense that they were called *black blizzards* and resulted in the deposition of fine silt and clay as far away as Chicago. In addition, the dust was so

Figure 18.26 Wall of windblown dust. Dust storms in the Great Plains, such as this one in Oklahoma, were common during the dust bowl era. These massive storms resulted from the combined impacts of drought, high winds, and plowed soils.

thick that as far east as Kansas City, streetlights were frequently left on during daylight hours. To give you a feel for just how bad it was in the region during this period, consider that from 1933 through the first nine months of 1937 there were 352 dust storms like the one pictured in Figure 18.26. Can you imagine what it must have been like to live there?

The overall result of the drought was to desertify the landscape in much of the region. From a cultural and economic standpoint, the impact was devastating, leaving many families literally on the verge of starvation. Although many people chose to stay and hope for rain, large numbers migrated to California in search of a better life. This migration is best depicted in John Steinbeck's famous novel, *The Grapes of Wrath*.

For those who stayed in the dust bowl region or moved into the area at a later time, the drought and resulting desertification demonstrated that any agricultural strategy must include conservation measures to mitigate future topsoil loss. As discussed in Chapter 15, one of the major changes that occurred in the region was the development of circle-pivot irrigation. This technology not only maintains soil wetness during drought conditions but leaves the portion of the field outside the cultivated circle available for stabilizing grass. Another soil conservation measure is a *windbreak* (Figure 18.27), which is a row of tall trees planted along a field that blocks the wind from blowing directly across the soil. Still another soil-conservation strategy is *conservation tillage*. In this method, the soil is not plowed as deeply

Figure 18.27 A windbreak. Windbreaks are rows of tall trees planted along fields that block the wind from blowing directly across the ground.

or as thoroughly, leaving at least 30% of the previous year's crop residue on the ground. This residue increases the surface roughness of the ground, which helps protect the soil from wind erosion and helps conserve water in the soil.

KEY CONCEPTS TO REMEMBER ABOUT DESERTIFICATION OF SEMI-ARID LANDS AND SAND DUNES AS INDICATORS OF CLIMATE CHANGE

1. Sand dunes are very sensitive to climate change. Thus, they often record cycles of drought and precipitation in their stratigraphy (vertical sequence of sediments).

2. The Sahara has been an arid desert for only the past few thousand years. Radar imagery of bedrock geology indicates that, before the desert period, the region was humid, with flowing rivers on the landscape.

3. Many dune fields on the Great Plains are less than 10,000 years old. Many dunes contain buried soils that formed during periods of relatively high precipitation and associated dense grassland vegetation that stabilized the dunes.

4. Desertification is the process through which marginal landscapes become desert. Two examples of human-induced desertification in the 20th century are the events that occurred in the African Sahel and the American Dust Bowl. Proper soil conservation can prevent desertification in many instances.

The Big Picture

In the next chapter, we finish our tour of the various geomorphic processes and associated landforms by discussing coastal landscapes. The coastal zone is a dynamic interface between the atmosphere, large water bodies, and landmasses. A variety of geomorphic processes occur along coastlines that are quite specific to a very narrow portion of the shore. This part of the Earth is shaped, often in dramatic ways, because waves contain a tremendous amount of energy that is continuously dispersed along the coast over time. If you happen to live by the shore or enjoy traveling to it, you'll learn a lot in this last chapter.

Summary of Key Concepts

1. Arid and semi-arid environments collectively comprise over 30% of the Earth's land surface. Arid landscapes are excellent places to study geomorphology because elements such as rock structure are easily seen. In places where the structure is horizontal, desert landscapes often contain a progression of erosional landforms, including plateaus, mesa, buttes, and pinnacles.

2. Eolian processes involve the shaping of the Earth due to the wind. The wind moves sediment in one of three ways: 1) fine grains such as silts and clays are carried great distances in suspension; 2) sand grains bounce along the ground in the process of saltation; and 3) heavier particles roll along the ground by creep.

3. The wind shapes the landscape through the combined processes of erosion and depositon. The formation of yardangs is an example of wind erosion. In contrast, sand dunes form when wind-blown sand is deposited and shaped.

4. Sand dunes come in many different shapes and sizes. The shapes of dunes depend largely on the supply of sand, the amount of vegetation, and prevailing wind speed and direction.

5. Loess is windblown silt carried a great distance by suspension. Thick deposits of loess occur in many places around the world, with most accumulating during glacial cycles.

6. Human-induced desertification is a major environmental issue facing the world today. Desertification occurs when the stabilizing vegetation on the landscape is reduced through human activity. This process exposes soils to the wind, which causes severe soil erosion and the transport of eolian sediment.

Check Your Understanding

1. What are the three large-scale environmental variables that contribute to the presence of deserts? What are the processes associated with each these factors and why do they produce arid environments?

2. Beginning with a plateau, describe the sequence of landform evolution where horizontal rocks of differing resistance influence the regional geomorphology.

3. What is a playa and where do they occur?

4. How does flowing air compare to flowing water? How is it similar and how is it different?

5. In the context of eolian processes, describe suspension, saltation, and creep. How do these various processes function and what kind of sediment is associated with each one?

6. What is the difference between deflation and abrasion?

7. Describe how ventifacts and yardangs form. How is a yardang somewhat similar to a glacial roche moutonée?

8. Sketch and describe a typical dune cross section and identify the three important components. How does sand move in these places and why do these processes collectively result in dune migration?

9. How do the factors of sand supply, wind (speed and direction), and vegetation influence the behavior and form of sand dunes? Provide at least three examples.

10. What is loess, what is its source, and how is it transported?

11. How do sand dunes often contain a stratigraphic record of climate change in a region?

12. Describe the process of desertification. Why are marginal landscapes most likely to experience human-induced desertification?

13. Name and describe two conservation techniques used to protect soils from wind erosion.

ANSWERS TO VISUAL CONCEPT CHECKS

Visual Concept Check 18.1

This landscape contains an example of desert pavement, which consists mostly of pebbles and gravel. Desert pavement forms when smaller sediment particles, such as sand and silt, are carried away by the wind, leaving the coarser particles behind. Over time the gravel pebbles and gravel coalesce to form a surface that resembles pavement.

Visual Concept Check 18.2

Since north is at the top of the photograph, the orientation of these barchan dunes indicates that prevailing winds are northeasterly, that is, from the north and east. These winds are moving the dunes in a slight by southwesterly direction toward the lower left of the photograph. The backslopes face the direction from which the winds are coming, whereas the slip faces lie in the shadows. The presence of the barchan dunes indicates that the available sand supply is relatively low, causing distinct dunes to form rather than a sand sea.

CHAPTER NINETEEN

COASTAL PROCESSES AND LANDFORMS

In the previous three chapters, we described fluvial, glacial, and eolian

processes and the distinctive landforms associated with them. Now we

examine the various processes and landforms that occur along coastlines.

Although these processes affect a very small part of the Earth's surface,

they create some of the most distinctive landscapes you can see.

Why should you be interested in coastal processes and landforms? One very

important reason is that there is a good chance you live along or very near a

Are you attracted to this landscape? If so, you're not alone. Coastlines are perhaps the most exotic landscapes on Earth, and people love to live near or visit them. This chapter focuses on the oceans of the world and the way they shape the landscape along coastlines.

coastline, as do a slim majority of people in the United States (and indeed, the world).

People are attracted to coastlines for a variety of economic, recreational, and lifestyle

reasons and are affected by the geomorphic processes that occur along these narrow

strips of land. In turn, people influence the behavior of coastlines in dramatic ways

through development. Thus, understanding the geomorphology of coastlines is

important, not only for helping you better appreciate this part of the landscape, but

also for helping you become a good steward for these areas.

Oceans and Seas on the Earth

Before we begin a discussion about various coastal processes, let's take a closer look at the character of the world's oceans and seas because it provides the context through which to view the shore. Figure 19.1 shows the names and locations of most of the notable large water bodies on Earth. The term **ocean** refers to the enormous body of salt water that covers about 71% of the Earth. There are five geographic divisions of the ocean. The largest is the Pacific Ocean, which encompasses a third of the Earth's surface. This ocean is significantly larger than the Earth's entire landmass, having an area of 179.7 million square kilometers (69.4 million mi²). The next largest ocean is the Atlantic, followed by the Indian Ocean, then the recently recognized (in 2000) Southern Ocean, and the Arctic Ocean. The Arctic Ocean covers an area of about 14,090,000 km² (5,440,000 mi²), which is slightly less than 1.5 times the size of the U.S. The Southern Ocean was recognized because it encircles Antarctica and has distinctive circulatory patterns.

We use the term "ocean" to refer to these large water bodies. The terminology for smaller areas of water more closely associated with land is inconsistently applied. In general, the next largest bodies of water are considered to be **seas,** which are subdivisions of oceans and are partially enclosed by land. For example, the Mediterranean Sea (Figure 19.1) is mostly surrounded by the European,

Middle Eastern, and African landmasses. In fact, it opens to the Atlantic only in its far western end.

After seas the next largest body of water is generally considered to be a **gulf,** which is a smaller arm of an ocean or sea that is also partially enclosed by land. Unfortunately, the use of the terms "gulf" or "sea" can be misleading because some named gulfs are actually larger than some seas. The Gulf of Mexico, for example, is much larger at 1.6 million km² (615,000 mi²), than the Black Sea in Eastern Europe, which is 422,000 km² (163,000 mi²). A distinctive characteristic of a gulf is that it usually has a recessed shoreline that opens outward to a larger body of water. Notice in Figure 19.1 how the shoreline of the Gulf of Mexico curves northward and that the Gulf opens to the Caribbean Sea east of Central America. On a still smaller level, a **bay** is a relatively small indentation on the coast that is directly connected to an ocean, sea, or gulf.

Although oceans, seas, and gulfs differ by size and their relationship to land, they are essentially interconnected and therefore have the same basic chemical composition. This chemistry is a result of complex interactions among several factors, including the atmosphere, seawater, minerals, sediments, and the myriad of organisms living in the water. Given that water is an excellent solvent, the world's oceans and seas contain a wide variety of dissolved solids, such as chlorine, sodium, and magnesium ions, among others. The term **salinity** refers to the concentration of dissolved solids in seawater and is most com-

---

Ocean *The entire body of salt water that covers about 71% of the Earth.*

Sea *A subdivision of an ocean that is partially enclosed by land.*

Gulf *A relatively small body of saltwater that is surrounded by land on three sides and opens to a sea or ocean.*

Bay *An indentation in the shoreline that is generally associated with an ocean, sea, or gulf.*

Salinity *Concentration of dissolved solids in water that is measured in parts per thousand (‰).*

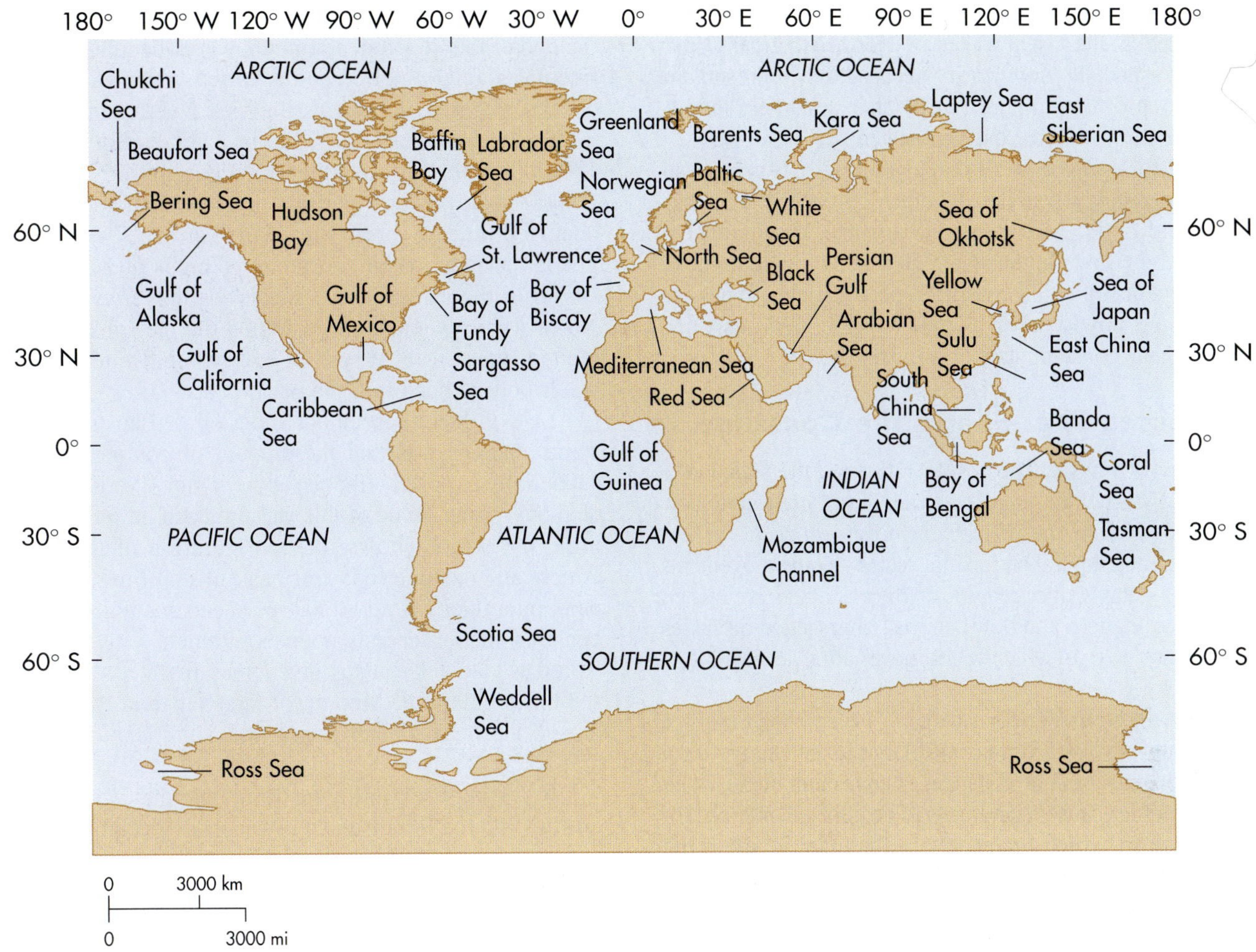

Map longitude labels: 180° 150° W 120° W 90° W 60° W 30° W 0° 30° E 60° E 90° E 120° E 150° E 180°

ARCTIC OCEAN   ARCTIC OCEAN

Chukchi Sea
Beaufort Sea
Bering Sea
Hudson Bay
Gulf of Alaska
Gulf of Mexico
Gulf of California
Caribbean Sea
Sargasso Sea
Baffin Bay
Labrador Sea
Gulf of St. Lawrence
Bay of Fundy
Bay of Biscay
Greenland Sea
Norwegian Sea
North Sea
Barents Sea
Kara Sea
Laptey Sea
East Siberian Sea
Baltic Sea
White Sea
Black Sea
Persian Gulf
Arabian Sea
Sea of Okhotsk
Yellow Sea
Sulu Sea
Sea of Japan
East China Sea
South China Sea
Banda Sea
Coral Sea
Tasman Sea
Mediterranean Sea
Red Sea
Gulf of Guinea
Mozambique Channel
Bay of Bengal
INDIAN OCEAN
PACIFIC OCEAN
ATLANTIC OCEAN
Scotia Sea
SOUTHERN OCEAN
Weddell Sea
Ross Sea   Ross Sea

Latitude labels: 60° N, 30° N, 0°, 30° S, 60° S

Scale: 0 — 3000 km ; 0 — 3000 mi

Figure 19.1 Global distribution of oceans, seas, and gulfs. There are five oceans on Earth. The largest ocean is the Pacific, followed by the Atlantic, Indian, Southern, and Arctic. Seas and gulfs are relatively small components of the oceanic network that are surrounded by land to a considerable extent.

monly expressed in parts per thousand. Overall, global salinity varies between 34 and 37‰. Water that exceeds a salinity of 35‰ is considered **brine,** whereas water with salinity less than 35‰ is called **brackish.**

The Nature of Coastlines: Intersection of Earth's Spheres

What makes a coastline distinctive from other landscapes? The primary distinguishing factor of coastlines is that they are places where large landmasses and large water bodies intersect (Figure 19.2), often for hundreds or even thousands of kilometers. Although coastlines are usually associated with oceans and seas, they also occur around large

Brine   *Water that has salinity greater than 35‰.*

Brackish   *Water that has salinity less than 35‰.*

Figure 19.2 A typical coastline in northern California. Coastlines such as this one at Big Sur are where the hydrosphere, lithosphere, and atmosphere meet in a very big way.

lakes such as the Great Lakes of North America. Coastlines are especially noteworthy because they represent the intersection of three major Earth spheres, namely the hydrosphere, lithosphere, and atmosphere.

Coastlines come in all shapes and sizes. Some coasts have wide, sandy beaches, whereas others feature rocky bluffs and cliffs. Along some coastlines, mountains extend straight down to the water. Because of wide range of environments in which coastlines occur, some are passive and change slowly, whereas others erode vigorously and are in a near-constant state of adjustment.

Processes That Shape the Coastline

A variety of physical processes influence the shape of a coastline. Most of these processes are directly related to how water moves along the shore. This movement occurs at several different temporal and spatial scales—some movements occur very slowly over centuries, whereas others are related to daily oscillations and flow patterns. This section discusses the primary ways in which water moves along a coastline.

Fluctuations in Water Level   Over long periods of time, water levels in oceans and large lakes change for a variety of reasons. For instance, changes can occur when tectonic forces cause a landmass to be uplifted or sink. Another way in which water levels fluctuate is when the amount of water in the ocean or lake varies due to adjustments in the hydrologic cycle. This form of water-level fluctuation is called **eustatic change.**

The best example of eustatic change in recent Earth history is the great change in sea level that occurred at

Eustatic change   *Fluctuations in sea level associated with adjustments in the hydrologic cycle.*

the end of the Pleistocene Epoch, when the massive continental glaciers melted and returned water back to the oceans. Figure 19.3 shows that sea level changed dramatically between the Illinoisan glaciation around 130,000 years ago and the Holocene period (the past 10,000 years). Major periods of low ocean water level occurred during both the Illinoisan and Wisconsin glacial periods, when enormous volumes of water were stored in the immense ice sheets that covered much of North America and Europe. When the ice subsequently melted, the sea level rose at a very rapid rate because water poured back into the ocean.

One of the important outcomes of the Pleistocene eustatic changes was that the amount of continental land subaerially exposed—that is, above water—varied a great deal over time. Some of this variation can be seen in Figure 19.4, which shows the configuration of the North American coastline today, at peak glacial time, and at a peak interglacial if all global ice sheets melted. Remember, this shape depends upon how much water is being stored as glacial ice at any given point in time. As you can see on the map, the amount of land subaerially exposed during the peak glacial period was significantly greater than today.

Referring back to Figure 19.3, note that water levels during both the Illinoisan and Wisconsin glaciations were about 125 m (410 ft) lower than today. At these times, large parts of the now-submerged continental shelf were subaerially exposed (see Figure 19.4 again) and rivers extended onto that platform. The evidence for these rivers still exists in the form of canyons that were cut into the shelf deposits during the major low ocean levels. An excellent example of such a canyon is the Hudson Canyon, which was eroded into the continental shelf by the ancestral Hudson River. Similar features also exist on the floor of Chesapeake Bay (Figure 19.5a), which is essentially a

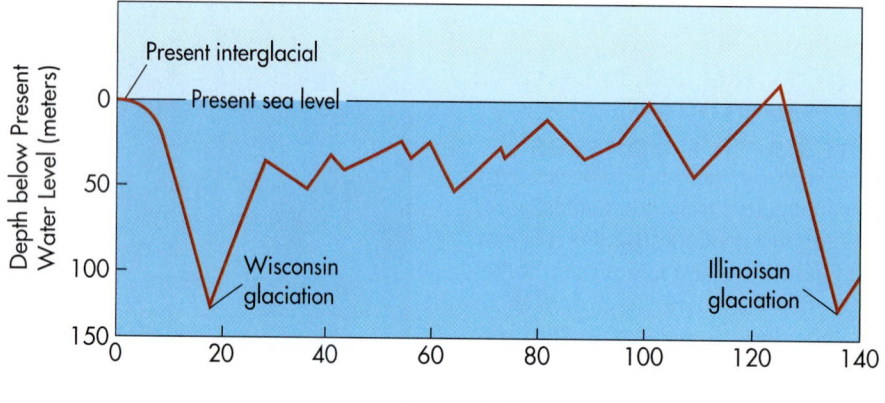

Age (thousands of years ago)

Figure 19.3 Eustatic sea level changes during the late Pleistocene Epoch at New Guinea. Dramatic low sea levels occurred during the Illinoisan and Wisconsin glacial maxima, with increases following these glaciations. (Source: Chappell, J., and Shackelton, N.J., 1986. Oxygen isotopes and sea level. Nature 324: 137–140)

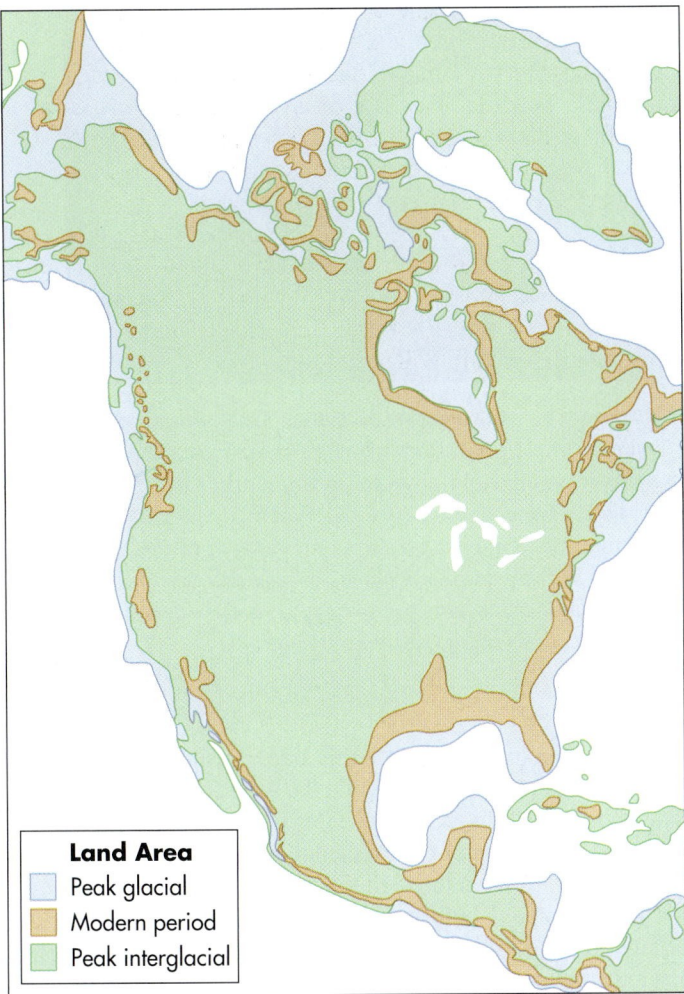

Land Area

☐ Peak glacial
☐ Modern period
☐ Peak interglacial

Figure 19.4 Configuration of the North American coastline during peak glacial, peak interglacial, and modern periods. Over time sea level has fluctuated greatly due to the amount of water tied up in ice sheets. Note the configuration of the coastline if all ice sheets melted.

large river valley, or **ria,** that was flooded when the sea level rose during the Holocene. Rias are common along many coastlines, with some of the most beautiful found in New Zealand (Figure 19.5b). The glacial counterpart of a ria is a **fjord,** which is an ice-formed trough that becomes flooded when sea level rises. Excellent examples of fjords exist in places such as Alaska, Norway, and the south island of New Zealand.

Ria   *A former river valley along the coast that is flooded by rising sea level.*

Fjord   *A former glaciated valley along the coast that is flooded by rising sea level.*

(a)

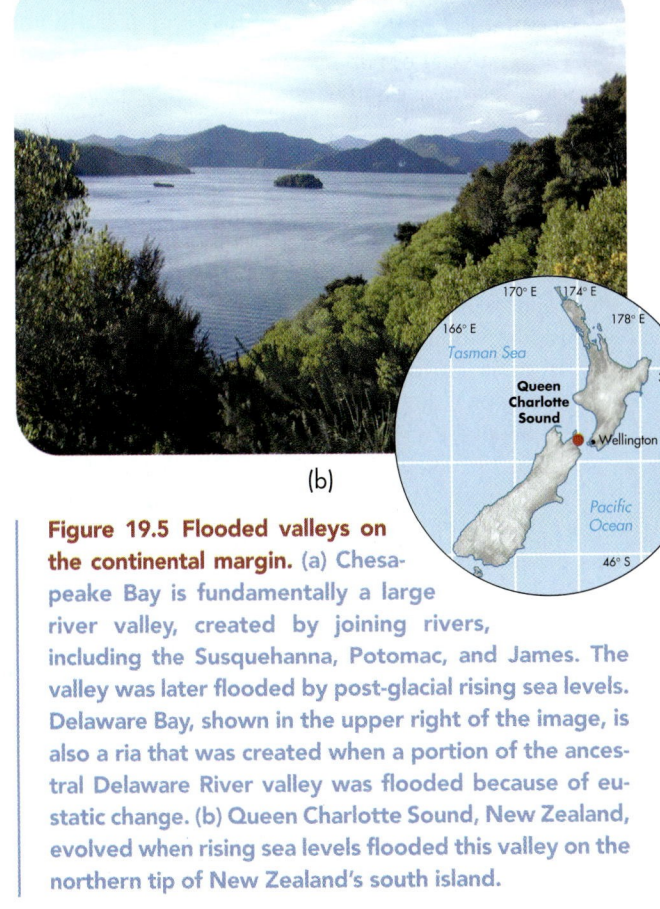

(b)

Figure 19.5 Flooded valleys on the continental margin. (a) Chesapeake Bay is fundamentally a large river valley, created by joining rivers, including the Susquehanna, Potomac, and James. The valley was later flooded by post-glacial rising sea levels. Delaware Bay, shown in the upper right of the image, is also a ria that was created when a portion of the ancestral Delaware River valley was flooded because of eustatic change. (b) Queen Charlotte Sound, New Zealand, evolved when rising sea levels flooded this valley on the northern tip of New Zealand's south island.

Figure 19.6 Tidal fluctuations in Chesapeake Bay. This image shows a boat dock at low tide. At high tide, the water reaches approximately half way up the support beams of the dock (arrow).

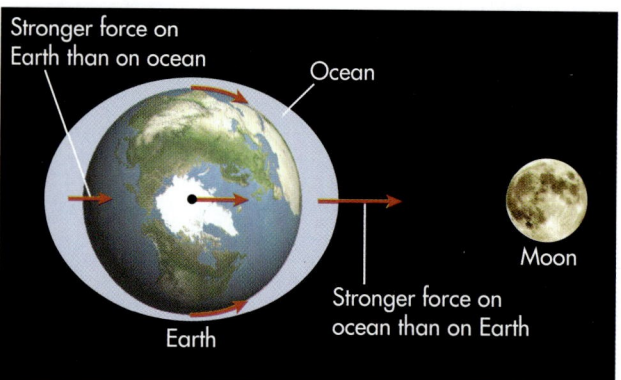

Figure 19.7 Formation of tides. Tides form mostly due to the gravitational pull of the Moon, which results in an ellipsoid of water in the world's oceans that extends from one side of the Earth to the other. Although tides may cause some movement of fine particles in tightly confined places where the tidal force is enhanced, tides are not usually associated with any large-scale shaping of the coastline.

Tides Tides are regular and predictable oscillations that occur with respect to the level of the world's oceans (Figure 19.6). These oscillations are related to Newton's law of gravitation, which states that every particle in the universe attracts every other particle in the universe. According to this law, both the Moon and Sun exert a gravitational force on the Earth. The most obvious effect of these gravitational forces that we can see on Earth are tides, which are basically the oceans being tugged back and forth by the Moon and Sun. Although it would seem, at first glance, that the Sun would have the biggest influence on tides due to its immense size, the Moon's gravitational effect is actually greater. This is because the strength of the force is inversely proportional to the distance: the closer the two bodies, the stronger the pull. Recall from Chapter 3 that the average distance from the Earth to the Sun is approximately 149,000,000 km (93,000,000 mi). In contrast, the Moon is only about 386,000 km (240,000 mi) away from the Earth. As a result, the Moon's gravitational pull is responsible for about 56% of the daily tides, whereas about 44% is related to the Sun's effect.

Figure 19.7 shows how the Moon's gravitational pull creates the tides. The Moon's gravitational effect on the ocean is greatest for the water that is closest to the Moon. This gravitational force progressively decreases across the Earth and is weakest on the side of the planet that faces away from the Moon. This difference in force causes the ocean facing the Moon to shift slightly closer to that body, forming a dome of relatively high water on that side of the Earth. This tide is called a *direct tide* because it results from the Moon pulling directly on the water. At the same time that a direct tide develops on one side of the Earth, a dome of relatively high water also exists on the other side of the planet. This tide is called an *opposite tide* and forms because the Moon pulls harder on

the Earth than on the water on this side, causing the Earth to move closer to the Moon while the water is left behind. The overall result is that water piles up in the form of an ellipsoid, with a long axis directed toward the Moon.

Midway between the high tides are the low tides. This cycle of alternating high and low tides lasts about 12 hours at any given location, resulting in two high tides and two low tides on most coasts. The highest high tides occur when the Earth, Moon, and Sun are in direct alignment and are called *spring tides*. In contrast, the lowest high tides occur when the Moon and Sun are at right angles to each other; these tides are called *neap tides*. In addition to the tidal variations that occur with respect to the geometric relationship of the Earth, Moon, and Sun, the overall tidal range can vary dramatically between places. For example, as you can see in Figure 19.8, the average tidal range at San Francisco, California is about 2.5 m (about 8.2 ft). The largest tidal range on Earth occurs in the Bay of Fundy, which lies between the Canadian provinces of New Brunswick and Nova Scotia in eastern North America (Figure 19.1). The shape of the coastline here has a funneling effect that produces a tidal range that varies from about 6 m (~20 ft) at the mouth of the bay to 16 m (53 ft) at the head of the bay!

Waves The most dynamic form of water movement along coastlines is associated with waves. **Waves** are oscillations in a body of water that form mostly due to the

Waves *Oscillations in a body of water that form mostly due to the frictional force generated by wind blowing across the surface of the water.*

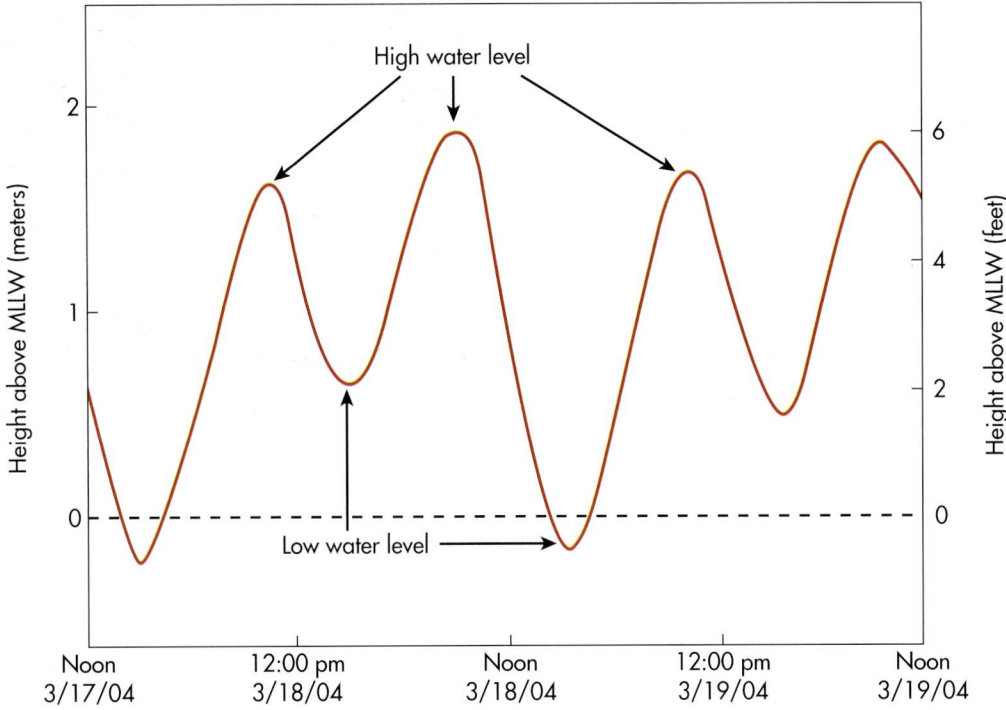

Figure 19.8 **Tidal range at San Francisco, California, over a two-day period.** The point of reference used here is the Mean Lower Low Water (MLLW), which is the average of the lower low water height of each day and is represented by the 0 baseline in the graph. Relative to this level, the average tidal range is about 2.0 m (about 6.5 ft). (Source: National Tide Center)

force of friction generated by wind blowing across the surface of the water. The specific wind factors involved in the formation of waves are: 1) wind strength, 2) wind duration, and 3) wind *fetch*—that is, the distance of unimpeded air flow. Although waves appear to be bodies of water that are moving horizontally, they are, in fact, nothing more than rising and falling disturbances of water masses. Very little water moves forward in a wave; instead, it is the wave energy that is transmitted forward through the elastic medium of the water. If you want to test this concept yourself, place a floatable object in the water the next time you're at the ocean and watch as the undulations pass beneath it. You'll see that, although many different waves pass by, the object moves only slightly.

Water waves are similar to the electromagnetic waves discussed in Chapter 4 because they consist of various measurable components. The *wave crest* is the high spot of an individual wave and forms when a vertical column of circularly rotating water particles pass through the water and cause the water to rise. Although the column of rotating water continues to some depth in the water, the diameter of individual oscillations progressively becomes less until they disappear at the *wave base*. After the wave crest passes, the water lowers into the *wave trough*. The horizontal distance between successive wave crests is the *wavelength*. *Wave amplitude* refers to the vertical distance between the wave crest and the level of the water if

it were still. *Wave height* is somewhat different from wave amplitude in that it measures the vertical distance between the wave crest and the base of the wave trough. In general, stronger winds result in greater wave heights and amplitudes. When waves move out of the area where the generating winds occur, they continue to move forward and are then known as *swells*.

As waves approach the coastline they begin to encounter shallower water because the sea floor gradually slopes upward to the shore (Figure 19.9). At some point the water depth becomes sufficiently shallow that the wave base and the ocean bottom intersect. From this point forward, the vertical space in which the rotating water has to move becomes progressively restricted, causing the overall wave height to increase and the overall speed of the waves to slow. At the same time that the forward speed of the waves is decreasing, from the point where the interaction of the wave base and ocean bottom begins, additional waves continue to advance into the area, resulting in compressed wavelength as more waves crowd into a smaller space. Ultimately, the height of each individual wave exceeds its vertical stability and the wave crashes in a feature called a **breaker.** This point typically occurs

Breaker *A wave that rises and crashes when the forward momentum of oscillations cannot be maintained.*

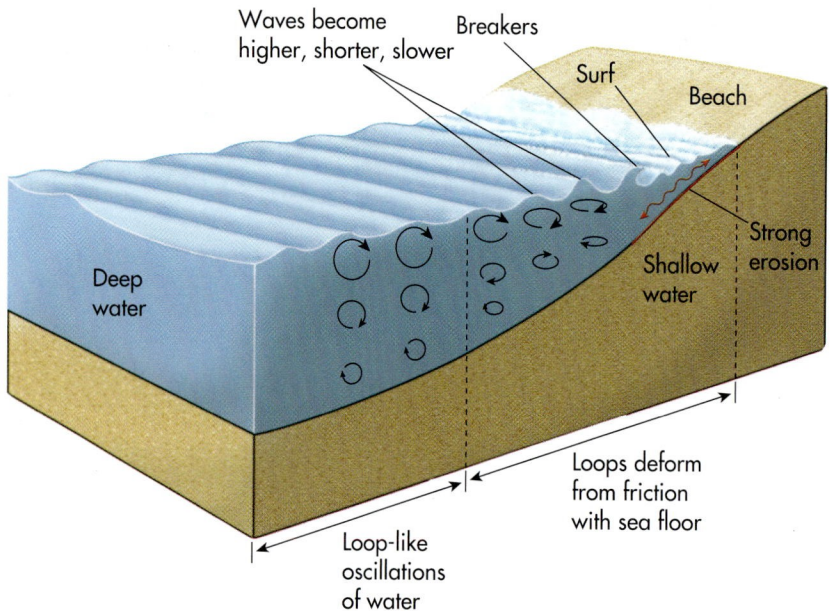

Waves become higher, shorter, slower

Breakers

Surf

Beach

Deep water

Shallow water

Strong erosion

Loops deform from friction with sea floor

Loop-like oscillations of water

Figure 19.9 Wave formation and components. Waves have a distinct structure that is driven by circular movements of particles of water. Waves crash at breakers when the wave height exceeds the vertical stability of the wave.

when the wave height is about seven times greater than the wavelength. From the time the wave breaks onward, the forward momentum of the wave causes water to rush onto the beach as *surf*, which is the only part of the wave system where water is actually moving forward, instead of up and down. Under normal conditions, crashing waves release a great deal of power that erodes the coast. This power is magnified many times when large storm waves are produced.

Although the vast majority of waves are caused by the wind, some waves also result from tidal surges and from major tectonic events such as an undersea earthquake, volcanic eruption, or major landslide. These waves are called *tsunamis* or *seismic sea waves* and form because water is rapidly displaced for some reason. To see how a tsunami forms from a seismic event, lower one of your hands—with your palm up and fingers touching—about 10 cm (4 in) into a filled bathtup. Once your hand is in position, quickly flick your fingers upward (mimicking the sudden rise of the sea floor) and watch how water is displaced vertically, which creates a small wave at the water surface that sweeps outward in every direction. In the open ocean, these waves are difficult if not impossible to see because they have low wave height, have wavelengths that can be tens of kilometers wide, and travel at several hundred kilometers per hour. However, when these waves reach shallow water, their rapid forward speed and encounter with the upward-sloping ocean bottom causes them to build quickly to considerable height, sometimes even 30

m (100 ft). As you can imagine, such a rapidly moving wave can be very destructive when it strikes the shore.

The devastating power of tsunamis was tragically illustrated in December 2004, when the Sumatra–Andaman earthquake occurred near Sumatra (part of Indonesia) in the eastern Indian Ocean where the India plate subducts beneath the Burma plate along the Sumatra fault (Figures 13.19 and 19.10a). As discussed in Chapter 13, this earthquake measured an incredible 9.1 on the Richter scale, making it the second strongest earthquake ever recorded on a seismograph (the strongest occurred in Chile in 1960 and measured 9.5 on the Richter scale). The Sumatra–Andaman earthquake caused a massive rupture along the fault that was over 1000 km (620 mi) long. In addition to the substantial lateral movement of the adjoining plates, the sea floor on the Burma plate was suddenly thrust up and over the India plate about 4 to 5 m (13 to 16 ft; Figure 19.10b). This thrust instantly displaced 30 km^3 (7.2 mi^3) of water, in a manner analogous to the flicking of your hand in water, resulting in a tsunami that rapidly radiated outward along the entire length of the rupture (Figure 19.10c).

Once in motion, the tsunami raced across the Indian Ocean at speeds ranging from 500 to 1000 km/h (310 to 620 mi/h). It reached the northwest coast of Sumatra to the east in about 15 minutes and the southern tip of South Africa, which is about 8500 km (5300 mi) to the southwest, in about 12 hours (Figure 19.11). Coincidentally, a French radar satellite happened to pass over the Indian Ocean two hours after the earthquake and measured the

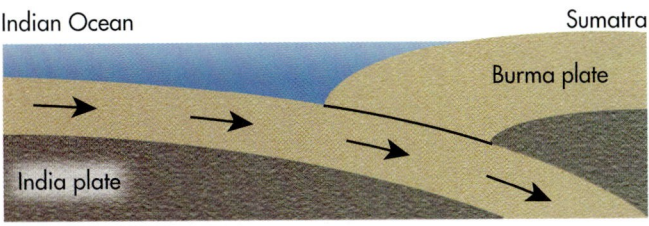

Indian Ocean — Sumatra

Burma plate

India plate

Stress gradually builds at subduction zone along India and Burma plate boundary.

(a) Before the earthquake

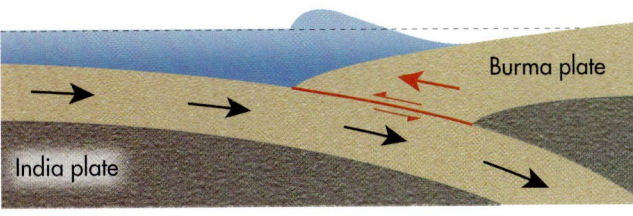

Burma plate

India plate

Earthquake causes Burma plate to thrust up and over the India plate several meters, causing massive displacement of overlying water.

(b) Earthquake strikes

Tsunami spreads

West ← → East

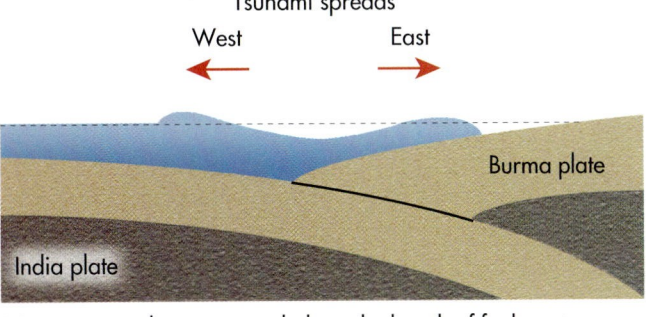

Burma plate

India plate

Tsunami spreads rapidly, reaching the nearby coast of Sumatra to the east in 15 minutes and the more distant shore of Somalia to the west in 7 hours.

(c) Tsunami radiates outward along the length of fault rupture

Figure 19.10 Evolution of the 2004 tsunami in the Indian Ocean. (a) Tectonic stress builds up where the India plate subducts beneath the Burma plate along the Sumatra fault. (Source: Geoscience Australia) **(b) Pent-up stress is suddenly released in a massive earth-quake. In addition to significant horizontal movement, the Burma plate is thrust up and over the India plate. This vertical movement displaces water in the overlying Indian Ocean. (c) After the water is initially displaced, waves rapidly radiate outward.**

tsunami wave heights as they moved across deep water. These observations indicated that the maximum wave height in the open ocean was only 60 cm (2 ft), which would have scarcely been noticed by ships in the area. As the waves approached the shore, however, they grew incredibly, reaching an estimated height of 24 m (80 ft) at Banda Aceh in the northwest coast of Sumatra. Some of the tsunami's energy spread into the Pacific Ocean, even creating a tsunami along the west coasts of North and South America that measured from 20 to 40 cm (7.9 to 15.7 in) in height.

Given the rare occurrence of tsunamis in the Indian Ocean, the waves came essentially without warning to the people in the region. The only precursors were the earthquake itself, which many people did not even feel or associate with a tsunami, and a temporary recession of the ocean shortly before the waves struck. In some places, the

ocean receded so much that the beach suddenly widened about 2.5 km (1.6 mi). This kind of rapid (and highly unusual) beach expansion is a clear tsunami warning to those familiar with these events.

An excellent example of why it's important to understand the behavior of Earth processes occurred during this tsunami. A 10-year-old British girl was vacationing with her parents in Thailand when the tsunami hit. Fortunately, she had just studied tsunamis in her grade-school class two weeks before the event so she understood that the rapid recession of water meant a tsunami was approaching. She warned her parents who, in turn, warned others nearby. The girl's awareness saved not only her own life but the lives of over 100 others as well. Unfortunately, however, to most people throughout the area the water recession and beach expansion was a source of curiosity because they were not

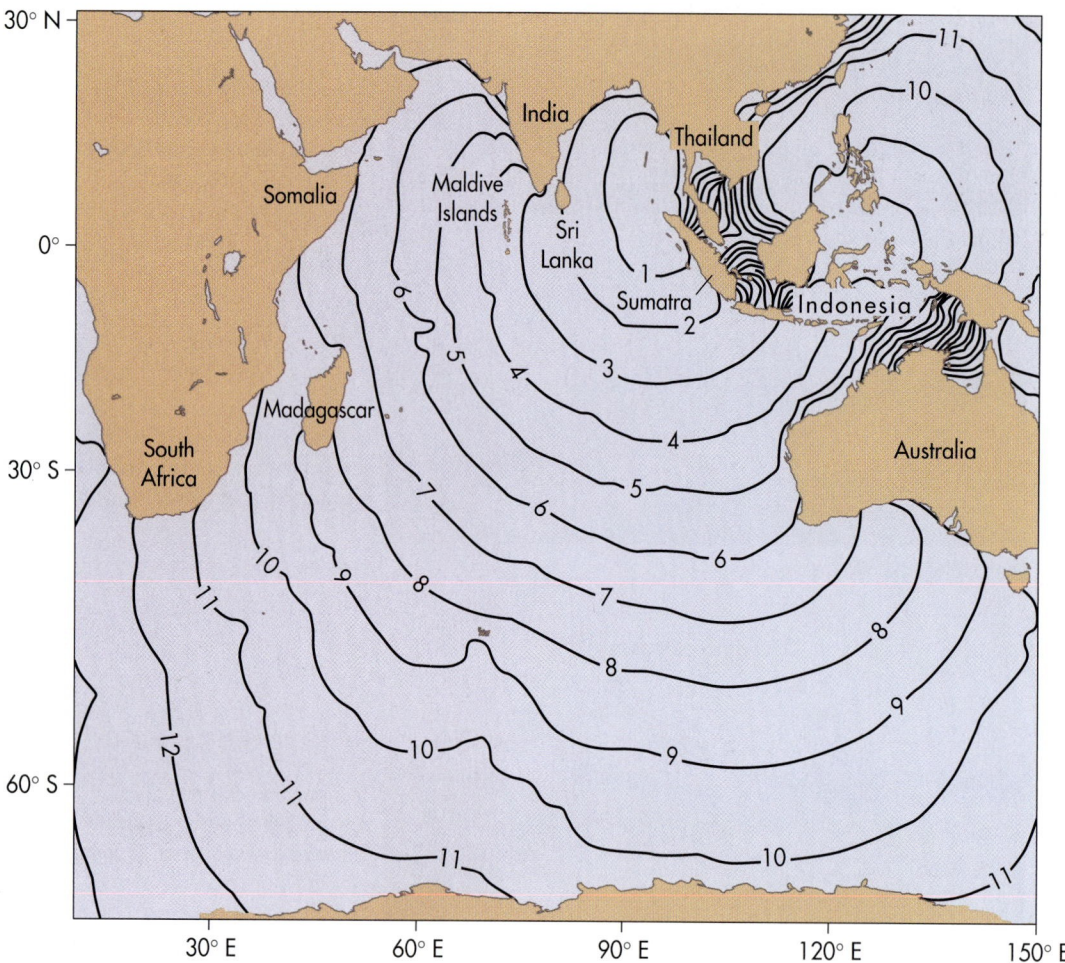

Figure 19.11 Travel times in hours for the 2004 tsunami in the Indian Ocean. The tsunami reached the shores of Indonesia and Thailand within minutes of the earthquake. In contrast, it reached the southern tip of South Africa in about 12 hours. (Source: NOAA)

aware of the danger signs. In fact, many people even ventured out onto the exposed sea floor to collect fish stranded there. Most of these people perished.

Amateur video of the tsunami in the hardest-hit areas of Indonesia and Sri Lanka illustrated the incredible power of moving water. You may have seen many of these scenes yourself. These videos showed not a single wave, but instead a series of massive water surges that engulfed coastal communities. In some places the water surged 2 km (1.2 mi) inland. The energy released by the tsunami is estimated to have been equivalent to about five megatons of TNT, which is more than all of the explosive energy used during World War II (including the two atomic bombs). This tremendous energy wreaked havoc along the coast (Figure 19.12a), causing billions of dollars in damage and killing at least 265,000 people. Many thousands more are missing and may never be found. The devastation was so extensive and widespread that it can clearly be seen in satellite imagery showing

the landscape before (Figure 19.12b) and after (Figure 19.12c) the tsunami.

The severity of the 2004 tsunami underscores the need for a warning system in the Indian Ocean. Such a warning system was established in the Pacific Ocean in 1949 because tsunamis are relatively common in this part of the world due to the extensive tectonic activity associated with the Pacific Ring of Fire. The Japanese, in particular, are well acquainted with tsunamis (hence the name: *tsu* = strong; *nami* = wave) because about 200 significant events have occurred in Japan alone in the past 1300 years. One of the most recent Pacific tsunamis was triggered by the massive 1964 Alaska earthquake, which created a wave that reached a height ranging from about 2 to 6.3 m (7 to 21 ft) in northern California and killed 11 people. In fact, a significant tsunami is considered by the United States Geological Survey to be statistically more likely along the U.S. west coast than in the Indian Ocean. Because of this concern, an extensive network of sensors

(a)

(b)

(c)

Figure 19.12 Devastation caused by the 2004 tsunami in the Indian Ocean. (a) Damage in coastal communities due to the surging water was catastrophic. (b) Satellite image of Banda Aceh before the tsunami. (c) Satellite image of Banda Aceh after the tsunami. Note the change in the configuration of the shoreline and reduction in island size.

was deployed throughout the Pacific basin in the 1950s and 1960s that monitors seismic events and associated sea-level fluctuations. Although this system is of little use during a sudden tsunami, it can provide warning within 15 minutes for events that originate far away. Such a system is now being considered for coastal communities along the Indian Ocean.

Littoral Processes In addition to being shaped by waves, coasts are also modified by **littoral processes,** which refer to the transport and deposition of sediment in the narrow shore zone. One way in which sediment moves along the littoral zone is in association with a feature called the **longshore current.** As the name implies, a longshore current consists of flowing water that moves along the coast. This current forms as a result of the way in which waves interact with the shore. Most of the time, waves approach the coastline from a distinct angle rather than in a perpendicular fashion (Figure 19.13). Because of this oblique approach, the force of the water is deflected downwind when it interacts with the shore. This deflection causes a current to develop that flows parallel to the shore, one that is capable of eroding sediment and transporting it down shore. This process is called **longshore drift.**

At the same time this current is flowing, water flows on and off the beach in an interesting way, related to wave run-up after waves break. Given the oblique approach of waves to the shore, the surf flows onto the beach at an angle; this surge is called *swash.* After the surf reaches its apex on the beach, it flows back to the ocean at a right angle to shore as *backwash.* Sediment is picked up in this fashion and is transported down the coast in a zigzag pattern known as **beach drift.** The next time you're on a large coastline, watch for the processes of swash and backwash and how they move pebbles down the shore.

Littoral processes *The processes through which sediment is transported and deposited in the shore zone.*

Longshore current *The current that develops parallel to the coast when waves approach a coast obliquely and forward momentum is deflected.*

Longshore drift *Transport of sediment by the longshore current.*

Beach drift *Sediment that is transported in the surf zone by swash and backwash, which form due to the oblique approach of waves.*

AMAZING PLACES

BIG WAVES IN HAWAII

Some of the largest waves on Earth occur at Waimea Bay, located on the north shore of Oahu in the Hawaiian Islands. During the winter months, mid-latitude cyclones pass far to the north of Hawaii. Nevertheless, these storms can pack strong winds, producing large swells that radiate outward from the center of the systems. Since Hawaii basically lies in the center of the Pacific Ocean, there is a tremendous fetch before the swells reach the islands, causing them to enlarge further.

Once the swells reach Waimea Bay, they encounter submerged shoals that are sufficiently high and far from the land to cause waves to rise up steeply. A great example of such waves occurred on January 28, 1998, when a strong mid-latitude cyclone passed 3000 km (1900 mi) north of Hawaii. This storm produced swells that were 9 m (28 ft) high in open water, which became waves that were 26 m (80 ft) tall when they finally broke at Waimea Bay! As you can imagine, only the best surfers took a shot at riding these waves. Similar big waves also occur at Cortes Bank, which is about 160 km (100 mi) west of San Diego. When conditions are ripe to produce big waves, surfers speed out to the area in a boat. Once there, they are then pulled by jet skis to catch up to a wave so they can ride it.

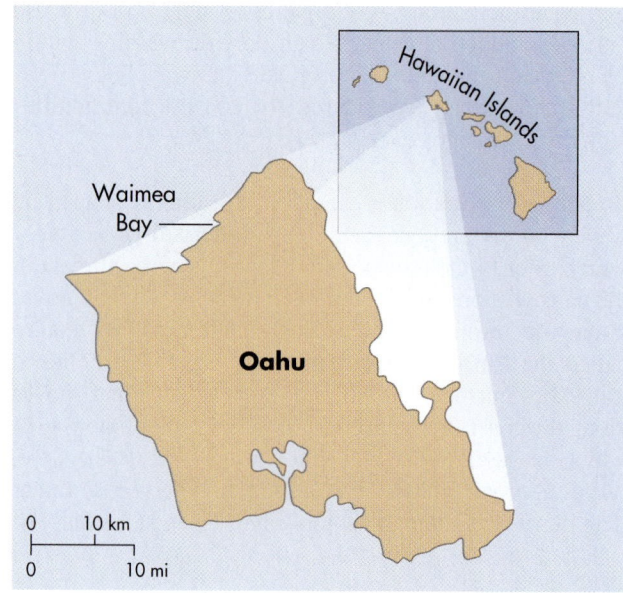

The big waves at Waimea Bay. This area produces some of the largest waves on Earth and, as a result, is a popular destination for surfers.

With your understanding of littoral processes in mind, examine this image of Boa Vigem, Brazil, and see if you can determine

a) The direction of prevailing winds

b) Which way the longshore current is flowing

c) Which way the beach drift is moving

Hint: North is to the right in the image.

These are among the more obvious geomorphic processes and are very easy to see. Taken together, the combined processes of longshore drift and beach drift are called **littoral drift.**

Swash moves sand up beach at an angle
Backwash moves sand down beach perpendicularly
Beach
Beach drift
Breaking waves
Longshore current
Wave crests move toward shore at an oblique angle

Figure 19.13 Littoral drift. Longshore current develops because waves rarely approach the coast in a perpendicular fashion; rather, they strike obliquely. Sediment subsequently moves down the shore as littoral drift. Littoral drift includes beach drift, which is the transport of sediment down shore due to the zig-zag pattern of swash and backwash, and longshore drift, which is the movement of sediment by the longshore current.

Littoral drift *Sediment that is transported through the combined processes of longshore drift and beach drift.*

KEY CONCEPTS TO REMEMBER ABOUT WATER MOVEMENT ALONG COASTLINES

1. Coastlines are the places on Earth where the hydrosphere, lithosphere, and atmosphere intersect over a large spatial extent.

2. The amount of water in the oceans fluctuates with time due to adjustments in the hydrologic cycle. These changes are called eustatic changes. Many coastlines show evidence of long-term eustatic change in the form of rias and fjords.

3. Tides are daily adjustments that occur with respect to the water level along any given coastline. Tidal processes are driven largely by the gravitational pull of the Moon and, to a lesser extent, the Sun.

4. Waves are the dominant factor that shapes coastlines. Waves are disturbances in bodies of water and form as a result of frictional forces related to wind. Waves generate rotating bodies of water that migrate obliquely toward the shore. At the point where the wave base interacts with the ocean bottom in shallow water, the oscillations become deformed, ultimately causing waves to crash.

5. Littoral processes refer to the way that sediment is transported and deposited in the shore zone. The longshore current develops along a coastline due to the deflected force of oblique waves striking the shore. In a related process, water runs up the beach and back into the ocean or sea in processes called swash and backwash, respectively.

Coastal Landforms

Let's now turn to the various landforms created along shorelines. We've seen that fluvial, glacial, and eolian landforms can be subdivided into erosional and depositional categories, and so too can coastal landforms.

Erosional Coastlines

Most coastal erosion is accomplished through the unceasing pounding of the shoreline by waves. Waves have tremendous power after they break, with the associated spray moving as fast as 113 km (70 mi) per hour. Given that water has a very high density, the rapidly moving water has a lot of power and is responsible for the majority of coastal erosional landforms.

Although wave erosion can occur along any part of the coast, it tends to be focused on a particular part of the landscape called a headland. A **headland** is a promontory that juts into the ocean or sea and thus is surrounded on three sides by water (Figure 19.14). Headlands tend to form along shorelines where bands of rock with alternating resistance run perpendicular to the coast. The headlands are associated with the more resistant rock such as limestone or granite. Where the rock is relatively soft, perhaps shale or sandstone, waves can erode sediment more effectively and a bay forms.

Once the pattern of headlands and intervening bays develops, the erosive power of waves begins to concentrate on the headlands. This change occurs because approaching waves initially begin to slow

Headland *A portion of the coast that extends outward into a large body of water.*

Figure 19.14 Headlands along the south coast of Tasmania in Australia. Headlands are prominent landforms that project out into the water. Note that at least three headlands appear in this image.

down in front of the headlands, where they first encounter shallow water. This friction causes the waves to pivot around the headlands in a process called **wave refraction.** As you can see in Figure 19.15a, b, waves bend around a headland, which results in the energy of the waves being expended on three sides of the landform. In this way, headlands are more vigorously eroded than the intervening bays. The bays become zones of deposition, forming features called *pocket beaches*, because sediment is funneled into them from the eroded headlands.

The basic coastal erosional landform is the wave-cut bluff, which is a vertical face that commonly forms along rocky shorelines. Excellent examples of large wave-cut bluffs occur on many parts of the coast of Scotland (Figure 19.16), as well as the coasts of England, Australia, and California. Wave-cut bluffs evolve in a predictable sequence of events as shown in Figure 19.17. This sequence begins with

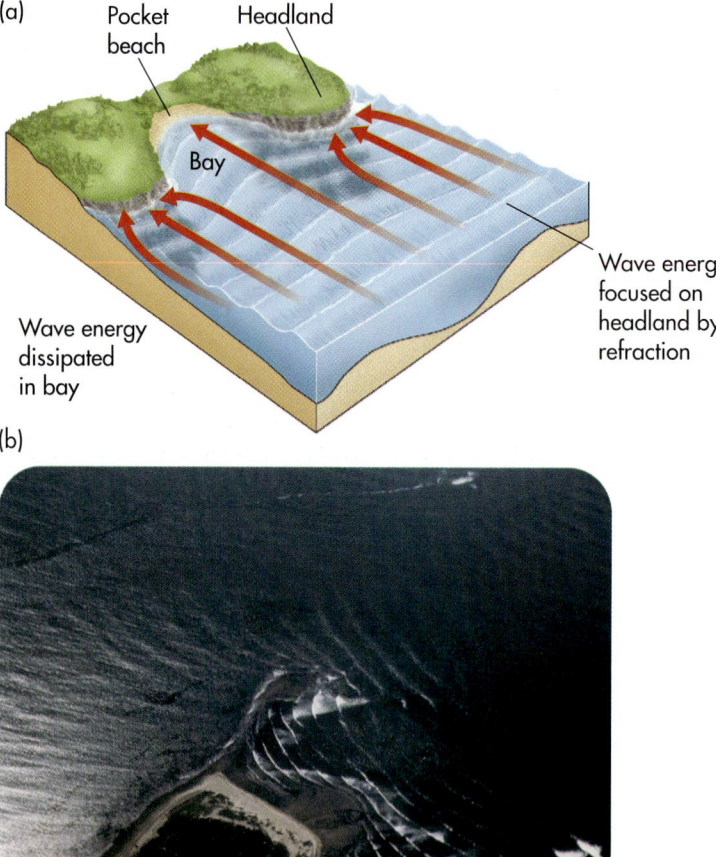

Figure 19.15 Headland erosion. (a) A model showing how waves refract around headlands, causing the formation of a pocket beach. (b) Wave refraction near Westport Point in Massachusetts.

Wave refraction *The process through which waves are focused and bent around headlands.*

Figure 19.16 Wave-cut cliff of Kilt Rock, Scotland. This massive wave-cut bluff was eroded into basalt cliffs over time. Note the waterfall in the foreground.

waves that are pounding against a non-eroded shoreline (Figure 19.17a). With time (19.17b), the waves cut a vertical face into the rock outcrop, and at the same time they plane a more horizontal surface called a *wave-cut platform* into the rocks in front of it. During higher water stages or strong storms, waves run up and cut a notch into the lower rocks of the bluff outcrop and further plane the nearshore platform (19.17c). If the notch is cut sufficiently deep, the rock overhang falls under the influence of gravity and the bluff retreats farther inland, leaving a longer wave-cut platform in front of it (19.17d). As the coastline is retreating overall due to erosion, it is said to be undergoing **retrogradation.**

As described earlier, one way that water moves along coastlines is through eustatic changes in sea level. In addition to the submerged river channels that occur on the continental shelf, one of the best lines of evidence for prehistoric eustatic sea-level changes is the presence of abandoned bluffs and marine terraces. Similar to a fluvial terrace, a marine terrace represents the former position of the active shoreline on the landscape. Figure 19.18 shows how this kind of landscape evolves. In Figure 19.18a, waves create a wave-cut bluff and

Retrogradation *The process through which a shoreline retreats through erosion.*

platform similar to that shown in Figure 19.17. Imagine what would happen if the landscape were subsequently uplifted or if sea level fell. If such a eustatic change occurred, a new bluff and platform would evolve below the abandoned one (Figure 19.18b), leaving a stair-stepped coastal landscape. Such landscapes are common along the active margin of the west coast of the United States (Figure 19.19).

In addition to wave-cut bluffs and terraces, additional evidence for coastal retrogradation comes in a variety of forms, which evolve in a continuum as the shoreline recedes. *Sea caves* are semicircular notches cut into the base of rock bluffs. If the rock is entirely cut through so that water passes through the notch, a *sea arch* results (Figure 19.20). With time and additional erosion, the top of the arch may collapse, leaving a detached fragment of the original bluff, called a *sea stack*, isolated in the ocean. These features are particularly common along the rocky coast of the Pacific Northwest. The net effect of shoreline retrogradation in places like this is that the coastline slowly transforms from one that has a number of prominent headlands to one that is relatively straight (Figure 19.21).

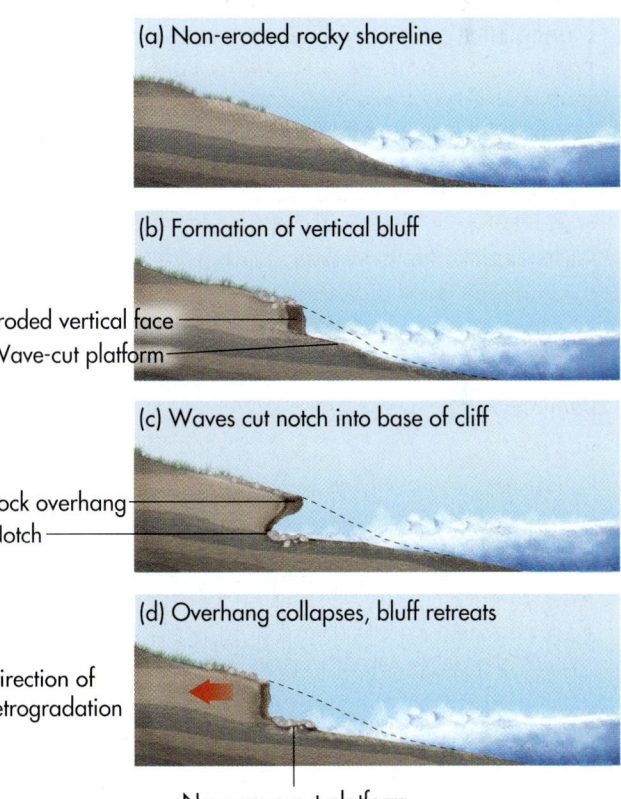

Figure 19.17 Bluff formation and retreat. (a) Waves attack a non-eroded rock outcrop. **(b)** With time, the waves cut a vertical bluff into the rock. **(c)** During strong storms or relatively high water, the waves cut a notch into the lowermost rocks of the bluff. **(d)** If the notch is cut sufficiently deep, the overhanging rocks in the bluff collapse, causing bluff retreat.

(a)

Wave-cut bluff

Notch   Beach   Wave-cut platform

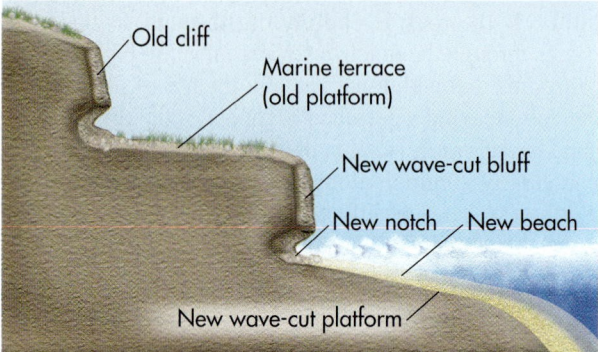

(b)

Old cliff

Marine terrace
(old platform)

New wave-cut bluff

New notch   New beach

New wave-cut platform

Figure 19.18 Evolution of a marine terrace. (a) Waves cut a bluff and platform. (b) Tectonic uplift or sea-level fall raises the previous wave-cut features relative to the position of current erosion, resulting in a marine terrace.

Figure 19.19 Marine terrace in California. This marine terrace formed sometime during the late Pleistocene, perhaps between 60,000 and 100,000 years ago, in association with tectonic uplift following a period of high sea level. Uplift rates in this area are about 40 cm (16 in) every 1000 years. Note the prominent bluff that stands between the terrace and the beach.

Figure 19.20 Erosional landforms along the coast of Wales. Sea arches develop when sea caves are entirely cut through, allowing free water passage beneath the arch. If the top of the arch collapses, an isolated sea stack remains, such as the one to the left. Note the tilted rock strata.

Imagine that you're driving down the Pacific coast in Oregon and you see a landscape that looks like the one pictured here. What conclusions can you make about the history of landscape evolution at this location? What were the processes involved? In short, how did this landscape form?

MARINE TERRACES ON SAN CLEMENTE ISLAND, CALIFORNIA

The elevation of marine terraces can provide a great deal of information about past ocean levels and tectonic uplift. This image shows marine terraces on San Clemente Island, which is 115 km (~72 mi) west of the southern coast of California. Each of these terraces formed during a prehistoric high sea level associated with an interglacial period, with the highest surface representing the oldest high-level sea stand. Sea level subsequently dropped during glacial periods when enormous volumes of water were stored as ice. At the same time, the old wave-cut bluff was slowly uplifted due to tectonic activity. With each successive interglacial, sea level rose again, creating new wave-cut bluffs that were progressively lower than the previous ones. Given that the rate of uplift is known, it is possible to reconstruct the height of the former sea-level high stands.

KEY CONCEPTS TO REMEMBER ABOUT EROSIONAL COASTLINES

1. A coastline that is essentially retreating through erosion is considered to be retrograding.

2. Erosional coastlines are shaped primarily by the process of erosion associated with strong waves. These coastlines tend to be rocky with high bluffs.

3. Headlands are prominent landforms that protrude into the ocean. These landforms are eroded more vigorously than other places along the shore because waves are refracted around them and erode on three sides.

4. Marine terraces represent ancient shorelines. The landforms indicate either variable water levels or uplift due to tectonic activity.

5. Other indicators of coastline recession include caves, sea arches, and sea stacks. These erosional features evolve in a continuum as a coastline erodes.

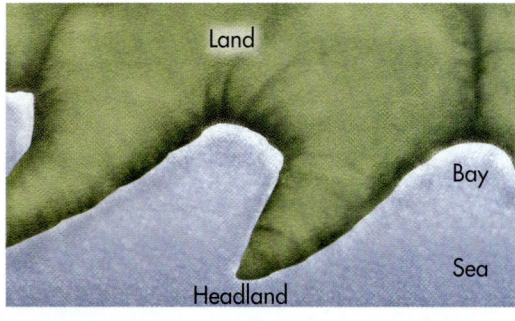

(a)

Labels: Land, Bay, Sea, Headland

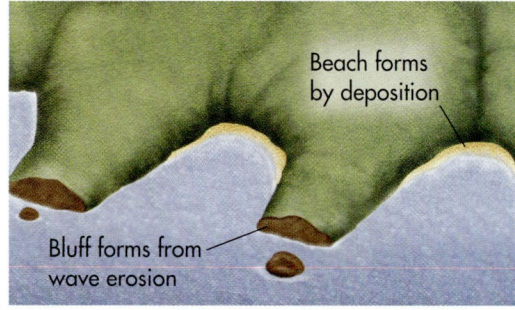

(b)

Labels: Beach forms by deposition, Bluff forms from wave erosion

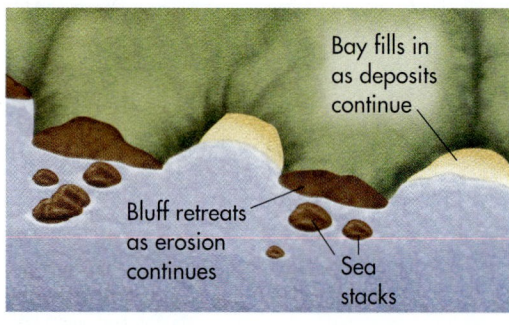

(c)

Labels: Bay fills in as deposits continue, Bluff retreats as erosion continues, Sea stacks

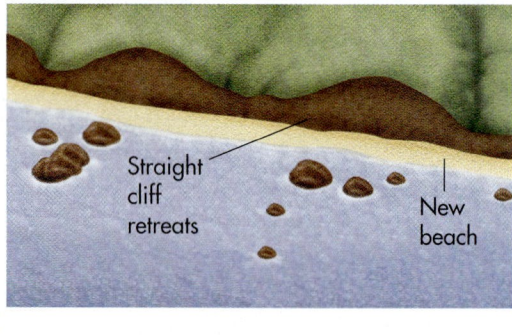

(d)

Labels: Straight cliff retreats, New beach

Figure 19.21 Evolution of a rocky coastline. Through the combined processes of erosion and deposition over time, rocky coastlines evolve in a predictable way that wears down headlands to form relatively straight beaches.

Depositional Coastlines

We've just seen how coastlines are shaped by erosion and some of the diagnostic landforms that result from this process. Now, let's turn to the coastal landforms created when sediment is deposited in various ways and places. Coastlines in the process of extending outward into the water through deposition are said to be undergoing **progradation.**

Beaches   The part of the coastal landscape you're probably most familiar with is the beach. Beaches are dynamic places where sediment is deposited through the combination of waves, beach drift, and wind. In some places, beaches are supplied with alluvial sands derived from land far inland. The beach is a transition between the water and the landmass and consists of exposed, unconsolidated sediments that usually range from sand to cobbles. Although finer silt and clay-sized particles are sometimes contained within a beach, they are usually carried away in suspension by the longshore current. Most beach growth occurs during the summer months in the Northern Hemisphere when the weather is relatively calm. In winter, however, beaches can be significantly eroded due to large waves created by strong storms. If you happen to live on or near a beach, notice how the shape of the beach can change between seasons.

Figure 19.22 shows how beach components are differentiated on the basis of their relative position to the water. The **offshore** is permanently submerged and is the zone where waves break and the surf is most active. Moving landward, the **foreshore** is the part of the beach that is influenced by the rise and fall of the tides; that is, it is regularly uncovered. Separating the offshore and foreshore is a submerged ridge called an **offshore bar,** which forms as a result of the interaction between breakers and the ocean bottom in nearshore environments, where wave velocity is relatively low. Another ridge, called a **beach ridge,** is present on the landward boundary of the foreshore and is cut at the high water line. Landward of the high water line is the **backshore,** which is the relatively flat part of the beach that is covered by water only during severe storms.

Notice in Figure 19.22 that dunes lie landward of the beach. Coastal dune fields are very common along wide, sandy beaches that provide a steady supply of sand blown by strong onshore winds. In places where the supply of sand is especially large and the winds are consistently

Progradation   *Outward extension of the shoreline through deposition of sediment.*

Offshore   *The nearshore zone that is permanently submerged and where waves break.*

Foreshore   *The nearshore zone that is regularly uncovered through the tidal fluctuations and movement of surf.*

Offshore bar   *A small ridge on the bottom of the ocean that separates the offshore and foreshore.*

Beach ridge   *An erosional ridge that marks the high water line and lies at the landward margin of the foreshore.*

Backshore   *The part of the beach that lies between the beach ridge and foredune and is covered by water only during strong storms.*

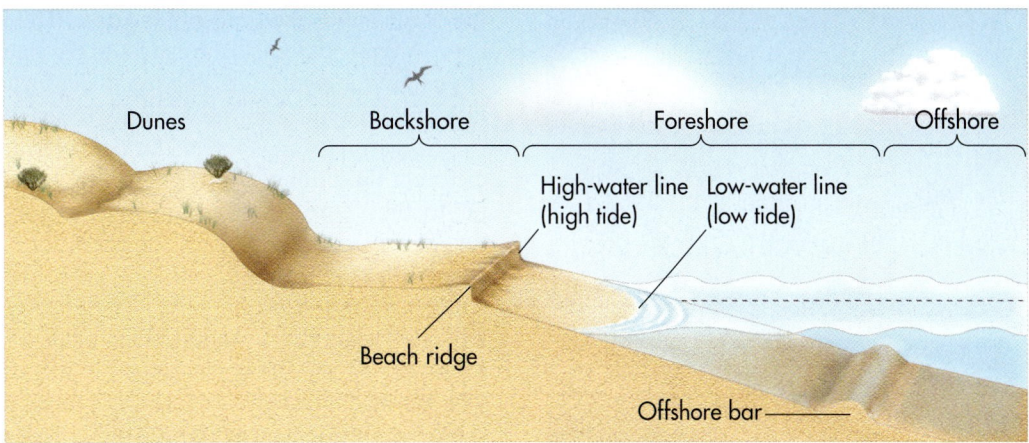

Figure 19.22 Generalized beach cross section. Beaches are divided into several regions, each associated with a different water level or the action of wind.

strong, massive *transgressive dune fields* form. These dune fields are so named because they transgress (advance) inland and bury older surfaces (Figure 19.23).

The most common type of individual dune associated with beaches is a **foredune.** Foredunes range in height from 1 m to perhaps 10 m (3.3 to 33 ft) and are parallel to the shore. These dunes form at times of low water level when sand is easily blown inland off the beach. Sometimes you can see rows of foredunes along the beach, which indicate fluctuating water levels have occurred.

Another type of dune that is common along the coast is the parabolic dune. As noted in Chapter 18, parabolic dunes are elongated deposits of eolian sand that form when blowouts are laterally enlarged at the same time that the dune arms are anchored by vegetation. Coastal parabolic dunes can grow quite large. Along the southern and eastern shores of Lake Michigan, for example, some coastal parabolic dunes reach heights of about 60 m (about 200 ft).

Spits and Baymouth Bars   As we discussed earlier, one of the ways in which water moves along the coast is in association with the longshore current (see Figure 19.13). This current is capable of eroding a great deal of fine sediment from the foreshore and offshore environments and transporting it down the beach (Figure 19.24). As long as the beach continues in an uninterrupted fashion along the shore, the current and its sediment load will flow progressively down the coast because of the interaction of waves and the landmass. Once the current encounters the mouth of a deep bay, however, its speed reduces in a manner consistent with a stream's loss of velocity when it meets the ocean. As with river deltas, sediment carried by the long-

shore current is deposited in a predictable fashion that produces distinctive landforms in bay areas (Figure 19.25).

The variety of landforms created by the deposition of longshore drift in bays forms a geomorphic continuum that evolves with time. This continuum begins with the formation of a **spit,** which is a linear bank of land that extends into a bay when longshore sediment is deposited (Figure 19.25). An important distinction to make here is that a spit is connected to land on the

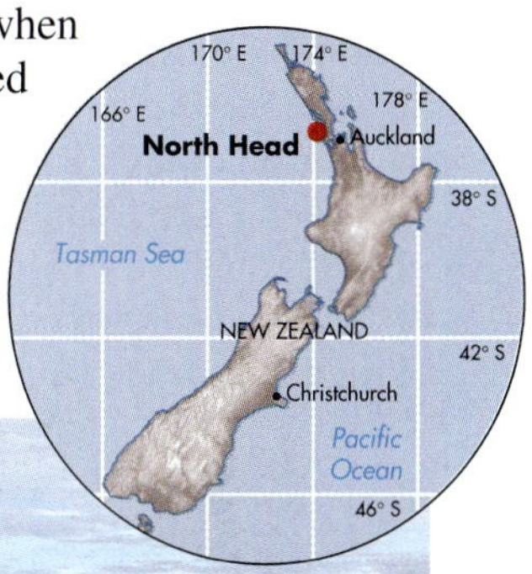

Figure 19.23 Transgressive dune field on North Head, New Zealand. Here, eolian sand has spread inland due to strong winds blowing off the Tasman Sea (left).

Foredune   *Small, linear dune that fronts the backshore and forms when winds transport eolian sediment from the backshore surface inland.*

Spit   *A linear bank of land that extends into a bay made by the deposition of longshore sediment.*

Figure 19.24 Longshore drift These plumes of sediment (arrows) are being carried along the coast due to the force of the longshore current.

the ocean and is then called a **lagoon.** The water in a lagoon is typically brackish relative to the open ocean. Lagoon environments are important ecosystems because they are often nurseries for a variety of sea life.

Sometimes longshore sediments converge from two different directions, causing a spit to grow perpendicular to the shore. If this kind of spit grows so that it reaches a large sea stack or island, it is called a **tombolo** (Figure 19.27).

Barrier Islands   Another type of depositional landform that occurs along coastlines is a **barrier island,** which is an elongated bar of sand that forms parallel to the shore for some distance (Figure 19.28). These islands tend to form on broad, gently sloping continental shelves. Although there is no firm agreement on how these islands develop, a leading hypothesis suggests that they are linked to rising sea levels following the most recent ice age. In this model, sediment deposited on the continental shelf by rivers during the Wisconsin glacial period was reworked

up-current side of the bay and extends into open water on the other side. A spit is usually straight for most of its length, but typically curves landward at its tip, forming a hook when viewed from the air above. This form develops because at the same time the longshore sediment is being deposited as the spit extends, incoming waves are moving the tip shoreward. An excellent example of a curved spit occurs on Cape Cod, Massachusetts (Figure 19.26). With time, the spit will extend entirely across the bay, closing it off with a landform called a **baymouth bar.** When this closing occurs the bay is no longer directly connected to

Baymouth bar   *A spit that entirely encloses a bay.*

Lagoon   *A brackish body of water that lies behind a baymouth bar.*

Tombolo   *A spit or sandbar that connects an island to the mainland.*

Barrier island   *An elongated bar of sand that forms parallel to the shore for some distance.*

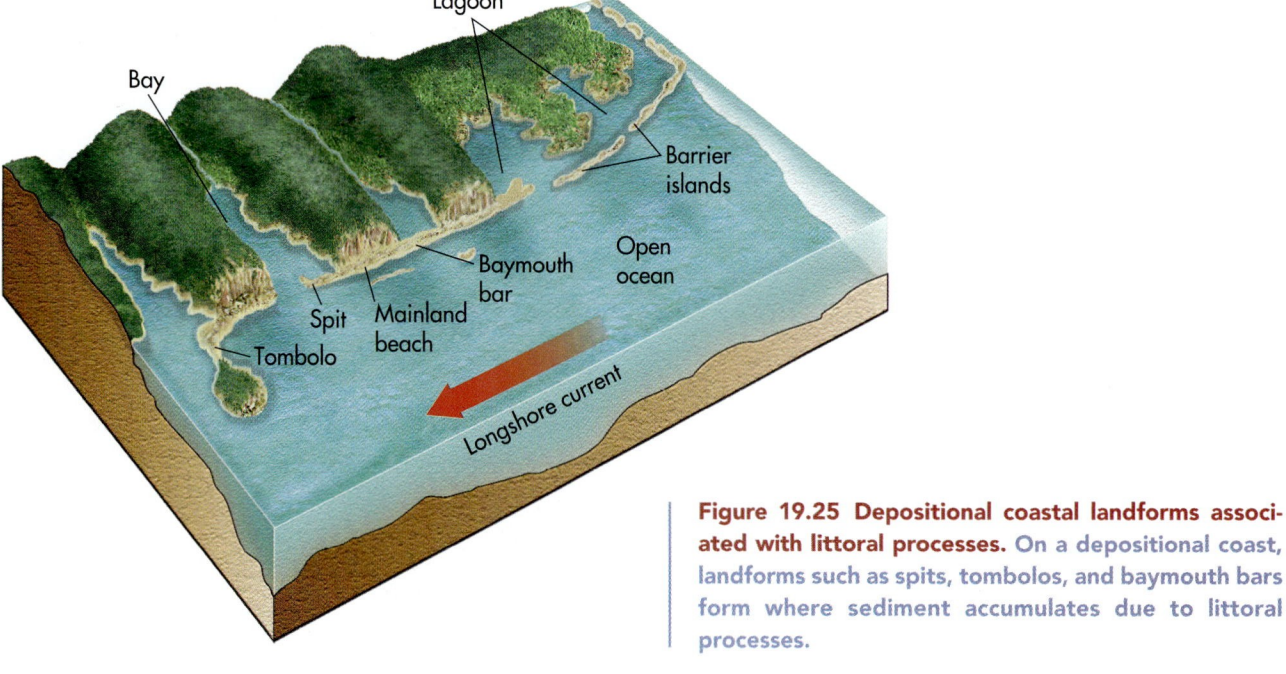

Figure 19.25 Depositional coastal landforms associated with littoral processes. On a depositional coast, landforms such as spits, tombolos, and baymouth bars form where sediment accumulates due to littoral processes.

Figure 19.26 Cape Cod, Massachusetts. This satellite image show a beautiful example of a curved spit, one that formed because the prevailing longshore drift flows to the north.

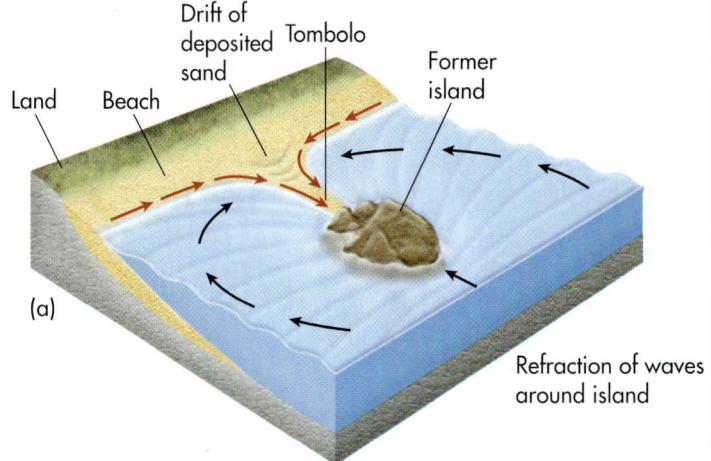

Land Beach Drift of deposited sand Tombolo Former island

(a)

Refraction of waves around island

(b)

Figure 19.27 Tombolos. (a) Tombolos form when longshore drift extends to a pre-existing island. (b) A beautiful example of a tombolo is Mont-Saint-Michel in France.

U.K. English Channel BELGIUM Paris **Mont-Saint-Michel** Bay of Biscay FRANCE ITALY SPAIN

VISUAL CONCEPT CHECK 19.3

As you view this image of a landform on the coast of South Carolina, can you describe how this landform evolved? Given that north is at the top of the image, what way is the longshore current flowing? Why is there a distinct curve at the end of the feature?

LONGSHORE PROCESSES AND DEPOSITIONAL COASTLINES

Now that we've discussed the basic processes associated with longshore drifting, we can view them in an animated format. To do so, go to the *GeoDiscoveries* website and access the module **Longshore Processes and Depositional Coastlines**. This module more thoroughly describes the way that water moves along a shoreline and the landforms that

result from this flow. Watch how oblique-approaching waves cause a current to develop parallel to the shore. Also see how this current moves sediment until it reaches bays where it is deposited to first form spits and then baymouth bars. Most importantly, watch how the coastal landscape is transformed through time and process. After you complete the animation, be sure to answer the questions at the end of the module to test your understanding of this concept.

and molded when sea level began to rise as the ice sheets melted. As the water level inched up, the shelf sediments were shaped by waves, tides, and the wind into the linear islands. At the same time as these islands began to form, they were gradually nudged toward their present position by the slowly rising water and storms crashing into them.

Given the high sand supply on barrier islands, dunes commonly form on them, providing additional relief to the landscape. Barrier islands lie in various proximities to the shoreline, with some being only several hundred meters away and others many kilometers from the coast. An excellent place to see barrier islands is along the Atlantic and Gulf coasts of the southeastern United States, where approximately 280 barrier islands are found. Figure 19.29 shows a satellite image of a reach of barrier islands along the coast of Texas. Notice that the barrier islands are separated by small gaps called *tidal inlets*. These gaps are pro-

duced when strong storms and hurricanes produce large waves that wash over the island in places. Given that barrier islands lie near sea level and are exposed to the open ocean, they are often the first place to be evacuated when a hurricane approaches the coast.

Like baymouth bars, brackish lagoons form behind barrier islands because they are not fully in contact with the ocean and thus have lower salinity. Once these lagoons become established, they become the focus of sediment deposition from three sources: 1) streams flowing off the mainland, 2) eolian sand blowing landward from the island, and 3) tides. Taken together, this deposition produces new land bodies called mudflats (Figure 19.30a),

Figure 19.28 A typical barrier island. Barrier islands are essentially long sand bars, with some sand dunes, that form parallel to the coast. These landforms may have formed when rising sea levels at the end of the Wisconsin glacial period reworked sediments on the continental shelf into these linear features.

Padre Island

Figure 19.29 Barrier islands along the Texas coast. Satellite image of a portion of Padre Island, a well-known barrier island. In this image, red indicates vegetation and blue represents water.

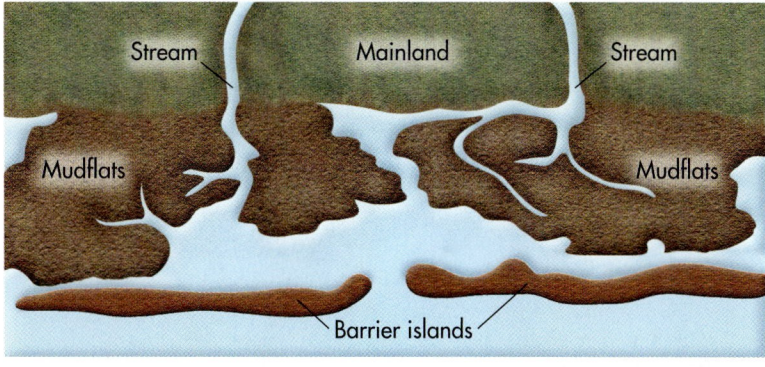

(a)

(b)

Figure 19.30 Infilling of a lagoon. (a) Because lagoons are water bodies of low energy, they are the focus of sediment deposition, forming mudflats. (b) With time, the lagoon will be transformed into a vegetated salt marsh that connects the mainland with the barrier island.

which, if vegetated, become salt marshes (Figure 19.30b). If plant colonization continues unchecked, due to the lack of strong storms or a thorough outlet to the ocean, the lagoon is slowly transformed into a continuous landmass connected to the barrier island.

Coral Reefs   Thus far our discussion of depositional coastlines has focused on landforms that develop due to the accumulation of sediment. Another way that coastal landforms evolve is through the growth of living organisms, specifically coral polyps and algae. In the course of their lives, these organisms secrete external skeletons of calcium carbonate that progressively accumulate as limestone deposits called **coral reefs** (Figure 19.31). These deposits grow larger with time because, when one generation of corals dies, a new one grows on top of it. Coral reefs grow best in tropical waters between about 30°N and 25° S that are warmer than 20° C (68° F), free of suspended sediment, and well aerated. Figure 19.32 shows the distribution of coral reefs on Earth.

Figure 19.31 Pillar coral in the Cayman Islands. Coral comes in a wide variety of colors, shapes, and species, but they all produce calcium carbonate external skeletons.

Coral reefs   *Resistant marine ridges or mounds consisting largely of compacted coral together with algal material and biochemically deposited calcium carbonates.*

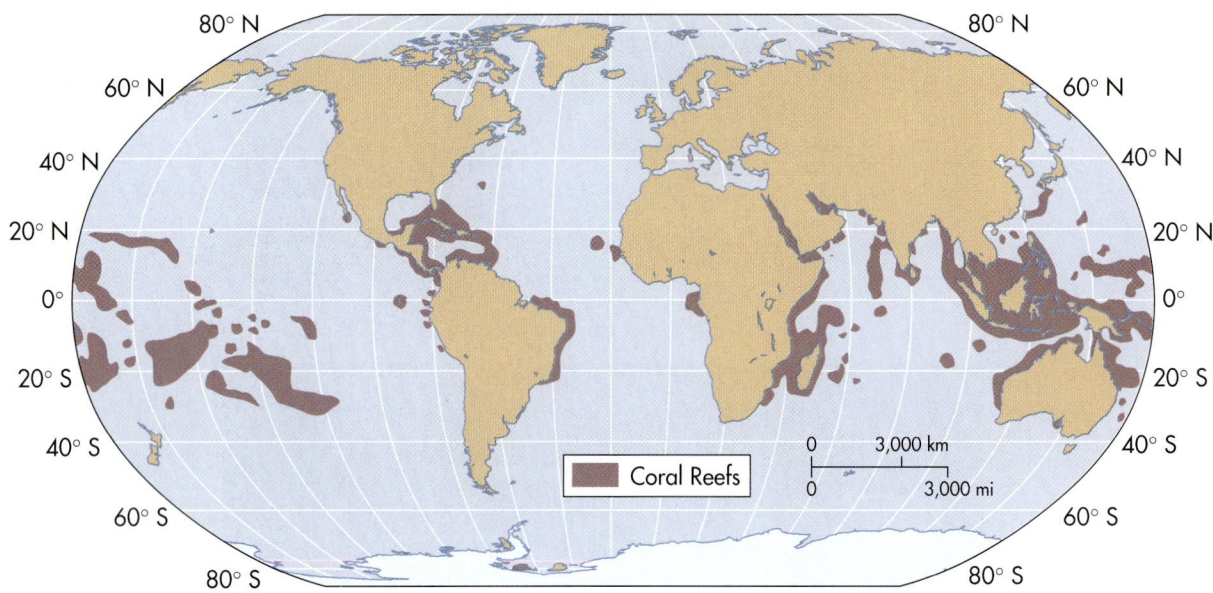

Figure 19.32 Global distribution of coral reefs. Coral reefs are very common in tropical areas.

Figure 19.33 Coral reefs. A fringing reef along the island of Moorea in Tahiti, South Pacific Ocean.

Coral reefs form in three distinct types of settings. A *fringing reef* forms on a shallow platform attached to the shore or island, such as the one shown in Figure 19.33. A second kind of reef is a *barrier reef*, which is a line of coral that is roughly parallel to the shore. Such reefs are separated from the coast either by a deep and wide lagoon or, in some cases, by several kilometers of open ocean that is much deeper than the water above the coral. The best known barrier reef, one you may be somewhat familiar with, is the Great Barrier Reef in Australia. This spectacular reef is actually a complex of more than 2000 individual reefs that lie about 50 km (30 mi) from the northeast Australian coast. This complex of reefs extends for about 2000 km (1250 mi), making it the longest reef in the world.

A third type of coral reef is an atoll. *Atolls* are semicircular reefs that are most closely associated with degradation of volcanic islands. These reefs evolve in a predictable pattern. In the first stage of development (Figure 19.34a), a fringing reef grows on the nearshore platform surrounding a relatively new volcanic island. In the second stage of development (b), the volcano is dormant and begins to wear down and become somewhat submerged due to erosion and subsidence. During this phase of reef development, the reef becomes more like a barrier reef because it is separated from the volcanic island. With the passage of additional time, the volcano disappears entirely (c), leaving an atoll reef ring isolated in the ocean. Atolls are particularly common in the South Pacific Ocean and, like reefs everywhere, are home to a wide variety of ocean animal species (Figure 19.35).

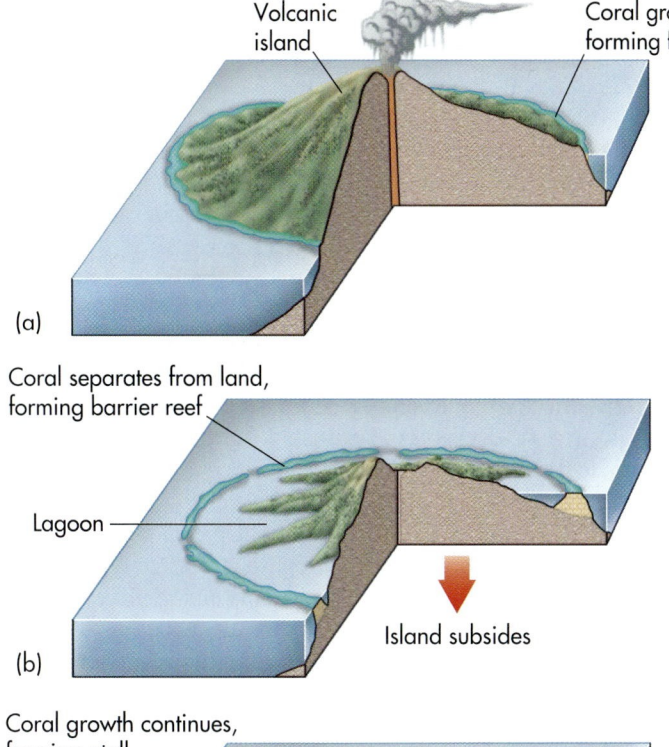

Volcanic island

Coral grows on shore, forming fringing reef

(a)

Coral separates from land, forming barrier reef

Lagoon

Island subsides

(b)

Coral growth continues, forming atoll

Shallow Lagoon

Island submerged

(c)

(d)

Figure 19.34 Three-stage evolution of an atoll. (a) During the first stage a fringing reef grows on the shallow platform surrounding a volcanic island. **(b)** In the second stage, the volcano begins to collapse on itself, which separates the reef from the landform, resulting in a barrier reef. **(c)** With time, the volcanic island is entirely submerged, leaving only the coral above water as an atoll. **(d)** This photograph of the Maldive Islands in the Indian Ocean illustrates the evolution of atolls, beginning with the fringing reef in the foreground.

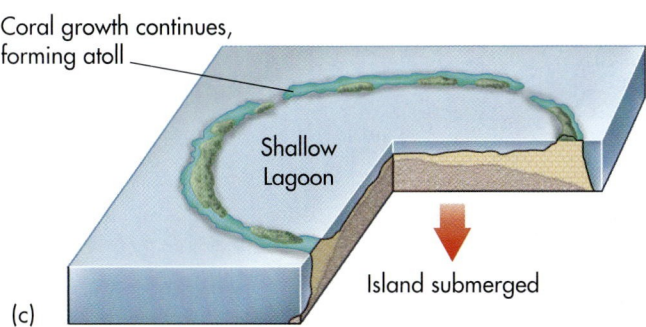

Figure 19.35 Coral reefs in the South Pacific Ocean. These shallow reefs in the Solomon Islands are home to a fascinating array of life.

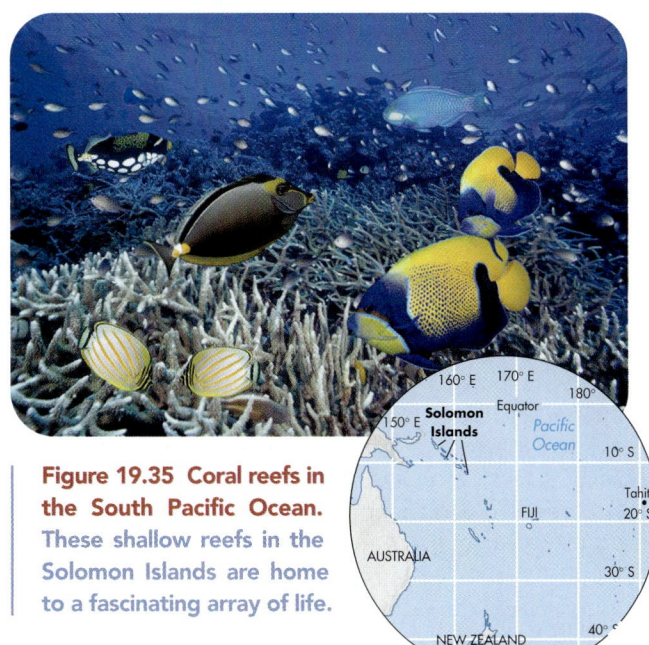

foreshore, backshore) on the basis of water depth and frequency and nature of submergence.

3. Sand dunes commonly occur in the lee of beaches where the supply of sand is high and strong winds frequently blow.

4. A range of depositional features is associated with the decreased velocity of the longshore current, including spits, baymouth bars, and tombolos.

5. A coral reef evolves through the growth of small marine organisms and can be categorized as a fringing reef, barrier reef, or atoll.

Human Impacts on Coastlines

As noted previously, coastlines are a favored place for human activities related to economic development, trade, housing, and recreation. Of the six billion people that live on the Earth, approximately 37% (2 billion) live within 100 km (60 mi) of a major coastline. Nearly 50% live within 200 km (120 mi) of a shore.

To see how these figures compare to the United States, examine Table 19.1. Overall, nearly 144 million people, almost 53% of people in the country, live near a coastline. For example, almost 63 million people (23% of U.S. population) live along the Atlantic Coast in cities such as New York, Boston, Philadelphia, and Miami. Another 17.3 million people (6.3% of U.S. population) live along the Gulf Coast in cities such as Tampa, New Orleans, and Houston. The Great Lakes region is home to almost 27 million people (9.8% of U.S. population), with cities such as Cleveland, Detroit, and Chicago being the major population clusters. Still another 37 million people (13.6% of U.S. population) reside along the Pacific Coast, with most living in cities such as Seattle, Portland, San Francisco, and Los Angeles. Given that most people in the United States (and the world) prefer to live along coastlines, the impact of human activities in these geomorphically sensitive regions is significant. In this section we examine a few of these impacts.

Coastal Engineering

One of the most obvious ways in which people impact the coastline is through engineered structures. Although these structures come in a wide variety of forms, they are generally built with the idea of 1) protecting the shore and shorefront homes from coastal hazards such as erosion and storm surges, 2) stabilizing and nourishing beaches, or 3) maintaining or improving the flow of trade or traffic into ports.

Shoreline Protection   Of the many people who live near U.S. coastlines, a large number choose to reside or work directly on the shore (Figure 19.36). The reasons for this lifestyle decision are probably obvious, as the shore provides a pleasurable landscape in which to live and work, with spectacular views and numerous recreational opportunities. However, coastlines are one of the more sensitive landscapes on Earth, with near-constant change occurring on a daily basis and major adjustments taking place when strong storms such as hurricanes strike. As discussed in Chapter 8, 2004 was particularly bad in Florida, with four

Figure 19.36 Coastal development on Padre Island, Texas. Much of this barrier island, which is shown in the satellite image in Figure 19.29, is heavily developed. This pattern of development is very common along the Gulf and Atlantic coasts of the U.S.

TABLE 19.1	U.S. Coastal Population Statistics Reported by the U.S. Department of Commerce					
Total U.S.	Atlantic Coast	Gulf of Mexico	Great Lakes	Pacific Coast	Total Coastal	Balance of U.S.
Population in Millions						
272.7	62.7	17.3	26.8	37.0	143.9	128.8
Percent of Total U.S. Population						
100.0	23.0	6.3	9.8	13.6	52.8	47.2

(Source: U.S. Department of Commerce, Bureau of the Census)

Figure 19.37 Effects of coastal erosion on personal property. Ongoing bluff erosion along Lake Michigan ultimately resulted in the collapse of this house.

Figure 19.38 Sea walls. This sea wall (arrow) at Palm Beach, Florida, protects a hotel complex from coastal erosion. Structures like this armor much of the shore in southeastern Florida.

major hurricanes (from first to last: Charley, Frances, Ivan, and Jeanne) striking the state. The ultimate amount of estimated damage from all four storms was $30 billion, including severe beach loss. In 2005, Hurricanes Katrina and Rita devastated the Gulf Coast. The combined cost of these storms is expected to be well over $100 billion.

Although extreme events such as hurricanes are indeed very dangerous, they are statistically pretty rare along any particular stretch of coast. As a result, people tend to be much more concerned about the erosion that occurs on a more regular basis, associated with less severe storms, waves, and high water level. Unchecked, the effects of shoreline erosion on people living along the coast can be disastrous, as you can see in Figure 19.37. In an effort to control the erosion that occurs along coastlines during a strong storm or period of high water level, people have devised a variety of protective structures.

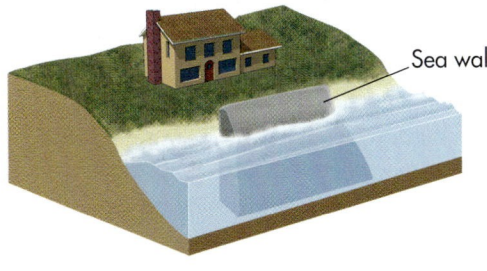

One way in which people protect their homes from high coastal waters is to raise them on stilts so that the waves roll under them. This construction method is common on barrier islands on the Atlantic and Gulf coasts and is also frequently seen in many other countries around the world, such as New Guinea and Vietnam.

Another way in which people attempt to mitigate coastal erosion and stabilize the shore is to build *sea walls* and *revetments* that armor the shore. As the name implies, a sea wall is a vertical, wall-like structure built along the coast, usually at the landward edge of a beach, where a bluff or dune may occur. Figure 19.38 shows a highly engineered structure at Palm Beach, Florida. Many protective structures are less elaborate, consisting of large rocks and sometimes even cement blocks from old roads. This kind of protective cover is called *rip-rap* and forms a revetment because it is usually applied to a preexisting slope, rather than being built vertically. Although sea walls and revetments work in the sense that they protect the property they front from erosion, they cause more extensive erosion up and down shore than would otherwise occur naturally because the erosive power of waves is transferred and concentrated on those locations (Figure 19.39).

Beach Nourishment   One of the primary reasons so many people like coastlines is that they are attracted to the beach and all the recreational and leisure opportunities it provides. Much of the nation's economy is directly tied to beaches, either through the myriad of tourist destinations along the coast or in the hundreds of thousands of people who live there. Given this attraction and important

Figure 19.39 Beach walls and coastal erosion. Although beach walls protect the property they front from erosion, they cause extensive erosion up and down shore because wave power is deflected.

economic status, people have a strong desire to maintain the physical integrity of their particular beach, whether it fronts a luxury hotel, condominiums, or a row of homes.

However, this desire to maintain a good recreational beach is frequently at odds with the natural behavior of the near-shore system because, as noted previously, beaches often consist of unconsolidated deposits of sand that are easily eroded, especially during the winter storm season. In an effort to circumvent the natural beach processes, people actively participate in a wide variety of beach nourishment programs. Basically, these programs involve bringing more sand to the beach in some way, usually at great cost. How much cost? Consider that in the very small state of Delaware alone, over 2,300,000 m^3 (3,008,286 yd^3) of sand has been used to nourish beaches since the early 1960s at a cost in 2002 dollars of over $47,000,000. In the tourist-dominated state of Florida, the cost to nourish beaches over the same length of time has been over $336,000,000 in 2002 dollars.

One way that humans nourish beaches is simply to bring fresh deposits of sand into a beach system to replenish sand lost through erosion. In this manner of beach maintenance, large earth-moving equipment transports sand from some other source, such as a quarry, to a heavily eroded beach. This form of beach nourishment usually occurs at the end of the storm season, when beaches are most vigorously eroded.

Another way in which people try to maintain a functional beach is somehow to limit the loss of sand. The most common method used to maintain beach sand supply is by building a simple structure called a **groin**. A groin is fundamentally a low wall built at a right angle from the shore outward a short distance into the water. This wall intercepts and slows the longshore current on the up-current side of the structure, causing deposition of sand at that locality (Figure 19.40).

Although this process results in beach widening on one side of the groin, the portion of the beach immediately on the down-current side of the structure becomes a zone of erosion because the longshore current is regenerated there due to unaltered waves striking the shore obliquely. Thus, it's possible for one landowner, say landowner A, to build a wide beach in front of his or her property by installing a groin that, in turn, causes erosion and beach loss at the adjacent property of landowner B down the coast. To avoid this loss, landowner B might also install a groin to intercept the sand eroded on his or her property caused by the groin on the property immediately up the coast. Given this downshore ripple effect, it is quite common to see a series of groins stretching along a coast in distinct *groin fields*.

Jetties   Another purpose for protecting beaches and coastlines is to maintain access between the open water and

Figure 19.40 Groins and shoreline processes. Groins such as these along Lake Michigan in Chicago are walls built perpendicular to the shore. These walls intercept the longshore current, causing deposition of sand on the up-current side and erosion on the down-current side of the structure. In Lake Michigan, the prevailing longshore current flows from north (at the top of the photo) to south. Note how sand has accumulated on the north side of these groins and has washed away on the south side.

interior water bodies such as lagoons, lakes, and rivers. This access is maintained with **jetties**, which are long stone or concrete structures used to create a permanent opening for a given channel by reinforcing both sides of the passageway (Figure 19.41). From the shore, jetties project several hundred meters out into the ocean and provide a zone of relatively quiet water for a ship or boat to approach the channel safely so it can navigate into the interior. Like groins, jetties cause deposition on the up-current side of the structure and erosion on the down-current side. To offset beach loss on the down-current side of jetties, beaches are often nourished either through direct deposition of sand or with groins.

Global Climate Change and the Impact on Coastlines

Recall from Chapter 9 that one of the major environmental issues facing people on Earth today is global climate change. Although this warming may be part of a natural

Groin   *A low wall built at a right angle from the shore out a short distance into the water.*

Jetties   *Long stone or concrete structures used to create a permanent opening for a channel by reinforcing both sides of the passageway.*

Figure 19.41 A typical jetty. This jetty on the Texas coast provides a reliable passageway into the ocean for boats. Given the variable beach thickness on either side of the jetty, can you tell which way the longshore current is flowing? It's moving from right (east) to left (west).

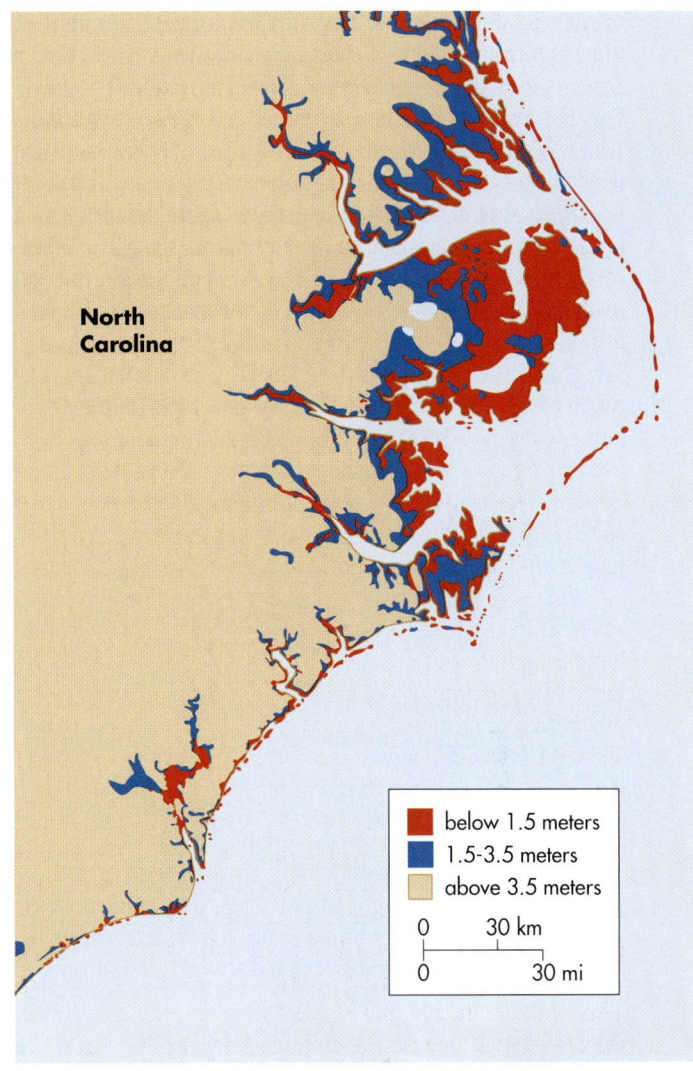

North Carolina

below 1.5 meters
1.5-3.5 meters
above 3.5 meters

0   30 km
0   30 mi

Figure 19.42 Coastal elevations in North Carolina. Much of the coastline lies less than 1.5 m (5 ft) above sea level and will likely be impacted to some extent by coastal flooding associated with global warming.

cycle, the majority of climatologists believe it is directly related to increased levels of atmospheric CO_2 associated with human industrial activity. As noted in that chapter, evidence is increasing that average global temperature may increase between 2° and 4° C (4° to 7° F) by the end of this century. Among other negative effects, this increased temperature will cause further melting of polar ice caps and alpine glaciers, as we discussed in Chapter 17. This meltwater will return to oceans, causing sea level to rise. The rising sea level caused by more ocean water will be magnified by the thermal expansion of water due to increased global temperature. According to current predictions, sea level will rise 15 to 90 cm (5.11 to 35.4 in) by 2100.

Naturally, the places that will be affected most by rising sea levels will be coastal zones around the world. On a global scale, it is estimated that perhaps 92 million people are at risk of coastal flooding. In the United States, significant regions are susceptible to coastal flooding. For example, Figure 19.42 shows the areas vulnerable to rising sea level along the coast of North Carolina. This map is color coded so that areas in red are coastal zones less than 1.5 m (5 ft) above sea level, whereas those in blue lie between 1.5 m (5 ft) and 3.5 m (11.5 ft) above sea level. As you can see, much of the eastern part of the state is very close to sea level. Of these regions, the extensive zone less than 1.5 m (5 ft) will likely be impacted somehow from coastal flooding associated with rising sea level. The likely response of environmental planners and government officials to this threat will be more funds devoted to coastal engineering and beach stabilization.

A landscape that will be particularly hard hit by rising sea levels is the Pacific Island Region, which contains hundreds of low-lying atolls and volcanic islands. Countries like the Marshall Islands, which consists of an archipelago of 34 atolls and coral reefs with a maximum elevation of less than 3 m (10 ft) above sea level, are already being affected by rising sea level. Several islands have been lost to erosion throughout the Pacific Island region, and plans are being developed in many places to relocate coastal inhabitants to more interior locations within the islands. In response to this regional crisis, people of the various island nations have formed the *Alliance of Small Island States* to lobby the United Nations about their particular plight, which, they argue, is impacting them disproportionately seriously given that they produce only 0.06% of all carbon emissions.

Coral-Reef Bleaching   Another coastal byproduct of global climate change is the impact that warmer ocean waters are having on coral-reef systems around the world. These systems are being affected by a process called *coral bleaching*, which occurs when coral reefs are stressed by warmer than normal waters. Coral bleaching can occur naturally as a result of short-term temperature changes, such as those that occurred in association with the El Niño in the early 1980s. Coral bleaching also occurs due to overexploitation of coral resources by people and other forms of local environmental degradation. Since the 1980s, however, increased stress caused by warmer oceans has been increasingly linked to coral bleaching. This stress loosens the algae that help feed the coral-building organisms from the reef. Because the algae give the corals their color, the starved corals look pale or bleached. Ultimately, continued bleaching kills corals. In the past decade, significant amounts of bleaching associated with above-normal ocean temperatures have been reported in a variety of Pacific islands, the Caribbean Sea, Indian Ocean, and along the coast of Australia.

> ### KEY CONCEPTS TO REMEMBER ABOUT HUMAN IMPACTS ON COASTLINES
>
> 1. Coastlines are favored places for people to work, live, and play. Given the inherent sensitivity of the coastal landscape, human impacts are significant and growing. Most human impacts along coastlines are associated with beach stabilization, protection, and access (to port) maintenance.
>
> 2. Beaches are stabilized predominantly in two ways: beach nourishment and groin construction. Beach nourishment is an active process involving the physical transport of sand to the beach, whereas groins take advantage of the natural behavior of the longshore current.
>
> 3. Beach and bluff protection is mostly accomplished through the installation of sea walls and revetments, which armor the shore. Although these features work in the sense that they protect the portion of shore they front, they transfer wave energy to adjacent localities where it is concentrated as an erosive force.
>
> 4. Jetties are long walls built along river channels and inlets to maintain a passageway from the open ocean to the interior. These features project some distance into the water to provide a safe coastal approach for ships.
>
> 5. Global warming appears to be causing a rise in sea level. In this context, significant parts of global coastlines are susceptible to increased erosion and, in some extreme cases, ultimate submergence. Warmer oceans are also contributing to coral bleaching.

Summary of Key Concepts

1. A coastline is a narrow zone where the hydrosphere, lithosphere, and atmosphere interact on a very large scale. These interactions result in very distinctive processes and landforms. Most coastlines are associated with the world's oceans, seas, and gulfs, but some prominent coastlines occur along very large lakes such as the North American Great Lakes.

2. The world's oceans collectively constitute the largest component of the hydrological cycle. Smaller bodies of water associated with oceans are seas and gulfs.

3. The most important agents of coastal change include water fluctuations, tides, waves, and littoral processes. The most extensive water fluctuations occurred when continental glaciers advanced and retreated during the Pleistocene. Tides are daily fluctuations caused largely by the gravitational pull of the Moon. Waves form when wind blows across the water, causing energy within the water to roll forward due to friction. Waves typically approach coastlines obliquely, resulting in the littoral processes of 1) longshore current in the water along the shore and 2) swash/backwash of water moving up and down, respectively, on the beach.

4. Erosional coastlines evolve when sediment is removed by coastal processes. Prominent erosional landforms along coastlines are headlands, bluffs, sea stacks, and marine terraces. Depositional coastlines develop where sediment accumulates. These landforms include beaches, spits, baymouth bars, tombolos, and barrier islands.

5. A slight majority of people on Earth live near coastlines. Given this population distribution, coastlines are impacted greatly by human behavior. Most of this impact is in the form of engineered structures, such as seawalls and jetties, that protect the shore and maintain the location of shipping channels.

6. It appears that global warming is having a major impact on coastlines around the world. These impacts include bleaching of coral reefs and rising sea levels, which are threatening coastal communities.

Check Your Understanding

1. What is the difference between an ocean and a sea? Name the oceans of the world and at least five major seas.

2. What does the term "salinity" mean and how is it related to the chemical composition of ocean water?

3. In the past approximately 150,000 years, what has been the primary cause of eustatic sea level change? Why have these changes occurred and what are some prominent landscapes that indicate such fluctuations?

4. With respect to tides, why is the gravitational pull of the Moon more significant than that of the Sun?

5. Describe the formation and movement of waves. What happens to waves when they approach the shore?

6. Describe the processes associated with littoral drift. How are they related to the approach of waves to the shore?

7. How does a marine terrace indicate that: 1) a period of coastal erosion occurred, and 2) sea level fell or tectonic activity occurred?

8. How are sea caves, sea arches, and sea stacks part of a geomorphic continuum that indicates shoreline erosion is occurring?

9. What is the difference between an erosional and a depositional coastline?

10. Name and describe the depositional landforms associated with the longshore current.

11. Why are coastlines a good place for dune fields to evolve?

12. How do coral reefs form and why is this process fundamentally different from other geomorphic processes?

13. What are the various ways in which people attempt to protect beaches and bluffs from erosion? Why do these approaches also cause problems in nearby parts of the coast?

14. How does a groin take advantage of the longshore current to cause beach growth? Why are groin fields common?

15. Name and describe the two ways that global warming may cause sea level rise.

ANSWERS TO VISUAL CONCEPT CHECKS

Visual Concept Check 19.1

Given that north is on the right side of the image, the prevailing winds in this area must be northerly or northeasterly. These winds are causing swells to migrate toward the southwest before they break as waves and strike the coast. This oblique approach causes the force of the water to be deflected toward the south, resulting in the formation of a longshore current flowing in that direction. In association with this current and swash/backwash of the surf, littoral drift is toward the south along the shore.

Visual Concept Check 19.2

Sea stacks and arches are indicative of a retrograding coastline. These landforms are bodies of relatively resistant rock that have, until now, withstood the force of erosion caused by waves.

Visual Concept Check 19.3

This landform indicates that the longshore current is flowing from north to south (the bottom of the photograph). The spit has formed because the longshore current encountered the relatively still water in the bay, causing deposition of sediment and progradation of the landform. The spit is curved because waves are moving sediment toward the west at the same time that it is being deposited at the tip of the landform.

A horizon The soil horizon that is enriched with humus. 313

Abrasion Erosion that occurs when particles grind against each other. 542

Absorption The assimilation and conversion of solar radiation into another form of energy by a medium such as water vapor. In this process, the temperature of the absorbing medium is raised. 82

Acid rain The precipitation by rain, fog, or snow of strong mineral acids, primarily sulfur dioxide and nitrogen oxides, that originated in factories. 421

Adiabatic processes Changes in temperature that occur due to variations in air pressure. 168

Adret slope The slope that faces the Sun most directly. 281

Advection fog Fog that develops when warm air flows over cooler air. 172

Advection The horizontal transfer of air. 124

Aerial photographs Photographs taken of the Earth's surface from the air. 36

Aggradation The progressive accumulation of sediment along or within a stream. 470

Air mass A large body of air in the lower atmosphere that has distinct temperature and humidity characteristics. 190

Air pressure The force that air molecules exert on a surface due to their weight. 120

Albedo The reflectivity of features on the Earth's surface or in the atmosphere. 82

Albic horizon A diagnostic horizon of podzolization from which clay and free iron oxides have been removed, resulting in a light-colored E horizon. 322

Alfisols Soils generally found in seasonal midlatitude regions that formed through podzolization and have an alkaline argillic horizon. 320

Alluvial fan A fan-shaped landform of low relief that forms where a stream flows out of an area of high relief into a broad, open plane where the gradient is less and deposition occurs. 482

Alluvial terrace A level, step-like landform that forms when a stream erodes its bed so that an essentially horizontal surface is raised relative to the channel. 481

Alluvium Sediment deposited by a stream. 481

Alpine glacier A glacier in mountainous regions that flows down pre-existing valleys. 501

Andisols Soils formed in parent material that is at least 50% volcanic ash. 334

Angle of incidence The angle at which the Sun strikes the Earth at any given place and time. 85

Anthropogenic Environmental changes caused by humans. 249

Anticlinal valley A topographic valley that occurs along the axis of a structural anticline. 385

Anticline A convex fold in rock in which rock layers are bent upward into an arch. 383

Anticyclones High-pressure systems. 123

Aphelion The point of the Earth's orbit where the distance between the Earth and Sun is greatest (~152 million km or 94.5 million mi). 52

Aquiclude An impermeable body of rock that may contain water but does not allow transmission of water through it. 438

Aquifer A geological formation that contains a suitable amount of water to be accessed for human use. 438

Argillic horizon A B horizon that is enriched in eluviated clay. 315

Aridisols Mineral soils that form in arid environments and thus are poorly developed. 328

Arroyo A deep, steep-sided gully that is cut into alluvium. 537

Artesian well A well in which water from a confined aquifer rises to the surface through natural pressure. 440

Artificial levee An engineered stucture along a river that effectively raises the height of the river bank and thus confines flood discharge. 488

Arête A sharp ridge that forms between two glacial cirques. 507

Asthenosphere The layer of very soft rock that occurs in the upper part of the upper mantle. This region is about 40 to 250 km (25 to 105 mi) below the surface of the Earth. The soft character of this rock allows isostatic adjustments to occur. 342

Atmosphere The gaseous shell that surrounds the Earth. 8

Avalanche A large mass of snow or rock that suddenly slides down a mountainside. 429

Axis The line around which the Earth rotates, extending through the poles. 52

B horizon The soil horizon that forms below the E horizon because translocated minerals recrystallize. 313

Backshore The part of the beach that lies between the beach ridge and foredune and is covered by water only during strong storms.   578

Backslope The gradual slope of a dune that faces the prevailing winds.   545

Backswamps Marsh floodplain landforms that develop behind natural levees in which fine-grained sediments settle after a flood.   479

Bankfull discharge The amount of discharge at which the stream channel is full.   464

Barrier island An elongated bar of sand that forms parallel to the shore for some distance.   580

Base flow The amount of stream discharge at any given place and time that is solely the product of groundwater seepage.   464

Base level The lowest level at which a stream can no longer lower its bed, because it flows into the ocean, a lake, or another stream.   472

Bay An indentation in the shoreline that is generally associated with an ocean, sea, or gulf.   562

Baymouth bar A spit that entirely encloses a bay.   580

Beach drift Sediment that is transported in the surf zone by swash and backwash, which form due to the oblique approach of waves.   571

Beach ridge An erosional ridge that marks the high water line and lies at the landward margin of the foreshore.   578

Big Bang theory The theory that the Universe originated about 14 billion years ago when all matter and energy erupted from a singular mass of extremely high density and temperature.   48

Biomass The amount of living matter in an area, including plants, large animals, and insects.   266

Biome The complex of living communities maintained by the climate of a region and characterized by a distinctive type of vegetation.   267

Biosphere The portion of the Earth and atmosphere that supports life.   8

Bioturbation The mixing of soil by plants or animals.   303

Brackish Water that has salinity less than 35‰.   563

Braided stream A network of converging and diverging stream channels within an individual stream system that are separated from each other by deposits of sand and gravel.   470

Breaker A wave that rises and crashes when the forward momentum of oscillations cannot be maintained.   567

Brine Water that has salinity greater than 35‰.   563

Butte A steep-sided hill or peak that is often a remnant of a plateau or mesa.   538

C horizon Unaltered soil parent material.   313

Calcic horizon A diagnostic soil horizon of calcification that is enriched in illuviated calcium carbonate.   328

Calcification A regional soil-forming process in which calcium carbonate is cycled within the soil.   327

Canyon A very steep-sided valley that is cut into bedrock.   538

Capillary action The force that causes water to rise in the small tubular conduits within the soil.   296

Capillary action The process through which water is able to move upward against the force of gravity.   154

Caprock A unit of relatively resistant rock that caps the top of a land form and thus protects underlying rocks from erosion.   412

Carbonation A type of chemical weathering caused by rainwater that has absorbed atmospheric carbon dioxide and formed a weak carbonic acid that slowly dissolves rock.   420

Cartography The design and production of maps.   24

Cations Positively charged ions, such as sodium, potassium, calcium, and magnesium.   311

Cation exchange capacity The total amount of exchangeable cations that a soil can absorb.   311

Cave A cavity in rock, produced by the dissolution of calcium carbonate, that is large enough for someone to enter.   448

Celestial dome A sphere that shows the Sun's arc, relative to the Earth, in the sky.   60

Chemical weathering The decomposition and alteration of rocks due to chemical actions of natural physical and biological processes.   418

Chinook winds Downslope air flow that results when a zone of high air pressure exists on one side of a mountain range and a zone of low pressure exists on the other.   144

Cinder-cone volcano A small, steep-sided volcano that consists of solidified magma fragments and rock debris that may form in only one eruption.   398

Circle of illumination The great circle on Earth that is the border between night and day.   53

Cirque A bowl-like depression that serves as a source area for some alpine glaciers.   502

Cirrus clouds Thin, wispy clouds that develop high in the troposphere.   173

Climate Average precipitation and temperature characteristics for a region that are based on long-term records.   220

Climax vegetation The vegetation within an area that has reached its ultimate complexity.   284

Climograph A graphical representation of climate that

shows average annual precipitation and temperature characteristics by month.   223

Coal A solid fossil fuel that consists of carbonized plants and animals.   354

Cold front A frontal boundary where cold air is advancing into relatively warm air. This front is typically associated with intense rain of short duration.   192

Colloids Very small (10 nanometers to 1 micrometer), evenly divided solids that do not settle in solution.   311

Colluvium Unconsolidated sediment that accumulates at the base of a slope.   424

Composite volcano A large, steep-sided volcano that grows through progressive volcanic eruptions, which are usually explosive, and consists of layers of volcanic debris.   399

Condensation The process through which water changes from the vapor to liquid phase.   156

Condensation nuclei Microscopic dust particles around which atmospheric water coalesces to form raindrops.   172

Conduction The transfer of heat energy from one substance to another by direct physical contact.   81

Cone of depression The cone-shaped depression of the water table that occurs around a well.   441

Confluence The place where two streams join together.   458

Conformal projection A map that maintains the correct shape of features on the Earth but distorts their relative size to one another.   27

Conglomorate Sandstone that contains a wide variety of particle sizes.   350

Constant gases Atmospheric gases such as nitrogen, oxygen, and argon that maintain relatively consistent levels in space and time.   71

Continental A place that is surrounded by a large body of land and that experiences a large annual range of temperature.   106

Continental crust Granitic part of the Earth's crust that makes up the continents. Continental crust is also called sial because it consists largely of silica and aluminum and averages about 40 km (25 mi) thick.   344

Continental drift The theory that the continents move relative to one another in association with plate tectonics.   374

Continental glacier An enormous body of ice that covers a significant part of a large landmass.   505

Continental shelf The part of the continental margin that extends from the lower water point seaward until the sea floor drops steeply.   365

Contours Isolines that connect points of equal elevation.   32

Convection A circular cell of moving matter that contains warm material moving up and cooler matter moving down.   81

Convectional uplift Uplift of air that occurs when bubbles of warm air rise within an unstable body of air.   179

Convergent uplift Uplift of air that occurs when large bodies of air meet in a central location.   180

Coral reefs Resistant marine ridges or mounds consisting largely of compacted coral together with algal material and biochemically deposited calcium carbonates.   583

Cordilleran Ice Sheet The ice cap that covered much of the mountains in the northwestern part of North America during the Pleistocene Epoch.   518

Coriolis force The force created by the Earth's rotation that causes winds to be deflected to the right in the Northern Hemisphere and to the left in the Southern Hemisphere.   128

Counter-radiation Longwave radiation that is emitted towards the Earth's surface from the atmosphere.   73

Crest The highest point of a dune.   545

Crevasse A deep crack in a glacier.   498

Cumulus clouds Individual puffy clouds that develop due to convection.   173

Cyclogenesis The sequence of atmospheric events that develops along the polar jet stream that produce midlatitude cyclones.   194

Cyclones Low-pressure systems.   123

Dam A barrier that blocks or restricts the downstream movement of a stream.   489

Debris flow A rapidly flowing and extremely powerful mass of water, rocks, sediment, boulders, and trees.   429

Debris slide A mass-wasting process in which slope failure occurs along a plane that is roughly parallel to the surface.   427

Deflation Removal of sediment from a surface by wind action.   541

Deflation hollow A depression created by wind erosion.   541

Deforestation The removal of trees for economic or agricultural purposes.   286

Degradation The topographic lowering of a stream channel by stream erosion.   470

Delta A low, level plain that develops where a stream flows into a relatively still body of water so that its velocity decreases and alluvial deposition occurs.   484

Dendrochronology The dating of past events and variations in the environment and climate by studying the annual growth rates of trees.   254

Deposition The process by which water vapor changes directly to ice without first becoming a liquid.   157

Desert pavement A resistant, pavement-like surface created when fine particles blow away and coarse sediment such as pebbles and gravel are left behind.   541

Desertification The process through which a formerly vegetated landscape gradually becomes desert-like.   553

Dew-point temperature The temperature at which condensation occurs in a definable body of air.   163

Direct radiation Solar radiation that flows directly to the surface of the Earth and is absorbed.   81

Disappearing stream A surface river or stream that flows into a sinkhole and subsequently moves into an underground river system.   449

Diurnal cycle A 24-hour cycle.   60

Dolomite Sedimentary rock that consists of over 50% calcium-magnesium carbonate ($CaMg[CO_3]_2$).   353

Downdraft A rapidly moving current of cool air that flows downward in a thunderstorm.   201

Downwelling current A current that sinks to great depths within the ocean because water temperature drops and salinity increases.   146

Drainage basin The geographical area that contributes groundwater and runoff to any particular stream.   458

Drainage density The measure of stream channel length per unit area of drainage basin.   460

Drainage divide An area of raised land that forms a separating rim between two adjacent drainage basins.   458

Drumlin A streamlined landform created when a glacier deforms previously deposited till.   513

Dry adiabatic lapse rate (DAR) The rate at which an unsaturated body of air cools while lifting or warms while descending. This rate is 10° C per 1000 m (5.5° F per 1000 ft).   168

E horizon The soil horizon that is progressively depleted of minerals through dissolution and translocation.   313

Earthflow A slow-to-rapid type of mass movement that involves soil and other loose sediments, some of which may be coarse.   429

Earthquake Shaking of the Earth's surface due to the instantaneous release of accumulated stress along a fault plane or from underground movements within a volcano.   390

Easterly wave A slow moving trough of low pressure that develops within the tropical latitudes.   209

Ecosystem A community of plants, animals, and micro-organisms that are linked by energy and nutrient flows.   267

Ecotone The transition area where two or more ecosystems merge.   268

Ecotone The transition between two distinct ecosystems that contains species of each area.   554

Electromagnetic spectrum The radiant energy produced by the Sun that is measured in progressive wavelengths.   68

Eluviation The dissolution and downward mobilization of minerals by water in the soil.   298

Emissivity The amount of electromagnetic energy released by some aspect of the Earth's surface.   38

Entisols Soils that are very weakly developed and thus have no distinct horizonation.   332

Environmental lapse rate The decrease in temperature that generally occurs with respect to altitude in the troposphere. This rate is 6.4° C per km or 3.5° F per 1000 feet (a negative lapse rate).   97

Eolian processes Geomorphic processes associated with the way that wind erodes, transports, and deposits sediment.   540

Epicenter The point on the Earth's surface that lies directly over the focus of an earthquake.   390

Epipedon The surface horizon of a soil profile.   315

Equator The great circle that lies halfway between the North and South Poles.   18

Equatorial trough Core of low pressure zone associated with the Intertropical Convergence Zone.   134

Equilibrium line The place on a glacier where snow accumulation and melting are in balance.   497

Equivalent projection A map projection that accurately portrays size features throughout the map.   28

Esker A winding ridge formed by a stream that flows beneath a glacier.   515

Eustatic change Fluctuations in sea level associated with adjustments in the hydrologic cycle.   564

Evaporation The process by which atoms and molecules of liquid water gain sufficient energy to enter the gaseous phase.   84

Evaporation The process through which water changes from the liquid to vapor phase.   156

Evaporites Surface salt residue that collects through the evaporation of water and the crystallization of sodium.   354

Evapotranspiration The combined processes of evaporation and transpiration.   167

Exfoliation Form of physical weathering in which sheets

of rock flake away due to seasonal temperature changes or by the expansion of the rock due to unloading.   417

Extrusive igneous rocks Rocks that form when magma cools on the Earth's surface.   346

Fall Equinox Assuming a Northern Hemisphere seasonal reference, the fall (or autumnal) equinox occurs on September 22 or 23, when the subsolar point is located at the Equator (0°).   57

Fault A crack in the Earth's crust that results in the displacement of one lithospheric plate or rock body relative to another.   390

Fault escarpment A step-like feature on the Earth's surface created by fault slippage.   393

Field capacity The amount of water remaining in the soil after the soil is completely drained of gravitational water.   437

Field capacity The maximum amount of water the soil can hold after gravitational water has moved away.   296

Firn The compact, granular substance that is the transition stage between snow and glacial ice.   496

Fjord A former glaciated valley along the coast that is flooded by rising sea level.   565

Flood stage The level at which stream discharge begins to spill out of the channel into the surrounding area.   464

Foredune Small, linear dune that fronts the backshore and forms when winds transport eolian sediment from the backshore surface inland.   579

Foreshore The nearshore zone that is regularly uncovered through the tidal fluctuations and movement of surf.   578

Fossil fuels Carbon-based energy sources, such as gasoline and coal, which are derived from ancient organisms.   250

Freezing The process through which water changes from the liquid to solid phase.   155

Frontal uplift Uplift of air that occurs along the boundary of contrasting bodies of air.   180

Frost heaving Upthrust of sediment or soil due to the freezing of wet soil beneath.   414

Frost wedging Expansion and contraction of water in rock cracks due to freezing and thawing.   412

Fumerole A steam vent that results because underlying groundwater is boiled away before reaching the surface.   406

Gelisols Soils in subarctic and arctic environments that contain permafrost within 2 meters of the surface.   325

Geologic time The period of time that encompasses all

of Earth history, from its formation to the present.   357

Geomorphology The branch of physical geography that investigates the form and evolution of the Earth's surface.   365

Geostationary orbit An orbit where satellites remain over the same place on Earth every day. This orbit is achieved because the satellite is placed very high above the Earth and travels at the same speed as the Earth's rotation   38

Geostrophic winds Air flow that moves parallel to isobars because of the combined effect of the pressure gradient force and Coriolis force.   129

Geyser A superheated fountain of water that suddenly sprays into the air on a periodic basis.   405

Glacial abrasion An erosional process caused by the grinding action of a glacier on rock.   506

Glacial drift Sediment deposited indirectly or directly by a glacier.   511

Glacial erratics Large boulders that have been plucked and transported a great distance before they are deposited.   507

Glacial grooves Deep furrows in rock produced by glacial abrasion.   506

Glacial outwash Sediment deposited by meltwater streams emanating from a glacier.   514

Glacial plucking An erosional process by which rocks are ripped out of the ground by a glacier.   507

Glacial striations Scratches in rock produced by glacial abrasion.   506

Glacial till Sediment deposited directly by a glacier.   511

Glacial trough A deep, U-shaped valley carved by an alpine glacier.   510

Glacier A slow-moving mass of dense ice.   496

Graben A downthrown block of rock that lies between two steeply inclined fault blocks.   395

Graded stream A stream that is capable of transporting the average sediment load provided to it over time.   470

Great circles Circles that pass through the center of the Earth that divide the planet into equal halves.   16

Greenhouse effect The process through which the lower part of the atmosphere is warmed because longwave radiation from the Earth is trapped by carbon dioxide (CO_2) and other greenhouse gases.   73

Groin A low wall built at a right angle from the shore out a short distance into the water.   588

Ground ice Distinct zones of frozen water that occur in permafrost regions.   529

Gulf A relatively small body of saltwater that is surrounded by land on three sides and opens to a sea or ocean.   562

Gyres Large oceanic circulatory systems that form because currents are deflected by landmasses.   145

Hadley cell Large-scale convection loop in the tropical latitudes that connects the Intertropical Convergence Zone (ITCZ) and the Subtropical High (STH).   136

Hanging valley An elevated U-shaped valley (with respect to a glacial trough) formed by a tributary alpine glacier.   510

Headland A portion of the coast that extends outward into a large body of water.   574

High latitudes The zone of latitude that lies from about 558 to 908 in both hemispheres.   20

High-pressure ridge An elongated area of elevated air pressure in the upper atmosphere that is typically associated with sunny skies and calm winds.   195

High-pressure system A rotating column of air that descends toward the surface of the Earth where it diverges. These systems spin clockwise in the Northern Hemisphere and counterclockwise in the Southern Hemisphere. Also called an anticyclone.   123

Histosols organic soils that form in cool, wet environments where organic carbon decomposes very slowly.   323

Hogback ridge A ridge underlain by gently tipped rock strata with a long, gradual slope on one side and a relative steep scarp or cliff on the other.   385

Hook echo The diagnostic feature in Doppler radar that indicates strong rotation is occurring within a thunderstorm and tornado development is thus possible.   206

Horn A mountain with three or more arêtes on its flanks.   510

Horst An upthrown block of rock that lies between two steeply inclined fault blocks.   395

Hotspot A stationary zone of magma upwelling that is associated with volcanism within the interior of a crustal plate.   404

Humidity A measure of how much water vapor is in the air. The ability of air to hold water vapor is dependent on temperature.   159

Humus Decomposed organic matter, typically dark, that is contained within the soil.   295

Hurricane A tropical circulatory system with maximum sustained winds greater than 63 knots (73 mph).   211

Hydrologic cycle A general model that illustrates the way that water is stored and moves on Earth from one reservoir to another.   158

Hydrolysis A chemical weathering process that results in the decomposition of silicate molecules within rock through the reaction of hydrogen and hydroxyl ions in water.   419

Hydrosphere The part of the Earth where water, in all its forms, flows and is stored.   8

Hydrosphere The water realm on Earth.   157

Hygroscopic water Soil water held so tightly by sediment grains that it is unavailable for plant use.   436

Ice cap A large ice mass in mountainous regions that is approximately circular in form.   501

Ice field A large ice mass in mountainous regions that is generally linear in form.   501

Igneous rocks Rocks that form when magma rises from the mantle and cools, either within the Earth's crust or on the surface.   346

Illuviation The recrystallization of minerals that occurs directly below the zone of eluviation.   298

Inceptisols Soils that have one or more weakly developed horizons due to some alteration and removal of soluble minerals.   333

Indirect radiation Radiation that reaches the Earth after it has been scattered or reflected.   83

Inner core The inner part of the Earth's core. This area is about 1220 km (760 mi) thick and consists of solid iron and nickel.   340

Insolation Amount of solar radiation measured in watts per meter (W/m^2) that strikes a surface perpendicular to the Sun's incoming rays.   79

Interfluves Topographic high points in a drainage basin that separate one tributary from another.   458

International Date Line This line generally occurs at 180° longitude, with some variations due to political boundaries, and marks the transition from one day to another on Earth.   54

Intertropical Convergence Zone (ITCZ) Band of low pressure, calm winds, and clouds in tropical latitudes where air converges from the Southern and Northern Hemispheres.   134

Intrusive igneous rocks Rocks that form when magma cools within the Earth's crust.   347

Isobars Isolines that connect points of equal atmospheric pressure.   32

Isohyets Isolines that connect points of equal precipitation.   32

Isolines Lines on a map that connect data points of equal value.   32

Isopachs Isolines that connect points of equal sediment or rock thickness.   32

Isotherms Isolines that connect points of equal temperature.   32

Jetties Long stone or concrete structures used to create a permanent opening for a channel by reinforcing both sides of the passageway.   588

Joint A crack or fissure along horizontal or vertical planes in a rock mass that divides the rock into large blocks.   412

Kame A large mound of sediment deposited along the front of a slowly melting or stationary glacier.   514

Karst topography Terrain that is generally underlain by soluble rocks, such as limestone and dolomite, where the landscape evolves largely through the dissolution of rock.   447

Katabatic winds Downslope air flow that evolves when pools of cold air develop over ice caps and subsequently descend into valleys.   144

Kettle lake A lake that forms when a block of ice falls off of the glacial front, is buried by glacial drift, and then melts, forming a depression that fills with water.   515

Kinetic energy The energy of motion in a body, measured as temperature, that is derived from movement of molecules within the body.   99

Lagoon A brackish body of water that lies behind a baymouth bar.   580

Land breeze Nighttime circulatory system along coasts where winds from a zone of high pressure over land flow to a zone of relatively low pressure over water.   144

Landform A natural feature, such as a hill or valley, on the surface of the Earth.   365

Landslide An instantaneous movement of soil and bedrock down a steep slope in response to gravity, triggered by an event such as an earthquake.   427

Large scale map A map that shows a relatively small geographic area with a relatively high level of detail.   30

Latent heat Heat stored in molecular bonds that cannot be measured.   84

Laterization A regional soil-forming process in tropical and subtropical environments that results in extensive eluviation of minerals except for iron and aluminum.   318

Latitude The part of the Earth's grid system that determines location north and south of the Equator.   17

Laurentide Ice Sheet The continental glacier that covered eastern Canada and parts of the northeastern United States during the Pleistocene Epoch.   517

Lava dome A steep-sided volcanic landform consisting of highly viscous lava that does not flow far from its point of origin before it solidifies.   400

Leeward side The side of a mountain range that faces away from prevailing winds.   144

Level of condensation The altitude at which water changes from the vapor to liquid phase.   169

Limestone Sedimentary rock that consists of over 50% calcium carbonate ($CaCO_3$).   353

Liquid outer core The outer part of the Earth's core. This area is about 2250 km (1400 mi) thick and consists of molten iron and nickel.   341

Lithification The process whereby sediments are cemented through compaction to form rock.   350

Lithosphere A layer of solid, brittle rock that comprises the outer 70 km (44 mi) of the Earth.   8

Lithosphere The outer, solid part of the Earth that is about 70 km (44 mi) thick and includes the uppermost part of the asthenosphere and the crust.   342

Littoral drift Sediment that is transported through the combined processes of longshore drift and beach drift.   573

Littoral processes The processes through which sediment is transported and deposited in the shore zone.   571

Loess Wind-blown silt.   548

Longitude The part of the Earth's grid system that determines location east and west of the Prime Meridian.   20

Longitudinal profile A graph that illustrates the change in stream gradient in cross section along a stream from its source to its mouth.   470

Longshore current The current that develops parallel to the coast when waves approach a coast obliquely and forward momentum is deflected.   571

Longshore drift Transport of sediment by the longshore current.   571

Longwave radiation The portion of the electromagnetic spectrum that includes thermal infrared radiation.   68

Low latitudes The zone of latitude that lies between about 35° N and 35° S.   19

Low-pressure system A rotating column of air where air converges at the surface and subsequently lifts. These systems spin counterclockwise in the Northern Hemisphere and clockwise in the Southern Hemisphere. Also called cyclones.   123

Low-pressure trough An elongated area of depressed air pressure in the upper atmosphere that is typically associated with cloudy skies and rain.   195

Magma Molten rock beneath the surface of the Earth.   346

Mantle The layer of the Earth's interior that lies between the liquid outer core and the crust. This area is about 2900 km (1800 mi) thick and consists largely of silicate rock.   341

Map projection The representation of the three-dimensional Earth on a two-dimensional surface.   26

Map scale The distance ratio that exists between features on a map and the real world.   29

Maritime vs. continental effect The difference in annual and daily temperature that exists between coastal locations and those that are surrounded by large bodies of land.   106

Maritime A place that is close to a large body of water that moderates temperature.   106

Mass wasting Movement of rock, sediment, and soil down slope due to the force of gravity.   424

Maximum humidity The maximum amount of water vapor that a definable body of air can hold at a given temperature.   159

Meandering stream A river or small stream that curves back and forth across its valley.   470

Mechanical weathering The breakup of a body of rock into smaller rocks of the same type.   412

Meridians Lines of longitude.   20

Meridional flow Jet stream pattern that develops when distinct Rossby waves exist and the polar front jet stream flows parallel to the meridians in many places.   138

Mesa A broad horizontal surface, smaller than a plateau, that is upheld by resistant caprock.   538

Mesocyclones Strong updrafts that are rotating within a supercell thunderstorm.   205

Mesopause The upper boundary of the mesosphere where temperature reaches its lowest point.   98

Mesosphere A layer of decreasing temperature in the atmosphere that occurs from about 50 to 80 km (~30 to 50 mi) in altitude.   98

Metamorphic rocks Rocks that form when igneous, sedimentary, or other metamorphic rocks are subjected to intense heat and pressure.   355

Microclimate Average temperature and precipitation characteristics within a small area within a larger climate region.   281

Mid-oceanic ridge A ridge-like feature that develops along a rift zone in the ocean due to magma upwelling.   380

Midlatitude cyclone A well-organized low-pressure system in the midlatitudes that contains warm and cold fronts.   193

Midlatitudes The zone of latitude that lies between about 35° and 55° in both hemispheres.   19

Milankovitch theory The theory that best explains Pleistocene glacial/interglacial cycles through long-term variations in the Earth's orbital eccentricity, tilt, and axial precession.   524

Minerals Naturally occurring substances with distinctive chemical configurations that usually manifest themselves in some kind of crystalline form.   345

Mohorovicic Discontinuity The boundary between the Earth's crust and the upper part of the asthenosphere;

seismic waves change speed at this boundary.   342

Mollisols Soils that form through calcification and have a mollic epipedon that overlies mineral matter that is more than 50% saturated with base ions.   326

Monocline A geologic landform in which rock beds are inclined in a single direction over a large distance.   383

Monsoon The seasonal change in wind direction that occurs in subtropical locations due to the migration of the Intertropical Convergence Zone (ITCZ) and Subtropical High (STH) pressure system.   140

Moraine A winding ridge-like feature that forms at the front or side of a glacier or between two glaciers.   512

Mountain breeze Downslope air flow that develops when mountain slopes cool off at night and relatively low pressure exists in valleys.   144

Mudflow A well saturated and highly fluid mass of fine-textured sediment.   429

Mudpot A bubbling mixture of gaseous mud and water at the Earth's surface that is associated with geothermal activity.   405

Multipath error Disruption of the GPS signal from satellites due to obstructions such as trees and buildings.   42

Natural gas Naturally occurring mixture of hydrocarbons that often occurs in association with petroleum and is found in porous geologic formations.   354

Natural levee A small ridge that develops along the channel of a stream through the deposition of relatively coarse sediment when flooding occurs.   479

Net radiation The difference between incoming and outgoing flows of radiation.   86

Normal fault A steeply inclined fault in which the hanging rock block moves relatively downward.   392

Northern Hemisphere The half of the Earth that lies north of the Equator.   18

O horizon The uppermost soil horizon that consists of undecomposed plant litter.   313

Occluded front The area where a cold front begins to overtake a warm front and thus lift warm surface air aloft.   198

Ocean The entire body of salt water that covers about 71% of the Earth.   562

Oceanic crust Basaltic part of the Earth's crust that makes up the ocean basins. Oceanic crust is also called sima because it consists largely of silica and magnesium and is about 8 km (about 5 mi) thick.   344

Offshore bar A small ridge on the bottom of the ocean that separates the offshore and foreshore.   578

Offshore The nearshore zone that is permanently submerged and where waves break.   578

Orogeny A period of mountain building, such as the Alleghany Orogeny.   385

Orographic uplift Uplift that occurs when a flowing body of air encounters a mountain range.   179

Outwash plain A broad landscape of limited relief created by the deposition of glacial outwash.   514

Overthrust fold A structural feature where one part of the rock mass is shoved up and over the other.   385

Overturned fold A structural feature in which the fold limb is tilted beyond vertical, which results in both limbs inclined in the same direction, but not at the same angle.   383

Oxbow lake A portion of an abandoned stream channel that is cut off from the rest of the stream by the meandering process and is filled with stagnant water.   477

Oxic horizon A diagnostic soil horizon in tropical and subtropical environments that is rich in iron oxides.   319

Oxidation A form of chemical weathering in which oxygen chemically combines with metallic iron to form iron oxides, resulting in the loss of electrons.   420

Oxisols Mineral soils in tropical and subtropical environments that formed through laterization and thus have an oxic horizon within 2 meters of the surface.   318

Oxygen isotope stages Periods of time that have distinct O-18/O-16 ratios, which are used to reconstruct prehistoric climate change.   519

Ozone hole The decrease in stratospheric ozone observed on a seasonal basis over Antarctica, and to a lesser extent over the Arctic.   76

Ozone layer The layer of the atmosphere that contains high concentrations of ozone, which protect the Earth from ultraviolet (UV) radiation.   76

Pacific Ring of Fire The chain of volcanoes that occurs along the edge of the Pacific lithospheric plate.   399

Pangea The hypothetical supercontinent, composed of all the present continents, which existed between 300 and 200 million years ago.   374

Parallels Lines of latitude.   19

Parent material The mineral or organic material in which the soil forms.   300

Passive plate margin A place where the continental crust and the oceanic crust are on the same tectonic plate and thus do not move relative to one another.   377

Pedogenic processes The natural processes of soil formation that involve additions, translocations, transformations, and losses.   297

Periglacial processes The suite of processes involving frost action, permafrost, and ground ice that occurs in arctic environments or along the margins of ice sheets.   528

Perihelion The point of the Earth's orbit where the distance between the Earth and Sun is least (~147 million km or 91.5 million mi).   51

Permafrost Ground that is permanently frozen.   528

Petroleum Naturally occurring, oily liquid that consists of ancient hydrocarbons.   354

pH The measure of acidity or alkalinity of a solution, ranging from 0 to 14, based on the activity of hydrogen ions (H^+).   310

Photosynthesis The conversion of solar radiation into chemical energy. Sugars and starches are produced from carbon dioxide and water, through the interaction of light and chlorophyll in green plants. The process releases oxygen into the atmosphere.   264

Physical geography Spatial analysis of the physical components and natural processes that combine to form the environment.   5

Pinnacle A steep-sided, narrow tower that is the final remnant of a plateau, mesa, or butte.   539

Pixel The smallest definable area of detail on an image; short for pixel element.   37

Plane of the ecliptic The flat plane on which the Earth travels as it revolves around the Sun.   51

Plant succession The natural changes that occur within a landscape over a period of time.   284

Plate tectonics The theory that the Earth's crust is divided into a small number of plates that move because they float on the asthenosphere.   374

Plateau A very broad, horizontal surface that is upheld by resistant caprock.   538

Playa A dried lake bed that forms when runoff collects in closed topographic depressions in arid regions.   539

Pluton An extremely large mass of intrusive igneous rock that forms within the Earth's crust.   347

Podzolization A regional soil-forming process in cool, humid environments that results in the eluviation of iron, aluminum, and organic acids to form well-developed E and Bs horizons.   321

Polar easterlies Band of easterly winds at high latitudes.   139

Polar front jet stream River of high-speed air in the upper atmosphere that flows along the polar front.   137

Polar front The contact in the midlatitudes between warm, tropical air and colder polar air.   136

Polar High Zone of high atmospheric pressure at high latitudes.   139

Potential natural vegetation The vegetation that would occur naturally within a specific area if no human influence occurred. 267

Pressure gradient force The difference in barometric pressure that exists between adjacent zones of low and high pressure that result in air flow. 126

Prime Meridian The arbitrary reference point for longitude that passes through Greenwich, England. 20

Process A naturally occurring series of events or reactions that can be measured and which result in predictable outcomes. 5

Progradation Outward extension of the shoreline through deposition of sediment. 578

Proxy data Indirect evidence of an event. For example, fossil pollen is a proxy indicator of climate change because vegetation reflects climate. 252

Pyroclastic material Fragmented rock materials resulting from a volcanic explosion or ejection from a volcanic vent. 399

R horizon Unweathered bedrock that underlies soil. 313

Radial dikes A long, wall-like feature of intrusive igneous rock that forms when magma is injected into thin cracks within older rock and cools. 364

Radiation Energy that is transmitted in the form of rays or waves. 79

Radiation budget The overall balance between incoming and outgoing radiation on Earth. 86

Radiation fog Fog that develops at night when a temperature inversion exists. 172

Radioactive isotopes Unstable isotopes that emit radioactivity as they decay from one element to another. 360

Rain shadow The body of land on the leeward side of a mountain range that is relatively dry and hot (compared to the windward side) due to adiabatic warming and drying. 184

Reflection The process through which solar radiation is returned directly to space without being absorbed by the Earth. 82

Regolith The fragmented and weathered rock material that overlies solid bedrock. 300

Relative humidity The ratio between the specific and maximum humidity of a definable body of air. 160

Relief The difference between the high and low elevation of an area. 304

Remote sensing The method through which information is gathered about the Earth from a distance. 36

Residual parent material Parent material that forms by the weathering of bedrock directly beneath it. 300

Retrogradation The process through which a shoreline retreats through erosion. 575

Reverse fault A steeply inclined fault in which the hanging rock block moves relatively upward. 393

Ria A former river valley along the coast that is flooded by rising sea level. 565

Richter scale The logarithmic scale used to measure the strength of an earthquake. 391

Rifting The spreading apart of the Earth's crust by magma rising between fractures in the Earth's plates. 380

Rills Small drainage channels that are cut into hill slopes by running water. 468

Riparian zone The strip of land, which borders a body of water, that supports plants and animals adapted to water systems. 285

Roche moutonnée A landform produced by glacial abrasion and plucking that has a shallow slope on one side and a steep slope on the other side. 507

Rock An amorphous mass of consolidated mineral matter. 346

Rock outcrop A place where rocks are exposed at the surface of the Earth. 346

Rock structure The internal arrangement of rock layers. 383

Rockfall A mass-wasting process in which rocks break free from cliff faces and rapidly tumble into the valley below. 425

Rossby waves Undulations that develop in the polar front jet stream when significant temperature differences exist between tropical and polar air masses. 137

Salic horizon The diagnostic horizon of salinization that forms due to the recrystallization of secondary salts. 330

Salinity Concentration of dissolved solids in water that is measured in parts per thousand (‰). 562

Salinization A regional soil-forming process in which soluble salts are cycled within the soil. 330

Salt-crystal growth A weathering process that involves the buildup of salts on rock surfaces through evaporation. These salts weaken the cement that bonds rock particles, allowing them later to be washed or blown away. 416

Sandstone A sedimentary rock created when individual sand grains are deposited in thick layers by wind or water and lithify. 350

Saturated zone The zone of rock below and including the water table where pore spaces are completely filled with water. 438

Scattering The redirection and deflection of solar radiation by atmospheric gases or particulates. 82

Sea A subdivision of an ocean that is partially enclosed by land. 562

Sea breeze Daytime circulatory system along coasts where winds flow from a zone of high pressure over water to a zone of relatively low pressure over land. 143

Sea fog Fog that develops when cool, marine air comes in direct contact with colder ocean water. 172

Sediment Solid fragments of rocks that are transported to some location and deposited by wind, water, or ice. 350

Sedimentary rocks Rocks that form through the deposition and lithification of small fragments or dissolved substances from other rocks, or, in some cases, marine animals. 350

Seismic waves Vibrations that travel through the Earth when stress is released in an earthquake. 340

Sensible heat Heat that can be felt and measured with a thermometer. 84

Shale Sedimentary rock that consists of lithified clay-sized sediments. 352

Shield volcano A very broad volcano with shallow slopes that forms in association with nonviscous lava flows. 403

Shortwave radiation The portion of the electromagnetic spectrum that includes gamma rays, X-rays, ultraviolet radiation, visible light, and near-infrared radiation. 68

Sinkhole A topographic depression that forms when underlying rock dissolves, causing the surface to collapse. 449

Slip face The steep slope that lies on the leeward side of a sand dune. 545

Slope The degree of steepness of a portion of the landscape. 281

Slump A mass-wasting process in which rock and sediment rotates and moves down the slope along a concave plane relative to the surface. 428

Small circles Circles that intersect the Earth's surface and that do not pass through the center of the planet. 17

Small scale map A map that shows a relatively large geographic area with a relatively low level of detail. 30

Soil The uppermost layer of the Earth's surface that forms by the influence of parent material, climate, relief, and chemical and biological agents. 294

Soil creep The gradual downhill movement of soil, trees, and rocks due to the force of gravity. 426

Soil horizons The distinct layers within a soil that result from pedogenesis. 312

Soil order A group of 12 distinctive soils differentiated at the most general level. 315

Soil profile A vertical exposure in which all soil components can be seen. 312

Soil science The study of soil as a natural resource through understanding of its physical, chemical, and biological properties. 315

Soil structure The way soil aggregates clump to form distinct physical characteristics. 309

Soil taxonomy The method of soil classification that is based on the physical, chemical, and biological properties of the soil. 315

Soil-forming factors The variables of climate, organisms, relief, parent material, and time that collectively influence the development of soil. 299

Soil-water budget The balance of soil water that involves the amount of precipitation, evapotranspiration, and water storage and loss. 296

Solar constant The average amount of solar radiation ($\sim$1370 W/m^2) received at the top of the atmosphere. 70

Solar noon The time of day when the Sun angle reaches its highest point as the Sun arcs across the sky. 60

Solifluction A form of soil creep that occurs in arctic environments where freeze-thaw processes result in lobes of soil moving gradually down slope. 427

Solum The A, E, and B horizons of a soil, which form through pedogenic processes. 313

Southern Hemisphere The half of the Earth that lies south of the Equator. 18

Spatial analysis A method of analyzing data that specifically includes information about the location of places and their defining characteristics. 4

Spatial resolution The area on the ground that can be viewed with detail from the air or space. 37

Specific humidity The measurable amount of water vapor that is in a definable body of air. 160

Spit A linear bank of land that extends into a bay made by the deposition of longshore sediment. 579

Spodic horizon A mineral soil horizon characterized by the illuvial accumulation of aluminum, iron, and organic carbon. 323

Spodosols Soils in cool, humid regions that form through podzolization and contain a spodic horizon enriched in eluviated iron, aluminum, and organic carbon. 323

Spring Equinox Assuming a Northern Hemisphere seasonal reference, the Spring (or vernal) Equinox occurs on March 20 or 21, when the subsolar point is located at the Equator (0°). 56

Stable air A body of air that has a relatively low environmental lapse rate compared to potential uplifting air; thus, strong convection cannot occur. 181

Stationary front A boundary where contrasting air masses are flowing parallel to one another.   192

Stratopause The upper boundary of the stratosphere where temperature reaches its highest point.   97

Stratosphere The layer of the atmosphere, between the troposphere and mesosphere, that ranges between about 12 km and 50 km (between ~7.5 mi and 31 mi) in altitude.   97

Stratus clouds Layered sheets of clouds that have a thick and dark appearance.   173

Stream hydrograph The graphical representation of stream discharge over a period of time.   464

Strike-slip fault A structural fault along which two lithospheric plates or rock blocks move horizontally in opposite directions and parallel to the fault line.   396

Subduction The process by which one lithospheric plate is forced beneath another. This usually happens when oceanic crust descends beneath continental crust.   388

Sublimation The process through which water changes directly from ice to the vapor phase.   157

Subsidence The settling or sinking of a surface as a result of the loss of support from underlying water, soils, or strata.   444

Subsolar point The point on Earth where the Sun angle is 90° and solar radiation strikes the surface most directly at any given point in time.   49

Subtropical High (STH) pressure system Band of high air pressure, calm winds, and clear skies that exists at about 25° to 30° N and S latitude.   135

Summer Solstice Assuming a Northern Hemisphere seasonal reference, the Summer Solstice occurs on June 20 or 21, when the subsolar point is located at the Tropic of Cancer (23.5° N).   57

Sun angle The angle at which the Sun's rays strike the Earth's surface at any given point and time. This angle is high at low latitudes and is progressively less at higher latitudes.   49

Sun-synchronous orbit A slightly inclined polar orbit that keeps pace with the Sun's westward progress as the Earth rotates, resulting in regular return intervals over every location on Earth.   37

Supercell thunderstorms Large thunderstorms that contain winds moving in opposing directions and are associated with strong winds, lightning, thunder, and sometimes hail and tornadoes.   205

Surface tension The contracting force that occurs when the water surface meets the air and acts like an elastic skin.   296

Synclinal valley A topographic valley that occurs along the axis of a structural syncline.   385

Syncline A concave fold in rock in which rock layers are bent downward to form a trough.   383

Talus A pile of rock fragments and boulders that accumulates below a cliff due to rockfall.   426

Tarn A small lake that forms within a glacial cirque.   507

Temperature inversion A layer of the atmosphere in which the air temperature increases, rather than cools, with altitude.   172

Temporal lag The difference in time between two events, such as when peak insolation and temperature occur.   105

Thermohaline circulation The global oceanic circulatory system that is driven by differences in salinity.   146

Thermosphere The upper layer of the atmosphere, which occurs between about 80 and 480 km (~50 and 300 mi) in altitude.   98

Thunderstorm A brief, but strong storm that contains strong winds, lightning, thunder, and perhaps hail.   192

Tombolo A spit or sandbar that connects an island to the mainland.   580

Topographic map A map that displays elevation data regarding the Earth's surface.   32

Topography The shape and configuration of the Earth's surface.   32

Trade winds The primary wind system in the tropics that flows toward the Intertropical Convergence Zone on the equatorial side of the Subtropical High pressure system. These winds flow to the southwest in the Northern Hemisphere and to the northwest in the Southern Hemisphere.   134

Transform plate margin A plate boundary where opposing plates move horizontally relative to each other.   380

Transpiration The passage of water from leaf pores to the atmosphere.   167

Transported parent material Parent material such as glacial or stream sediments that has recently been deposited and in which soil forms.   300

Treeline The line that represents the upper limit in mountains and high latitudes where environmental conditions support the growth of trees.   283

Tributary A stream or river that flows into a larger stream or river.   458

Tropical depression A tropical low-pressure system with central sustained winds ranging between 20 to 34 knots (23 to 39 mph).   210

Tropical easterlies Band of easterly winds that exist where northern and southern trade winds converge.   134

Tropical storm A tropical low-pressure system with maximum sustained winds between 35 to 63 knots (39 to 73 mph).   210

Tropopause The top part of the troposphere, which is identified by where the air temperature is –57° C (–70° F).   97

Troposphere The lowermost layer of the atmosphere, which lies between the Earth's surface and an altitude of about 12 km (~7.5 mi).   97

Trunk stream The primary stream of a drainage basin. 458

Ubac slope The slope that faces away from the Sun. 281

Ultisols Mineral soils in subtropical environments that formed through laterization and thus are depleted of calcium and have an argillic horizon. 319

Unsaturated zone The area between the soil water belt and the water table where pore spaces are not saturated with water. 437

Unstable air A body of air that has a relatively high environmental lapse rate compared to uplifting air within it; thus strong convection can occur. 181

Updrafts An area of rapidly flowing air that is moving upward within a thunderstorm. 204

Upwelling current A current that ascends to the surface of the ocean because water temperature warms and salinity decreases. 146

Urban heat island The relatively warm temperatures associated with cities that occur because paved surfaces and urban structures absorb and release radiation differently than the surrounding countryside. 109

Valley breeze Upslope air flow that develops when mountain slopes heat up due to re-radiation and conduction over the course of the day. 144

Variable gases Atmospheric gases such as carbon dioxide, water vapor, and ozone that vary in concentration in space and time. 72

Ventifact An individual rock that is pitted, grooved, or streamlined through wind abrasion. 542

Vertical zonation The change in environmental characteristics that occur with respect to altitude. 282

Vertisols Soils that contain an abundance of expandable clay and thus swell and shrink during wet and dry cycles, respectively. 319

Volcano A mountain or large hill containing a conduit that extends down into the upper mantle, through which magma, ash, and gases are periodically ejected onto the surface of the Earth or into the atmosphere. 398

Warm front A frontal boundary where warm air is advancing into relatively cool air. This front is typically associated with slow, steady precipitation. 192

Water table The top of the saturated zone. 438

Wave amplitude The overall height of any given wave as measured from the wave trough to the wave crest. 68

Wave refraction The process through which waves are focused and bent around headlands. 574

Wavelength The distance between adjacent wave crests or wave troughs. 68

Waves Oscillations in a body of water that form mostly due to the frictional force generated by wind blowing across the surface of the water. 566

Weather Day to day changes that occur with respect to temperature and precipitation. 220

Weathering Physical or chemical modification of rock or sediment that occurs over time. 412

Westerlies Midlatitude winds that generally flow from west to east. 136

Wet adiabatic lapse rate (WAR) The rate at which a saturated body of air cools as it lifts. The average rate is about 5° C per 1000 m (2.7° F per 1000 ft). 170

Wilting point The threshold amount of soil water below which plants can no longer transpire water. 296

Windward side The side of a mountain range that faces oncoming winds. 144

Winter Solstice Assuming a Northern Hemisphere seasonal reference, the Winter Solstice occurs on December 21 or 22, when the subsolar point is at the Tropic of Capricorn (23.5° S) 57

Xerophytic plants Plants that live in very dry places that have a number of survival mechanisms in response to prolonged periods of drought. 279

Yardangs Ridges that are sculpted and streamlined by wind abrasion and deflation. 542

Zonal flow Jet stream pattern that is tightly confined to the high latitudes and is thus circular to semicircular in polar view. 138

Zone of ablation The part of a glacier where melting exceeds snow accumulation. 497

Zone of accumulation The geographical region where snow accumulates and feeds the growth of a glacier. 497

PHOTO CREDITS

CHAPTER 1

CO1 Craig Tuttle/Corbis Images; **1.2a** Eric Nguyen/Jim Reed Photography/Photo Researchers, Inc.; **1-2b** Neil Rabinowitz/Corbis Images; **1-2c** Iconica/Getty Images; **1-2d** James Randklev/Stone/Getty Images; **1-3a** Roger Harris/Photo Researchers, Inc.; **1-3b** Science Photo Library/Photo Researchers; **1-3c** Alan Arbogast; **Fig. 1-4** USGS; **1-3d** Lynn Betts; **1-6a** NASA Media Services; **1-6b** Alan Arbogast; **1-6c** Felix Stampfli; **1-6d** PhotoDisc, Inc./Getty Images; **p. 13** NASA/GSFC/MITI/ERSDAC/JAROS, and U.S./Japan ASTER Science Team

CHAPTER 2

CO2 NASA/JPL/NIMA; **2.6** National Maritime Museum Publications; **p. 22** (*top*) Michael Busselle/Corbis Images; **p. 22** (*bottom left*) Aurora & Quanta Productions; **p. 22** (*bottom right*) Hubert Stadler/Corbis Images; **p. 24** Art Wolfe/Stone/Getty Images; **p. 25** (*top right*) Kevin Schafer/Age Fotostock America, Inc. ; **p. 25** (*bottom right*) Harvey Lloyd/Taxi/Getty Images; **2.10a** Age Fotostock America, Inc. **2.10b** Nasa Jet Propulsion Laboratory; **2.12** Alan Arbogast; **2.19a** USGS; **2.19b** Richard Price/Taxi/Getty Images; **2.2** David Zimmerman/Corbis Images; **2.22** NASA GSFC; **2.23** Landsat.org/TRFIC NASA ESIP, Michigan State University.; **2.24a** NASA/GSFC/MITI/ERSDAC/JAROS, and U.S./Japan ASTER Science Team; **2.24b** NASA; **p. 40** NASA GSFC, MITI, ERSDAC, JAROS, and U.S./Japan ASTER Science Team; **Fig. 2-26** Nasa Jet Propulsion Laboratory; **Fig. 2-28** Floris Leeuwenberg/Corbis Images; **Fig. 2-30** Landsat 7, GSFC, NASA; **p. 44** Landsat/GSFC/NASA

CHAPTER 3

CO3 Arnulf Husmo; /Stone/Getty Images; **p. 50** SOHO Extreme Ultraviolet Imaging Telescope/Photo Researchers; **p. 51** Swedish One-metre Solar Telescope/Photo Researchers; **p. 63** Alan F. Arbogast; **p. 64** Andrea N. Hahmann, Ph.D. Research Applications Laboratory, National Center for Atmospheric Research

CHAPTER 4

CO4 NASA / Johnson Space Center; **p. 70** SOHO s Extreme Ultraviolet Imaging Telescope/Science Photo Library/Photo Researchers; **4.4** NASA Johnson Space Center; **4.6** NOAA; **p. 74** Raymond Gehman/Corbis Images; **4.11** Dean Conger/Corbis Images; **4.13a** NASA/Goddard Space Flight Center Scientific Visualization Studio; **4.13b** NASA/Goddard Space Flight Center Scientific Visualization Studio; **4.14** AP/Wide World Photos; **p. 80** Corbis Images; **4.18a** R. Spoenlein/Corbis Images; **4-18c** Onne van der Wal/Corbis Images; **Figs. 4-20, 4-22, 4-23, 4-25** Andrea N. Hahmann, PhD Research Applications Laboratory Natl. Center for Atmospheric Research; **p. 91** Suomi Virtual Museum; **p. 92** (*top*) Frans Lemmens/Getty Images; **p. 92** (*bottom*) Frans Lanting/Minden Pictures, Inc.

CHAPTER 5

CO5 NASA Media Services; **p. 98** George Hall/Corbis Images; **p.100** Kevin Schafer/The Image Bank/Getty Images; **p.101** NASA/Science Photo Library/Photo Researchers, Inc.; **5.3b** Science Photo Library/Chris Priest/Photo Researchers, Inc.; **5.3c** David Joel/Stone/Getty Images; **5.3d** Jody Doyle/The Image Bank/Digital Vision; **5.3e** Digital Vision/Getty Images; **p. 104** Frans Lemmens/The Image Bank/Getty; **p. 105** Frans Lanting/Minden Pictures, Inc.; **5.7** PODAAC; **p. 110** (*top*) Courtesy of Dept. Geography, University of Oregon; **p. 110** (*bottom*) North Light Images/Age Fotostock America, Inc. ; **5.9a** Digital Vision/Getty Images; **5.9b** Pete Seaward/Stone/Getty Images; **5.10** Courtesy NASA/EPA. Provided by Dr. Dale Quattrochi, Marshall Space Flight Plan; **p. 115** (*bottom*) Science Photo Library/Photo Researchers, Inc.

CHAPTER 6

CO6 Peter Lillie;Gallo Images/Corbis Images; **6.1** MODIS/NASA; **p. 122** Age Fotostock America, Inc.; **6.8** Science Photo Library/Photo Researchers, Inc.; **p. 126** Provided by the SeaWiFS Project, NASA/Goddard Space Flight Center, and ORBIMAGE; **p. 127** NOAA/NESDIS/NCDC; **6.17** Image Courtesy GOES Project Science Office; **6.18a** Image Courtesy MODIS Land Group/Vegetation Indices, Alfredo Huete, Principal Investigator, and Kamel Didan, University of Arizona; **6.18b** Martin Zwick/WWI/Peter Arnold, Inc.; **p. 136** GOES; **6-20b** Science Photo Library/NASA/Photo Researchers, Inc.; **p. 140** Stanley Sadkowski Photography. www.tucman.com; **p. 141** Christopher J. Morris/Corbis Images; **6.29** NASA, Courtesy of Otis B. Brown, Robert Evans, and M. Carl, University of Miami, Rosentiel School of Marine and Atmospheric Science; **p. 149** Digital Vision/Getty Images

CHAPTER 7

CO7 Thomas Wiewandt;Visions of America/Corbis Images; **7.2a** Corbis Images; **7.2b** Age Fotostock America, Inc.; **7.2c** Digital Vision/Getty Images; **p. 158** Ric Ergenbright/Corbis Images; **7.8** Joseph M. Moran, American Meteorological Society; **7.9** Dr. Timothy Miller Deputy Chief, Earth and Planetary Science Brach NASA Marshall Space Flight; **p. 164** David Muench/Corbis Images; **7.11** Corbis Images; **7.17** NOAA; **7.18a** Alan Arbogast; **7.18b** Paul Harris/Stone/Getty Images; **7.19a** John Howard/SPL/Photo Researchers; **7.19b** S. J. Krasemann/Peter Arnold, Inc.; **7.19c** Punchstock; **7.19d** Jim Reed/SPL//Photo Researchers, Inc.; **7.19e** MedioImages/Getty Images, Inc.; **7.19f** Sheila Terry/SPL/Photo Researchers, Inc.; **7.19g** John Sanford/Science Photo Library/Photo Researchers, Inc.; **7.19h** John Meade/Science Photo Library/Photo Researchers, Inc.; **7.19i** David Parker/Science Photo Library/Photo Researchers, Inc.; **7.19j** Jim Wark/Peter Arnold, Inc.; **p. 173** Anthony Jay West/Corbis Images; **p. 176** Art Wolfe/The Image Bank/Getty Images; **p. 177** Pekka Parviainen/Science Photo Library/Photo Researchers, Inc.; **7.22b** Alan Arbogast; **p. 180** NOAA;

GOES East, 38
GOES West, 38
Goosenecks of the San Juan River, 480
GPS, 40–42
GPS triangulation, 42
Graben, 395
Graded stream, 469–474, 470
Graded stream model, 481–482, 483
Grand Canyon, 338–339, 362–363
Grand Unconformity, 362
Granite, 347, 349
Granular structure, 309
Grapes of Wrath, The (Steinbeck), 557
Graphic scale, 30
Grassland biomes, 276–278
Grazing, 288–289, 290
Great Barrier Reef, 584
Great circles, 16–17
Great Lakes, 520, 521
Great Lakes coastal dunes, 285
Great Plains, 278, 439–443, 555
Great Rift system, 381
Great Sand Dunes National Monument, 548–549
Greenhouse effect, 73, 249. *See also* Global climate change
Greenland ice sheet, 506, 527, 528
Grimsvötn volcano, 503
Groin, 588
Ground ice, 531
Ground-level ozone, 76
Ground moraine, 513
Groundwater, 436–447
    aquiclude, 438
    aquifer, 438
    artesian well, 440, 441
    contamination, 444–446
    depletion, 441–442
    extraction, 442
    High Plains Aquifer, 438–443
    spring, 441
    subsidence, 443–444
    water table, 438
Groundwater contamination, 444–446
Groundwater depletion, 442
Groundwater extraction, 442
Groundwater model, 436
Gulf, 562, 563
Gully, 467, 468
Gust front, 201, 202
Gyres, 145

H climates, 246–249
Hadley cell, 136
Hail, 179, 204
Half Dome mountain, 420
Half-life, 360
Hand-held GPS receiver, 42
Hanging valley, 510
Hawaiian hotspot, 404, 405
Headland, 574
Heat index, 101, 103
Heat transfer, 79–81
High clouds, 173
High latitudes, 20
High Plains Aquifer, 438–443
High-pressure ridge, 195
High-pressure systems, 123–124
Highland *(H)* climates, 246–249
Hill slopes, 468
Himalaya Mountains, 386, 387
Histosols, 323, 325

Hogback ridge, 385
Holocene Epoch, 358, 360, 517
Hook echo, 206
Horn, 510
Horst, 395
Hot and dry desert biome, 279, 280
Hot low-latitude desert climate *(BWh)*, 228–230
Hot low-latitude steppe climate *(BSh)*, 230–231
Hotspot, 404–406
Human interaction
    coastal processes, 586–590
    eolian processes, 553–557
    global climate change. *See* Global climate change
    streams, 487–491
Humboldt Current, 145
Humid continental hot-summer climates *(Dfa, Dwa)*, 240–241, 242
Humid continental mild-summer climates *(Dfb, Dwb)*, 241, 243
Humid subtropical hot-summer climate *(Cfa, Cwa)*, 235–237
Humidity, 159–163, 165–166
Humidity maps, 162
Humus, 295
Hurricane, 211–215
Hurricane Katrina, 10, 188–189, 214, 215
Hurricane weather map, 211
Hurricane Wilma, 214
Hydraulic geometry, 462–466
Hydraulic variables, 463
Hydroelectricity, 490
Hydrogen bonding, 154
Hydrograph, 489, 490
Hydrologic cycle, 158
Hydrolysis, 419–420
Hydrosphere, 8, 157
Hygroscopic water, 436, 437

Ice, 155
Ice age, 517
Ice cap, 501, 502
Ice-cap climate *(EF)*, 244, 245
Ice core, 519
Ice crystallization, 178
Ice field, 501, 502
Ice sheets, 505
Ice wedge, 531
Iceberg, 502
Igneous rocks, 346–350
Iguacu Falls, 472
Illinoisan glaciation, 517
Illuviation, 298, 303
Inceptisols, 333–334
Inconceivable tornado, 207
Incredible tornado, 207
Indirect radiation, 83
Infrared satellite images, 39, 40
Inner core, 340
Insolation, 79, 90, 91
Interactive tool. *See* GeoDiscoveries
Interfluves, 458
Interglacial, 516
International Date Line, 54
Interpolation, 32
Intertropical convergence zone (ITCZ), 134, 140, 141
Intrusive ice, 531
Intrusive igneous rocks, 347–349
Isobars, 32
Isohyets, 32
Isoline, 32, 33
Isopachs, 32
Isostatic adjustment, 343, 344

Isostatic depression, 505
Isotherms, 32
Isotopic decay, 360
Itaipu Dam, 491
ITCZ, 134, 140, 141

Jet stream, 137
Jetties, 588, 589
Joint, 412–413
Jokulhlaup, 503
Journey to the Center of the Earth (Verne), 340
Jurassic Period, 358, 359

K2, 24
Kame, 514
Kansan glaciation, 517
Karst landforms and landscapes, 447–450
    caves and caverns, 448–449
    disappearing stream, 449
    sinkhole, 449
    tower karst, 450
Karst topography, 447, 449
Katabatic winds, 144
Kelvin, Lord, 100
Kelvin scale, 100, 102
Kettle lake, 515
Kilt Rock, Scotland, 575
Kinetic energy, 99
Knickpoint, 471
Köppen climate classification system, 221–225
Kyoto Protocol, 259

La Conchita debris flow, 429, 430
Laccolith, 346, 348
Lagoon, 580, 582, 583
Lahar, 401
Lake Victoria, 457
Land breeze, 144
Landfills, 446
Landforms, 365. *See also* Geomorphology of continents/ocean
    basins
Landsat systems, 37
Landsat TM, 37
Landscape, 365
Landscape denudation, 411
Landslide, 427–429
Large scale map, 30
Larsen B Ice Shelf, 527, 529
Latent heat, 84
Latent heat of condensation, 156
Latent heat of freezing, 155
Latent heat of melting, 156
Latent heat of sublimation, 157
Latent heat of vaporization, 156
Lateral moraine, 513
Laterization, 318
Latitude, 17–20, 104
Laurentide Ice Sheet, 517
Lava dome, 400
Leader, 204
Leap year, 51
Leeward side, 144
Length of day, 61, 104
Lenticular cloud, 176–177
Levee, 479, 488
Level of condensation, 169
Lianas, 270
Lightning, 201–204
Limb, 383
Limestone, 353

Liquid outer core, 341
Liquid particulates, 78
Lithification, 350
Lithosphere, 8, 342
Lithospheric plates, 374–375, 376
Little Ice Age, 256
Littoral drift, 573
Littoral processes, 571–573
Loamy sand, 307
Local relief, 424
Local wind systems, 143–145
Lodgement till, 511
Loess, 548–552
Longitude, 20–21
Longitudinal dune, 546, 547
Longitudinal profile, 470
Longshore current, 571
Longshore drift, 571, 580
Longwave radiation, 68
Low clouds, 173
Low latitudes, 19
Low-pressure ridge, 195
Low-pressure systems, 123
Lower mantle, 341
Lower soil depletions, 298
Lungs of the Earth, 288

Madison slide, 427, 428
Mafic rocks, 347, 348
Magma, 346
Manilla, Philippines, 228
Mantle, 341–342
Map, 23–25
Map projection, 26–29
Map scale, 29–31
Marble, 355
Marine terrace, 575–577
Marine west-coast climates *(Cfb, Cfc)*, 237–239
Maritime, 106
Maritime Polar (mP), 191
Maritime Tropical (mT), 191
Maritime *vs.* continental effect, 106–109
Maroon Bells, 34
Mars, 73–74
Marshall Islands, 589
Marshland forest (Amazon River), 262–263
Mass wasting, 424–431
    avalanche, 429–430, 431
    defined, 424
    flows, 429
    landslide, 427–429
    overview, 425
    rockfall, 425–426
    soil creep, 426–427
Matterhorn, 510
Mature stage (thunderstorm), 201, 202
Maximum humidity, 159–160
mb, 121
Mbandaka, Democratic Republic of Congo, 225, 227
McMurdo Station, Antarctica, 244, 246
Mean annual albedo, 85
Mean annual net radiation, 88
Mean annual outgoing longwave radiation, 87
Meander scar, 479
Meandering stream, 470
Mechanical weathering, 412–418
    defined, 412
    exfoliation, 417–418
    frost heaving, 414–415
    frost wedging, 412–414, 416

Sumatra-Andaman earthquake, 392, 568568
Summer monsoon, 142
Summer solstice, 57
Sun, 50–51
Sun angle, 49, 60–61
Sun-synchronous orbit, 37
Sunrise (Pacific Ocean), 66–67
Sunspots, 51
Supercell thunderstorm, 205
Surf, 568
Surface temperature, 99, 101–112. *See also* Global temperature patterns
Surface tension, 296
Surging, 498
Suspended load, 469
Swash, 571
Swells, 567
Synclinal valley, 385
Syncline, 383, 384
Synthetic aperture radar (SAR), 41

Taconic Orogeny, 385
Taiga, 275
Talik, 531
Talus, 426
Talus cones, 426
Tarn, 507, 510
Temperate rainforest, 273
Temperature. *See* Global temperature patterns
Temperature inversion, 172
Temperature scales, 99–100, 102
Temporal lag, 105
Terminal moraine, 513
Termite mounds, 304
Terrace formation, 481
Tertiary Period, 358, 359
Teton Mountains, 366, 367
Textural triangle of soils, 307, 309
Thalweg, 477
The Snows of Kilimanjaro (Hemingway), 527
Thematic map, 23
Thermohaline circulation, 146
Thermometer, 102
Thermosphere, 98–99
Third-order stream, 462
Thomson, William (Lord Kelvin), 100
Three Gorges Dam, 491
Threshold point, 424
Thunder, 204
Thunderstorm, 193, 201–204
Tides, 566
Tidewater glaciers, 501, 504
Tilt obliquity, 524–535
Time of day, 104–105
Time zones, 53, 54
Tombolo, 580, 581
Tools. *See* Geographers' tools
Topeka Tornado, 208, 209
Topographic map, 32, 34
Topographic relief, 366–367
Topographic winds, 144–145
Topography, 32
Tornado, 6, 205–209
Tornado Alley, 208–209
Tornado outbreak, 209
Tornado warning, 205
Tornado watch, 205
Tower karst, 450
Track, 430
Trade easterlies, 134

Trade winds, 134
Transform plate margins, 380
Transformations, 298, 299
Transgressive dune fields, 579
Translocations, 298
Transpiration, 167
Transported parent material, 300, 301
Transverse ridges, 546, 547
Tree rings, 253–254
Treeline, 283
Trellis drainage pattern, 460
Triangulation, 42
Tributary, 458, 459
Triple junction, 380
Tropic of cancer, 57
Tropic of Capricorn, 57
Tropical *(A)* climates, 225–228
Tropical circulation, 134–136
Tropical cyclone, 209–211
Tropical deciduous forest and scrub biome, 271
Tropical depression, 210
Tropical forest biome, 268–271
Tropical monsoon climate *(Am)*, 225–228
Tropical rainforest biome, 268–271
Tropical rainforest climate *(Af)*, 225
Tropical savanna biome, 277–278
Tropical savanna climate *(Aw)*, 228
Tropical storm, 210
Tropical wave, 210
Tropopause, 97
Troposphere, 97
Trunk stream, 458
Tsunami, 568–571
Tuckerman's Ravine, 141
Tundra biome, 280
Tundra climate *(ET)*, 244, 245
Typhoon, 211

Ubac slope, 281, 282
Uluru, 25
Understory, 270
Unequal heating of land surfaces, 125–126
Universal Time Coordinated (UTC), 53
Universe, 48
Unloading, 417
Unpaired terraces, 481
Unsaturated zone, 437
Unstable air, 181, 182
Updraft, 204
Upper mantle, 342
Upper soil depletions, 298
Upwelling current, 146
Urban effect on temperature, 109–111
Urban heat island, 109, 111
Urbanization, 487–488
UTC, 53
Utisols, 319

Valley breeze, 144
Valley glaciers, 501, 504
Van Allen belts, 101
Variable gases, 72–78
Vatnajökull ice cap, 502, 503
Vegetation. *See* Plant geography
Ventifact, 542, 543
Venus, 74
Vernal equinox, 56–57
Vertical zonation, 282–284
Vertisols, 319–320
Victoria Falls, 473